T0290975

Applications of Data Assimilation and Inverse Problems in the Earth Sciences

Many contemporary problems faced by Earth sciences and society are complex, for example, climate change, disaster risk, energy and water security, and preservation of oceans. Studies of these challenges require an interdisciplinary approach and common knowledge. This book contributes to closing the gap between Earth science disciplines and assists in utilisation of the growing amount of data from observations and experiments using modern techniques on data assimilation and inversions developed within the same/another discipline or across the disciplines.

This book covers basic knowledge about geophysical inversions and data assimilation and discusses a range of important research issues and applications in atmospheric and cryospheric sciences, hydrology, geochronology, geodesy, geodynamics, geomagnetism, gravity, near-Earth electron radiation, seismology, and volcanology. This book creates an opportunity to inspire more researchers to focus on data assimilation and inverse problems in Earth sciences and provides a useful theoretical reference and practical applications of data assimilation and inverse problems. This is a key resource for academic researchers and graduate students in a wide range of Earth and environmental science disciplines, who specialise in data assimilation and inverse problems.

ALIK ISMAIL-ZADEH is Senior Research Fellow at the Institute of Applied Geosciences of the Karlsruhe Institute of Technology, Germany. His research interests cover dynamics of the lithosphere/mantle, seismology, volcanology, and sedimentary geology, as well as natural hazards and disaster risks. His research methods span from theoretical analysis and numerical modelling of direct and inverse problems and data assimilation to interdisciplinary synthesis. He is an author/editor of *Computational Methods for Geodynamics* and *Extreme Natural Hazards, Disaster Risks, and Societal Implications* (Cambridge University Press), and *Data-Driven Numerical Modeling in Geodynamics*. He is a recipient of several awards, Member of Academia Europaea, Fellow of the American Geophysical Union, the International Science Council, and the International Union of Geodesy and Geophysics, and Honorary Fellow of the Royal Astronomical Society.

FABIO CASTELLI is Professor of Hydraulic Constructions at the University of Firenze, Italy. His research interests cover fluid-dynamical modelling of atmospheric fronts and frontal precipitation; physically based stochastic models of rainfall; stochastic modelling of infiltration into heterogeneous soils; spatial variability and scaling properties of soil moisture; land-atmosphere interaction and soil moisture feedback on precipitation; distributed hydrologic modelling for real-time flood forecasting; use of quantitative precipitation forecast in flood risk management; flood risk assessment in urban environments; sustainable water supply in arid regions; remote sensing of vegetation for hydraulic and hydrologic applications; climate change impact on eco-hydrology and water resources; and the dynamic data assimilation in large non-linear systems.

DYLAN JONES is Professor of Physics at the University of Toronto, Canada. He held a Tier II Canada Research Chair from 2004 to 2014. His research uses chemical data assimilation and inverse modelling techniques to integrate measurements of atmospheric composition with global three-dimensional models of chemistry and transport to develop a better understanding of how pollution influences the chemical and dynamical state of the atmosphere. He is a member of the steering committees for the GEOS-Chem model and for the Analysis of eMIssions usinG Observations (AMIGO) project, which is a sponsored activity of the International Global Atmospheric Chemistry (IGAC) project operating under the umbrella of Future Earth and jointly sponsored by iCACGP of the IAMAS-IUGG. He is the co-chair for the GEOS-Chem Carbon Gases and the Stratosphere Working Groups.

SABRINA SANCHEZ is an early career scientist with primary interest in geomagnetism and dynamo simulations. She has been a postdoc at the Max Planck Institute for Solar System Research, Department of Planets and Comets in Göttingen (Germany) and is currently at Institut de Physique du Globe de Paris (France). Her research covers assimilation of geomagnetic data, modelling of the geomagnetic field evolution, archaeomagnetic field from dynamo simulations, and solar dynamo modelling. She is a recipient of the American Geophysical Union's Max Hammond Award for young researchers given by the Space Physics and Aeronomy (SPA) Section.

SERIES EDITORS
Christophe Cudennec, *IAHS, France*
Hermann Drewes, *Deutsches Geodätisches Forschungsinstitut, München, Germany*
Alik Ismail-Zadeh (Editor-in-Chief), *Karlsruhe Institute of Technology, Germany*
Andrew Mackintosh, Mioara Mandea, *Institut de Physique du Globe de Paris, France*
Joan Martí, *Instituto de Ciencias de la Tierra "Jaume Almera", Barcelona, Spain*
Johan Rodhe, *University of Gothenburg, Sweden*
Peter Suhadolc, *University of Trieste, Italy*
Hans Volkert, *Deutsches Zentrum für Luft- und Raumfahrt, Oberpfaffenhofen, German*

ADVISORY BOARD
Ian Allison, *Australian Antarctic Division*
Isabelle Ansorge, *University of Cape Town, South Africa*
Tom Beer, *Commonwealth Scientific and Industrial Research Organisation, Australia*
Claudio Brunini, *Universidad Nacional de La Plata, Argentina*
Jeff Freymuller, *University of Alaska, Fairbanks*
Harsh Gupta, *National Geophysical Research Institute, Hyderabad, India*
David Jackson, *University of California, Los Angeles*
Setsuya Nakada, *University of Tokyo, Japan*
Guoxiang Wu, *Chinese Academy of Sciences, Beijing, China*
Gordon Young, *Wilfrid Laurier University, Waterloo, Canada*

TITLES IN PRINT IN THIS SERIES
A. Ismail-Zadeh, J. Urrutia-Fucugauchi, A. Kijko, K. Takeuchi, and I. Zaliapin, *Extreme Natural Hazards, Disaster Risks and Societal Implications*
J. Li, R. Swinbank, R. Grotjahn, and H. Volkert, *Dynamics and Predictability of Large-Scale, High-Impact Weather and Climate Events*
T. Beer, J. Li, and K. Alverson, *Global Change and Future Earth: The Geoscience Perspective*
M. Mandea, M. Korte, A. W. Yau, and E. Petrovsky, *Geomagnetism, Aeronomy and Space Weather: A Journey from the Earth's Core to the Sun*

Applications of Data Assimilation and Inverse Problems in the Earth Sciences

EDITED BY

ALIK ISMAIL-ZADEH
Karlsruhe Institute of Technology

FABIO CASTELLI
University of Florence

DYLAN JONES
University of Toronto

SABRINA SANCHEZ
Institut de Physique du Globe de Paris

CAMBRIDGE
UNIVERSITY PRESS

Shaftesbury Road, Cambridge CB2 8EA, United Kingdom

One Liberty Plaza, 20th Floor, New York, NY 10006, USA

477 Williamstown Road, Port Melbourne, VIC 3207, Australia

314–321, 3rd Floor, Plot 3, Splendor Forum, Jasola District Centre, New Delhi – 110025, India

103 Penang Road, #05–06/07, Visioncrest Commercial, Singapore 238467

Cambridge University Press is part of Cambridge University Press and Assessment, a department of the University of Cambridge.

We share the University's mission to contribute to society through the pursuit of education, learning and research at the highest international levels of excellence.

www.cambridge.org
Information on this title: www.cambridge.org/9781009180405

DOI: 10.1017/9781009180412

© Cambridge University Press & Assessment 2023

This publication is in copyright. Subject to statutory exception and to the provisions of relevant collective licensing agreements, no reproduction of any part may take place without the written permission of Cambridge University Press and Assessment.

First published 2023

Printed in the United Kingdom by TJ Books Limited, Padstow Cornwall

A catalogue record for this publication is available from the British Library.

ISBN 978-1-009-18040-5 Hardback

Cambridge University Press & Assessment has no responsibility for the persistence or accuracy of URLs for external or third-party internet websites referred to in this publication and does not guarantee that any content on such websites is, or will remain, accurate or appropriate.

Contents

Contributors

Nikita Aseev
Helmholtz Centre Potsdam – GFZ German Research Centre for Geosciences, Potsdam, Germany

Marc Bocquet
École des Ponts ParisTech, Champs-sur-Marne, France

Hans-Peter Bunge
Ludwig Maximilian University of Munich, Munich, Germany

Dan G. Cacuci
University of South Carolina, Columbia, SC, USA

Fabio Castelli
University of Florence, Florence, Italy

Angelica M. Castillo
Helmholtz Centre Potsdam – GFZ German Research Centre for Geosciences, Potsdam, Germany

University of Potsdam, Potsdam, Germany

Sebastian Cervantes
Helmholtz Centre Potsdam – GFZ German Research Centre for Geosciences, Potsdam, Germany

Now at University of Cologne, Cologne, Germany

Lorenzo Colli
University of Houston, Houston, TX, USA

Richard L. H. Essery
University of Edinburgh, Edinburgh, UK

Andreas Fichtner
ETH Zürich, Zürich, Switzerland

Arnau Folch
Geosciences Barcelona, Barcelona, Spain

Kerry Gallagher
University of Rennes, Rennes, France

Lars Gebraad
ETH Zürich, Zürich, Switzerland

Siavash Ghelichkhan
Australian National University, Canberra, Australia

Manuela Girotto
University of California, Berkeley, CA, USA

Daniel Goldberg
University of Edinburgh, Edinburgh, UK

Colin Grudzien
University of California–San Diego, San Diego, CA, USA

University of Nevada–Reno, Reno, NV, USA

Kyle Gwirtz
University of California, San Diego, CA, USA

Now at Oak Ridge Associated Universities, Oak Ridge, TN, USA

Jorge N. Hayek
Ludwig Maximilian University of Munich, Munich, Germany

Andre Horbach
Technical University of Munich, Munich, Germany

Mitsuyuki Hoshiba
Japan Meteorological Agency, Tsukuba, Japan

Alik Ismail-Zadeh
Karlsruhe Institute of Technology, Karlsruhe, Germany

Shuanggen Jin
Shanghai Astronomical Observatory, Chinese Academy of Sciences, Shanghai, China

Nanjing University of Information Science and Technology, Nanjing, China

Dylan Jones
University of Toronto, Toronto, Canada

Alexander Korotkii
Institute of Mathematics and Mechanics, Russian Academy of Sciences, Yekaterinburg, Russia

Weijia Kuang
Goddard Space Flight Center, National Aeronautics and Space Administration, USA

Oleg Melnik
Lomonosov Moscow State University, Moscow, Russia
Now at Université Grenoble Alpes, Grenoble, France

Ingo Michaelis
Helmholtz Centre Potsdam – GFZ German Research Centre for Geosciences, Potsdam, Germany

Leonardo Mingari
Barcelona Supercomputing Centre, Barcelona, Spain

Max Moorkamp
Ludwig Maximilian University of Munich, Munich, Germany

Mathieu Morlighem
Dartmouth College, Hanover, NH, USA

University of California–Irvine, Irvine, CA, USA

Mathias Mortzfeld
University of California, San Diego, CA, USA

Keith N. Musselman
University of Colorado, Boulder, CO, USA

Barbara Romanowicz
University of California, Berkeley, CA, USA

Collège de France, Paris, France

Malcolm Sambridge
Australian National University, Canberra, Australia

Sabrina Sanchez
Institut de Physique du Globe de Paris, Paris, France

Yuri Y. Shprits
Helmholtz Centre Potsdam – GFZ German Research Centre for Geosciences, Potsdam, Germany

University of Potsdam, Potsdam, Germany

University of California, Los Angeles, CA, USA

Artem Smirnov
Helmholtz Centre Potsdam – GFZ German Research Centre for Geosciences, Potsdam, Germany

University of Potsdam, Potsdam, Germany

Ilya Starodubtsev
Ural Federal University, Yekaterinburg, Russia

Institute of Mathematics and Mechanics, Russian Academy of Sciences, Yekaterinburg, Russia

Yulia Starodubtseva
Institute of Mathematics and Mechanics, Russian Academy of Sciences, Yekaterinburg, Russia

Grigory M. Steblov
Institute of Earthquake Prediction Theory and Mathematical Geophysics, Russian Academy of Sciences, Moscow, Russia

Andrew Tangborn
University of Maryland, Baltimore County, Baltimore, MD, USA

Solvi Thrastarson
ETH Zürich, Zürich, Switzerland

Igor Tsepelev
Institute of Mathematics and Mechanics, Russian Academy of Sciences, Yekaterinburg, Russia

Andrew P. Valentine
Durham University, Durham, UK

Dirk-Philip van Herwaarden
ETH Zürich, Zürich, Switzerland

Berta Vilacís
Ludwig Maximilian University of Munich, Munich, Germany

Irina S. Vladimirova
Shirshov Institute of Oceanology, Russian Academy of Sciences, Moscow, Russia

Dedong Wang
Helmholtz Centre Potsdam – GFZ German Research Centre for Geosciences, Potsdam, Germany

Songbai Xuan
Shanghai Astronomical Observatory, Chinese Academy of Sciences, Shanghai, China

Natalya Zeinalova
Karlsruhe Institute of Technology, Karlsruhe, Germany

Irina Zhelavskaya
Helmholtz Centre Potsdam – GFZ German Research Centre for Geosciences, Potsdam, Germany

Preface

At the end of the last century, Stephen Hawking mentioned that 'the next century will be the century of complexity'. Indeed, many contemporary problems faced by Earth sciences and society are complex (e.g. climate change, disaster risk, energy and water security, and preservation of oceans). Studies of these challenges require an interdisciplinary approach and common knowledge. Ability to utilise data within and across geophysical disciplines remains limited as the knowledge of the methodologies for data analyses and data assimilation developed in one discipline is limited, if not unknown, within the scientific communities of other geoscience disciplines.

For years, conferences and symposia of the International Union of Geodesy and Geophysics (IUGG), organised by the Union Commission on Mathematical Geophysics, featured data assimilation and inverse problems related to a specific geophysical field or to an interdisciplinary field. The idea of a book on data assimilation and inversions in Earth sciences matured during a series of scientific meetings since the 2015 IUGG General Assembly in Prague. The meetings brought together prominent experts in different fields of geoscience to address recent developments, challenges, and perspectives in data assimilation and geophysical inversions.

The book *Applications of Data Assimilation and Inverse Problems in the Earth Sciences* contributes to closing the gap between Earth science disciplines and assists in utilisation of the growing amount of data from observations and experiments using modern techniques on data assimilation and inversions developed within the same/another discipline or across the disciplines. This book sets out basic principles of inverse problems and data assimilation and presents applications of data assimilation and inverse problems in many geoscience disciplines. The book's goal is to highlight the importance of research in data assimilation and geophysical inversions for predictability and for understanding dynamic processes of the Earth and its space environment. The book summarises new advances in the field of data assimilation and inverse problems related to different geoscience fields. Data assimilation and geophysical inversions assist in

scientific understanding and forecasting natural hazard events, such as atmospheric pollution, floods, earthquakes, and volcanoes. This interdisciplinary book provides guidance on future research directions to experts, early career scientists, and graduate students.

This book covers a range of important research issues, consists of twenty-two peer-reviewed chapters, and is organised into the following parts: basic knowledge about inverse problems and data assimilation (Part I); and applications of the techniques of data assimilation and geophysical inversions to problems related to cryosphere, hydrology, atmospheric chemistry, volcanic cloud propagation, and near-Earth electron radiation (Part II); and to problems related to geochronology, lava dynamics, ground shaking due to earthquakes, seismic tomography, gravity and geodetic inversions, geodynamics, geomagnetism, and joint geophysical inversions (Part III). We hope that this book will inspire more researchers to focus on data assimilation and inverse problems in Earth sciences and provide a useful theoretical reference and practical applications of data assimilation and inverse problems.

The book is intended for academic researchers and graduate students from a broad spectrum of Earth and environmental science disciplines interested in data assimilation and inverse problems. We believe that this book will complement other publications on the topic in terms of the coverage of geoscience topics, and hope that the book will form an important reference for researchers dealing with data assimilation and geophysical inversions.

We apologise that the book does not contain all methods for data assimilation and inversions, but only those frequently used in the geosciences. However, we believe that the methods and the applications described here will be helpful for understanding how Earth observations and data can be utilised to quantitatively resolve some problems in dynamics of the Earth and its space environment. We hope that readers will enjoy reading this book to learn more about inverse problems and assimilation of data in geosciences.

Acknowledgements

The book could not be written without the scientific and financial support of the International Union of Geodesy and Geophysics (IUGG). We are grateful to the IUGG Commission on Mathematical Geophysics for providing advice on the topics and suggesting potential authors for the book's chapters. We thank very much all authors of the book for their outstanding contributions.

We are very thankful to our colleagues listed here for their in-depth and constructive reviews of the book's chapters, which improved the original manuscripts and resulted in the volume that you have in your hands. The reviewers of the book are:

Abedeh Abdolghafoorian, George Mason University, Fairfax, USA

Julien Baerenzung, University of Potsdam, Potsdam, Germany

Mohammad Bagherbandi, University of Gävle, Gävle, Sweden

Ciaran Beggan, British Geological Survey, Edinburgh, UK

Patrick Blaise, French Alternative Energies and Atomic Energy Commission, Paris, France

Ebru Bozdag, Colorado School of Mines, Golden, USA

Stephen Cornford, Swansea University, Swansea, UK

Sarah Dance, University of Reading, Reading, UK

Huw Davies, Cardiff University, UK

Andreas Fichtner, ETH Zurich, Switzerland

Jeff Freymueller, Michigan State University, East Lansing, USA

Petar Glisovic, Université du Québec à Montréal, Montreal, Canada

Kyle Gwirtz, University of California, San Diego, USA

Takao Kagawa, Tottori University, Tottori, Japan

Jochen Kamm, Geological Survey of Finland, Espoo, Finland

Brenhin Keller, Dartmouth College, Hanover, USA

Nina Kristiansen, Met Office, Exeter, UK

Daniel Martin, Lawrence Berkeley National Laboratory, Berkeley, USA

Catherine Meriaux, University of Rwanda, Kigali, Rwanda

Klaus Mosegaard, University of Copenhagen, Copenhagen, Denmark

Mauro Perego, Sandia National Laboratories, Albuquerque, USA

Saroja Polavarapu, Environment and Climate Change Canada, Toronto, Canada

Fred Pollitz, United States Geological Survey, Menlo Park, USA

Andrew Prata, University of Oxford, Oxford, UK

Chris Rizos, University of New South Wales, Sydney, Australia

Lassi Roininen, LUT University, Lappeenranta, Finland

Quentin Schiller, Space Science Institute, Boulder, USA

Frederik J. Simons, Princeton University, Princeton, USA

Eugene Shwageraus, University of Cambridge, Cambridge, UK

Polly Smith, University of Reading, Reading, UK

Jiajia Sun, University of Houston, Houston, USA

Helen Worden, National Center for Atmospheric Research, Boulder, USA

Zhang Xin, University of Edinburgh, Edinburgh, UK

This book would not perhaps have been written without the warm encouragement received from Susan Francis of Cambridge University Press, whom we thank for her kind assistance at the initial stage of the book proposal. We are grateful to Sarah Lambert for support and excellent technical assistance and to Ursula Acton for her thorough technical editing of the book.

We would be very grateful for comments, inquiries, and complaints on the content of the book sent to us by e-mail (alik.ismail-zadeh@kit.edu).

PART I

Introduction

1

Inverse Problems and Data Assimilation in Earth Sciences

Alik Ismail-Zadeh, Fabio Castelli, Dylan Jones, and Sabrina Sanchez

Abstract: We introduce direct and inverse problems, which describe dynamical processes causing change in the Earth system and its space environment. A well-posedness of the problems is defined in the sense of Hadamard and in the sense of Tikhonov, and it is linked to the existence, uniqueness, and stability of the problem solution. Some examples of ill- and well-posed problems are considered. Basic knowledge and approaches in data assimilation and solving inverse problems are discussed along with errors and uncertainties in data and model parameters as well as sensitivities of model results. Finally, we briefly review the book's chapters which present state-of-the-art knowledge in data assimilation and geophysical inversions and applications in many disciplines of the Earth sciences: from the Earth's core to the near-Earth environment.

1.1 Introduction

Many problems in Earth sciences are related to dynamic processes within the planet, on its surface, and in its space environment. Geoscientists study the processes using observations and measurements of their manifestations. Each process can be presented by a model described by physical and/or chemical laws and a set of relevant parameters. The model, in its turn, can be represented by a mathematical model; that is, a set of partial differential equations or ordinary differential equations with boundary and/or initial conditions defined in a specific domain. The mathematical model links its parameters and variables with a set of data from observations and measurements and provides a connection between the causal characteristics of the dynamic process and its effects. The causal characteristics include, for example, physical parameters (such as velocity, temperature, pressure), parameters of the initial and boundary conditions, and geometrical parameters of a model domain.

The aim of the direct mathematical problem is to determine the effects of a dynamic model process based on the knowledge of its causes, and hence to find a solution to the mathematical problem for a given set of model parameters. An inverse problem is the opposite of a direct problem. An inverse problem is considered when there is a lack of information on the causal characteristics but information on the effects of the dynamic process exists (e.g. Kirsch, 1996; Kabanikhin, 2011; Ismail-Zadeh et al., 2016). For example, the seismic wave velocities inferred from seismograph's measurements on the Earth's surface are related to a fault rupture in the lithosphere; the rupture process and the wave propagation are described mathematically by the wave equation, which relates causal characteristics (the velocity, density, and elastic properties) with their effects. The heat flow measured at the Earth's surface and inferred temperature can be related to the heat equation linking temperature, thermal conductivity, density, and specific heat.

For centuries, physicists searched for and discovered the causes of the effects of the geophysical processes they observed, so that they solved simplified inverse problems. Inverse problems, as formalised mathematical studies, have been initiated in the twentieth century (e.g. Weyl, 1911). These problems allow for determining model parameters or specific model conditions that cannot be directly observed. Inverse problems can be subdivided into time-reverse or retrospective problems (e.g. to determine initial conditions in the past and/or to restore the development of a dynamic process), coefficient problems (e.g. to determine the coefficients of the model and/or boundary conditions), geometrical problems (e.g. to determine the location of heat sources in a model domain or the geometry of the model boundary), and some others.

1.2 Inverse Problems and Well-Posedness

The idea of well- (and ill-)posedness in the theory of partial differential equations was introduced by Hadamard (1902). A mathematical model of a geophysical problem is *well-posed* if (i) a solution to this problem exists, and the solution is (ii) unique and (iii) stable. Problems for which at least one of these three properties does not hold are called *ill-posed*. Existence of the problem's solution is normally proven by mathematicians, at least in the simplest cases. Meanwhile, if the solution exists, it may not be unique, allowing for multiple theoretical solutions, as in the case of potential-field inter-pretation. The non-uniqueness of potential-field studies is associated with the general topic of scientific uncertainty in the Earth sciences, because problems are generally addressed with incomplete and imprecise data (Saltus and Blakely, 2011). The requirement of stability is the most important in numerical modelling. If a problem lacks the property of stability, then its solution is almost impossible to compute, because computations are polluted by unavoidable errors. If the solution of a problem does not depend continuously on the initial data, then, in general, the computed solution may have nothing to do with the true solution.

Inverse problems are often ill-posed. For example, the retro-spective (inverse) problem of thermal convection is an ill-posed problem, since the backward heat problem, describing both heat advection and conduction backwards in time, pos-sesses the property of instability (e.g. Kirsch, 1996). In particu-lar, the solution to the problem does not depend continuously on the initial data. This means that small changes in the present-day temperature field may result in large changes of

predicted temperatures in the past. Following Ismail-Zadeh et al. (2016), this statement is explained using a simple problem related to the one-dimensional (1-D) diffusion equation (Example 1).

Although inverse problems are quite often unstable and hence ill-posed, there are some methods for solving them. The idea of conditionally well-posed problems and the regularisation method were introduced by Tikhonov (1963). According to Tikhonov, a class of admissible solutions to conditionally ill-posed problems should be selected to satisfy the following conditions: (i) a solution exists in this class, (ii) the solution is unique in the same class, and (iii) the solution depends continu-ously on the input data (i.e. the solution is stable). The Tikhonov regularisation is essentially a trade-off between fitting the observations and reducing a norm of the solution to the mathematical model of a geophysical problem. We show the differences between the Hadamard's and Tikhonov's approaches to ill-posed problems in Example 2.

Example 1

Consider the following problem for the 1-D backward diffusion equation

$$\partial u(t,x)/\partial t = \partial^2 u(t,x)/\partial x^2,\ 0 \le x \le \pi,\ t < 0, \tag{1.1}$$

with the following boundary and initial conditions

$$u(t,0) = 0 = u(t,\pi),\ t \le 0,\ u(0,x) = \phi_n(x),\ 0 \le x \le \pi. \tag{1.2}$$

At the initial time, the function $\phi_n(x)$ is assumed to take the following two forms:

$$\phi_n(x) = \frac{\sin((4n+1)x)}{4n+1}\ \text{and}\ \phi_0(x) \equiv 0. \tag{1.3}$$

We note that

$$\max_{0 \le x \le \pi} |\phi_n(x) - \phi_0(x)| \le \frac{1}{4n+1} \to 0\ \text{at}\ n \to \infty. \tag{1.4}$$

The two solutions of the problem

$$u_n(t,x) = \frac{\sin((4n+1)x)}{4n+1}\exp(-(4n+1)^2 t)\ \text{at}\ \phi_n(x) = \phi_n\ \text{and} \tag{1.5}$$

$$u_0(t,x) \equiv 0\ \text{at}\ \phi_n(x) = \phi_0 \tag{1.6}$$

correspond to the two chosen functions of $\phi_n(x)$, respectively. At $t = -1$ and $x = \pi/2$

$$u_n\left(-1,\frac{\pi}{2}\right) - u_0\left(-1,\frac{\pi}{2}\right) = \frac{1}{4n+1}\exp((4n+1)^2) \to \infty\ \text{at}\ n \to \infty. \tag{1.7}$$

At large n, two closely set initial functions ϕ_n and ϕ_0 are associated with the two strongly different solutions at $t = -1$ and $x = \pi/2$. Hence, a small error in the initial data (1.4) can result in very large errors in the solution to the backward problem (1.7), and therefore the solution is unstable, and the problem is ill-posed in the sense of Hadamard.

Example 2

Consider the problem for the 1-D backward diffusion equation (like the problem presented in Example 1)

$$\partial u(t,x)/\partial t = \partial^2 u(t,x)/\partial x^2,\ 0 \le x \le \pi,\ -T \le t < 0, \tag{1.11}$$

with the boundary and initial conditions (1.2). The solution of the problem satisfies the inequality

$$\|u(t,x)\| \le \|u(T,x)\|^{-t/T}\|u(0,x)\|^{1+t/T}, \tag{1.12}$$

where the norm is presented as $\|u(t,x)\|^2 \equiv \int_0^\pi u^2(t,x)dx$. We note that the inequality

$$\|u(t,x)\| \le M^{-t/T}\|u_0\|^{1+t/T} \tag{1.13}$$

is valid in the class of functions $\|u(t,x)\| \le M = const$ (Samarskii and Vabischevich, 2007). Inequality (1.13) yields a continuous dependence of the problem's solution on the initial conditions, and hence to well-posedness of the problem in the sense of Tikhonov. Therefore, the Tikhonov approach allows for developing methods for regularisation of the numerical solution of unstable problems.

1.3 Data Assimilation

With a growth of data related to Earth observations and laboratory measurements, the enhancement of data quality and instrumental accuracy, as well as the sophistication of mathematical and numerical models, the assimilation of available information into the models to determine specific states of geophysical/geochemical dynamic processes as accurately as possible becomes an essential tool in solving inverse problems. Data assimilation can be defined as the incorporation of observations and initial/boundary conditions in an explicit dynamic model to provide time continuity and coupling among the physical characteristics of the dynamic model (e.g. Kalnay, 2003; Lahoz et al., 2010; Law et al., 2015; Asch et al., 2016; Ismail-Zadeh et al., 2016; Fletcher, 2017).

Data assimilation has been pioneered by meteorologists and successfully applied to improve operational weather forecasts (e.g. Ghil and Malanotte-Rizzoli, 1991; Kalnay, 2003). To produce forecasts, initial conditions are required for the weather prediction models resembling the current state of the atmosphere. Data assimilation starts with 'unknown' forecasts and applies corrections to the forecast based on observations and estimated errors in the observations and in the forecasts. The difference between the forecast and the observed data at a certain time is assessed using different methods to make new forecast to better fit the observations.

Data assimilation and geophysical inversions have also been widely used in oceanography (e.g. Ghil and Malanotte-Rizzoli, 1991; Bennett, 1992), hydrology (e.g. McLaughlin, 2002), seismology (e.g. Backus and Gilbert, 1968), geodynamics (e.g. Bunge et al., 2003; Ismail-Zadeh et al., 2003; 2004; 2016), geomagnetism (Fournier et al., 2007; Liu et al., 2007), and other Earth science disciplines (e.g. Park and Xu, 2009; Lahoz et al., 2010; Blayo et al., 2014). We note, that depending on the geoscience discipline, data assimilation is also referred to as state estimation, history matching, and data-driven analysis.

1.4 Data Assimilation and Inversions: Basic Approaches and Sensitivity Analysis

Part I of the book introduces basic knowledge and approaches in data assimilation and inversions and presents a high-order sensitivity analysis to obtain best estimate results with reduced uncertainties.

There are two basic approaches to solve inverse problems: classical and Bayesian. The classical approach considers a mathematical model as a true model describing the physical process under study, and geoscientific data as the only available data set with some measurement errors. The goal

of this approach is to recover the true model (e.g. initial or boundary conditions). Another way to treat a mathematical model is the Bayesian approach, where the model is considered as a random variable, and the solution is a probability distribution for the model parameters (e.g. Aster et al., 2005).

Chapter 2 discusses these approaches in more detail. This chapter also provides an accessible general introduction to the breadth of geophysical inversions and presents similarities and connections between different approaches (Valentine and Sambridge, this volume). Chapter 3 introduces the Bayesian data assimilation providing a history of geophysical data assimilation and its current directions (Grudzien and Bocquet, this volume).

All variables in data assimilation models (e.g. state variables describing physical properties, such as velocity, pressure, or temperature; initial and/or boundary conditions; and parameters such as viscosity or thermal diffusivity) can be polluted by errors. The source of errors comes from imperfect measurements and computations. Experimental or calibration standard errors result in measurement errors. Systematic errors in numerical modelling are associated with a mathematical model, its discretisation, and iteration errors. Model errors are associated with the idealisation of Earth system dynamics by a set of conservation equations governing the dynamics. Model errors can be defined as the difference between the actual Earth system dynamics and the exact solution of the mathematical model. Discretisation errors are associated with the difference between the exact solution of the conservation equations and the exact solution of the algebraic system of equations obtained by discretising these equations. And iteration errors are defined as the difference between the iterative and exact solutions of the algebraic system of equations (Ismail-Zadeh and Tackley, 2010). Also, errors can stem from imperfectly known physical processes or geometry. Determining the changes in computed model responses that are induced by variations in the model parameters (e.g. due to errors) is the scope of sensitivity analysis, which is linked to the stability of systems to small errors.

Sensitivity analysis assists in understanding the stability of the model solution to small perturbations in input variables or parameters. For instance, consider the thermal convection in the Earth's mantle. If the temperature in the geological past is determined from the solution of the backward thermal convection problem using present mantle temperature assimilated to the past, the following question arises: what will be the temperature variation due to small perturbations of the present temperature data? The gradient of the objective functional (representing the misfit between the model and measured temperature) with respect to the present temperature in a variational data assimilation gives the first-order sensitivity coefficients (e.g. Hier-Majumder et al., 2006). A second-order adjoint sensitivity analysis presents some challenges associated with cumbersome computations of the product of the Hessian matrix of the objective functional with a vector (Le Dimet et al., 2002).

Chapter 4 discusses higher-order sensitivity and uncertainty analysis to obtain best estimates with reduced uncertainties. The analysis is applied to an inverse radiation transmission problem, to an oscillatory dynamical model, and to a large-scale computational model involving thousands of uncertain parameters. The examples illustrate impacts of the first-, second-, and third-order response sensitivities to parameters on the expectation, variance, and skewness of the respective model responses (Cacuci, this volume).

1.5 Data Assimilation and Inverse Problems in 'Fluid' Earth Sciences

Part II of the book is dedicated to applications of data assimilation and inversions to problems related to the cryosphere, hydrosphere, atmosphere, and near-Earth environment ('fluid' Earth spheres).

Estimates of seasonal snow often bear significant uncertainties (e.g. due to error-prone forcing data and parameter estimates), and data assimilation becomes a useful tool to minimise inherent limitations that result from the uncertainty. Chapter 5 reviews current snow models, snow remote sensing methods, and data assimilation techniques that can reduce uncertainties in the characterisation of seasonal snow (Girotto et al., this volume). Although some properties at the surface of glaciers and ice sheets can be measured from remote sensing or in-situ observations, other characteristics, such as englacial and basal properties, or past climate conditions, remain difficult or impossible to observe. Data assimilation in glaciology assists in inferring unknown properties and boundary conditions to be employed in numerical models (Morlighem and Goldberg, this volume). Chapter 6 presents common applications of data assimilation in glaciology, and some of the new directions that are currently being developed.

Data assimilation in many hydrological problems shares distinct peculiarities: scarce or indirect observation of important state variables (e.g. soil moisture, river discharge, groundwater level), very incomplete or largely conceptual modelling, extreme spatial heterogeneity, and uncertainty on controlling physical parameters (Castelli, this volume). Chapter 7 discusses the peculiarities of data assimilation for state estimation and model inversion in hydrology related to the following applications: soil moisture, runoff for flood and inundation prediction, static geophysical inversion in

groundwater modelling, and dynamic geophysical inversion in coupled surface water and energy balance.

Robust estimates of trace gas emissions that impact air quality and climate provide important knowledge for informed decision-making. Better monitoring and increasing data availability due to expanding observing networks provide information on the changing composition of the atmosphere. Chapter 8 discusses the use of various inverse modelling approaches to quantify emissions of environmentally important trace gases, with a focus on the use of satellite observations. It presents the inverse problem of retrieving the atmospheric trace gas information from the satellite measurements, and the subsequent use of these satellite data for inverse modelling of sources and sinks of the trace gases (Jones, this volume).

Models of volcanic cloud propagation due to volcanic eruptions assist in operational forecasts and provide invaluable information for civil protection agencies and aviation authorities during volcanic crises. Quantitative operational forecasts are challenging due to the large uncertainties that typically exist when characterising volcanic emissions in real time, and data assimilation assists in reduction of quantitative forecast errors (Folch and Mingari, this volume). Chapter 9 reviews state-of-the-art in data assimilation of volcanic clouds and its use in operational forecasts.

Energetic charged particles trapped by the Earth magnetic field present a significant hazard for Earth orbiting satellites and humans in space, and data assimilation helps to reconstruct the global state of the radiation particle environment from observations (Shprits et al. this volume). Chapter 10 describes recent studies related to data assimilation in the near-Earth electron radiation environment. Applications to the reanalysis of the radiation belts and ring current, real-time predictions, and analysis of the missing physical processes are discussed in the chapter.

1.6 Data Assimilation and Inverse Problems in Solid Earth Sciences

Part III presents methods and applications of data assimilation and inversions in problems of the solid Earth sciences: geochronology, volcanology, seismology, gravity, geodesy, geodynamics, and geomagnetism.

Chapter 11 presents applications of inverse methods, namely, trans-dimensional Markov chain Monte Carlo, to geochronological and thermochronological data to identify the number of potential source components for detrital material in sedimentary basins and to extract temperature histories of rocks over geological time (Gallagher, this volume).

Lava dome growth and lava flow are two main manifestations of effusive volcanic eruptions. Chapter 12 discusses inverse problems related to lava dynamics. One problem is related to a determination of the thermal state of a lava flow from thermal measurements at its surface using a variational data assimilation method. Another problem aims to determine magma viscosity by comparison of observed and simulated lava domes employing artificial intelligence methods (Ismail-Zadeh et al., this volume).

Chapter 13 deals with data assimilation for real-time shake-mapping and ground shaking forecasts to assist in earthquake early warning systems. The current seismic wavefield is rapidly estimated using data assimilation, and then the future wavefield is predicted based on the physics of wave propagation (Hoshiba, this volume).

Global seismic tomography using time domain waveform inversion is overviewed in Chapter 14 in the context of imaging the Earth's whole mantle at the global scale. The chapter discusses how the tomography problem is addressed, data selection approaches, definitions of the misfit function, and computation of kernels for the inverse step of the imaging procedure, as well as the choice of optimisation method (Romanowicz, this volume). The diversity of seismic inverse problems – in terms of scientific scope, spatial scale, nature of the data, and available resources – precludes the existence of a silver bullet to solve the problems. Chapter 15 describes smart methods for solving the inverse problems, which increase computational efficiency and usable data volumes, sometimes by orders of magnitude (Gebraad et al., this volume).

Chapter 16 deals with joint inversions as a hypothesis testing tool to study the Earth's subsurface. It presents an application of joint inversions of gravity and magnetic data with seismic constraints in the western United States. As a result of the joint inversions, high velocity structures in the crust are found to be associated with relatively low-density anomalies, potentially indicating the presence of melt in a rock matrix (Moorkamp, this volume). An application of gravity inversion of Bouguer anomalies is presented in Chapter 17 focusing on the Moho depth and crustal density structure in the Tibetan Plateau. The inversion results clearly recognise a thick Tibetan crust and Moho depths of more than 60 km (Jin and Xuan, this volume).

Chapter 18 describes geodetic inversions and applications in geodynamics. Rapid development of the Global Navigation Satellite Systems (GNSS) allows enhanced geodynamic studies providing information about global-scale plate motions, plate boundary deformation, seismo-tectonic deformation, volcanology, postglacial isostatic rebound, ice flow, and water mass flow. A geophysical interpretation of GNSS observations is based on rheological models used to predict surface motions related to various tectonic processes and the corresponding inversion technique permitting us to separate the processes and to evaluate their parameters (Steblov and Vladimirova, this volume).

In Chapter 19, basic methods for data assimilation used in geodynamic modelling are described: backward advection method, variational (adjoint) method, and quasi-reversibility method. To demonstrate the applicability of the methods, two models are considered: a model of restoring prominent mantle plumes from their diffused stage, and a model of reconstruction of the thermal state of the mantle beneath the Japanese islands and their surroundings during forty million years. Also, this chapter discusses challenges, advantages, and disadvantages of the data assimilation methods (Ismail-Zadeh, Tsepelev, and Korotkii, this volume). Chapter 20 deals with global mantle convection in the Earth. Variational data assimilation allows for retrodicting past states of the Earth's mantle as optimal flow histories relative to the current state. Poorly known mantle flow parameters, such as rheology and composition, can be then tested against observations extracting information from the geologic records (Bunge et al., this volume).

Chapter 21 presents geomagnetic data assimilation, which aims to optimally combine geomagnetic observations and numerical geodynamo models to better estimate the dynamic state of the Earth's outer core and to predict geomagnetic secular variation. It provides a comprehensive overview of recent advances in the field, as well as some of the immediate challenges of geomagnetic data assimilation, possible solutions, and pathways to move forward (Kuang et al., this volume). Chapter 22 introduces main characteristics of geomagnetic data and magnetic field models and explores the role of model and observation covariances and localisation in typical assimilation setups, focusing on the use of three-dimensional dynamo simulations as the background model (Sanchez, this volume).

Conclusion

Inverse problems and data assimilation in Earth sciences provide many benefits to science and society. Mathematical and numerical models and methods involved in solving inverse problems and in assimilating data assist in utilisation of Earth observations and add value to the observations, for example, providing insight into physical/chemical processes and their observed manifestations. At the same time, observed and measured data help to constrain and sophisticate models and hence provide more reliable model estimates and forecasts. Society benefits from the knowledge obtained by using scientific products such as weather, air quality, space weather, and other forecasts. Applications of inverse problems and data assimilation in Earth sciences are broad, and this book covers only a part of them, including applications in atmospheric, cryospheric, geochronological, geodetical, geomagnetic, hydrological, seismological, and volcanological sciences. It presents the basics of modern theory and how theoretical methods works to decipher the puzzles of nature.

References

Asch, M., Bocquet, M., and Nodet, M. (2016). *Data Assimilation: Methods, Algorithms, and Applications*. Philadelphia, PA: Society of Industrial and Applied Mathematics.

Aster, R. C., Borchers, B., and Thurber C. H. (2005). *Parameter Estimation and Inverse Problems*, vol. 90. San Diego, CA: Elsevier.

Backus, G. E., and Gilbert, F. (1968). The resolving power of gross Earth data. *Geophysical Journal of the Royal Astronomical Society*, 16, 169–205.

Bennett, A. F. (1992). *Inverse Methods in Physical Oceanography*. Cambridge: Cambridge University Press.

Blayo, E., Bocquet, M., Cosme, E., and Cugliandolo, L. F., eds. (2014). *Advanced Data Assimilation for Geosciences*. Oxford: Oxford University Press.

Bunge, H.-P., Hagelberg, C. R., and Travis, B. J. (2003). Mantle circulation models with variational data assimilation: Inferring past mantle flow and structure from plate motion histories and seismic tomography. *Geophysical Journal International*, 152, 280–301.

Fletcher, S. J. (2017). *Data Assimilation for the Geosciences: From Theory to Application*. Amsterdam: Elsevier.

Fournier, A., Eymin, C., and Alboussiere, T. (2007). A case for variational geomagnetic data assimilation: Insights from a one-dimensional, non-linear, and sparsely observed MHD system. *Nonlinear Processes in Geophysics*, 14(3), 163–80.

Ghil, M., and Malanotte-Rizzoli, P. (1991). Data assimilation in meteorology and oceanography. *Advances in Geophysics*, 33, 141–266.

Hadamard, J. (1902). Sur les problèmes aux dérivées partielles et leur signification physique. *Princeton University Bulletin*, XIII (4), 49–52.

Hier-Majumder, C. A., Travis, B. J., Belanger, E. et al. (2006). Efficient sensitivity analysis for flow and transport in the Earth's crust and mantle. *Geophysical Journal International*, 166, 907–22.

Ismail-Zadeh, A., and Tackley, P. (2010). *Computational Methods for Geodynamics*. Cambridge: Cambridge University Press.

Ismail-Zadeh, A. T., Korotkii, A. I., Naimark, B. M., and Tsepelev, I. A. (2003). Three-dimensional numerical simulation of the inverse problem of thermal convection. *Computational Mathematics and Mathematical Physics*, 43(4), 587–99.

Ismail-Zadeh, A., Schubert, G., Tsepelev, I., and Korotkii, A. (2004). Inverse problem of thermal convection: Numerical approach and application to mantle plume restoration. *Physics of the Earth and Planetary Interiors*, 145, 99–114.

Ismail-Zadeh, A., Korotkii, A., and Tsepelev, I. (2016). *Data-Driven Numerical Modeling in Geodynamics: Methods and Applications*. Heidelberg: Springer.

Kabanikhin, S.I. (2011). *Inverse and Ill-Posed Problems. Theory and Applications*. Berlin: De Gruyter.

Kalnay, E. (2003). *Atmospheric Modeling, Data Assimilation and Predictability*. Cambridge: Cambridge University Press.

Kirsch, A. (1996). *An Introduction to the Mathematical Theory of Inverse Problems*. New York: Springer.

Lahoz, W., Khattatov, B., and Menard R., eds. (2010). *Data Assimilation: Making Sense of Observations*. Berlin: Springer.

Law, K., Stuart, A., and Zygalakis, K. (2015). *Data Assimilation: A Mathematical Introduction*. Berlin: Springer.

Le Dimet, F.-X., Navon, I. M., and Daescu D. N. (2002). Second-order information in data assimilation. *Monthly Weather Review*, 130, 629–48.

Liu, D., Tangborn, A., and Kuang, W. 2007. Observing system simulation experiments in geomagnetic data assimilation. *Journal of Geophysical Research*, 112. https://doi.org/10.1029/2006JB004691.

McLaughlin, D. (2002). An integrated approach to hydrologic data assimilation: Interpolation, smoothing, and forecasting. *Advances in Water Research*, 25, 1275–86.

Park, S. K., and Xu, L., eds. (2009). *Data Assimilation for Atmospheric, Oceanic and Hydrologic Applications*. Berlin: Springer.

Saltus, R. W., and Blakely, R. J. (2011). Unique geologic insights from 'non-unique' gravity and magnetic interpretation. *GSA Today*, 21(12), 4–11. https://doi.org/10.1130/G136A.1

Samarskii, A. A., and Vabishchevich, P. N. (2007). *Numerical Methods for Solving Inverse Problems of Mathematical Physics*. Berlin: De Gruyter.

Tikhonov, A. N. (1963). Solution of incorrectly formulated problems and the regularization method. *Doklady Akademii Nauk SSSR*, 151, 501–4 (Engl. transl.: Soviet Math. Dokl. 4, 1035–8).

Weyl, H. (1911). Über die asymptotische Verteilung der Eigenwerte: Nachrichten der Königlichen Gesellschaft der Wissenschaften zu Göttingen. Mathematisch-Naturwissenschaftliche Klasse, 110–17.

2
Emerging Directions in Geophysical Inversion

Andrew P. Valentine and Malcolm Sambridge

Abstract: In this chapter, we survey some recent developments in the field of geophysical inversion. We aim to provide an accessible general introduction to the breadth of current research, rather than focusing in depth on particular topics. We hope to give the reader an appreciation for the similarities and connections between different approaches, and their relative strengths and weaknesses.

2.1 Introduction

Geophysics is built upon indirect information. We cannot travel deep into the Earth to directly measure rheological properties, nor journey back through geological time to record the planet's tectonic evolution. Instead, we must draw inferences from whatever observations we can make, constrained as we are to the Earth's surface and the present day. Inevitably, such datasets are sparse, incomplete, and contaminated with signals from many unknown events and processes. We therefore rely on a variety of mathematical, statistical, and computational techniques designed to help us learn from available data. Collectively, these are the tools of 'geophysical inversion', and they lie at the heart of all progress in geophysics.

To achieve this progress, geophysicists have long pioneered – and indeed driven – developments in the mathematical and statistical theory that underpins inference. The acclaimed French mathematician Pierre-Simon Laplace played a central role in our understanding of tidal forcing, developing the theory of spherical harmonics along the way. He is also credited (along with Gauss and Legendre) with the development of the least-squares algorithm and the underpinnings of modern Bayesian statistics – an approach which was subsequently extended and popularised within the physical sciences by Sir Harold Jeffreys (1931, 1939), who is, of course, also well known for his contributions to seismology and solid-earth geophysics (see, e.g., Cook, 1990). Technological developments have also been significant, with (for example) the challenges of handling and processing the huge volumes of data obtained from continuously operating terrestrial and satellite sensor systems stimulating innovation in computational science.

In this chapter, we discuss some current and emerging ideas that we believe to have significance for the broad field of geophysical inversion. In doing so, we aim to not just highlight novelty, but also demonstrate how such 'new' ideas can be connected into the canon of established techniques and methods. We hope that this can help provide insight into the potential strengths and weaknesses of different strategies, and support the interpretation and integration of results obtained using different approaches. Inevitably, constraints of time and space mean that our discussion here remains far from comprehensive; much interesting and important work must be omitted, and our account is undoubtedly biased by our own perspectives and interests. Nevertheless, we hope that the reader is able to gain some appreciation for the current state of progress in geophysical inversion.

In order to frame our discussion, and to enable us to clearly define notation and terminology, we begin with a brief account of the basic concepts of geophysical inversion. For a more in-depth account, readers are encouraged to refer to one of the many textbooks and monographs covering the subject, such as those by Menke (1989), Parker (1994), Tarantola (2005), or Aster et al. (2013).

2.2 Fundamentals

The starting point for any geophysical inversion must be a mathematical description of the earth system of interest. In practical terms, this amounts to specifying some relationship of the form

$$\mathcal{F}[m(\mathbf{x}, t), u(\mathbf{x}, t)] = 0 \qquad (2.1)$$

where $m(\mathbf{x}, t)$ represents some property (or collection of properties) of the Earth with unknown value that may vary across space, $\mathbf{x}$, and/or time, t; and where $u(\mathbf{x}, t)$ represents some quantity (or collection of quantities) that can – at least in principle – be measured or observed. Most commonly in geophysics, $\mathcal{F}$ has the form of an integrodifferential operator. Underpinning Eq. (2.1) will be some set of assumptions, $\mathcal{A}$, although these may not always be clearly or completely enunciated.

2.2.1 The Forward Problem

The fundamental physical theory embodied by Eq. (2.1) may then be used to develop predictions, often via a computational simulation. This invariably involves introducing additional assumptions, $\mathcal{B}$. In particular, it is common to place restrictions on the function m, so that it may be assumed to have properties amenable to efficient computation. For example, it is very common to assert that the function must lie within the span of a finite set of basis functions, $\psi_1, \ldots, \psi_M$, allowing it to be fully represented by a set of M expansion coefficients,

$$m(\mathbf{x}, t) = \sum_{i=1}^{M} m_i \psi_i(\mathbf{x}, t). \qquad (2.2)$$

It is important to recognise that such restrictions are primarily motivated by computational considerations, but may impose certain characteristics – such as a minimum length-scale, or smoothness properties – upon the physical systems that can be represented. Nevertheless, by doing so, we enable Eq. (2.1) to be expressed, and implemented, as a 'forward model'

$$u(\mathbf{x}, t) = \mathcal{G}(\mathbf{x}, t, m), \qquad (2.3)$$

which computes simulated observables for any 'input model' conforming to the requisite assumptions. Typically, the function $\mathcal{G}$ exists only in the form of a numerical computer code, and not as an analytical expression in any meaningful sense. As a result, we often have little concrete understanding of the function's global behaviour or properties, and the computational cost associated with each function evaluation may be high.

2.2.2 Observational Data and the Inverse Problem

We use $\mathbf{d}$ to represent a data vector, with each element d_i representing an observation made at a known location in space and time, $(\mathbf{x}_i, t_i)$. This is assumed to correspond to $u(\mathbf{x}_i, t_i)$, corrupted by 'noise' (essentially all processes not captured within our modelling assumptions, $\mathcal{A} \cup \mathcal{B}$), any limitations of the measurement system itself, and any preprocessing (e.g. filtering) that has been applied to the dataset. We address the latter two factors by applying transformations (e.g. equivalent preprocessing and filters designed to mimic instrument responses) to the output of our forward model; mathematically, this amounts to composing $\mathcal{G}$ with some transfer function $\mathcal{T}$. For notational convenience, we define a new function, $\mathbf{g}$, which synthesises the entire dataset $\mathbf{d}$: $[\mathbf{g}(m)]_i = \mathcal{T} \circ \mathcal{G}(\mathbf{x}_i, t_i, m)$. We also introduce the concept of a data covariance matrix, $\mathbf{C_d}$, which encapsulates our assumptions about the uncertainties and covariances within the dataset. The fundamental goal of inversion is then to find – or somehow characterise – m such that $\mathbf{g}(m)$ matches or explains $\mathbf{d}$.

Since $\mathbf{d}$ contains noise, we do not expect any model to be able to reproduce the data perfectly. Moreover, the forward problem may be fundamentally non-unique: it may generate identical predictions for two distinct models. As such, there will typically be a range of models that could be taken to 'agree with' observations. We must therefore make a fundamental decision regarding the approach we wish to take. We may:

1. Seek a single model, chosen to yield predictions that are 'as close as possible' to the data, usually with additional requirements that impose characteristics we deem desirable and ensure that a unique solution exists to be found, e.g. that the model be 'as smooth as possible';
2. Seek a collection or ensemble of models, chosen to represent the spectrum of possibilities that are compatible with observations – again, perhaps tempered by additional preferences;
3. Disavow the idea of recovering a complete model, and instead focus on identifying specific characteristics or properties that must be satisfied by any plausible model.

In the context of this chapter, we have deliberately framed these three categories to be quite general in scope. Nevertheless, readers may appreciate some specific examples: the first category includes methods based upon numerical optimisation of an objective function, including the familiar least-squares algorithm (e.g. Nocedal and Wright, 1999), while Markov chain Monte Carlo and other Bayesian methods fall within the second (e.g. Sambridge and Mosegaard, 2002); Backus–Gilbert theory (e.g. Backus and Gilbert, 1968) lies within the third. Each of these groups is quite distinct – at least in philosophy – from the others, and in the remainder of this chapter we address each in turn.

2.3 Single Models

Before we can set out to find the model that 'best' explains the data, we must introduce some measure of the agreement between observations and predictions. This 'misfit function' or 'objective function' is of fundamental importance in determining the properties of the recovered model and the efficiency of the solution algorithms that may be available to us. In general, misfit functions take the form

$$\phi(m) = \phi_d(\mathbf{d}, \mathbf{g}(m)) + \phi_m(m), \qquad (2.4)$$

where ϕ_d is a metric defined in the 'data space', measuring how far a model's predictions are from observations, and ϕ_m is a 'regularisation' or 'penalty' term (see Fig. 2.1). This encapsulates any preferences we may have regarding the solution, and aims to ensure that the function ϕ has a unique minimum.

Once a misfit function has been defined, it is conceptually straightforward to search for the model that minimises $\phi(m)$. However, it is often challenging to achieve this in practice. The most complete characterisation of ϕ comes from a grid-search strategy, with systematic evaluation of the function throughout a discretised 'model space' (typically following Eq. 2.2). This is viable for small problems, and is commonly encountered in the geophysical literature (e.g. Sambridge and Kennett, 1986; Dinh and Van der Baan, 2019; Hejrani and Tkalčić, 2020), but the computational costs of evaluating the forward model,

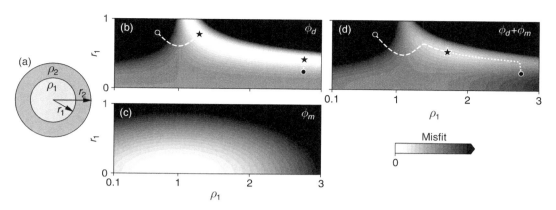

Figure 2.1 Misfit functions for a simple inverse problem (after Valentine and Sambridge, 2020a). (a) A planet is modelled as comprising two spherical layers: a core of radius r_1 and density ρ_1, and an outer unit of density ρ_2 extending to radius r_2. Defining units such that $r_2 = 1$ and $\rho_2 = 1$, we find the overall mass of the planet to be $M = 4.76 \pm 0.25$ units. What can be said about r_1 and ρ_1? (b) The data misfit, $\phi_d(d, g(r_1, \rho_1))$, as in Eq. (2.5), highlighting non-linear behaviours. Two gradient-based optimisation trajectories are shown for different starting points (circles), with convergence to distinct solutions (stars). The inverse problem is inherently non-unique. (c) Penalty term, $\phi_m(r_1, \rho_1)$, expressing a preference for a small core with density similar to that of the surface layer. (d) Combined (regularised) misfit, $\phi_d(d, g(r_1, \rho_1)) + \phi_m(r_1, \rho_1)$. Both optimisation trajectories now converge to the same point.

combined with the 'curse of dimensionality' (Curtis and Lomax, 2001; Fernández-Martínez and Fernández-Muñiz, 2020), rapidly become prohibitive. However, in many cases, it is possible to obtain Fréchet derivatives of the forward problem (Eq. 2.3) with respect to the model, $\delta\mathcal{G}/\delta m$, and this information can be used to guide a search towards the minimum of $\phi(m)$.

2.3.1 Euclidean Data Metrics

Overwhelmingly, the conventional choice for ϕ_d is the squared L_2, or Euclidean, norm of the residuals weighted using the data covariance matrix, $\mathbf{C_d}$,

$$\phi_d(\mathbf{d}, \mathbf{g}(m)) = \left\| \mathbf{C_d}^{-\frac{1}{2}}(\mathbf{d} - \mathbf{g}(m)) \right\|_2^2$$
$$= (\mathbf{d} - \mathbf{g}(m))^{\mathrm{T}} \mathbf{C_d}^{-1}(\mathbf{d} - \mathbf{g}(m)). \quad (2.5)$$

Relying on the Fréchet derivatives is essentially an assumption that $g(m)$ is (locally) linear. For the usual case, where the model has been discretised as in Eq. (2.2) and can be represented as a vector of coefficients, $\mathbf{m}$, we have $\mathbf{g(m)} = \mathbf{g(m_0)} + \mathbf{G(m - m_0)}$, where $\mathbf{m_0}$ is the linearisation point and $[\mathbf{G}]_{ij} = \partial[\mathbf{g(m)}]_i/\partial m_j|_{\mathbf{m}=\mathbf{m_0}}$. We therefore find

$$\frac{\partial \phi}{\partial \mathbf{m}} = 2\mathbf{G^T C_d^{-1}}[\mathbf{G(m - m_0)} - (\mathbf{d} - \mathbf{g(m_0)})] + \frac{\partial \phi_m}{\partial \mathbf{m}}. \quad (2.6)$$

This can be used to define an update to the model, following a range of different strategies. Setting $\mathbf{m} = \mathbf{m_0}$, we obtain the gradient of ϕ with respect to each coordinate direction at the point of linearisation: this information may then be used to take a step towards the optimum, using techniques such as conjugate-gradient methods (as in Bozdağ et al., 2016) or the L-BFGS algorithm of Liu and Nocedal (1989), as

employed by Lei et al. (2020). Alternatively, we can exploit the fact that, at the optimum, the gradient should be zero: for a suitable choice of ϕ_m, it is possible to solve Eq. (2.6) directly for the $\mathbf{m}$ that should minimise the misfit within the linearised regime. This is 'the' least-squares algorithm, employed by many studies (e.g. Wiggins, 1972; Dziewonski et al., 1981; Woodhouse and Dziewonski, 1984). Few interesting problems are truly linear, and so it is usually necessary to adopt an iterative approach, computing a new linear approximation at each step.

2.3.1.1 Stochastic Algorithms

Since the fundamental task of optimising an objective function is also central to modern machine learning efforts, recent geophysical studies have also sought to exploit advances from that sphere. In particular, methods based on 'stochastic gradient descent' have attracted some attention (e.g. van Herwaarden et al., 2020; Bernal-Romero and Iturrarán-Viveros, 2021). These exploit the intuitive idea that the gradient obtained using all available data can be approximated by a gradient obtained using only a subset of the dataset – and that by using different randomly chosen subsets on successive iterations of gradient descent, one may reach a point close to the overall optimum. In appropriate problems, this can yield a substantial reduction in the overall computational effort expended on gradient calculations. It should be noted that the success of this approach relies on constructing approximate gradients that are, on average, unbiased; as discussed in Valentine and Trampert (2016), approximations that induce systematic errors into the gradient operator will lead to erroneous results.

2.3.2 Sparsity

As has been discussed, we commonly assume that a model can be discretised in terms of some finite set of basis functions. Usually, these are chosen for computational convenience, and inevitably there will be features in the real earth system that cannot be represented within our chosen basis. This leads to the problem of 'spectral leakage' (Trampert and Snieder, 1996): features that are unrepresentable create artefacts within the recovered model.

In digital signal processing, the conditions for complete and accurate recovery of a signal are well known. According to Nyquist's theorem, the signal must be band limited and sampled at a rate at least twice that of the highest frequency component present (Nyquist, 1928). Failure to observe this leads to spurious features in the reconstructed signal, known as aliasing – essentially the same issue as spectral leakage. This has far-reaching consequences, heavily influencing instrument design, data collection, and subsequent processing and analysis.

However, recent work has led to the concept of 'compressed sensing' (Donoho, 2006; Candès and Wakin, 2008). Most real-world signals are, in some sense, sparse: when expanded in terms of an appropriately chosen basis (as per Eq. 2.2), only a few non-zero coefficients are required. If data is collected by random sampling, and in a manner designed to be incoherent with the signal basis, exploiting this sparsity allows the signal to be reconstructed from far fewer observations than Nyquist would suggest. The essential intuition here is that incoherence ensures that each observation is sensitive to many (ideally: all) coefficients within the basis function expansion; the principle of sparsity then allows us to assign the resulting information across the smallest number of coefficients possible.

In theory, imposing sparsity should require us to use a penalty term that counts the number of non-zero model coefficients: $\phi_m(\mathbf{m}) = \alpha^2\|\mathbf{m}\|_0$. However, this does not lead to a tractable computational problem. Instead, Donoho (2006) has shown that it is sufficient to penalise the L_1 norm of the model vector, $\phi_m(\mathbf{m}) = \alpha^2\|\mathbf{m}\|_1 = \alpha^2 \sum_i |m_i|$.

This can be implemented using a variety of algorithms, including quadratic programming techniques and the Lasso (Tibshirani, 1996). Costs are markedly higher than for L_2-based penalty functions, but remain tolerable.

Sparsity-promoting algorithms have significant potential: they open up new paradigms for data collection, offering the opportunity to substantially reduce the burden of storing, transmitting, and handling datasets. The success of compressed sensing also suggests that the data misfit $\phi_d(\mathbf{d}, \mathbf{g}(m))$ may be accurately estimated using only a small number of randomly chosen samples: for certain classes of forward model, this may offer a route to substantially reduced computational costs. Again, work is ongoing to explore the variety of ways in which concepts of sparsity can be applied and exploited within the context of geophysical inversion (e.g. Herrmann et al., 2009; Wang et al., 2011; Simons et al., 2011; Bianco and Gerstoft, 2018; Muir and Zhang, 2021).

2.3.3 Non-Euclidean Data Metrics

A common challenge for gradient-based methods is convergence to a local – rather than global – minimum. This situation is difficult to identify or robustly avoid, since doing so would require knowledge of the global behaviour of the forward model. In this context, a particular downside to the use of a Euclidean data norm is that it treats each element of the data vector (i.e. each individual digitised data point) independently. For geophysical datasets, this is often undesirable: the spatial and temporal relationships connecting distinct data points are physically meaningful, and a model that misplaces a data feature (such as a seismic arrival) in time or space is often preferable to one that fails to predict it at all. This problem is particularly familiar in waveform-fitting tasks, where the Euclidean norm is unduly sensitive to any phase differences between data and synthetics. From an optimisation perspective, this can manifest as 'cycle-skipping', where waveforms end up misaligned by one or more complete periods.

As a result, there is interest – and perhaps significant value – in exploring alternative metrics for quantifying the agreement between real and observed data sets. A particular focus of current research is measures built upon the theory of Optimal Transport (e.g. Ambrosio, 2003; Santambrogio, 2015). This focuses on quantifying the 'work' (appropriately defined) required to transform one object into another, and the most efficient path between the two states. In particular, the p-Wasserstein distance between two densities, $f(x)$ and $g(x)$ may be defined

$$W_p(f,g) = \left[\inf_{T \in \mathcal{T}} \int c(x, T(x))^p f(x)\, dx\right]^{1/p}, \qquad (2.7)$$

where $\mathcal{T}$ is the set of all 'transport plans' $T(x)$ that satisfy

$$f(x) = g(T(x))|\nabla T(x)|, \qquad (2.8)$$

and $c(x,y)$ is a measure of the distance between points x and y. The resulting metric provides a much more intuitive measure of the difference between two datasets, and perhaps offers a principled route to combining information from multiple distinct data types (sometimes known as 'joint inversion').

Pioneered in geophysics by Engquist and Froese (2014), this has subsequently been employed for numerous studies, including the work of Métivier et al. (2016a,b,c,d) and others (e.g. He et al., 2019; Hedjazian et al., 2019; Huang et al., 2019). However, numerous challenges remain to be fully overcome. Since Optimal Transport is conceived around density functions – which are inherently positive – signed datasets such as waveforms require special treatment. In addition, since computing the Wasserstein distance between two functions is itself an optimisation problem, there are practical challenges associated with employing it in large-scale inversion problems, and these are the focus of current work.

2.4 Ensemble-Based Methods

We now switch focus, and consider the second fundamental approach to geophysical inversion: instead of seeking a single model that explains the data, we now aim to characterise the collection, or ensemble, of models that are compatible with observations. Clearly, this has potential to be more informative, providing insight into uncertainties and trade-offs; however, it also brings new challenges. Computational costs may be high, and interpretation and decision-making may be complicated without the (illusion of) certainty promised by single-model strategies.

There are many different ways in which one might frame an ensemble-based inversion strategy: at the simplest, one might adapt the grid-search strategy of Section 2.3 so that the 'ensemble' is the set of all grid nodes for which $\phi(m)$ is below some threshold. This approach, with models generated randomly rather than on a grid, underpinned some of the earliest ensemble-based studies in geophysics (e.g. Press, 1970; Anderssen et al., 1972; Worthington et al., 1972). However, it is not particularly convenient from a computational perspective, since such an ensemble has little structure that can be exploited for efficiency or ease of analysis. Techniques exist that seek to address this (e.g. Sambridge, 1998), but the most common strategy is to adopt a probabilistic – and typically Bayesian – perspective. This involves a subtle, but important, change of philosophy: rather than seeking to determine the Earth structure directly, Bayesian inversion aims to quantify our state of knowledge (or 'degree of belief') about that structure (for more discussion, see, e.g., Scales and Snieder, 1997).

The hallmark of Bayesian methods is that the posterior distribution – $\mathcal{P}(m|\mathbf{d})$, the probability of a model m given the observations $\mathbf{d}$ – is obtained by taking the prior distribution ($\mathcal{P}(m)$, our state of knowledge before making any observations), and weighting it by the likelihood, $\mathcal{P}(\mathbf{d}|m)$, which encapsulates the extent to which the data support any given model (see Fig. 2.2a–c). When normalised to give a valid probability distribution, we obtain

$$\mathcal{P}(m|\mathbf{d}) = \frac{\mathcal{P}(\mathbf{d}|m)\,\mathcal{P}(m)}{\mathcal{P}(\mathbf{d})}, \qquad (2.9)$$

which is well known as Bayes' Theorem (Bayes, 1763). We take this opportunity to remark that whereas a misfit function may be chosen in rather ad hoc fashion to exhibit whatever sensitivity is desired, a likelihood has inherent

meaning as 'the probability that the observations arose from a given model', and ought to be defined by reference to the expected noise characteristics of the data. We also highlight the work of Allmaras et al. (2013), which provides a comprehensive but accessible account of the practical application of Bayes' Theorem to an experimental inference problem. However, it is usually challenging to employ Eq. (2.9) directly, since evaluating the 'evidence', $\mathcal{P}(\mathbf{d})$, requires an integral over the space of all allowable models, $\mathcal{M}$,

$$\mathcal{P}(\mathbf{d}) = \int_{\mathcal{M}} \mathcal{P}(\mathbf{d}|m)\mathcal{P}(m)\,\mathrm{d}m, \qquad (2.10)$$

which is not computationally tractable for arbitrary large-scale problems. Instead, most Bayesian studies either make additional assumptions that enable analytic or semi-analytic evaluation of the evidence, or they exploit the fact that the ratio $\mathcal{P}(m_A|\mathbf{d})/\mathcal{P}(m_B|\mathbf{d})$ can be evaluated without knowledge of the evidence to obtain information about the *relative* probability of different models.

2.4.1 Bayesian Least Squares

The choice of prior is central to the success of any Bayesian approach – and also lies at the heart of many controversies and interpretational challenges, largely due to the impossibility of representing the state of no information (e.g. Backus, 1988). It is therefore apparent that within a Bayesian framework all inference is considered relative to a known prior. In principle, the prior should be chosen based on a careful consideration of what is known about the problem of interest; in practice, this is often tempered by computational pragmatism, and a distribution with useful analytic properties is adopted.

2.4.1.1 Gaussian Process Priors

A convenient choice when dealing with an unknown model function $m(\mathbf{x}, t)$ is a Gaussian Process prior,

$$m(\mathbf{x}, t) \sim \mathcal{GP}\Big(\mu(\mathbf{x}, t), k(\mathbf{x}, t, \mathbf{x}', t')\Big). \qquad (2.11)$$

This is essentially the extension of the familiar normal distribution into function space, with our knowledge at any given point, $(\mathbf{x}, t)$, quantified by a mean $\mu(\mathbf{x}, t)$ and standard deviation $k(\mathbf{x}, t; \mathbf{x}, t)^{1/2}$; however, the covariance function k also quantifies our knowledge (or assumptions) about the expected covariances if m were to be measured at multiple distinct points. A comprehensive introduction to the theory of Gaussian Processes may be found in, for example, Rasmussen and Williams (2006).

In some geophysical problems, the data–model relationship is – or can usefully be approximated as – linear (see also Section 2.3.1), and so can be expressed in the form

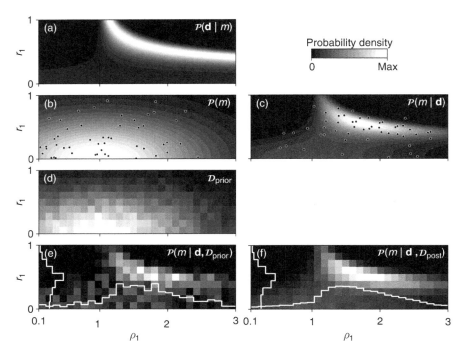

Figure 2.2 Bayesian analysis for the simple inverse problem introduced in Fig. 2.1. (a) The likelihood, $\mathcal{P}(\mathbf{d}|m)$, quantifies the extent to which any given choice of model can explain the data. (b) The prior distribution, $\mathcal{P}(m)$, encapsulates our beliefs before observing any data, and can be 'sampled' to generate a collection of candidate models ($\mathcal{D}_{\mathrm{prior}}$; dots; 50 shown). (c) The posterior distribution, $\mathcal{P}(m|\mathbf{d})$ combines prior and likelihood (Eq. 2.9) to encapsulate our state of knowledge *after* taking account of the data. In realistic problems, visualising the posterior is intractable, but we can generate samples from it ($\mathcal{D}_{\mathrm{post}}$; 50 shown). (d) We can evaluate the forward model $g(m)$ for each example within an ensemble of prior samples, and additionally simulate the effects of noise processes. This can be completed without reference to any data. The information can be stored in many forms, including as a machine learning model. (e) Once data becomes available, this information can be queried to identify regions of parameter space that may explain observations; see Section 2.4.2. This provides an approximation to the posterior; we additionally show 1-D marginals for each model parameter. (f) A similarly sized set of posterior samples provides a much better approximation to the true posterior, as it is targeted towards explaining one specific set of observations: see Section 2.4.3. However, computational costs may be prohibitive for some applications.

$$d_i = \int_0^T \int_{\mathcal{X}} q_i(\mathbf{x}, t)\, m(\mathbf{x}, t)\, \mathrm{d}\mathbf{x}\, \mathrm{d}t, \qquad (2.12)$$

where $q_i(\mathbf{x}, t)$ is some 'data kernel', and where $\mathcal{X}$ represents the domain upon which the model is defined. Moreover, we assume that the noise process represented by $\mathbf{C_d}$ is explicitly Gaussian. These assumptions permit analytic evaluation of the evidence, and the posterior distribution can be written in the form (Valentine and Sambridge, 2020a)

$$\widetilde{m}(\mathbf{x}, t) \sim \mathcal{GP}\left(\widetilde{\mu}(\mathbf{x}, t), \widetilde{k}(\mathbf{x}, t; \mathbf{x}', t')\right), \qquad (2.13)$$

where we use a tilde to denote a posterior quantity, and where

$$\widetilde{\mu}(\mathbf{x}, t) = \mu(\mathbf{x}, t) + \sum_{ij} w_i(\mathbf{x}, t)\left[(\mathbf{W} + \mathbf{C_d})^{-1}\right]_{ij}(d_j - \omega_j),$$
$$(2.14)$$

$$\widetilde{k}(\mathbf{x}, t; \mathbf{x}', t') = k(\mathbf{x}, t; \mathbf{x}', t')$$
$$- \sum_{ij} w_i(\mathbf{x}, t)\left[(\mathbf{W} + \mathbf{C_d})^{-1}\right]_{ij} w_j(\mathbf{x}', t') \quad (2.15)$$

with

$$w_i(\mathbf{x}, t) = \int_0^T \int_{\mathcal{X}} k(\mathbf{x}, t; \mathbf{x}', t') q_i(\mathbf{x}', t') \mathrm{d}\mathbf{x}'\, \mathrm{d}t', \qquad (2.16)$$

$$W_{ij} = \int_0^T \int_0^T \iint_{\mathcal{X}^2} q_i(\mathbf{x}, t)\, k\,(\mathbf{x}, t; \mathbf{x}', t') q_j(\mathbf{x}', t')\, \mathrm{d}\mathbf{x}\, \mathrm{d}\mathbf{x}'\, \mathrm{d}t\, \mathrm{d}t',$$
$$(2.17)$$

$$\omega_i = \int_0^T \int_{\mathcal{X}} \mu(\mathbf{x}, t) q_i(\mathbf{x}, t) \mathrm{d}\mathbf{x}\, \mathrm{d}t. \qquad (2.18)$$

This approach has formed the basis for a variety of geophysical studies (e.g. Tarantola and Nercessian, 1984; Montagner and Tanimoto, 1990, 1991; Valentine and Davies, 2020) and has the attractive property that the inference problem is posed directly in a function space, avoiding some of the difficulties associated with discretisation (such as spectral leakage).

2.4.1.2 Discretised Form

Nevertheless, if one chooses to introduce a finite set of basis functions, as in Eq. (2.2), it is possible to express Eqs. (2.11–2.18) in discretised form (for full discussion, see Valentine and Sambridge, 2020b). The prior distribution on the expansion coefficients becomes

$$\mathbf{m} \sim \mathcal{N}(\mathbf{m_p}, \mathbf{C_m}) \tag{2.19}$$

and the linear data–model relationship is expressed in the form $\mathbf{g(m)} = \mathbf{Gm}$. The posterior distribution may be written in a variety of forms, including

$$\mathbf{m} \sim \mathcal{N}(\widetilde{\mathbf{m}}, \widetilde{\mathbf{C}}_\mathbf{m}), \tag{2.20}$$

where

$$\widetilde{\mathbf{m}} = \mathbf{m_p} + (\mathbf{G^T C_d^{-1} G} + \mathbf{C_m^{-1}})^{-1} \mathbf{G^T C_d^{-1}} (\mathbf{d} - \mathbf{Gm_p}) \tag{2.21}$$

$$\widetilde{\mathbf{C}}_\mathbf{m} = (\mathbf{G^T C_d^{-1} G} + \mathbf{C_m^{-1}})^{-1}. \tag{2.22}$$

This well-known result, found in Tarantola and Valette (1982), has formed the basis of much work in geophysics. The expression for $\widetilde{\mathbf{m}}$ is also often applied in non-Bayesian guise – compare with the discussion in Section 2.3.1 – with the prior covariance matrix $\mathbf{C_m}$ regarded as a generic 'regularisation matrix' without probabilistic interpretation.

2.4.2 Prior Sampling

The results of Section 2.4.1 are built upon assumptions that our prior knowledge is Gaussian and the forward model is linear. This is computationally convenient, but will rarely be an accurate representation of the true state of affairs. Unfortunately, more general assumptions tend not to support analytic expressions for the posterior, and hence it becomes necessary to adopt 'sampling-based methods'. These rely on evaluating the forward problem for a large number of models, in order to accumulate information about the relationship between model and data. Various strategies exist, which can be characterised by the manner in which sampling is performed.

The first group of strategies are those where candidate models are generated according to the prior distribution, and predicted data (potentially including simulated 'noise') is computed for each. This provides a set of samples

$$\mathcal{D}_{\mathrm{prior}} = \{(\mathbf{m}_i, \mathbf{g}(\mathbf{m}_i)), \quad i = 1, ..., N\}, \tag{2.23}$$

which may then be interpolated as necessary to address inversion questions (see Fig. 2.2d–e). This family of approaches is known as 'prior sampling' (Käufl et al., 2016a), with different examples characterised by differing approaches to interpolation. Many of the recent studies that exploit machine learning to perform inversion may be seen within the prior sampling framework, although not all are explicitly Bayesian in design.

2.4.2.1 Mixture Density Networks

If we *do* take a Bayesian approach, then we may note that the density of samples within $\mathcal{D}_{\mathrm{prior}}$ approximates – by construction – the joint probability density, $\mathcal{P}(\mathbf{m}, \mathbf{d})$. If we can fit an appropriate parametric density function to the samples, it is then straightforward to interpolate to obtain the conditional density $\mathcal{P}(\mathbf{m}|\mathbf{d})$ corresponding to observations (which we recognise to be the posterior distribution). One currently popular way to achieve this is to employ Mixture Density Networks (MDNs; Bishop, 1995), which involve an assumption that the conditional distribution can be written as a Gaussian Mixture Model (GMM),

$$\mathcal{P}(m_j|\mathbf{d}) \approx \sum_{k=1}^{K} \frac{w_k(\mathbf{d})}{\sqrt{2\pi\sigma_k^2(\mathbf{d})}} \exp\left(-\frac{\left(m_j - \mu_k(\mathbf{d})\right)^2}{2\sigma_k^2(\mathbf{d})}\right), \tag{2.24}$$

where the weights w_k (which are subject to an additional constraint, $\Sigma_k w_k = 1$), means μ_k, and standard deviations σ_k that define the GMM are assumed to be functions of the data. These relationships may in turn be represented by a neural network. The set of prior samples, $\mathcal{D}_{\mathrm{prior}}$, is then used to optimise the neural network parameters, such that the expected value

$$\mathbb{E}_{\mathcal{D}_{\mathrm{prior}}}\{\mathcal{P}(m_j|\mathbf{d})\} = \frac{1}{N} \sum_{i=1}^{N} \mathcal{P}\left([\mathbf{m_i}]_j | \mathbf{g(m_i)}\right) \tag{2.25}$$

is maximised. This approach has been applied to a variety of geophysical problems, including structural studies at global (e.g. Meier et al., 2007; de Wit et al., 2014) and local (e.g. Earp et al., 2020; Mosher et al., 2021) scales, seismic source characterisation (Käufl et al., 2014), and mineral physics (e.g. Rijal et al., 2021).

2.4.2.2 Challenges and Opportunities

The principal downside to prior sampling – discussed in detail by Käufl et al. (2016a) in the context of MDNs, but applicable more broadly – is the fact that only a few of the samples within $\mathcal{D}_{\mathrm{prior}}$ will provide useful information about any given set of observations. In realistic problems, the range of models encompassed by the prior is large in comparison to the range encompassed by the posterior, and much computational effort is expended on generating predictions that turn out to have little similarity to observations. This is exacerbated by issues associated with the 'curse of dimensionality', motivating the common choice (implicit in our notation for Eq. 2.24) to use prior sampling to infer low or uni-dimensional marginal distributions rather than the full posterior. Overall, the consequence is that prior sampling tends to yield rather broad posteriors, representing 'our state of knowledge in the light of the simulations we have performed', rather than 'the most we can hope to learn from the available data'. We also emphasise that results are wholly dependent on the choice of prior, and will be meaningless if this does not encompass the real earth system. This is perhaps obvious in an explicitly Bayesian context, but may be lost when studies are framed primarily from the perspective of machine learning.

The great benefit of prior sampling is that nearly all of the computational costs are incurred *before* any knowledge of observed data is required. As a result, it may be effective in situations where it is desirable to obtain results as rapidly or cheaply as possible following data collection – for example, to enable expensive numerical wave propagation simulations to be employed for earthquake early warning (Käufl et al., 2016b). We note parallels here to the use of scenario-matching approaches in the field of tsunami early warning (e.g. Steinmetz et al., 2010). It is also well-suited to applications where the same fundamental inverse problem must be solved many times for distinct datasets, perhaps representing observations repeated over time, or at many localities throughout a region.

Prior sampling may also be effective in settings requiring what we term 'indirect' inference, where the primary goal is to understand some quantity derived from the model, rather than the model itself. For example, in an earthquake early warning setting, one might seek to determine seismic source information with a view to then using this to predict tsunami run-up, or the peak ground acceleration at critical infrastructure sites (Käufl, 2015). In a prior sampling setting, one may augment $\mathcal{D}_{\text{prior}}$ to incorporate a diverse suite of predictions, $\mathcal{D}_{\text{prior}} = \{(\mathbf{m}_i, \mathbf{g_1}(\mathbf{m}_i), \mathbf{g_2}(\mathbf{m}_i), \ldots), \ i = 1, \ldots, N\}$, and then employ some interpolation framework to use observations of the process associated with (say) $\mathbf{g_1}$ to make inferences about $\mathbf{g_2}$. From a Bayesian perspective, this can be seen as a process of marginalisation over the model parameters themselves.

2.4.3 Posterior Sampling

As an alternative to prior sampling, one may set out to generate a suite of samples, $\mathcal{D}_{\text{post}}$, distributed according to the posterior (see Fig. 2.2f). Again, there are a variety of ways this can be achieved – for example, a simple (but inefficient) approach might involve rejection sampling. More commonly, Markov chain Monte Carlo (McMC) methods are employed, with the posterior forming the equilibrium distribution of a random walk. Encompassed within the term McMC lie a broad swathe of algorithms, of which

the Metropolis–Hastings is probably most familiar, and the field is continually the subject of much development. We do not attempt to survey these advances, but instead direct the reader to one of the many recent reviews or tutorials on the topic (e.g. Brooks et al., 2011; Hogg and Foreman-Mackey, 2018; Luengo et al., 2020).

As set out in Käufl et al. (2016a), prior and posterior sampling procedures generate identical results in the theoretical limit. However, in practical settings they are suited to different classes of problems. Posterior sampling approaches are directed towards explaining a specific dataset: this allows computational resources to be targeted towards learning the specifics of the problem at hand, but prevents expensive simulations from being 'recycled' in conjunction with other datasets. It should also be noted that the 'solution' obtained via posterior sampling takes the form of an ensemble of discrete samples. This can be challenging to store, represent, and interrogate in a meaningful way: many studies resort to reducing the ensemble to a single maximum-likelihood or mean model, and perhaps some statistics about the (co)variances associated with different parameters, and thereby neglect much of the power of McMC methods. Effective solutions to this issue may be somewhat problem dependent, but remain the focus of much work.

2.4.3.1 *Improving Acceptance Ratios*

Generation of ensembles of posterior samples is inherently wasteful: by definition, one does not know in advance where samples should be placed, and hence for every 'useful' sample, a large numbers of candidate models must be tested (i.e. we must evaluate the forward problem) and rejected. This is exacerbated by requirements for 'burn-in' (so that the chain is independent of the arbitrary starting point) and 'chain thinning' (to reduce correlations between consecutive samples), which also cause substantial numbers of samples to be discarded. Much effort is therefore expended on developing a variety of strategies to improve 'acceptance ratios' (i.e. the proportion of all tested models that end up retained within the final ensemble).

One route forward involves improving the 'proposal distribution', that is, the manner in which samples are generated for testing. Ideally, we wish to make the proposal distribution as close as possible to the posterior, so that nearly all samples may be retained. Of course, the difficulty in doing so is that the posterior is not known in advance. An avenue currently attracting considerable interest is Hamiltonian McMC (HMC) methods, which exploit analogies with Hamiltonian dynamics to guide the random walk process towards 'acceptable' samples (see, e.g., Neal, 2011; Betancourt, 2017). In order to do so, HMC methods require, and exploit, knowledge of the gradient of the likelihood with respect to the model parameters at each sampling point. This provides additional information about the underlying physical problem, enabling extrapolation away from the sample point, and the identification of 'useful' directions for exploration. To apply this idea, we must be able to compute the required gradients efficiently; early applications in geophysics have included seismic exploration and full-waveform inversion (e.g. Sen and Biswas, 2017; Fichtner et al., 2019; Aleardi and Salusti, 2020).

In many cases, the fundamental physical problem of Eq. (2.1) is amenable to implementation (Eq. 2.3) in a variety of ways, depending on the assumptions made ($\mathcal{B}$). Usually, simplified assumptions lead to implementations with lower computational costs, at the expense of introducing systematic biases into predictions. Recently, Koshkholgh et al. (2021) has exploited this to accelerate McMC sampling, by using a low-cost physical approximation to help define a proposal distribution. Likelihood evaluations continue to rely on a more complex physical model, so that accuracy is preserved within the solution to the inverse problem – but the physically motivated proposal distribution improves the acceptance rate and reduces overall computational costs. This is an attractive strategy, and seems likely to underpin future theoretical developments.

2.4.3.2 Transdimensional Inference

In practice, McMC studies typically assume a discretised model, expressed relative to some set of basis functions in as in Eq. (2.2), and the choice of basis functions is influential in determining the characteristics of the solution. In particular, the number of terms in the basis function expansion typically governs the flexibility of the solution, and the scale-lengths that can be represented. However, it also governs the dimension of the search space: as the number of free parameters in the model grows, so does the complexity (and hence computational cost) of the Monte Carlo procedure. Transdimensional approaches arise as an attempt to strike a balance between these two competing considerations: both basis set and expansion coefficients are allowed to evolve during the random walk process (Green, 1995; Sambridge et al., 2006; Bodin and Sambridge, 2009; Sambridge et al., 2012).

The transdimensional idea has been applied to a wide variety of geoscience problems, including source (e.g. Dettmer et al., 2014) and structural (Burdick and Lekić, 2017; Galetti et al., 2017; Guo et al., 2020) studies using seismic data, in geomagnetism (Livermore et al., 2018), and in hydrology (Enemark et al., 2019). It can be particularly effective in settings where basis functions form a natural hierarchy of scale lengths, such as with wavelets and spherical harmonics, although keeping track of information creates computational challenges (Hawkins and Sambridge, 2015). We note that model complexity is not confined only to length-scales: one can also employ a transdimensional approach to the physical theory, perhaps to assess whether mechanisms such as anisotropy are truly mandated by available data. The approach can also be employed to identify change-points or discontinuities within a function (e.g. Gallagher et al., 2011), and used in combination with other techniques such as Gaussian Processes (Ray and Myer, 2019; Ray, 2021).

2.4.4 Variational Methods

One of the drawbacks of posterior sampling is the fact that the sampling procedure must achieve two purposes: it not only 'discovers' the form of the posterior distribution, but also acts as our mechanism for representing the solution (which takes the form of a collection of appropriately distributed samples). Large numbers of samples are often required to ensure stable statistics and 'convincing' figures, even if the underlying problem itself is rather simple. To address this, one may introduce a parametric representation of the posterior distribution, and frame the inference task as determination of the optimal values for the free parameters – much as with the Mixture Density Network (Section 2.4.2.1). This approach, often known as Variational Inference (e.g. Blei et al., 2017), transforms inference for ensembles into an optimisation problem, and offers potentially large efficiency gains.

We sketch the basic concept here, noting that a galaxy of subtly different strategies can be found in recent literature (see, e.g. Zhang et al., 2019, for a review). As usual, our goal is to determine the posterior distribution, $\mathcal{P}(m \,|\, \mathbf{d})$. To approximate this, we introduce a distribution function $\mathcal{Q}(m \,|\, \boldsymbol{\theta})$, that has known form, parameterised by some set of variables $\boldsymbol{\theta}$ – for example, we might decide that Q should be a Gaussian mixture model, in which case $\boldsymbol{\theta}$ would encapsulate the weights, means, and variances for each mixture component. Our basic goal is then to optimise the parameters $\boldsymbol{\theta}$ such that $\mathcal{Q}(m \,|\, \boldsymbol{\theta}) \approx \mathcal{P}(m \,|\, \mathbf{d})$.

To make this meaningful, we must – much as in Section 2.3 – first define some measure of the difference between the two distributions. In Variational Inference, the usual choice is the Kullback–Leibler divergence (Kullback and Leibler, 1951),

$$D_{\mathrm{KL}}(\mathcal{Q}\|\mathcal{P}) = \int \mathcal{Q}(m \,|\, \boldsymbol{\theta}) \log \frac{\mathcal{Q}(m \,|\, \boldsymbol{\theta})}{\mathcal{P}(m \,|\, \mathbf{d})} \, \mathrm{d}m$$

$$= \mathbb{E}_{\mathcal{Q}(m \,|\, \theta)} \left\{ \log \frac{\mathcal{Q}(m \,|\, \boldsymbol{\theta})}{\mathcal{P}(m \,|\, \mathbf{d})} \right\}, \tag{2.26}$$

where the notation $\mathbb{E}_{\mathcal{Q}(m)}\{f(m)\}$ signifies 'the expected value of $f(m)$ when m m is distributed according to $\mathcal{Q}$'. Exploiting the properties of logarithms, and applying Bayes' Theorem, we can rewrite this in the form

$$D_{\mathrm{KL}}(\mathcal{Q}\|\mathcal{P}) = \log \mathcal{P}(\mathbf{d}) + \mathbb{E}_{\mathcal{Q}(m \,|\, \theta)} \{ \log \mathcal{Q}(m | \boldsymbol{\theta}) - \log \mathcal{P}(\mathbf{d} | m) $$
$$ - \log \mathcal{P}(m) \}, \tag{2.27}$$

where $\mathcal{P}(\mathbf{d})$ has been moved outside the expectation since it is independent of m. While this quantity is unknown, it is also constant – and so can be neglected from the perspective of determining the value of θ at which D_{KL} is minimised. The quantity $\mathcal{P}(\mathbf{d}) - D_{KL}(Q\|\mathcal{P})$ is known as the 'evidence lower bound' (ELBO), and maximisation of this is equivalent to minimising the Kullback–Leibler divergence. Because the variational family $Q(m\,|\,\theta)$ has a known form, the ELBO can be evaluated, as can the derivatives $\partial D_{KL}/\partial \theta_i$. Thus, it is conceptually straightforward to apply any gradient-based optimisation scheme to determine the parameters such that Q best approximates the posterior distribution.

2.4.4.1 A Gaussian Approximation
To illustrate this procedure, and to highlight connections to other approaches, we consider an inverse problem where: (i) the model is discretised, as in Eq. (2.2), so that we seek an M-component model vector $\mathbf{m}$; (ii) the prior distribution on those model coefficients is Gaussian with mean $\mathbf{m_p}$ and covariance $\mathbf{C_m}$; and (iii) the likelihood takes the form $\mathcal{P}(\mathbf{m}\,|\,\mathbf{d}) = k\,\exp(-\frac{1}{2}\phi(\mathbf{m}))$ for some appropriate function ϕ. We choose to assert that the solution can be approximated by a Gaussian of mean $\boldsymbol{\mu}$ and covariance matrix $\boldsymbol{\Sigma}$, and seek the optimal values of these quantities. Thus, we choose

$$Q(\mathbf{m}\,|\,\boldsymbol{\mu}, \boldsymbol{\Sigma}) = \frac{1}{(2\pi)^{M/2}(\det \boldsymbol{\Sigma})^{1/2}} \exp\{-(\mathbf{m}-\boldsymbol{\mu})^{\mathsf{T}}\boldsymbol{\Sigma}^{-1}(\mathbf{m}-\boldsymbol{\mu})\}. \tag{2.28}$$

To proceed, we need to determine the expectation of various functions of $\mathbf{m}$ under this distribution – and their gradients with respect to $\boldsymbol{\mu}$ and $\boldsymbol{\Sigma}$.

A number of useful analytical results and expressions can be found in Petersen and Pedersen (2012). It is straightforward to determine that

$$\frac{\partial}{\partial \boldsymbol{\mu}} D_{KL}(Q\|\mathcal{P}) = \mathbf{C_m^{-1}}(\boldsymbol{\mu}-\mathbf{m_p}) + \frac{1}{2}\frac{\partial}{\partial \boldsymbol{\mu}}\mathbb{E}_Q\{\phi(\mathbf{m})\}, \tag{2.29}$$

$$\frac{\partial}{\partial \boldsymbol{\Sigma}} D_{KL}(Q\|\mathcal{P}) = -\frac{1}{2}(\boldsymbol{\Sigma}^{-1}-\mathbf{C_m^{-1}}) + \frac{1}{2}\frac{\partial}{\partial \boldsymbol{\Sigma}}\mathbb{E}_Q\{\phi(\mathbf{m})\}. \tag{2.30}$$

These expressions can be used to drive an iterative optimisation procedure to determine the optimal variational parameters. In implementing this, the result

$$\frac{\partial}{\partial \theta_i}\mathbb{E}_{Q(m\,|\,\theta)}\{f[m]\} = \mathbb{E}_{Q(m\,|\,\theta)}\left\{f(m)\frac{\partial}{\partial \theta_i}\log Q(m\,|\,\theta)\right\} \tag{2.31}$$

may be useful.

In the case where $\mathbf{g}(\mathbf{m})$ is (or is assumed to be) linear, and where the function ϕ is defined as the L_2 norm of the residuals, the expected value can be evaluated analytically. The misfit is quadratic in form,

$$\phi(\mathbf{m}) = \mathbf{d^{\mathsf{T}}C_d^{-1}d} - 2\mathbf{d^{\mathsf{T}}C_d^{-1}Gm} + \mathbf{m^{\mathsf{T}}G^{\mathsf{T}}C_d^{-1}Gm}, \tag{2.32}$$

and the Gaussian Q, can be determined, along with its derivatives, as in Section 2.4.1.1. Hence the expected value, given m is distributed according to the Gaussian Q, can be determined, along with its derivatives

$$\mathbb{E}_Q\{\phi(\mathbf{m})\} = -2\mathbf{d^{\mathsf{T}}C_d^{-1}G\boldsymbol{\mu}} + \mathrm{Tr}(\mathbf{G^{\mathsf{T}}C_d^{-1}G\boldsymbol{\Sigma}}) + \boldsymbol{\mu}^{\mathsf{T}}\mathbf{G^{\mathsf{T}}C_d^{-1}G\boldsymbol{\mu}} \tag{2.33}$$

$$\frac{\partial}{\partial \boldsymbol{\mu}}\mathbb{E}_Q\{\phi(\mathbf{m})\} = -2\mathbf{G^{\mathsf{T}}C_d^{-1}d} + 2\mathbf{G^{\mathsf{T}}C_d^{-1}G\boldsymbol{\mu}}, \tag{2.34}$$

$$\frac{\partial}{\partial \boldsymbol{\Sigma}}\mathbb{E}_Q\{\phi(\mathbf{m})\} = \mathbf{G^{\mathsf{T}}C_d^{-1}G}. \tag{2.35}$$

Substituting these expressions into Eqs. (2.29–2.30), and solving for the $\boldsymbol{\mu}$ and $\boldsymbol{\Sigma}$ such that the gradients of D_{KL} are zero (as is required at a minimum), we find that the optimal distribution Q is identical to the posterior distribution obtained in Eq. (2.20). This is unsurprising, since our underlying assumptions are also identical, but demonstrates the self-consistency of, and connections between, the different approaches. Again, we also highlight the similarity with the expressions obtained in Section 2.3.1, although the underlying philosophy differs.

2.4.4.2 Geophysical Applications
Variational methods offer a promising route to flexible but tractable inference. As the preceding example illustrates, they provide opportunities to balance the (assumed) complexity and expressivity of the solution against computational costs. A number of recent studies have therefore explicitly sought to explore their potential in particular applications, including for earthquake hypocentre determination (Smith et al., 2022), seismic tomography (Zhang and Curtis, 2020; Siahkoohi and Herrman, 2021; Zhao et al., 2022), and hydrogeology (Ramgraber et al., 2021). However, given the fairly broad ambit of variational inference, many past studies could also be seen as falling under this umbrella.

2.4.5 Generative Models
Many of the methods discussed so far rely on strong assumptions about the form of prior and/or posterior distributions: we suppose that these belong to some relatively simple family, with properties that we can then exploit for efficient calculations. However, such assertions are typically justified by their convenience – perhaps aided by an appeal to the principle known as Occam's Razor – and not through any fundamental physical reasoning (see, e.g., Constable

et al., 1987). This is unsatisfactory, and may contribute substantial unquantifiable errors into solutions and their associated uncertainty estimates.

Recently, a number of techniques have emerged that allow representation of, and computation with, relatively general probability distributions. In broad terms, these are built upon the idea that arbitrarily complex probability distributions can be constructed via transformations of simpler distributions. This is familiar territory: whenever we need to generate normally distributed random numbers, a technique such as the Box–Muller transform (Box and Muller, 1958) is applied to the uniformly distributed output of a pseudorandom number generator. However, the versatility of such approaches is vastly increased in conjunction with the tools and techniques of modern machine learning.

This is an area that is currently the focus of rapid development; recent reviews include those of Bond-Taylor et al. (2022) and Ruthotto and Haber (2021). Clearly, the concept is closely connected to the idea of variational inference, as discussed in Section 2.4.4. Several major techniques have emerged, including 'generative adversarial networks' (GANs) (e.g. Goodfellow et al., 2014; Creswell et al., 2018), 'variational autoencoders' (Kingma and Welling, 2014), and 'normalising flows' (Rezende and Mohamed, 2015; Kobyzev et al., 2021). A variety of recent studies have explored diverse applications of these concepts within the context of geophysical inversion: examples include Mosser et al. (2020), Lopez-Alvis et al. (2021), Scheiter et al. (2022), and Zhao et al. (2022). We have no doubt that this area will lead to influential developments, although the precise scope of these is not yet clear.

2.5 Model Properties

The third fundamental approach builds on the work of Backus and Gilbert (1968) and Backus (1970a,b,c), and we sketch it briefly for completeness. For certain classes of problem, as in Eq. (2.12), each of the observables d_i can be regarded as representing an average of the model function weighted by some data kernel $q_i(\mathbf{x}, t)$. It is then straightforward to write down a weighted sum of the observations,

$$D_\alpha = \sum_i \alpha_i d_i = \int_0^T \int_{\mathcal{X}} Q(\boldsymbol{\alpha}, \mathbf{x}, t) m(\mathbf{x}, t) \mathrm{d}\mathbf{x} \, \mathrm{d}t, \quad (2.36)$$

where $Q(\boldsymbol{\alpha}, \mathbf{x}, t) = \sum_i \alpha_i q_i(\mathbf{x}, t)$, and $\boldsymbol{\alpha} = (\alpha_1, \alpha_2 \ldots)$ represents some set of tunable weights. By adjusting these, one may vary the form of the averaging kernel Q, and frame a functional optimisation problem to determine the $\boldsymbol{\alpha}$ that brings Q as close as possible to some desired form. In this way, the value of the average that is sought can be estimated as a linear combination of the observed data.

Backus–Gilbert theory has an inherent honesty: it is data led, with a focus on understanding what the available data can – or cannot – constrain within the system. On the other hand, this can be seen as a downside: it is not usually possible to use the results of a Backus–Gilbert style analysis as the foundation for further simulations. Moreover, interpretation can be challenging in large-scale applications, as the 'meaning' of each result must be considered in the light of the particular averaging kernel found. Perhaps for this reason – and because it is designed for strictly linear problems (although we note the work of Snieder, 1991) – the method is well known but has found comparatively little use. Notable early examples include Green (1975), Chou and Booker (1979), and Tanimoto (1985, 1986). More recently, it has been adopted by the helioseismology community (Pijpers and Thompson, 1992), and applied to global tomography (Zaroli, 2016) and to constrain mantle discontinuities (Lau and Romanowicz, 2021). Concepts from Backus–Gilbert theory are also sometimes used to support interpretation of models produced using other approaches: for example, Ritsema et al. (1999) presents Backus–Gilbert kernels to illustrate the resolution of a model obtained by least-squares inversion.

2.6 Miscellanea

The preceding sections have focused on the range of different philosophies, and associated techniques, by which geophysical inversion can be framed. We now turn to consider some additional concepts and developments that are not themselves designed to solve inverse problems, but which can potentially be employed in conjunction with one or other of the approaches described in this chapter.

2.6.1 Approximate Forward Models
One of the major limiting factors in any geophysical inversion is computational cost. High-fidelity numerical models tend to be computationally expensive, and costs may reach hundreds or even thousands of CPU-hours per simulation. In such cases, resource availability may severely constrain the number of simulations that may be performed, rendering certain approaches infeasible. There is therefore considerable potential value in any technique that may lower the burden of simulation.

2.6.1.1 Surrogate Modelling
One possible solution to this lies in 'surrogate modelling': using techniques of machine learning to mimic the behaviour of an expensive forward model, but at much lower computational cost. This is an idea that has its origins in engineering design (see, e.g., Quiepo et al., 2005; Forrester

and Keane, 2009), and typically involves tuning the free parameters of a neural network or other approximator to match a database of examples obtained via expensive computations (or, indeed, physical experiments). The approximate function can then be interrogated to provide insights, or to serve as a drop-in replacement for the numerical code.

Although the term 'surrogate modelling' only appears relatively recently in the geophysics literature, the underlying idea has a long history. For example, seismologists have long recognised that travel times of seismic arrivals from known sources can be interpolated, and the resulting travel-time curves used to assist in the location of new events (e.g. Jeffreys and Bullen, 1940; Kennett and Engdahl, 1991; Nicholson et al., 2004). One may also regard the Neighbourhood Algorithm (Sambridge, 1999a,b) within this framework: it uses computational geometry to assemble a surrogate approximation to evaluation of (typically) the likelihood for any given model. By employing and refining this within a Markov chain, it is possible to substantially reduce the computational costs of McMC-based inference. In doing so, we exploit the fact that the mapping from models to likelihood (a scalar quantity) is, typically, much simpler than the mapping from models to data. Closely related is the field of 'Bayesian optimisation', which relies on a surrogate (often a Gaussian Process) to encapsulate incomplete knowledge of an objective function, and takes this uncertainty into account within the optimisation procedure (e.g. Shahriari et al., 2016; Wang et al., 2016).

Latterly, surrogate models (also known as emulators) have been explicitly adopted for geophysical studies. Similar to the Neighbourhood Algorithm, Chandra et al. (2020) employed a neural network-based surrogate to replace likelihood calculations within a landscape evolution model; on the other hand, Das et al. (2018) and Spurio Mancini et al. (2021) both developed a surrogate that directly replaces a forward model and outputs synthetic seismograms. Other geophysical examples include modelling of climate and weather (e.g. Field et al., 2011; Castruccio et al., 2014), and applications in hydrology (Hussain et al., 2015) and planetary geophysics (Agarwal et al., 2020).

2.6.1.2 *Physics-Informed Neural Networks*
A number of recent studies have also explored the concept and applications of 'physics informed neural networks' (PINNs; see, e.g., Raissi et al., 2019; Karniadakis et al., 2021). As with surrogate models, these exploit machine learning techniques to provide a version of the forward model that has significantly lower computational cost than 'conventional' implementations. However, whereas a surrogate is constructed using a suite of examples obtained by running the conventional model (at substantial expense), a PINN is directly trained to satisfy the physical constraints. Typically, this amounts to defining a neural network to represent the observable function,

$u(\mathbf{x}, t)$, and then employing a training procedure to minimise the deviation from Eq. (2.1). This is potentially a more efficient approach, and provides the researcher with greater oversight of the behaviour and limitations of the learned model.

A number of recent examples may be found, particularly in the seismological literature. Moseley et al. (2020), Song et al. (2021), and Smith et al. (2020) all use PINNs to solve problems related to the wave equation, with the latter underpinning the variational inference approach of Smith et al. (2022). A range of potential applications in climate science and meteorology are discussed in Kashinath et al. (2021), while He and Tartakovsky (2021) consider hydrological problems. Again, it is clear that PINNs present a promising opportunity that is likely to bring substantial benefits for geophysics, but it is not yet clear how the field will evolve.

2.6.1.3 Conventional Approximations
Surrogate models and PINNs both rely on machine learning, and their 'approximate' nature arises from this: they are constructed to give good average performance for a particular task, but there are few hard constraints on their accuracy in any specific case. In many geophysical problems, an alternate route exists, and has long been exploited: rather than seeking an approximate solution to a complex physical problem, we can use conventional methods to obtain an accurate solution for a simplified physical system (i.e. adopting a more restrictive set of assumptions, $\mathcal{A} \cup \mathcal{B}$). Thus, for example, seismic waves might be modelled under the assumption that propagation is only affected by structure in the great-circle plane between source and receiver (Woodhouse and Dziewonski, 1984) at far lower cost than (almost) physically complete simulation (e.g. Komatitsch et al., 2002). Depending on circumstances, it may be beneficial to exploit a known approximation of this kind, where impacts can be understood and interpretations adjusted accordingly. We also highlight that it may be desirable to vary the level of approximation used for forward simulations within an inversion framework, using a fast approximate technique for initial characterisation, and increasing accuracy as solutions are approached. In the ideal case, one might envisage a forward model where the level of approximation is itself a tuneable parameter (e.g. via the coupling band-width in a normal-mode–based solver, Woodhouse, 1980), enabling a smooth transition from simplified to complete modelling as a solution is approached.

2.6.2 *Computational Advances*
Modern geophysics is computationally intensive, and – as we have seen – the feasibility of various inversion strategies is directly linked to the available resources. As such,

computational developments are often important in driving the development and adoption of novel inference approaches. In particular, current progress leverages a number of technological advances that have been stimulated by the rapid growth of 'machine learning' applications across society. This includes general-purpose computational libraries such as Tensorflow (Abadi et al., 2016) and Pytorch (Paszke et al., 2019), along with more specialist tools such as Edward (Tran et al., 2016). A key feature of these libraries is native support for auto-differentiation, making it easy to exploit gradient-based optimisation strategies. This is an area that has previously been highlighted as ripe for exploitation in geophysics (Sambridge et al., 2007), although its use is not yet widespread. Another interesting development is the rise of packages such as FEniCS (Logg et al., 2012), which aim to automatically generate forward models from a statement of the relevant physical equations (e.g. Reuber and Simons, 2020). This has the potential to greatly expand the range of problems that it is feasible to address.

2.6.3 Novel Data – Novel Strategies

An ongoing theme of geophysics is the growth in data quantity. This is often driven by concerted efforts to collect high-resolution datasets: examples include high-quality satellite gravity measurements (e.g. Kornfeld et al., 2019), and systematic continental scale surveys such as USArray (Meltzer et al., 1999) or AusAEM (Ley-Cooper et al., 2020). Handling and processing such massive datasets has necessitated new tools and standards designed to enable easy exploitation of high-performance computing (e.g. Krischer et al., 2016; Hassan et al., 2020). On the other hand, we have also seen exciting recent developments in planetary seismology, with the recent breakthrough analysis of Martian seismic data from the InSight mission (Khan et al., 2021; Knapmeyer-Endrun et al., 2021; Stähler et al., 2021). In this context, the available dataset is very limited: we must work with a single instrument, limited capacity for data transmission, and with data characteristics quite different from those of Earth. Undoubtedly techniques will need to develop accordingly.

Another driver for innovation in geophysical inversion is innovation in data collection. Recent advances in sensor technology include the growth of distributed acoustic sensing (e.g. Daley et al., 2013; Parker et al., 2014), which uses fibre-optic cables to measure strain rates, and nodal seismic acquisition systems (Dean et al., 2018), which enable dense deployments of semi-autonomous instruments. Fully exploiting these technologies within an inversion context will doubtless motivate a new generation of analysis techniques (e.g. Lythgoe et al., 2021; Muir and Zhang, 2021), and ongoing innovation in the field of geophysical inversion.

2.7 Concluding Remarks

Athanasius Kircher published his *Mundus Subterraneus* in 1665, with his now-famous images of fiery chambers crisscrossing the Earth's interior to feed its volcanoes. What was his evidence for this structure? He acknowledges: '*sive ea jam hoc modo, sive alio*' – 'either like this, or something else'. As Waddell (2006) writes, this

> makes very clear that Kircher was not interested in whether his images had managed to capture exactly the subterranean structure of the Earth. Such large and detailed copper engravings must have been extremely expensive to commission and print, suggesting that Kircher did believe them to be important. But their value lay in their ability to encourage speculation and consideration.

Some 350 years later, geophysical images are produced with more emphasis on rigour – but otherwise, perhaps little has changed.

In this chapter, we have sought to survey and summarise the state of the art of geophysical inversion, and to highlight some of the theoretical and conceptual connections between different approaches. As we hope is clear, the field continues to develop at pace: driven by the need to better address geoscience questions; drawn on towards exciting horizons across mathematics, statistics, and computation. In particular, the growth of machine learning has focused much attention on techniques of regression, model building, and statistical inference, and the fruits of this have been evident throughout our discussion. We have no doubt that geophysical inversion will continue to produce images and models that can inspire and stimulate geoscientists for many years to come.

Acknowledgements. We are grateful to the many students, colleagues, and collaborators who have contributed to our understanding of the topics discussed in this chapter. We also thank several colleagues, and an anonymous reviewer, for helpful comments and suggestions on a draft of this work. We acknowledge financial support from the CSIRO Future Science Platform in Deep Earth Imaging, and from the Australian Research Council under grant numbers DP180100040 and DP200100053.

References

Abadi, M., Barham, P., Chen, J. et al. 2016. Tensorflow: A system for large-scale machine learning. *12th USENIX Symposium on Operating Systems Design and Implementation (OSDI 16)*, pp. 265–83, USENIX Association, Savannah, GA.

Agarwal, S., Tosi, N., Breuer, D. et al. 2020. A machine-learning-based surrogate model of Mars' thermal evolution. *Geophysical Journal International*, 222, 1656–70.

Aleardi, M., and Salusti, A. 2020. Hamiltonian Monte Carlo algorithms for target and interval-oriented amplitude versus angle inversions. *Geophysics*, 85, R177–R194.

Allmaras, M., Bangerth, W., Linhart, J. et al. 2013. Estimating parameters in physical models through Bayesian inversion: A complete example. *SIAM Review*, 55, 149–67.

Ambrosio, L. 2003. Lecture notes on optimal transport problems. In L. Ambriosio, K. Deckelnick, G. Dziuk, M. Mimura, V. Solonnikov, and H. Soner, eds., *Mathematical Aspects of Evolving Interfaces*. Heidelberg: Springer, pp. 1–52.

Anderssen, R., Worthington, M., and Cleary, J. 1972. Density modelling by Monte Carlo inversion – I. Methodology. *Geophysical Journal of the Royal Astronomical Society*, 29, 433–44.

Aster, R., Borchers, B., and Thurber, C. 2013. *Parameter estimation and inverse problems*. Amsterdam: Academic Press.

Backus, G. 1970a. Inference from inadequate and inaccurate data, i. *Proceedings of the National Academy of Sciences*, 65, 1–7.

Backus, G. 1970b. Inference from inadequate and inaccurate data, ii. *Proceedings of the National Academy of Sciences*, 65, 281–7.

Backus, G. 1970c. Inference from inadequate and inaccurate data, iii. *Proceedings of the National Academy of Sciences*, 67, 282–9.

Backus, G. 1988. Bayesian inference in geomagnetism. *Geophysical Journal*, 92, 125–42.

Backus, G., and Gilbert, F. 1968. The resolving power of gross Earth data. *Geophysical Journal of the Royal Astronomical Society*, 16, 169–205.

Bayes, T. 1763. An essay towards solving a problem in the doctrine of chances. *Philosophical Transactions*, 53, 370–418.

Bernal-Romero, M., and Iturrarán-Viveros, U. 2021. Accelerating full-waveform inversion through adaptive gradient optimization methods and dynamic simultaneous sources. *Geophysical Journal International*, 225, 97–126.

Betancourt, M. 2017. A conceptual introduction to Hamiltonian Monte Carlo. arXiv:1701.02434v1.

Bianco, M., and Gerstoft, P. 2018. Travel time tomography with adaptive dictionaries. *IEEE Transactions on Computational Imaging*, 4, 499–511.

Bishop, C. 1995. *Neural Networks for Pattern Recognition*. Oxford: Oxford University Press.

Blei, D., Kucukelbir, A., and McAuliffe, J. 2017. Variational inference: A review for statisticians. *Journal of the American Statistical Association*, 112, 859–77.

Bodin, T., and Sambridge, M. 2009. Seismic tomography with the reversible jump algorithm. *Geophysical Journal International*, 178, 1411–36.

Bond-Taylor, S., Leach, A., Long, Y., and Willcocks, C. 2022. Deep generative modelling: A comparative review of VAEs, GANs, normalizing flows, energy-based and autoregressive models. *IEEE Transactions on Pattern Analysis and Machine Intelligence*, 44(11), 7327–47. https://doi.org/10.1109/TPAMI.2021.3116668.

Box, G., and Muller, M. 1958. A note on the generation of random normal deviates. *Annals of Mathematical Statistics*, 29, 610–611.

Bozdağ, E., Peter, D., Lefebvre, M. et al. 2016. Global adjoint tomography: First-generation model. *Geophysical Journal International*, 207, 1739–66.

Brooks, S., Gelman, A., Jones, G., and Meng, X.-L. 2011. *Handbook of Markov Chain Monte Carlo*. Boca Raton, FL: CRC Press.

Burdick, S., and Lekić, V. 2017. Velocity variations and uncertainty from trans dimensional *P*-wave tomography of North America. *Geophysical Journal International*, 209, 1337–51.

Candès, E., and Wakin, B. 2008. An introduction to compressive sampling. *IEEE Signal Processing Magazine*, 25, 21–30.

Castruccio, S., McInerney, D., Stein, M. et al. 2014. Statistical emulation of climate model projections based on precomputed GCM runs. *Journal of Climate*, 27, 1829–44.

Chandra, R., Azam, D., Kapoor, A., and Müller, R. 2020. Surrogate-assisted Bayesian inversion for landscape and basin evolution models. *Geoscientific Model Development*, 13, 2959–79.

Chou, C., and Booker, J. 1979. A Backus–Gilbert approach to inversion of travel-time data for three-dimensional velocity structure. *Geophysical Journal of the Royal Astronomical Society*, 59, 325–44.

Constable, S., Parker, R., and Constable, C. 1987. Occam's inversion: A practical algorithm for generating smooth models from electromagnetic sounding data. *Geophysics*, 52, 289–300.

Cook, A. 1990. Sir Harold Jeffreys. *Biographical Memoirs of Fellows of the Royal Society*, 36, 303–33.

Creswell, A., White, T., Dumoulin, V. et al. 2018. Generative adversarial networks: An overview. *IEEE Signal Processing Magazine*, 35, 53–65.

Curtis, A., and Lomax, A. 2001. Prior information, sampling distributions, and the curse of dimensionality. *Geophysics*, 66, 372–78.

Daley, T., Freifeld, B., Ajo-Franklin, J. et al. 2013. Field testing of fiberoptic distributed acoustic sensing (DAS) for subsurface seismic monitoring. *The Leading Edge*, 32, 699–706.

Das, S., Chen, X., Hobson, M. et al. 2018. Surrogate regression modelling for fast seismogram generation and detection of microseismic events in heterogeneous velocity models. *Geophysical Journal International*, 215, 1257–90.

de Wit, R., Käufl, P., Valentine, A., and Trampert, J. 2014. Bayesian inversion of free oscillations for Earth's radial (an) elastic structure. *Physics of the Earth and Planetary Interiors*, 237, 1–17.

Dean, T., Tulett, J., and Barwell, R. 2018. Nodal land seismic acquisition: The next generation. *First Break*, 36, 47–52.

Dettmer, J., Benavente, R., Cummins, P., and Sambridge, M. 2014. Trans-dimensional finite-fault inversion. *Geophysical Journal International*, 199, 735–51.

Dinh, H., and Van der Baan, M. 2019. A grid-search approach for 4d pressure-saturation discrimination. *Geophysics*, 84, IM47–IM62.

Donoho, D. 2006. Compressed sensing. *IEEE Transactions on Information Theory*, 52, 1289–306.

Dziewonski, A., Chou, T.-A., and Woodhouse, J. 1981. Determination of earthquake source parameters from waveform data for studies of global and regional seismicity. *Journal of Geophysical Research*, 86, 2825–52.

Earp, S., Curtis, A., Zhang, X., and Hansteen, F. 2020. Probabilistic neural network tomography across Grane field (North Sea) from surface wave dispersion data. *Geophysical Journal International*, 223, 1741–57.

Enemark, T., Peeters, L., Mallants, D. et al. 2019. Hydrogeological Bayesian hypothesis testing through trans-dimensional sampling of a stochastic water balance model. *Water*, 11(7), 1463. https://doi.org/10.3390/w11071463.

Engquist, B., and Froese, B. 2014. Application of the Wasserstein metric to seismic signals. *Communications in Mathematical Sciences*, 12, 979–88.

Fernández-Martínez, J. L., and Fernández-Muñiz, Z. 2020. The curse of dimensionality in inverse problems. *Journal of Computational and Applied Mathematics*, 369, 112571. https://doi.org/10.1016/j.cam.2019.112571.

Fichtner, A., Zunino, A., and Gebraad, L. 2019. Hamiltonian Monte Carlo solution of tomographic inverse problems. *Geophysical Journal International*, 216, 1344–63.

Field, R., Constantine, P., and Boslough, M. 2011. Statistical surrogate models for prediction of high-consequence climate change, Tech. Rep. SAND2011-6496, Sandia National Laboratories.

Forrester, A., and Keane, A. 2009. Recent advances in surrogate-based optimization. *Progress in Aerospace Sciences*, 45, 50–79.

Galetti, E., Curtis, A., Baptie, B., Jenkins, D., and Nicolson, H. 2017. Transdimensional Love-wave tomography of the British Isles and shear-velocity structure of the East Irish Sea Basin from ambient-noise interferometry. *Geophysical Journal International*, 208, 35–58.

Gallagher, K., Bodin, T., Sambridge, M. et al. 2011. Inference of abrupt changes in noisy geochemical records using transdimensional changepoint models. *Earth and Planetary Science Letters*, 311, 182–94.

Goodfellow, I., Pouget-Abadie, J., Mirza, M. et al. 2014. Generative adversarial nets. In Z. Ghahramani, M. Welling, C. Cortes, N. Lawrence, and K. Q. Weinberger, eds., *Advances in Neural Information Processing Systems*, vol. 27. Red Hook, NY: Curran Associates. https://proceedings.neurips.cc/paper/2014/file/5ca3e9b122f61f8f06494c97b1afccf3-Paper.pdf.

Green, P. 1995. Reversible jump Markov chain Monte Carlo computation and Bayesian model determination. *Biometrika*, 82, 711–32.

Green, W. 1975. Inversion of gravity profiles by a Backus–Gilbert approach. *Geophysics*, 40, 763–72.

Guo, P., Visser, G., and Saygin, E. 2020. Bayesian trans-dimensional full waveform inversion: Synthetic and field data application. *Geophysical Journal International*, 222, 610–27.

Hassan, R., Hejrani, B., Medlin, A., Gorbatov, A., and Zhang, F. 2020. High-performance seismological tools (HiPerSeis). In K. Czarnota, I. Roach, S. Abbott, M. Haynes, N. Kositcin, A. Ray, and E. Slatter, eds., *Exploring for the Future: Extended Abstracts*. Canberra: Geoscience Australia, pp. 1–4.

Hawkins, R., and Sambridge, M. 2015. Geophysical imaging using trans-dimensional trees. *Geophysical Journal International*, 203, 972–1000.

He, Q., and Tartakovsky, A. 2021. Physics-informed neural network method for forward and backward advection-dispersion equations. *Water Resources Research*, 57, e2020WR029479.

He, W., Brossier, R., Métivier, L., and Plessix, R.-E. 2019. Land seismic multiparameter full waveform inversion in elastic VTI media by simultaneously interpreting body waves and surface waves with an optimal transport based objective function. *Geophysical Journal International*, 219, 1970–88.

Hedjazian, N., Bodin, T., and Métivier, L. 2019. An optimal transport approach to linearized inversion of receiver functions. *Geophysical Journal International*, 216, 130–47.

Hejrani, B., and Tkalčić, H. 2020. Resolvability of the centroid-moment-tensors for shallow seismic sources and improvements from modeling high-frequency waveforms. *Journal of Geophysical Research*, 125, e2020JB019643.

Herrmann, F., Erlangga, Y., and Lin, T. 2009. Compressive simultaneous full-waveform simulation. *Geophysics*, 74, A35–A40.

Hogg, D., and Foreman-Mackey, D. 2018. Data analysis recipes: Using Markov chain Monte Carlo. *The Astrophysical Journal Supplement Series*, 236:11 (18 pp.). https://doi.org/10.3847/1538-4365/aab76e.

Huang, G., Zhang, X., and Qian, J. 2019. Kantorovich–Rubinstein misfit for inverting gravity-gradient data by the level-set method. *Geophysics*, 84, 1–115.

Hussain, M., Javadi, A., Ahangar-Asr, A., and Farmani, R. 2015. A surrogate model for simulation-optimization of aquifer systems subjected to seawater intrusion. *Journal of Hydrology*, 523, 542–554.

Jeffreys, H. 1931. *Scientific Inference*. Cambridge: Cambridge University Press.

Jeffreys, H. 1939. *The Theory of Probability*. Oxford: Oxford University Press.

Jeffreys, H., and Bullen, K. 1940. *Seismological Tables*. London: British Association for the Advancement of Science.

Karniadakis, G., Kevrekidis, I., Lu, L. et al. 2021. Physics-informed machine learning. *Nature Reviews Physics*, 3, 422–40.

Kashinath, K., Mustafa, M., Albert, A. et al. 2021. Physics-informed machine learning: Case studies for weather and climate modelling. *Philosophical Transactions*, 379, 20200093.

Käufl, P. 2015. Rapid probabilistic source inversion using pattern recognition. Ph.D. thesis, University of Utrecht.

Käufl, P., Valentine, A., O'Toole, T., and Trampert, J. 2014. A framework for fast probabilistic centroid – moment-tensor determination – inversion of regional static displacement measurements. *Geophysical Journal International*, 196, 1676–93.

Käufl, P., Valentine, A., de Wit, R., and Trampert, J. 2016a. Solving probabilistic inverse problems rapidly with prior samples. *Geophysical Journal International*, 205, 1710–28.

Käufl, P., Valentine, A., and Trampert, J. 2016b. Probabilistic point source inversion of strong-motion data in 3-D media using pattern recognition: A case study for the 2008 *Mw* 5.4 Chino Hills earthquake. *Geophysical Research Letters*, 43, 8492–8.

Kennett, B., and Engdahl, E. 1991. Traveltimes for global earthquake location and phase identification. *Geophysical Journal International*, 105, 429–65.

Khan, A., Ceylan, S., van Driel, M. et al. 2021. Upper mantle structure of Mars from InSight seismic data. *Science*, 373, 434–8.

Kingma, D., and Welling, M. 2014. Auto-encoding variational Bayes, in *2nd International Conference on Learning Representations*, ICLR 2014, Banff, AB, Canada, 14–16 April 2014, Conference Track Proceedings. https://arxiv.org/abs/1312.6114.

Kircher, A. 1665. *Mundus subterraneus*. Amsterdam: Joannem Janssonium & EliziumWegerstraten.

Knapmeyer-Endrun, B., Panning, M. P., Bissig, F. et al. 2021. Thickness and structure of the martian crust from In Sight seismic data. *Science*, 373, 438–43.

Kobyzev, I., Prince, S., and Brubaker, M. 2021. Normalizing flows: An introduction and review of current methods. *IEEE Transactions on Pattern Analysis and Machine Intelligence*, 43, 3964–79.

Komatitsch, D., Ritsema, J., and Tromp, J. 2002. The spectral-element method, Beowulf computing, and global seismology. *Science*, 298, 1737–42.

Kornfeld, R., Arnold, B., Gross, M. et al. 2019. GRACE-FO: The gravity recovery and climate experiment follow-on mission. *Journal of Spacecraft and Rockets*, 56, 931–51.

Koshkholgh, S., Zunino, A., and Mosegaard, K. 2021. Informed proposal Monte Carlo. *Geophysical Journal International*, 226, 1239–48.

Krischer, L., Smith, J., Lei, W. et al. 2016. An adaptable seismic data format. *Geophysical Journal International*, 207, 1003–11.

Kullback, S., and Leibler, R. 1951. On information and sufficiency. *The Annals of Mathematical Statistics*, 22, 79–86.

Lau, H., and Romanowicz, B. 2021. Constraining jumps in density and elastic properties at the 660 km discontinuity using normal mode data via the Backus-Gilbert method. *Geophysical Research Letters*, 48, e2020GL092217.

Lei, W., Ruan, Y., Bozdağ, E. et al. 2020. Global adjoint tomography: Model GLAD-M25. *Geophysical Journal International*, 223, 1–21.

Ley-Cooper, A., Brodie, R., and Richardson, M. 2020. AusAEM: Australia's airborne electromagnetic continental-scale acquisition program. *Exploration Geophysics*, 51, 193–202.

Liu, D., and Nocedal, J. 1989. On the limited memory BFGS method for large-scale optimization. *Mathematical Programming*, 45, 503–28.

Livermore, P., Fournier, A., Gallet, Y., and Bodin, T. 2018. Transdimensional inference of archeomagnetic intensity change. *Geophysical Journal International*, 215, 2008–34.

Logg, A., Mardal, K.-A., and Wells, G. N. 2012. *Automated Solution of Differential Equations by the Finite Element Method*. Heidelberg: Springer.

Lopez-Alvis, J., Laloy, E., Nguyen, F., and Hermans, T. 2021. Deep generative models in inversion: The impact of the generator's nonlinearity and development of a new approach based on a variational autoencoder. *Computers & Geosciences*, 152, 104762.

Luengo, D., Martino, L., Bugallo, M., Elvira, V., and Särkkä, S. 2020. A survey of Monte Carlo methods for parameter estimation. *EURASIP Journal on Advances in Signal Processing*, 2020, 25. https://doi.org/10.1186/s13634-020-00675-6.

Lythgoe, K., Loasby, A., Hidayat, D., and Wei, S. 2021. Seismic event detection in urban Singapore using a nodal array and frequency domain array detector: Earthquakes, blasts and thunderquakes. *Geophysical Journal International*, 226, 1542–57.

Meier, U., Curtis, A., and Trampert, J. 2007. Global crustal thickness from neural network inversion of surface wave data. *Geophysical Journal International*, 169, 706–722.

Meltzer, A., Rudnick, R., Zeitler, P. et al. 1999. USArray initiative. *GSA Today*, 11, 8–10.

Menke, W. 1989. *Geophysical Data Analysis: Discrete Inverse Theory*. New York: Academic Press.

Métivier, L., Brossier, R., Mérigot, Q., Oudet, E., and Virieux, J. 2016a. Increasing the robustness and applicability of full-waveform inversion: An optimal transport distance strategy. *The Leading Edge*, 35, 1060–7.

Métivier, L., Brossier, R., Mérigot, Q., Oudet, E., and Virieux, J. 2016b. Measuring the misfit between seismograms using an optimal transport distance: Application to full waveform inversion. *Geophysical Journal International*, 205, 345–77.

Métivier, L., Brossier, R., Mérigot, Q., Oudet, e., and Virieux, J. 2016c. An optimal transport approach for seismic tomography: Application to 3D full waveform inversion. *Inverse Problems*, 32, 115008.

Métivier, L., Brossier, R., Oudet, E., Mérigot, Q., and Virieux, J. 2016d. An optimal transport distance for full-waveform inversion: Application to the 2014 chevron benchmark data set. In C. Sicking and J. Ferguson *SEG Technical Program Expanded Abstracts*. Tulsa, OK: Society of Exploration Geophysicists, pp. 1278–83.

Montagner, J.-P., and Tanimoto, T. 1990. Global anisotropy in the upper mantle inferred from the regionalization of phase velocities. *Journal of Geophysical Research*, 95, 4797–819.

Montagner, J.-P., and Tanimoto, T. 1991. Global upper mantle tomography of seismic velocities and anisotropies. *Journal of Geophysical Research*, 96, 20337–51.

Moseley, B., Nissen-Meyer, T., and Markham, A. 2020. Deep learning for fast simulation of seismic waves in complex media. *Solid Earth*, 11, 1527–49.

Mosher, S., Eilon, Z., Janiszewski, H., and Audet, P. 2021. Probabilistic inversion of seafloor compliance for oceanic crustal shear velocity structure using mixture density networks. *Geophysical Journal International*, 227, 1879–92.

Mosser, L., Dubrule, O., and Blunt, M. 2020. Stochastic seismic waveform inversion using generative adversarial networks as a geological prior. *Mathematical Geosciences*, 52, 53–79.

Muir, J., and Zhang, Z. 2021. Seismic wavefield reconstruction using a pre-conditioned wavelet-curvelet compressive sensing approach. *Geophysical Journal International*, 227, 303–15.

Neal, R. 2011. MCMC using Hamiltonian dynamics. In S. Brooks, A. Gelman, G. Jones, and X.-L. Meng, eds., *Handbook of Markov Chain Monte Carlo*. Boca Raton, FL: CRC Press.

Nicholson, T., Sambridge, M., and Gudmundsson, O. 2004. Three-dimensional empirical traveltimes: Construction and applications. *Geophysical Journal International*, 156, 307–28.

Nocedal, J., and Wright, S. 1999. *Numerical Optimization*. New York: Springer.

Nyquist, H. 1928. Certain topics in telegraph transmission theory. *Transactions of the American Institute of Electrical Engineers*, 47, 617–44.

Parker, R. 1994. *Geophysical Inverse Theory*. Princeton, NJ: Princeton University Press.

Parker, T., Shatalin, S., and Farhadiroushan, M. 2014. Distributed acoustic sensing – a new tool for seismic applications. *First Break*, 32, 61–69.

Paszke, A., Gross, S., Massa, F. et al. 2019. Pytorch: An imperative style, high-performance deep learning library. In H. M. Wallach, H. Larochelle, A. Beygelzimer, F. d'Alché-Buc, E. A. Fox, and R. Garnett, eds. *Advances in Neural Information Processing Systems*, vol. 32. Red Hook, NY: Curran Associates. https://proceedings.neurips.cc/paper/2019/file/bdbca288fee7f92f2bfa9f7012727740-Paper.pdf.

Petersen, K., and Pedersen, M. 2012. The matrix cookbook, Tech. rep. Technical University of Denmark, Kongens Lyngby. 72 p.

Pijpers, F., and Thompson, M. 1992. Faster formulations of the optimally localized averages method for helioseismic inversions. *Astronomy and Astrophysics*, 262, L33–L36.

Press, F. 1970. Earth models consistent with geophysical data. *Physics of the Earth and Planetary Interiors*, 3, 3–22.

Quiepo, N., Haftka, R., Shyy, W. et al. 2005. Surrogate-based analysis and optimization. *Progress in Aerospace Sciences*, 41, 1–28.

Raissi, M., Perdikaris, P., and Karniadakis, G. 2019. Physics-informed neural networks: A deep learning framework for solving forward and inverse problems involving nonlinear partial differential equations. *Journal of Computational Physics*, 378, 686–707.

Ramgraber, M., Weatherl, R., Blumensaat, F., and Schirmer, M. 2021. Non-Gaussian parameter inference for hydrogeological models using Stein variational gradient descent. *Water Resources Research*, 57, e2020WR029339.

Rasmussen, C., and Williams, C. 2006. *Gaussian Processes for Machine Learning*. Cambridge, MA: MIT Press.

Ray, A. 2021. Bayesian inversion using nested trans-dimensional Gaussian processes. *Geophysical Journal International*, 226, 302–26.

Ray, A., and Myer, D. 2019. Bayesian geophysical inversion with trans-dimensional Gaussian process machine learning. *Geophysical Journal International*, 217, 1706–26.

Reuber, G., and Simons, F. 2020. Multi-physics adjoint modeling of Earth structure: combining gravimetric, seismic and geo-dynamic inversions. *International Journal on Geomathematics*, 11(30), 1–38.

Rezende, D., and Mohamed, S. 2015. Variational inference with normalizing flows. In F. Bach and D. Blei, eds., *Proceedings of the 32nd International Conference on Machine Learning*, vol. 37; International Conference on Machine Learning, 7–9 July 2015, Lille, France, pp.1530–8. http://proceedings.mlr.press/v37/rezende15.pdf.

Rijal, A., Cobden, L., Trampert, J., Jackson, J., and Valentine, A. 2021. Inferring equations of state of the lower mantle minerals using mixture density networks. *Physics of the Earth and Planetary Interiors*, 319, 106784.

Ritsema, J., van Heijst, H., and Woodhouse, J. 1999. Complex shear wave velocity structure imaged beneath Africa and Iceland., *Science*, 286, 1925–31.

Ruthotto, L., and Haber, E. 2021. An introduction to deep generative modelling. *GAMM-Mitteilungen*, 44, e202100008.

Sambridge, M. 1998. Exploring multidimensional landscapes without a map. *Inverse Problems*, 14, 427–40.

Sambridge, M. 1999a. Geophysical inversion with a neighbourhood algorithm – I. Searching a parameter space. *Geophysical Journal International*, 138, 479–94.

Sambridge, M. 1999b. Geophysical inversion with a neighbourhood algorithm – II. Appraising the ensemble *Geophysical Journal International*, 138, 727–46.

Sambridge, M., and Kennett, B. 1986. A novel method of hypo-centre location. *Geophysical Journal International*, 87, 679–97.

Sambridge, M., and Mosegaard, K. 2002. Monte Carlo methods in geophysical inverse problems. *Reviews of Geophysics*, 40, 3-1–3-29. https://doi.org/10.1029/2000RG000089.

Sambridge, M., Gallagher, K., Jackson, A., and Rickwood, P. 2006. Trans-dimensional inverse problems, model comparison and the evidence. *Geophysical Journal International*, 167, 528–42.

Sambridge, M., Rickwood, P., Rawlinson, N., and Sommacal, S. 2007. Automatic differentiation in geophysical inverse problems. *Geophysical Journal International*, 170, 1–8.

Sambridge, M., Bodin, T., Gallagher, K., and Tkalcic, H. 2012. Transdimensional inference in the geosciences. *Philosophical Transactions of the Royal Society*, 371.

Santambrogio, F. 2015. *Optimal Transport for Applied Mathematicians*. Basel: Birkhäuser.

Scales, J., and Snieder, R. 1997. To Bayes or not to Bayes. *Geophysics*, 62, 1045–46.

Scheiter, M., Valentine, A., and Sambridge, M. 2022. Upscaling and downscaling Monte Carlo ensembles with generative models. *Geophysical Journal International*, 230(2), 916–31. https://doi.org/10.1093/gji/ggac100.

Sen, M., and Biswas, R. 2017. Transdimensional seismic inversion using the reversible jump Hamiltonian Monte Carlo algorithm. *Geophysics*, 82, R119–R134.

Shahriari, B., Swersky, K., Wang, Z., Adams, R., and de Freitas, N. 2016. Taking the human out of the loop: A review of Bayesian optimization. *Proceedings of the IEEE*, 104, 148–75.

Siahkoohi, A., and Herrman, F. 2021. Learning by example: Fast reliability-aware seismic imaging with normalizing flows. arXiv:2104.06255v1. https://arxiv.org/abs/2104.06255

Simons, F., Loris, I., Nolet, G. et al. 2011. Solving or resolving global tomographic models with spherical wavelets, and the scale and sparsity of seismic heterogeneity. *Geophysical Journal International*, 187, 969–88.

Smith, J., Azizzadenesheli, K., and Ross, Z. 2020. EikoNet: Solving the eikonal equation with deep neural networks. *IEEE Transactions on Geoscience and Remote Sensing*, 59 (12), 10685–96. https://doi.org/10.1109/TGRS.2020.3039165.

Smith, J., Ross, Z., Azizzadenesheli, K., and Muir, J. 2022. HypoSVI: Hypocenter inversion with Stein variational inference and physics informed neural networks. *Geophysical Journal International*, 228, 698–710.

Snieder, R. 1991. An extension of Backus–Gilbert theory to non-linear inverse problems. *Inverse Problems*, 7, 409–433.

Song, C., Alkhalifah, T., and Bin Waheed, U. 2021. Solving the frequency-domain acoustic VTI wave equation using physics-informed neural networks. *Geophysical Journal International*, 225, 846–59.

Spurio Mancini, A., Piras, D., Ferreira, A., Hobson, M., and Joachimi, B. 2021. Accelerating Bayesian microseismic event location with deep learning. *Solid Earth*, 12, 1683–705.

Stähler, S., Khan, A., Banerdt, W. et al. 2021. Seismic detection of the Martian core. *Science*, 373, 443–48.

Steinmetz, T., Raape, U., Teßmann, S. et al. 2010. Tsunami early warning and decision support. *Natural Hazards and Earth System Sciences*, 10, 1839–50.

Tanimoto, T. 1985. The Backus–Gilbert approach to the three-dimensional structure in the upper mantle – I. Lateral variation of surface wave phase velocity with its error and resolution. *Geophysical Journal International*, 82, 105–23.

Tanimoto, T. 1986. The Backus–Gilbert approach to the three-dimensional structure in the upper mantle – II. *SH* and *SV* velocity. *Geophysical Journal International*, 84, 49–69.

Tarantola, A. 2005. *Inverse Problem Theory and Methods for Model Parameter Estimation*. Philadelphia, PA: Society of Industrial and Applied Mathematics.

Tarantola, A., and Nercessian, A. 1984. Three-dimensional inversion without blocks. *Geophysical Journal of the Royal Astronomical Society*, 76, 299–306.

Tarantola, A., and Valette, B. 1982. Generalized nonlinear inverse problems solved using the least squares criterion. *Reviews of Geophysics and Space Physics*, 20, 219–32.

Tibshirani, R. 1996. Regression shrinkage and selection via the lasso. *Journal of the Royal Statistical Society B*, 58, 267–88.

Trampert, J., and Snieder, R. 1996. Model estimations biased by truncated expansions: Possible artifacts in seismic tomography. *Science*, 271, 1257–60.

Tran, D., Kucukelbir, A., Dieng, A. B. et al. 2016. Edward: A library for probabilistic modeling, inference, and criticism. arXiv:1610.09787.

Valentine, A., and Davies, D. 2020. Global models from sparse data: A robust estimate of earth's residual topography spectrum. *Geochemistry, Geophysics, Geosystems*, e2020GC009240.

Valentine, A., and Sambridge, M. 2020a. Gaussian process models – I. A framework for probabilistic continuous inverse theory. *Geophysical Journal International*, 220, 1632–47.

Valentine, A., and Sambridge, M. 2020b. Gaussian process models – II. Lessons for discrete inversion. *Geophysical Journal International*, 220, 1648–56.

Valentine, A., and Trampert, J. 2016. The impact of approximations and arbitrary choices on geophysical images. *Geophysical Journal International*, 204, 59–73.

van Herwaarden, D. P., Boehm, C., Afansiev, M. et al. 2020. Accelerated full-waveform inversion using dynamic mini-batches. *Geophysical Journal International*, 221, 1427–38.

Waddell, M. 2006. The world, as it might be: Iconography and probabalism in the Mundus subterraneus of Athanasius Kircher. *Centaurius*, 48, 3–22. https://doi.org/10.1111/j.1600-0498.2006.00038.x.

Wang, Y., Cao, J., and Yang, C. 2011. Recovery of seismic wavefields based on compressive sensing by an l_1-norm constrained trust region method and the piecewise random subsampling. *Geophysical Journal International*, 187, 199–213.

Wang, Z., Hutter, F., Zoghi, M., Matheson, D., and de Freitas, N. 2016. Bayesian optimization in a billion dimensions via random embeddings. *Journal of Artificial Intelligence Research*, 55, 361–87.

Wiggins, R. 1972. The general linear inverse problem: Implication of surface waves and free oscillations for Earth structure. *Reviews of Geophysics and Space Physics*, 10, 251–85.

Woodhouse, J. 1980. The coupling and attenuation of nearly resonant multiplets in the Earth's free oscillation spectrum. *Geophysical Journal of the Royal Astronomical Society*, 61, 261–83.

Woodhouse, J., and Dziewonski, A. 1984. Mapping the upper mantle: Three-dimensional modelling of Earth structure by inversion of seismic waveforms. *Journal of Geophysical Research*, 89, 5953–86.

Worthington, M., Cleary, J., and Anderssen, R. 1972. Density modelling by Monte Carlo inversion – II. Comparison of recent Earth models. *Geophysical Journal of the Royal Astronomical Society*, 29, 445–57.

Zaroli, C. 2016. Global seismic tomography using Backus–Gilbert inversion. *Geophysical Journal International*. 207, 876–88.

Zhang, C., Bütepage, J., Kjellström, H., and Mandt, S. 2019. Advances in variational inference. *IEEE Transactions on Pattern Analysis and Machine Intelligence*, 41, 2008–26.

Zhang, X., and Curtis, A. 2020. Seismic tomography using variational inference methods. *Journal of Geophysical Research*, 125, e2019JB018589.

Zhao, X., Curtis, A., and Zhang, X. 2022. Bayesian seismic tomography using normalizing flows. *Geophysical Journal International*, 228, 213–39.

3

A Tutorial on Bayesian Data Assimilation

Colin Grudzien and Marc Bocquet

Abstract: This chapter provides a broad introduction to Bayesian data assimilation that will be useful to practitioners in interpreting algorithms and results, and for theoretical studies developing novel schemes with an understanding of the rich history of geophysical data assimilation and its current directions. The simple case of data assimilation in a 'perfect' model is primarily discussed for pedagogical purposes. Some mathematical results are derived at a high level in order to illustrate key ideas about different estimators. However, the focus of this chapter is on the intuition behind these methods, where more formal and detailed treatments of the data assimilation problem can be found in the various references. In surveying a variety of widely used data assimilation schemes, the key message of this chapter is how the Bayesian analysis provides a consistent framework for the estimation problem and how this allows one to formulate its solution in a variety of ways to exploit the operational challenges in the geosciences.

3.1 Introduction

In applications such as short- to medium-range weather prediction, *data assimilation* (DA) provides a means to sequentially and recursively update forecasts of a time-varying physical process with newly incoming information (Daley, 1991; Kalnay, 2003; Asch et al., 2016), typically Earth observations. The Bayesian approach to DA is widely adopted (Lorenc, 1986) because it provides a unified treatment of tools from statistical estimation, non-linear optimisation, and machine learning for handling such a problem. This chapter illustrates how this approach can be utilised to develop and interpret a variety of widely used DA algorithms, both classical and those at the current state-of-the-art.

Suppose that the time-dependent physical states to be modelled can be written as a vector, $x_k \in \mathbb{R}^{N_x}$, where k labels some time t_k. Formally, the time-evolution of these states is represented with the non-linear map $\mathcal{M}$,

$$x_k = \mathcal{M}_k(x_{k-1}, \lambda) + \eta_k, \tag{3.1}$$

where: (i) x_{k-1} is the state variable vector at an earlier time t_{k-1}; (ii) λ is a vector of uncertain static physical parameters but on which the time evolution depends; and (iii) η_k is an additive (for simplicity but other choices are possible), stochastic noise term, representing errors in our model for the physical process. Define $\Delta_t := t_k - t_{k-1}$ to be a fixed-length forecast horizon in this chapter, though none of the following results require this to be fixed in practice.

The basic goal of sequential DA is to estimate the random state vector x_k, given a prior distribution on (x_{k-1}, λ), and given knowledge of $\mathcal{M}_k$ and knowledge of how η_k is statistically distributed. At time t_{k-1}, a forecast is made for the distribution of x_k utilising the prior knowledge, which includes the physics-based model. For simplicity, most of this chapter is restricted to the case where λ is a known constant, and the forecast model is *perfect*, that is,

$$x_k = \mathcal{M}_k(x_{k-1}). \tag{3.2}$$

However, a strength of Bayesian analysis is how it easily extends to include a general treatment of model errors and the estimation of model parameters (see, e.g., Asch et al. (2016) for a more general introduction).

While a forecast is made, one acquires a collection of observations of the real-world process. This is written as the observation vector $y_k \in \mathbb{R}^{N_y}$, which is related to the state vector by

$$y_k = \mathcal{H}_k(x_k) + \epsilon_k. \tag{3.3}$$

The (possibly non-linear) map $\mathcal{H}_k : \mathbb{R}^{N_x} \to \mathbb{R}^{N_y}$ relates the physical states being modelled, x_k, to the values that are actually observed, y_k. Typically, in geophysical applications, observations are not $1 : 1$ with the state variables; while the data dimension can be extremely large, $N_y \ll N_x$, so this information is sparse relative to the model state dimension. The term ϵ_k in Eq. (3.3) is an additive, stochastic noise term representing errors in the measurements, or a mismatch between the state variable representation and the observation (the representation error, Janjić et al., 2018).

Therefore, sometime after the real-life physical system has reached time t_k, one has a forecast distribution for the states x_k, generated by the prior on $x_k - 1$ and the physics-based model $\mathcal{M}$, and the observations y_k with some associated uncertainty. The goal of Bayesian DA is to estimate the posterior distribution for x_k conditioned on y_k, or some statistics of this distribution.

3.2 Hidden Markov Models and Bayesian Analysis

3.2.1 The Observation-Analysis-Forecast Cycle

Recursive estimation of the distribution for x_k conditional on y_k (i.e. assuming y_k is known) can be described as an observation-analysis-forecast cycle (Trevisan and Uboldi, 2004). Given the forecast-prior for the model state, and the likelihood function for the observation, Bayes' law updates the prior for the modelled state to the posterior conditioned on the observation. Bayes' law is a simple re-arrangement of *conditional probability*, defined by Kolmogorov (2018) as

$$\mathcal{P}(A|B) := \frac{\mathcal{P}(A,\, B)}{\mathcal{P}(B)}, \qquad (3.4)$$

for two events A and B where the probability of B is non-zero. Intuitively, this says that the probability of an event A, given knowledge that an event B occurs, is equal to the probability of A occurring relative to a sample space restricted to the event B. Using the symmetry in the joint event $\mathcal{P}(A,B) = \mathcal{P}(B,A)$, Bayes' law is written

$$\mathcal{P}(A|B) = \frac{\mathcal{P}(B|A)\mathcal{P}(A)}{\mathcal{P}(B)}. \qquad (3.5)$$

In the observation-analysis-forecast cycle, A is identified with the state vector (seen as a random vector) taking its value in a neighbourhood of x_k, and B is identified with the observation vector (seen as a random vector) taking its value in a neighbourhood of y_k. The power of this statement is in how it describes an 'inverse' probability – while the *posterior*, $\mathcal{P}(A|B)$, on the left-hand-side may not be directly accessible, often the *likelihood* $\mathcal{P}(B|A)$ and the *prior* $\mathcal{P}(A)$ are easy to compute, and this is sufficient to develop a variety of probabilistic DA techniques.

A conceptual diagram of this process is pictured in Fig. 3.1. Given the initial first prior (represented as the first 'posterior' at time t_0), a forecast prior for the model state at t_1, $\mathcal{P}(A)$, is produced with the numerical model. At the update time, there is an (possibly indirect and noisy) observation of the physical state with an associated likelihood $\mathcal{P}(B|A)$. The posterior for the model state conditioned on this observation $\mathcal{P}(A|B)$, commonly denoted the *analysis* in geophysical DA, is used to initialise the subsequent

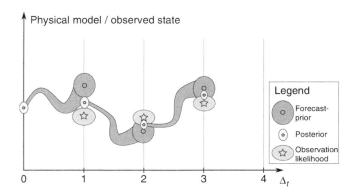

Figure 3.1 Conceptual diagram of the observation-analysis-forecast cycle. The y-axis represents a variable of the state and observation vectors; the x-axis represents time. Ellipses represent the spread of the observation/posterior/forecast errors. Original figure, adapted from Carrassi et al. (2018).

numerical forecast. Recursive estimates of the current modelled state can be performed in this fashion, but a related question regards the past states. The newly received observation gives information about the model states at past times and this allows one to produce a retrospective posterior estimate for the past states. Recursive estimation of the present state using these incoming observations is commonly known as *filtering*. Conditional estimation of a past state given a time series including future observations is commonly known as *smoothing*.

It is important to recognise that the filtering *probability density function* (pdf) for the current time $p(x_k|y_k, y_{k-1}, y_{k-2}, \ldots)$ is actually just a marginal of the joint posterior pdf over all states in the current *data assimilation window* (DAW) (i.e. the window of lagged past and current states being estimated). In Fig. 3.1, the DAW is the time window $\{t_1,\ t_2,\ t_3\}$. The conditional pdf for the model state at time t_3, given observations in the DAW, is written in terms of the joint posterior as

$$p(x_3|y_3, y_2, y_1) = \iiint p(x_3, x_2, x_1, x_0|y_3, y_2, y_1)\, \mathrm{d}x_2 \mathrm{d}x_1 \mathrm{d}x_0, \qquad (3.6)$$

by averaging out the past history of the model state from the joint posterior in the integrand. A smoothing estimate may be produced in a variety of ways, exploiting different formulations of the Bayesian problem. One may estimate only a *marginal pdf* as on the left-hand-side of Eq. (3.6), or the entire *joint posterior pdf* as in the integrand in Eq. 3.6 (Anderson and Moore, 1979; Cohn et al., 1994; Cosme et al., 2012). The approach chosen may strongly depend on whether the DAW is static or is advanced in time. Particularly, if one produces a smoothing estimate for the state at times t_1 through t_3, one may subsequently shift the DAW so that, in the new cycle, the posterior for times t_2

through t_4 is estimated: this type of analysis is known as *fixed-lag smoothing*. This chapter considers how one can utilise a Bayesian *maximum a posteriori* (MAP) formalism to efficiently solve the filtering and smoothing problems, using the various tools of statistical estimation, non-linear optimisation, and machine learning.

3.2.2 A Generic Hidden Markov Model

Recall the perfect physical process model and the noisy observation model,

$$\boldsymbol{x}_k = \mathcal{M}_k(\boldsymbol{x}_{k-1}), \tag{3.7a}$$

$$\boldsymbol{y}_k = \mathcal{H}_k(\boldsymbol{x}_k) + \boldsymbol{\epsilon}_k. \tag{3.7b}$$

Denote the sequence of the process model states and observation model states between time t_k and time t_l, for $k < l$, as

$$\boldsymbol{x}_{l:k} := \{\boldsymbol{x}_l, \boldsymbol{x}_{l-1}, \dots, \boldsymbol{x}_k\}, \quad \boldsymbol{y}_{l:k} := \{\boldsymbol{y}_l, \boldsymbol{y}_{l-1}, \dots, \boldsymbol{y}_k\}. \tag{3.8}$$

For arbitrary $l \in \mathbb{N}$, assume that the sequence of observation error

$$\{\boldsymbol{\epsilon}_l, \boldsymbol{\epsilon}_{l-1}, \dots, \boldsymbol{\epsilon}_{-1}\} \tag{3.9}$$

is independent-in-time (i.e. a white process).

This formulation is a type of *hidden Markov model*, where the dynamic state variables $\boldsymbol{x}_k$ are known as the hidden variables because they are not directly observed. A Markov model is a type of 'memoryless' process, described this way because of how the conditional probability for the state is represented between different times (Ross, 2014, chapter 4). Particularly, if $\boldsymbol{x}_{k:1}$ is a Markov process, the Markov property is defined as

$$p(\boldsymbol{x}_k|\boldsymbol{x}_{k-1:0}) = p(\boldsymbol{x}_k|\boldsymbol{x}_{k-1}). \tag{3.10}$$

Equation (3.10) says that, given knowledge of the state $\boldsymbol{x}_{k-1}$, the conditional probability for $\boldsymbol{x}_k$ is independent of the past history of the state before time t_{k-1}, representing the probabilistic analogue of an initial value problem.

Applying the Markov property recursively with the definition of the conditional pdf yields

$$p(\boldsymbol{x}_{L:0}) = p(\boldsymbol{x}_0) \prod_{k=1}^{L} p(\boldsymbol{x}_k|\boldsymbol{x}_{k-1}). \tag{3.11}$$

Therefore, the joint pdf for a forecast of the model state can be written as the product of the first prior at time t_0, representing the uncertainty of the data used to initialise the model forecast, and the product of the *Markov transition pdfs*, describing the evolution of the state between discrete times.

With the perfect state model, as in Eq. (3.7a), the transition probability for some subset $\mathrm{d}\boldsymbol{x} \subset \mathbb{R}^{N_x}$ is written

$$\mathcal{P}(\boldsymbol{x}_k \in \mathrm{d}\boldsymbol{x}|\boldsymbol{x}_{k-1}) = \delta_{\mathcal{M}_k(\boldsymbol{x}_{k-1})}(\mathrm{d}\boldsymbol{x}), \tag{3.12}$$

with δ_v referring to the *Dirac measure* at $v \in \mathbb{R}^{N_x}$. The Dirac measure satisfies

$$\int f(\boldsymbol{x})\delta_v(\mathrm{d}\boldsymbol{x}) = f(v), \tag{3.13}$$

where this is a singular measure, to be understood by the integral equation. Accordingly, the transition pdf is often written proportional as

$$p(\boldsymbol{x}_k|\boldsymbol{x}_{k-1}) \propto \delta\{\boldsymbol{x}_k - \mathcal{M}_k(\boldsymbol{x}_{k-1})\}, \tag{3.14}$$

where δ represents the *Dirac distribution*. Heuristically, this is known as the 'function' defined by the property

$$\delta_\epsilon(\boldsymbol{x}) = \begin{cases} \dfrac{1}{\epsilon} & \boldsymbol{x} \in [-\epsilon, +\epsilon] \\ 0 & \text{else} \end{cases}, \quad \delta(\boldsymbol{x}) = \lim_{\epsilon \to 0^+} \delta_\epsilon(\boldsymbol{x}). \tag{3.15}$$

However, this is just a convenient abuse of notations, as the Dirac measure does not have a pdf with respect to the standard Lebesgue measure. Rather, the Dirac distribution is understood through the generalised function theory of distributions (Taylor, 1996, section 3.4) as a type of integral kernel satisfying

$$\int f(\boldsymbol{x}_k)\delta\{\boldsymbol{x}_k - \mathcal{M}_k(\boldsymbol{x}_{k-1})\}\mathrm{d}\boldsymbol{x}_k = f(\mathcal{M}_k(\boldsymbol{x}_{k-1})). \tag{3.16}$$

Equation (3.16) represents the existence and uniqueness of the solution to an initial value problem in deterministic systems of ordinary and partial differential equations, where the knowledge of the state $\boldsymbol{x}_{k-1}$ completely determines the state $\boldsymbol{x}_k$ via the perfect forecast model $\mathcal{M}_k$. Particularly, there is probability 1 of the state following the unique solution to the time evolution, and probability 0 of all other outcomes. Each Markov transition pdf represents the evolution of an initial condition with respect to the dynamical model, conditional ultimately on an uncertain outcome of the initial data from the first prior. While the perfect model assumption is used for simplicity, the decomposition of the forecast pdf can be derived for erroneous models under the additional assumption that the model errors are independent-in-time. Like the decomposition of the forecast pdf, for the observation likelihood we can write

$$p(\boldsymbol{y}_k|\boldsymbol{x}_k, \boldsymbol{y}_{k-1:1}) = p(\boldsymbol{y}_k|\boldsymbol{x}_k), \tag{3.17}$$

due to the independence assumption on the observation errors, and the relationship between $\boldsymbol{x}_k$ and $\boldsymbol{y}_k$. This says that the knowledge of the physical state $\boldsymbol{x}_k$ completely determines the likelihood of the observation $\boldsymbol{y}_k$. Indeed,

$$\boldsymbol{\epsilon}_k = \boldsymbol{y}_k - \mathcal{H}_k(\boldsymbol{x}_k), \tag{3.18}$$

which follows a known distribution and is independent of the other observation error outcomes by assumption.

Consider thus how to estimate the filtering pdf $p(\boldsymbol{x}_k|\boldsymbol{y}_{k:1})$. Using the definition of the conditional pdf, one has

$$p(\boldsymbol{x}_k | \boldsymbol{y}_{k:1}) = \frac{p(\boldsymbol{y}_{k:1}, \boldsymbol{x}_k)}{p(\boldsymbol{y}_{k:1})}. \tag{3.19}$$

Rewriting these pdfs as conditional pdfs, and by using the independence assumption,

$$p(\boldsymbol{x}_k | \boldsymbol{y}_{k:1}) = \frac{p(\boldsymbol{y}_k, (\boldsymbol{x}_k, \boldsymbol{y}_{k-1:1}))}{p(\boldsymbol{y}_{k:1})} \tag{3.20a}$$

$$= \frac{p(\boldsymbol{y}_k | \boldsymbol{x}_k, \boldsymbol{y}_{k-1:1}) p(\boldsymbol{x}_k, \boldsymbol{y}_{k-1:1})}{p(\boldsymbol{y}_{k:1})}$$

$$= \frac{p(\boldsymbol{y}_k | \boldsymbol{x}_k) p(\boldsymbol{x}_k, \boldsymbol{y}_{k-1:1})}{p(\boldsymbol{y}_{k:1})}. \tag{3.20b}$$

Writing the joint pdfs again in terms of conditional pdfs

$$p(\boldsymbol{x}_k | \boldsymbol{y}_{k:1}) = \frac{p(\boldsymbol{y}_k | \boldsymbol{x}_k) p(\boldsymbol{x}_k | \boldsymbol{y}_{k-1:1}) p(\boldsymbol{y}_{k-1:1})}{p(\boldsymbol{y}_k | \boldsymbol{y}_{k-1:1}) p(\boldsymbol{y}_{k-1:1})} \tag{3.21a}$$

$$= \frac{p(\boldsymbol{y}_k | \boldsymbol{x}_k) p(\boldsymbol{x}_k | \boldsymbol{y}_{k-1:1})}{p(\boldsymbol{y}_k | \boldsymbol{y}_{k-1:1})}. \tag{3.21b}$$

Now, suppose that the posterior pdf $p(\boldsymbol{x}_{k-1} | \boldsymbol{y}_{k-1:1})$ at the last observation time t_{k-1} is already computed – then the model forecast of this pdf is given by averaging over the state at time t_{k-1} with respect to the Markov transition pdf,

$$p(\boldsymbol{x}_k | \boldsymbol{y}_{k-1:1}) = \int p(\boldsymbol{x}_k | \boldsymbol{x}_{k-1}) p(\boldsymbol{x}_{k-1} | \boldsymbol{y}_{k-1:1}) \, \mathrm{d}\boldsymbol{x}_{k-1}, \tag{3.22}$$

yielding the forecast-prior. The filtering pdf, on the left-hand-side of Eq. (3.21) is written in terms of: (i) the likelihood of the observed data given the model forecast, $p(\boldsymbol{y}_k | \boldsymbol{x}_k)$; (ii) the forecast-prior given the last best estimate of the state, $p(\boldsymbol{x}_k | \boldsymbol{y}_{k-1:1})$; and (iii) the marginal of the joint pdf $p(\boldsymbol{y}_k, \boldsymbol{x}_k | \boldsymbol{y}_{k-1:1})$, integrating out the hidden variables,

$$p(\boldsymbol{y}_k | \boldsymbol{y}_{k-1:1}) = \int p(\boldsymbol{y}_k | \boldsymbol{x}_k) p(\boldsymbol{x}_k | \boldsymbol{y}_{k-1:1}) \, \mathrm{d}\boldsymbol{x}_k. \tag{3.23}$$

This type of pdf, only depending on the observations, is called an *evidence* (e.g. Carrassi et al., 2017).

Typically, the pdf in the denominator of Eq. (3.21) is mathematically intractable. However, the denominator is independent of the hidden variable $\boldsymbol{x}_k$ by construction – the free argument in the pdf on the left-hand-side is the model state $\boldsymbol{x}_k$ and the purpose of the denominator on the right-hand-side is only to normalise the integral of the posterior pdf to 1. Instead, as a proportionality statement,

$$p(\boldsymbol{x}_k | \boldsymbol{y}_{k:1}) \propto p(\boldsymbol{y}_k | \boldsymbol{x}_k) p(\boldsymbol{x}_k | \boldsymbol{y}_{k-1:1}), \tag{3.24}$$

one can devise the Bayesian MAP estimate as the choice of $\bar{\boldsymbol{x}}_k$ that maximises the posterior pdf, but written in terms of the two right-hand-side components in Eq. (3.24). For the purpose of maximising the posterior

pdf, the denominator leads to insignificant constants that can be discarded. Thus, in order to compute the MAP sequentially and recursively in time, one can develop a recursion in proportionality. However, note that the evidence can be estimated and used for other significant purposes still within a Bayesian framework (Carrassi et al., 2017).

3.2.3 Linear-Gaussian Models

Generally, the filtering pdf $p(\boldsymbol{x}_k | \boldsymbol{y}_{k:1})$ has no analytical solution (i.e. no explicit expression). However, when the models are linear, that is, both the state and observation models are written as matrix actions

$$\boldsymbol{x}_k = \mathbf{M}_k \boldsymbol{x}_{k-1}, \tag{3.25a}$$

$$\boldsymbol{y}_k = \mathbf{H}_k \boldsymbol{x}_k + \boldsymbol{\epsilon}_k, \tag{3.25b}$$

and the error pdfs are *Gaussian*:

$$p(\boldsymbol{x}_0) = n(\boldsymbol{x}_0 | \bar{\boldsymbol{x}}_0, \mathbf{B}_0), \ p(\boldsymbol{y}_k | \boldsymbol{x}_k) = n(\boldsymbol{y}_k | \mathbf{H}_k \boldsymbol{x}_k, \mathbf{R}_k), \tag{3.26}$$

$$n(\boldsymbol{z} | \bar{\boldsymbol{z}}, \mathbf{B}) := \frac{1}{\sqrt{(2\pi)^{N_z} \det(\mathbf{B})}} \exp\left\{ -\frac{1}{2} (\bar{\boldsymbol{z}} - \boldsymbol{z})^{\mathrm{T}} \mathbf{B}^{-1} (\bar{\boldsymbol{z}} - \boldsymbol{z}) \right\}, \tag{3.27}$$

then the forecast and posterior pdfs are Gaussian at all times, and are parametrised in terms of their mean and covariance. In particular, Gaussian distributions are closed under affine transformations, that is, maps of the form

$$\boldsymbol{f}(\boldsymbol{x}) = \mathbf{A}\boldsymbol{x} + \boldsymbol{b}, \tag{3.28}$$

corresponding to a linear transformation when $\boldsymbol{b}$ is a vector of zeros, such that $\boldsymbol{b} = \mathbf{0}$ (Tong, 2012, see theorem 3.3.3). If $\boldsymbol{x}$ is distributed with pdf $p(\boldsymbol{x}) = n(\boldsymbol{x} | \bar{\boldsymbol{x}}, \mathbf{B})$, then the random vector $\boldsymbol{y} := \mathbf{A}\boldsymbol{x} + \boldsymbol{b}$ is distributed with pdf

$$p(\boldsymbol{y}) = n(\boldsymbol{y} | \mathbf{A}\bar{\boldsymbol{x}} + \boldsymbol{b}, \mathbf{A}\mathbf{B}\mathbf{A}^{\mathrm{T}}). \tag{3.29}$$

Suppose that the last analysis pdf is given as

$$p(\boldsymbol{x}_{k-1} | \boldsymbol{y}_{k-1:1}) = n(\boldsymbol{x}_{k-1} | \bar{\boldsymbol{x}}_{k-1}^{\mathrm{a}}, \mathbf{B}_{k-1}^{\mathrm{a}}), \tag{3.30}$$

parametrised in terms of the analysis mean $\bar{\boldsymbol{x}}_{k-1}^{\mathrm{a}}$ and analysis error covariance $\mathbf{B}_{k-1}^{\mathrm{a}}$. Then, the forecast-prior pdf is written as

$$p(\boldsymbol{x}_k | \boldsymbol{y}_{k-1:1}) = n(\boldsymbol{x}_k | \bar{\boldsymbol{x}}_{k-1}^{\mathrm{f}}, \mathbf{B}_k^{\mathrm{f}}), \tag{3.31}$$

where the forecast mean and forecast error covariance are defined by

$$\bar{\boldsymbol{x}}_k^{\mathrm{f}} := \mathbf{M}_k \bar{\boldsymbol{x}}_{k-1}^{\mathrm{a}}, \tag{3.32a}$$

$$\mathbf{B}_k^{\mathrm{f}} := \mathbf{M}_k \mathbf{B}_{k-1}^{\mathrm{a}} \mathbf{M}_k^{\mathrm{T}}, \tag{3.32b}$$

respectively.

Similarly, the conditional and marginal distributions of a Gaussian random vector are also Gaussian and their pdf has an analytical form. Suppose that $z \in \mathbb{R}^{N_z}$ is an arbitrary Gaussian random vector, partitioned as

$$z := \begin{pmatrix} x \\ y \end{pmatrix} \quad p(z) := n\left(z \middle| \begin{pmatrix} \bar{x} \\ \bar{y} \end{pmatrix}, \begin{pmatrix} \Sigma_{xx} & \Sigma_{xy} \\ \Sigma_{yx} & \Sigma_{yy} \end{pmatrix}\right), \quad (3.33)$$

with the dimensions given as $N_z = N_x + N_y$ and

$$x, \bar{x} \in \mathbb{R}^{N_x}, \quad y, \bar{y} \in \mathbb{R}^{N_y}, \quad \Sigma_{xx} \in \mathbb{R}^{N_x \times N_x}, \quad (3.34a)$$

$$\Sigma_{xy} = \Sigma_{yx}^{\mathrm{T}} \in \mathbb{R}^{N_x \times N_y}, \quad \Sigma_{yy} \in \mathbb{R}^{N_y \times N_y}. \quad (3.34b)$$

Then, the general form of the pdf for x conditioned on the outcome of y is given by the Gaussian

$$p(x|y) = n\left(x \middle| \bar{x} + \Sigma_{xy}\Sigma_{yy}^{-1}(y - \bar{y}), \Sigma_{xx} - \Sigma_{xy}\Sigma_{yy}^{-1}\Sigma_{yx}\right), \quad (3.35)$$

where the covariance matrix $\Sigma_{xx} - \Sigma_{xy}\Sigma_{yy}^{-1}\Sigma_{yx}$ is called the *Schur complement* (see, e.g., theorem 3.3.4 of Tong, 2012). Noting the form of the linear observation model in Eq. (3.25), and the independence of the observation errors, it is easy to see that the vector composed of $\begin{pmatrix} x_k^{\mathrm{T}} & y_k^{\mathrm{T}} \end{pmatrix}^{\mathrm{T}}$ is jointly Gaussian. Relying on the identifications

$$\Sigma_{xy} = \mathbf{B}_k^{\mathrm{f}} \mathbf{H}_k^{\mathrm{T}}, \quad \Sigma_{yy} = \mathbf{H}_k \mathbf{B}_k^{\mathrm{f}} \mathbf{H}_k^{\mathrm{T}} + \mathbf{R}_k, \quad (3.36)$$

the posterior pdf at time t_k is derived as the Gaussian with analysis mean and analysis error covariance given by

$$\bar{x}_k^{\mathrm{a}} := \bar{x}_k^{\mathrm{f}} + \mathbf{B}_k^{\mathrm{f}} \mathbf{H}_k^{\mathrm{T}} \left(\mathbf{H}_k \mathbf{B}_k^{\mathrm{f}} \mathbf{H}_k^{\mathrm{T}} + \mathbf{R}_k\right)^{-1} (y_k - \mathbf{H}_k \bar{x}_k^{\mathrm{f}}), \quad (3.37a)$$

$$\mathbf{B}_k^{\mathrm{a}} := \mathbf{B}_k^{\mathrm{f}} - \mathbf{B}_k^{\mathrm{f}} \mathbf{H}_k^{\mathrm{T}} \left(\mathbf{H}_k \mathbf{B}_k^{\mathrm{f}} \mathbf{H}_k^{\mathrm{T}} + \mathbf{R}_k\right)^{-1} \mathbf{H}_k \mathbf{B}_k^{\mathrm{f}}. \quad (3.37b)$$

Defining $\mathbf{K}_k := \mathbf{B}_k^{\mathrm{f}} \mathbf{H}_k^{\mathrm{T}} \left(\mathbf{H}_k \mathbf{B}_k^{\mathrm{f}} \mathbf{H}_k^{\mathrm{T}} + \mathbf{R}_k\right)^{-1}$ as the Kalman gain, Eqs. (3.37a,b) yield the classical *Kalman filter* (KF) update, and this derivation inductively defines the forecast and posterior distribution for x_k at all times. The KF is recognised thus as the parametric representation of a linear-Gaussian hidden Markov model, providing a recursion on the first two moments for the forecast and posterior.

This analysis extends, as with the classical KF, easily to incorporate additive model errors as in Eq. (3.1); see, for example, Anderson and Moore (1979). However, there are many ways to formulate the KF and there are some drawbacks of this approach. Even when the model forecast equations themselves are linear, if they functionally depend on an uncertain parameter vector, defined $\mathbf{M}_k(\lambda)$, the joint estimation problem of (x_k, λ) can become highly non-linear and an iterative approach to the joint estimation may be favourable. Therefore, this chapter develops the subsequent extensions to the KF with the MAP approach, which coincides with the development in least-squares and non-linear optimisation.

3.3 Least-Squares and Non-linear Optimisation

3.3.1 The 3-D Cost Function from Gaussian Statistics

As seen in the previous section, the linear-Gaussian analysis admits an analytical solution for the forecast and posterior pdf at all times. However, an alternative approach for deriving the KF is formed using the MAP estimation in proportionality. Consider that the *natural logarithm* (log) is monotonic, so that an increase in the input of the argument corresponds identically to an increase in the output of the function. Therefore, maximising the posterior pdf as in Eq. (3.24) is equivalent to maximising

$$\log(p(y_k|x_k)p(x_k|y_{k-1:1})) = \log(p(y_k|x_k)) + \log(p(x_k|y_{k-1:1})). \quad (3.38)$$

Given the form of the multivariate Gaussian pdf in Eq. (3.27), maximising the log-posterior Eq. (3.38) is equivalent to minimising the following least-squares cost function, derived in proportionality to the minus-log-posterior:

$$\mathcal{J}_{\mathrm{KF}}(x_k) = \frac{1}{2}\|\bar{x}_k^{\mathrm{f}} - x_k\|_{\mathbf{B}_k^{\mathrm{f}}}^2 + \frac{1}{2}\|y_k - \mathbf{H}_k x_k\|_{\mathbf{R}_k}^2. \quad (3.39)$$

For an arbitrary positive definite matrix $\mathbf{A}$, the *Mahalanobis distance* (Mahalanobis, 1936) with respect to $\mathbf{A}$ is defined as

$$\|v\|_{\mathbf{A}} := \sqrt{v^{\mathrm{T}} \mathbf{A}^{-1} v}. \quad (3.40)$$

These distances are weighted Euclidean norms with: (i) $\| \circ \|_{\mathbf{B}_k^{\mathrm{f}}}$ weighting relative to the forecast spread; and (ii) $\| \circ \|_{\mathbf{R}_k}$ weighting relative to the observation imprecision. The MAP state thus interpolates the forecast mean and the observation relative to the uncertainty in each piece of data. Due to the unimodality of the Gaussian, and its symmetry about its mean, it is clear that the conditional mean $\bar{x}_k^{\mathrm{a}}$ is also the MAP state.

While this cost function analysis provides the solution to finding the first moment of the Gaussian posterior, this does not yet address how to find the posterior error covariance. In this linear-Gaussian setting, it is easily shown that the analysis error covariance is actually given by

$$\mathbf{B}_k^{\mathrm{a}} := \left[(\mathbf{B}_k^{\mathrm{f}})^{-1} + \mathbf{H}_k^{\mathrm{T}} \mathbf{R}_k^{-1} \mathbf{H}_k\right]^{-1} = \Xi_{\mathcal{J}_{\mathrm{KF}}}^{-1}, \quad (3.41)$$

where $\Xi_{\mathcal{J}_{\mathrm{KF}}}$ refers to the *Hessian* of the least-squares cost function in Eq. (3.39), that is, the matrix of its mixed second partial derivatives in the state vector variables. This is a fundamental result that links the recursive analysis in time to the MAP estimation performed with linear least-squares; see, for example, section 6.2 of Reich and Cotter (2015) for further details.

A more general result from maximum likelihood estimation extends this analysis as an approximation to non-linear state and observation models, and to non-Gaussian error distributions. Define $\hat{\theta}$ to be the *maximum likelihood estimator* (MLE) for some unknown parameter vector $\theta \in \mathbb{R}^{N_\theta}$, where $\hat{\boldsymbol{\theta}}$ depends on the realisation of some arbitrarily distributed random sample of observed data $\{z_i\}_{i=1}^N$. Then, under fairly general regularity conditions, the MLE satisfies

$$\mathcal{I}(\boldsymbol{\theta})^{-\frac{1}{2}}(\hat{\boldsymbol{\theta}} - \boldsymbol{\theta}) \to_d \mathcal{N}(\mathbf{0}, \mathbf{I}_{N_\theta}), \qquad (3.42)$$

where: (i) Eq. (3.42) refers to convergence in distribution of the random vector $\mathcal{I}(\boldsymbol{\theta})^{-\frac{1}{2}}(\hat{\boldsymbol{\theta}} - \boldsymbol{\theta})$ as the sample size $N \to \infty$; (ii) $\mathbf{I}_{N_\theta}$ is the identity matrix in the dimension of $\boldsymbol{\theta}$; (iii) $\mathcal{N}(\mathbf{0}, \mathbf{I}_{N_\theta})$ is the multivariate Gaussian distribution with mean zero and covariance equal to the identity matrix, that is, with pdf $n(\boldsymbol{x}|\mathbf{0}, \mathbf{I}_{N_\theta})$; and (iv) $\mathcal{I}(\boldsymbol{\theta})$ is the *Fisher information matrix*. The Fisher information matrix is defined as the expected value of the Hessian of the minus-log-likelihood, taken over the realisations of the observed data $\{z_i\}_{i=1}^N$, and with respect to the true parameter vector $\boldsymbol{\theta}$. It is common to approximate the distribution of the MLE as

$$\hat{\boldsymbol{\theta}} \sim \mathcal{N}(\boldsymbol{\theta}, I(\hat{\boldsymbol{\theta}})), \qquad (3.43)$$

where $I(\hat{\boldsymbol{\theta}})$ is the observed Fisher information (i.e. the realisation of the Hessian of the minus-log-likelihood given the observed sample data). This approximation improves in large sample sizes like the central limit theorem; see, for example, chapter 9 of Pawitan (2001) for the details of this discussion. In non-linear optimisation, the approximate distribution of the MLE therefore relates the geometry in the neighbourhood of a local minimiser to the variation in the optimal estimate. Particularly, the curvature of the cost function level contours around the local minimum is described by the Hessian while the spread of the distribution of the MLE is described by the analysis error covariance.

3.3.2 3D-VAR and the Extended Kalman Filter

One of the benefits of the aforementioned cost function approach is that this immediately extends to handle a non-linear observation operator $\mathcal{H}_k$ as a problem of non-linear least-squares. When the observations are related non-linearly to the state vector, the Bayesian posterior is no longer generally Gaussian, but the interpretation of the analysis state $\bar{\boldsymbol{x}}_k^{\mathrm{a}}$ interpolating between a background proposal $\bar{\boldsymbol{x}}_k^{\mathrm{f}}$ and the observed data $\boldsymbol{y}_k$ relative to their respective uncertainties remains valid. The use of the non-linear least-squares cost function can be considered as making a Gaussian approximation, similar to the large sample theory in maximum likelihood estimation. However, the forecast and analysis background error covariances, $\mathbf{B}_k^{\mathrm{f/a}}$, take on different interpretations depending on the DA scheme.

In full-scale geophysical models, the computation and storage of the background error covariance $\mathbf{B}_k^{\mathrm{f/a}}$ is rarely feasible due to its large size; in practice, this is usually treated abstractly in its action as an operator by preconditioning the optimisation (Tabeart et al., 2018). One traditional approach to handle this reduced representation is to use a static-in-time background for the cost function in Eq.

(3.39), rendering the *three-dimensional, variational* (3D-VAR) cost function

$$\mathcal{J}_{\mathrm{3D-VAR}}(\boldsymbol{x}) := \frac{1}{2}\|\bar{\boldsymbol{x}}_k^{\mathrm{f}} - \boldsymbol{x}\|_{\mathbf{B}_{\mathrm{3D-VAR}}}^2 + \frac{1}{2}\|\boldsymbol{y}_k - \mathcal{H}_k(\boldsymbol{x})\|_{\mathbf{R}_k}^2. \qquad (3.44)$$

The 3D-VAR background error covariance is typically defined as a posterior, 'climatological' covariance, taken with respect to a long-time average over the modelled state. One way that this is roughly estimated in meteorology is by averaging over a *reanalysis data set*, which is equivalent to averaging with respect to a joint-smoothing posterior over a long history for the system (Kalnay, 2003, and references therein). If one assumes that the DA cycle represents a stationary, ergodic process in its posterior statistics, and that this admits an invariant, ergodic measure, π^*, one would define the 3D-VAR background error covariance matrix with the expected value with respect to this measure as

$$\bar{\boldsymbol{x}}^* := \mathbb{E}_{\pi^*}[\boldsymbol{x}], \qquad (3.45a)$$

$$\mathbf{B}_{\mathrm{3D-VAR}} := \mathbb{E}_{\pi^*}\left[(\boldsymbol{x} - \bar{\boldsymbol{x}}^*)(\boldsymbol{x} - \bar{\boldsymbol{x}}^*)^\top\right]. \qquad (3.45b)$$

The 3D-VAR cycle is defined where $\bar{\boldsymbol{x}}_k^{\mathrm{a}}$ is the *minimising argument* (argmin) of Eq. (3.44) and subsequently $\bar{\boldsymbol{x}}_{k+1}^{\mathrm{f}} := \mathcal{M}_{k+1}(\bar{\boldsymbol{x}}_k^{\mathrm{a}})$, meaning, the forecast mean is estimated with the background control trajectory propagated through the fully non-linear model. Therefore, at every iteration of the algorithm, 3D-VAR can be understood to treat the proposal $\bar{\boldsymbol{x}}_k^{\mathrm{f}}$ as a random draw from the invariant, climatological-posterior measure of the system; the optimised solution $\bar{\boldsymbol{x}}_k^{\mathrm{a}}$ is thus the state that interpolates the forecast mean and the currently observed data as if the forecast mean was drawn randomly from the climatological-posterior.

For illustration, consider a cost-effective, explicit formulation of 3D-VAR where the background error covariance matrix is represented as a Cholesky factorisation. Here, one performs the optimisation by writing the modelled state as a perturbation of the forecast mean,

$$\boldsymbol{x}_k := \bar{\boldsymbol{x}}_k^{\mathrm{f}} + \Sigma\boldsymbol{w}, \qquad (3.46)$$

where $\mathbf{B}_{\mathrm{3D-VAR}} := \Sigma\Sigma^\mathrm{T}$. The weight vector $\boldsymbol{w}$ gives the linear combination of the columns of the matrix factor that describe $\boldsymbol{x}_k$ as a perturbation. One can iteratively optimise the 3D-VAR cost function as such with a locally quadratic approximation. Let $\bar{\boldsymbol{x}}_k^i$ be the i-th iterate for the proposal value at time t_k, defining $\bar{\boldsymbol{x}}_k^0 := \bar{\boldsymbol{x}}_k^{\mathrm{f}}$. The $i + 1$-st iteration is defined (up to a subtlety pointed out by Ménétrier and Auligné (2015)) as

$$\bar{\boldsymbol{x}}_k^{i+1} = \bar{\boldsymbol{x}}_k^i + \Sigma\overline{\boldsymbol{w}}^{i+1}, \qquad (3.47)$$

where $\overline{\boldsymbol{w}}^{i+1}$ is the argmin of the following *incremental cost function*

$$\mathcal{J}_{\text{3D-VARI}}(\boldsymbol{w}) := \frac{1}{2}\|\boldsymbol{w}\|^2 + \frac{1}{2}\|\boldsymbol{y}_k - \mathcal{H}_k(\overline{\boldsymbol{x}}_k^{\text{i}}) - \mathbf{H}_k\boldsymbol{\Sigma}\boldsymbol{w}\|_{\mathbf{R}_k}^2.$$

(3.48)

This approximate cost function follows from truncating the Taylor expansion

$$\mathcal{H}_k(\boldsymbol{x}) = \mathcal{H}_k(\overline{\boldsymbol{x}}_k^{\text{i}}) + \mathbf{H}_k\boldsymbol{\Sigma}\boldsymbol{w} + \mathcal{O}(\|\boldsymbol{w}\|^2) \qquad (3.49)$$

and the definition of the Mahalanobis distance. Notice that $\overline{\boldsymbol{w}}^{i+1}$ is defined uniquely as Eq. (3.48) is quadratic with respect to the weights, allowing one to use fast and adequate methods such as the conjugate gradient (Nocedal and Wright, 2006, chapter 5). The iterations are set to terminate when $\|\overline{\boldsymbol{w}}^i\|$ is sufficiently small, representing a negligible change from the last proposal. This approach to non-linear optimisation is known as the incremental approach, and corresponds mathematically to the Gauss-Newton optimisation method, an efficient simplification of Newton's descent method for non-linear least-squares (Nocedal and Wright, 2006, chapter 10). As described, it consists of an iterative linearisation of the non-linear optimisation problem, which is a concept utilised throughout this chapter.

The 3D-VAR approach described here played an important role in early DA methodology, and aspects of this technique are still widely used in ensemble-based DA when using the regularisation technique known as covariance hybridisation (Hamill and Snyder, 2000; Lorenc, 2003; Penny, 2017). However, optimising the state with the climatological-posterior background neglects important information in the time-dependent features of the forecast spread, described sometimes as the 'errors of the day' (Corazza et al., 2003). The *extended Kalman filter* (EKF) can be interpreted to generalise the 3D-VAR cost function by including a time-dependent background, like the KF, by approximating its time-evolution at first order with the tangent linear model as follows.

Suppose that the equations of motion are generated by a non-linear function, independent of time for simplicity,

$$\frac{\text{d}}{\text{d}t}\boldsymbol{x} := \boldsymbol{f}(\boldsymbol{x}), \quad \boldsymbol{x}_k = \mathcal{M}_k(\boldsymbol{x}_{k-1}) := \int_{t_{k-1}}^{t_k} \boldsymbol{f}(\boldsymbol{x})\,\text{d}t + \boldsymbol{x}_{k-1}. \quad (3.50)$$

One can extend the linear-Gaussian approximation for the forecast pdf by modelling the state as a perturbation of the analysis mean,

$$\boldsymbol{x}_{k-1} := \overline{\boldsymbol{x}}_{k-1}^{\text{a}} + \boldsymbol{\delta}_{k-1} \sim \mathcal{N}(\overline{\boldsymbol{x}}_{k-1}^{\text{a}}, \mathbf{B}_{k-1}^{\text{a}}), \qquad (3.51\text{a})$$

$$\Leftrightarrow \boldsymbol{\delta}_{k-1} \sim \mathcal{N}(\mathbf{0}, \mathbf{B}_{k-1}^{\text{a}}). \qquad (3.51\text{b})$$

The evolution of the perturbation $\boldsymbol{\delta}_{k-1}$ is written via Taylor's theorem as

$$\frac{\text{d}}{\text{d}t}\boldsymbol{\delta}_{k-1} := \frac{\text{d}}{\text{d}t}(\boldsymbol{x}_{k-1} - \overline{\boldsymbol{x}}_{k-1}^{\text{a}}) \qquad (3.52\text{a})$$

$$= \boldsymbol{f}(\boldsymbol{x}_{k-1}) - \boldsymbol{f}(\overline{\boldsymbol{x}}_{k-1}^{\text{a}}) \qquad (3.52\text{b})$$

$$= \nabla_{\boldsymbol{x}}\boldsymbol{f}(\overline{\boldsymbol{x}}_{k-1}^{\text{a}})\boldsymbol{\delta}_{k-1} + \mathcal{O}(\|\boldsymbol{\delta}_{k-1}\|^2), \qquad (3.52\text{c})$$

where $\nabla_{\boldsymbol{x}}\boldsymbol{f}(\overline{\boldsymbol{x}}_{k-1}^{\text{a}})$ is the Jacobian equation with dependence on the underlying analysis mean state. The linear evolution defined by the truncated Taylor expansion about the underlying reference solution $\overline{\boldsymbol{x}}$,

$$\frac{\text{d}}{\text{d}t}\boldsymbol{\delta} := \nabla_{\boldsymbol{x}}\boldsymbol{f}(\overline{\boldsymbol{x}}) \cdot \boldsymbol{\delta}, \qquad (3.53)$$

is known as the *tangent linear model*.

Making the approximation of the tangent linear model for the first order evolution of the modelled state, we obtain

$$\frac{\text{d}}{\text{d}t}\boldsymbol{x} \approx \boldsymbol{f}(\overline{\boldsymbol{x}}) + \nabla_{\boldsymbol{x}}\boldsymbol{f}(\overline{\boldsymbol{x}}) \cdot \boldsymbol{\delta} \qquad (3.54\text{a})$$

$$\Rightarrow \int_{t_{k-1}}^{t_k} \frac{\text{d}}{\text{d}t}\boldsymbol{x}\,\text{d}t \approx \int_{t_{k-1}}^{t_k} \boldsymbol{f}(\overline{\boldsymbol{x}})\,\text{d}t + \int_{t_{k-1}}^{t_k} \nabla_{\boldsymbol{x}}\boldsymbol{f}(\overline{\boldsymbol{x}}) \cdot \boldsymbol{\delta}\,\text{d}t \qquad (3.54\text{b})$$

$$\Rightarrow \boldsymbol{x}_k \approx \mathcal{M}_k(\overline{\boldsymbol{x}}_{k-1}) + \mathbf{M}_k\boldsymbol{\delta}_{k-1}, \qquad (3.54\text{c})$$

where $\mathbf{M}_k$ is the resolvent of the tangent linear model. Given that Gaussians are closed under affine transformations, the EKF approximation for the (perfect) evolution of the state vector is defined as

$$p(\boldsymbol{x}_k|\boldsymbol{y}_{k-1:1}) \approx n\left(\boldsymbol{x}_k|\mathcal{M}_k(\overline{\boldsymbol{x}}_{k-1}^{\text{a}}), \mathbf{M}_k\mathbf{B}_{k-1}^{\text{a}}\mathbf{M}_k^{\top}\right). \qquad (3.55)$$

Respectively, define the EKF analysis cost function as

$$\mathcal{J}_{\text{EKF}}(\boldsymbol{x}) := \frac{1}{2}\|\overline{\boldsymbol{x}}_k^{\text{f}} - \boldsymbol{x}\|_{\mathbf{B}_k^{\text{f}}}^2 + \frac{1}{2}\|\boldsymbol{y}_k - \mathcal{H}_k(\boldsymbol{x})\|_{\mathbf{R}_k}^2, \qquad (3.56)$$

where

$$\overline{\boldsymbol{x}}_k^{\text{f}} := \mathcal{M}_k(\overline{\boldsymbol{x}}_{k-1}^{\text{a}}), \qquad (3.57\text{a})$$

$$\mathbf{B}_k^{\text{f}} := \mathbf{M}_k\mathbf{B}_{k-1}^{\text{a}}\mathbf{M}_k^{\top}. \qquad (3.57\text{b})$$

A similar, locally quadratic, weight–space optimisation, can be performed versus the non-linear observation operator using a matrix decomposition for the time-varying background forecast error covariance, defining the state as a perturbation using this matrix factor similar to Eq. (3.46). This gives the square root EKF, which was used historically to improve the stability over the direct approach (Tippett et al., 2003).

The accuracy and stability of this EKF depends strongly on the length of the forecast horizon Δ_t (Miller et al., 1994). For short-range forecasting, the perturbation dynamics of the tangent linear model can be an adequate approximation, and is an underlying approximation for most operational

DA (Carrassi et al., 2018). However, the explicit linear approximation can degrade quickly, and especially when the mean state is not accurately known. The perturbation dynamics in the linearisation about the approximate mean can differ substantially from the non-linear dynamics of the true system as the approximate mean state diverges from the modelled system (Grewal and Andrews, 2014, see chapters 7 and 8). These stability issues, and the computational cost of explicitly representing the evolution of the background covariance in the tangent linear model, limit the use of the EKF for geophysical DA. Nonetheless, this technique provides important intuition that is developed later in this chapter.

3.3.3 The Gaussian 4-D Cost Function

Consider again the perfect, linear-Gaussian model represented by Eq. (3.25). Using the Markov assumption, Eq. (3.10), and the independence of observation errors, Eq. (3.17), recursively for the hidden Markov model, the joint posterior decomposes as

$$p(\boldsymbol{x}_{L:0}|\boldsymbol{y}_{L:1}) \propto \underbrace{p(\boldsymbol{x}_0)}_{(i)} \underbrace{\left[\prod_{k=1}^{L} p(\boldsymbol{x}_k|\boldsymbol{x}_{k-1})\right]}_{(ii)} \underbrace{\left[\prod_{k=1}^{L} p(\boldsymbol{y}_k|\boldsymbol{x}_k)\right]}_{(iii)},$$

(3.58)

where (i) is the prior for the initial state $\boldsymbol{x}_0$; (ii) is the free-forecast with the perfect model $\mathbf{M}_k$, depending on some outcome drawn from the prior; and (iii) is the joint likelihood of the observations in the DAW, given the background forecast.

Define the composition of the linear model forecast from time t_{k-1} to t_l as

$$\mathbf{M}_{l:k} := \mathbf{M}_l \cdots \mathbf{M}_k, \qquad \mathbf{M}_{k:k} := \mathbf{I}_{N_x}. \quad (3.59)$$

Using the perfect, linear model hypothesis, note that

$$\boldsymbol{x}_k := \mathbf{M}_{k:1}\boldsymbol{x}_0 \quad (3.60)$$

for every k. Therefore, the transition pdfs in Eq. (3.58) are reduced to a trivial condition by re-writing

$$p(\boldsymbol{x}_{L:0}|\boldsymbol{y}_{L:1}) \propto p(\boldsymbol{x}_0)\left[\prod_{k=1}^{L} p(\boldsymbol{x}_k|\mathbf{M}_{k-1:1}\boldsymbol{x}_0)\right]\left[\prod_{k=1}^{L} p(\boldsymbol{y}_k|\boldsymbol{x}_k)\right]$$

(3.61a)

$$\propto p(\boldsymbol{x}_0)\left[\prod_{k=1}^{L} p(\boldsymbol{y}_k|\mathbf{M}_{k:1}\boldsymbol{x}_0)\right], \quad (3.61b)$$

as this pdf evaluates to zero whenever $\boldsymbol{x}_k \neq \mathbf{M}_{k:1}\boldsymbol{x}_0$. Given a Gaussian prior, and Gaussian observation error distributions, the minus-log-posterior *four-dimensional* (4-D) cost function is derived as

$$\mathcal{J}_{4D}(\boldsymbol{x}_0) := \frac{1}{2}\|\bar{\boldsymbol{x}}_0 - \boldsymbol{x}_0\|_{\mathbf{B}_0}^2 + \frac{1}{2}\sum_{k=1}^{L}\|\boldsymbol{y}_k - \mathbf{H}_k\mathbf{M}_{k:1}\boldsymbol{x}_0\|_{\mathbf{R}_k}^2.$$

(3.62)

Notice that the 4-D cost function in Eq. (3.62) is actually quadratic with respect to the initial condition $\boldsymbol{x}_0$. Therefore, the smoothing problem in the perfect, linear-Gaussian model has a unique, optimal initial condition that minimises the sum-of-square deviations from the prior mean, with distance weighted with respect to the prior covariance, and the observations, with distance weighted with respect to the observation error covariances. The optimal, smoothed initial condition using observation information up to time t_L is denoted $\bar{\boldsymbol{x}}_{0|L}^s$; this gives the smoothed model states at subsequent times by the perfect model evolution,

$$\bar{\boldsymbol{x}}_{k|L}^s := \mathbf{M}_{k:1}\bar{\boldsymbol{x}}_{0|L}^s. \quad (3.63)$$

This derivation is formulated for a smoothing problem with a static-in-time DAW, in which an entire time series of observations $\boldsymbol{y}_{L:1}$ is available, and for which the estimation may be performed 'offline'. However, a simple extension of this analysis accommodates DAWs that are sequentially shifted in time, allowing an 'online' smoothing estimate to be formed analogously to sequential filtering.

Fixed-lag smoothing sets the length of the DAW, L, to be fixed for all time, while the underlying states are cycled through this window as it shifts forward in time. Given the *lag* of length L, suppose that a *shift* S is defined for which $1 \leq S \leq L$. It is convenient to consider an algorithmically stationary DAW, referring to the time indices $\{t_1, \cdots, t_L\}$. In a given cycle, the joint posterior $p(\boldsymbol{x}_{L:1}|\boldsymbol{y}_{L:1})$ is estimated. After the estimate is produced, the DAW is subsequently shifted in time by $S \times \Delta_t$ and all states are re-indexed by $t_k \leftarrow t_{k+S}$ to begin the next cycle. For a lag of L and a shift of S, the observation vectors at times $\{t_{L-S+1}, \cdots, t_L\}$ correspond to observations newly entering the DAW for the analysis performed up to time t_L. When $S = L$, the DAWs are disconnected and adjacent in time, whereas for $S < L$ there is an overlap between the estimated states in sequential DAWs. Figure 3.2 provides a schematic of how the DAW is shifted for a lag of $L = 5$ and shift $S = 2$. Following the common convention in DA that there is no observation at time zero, in addition to the DAW, $\{t_1, \cdots, t_L\}$, states at time t_0 may be estimated or utilised in order to connect estimates between adjacent/overlapping DAWs.

Consider the algorithmically stationary DAW, and suppose that the current analysis time is t_L, where the joint posterior pdf $p(\boldsymbol{x}_{L-S:1-S}|\boldsymbol{y}_{L-S:1-S})$ is available from the last fixed-lag smoothing cycle at analysis time t_{L-S}. Using the independence of observation errors and the Markov assumption recursively,

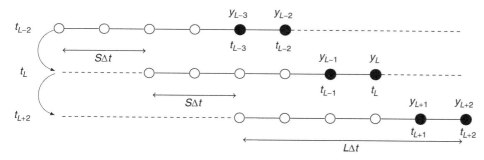

Figure 3.2 Three cycles of a shift $S = 2$, lag $L = 5$ fixed-lag smoother, cycle number is increasing top to bottom. Time indices on the left-hand margin indicate the current time for the associated cycle of the algorithm. New observations entering the current DAW are shaded black. Source: Grudzien and Bocquet (2021), adapted from Asch et al. (2016).

$$p(\boldsymbol{x}_{L:1} | \boldsymbol{y}_{L:1-S}) \propto$$

$$\int \underbrace{\mathrm{d}\boldsymbol{x}_0\, p(\boldsymbol{x}_0 | \boldsymbol{y}_{L-S:1-S})}_{(i)} \underbrace{\left[\prod_{k=1}^{L} p(\boldsymbol{x}_k | \boldsymbol{x}_{k-1})\right]}_{(ii)} \underbrace{\left[\prod_{k=L-S+1}^{L} p(\boldsymbol{y}_k | \boldsymbol{x}_k)\right]}_{(iii)},$$

$$(3.64)$$

where: (i) now represents averaging out the initial condition at time t_0 with respect to the marginal smoothing pdf for $\boldsymbol{x}^{\mathrm{s}}_{0|L-S}$ over the last DAW; (ii) represents the free forecast of the smoothed estimate for $\boldsymbol{x}^{\mathrm{s}}_{0|L-S}$; and term (iii) represents the joint likelihood of the newly incoming observations to the DAW given the forecasted model state. Noting that $p(\boldsymbol{x}_{L:1} | \boldsymbol{y}_{L:1}) \propto p(\boldsymbol{x}_{L:1} | \boldsymbol{y}_{L:1-S})$, this provides a recursive form of the 4-D cost function, shifted sequentially in time,

$$\mathcal{J}_{\mathrm{4D-seq}}(\boldsymbol{x}_0) := \frac{1}{2} \|\bar{\boldsymbol{x}}^{\mathrm{s}}_{0|L-S} - \boldsymbol{x}_0\|^2_{\mathbf{B}^{\mathrm{s}}_{0|L-S}}$$

$$+ \frac{1}{2} \sum_{k=L-S+1}^{L} \|\boldsymbol{y}_k - \mathbf{H}_k \mathbf{M}_{k:1} \boldsymbol{x}_0\|^2_{\mathbf{R}_k}, \qquad (3.65)$$

where $\mathbf{B}^{\mathrm{s}}_{0|L-S}$ refers to the smoothed error covariance from the last DA cycle. As with the KF cost function, the posterior error covariance, conditioning on observations up to time t_L, is identified with

$$\mathbf{B}^{\mathrm{s}}_{0|L} := \boldsymbol{\Xi}^{-1}_{\mathcal{J}_{\mathrm{4D-seq}}}. \qquad (3.66)$$

Given the perfect, linear-Gaussian model assumption, the mean and covariance are propagated to time t_S via

$$\bar{\boldsymbol{x}}^{\mathrm{s}}_{S|L} := \mathbf{M}_{S:1} \bar{\boldsymbol{x}}^{\mathrm{s}}_{0|L}, \qquad (3.67a)$$

$$\mathbf{B}^{\mathrm{s}}_{S|L} := \mathbf{M}_{S:1} \mathbf{B}^{\mathrm{s}}_{0|L} \mathbf{M}^{\mathsf{T}}_{S:1}, \qquad (3.67b)$$

and states are re-indexed as $t_{k+S} \leftarrow t_k$ to initialise the next cycle. This provides a 4-D derivation of the *Kalman smoother* (KS), assuming a perfect model, though not all developments take this approach, using a global analysis over all observations in the current DAW at once. Other formulations use an alternating (i) forward sequential filtering pass to update the current

forecast; and (ii) backward-in-time sequential filtering pass over the DAW to condition lagged states on the new information. Examples of smoothers that follow this alternating analysis include the *ensemble Kalman smoother* (EnKS) (Evensen and Leeuwen, 2000) and the ensemble *Rauch–Tung–Striebel* (RTS) smoother (Raanes, 2016), with the EnKS to be discussed in Section 3.4.3.

3.3.4 Incremental 4D-VAR

The method of incremental *four-dimensional, variational* (4D-VAR) data assimilation (Le Dimet and Talagrand, 1986; Talagrand and Courtier, 1987; Courtier et al., 1994) is a classical and widely used DA technique that extends the linear-Gaussian 4-D analysis to non-linear settings, both for fixed and sequential DAWs. Modern formulations of the 4D-VAR analysis furthermore handle model errors, as in weak-constraint 4D-VAR (Trémolet, 2006; Desroziers et al., 2014; Laloyaux et al., 2020), and may include time-varying background error covariances by combining an ensemble of 4D-VAR (Bonavita et al., 2012), which is known as *ensemble of data assimilation* (EDA).

Assuming now that the state and observation models are non-linear, as in Eq. (3.7), denote the composition of the non-linear model forecast from time t_{k-1} to time t_l as

$$\mathcal{M}_{l:k} := \mathcal{M}_l \circ \cdots \circ \mathcal{M}_k, \quad \mathcal{M}_{k:k} := \mathbf{I}_{N_x}. \qquad (3.68)$$

Then, the linear-Gaussian 4-D cost function is formally extended as a Gaussian approximation in 4D-VAR with

$$\mathcal{J}_{\mathrm{4D-VAR}}(\boldsymbol{x}_0) := \frac{1}{2} \|\bar{\boldsymbol{x}}_0 - \boldsymbol{x}_0\|^2_{\mathbf{B}_{\mathrm{4D-VAR}}}$$

$$+ \frac{1}{2} \sum_{k=1}^{L} \|\boldsymbol{y}_k - \mathcal{H}_k \circ \mathcal{M}_{k:1}(\boldsymbol{x}_0)\|^2_{\mathbf{R}_k}, \qquad (3.69)$$

where $\mathbf{B}_{\mathrm{4D-VAR}} \leftarrow \mathbf{B}_{\mathrm{3D-VAR}}$. A similar indexing to equation to Eq. (3.65) gives the sequential form over newly incoming observations.

The incremental linearisation that was performed in 3D-VAR in Eq. (3.48) forms the basis for the classical technique of incremental 4D-VAR. Suppose again that an explicit Cholesky factor for the background covariance is given $\mathbf{B}_{\text{4D-VAR}} := \Sigma\Sigma^\top$, where the state is written as a perturbation of the i-th iterate as

$$\boldsymbol{x}_0 := \overline{\boldsymbol{x}}_0^i + \Sigma\boldsymbol{w}. \tag{3.70}$$

Taking a Taylor expansion of the composition of the non-linear observation model and the non-linear state model at the i-th iterate,

$$\mathcal{H}_k \circ \mathcal{M}_{k:1}(\boldsymbol{x}_0) = \mathcal{H} \circ \mathcal{M}_{k:1}(\overline{\boldsymbol{x}}_0^i) + \mathbf{H}_k\mathbf{M}_{k:1}\Sigma\boldsymbol{w} + \mathcal{O}(\|\boldsymbol{w}\|^2), \tag{3.71}$$

the incremental cost function is rendered

$$\mathcal{J}_{\text{4D-VARI}}(\boldsymbol{w}) = \frac{1}{2}\|\boldsymbol{w}\|^2 + \frac{1}{2}\sum_{k=1}^{L}\|\boldsymbol{y}_k - \mathcal{H}_k \circ \mathcal{M}_{k:1}(\overline{\boldsymbol{x}}_0^i) \\ - \mathbf{H}_k\mathbf{M}_{k:1}\Sigma\boldsymbol{w}\|^2_{\mathbf{R}_k}. \tag{3.72}$$

The argmin of Eq. (3.72) is defined as $\overline{\boldsymbol{w}}^i$, where $\overline{\boldsymbol{x}}^{i+1}$ is defined as in Eq. (3.47).

It is important to remember that in Eq. (3.72), the term $\mathbf{M}_{k:1}$ involves the calculation of the tangent linear model with reference to the underlying non-linear solution $\mathcal{M}_{L:1}(\overline{\boldsymbol{x}}_0^i)$, simulated freely over the entire DAW. A key aspect to the efficiency of the incremental 4D-VAR approach is in the adjoint model computation of the gradient to this cost function, commonly known as *backpropagation* in statistical and machine learning (Rojas, 2013, see chapter 7). The *adjoint model* with respect to the proposal is defined as

$$\frac{\mathrm{d}}{\mathrm{d}t}\widetilde{\boldsymbol{\delta}} = \left(\nabla_x\boldsymbol{f}(\overline{\boldsymbol{x}}^i)\right)^\top\widetilde{\boldsymbol{\delta}}. \tag{3.73}$$

The solution to the linear adjoint model then defines the adjoint resolvent matrix $\mathbf{M}_k^\top$ (i.e. the transpose of the tangent linear resolvent). Notice that for the composition of the tangent linear model forecasts with $k \leq l$, the adjoint is given as

$$\mathbf{M}_{l:k}^\top := (\mathbf{M}_l\mathbf{M}_{l-1}\cdots\mathbf{M}_{k+1}\mathbf{M}_k)^\top = \mathbf{M}_k^\top\mathbf{M}_{k+1}^\top\cdots\mathbf{M}_{l-1}^\top\mathbf{M}_l^\top. \tag{3.74}$$

This means that the adjoint model variables $\widetilde{\delta l}$ are propagated by the linear resolvents of the adjoint model, but applied in reverse chronological order from the tangent linear model from last time t_l to the initial time t_{k-1},

$$\widetilde{\boldsymbol{\delta}}_{k-1} := \mathbf{M}_{l:k}^\top\widetilde{\boldsymbol{\delta}}_l \tag{3.75}$$

transmitting the sensitivity from a future time back to a perturbation of (Kalnay, 2003, see section 6.3).

Define the *innovation* vector of the i-th iterate and the k-th observation vector as

$$\overline{\boldsymbol{\delta}}_k^i := \boldsymbol{y}_k - \mathcal{H}_k \circ \mathcal{M}_{k:1}(\overline{\boldsymbol{x}}_0^i). \tag{3.76}$$

Then the gradient of Eq. (3.72) with respect to the weight vector is written

$$\nabla_w\mathcal{J}_{\text{4D-VARI}} = \boldsymbol{w} - \sum_{k=1}^{L}\Sigma^\top\mathbf{M}_{k:1}^\top\mathbf{H}_k^\top\mathbf{R}_k^{-1}\left[\overline{\boldsymbol{\delta}}_k^i - \mathbf{H}_k\mathbf{M}_{k:1}\Sigma\boldsymbol{w}\right]. \tag{3.77}$$

Making the substitution

$$\Delta_k := \mathbf{R}_k^{-1}\left[\overline{\boldsymbol{\delta}}_k^i - \mathbf{H}_k\mathbf{M}_{k:1}\Sigma\boldsymbol{w}\right], \tag{3.78}$$

notice that the gradient is written in Horner factorisation as

$$\nabla_w\mathcal{J}_{\text{4D-VARI}} = \boldsymbol{w} - \Sigma^\top\mathbf{M}_1^\top[\mathbf{H}_1^\top\Delta_1 + \mathbf{M}_2^\top[\mathbf{H}_2^\top\Delta_2 + \cdots \\ + [\mathbf{M}_L^\top\mathbf{H}_L^\top\Delta_L]]]. \tag{3.79}$$

Defining the adjoint variable as $\widetilde{\boldsymbol{\delta}}_L := \mathbf{H}_L^\top\Delta_L$, with recursion in t_k,

$$\widetilde{\boldsymbol{\delta}}_k := \mathbf{H}_k^\top\Delta_k + \mathbf{M}_{k+1}^\top\widetilde{\boldsymbol{\delta}}_{k+1}, \tag{3.80}$$

the gradient of the cost function is written as

$$\nabla_w\mathcal{J}_{\text{4D-VARI}} := \boldsymbol{w} - \Sigma\widetilde{\boldsymbol{\delta}}_0. \tag{3.81}$$

Using these definitions, the gradient of the incremental cost function is computed from the following steps: (i) a forward pass of the free evolution of $\boldsymbol{x}_0^i$ under the non-linear forecast model, computing the innovations as in Eq. (3.76); (ii) the propagation of the perturbation $\Sigma\boldsymbol{w}$ in the tangent linear model and the linearised observation operator with respect to the proposal, $\overline{\boldsymbol{x}}_k^i$, in order to compute the terms in Eq. (3.78); and (iii) the back propagation of the sensitivities in the adjoint model, Eq. (3.73), to obtain the adjoint variables recursively as in Eq. (3.80) and the gradient from Eq. (3.81).

The benefit of this approach is that it gives an extremely efficient calculation of the gradient, provided that the tangent linear and adjoint models are available, which, in turn, is key to very efficient numerical optimisations of cost functions. However, the trade-off is that the tangent linear and adjoint models of a full-scale geophysical state model, and of the observation model, require considerable development time and expertise. Increasingly, these models can be computed abstractly by differentiating a computer program alone, in what is known as *automatic differentiation* of code (Griewank, 1989; Griewank and Walther, 2003; Hascöet, 2014; Baydin et al., 2018). When tangent linear and adjoint models are not available, one alternative is to use ensemble sampling techniques in the fully non-linear model $\mathcal{M}$ alone. The ensemble-based analysis provides a complementary approach to the explicit tangent linear and adjoint model analysis – this approach can be developed independently, or hybridised with the use of the tangent linear and adjoint models as in various flavours of hybrid *ensemble-variational* (EnVAR) techniques (Asch et al., 2016; Bannister, 2017). This chapter focuses on the independent development of EnVAR estimators in the following section.

3.4 Bayesian Ensemble-Variational Estimators

3.4.1 The Ensemble Transform Kalman Filter

Consider once again the perfect, linear-Gaussian model in Eq. (3.25). Rather than explicitly computing the evolution of the background mean and error covariance in the linear model as in Eq. (3.32) and the KF, one can alternatively estimate the state mean and the error covariances using a statistical sampling approach. Let $\left\{ x_{k,i}^{f/a} \right\}_{i=1}^{N_e}$ be replicates of the model state, *independently and identically distributed* (iid) according to the distribution

$$x_{k,i}^{f/a} \sim \mathcal{N}(\bar{x}_k^{f/a}, \mathbf{B}_k^{f/a}). \tag{3.82}$$

Given the iid assumption, the ensemble-based mean and the ensemble-based covariance

$$\hat{x}_k^{f/a} := \frac{1}{N_e} \sum_{i=1}^{N_e} x_{k,i}^{f/a}, \tag{3.83a}$$

$$\mathbf{P}_k^{f/a} := \frac{1}{N_e - 1} \sum_{i=1}^{N_e} \left(x_{k,i}^{f/a} - \hat{x}_k^{f/a} \right) \left(x_{k,i}^{f/a} - \hat{x}_k^{f/a} \right)^{\mathsf{T}}, \tag{3.83b}$$

are (asymptotically) consistent estimators of the background, that is,

$$\mathbb{E}\left[\hat{x}_k^{f/a} \right] = \bar{x}_k^{f/a}, \qquad \lim_{N_e \to \infty} \mathbb{E}\left[\mathbf{P}_k^{f/a} \right] = \mathbf{B}_k^{f/a}, \tag{3.84}$$

where the expectation is over the possible realisations of the random sample. Particularly, the multivariate central limit theorem gives that

$$\left(\mathbf{P}_k^{f/a} \right)^{-\frac{1}{2}} \left(x_k^{f/a} - \hat{x}_k^{f/a} \right) \to_d \mathcal{N}(\mathbf{0}, \mathbf{I}_{N_x}), \tag{3.85}$$

referring to convergence in distribution as $N_e \to \infty$ (Härdle et al., 2017, see section 6.2).

Using the relationships in Eq. (3.83), these estimators are efficiently encoded as linear operations on the ensemble matrix. Define the *ensemble matrix* and the *perturbation matrix* as

$$\mathbf{E}_k^{f/a} := \left(x_{k,1}^{f/a} \quad \cdots \quad x_{k,N_e}^{f/a} \right) \in \mathbb{R}^{N_x \times N_e}, \tag{3.86}$$

$$\mathbf{X}_k^{f/a} := \left(x_{k,1}^{f/a} - \hat{x}_k^{f/a} \quad \cdots \quad x_{k,N_e}^{f/a} - \hat{x}_k^{f/a} \right) \in \mathbb{R}^{N_x \times N_e}, \tag{3.87}$$

that is, as the arrays with the columns given as the ordered replicates of the model state and their deviations from the ensemble mean respectively. For $\mathbf{1}$, defined as the vector composed entirely of ones, define the following linear operations conformally in their dimensions

$$\hat{x}_k^{f/a} = \mathbf{E}_k^{f/a} \mathbf{1} / N_e, \tag{3.88a}$$

$$\mathbf{X}_k^{f/a} = \mathbf{E}_k^{f/a} (\mathbf{I}_{N_e} - \mathbf{1}\mathbf{1}^{\mathsf{T}} / N_e), \tag{3.88b}$$

$$\mathbf{P}_k^{f/a} = \left(\mathbf{X}_k^{f/a} \right) \left(\mathbf{X}_k^{f/a} \right)^{\mathsf{T}} / (N_e - 1). \tag{3.88c}$$

The operator $\mathbf{I}_{N_e} - \mathbf{1}\mathbf{1}^{\mathsf{T}} / N_e$ is the orthogonal complementary projection operator to the span of the vector of ones, known as the *centring operator* in statistics (Härdle et al., 2017, see section 6.1); this has the effect of transforming the ensemble to mean zero.

Notice, when $N_e \le N_x$, the ensemble-based error covariance has rank of at most $N_e - 1$, irrespective of the rank of the background error covariance, corresponding to the one degree of freedom lost in computing the ensemble mean. Therefore, to utilise the ensemble error covariance in the least-squares optimisation as with the KF in Eq. (3.39), a new construction is necessary. For a generic matrix $\mathbf{A} \in \mathbb{R}^{N \times M}$ with full column rank M, define the (left Moore–Penrose) *pseudo-inverse* (Meyer, 2000, see page 423)

$$\mathbf{A}^{\dagger} := (\mathbf{A}^{\top} \mathbf{A})^{-1} \mathbf{A}^{\top}. \tag{3.89}$$

In particular, $\mathbf{A}^{\dagger}\mathbf{A} = \mathbf{I}_M$ (and the orthogonal projector into the column span of $\mathbf{A}$) is defined by $\mathbf{A}\mathbf{A}^{\dagger}$. When $\mathbf{A}$ has full column rank as here, define the Mahalanobis 'distance' with respect to $\mathbf{G} = \mathbf{A}\mathbf{A}^{\top}$ as

$$\|v\|_{\mathbf{G}} := \sqrt{(\mathbf{A}^{\dagger} v)^{\top} (\mathbf{A}^{\dagger} v)}. \tag{3.90}$$

Note that in the case that $\mathbf{G}$ does not have full column rank (i.e. $N > M$), this is not a true norm on $\mathbb{R}^N$ as it is degenerate in the null space of $\mathbf{A}^{\dagger}$. This instead represents a lift of a non-degenerate norm in the column span of $\mathbf{A}$ to $\mathbf{R}^N$. In the case that v is in the column span of $\mathbf{A}$, one can equivalently write

$$v = \mathbf{A}w, \tag{3.91a}$$

$$\|v\|_{\mathbf{G}} = \|w\|, \tag{3.91b}$$

for a vector of weights $w \in \mathbf{R}^M$.

The *ensemble Kalman filter* (EnKF) cost function for the linear-Gaussian model is defined

$$\mathcal{J}_{\text{EnKF}}(x_k) := \frac{1}{2} \|\hat{x}_k^f - x_k\|_{\mathbf{P}_k^f}^2 + \frac{1}{2} \|y_k - \mathbf{H}_k x_k\|_{\mathbf{R}_k}^2 \tag{3.92a}$$

$$\Leftrightarrow \mathcal{J}_{\text{EnKF}}(w) := \frac{1}{2} (N_e - 1) \|w\|^2 + \frac{1}{2} \|y_k - \mathbf{H}_k \hat{x}_k^f - \mathbf{H}_k \mathbf{X}_k^f w\|_{\mathbf{R}_k}^2, \tag{3.92b}$$

where the model state is written as a perturbation of the ensemble mean

$$\boldsymbol{x}_k = \hat{\boldsymbol{x}}_k^{\mathrm{f}} + \mathbf{X}_k^{\mathrm{f}} \boldsymbol{w}. \tag{3.93}$$

Notice, $\boldsymbol{w} \in \mathbb{R}^{N_e}$, giving the linear combination of the ensemble perturbations used to represent the model state.

Define $\hat{\boldsymbol{w}}$ to be the argmin of the cost function in Eq. (3.92b). Hunt et al. (2007) and Bocquet (2011) demonstrate that, up to a gauge transformation, $\hat{\boldsymbol{w}}$ yields the argmin of the state-space cost function, Eq. (3.92a), when the estimate is restricted to the ensemble span. Equation (3.92b) is quadratic in $\boldsymbol{w}$ and can be solved to render

$$\hat{\boldsymbol{w}} := \boldsymbol{0} - \boldsymbol{\Xi}_{\mathcal{J}_{\mathrm{EnKF}}}^{-1} \nabla_{\boldsymbol{w}} \mathcal{J}_{\mathrm{EnKF}}|_{\boldsymbol{w}=0}, \tag{3.94a}$$

$$\mathrm{T} := \boldsymbol{\Xi}_{\mathcal{J}_{\mathrm{EnKF}}}^{-\frac{1}{2}}, \tag{3.94b}$$

$$\mathbf{P}_k^{\mathrm{a}} = (\mathbf{X}_k^{\mathrm{f}}\mathrm{T})(\mathbf{X}_k^{\mathrm{f}}\mathrm{T})^\top/(N_e - 1), \tag{3.94c}$$

corresponding to a single iteration of Newton's descent algorithm in Eq. (3.94a), initialised with the ensemble mean, to find the optimal weights.

The linear *ensemble transform Kalman filter* (ETKF) equations (Bishop et al., 2001; Hunt et al., 2007) are then given by

$$\mathbf{E}_k^{\mathrm{f}} = \mathbf{M}_k \mathbf{E}_{k-1}^{\mathrm{a}}, \tag{3.95a}$$

$$\mathbf{E}_k^{\mathrm{a}} = \hat{\boldsymbol{x}}_k^{\mathrm{f}} \mathbf{1}^\top + \mathbf{X}_k^{\mathrm{f}}\left(\hat{\boldsymbol{w}}\mathbf{1}^\top + \sqrt{N_e - 1}\mathrm{T}\mathbf{U}\right), \tag{3.95b}$$

where $\mathbf{U} \in \mathbb{R}^{N_e \times N_e}$ can be any mean-preserving, orthogonal transformation (i.e. $\mathbf{U}\mathbf{1} = \mathbf{1}$). The simple choice of $\mathbf{U} := \mathbf{I}_{N_e}$ is sufficient, but it has been demonstrated that choosing a random, mean-preserving orthogonal transformation at each analysis can improve the accuracy and robustness of the ETKF, smoothing out higher-order artefacts in the empirical covariance estimate (Sakov and Oke, 2008).

Notice that Eq. (3.95b) is written equivalently as a single right ensemble transformation:

$$\mathbf{E}_k^{\mathrm{a}} = \mathbf{E}_k^{\mathrm{f}} \boldsymbol{\Psi}_k, \tag{3.96a}$$

$$\boldsymbol{\Psi}_k := \mathbf{1}\mathbf{1}^\top/N_e + (\mathbf{I}_{N_e} - \mathbf{1}\mathbf{1}^\top/N_e)\left(\hat{\boldsymbol{w}}\mathbf{1}^\top + \sqrt{N_e - 1}\mathrm{T}\mathbf{U}\right), \tag{3.96b}$$

where the columns are approximately distributed as in Eq. (3.82), with the (asymptotic) consistency as with the central limit theorem, Eq. (3.85). However, in the small feasible sample sizes for realistic geophysical models, this approximation can lead to a systematic underestimation of the uncertainty of the analysis state, where the ensemble-based error covariance can become overly confident in its own estimate with artificially small variances. Covariance inflation is a technique that is widely used to regularise the ensemble-based error covariance by increasing the empirical variances of the estimate. This can be used to handle the inaccuracy of the estimator due to the finite sample size

approximation of the background mean and error covariance as in Eq. (3.85), as well as inaccuracies due to a variety of other sources of error (Carrassi et al., 2018; Raanes et al., 2019a; Tandeo et al., 2020).

A Bayesian hierarchical approach can model the inaccuracy in the approximation error due to the finite sample size by including a prior additionally on the background mean and error covariance $p(\overline{\boldsymbol{x}}_k^{\mathrm{f}}, \mathbf{B}_k^{\mathrm{f}})$, as in the finite-size ensemble Kalman filter formalism of Bocquet (2011), Bocquet and Sakov (2012), and Bocquet et al. (2015). Mathematical results demonstrate that covariance inflation can ameliorate the systematic underestimation of the variances due to model error in the presence of a low-rank ensemble (Grudzien et al., 2018). In the presence of significant model error, the finite-size analysis is extended by the variant developed by Raanes et al. (2019a).

The linear transform of the ensemble matrix in Eq. (3.96) is key to the efficiency of the (deterministic) EnKF presented in this section (Sect. 3.4.1). The ensemble-based cost function Hessian $\boldsymbol{\Xi}_{\mathcal{J}_{\mathrm{EnKF}}} \in \mathbb{R}^{N_e \times N_e}$, where $N_e \ll N_x$ for typical geophysical models. Particularly, the cost of computing the optimal weights, as in Eq. (3.94a), and computing the transform matrix $\mathbf{T}$ in Eq. (3.94b) are both subordinate to the cost of the eigenvalue decomposition of the Hessian at $\mathcal{O}(N_e^3)$ *floating point operations* (flops), or to a randomised singular value decomposition (Farchi and Bocquet, 2019). However, the extremely low ensemble size means that the correction to the forecast is restricted to the low-dimensional ensemble span, which may fail to correct directions of rapidly growing errors due to the rank deficiency. While chaotic, dissipative dynamics implies that the background covariance $\mathbf{B}_k^{\mathrm{f/a}}$ has spectrum concentrated on a reduced rank subspace (Carrassi et al., 2022, and referenced therein), covariance hybridisation (Penny, 2017) or localisation (Sakov and Bertino, 2011) are used in practice to regularise the estimator's extreme rank deficiency, and the spurious empirical correlations that occur as a result of the degenerate sample size.

When extended to non-linear dynamics, as in Eq. (3.7a), the EnKF can be seen to make an ensemble-based approximation to the EKF cost function, where the forecast ensemble is defined by

$$\boldsymbol{x}_{k,i}^{\mathrm{f}} := \mathscr{M}_{k-1}(\boldsymbol{x}_{k-1,i}^{\mathrm{a}}), \quad \mathbf{E}_k^{\mathrm{f}} := (\boldsymbol{x}_{k,1}^{\mathrm{f}} \cdots \boldsymbol{x}_{k,N_e}^{\mathrm{f}}). \tag{3.97}$$

The accuracy of the linear-Gaussian approximation to the dynamics of the non-linear evolution of the ensemble, like the approximation of the EKF, depends strongly on the length of the forecast horizon Δ_t. When the ensemble mean is a sufficiently accurate approximation of the mean state, and if the ensemble spread is of the same order as the error in the mean estimate (Whitaker and Loughe, 1998), a similar approximation can be made for the ensemble evolution at first order, as with Eq. (3.53), but linearised about the ensemble mean as discussed later in Eq. (3.112). Despite the similarity to the EKF, in the moderately non-linear dynamics present in medium- to longer-range forecast horizons, the EnKF does not suffer from the same inaccuracy as the EKF in truncating the time-evolution at first order (Evensen, 2003). The forecast ensemble members themselves evolve fully non-linearly, tracking the higher-order dynamics, but the analysis update based on the Gaussian approximation becomes increasingly biased and fails to discriminate features like multimodality of the posterior, even though this is occasionally an asset with higher-order statistical artefacts (see the discussion in Lawson and Hansen, 2004).

3.4.2 The Maximum Likelihood Ensemble Filter

The EnKF filter analysis automatically accommodates weak non-linearity in the state model without using the tangent linear model, as the filtering cost function has no underlying dependence on the state model. However, the EnKF analysis must be adjusted to account for non-linearity in the observation model. When a non-linear observation operator is introduced, as in Eq. (3.7b), the EnKF cost function can be re-written in the incremental analysis as

$$\mathcal{J}_{\text{EnKFI}}(\boldsymbol{w}) := \frac{1}{2}(N_e - 1)\|\boldsymbol{w}\|^2 + \frac{1}{2}\|\boldsymbol{y}_k - \mathcal{H}_k\left(\hat{\boldsymbol{x}}_k^{i,\text{f}}\right)$$
$$- \mathbf{H}_k \mathbf{X}_k^{\text{f}} \boldsymbol{w}\|_{\mathbf{R}_k}^2, \qquad (3.98)$$

where $\hat{\boldsymbol{x}}_k^{i,\text{f}}$ refers to the i-th iteration for the forecast mean. Note that this does not refer to an estimate derived by an iterated ensemble forecast through the non-linear state model – rather, this is an iteration only with respect to the estimate of the optimal weights for the forecast perturbations. If $\hat{\boldsymbol{w}}^i$ is defined as the argmin of the cost function in Eq. (3.98), then the iterations of the ensemble mean are given as

$$\hat{\boldsymbol{x}}_k^{i+1,\text{f}} := \hat{\boldsymbol{x}}_k^{i,\text{f}} + \mathbf{X}_k^{\text{f}} \hat{\boldsymbol{w}}^i. \qquad (3.99)$$

When $\|\hat{\boldsymbol{w}}^i\|$ is sufficiently small, the optimisation terminates and the transform and the ensemble update can be performed as with the ETKF as in Eqs. (3.94b) and (3.96). However, the direct, incremental approach here used the computation of the Jacobian of the observation operator $\mathbf{H}_k$.

The *maximum likelihood ensemble filter* (MLEF) of Zupanski (2005) and Zupanski et al. (2008) is an estimator designed to perform incremental analysis in the ETKF formalism, but without taking an explicit Taylor expansion for the observation operator. The method is designed to approximate the directional derivative of the non-linear observation operator with respect to the ensemble perturbations,

$$\mathbf{H}_k \mathbf{X}_k^{\text{f}} := \nabla|_{\hat{\boldsymbol{x}}_k^{i,\text{f}}} [\mathcal{H}_k] \mathbf{X}_k^{\text{f}}, \qquad (3.100)$$

equivalent to computing the ensemble sensitivities of the map linearised about the ensemble-based mean. This is often performed with an explicit finite differences approximation between the ensemble members and the ensemble mean, mapped to the observation space. Particularly, one may write

$$\mathcal{H}_k\left(\boldsymbol{x}_k^{i,\text{f}}\right) \approx \hat{\boldsymbol{y}}_k^i := \mathcal{H}_k\left(\hat{\boldsymbol{x}}_k^{i,\text{f}} \mathbf{1}^\top + \epsilon \mathbf{X}_k^{\text{f}}\right) \mathbf{1}/N_e, \qquad (3.101a)$$

$$\mathbf{H}_k \mathbf{X}_k^{\text{f}} \approx \widetilde{\mathbf{Y}}_k := \frac{1}{\epsilon} \mathcal{H}_k\left(\hat{\boldsymbol{x}}_k^{i,\text{f}} \mathbf{1}^\top + \epsilon \mathbf{X}_k^{\text{f}}\right)(\mathbf{I}_{N_e} - \mathbf{1}\mathbf{1}^\top/N_e), \qquad (3.101b)$$

where ϵ is a small constant that re-scales the ensemble perturbations to approximate infinitesimals about the mean, and rescales the finite differences about the ensemble mean in the observation space. This technique is used, for example, in a modern form of the MLEF algorithm, based on the analysis of the iterative ensemble Kalman filter and smoother (Asch et al., 2016, see section 6.7.2.1). The approximation in Asch et al. (2016) easily generalises to a 4-D analysis, taking the directional derivative with respect to the state and observation models simultaneously, as in Eq. (3.111). However, in the smoothing problem, one can also extend the analysis in terms of an alternating forward-filtering pass and backward-filtering pass to estimate the joint posterior. This chapter returns to the 4-D analysis with the iterative ensemble Kalman filter and smoother in Section 3.4.4, after an interlude on the retrospective analysis of the EnKS in the following section.

3.4.3 The Ensemble Transform Kalman Smoother

The EnKS extends the filter analysis in the ETKF over the smoothing DAW by sequentially reanalysing past states with future observations with an additional filtering pass over the DAW backward-in-time. This analysis is performed retrospectively in the sense that the filter cycle of the ETKF is left unchanged, while an additional inner-loop of the DA cycle performs an update on the estimated lagged state ensembles within the DAW, stored in memory. This can be formulated both for a fixed DAW and for fixed-lag smoothing, where only minor modifications are needed. Consider here the algorithmically stationary DAW $\{t_1, \ldots, t_L\}$ of fixed-lag smoothing, with a shift S and lag L, and where it is assumed that $S = 1 \le L$. The fixed-lag smoothing cycle of the EnKS begins by estimating the joint posterior pdf $p(\boldsymbol{x}_{L:1}|\boldsymbol{y}_{L:1})$ recursively, given the joint posterior estimate over the last DAW $p(\boldsymbol{x}_{L-1:0}|\boldsymbol{y}_{L-1:0})$.

Given $p(\boldsymbol{x}_{L-1:0}, \boldsymbol{y}_{L-1:0})$, one can write the filtering pdf up to proportionality:

$$p(\boldsymbol{x}_L|\boldsymbol{y}_{L:0}) \propto p(\boldsymbol{y}_L|\boldsymbol{x}_L, \boldsymbol{y}_{L-1:0}) \, p(\boldsymbol{x}_L, \boldsymbol{y}_{L-1:0}) \qquad (3.102a)$$

$$\propto \underbrace{p(\boldsymbol{y}_L|\boldsymbol{x}_L)}_{(i)} \underbrace{\int p(\boldsymbol{x}_L|\boldsymbol{x}_{L-1}) p(\boldsymbol{x}_{L-1:0}|\boldsymbol{y}_{L-1:0}) \mathrm{d}\boldsymbol{x}_{L-1:0}}_{(ii)}, \quad (3.102b)$$

as the product of (i) the likelihood of the observation $\boldsymbol{y}_L$ given $\boldsymbol{x}_L$; and (ii) the forecast for $\boldsymbol{x}_L$ using the transition pdf on the last joint posterior estimate, marginalising out the past history of the model state $\boldsymbol{x}_{L-1:0}$. Recalling that $p(\boldsymbol{x}_L|\boldsymbol{y}_{L:1}) \propto p(\boldsymbol{x}_L|\boldsymbol{y}_{L:0})$, this provides a means to estimate the filter marginal of the joint posterior. An alternating filtering pass, backward-in-time, completes the smoothing cycle by estimating the joint posterior pdf $p(\boldsymbol{x}_{L:1}, \boldsymbol{y}_{L:1})$.

Consider that the marginal smoother pdf is proportional to

$$p(\boldsymbol{x}_{L-1}|\boldsymbol{y}_{L:0}) \propto p(\boldsymbol{y}_L|\boldsymbol{x}_{L-1}, \boldsymbol{y}_{L-1:0}) p(\boldsymbol{x}_{L-1}, \boldsymbol{y}_{L-1:0}) \quad (3.103a)$$

$$\propto \underbrace{p(\boldsymbol{y}_L|\boldsymbol{x}_{L-1})}_{(i)} \underbrace{p(\boldsymbol{x}_{L-1}|\boldsymbol{y}_{L-1:0})}_{(ii)}, \quad (3.103b)$$

where (i) is the likelihood of the observation $\boldsymbol{y}_L$ given the past state $\boldsymbol{x}_{L-1}$, and (ii) is the marginal pdf for $\boldsymbol{x}_{L-1}$ from the last joint posterior. The corresponding linear-Gaussian Bayesian MAP cost function is given for the retrospective analysis of the KS as

$$\mathcal{J}_{\mathrm{KS}}(\boldsymbol{x}_{L-1}) = \frac{1}{2} \|\boldsymbol{x}_{L-1} - \overline{\boldsymbol{x}}^{\mathrm{s}}_{L-1|L-1}\|^2_{\mathbf{B}^{\mathrm{s}}_{L-1|L-1}} + \frac{1}{2}\|\boldsymbol{y}_L$$
$$- \mathbf{H}_L \mathbf{M}_L \boldsymbol{x}_{L-1}\|^2_{\mathbf{R}_L}, \quad (3.104)$$

where $\overline{\boldsymbol{x}}^{\mathrm{s}}_{L-1|L-1}$ and $\mathbf{B}^{\mathrm{s}}_{L-1|L-1}$ are the mean and covariance of the marginal smoother pdf $p(\boldsymbol{x}_{L-1}|\boldsymbol{y}_{L-1:0})$. Define the matrix decomposition with the factorisation (e.g. a Cholesky decomposition):

$$\mathbf{B}^{\mathrm{s}}_{L-1|L-1} = \boldsymbol{\Sigma}^{\mathrm{s}}_{L-1|L-1} (\boldsymbol{\Sigma}^{\mathrm{s}}_{L-1|L-1})^\top, \quad (3.105)$$

and write $\boldsymbol{x}_{L-1} = \overline{\boldsymbol{x}}^{\mathrm{s}}_{L-1|L-1} + \boldsymbol{\Sigma}^{\mathrm{s}}_{L-1|L-1} \boldsymbol{w}$, rendering the cost function as

$$\mathcal{J}_{\mathrm{KS}}(\boldsymbol{w}) = \frac{1}{2}\|\boldsymbol{w}\|^2 + \frac{1}{2}\|\boldsymbol{y}_L - \mathbf{H}_L \mathbf{M}_L (\overline{\boldsymbol{x}}^{\mathrm{s}}_{L-1|L-1} + \boldsymbol{\Sigma}^{\mathrm{s}}_{L-1|L-1} \boldsymbol{w})\|^2_{\mathbf{R}_L} \quad (3.106a)$$

$$= \frac{1}{2}\|\boldsymbol{w}\|^2 + \frac{1}{2}\|\boldsymbol{y}_L - \mathbf{H}_L \overline{\boldsymbol{x}}^{\mathrm{f}}_L - \mathbf{H}_L \boldsymbol{\Sigma}^{\mathrm{f}}_L \boldsymbol{w}\|^2_{\mathbf{R}_L}. \quad (3.106b)$$

Let $\overline{\boldsymbol{w}}$ now denote the argmin of Eq. (3.106). It is important to recognise that

$$\boldsymbol{x}_L := \mathbf{M}_L \left(\overline{\boldsymbol{x}}^{\mathrm{s}}_{L-1|L-1} + \boldsymbol{\Sigma}^{\mathrm{s}}_{L-1|L-1} \boldsymbol{w} \right) \quad (3.107a)$$

$$= \overline{\boldsymbol{x}}^{\mathrm{f}}_L + \boldsymbol{\Sigma}^{\mathrm{f}}_L \boldsymbol{w}, \quad (3.107b)$$

such that the argmin for the smoothing problem $\overline{\boldsymbol{w}}$ is also the argmin for the filtering MAP analysis.

The ensemble-based approximation,

$$\boldsymbol{x}_{L-1} = \hat{\boldsymbol{x}}^{\mathrm{s}}_{L-1|L-1} + \mathbf{X}^{\mathrm{s}}_{L-1|L-1} \boldsymbol{w}, \quad (3.108a)$$

$$\mathcal{J}_{\mathrm{EnKS}}(\boldsymbol{w}) = \frac{1}{2}(N_e - 1)\|\boldsymbol{w}\|^2 + \frac{1}{2}\|\boldsymbol{y}_L - \mathbf{H}_L \hat{\boldsymbol{x}}^{\mathrm{f}}_L - \mathbf{H}_L \mathbf{X}^{\mathrm{f}}_L \boldsymbol{w}\|^2, \quad (3.108b)$$

to the exact smoother cost function in Eq. (3.106), yields the retrospective analysis of the EnKS as

$$\hat{\boldsymbol{w}} := \boldsymbol{0} - \boldsymbol{\Xi}^{-1}_{\mathcal{J}_{\mathrm{EnKS}}} \nabla \mathcal{J}_{\mathrm{EnKS}}|_{\boldsymbol{w}=\boldsymbol{0}}, \quad (3.109a)$$

$$\mathbf{T} := \boldsymbol{\Xi}^{-\frac{1}{2}}_{\mathcal{J}_{\mathrm{EnKS}}}, \quad (3.109b)$$

$$\mathbf{E}^{\mathrm{s}}_{L-1|L} = \hat{\boldsymbol{x}}^{\mathrm{s}}_{L-1|L-1} \mathbf{1}^\top + \mathbf{X}^{\mathrm{s}}_{L-1|L-1} \left(\hat{\boldsymbol{w}} \mathbf{1}^\top + \sqrt{N_e - 1} \mathsf{T} \mathbf{U} \right),$$
$$= \mathbf{E}^{\mathrm{s}}_{L-1|L-1} \boldsymbol{\Psi}_L, \quad (3.109c)$$

where $\boldsymbol{\Psi}_L$ is the ensemble transform as defined for the filtering update as in Eq. (3.96).

These equations generalise for arbitrary indices $k|L$ over the DAW, providing the complete description of the inner-loop between each filter cycle of the EnKS. After each new observation is assimilated with the ETKF analysis step, a smoother inner-loop makes a backward pass over the DAW applying the transform and the weights of the ETKF filter update to each past ensemble state stored in memory. This analysis easily generalises to the case where there is a shift of the DAW with $S > 1$, though the EnKS alternating forward- and backward-filtering passes must be performed in sequence over the observations, ordered-in-time, rather than making a global analysis over $\boldsymbol{y}_{L:L-S+1}$. Finally, this easily extends to accommodate a non-linear observation model by using the MLEF filtering step to obtain the optimal ensemble transform, and by applying this recursively backward-in-time to the lagged ensemble states.

A schematic of the EnKS cycle for a lag of $L = 4$ and a shift of $S = 1$ is pictured in Fig. 3.3. Time moves forward from left to right in the horizontal axis with a step size of Δ_t. At each analysis time, the ensemble forecast from the last filter pdf is combined with the observation to produce the ensemble transform update. This transform is then utilised to produce the posterior estimate for all lagged ensemble states, conditioned on the new observation. The information in the posterior estimate thus flows in reverse time to the lagged states stored in memory, but the information flow is unidirectional in this scheme. This type of retrospective analysis maintains the computational cost of the fixed-lag EnKS described here at a comparable level to the EnKF, with the only significant additional cost being the storage of the ensemble at lagged times to be reanalysed.

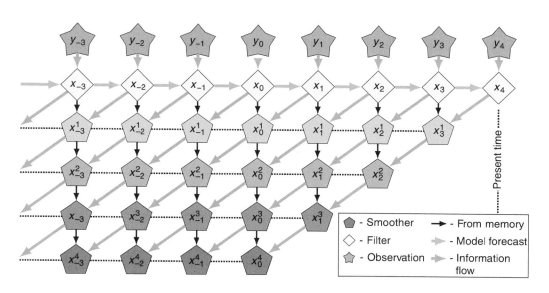

Figure 3.3 $L = 4$ (lag), $S = 1$ (shift) EnKS. Observations are assimilated sequentially via the filter cost function and a retrospective reanalysis is applied to all ensemble states within the lag window stored in memory. Source: Grudzien and Bocquet (2021), adapted from Asch et al. (2016).

3.4.4 The Iterative Ensemble Kalman Filter and Smoother

While the EnKS is computationally efficient, its retrospective analysis has some drawbacks compared to the 4-D analysis in terms of forecast accuracy. The 4-D fixed-lag smoothing analysis re-initialises each cycle with a reanalysed estimate for the initial data, transmitting the observations' information forward-in-time through the non-linear dynamics. The challenge with the 4-D analysis in the absence of the tangent linear and adjoint models is in devising how to efficiently and accurately compute the gradient of the 4-D cost function. Building on Zupanski (2005) and Liu et al. (2008), the analysis of the *iterative ensemble Kalman filter* (IEnKF) and the *iterative ensemble Kalman smoother* (IEnKS) extends the ensemble transform method of the ETKF to iteratively optimise the 4-D cost function.

Recall the quadratic cost function at the basis of incremental 4D-VAR, Eq. (3.72) – making an ensemble-based approximation for the background mean and error covariance, and writing the model state as an ensemble perturbation of the i-th proposal for the ensemble mean, one has

$$\mathcal{J}_{\text{IEnKS}}(\boldsymbol{w}) := \frac{1}{2}\|\hat{\boldsymbol{x}}_0^{\text{i}} - \hat{\boldsymbol{x}}_0^{\text{i}} - \mathbf{X}_0\boldsymbol{w}\|_{\mathbf{P}_0}^2$$
$$+ \frac{1}{2}\sum_{k=1}^{L}\|\boldsymbol{y}_k - \mathcal{H}_k \circ \mathcal{M}_{k:1}(\hat{\boldsymbol{x}}_0^i) - \mathbf{H}_k\mathbf{M}_{k:1}\mathbf{X}_0\boldsymbol{w}\|_{\mathbf{R}_k}^2$$

$$(3.110\text{a})$$

$$= \frac{1}{2}(N_e - 1)\|\boldsymbol{w}\|^2$$
$$+ \frac{1}{2}\sum_{k=1}^{L}\|\boldsymbol{y}_k - \mathcal{H}_k \circ \mathcal{M}_{k:1}(\hat{\boldsymbol{x}}_0^i) - \mathbf{H}_k\mathbf{M}_{k:1}\mathbf{X}_0\boldsymbol{w}\|_{\mathbf{R}_k}^2.$$

$$(3.110\text{b})$$

Applying the finite differences approximation as given in Eq. (3.101a,b), but with respect to the composition of the non-linear observation operator with the non-linear state model

$$\mathcal{H}_k \circ \mathcal{M}_{k:1}(\hat{\boldsymbol{x}}_0^i) \approx \hat{\boldsymbol{y}}_k := \mathcal{H}_k \circ \mathcal{M}_{k:1}(\hat{\boldsymbol{x}}_0^i + \mathbf{X}_0\epsilon)\mathbf{1}/N_e,$$

$$(3.111\text{a})$$

$$\mathbf{H}_k\mathbf{M}_{k:1}\mathbf{X}_0 \approx \widetilde{\mathbf{Y}}_k := \frac{1}{\epsilon}\mathcal{H}_k \circ \mathcal{M}_{k:1}(\hat{\boldsymbol{x}}_0^i + \mathbf{X}_0\epsilon)(\mathbf{I}_{N_e} - \mathbf{1}\mathbf{1}^{\text{T}}/N_e),$$

$$(3.111\text{b})$$

this sketches the 'bundle' formulation of the IEnKS (Bocquet and Sakov, 2013; Bocquet and Sakov, 2014). This indexing refers to the case where the smoothing problem is performed offline, with a fixed DAW, though a similar indexing to Eq. (3.65) gives the sequential form over newly incoming observations.

The sequential form of the IEnKS can be treated as a non-linear sequential filter, which is the purpose that the method was originally devised for. Indeed, setting the number of lagged states $L = 1$, this provides a direct extension of the EnKF cost function, Eq. (3.92b), but where there is an additional dependence on the initial conditions for the ensemble forecast. This lag-1 iterative filtering scheme is called the IEnKF (Bocquet and Sakov, 2012; Sakov et al., 2012), which formed the original basis for the IEnKS. Modern forms of the IEnKF/S analysis furthermore include the treatment of model errors, like weak-constraint 4D-VAR (Sakov and Bocquet, 2018; Sakov et al., 2018; Fillion et al., 2020). Alternative formulations of this analysis, based on the original stochastic, perturbed observation EnKF (Evensen, 1994; Burgers et al., 1998), also have a parallel development in the *ensemble randomised maximum likelihood method* (EnRML) of Gu and Oliver (2007), Chen and Oliver (2012), and Raanes et al. (2019b). A similar ensemble-variational estimator based on the EnKF analysis is the *ensemble Kalman inversion* (EKI) of Iglesias et al. (2013), Schillings and Stuart (2018), and Kovachki and Stuart (2019).

3.4.5 The Single-Iteration Ensemble Kalman Smoother

The sequential smoothing analysis so far has presented two classical approaches: (i) a 4-D approach using the global analysis of all new observations available within a DAW at once, optimising an initial condition for the lagged model state; and (ii) the 3-D approach based upon alternating forward- and backward-filtering passes over the DAW. Each of these approaches has strengths and weaknesses in a computational cost/forecast-accuracy trade-off. Particularly, the 4-D approach, as in the IEnKS, benefits from the improved estimate of the initial condition when producing the subsequent forecast statistics in a shifted DAW; however, each step of the iterative optimisation comes at the cost of simulating the ensemble forecast over the entire DAW in the fully nonlinear state model, which is typically the greatest numerical expense in geophysical DA. On the other hand, the 3-D approach of the classical EnKS benefits from a low computational cost, requiring only a single ensemble simulation of the non-linear forecast model over the DAW, while the retrospective analysis of lagged states is performed with the filtering transform without requiring additional model simulations (though at a potentially large memory storage cost). It should be noted that in the perfect, linear-Gaussian model Bayesian analysis, both approaches produce equivalent estimates of the joint posterior. However, when non-linearity is present in the DA cycle, the approaches produce distinct estimates: for this reason, the source of the non-linearity in the DA cycle is of an important practical concern.

Consider the situation in which the forecast horizon Δ_t is short, so that the ensemble time-evolution is weakly non-linear, and where the ensemble mean is a good approximation for the mean of the filtered distribution. In this case, the model forecast dynamics are well-approximated by the tangent linear evolution of a perturbation about the ensemble mean, that is,

$$\boldsymbol{x}_k = \mathcal{M}_k(\boldsymbol{x}_{k-1}) \approx \mathcal{M}_k(\hat{\boldsymbol{x}}_{k-1}) + \mathbf{M}_k \boldsymbol{\delta}_{k-1}. \tag{3.112}$$

For such short-range forecasts, non-linearity in the observation-analysis-forecast cycle may instead be dominated by the non-linearity in the observation operator $\mathcal{H}_k$, or in the optimisation of hyper-parameters of the filtering cost function, and not by the model forecast itself. In this situation, an iterative simulation of the forecast dynamics as in the 4-D approach may not produce a cost-effective reduction in the analysis error as compared to, for example, the MLEF filter analysis optimising the filtering cost function alone. However, one may still obtain the benefits of re-initialisation of the forecast with a reanalysed prior by using a simple hybridisation of the 3-D and 4-D smoothing analyses.

Specifically, in a given fixed-lag smoothing cycle, one may iteratively optimise the sequential filtering cost functions for a given DAW corresponding to the new observations at times $\{t_{L-S+1}, \ldots, t_L\}$ as with the MLEF. These filtering ensemble transforms not only condition the ensemble at the corresponding observation time instance, but also produce a retrospective analysis of a lagged state as in Eq. (3.109c) with the EnKS. However, when the DAW itself is shifted, one does not need to produce a forecast from the latest filtered ensemble – instead one can initialise the next ensemble forecast with the lagged, retrospectively smoothed ensemble at the beginning of the last DAW. This hybrid approach utilising the retrospective analysis, as in the classical EnKS, and the ensemble simulation over the lagged states while shifting the DAW, as in the 4-D analysis of the IEnKS, was recently developed by Grudzien and Bocquet (2022), and is called the *single-iteration ensemble Kalman smoother* (SIEnKS). The SIEnKS is named as such because it produces its forecast, filter, and reanalysed smoother statistics with a single iteration of the ensemble simulation over the DAW in a fully consistent Bayesian analysis. By doing so, it seeks to minimise the leading order cost of EnVAR smoothing (i.e. the ensemble simulation in the non-linear forecast model). However, the estimator is free to iteratively optimise the filter cost function for any single observation vector without additional iterations of the ensemble simulation. In observation-analysis-forecast cycles in which the forecast error dynamics are weakly non-linear, yet other aspects of the cycle are moderately to strongly non-linear, this scheme is shown to produce a forecast accuracy comparable to, and at times better than, the 4-D approach but with an overall lower leading-order computational burden (Grudzien and Bocquet, 2022).

A schematic of the SIEnKS cycle for a lag of $L = 4$ and a shift of $S = 2$ is pictured in Fig. 3.4. This demonstrates how the sequential analysis of the filter cost function and sequential, retrospective reanalysis for each incoming observation differs from the global analysis of the 4-D approach of the IEnKS. Other well-known DA schemes combining a retrospective reanalysis and re-initialisation of the ensemble forecast include the *running in place* (RIP) smoother of Kalnay and Yang (2010) and Yang et al. (2013) and the *one step ahead* (OSA) smoother of Desbouvries et al. (2011) and Ait-El-Fquih and Hoteit (2022). It can be shown that, with an ETKF-style filter analysis, a single iteration of the ensemble over the DAW, a perfect model assumption, and a lag of $L = S = 1$, the SIEnKS, RIP, and OSA smoothers all coincide (Grudzien and Bocquet, 2022).

3.5 Machine Learning and Data Assimilation in a Bayesian Perspective

This final section demonstrates how the Bayesian DA framework can be extended to estimate more than the state vector, including key parameters of the model and the observation-analysis-forecast cycle. In particular, the Bayesian framework can be used to formulate techniques

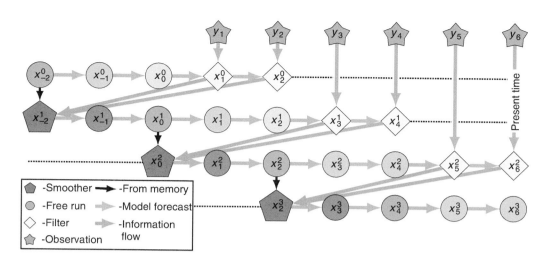

Figure 3.4 $L = 4$ (lag), $S = 2$ (shift) SIEnKS diagram. An initial condition from the last smoothing cycle initialises a forecast simulation over the current DAW of $L = 4$ states. New observations entering the DAW are assimilated sequentially via the filter cost function. After each filter analysis, a retrospective reanalysis is applied to the initial ensemble, and this reanalysed initial condition is evolved via the model S analysis times forward to begin the next cycle. Source: Grudzien and Bocquet (2021).

for learning both the state vector and part of, if not the full, dynamical model. If this objective was always in the scope of classical DA, it was actually made possible, significantly beyond linear regression, by the introduction of *machine learning* (ML) techniques to supplement traditional DA schemes.

Incorporating ML into DA algorithms was suggested quite early by Hsieh and Tang (1998), who clearly advocated the use of *neural networks* (NNs) in a variational DA framework. More recently this was put forward and illustrated by Abarbanel et al. (2018) and Bocquet et al. (2019) with a derivation of the generalised cost function from Bayes' law, showing that the classical ML cost function for learning a surrogate model of the dynamical model is a limiting case of the DA framework, as in Eq. (3.117). This formalism was further generalised by Bocquet et al. (2020), using classical DA notation, for learning the state vector, the dynamical model, and the error statistics attached to this full retrieval. This section follows this latest paper, showing how to derive the cost function of the generalised estimation problem. Note that going even beyond this approach, attempting to learn the optimisation scheme of the cost function, or even the full DA procedure, an approach called *end-to-end* in ML, is a subject of active investigations (Fablet et al., 2021; Peyron et al., 2021).

3.5.1 Prior Error Statistics
For the sake of simplicity, again assume Gaussian statistics for the observation errors $p(\boldsymbol{y}_k|\boldsymbol{x}_k) = n(\boldsymbol{y}_k|\boldsymbol{x}_k, \mathbf{R}_k)$, where the observation error covariance matrices $\mathbf{R}_{L:1} := \{\mathbf{R}_L, \mathbf{R}_{L-1}, \ldots, \mathbf{R}_1\}$ are supposed to be known. The dynamical model is meant to be learned or approximated and thus stands as a surrogate model for the unknown true physical dynamics. Assuming that the model does not explicitly depend on time, its resolvent is defined by

$$\boldsymbol{x}_k := \mathbf{F}_{\mathbf{A}}^k(\boldsymbol{x}_{k-1}) + \boldsymbol{\eta}_k, \qquad (3.113)$$

depending on a (possibly very large) set of parameters $\mathbf{A}$. Prototypically, $\mathbf{A}$ represents the weights and biases of an NN, which are learned from the observations alongside the state vectors within the DAW. The distribution for model error, such as $\boldsymbol{\eta}_k$ in Eq. (3.113), is also assumed Gaussian such that

$$p(\boldsymbol{x}_k|\boldsymbol{x}_{k-1}, \mathbf{A}, \mathbf{Q}_k) := n\left(\boldsymbol{x}_k|\mathbf{F}_{\mathbf{A}}^k(\boldsymbol{x}_{k-1}), \mathbf{Q}_k\right), \qquad (3.114)$$

where $\mathbf{Q}_{L:1} := \{\mathbf{Q}_L, \mathbf{Q}_{L-1}, \ldots, \mathbf{Q}_1\}$ are not necessarily known. Further assume that these Gaussian errors are white-in-time and that the observation and model errors are mutually independent.

Note the intriguing status of $\mathbf{Q}_{L:1}$ since it depends, *a posteriori*, on how well the surrogate model is estimated. This calls for an adaptive estimation of the model error statistics $\mathbf{Q}_{L:1}$, as the surrogate model, parametrised by $\mathbf{A}$, is better approximated.

3.5.2 Joint Estimation of the Model, Its Error Statistics, and the State Trajectory
Following the Bayesian formalism, one may form a MAP estimate for the joint pdf in $\mathbf{A}$ and $\boldsymbol{x}_{L:0}$ conditioned on the observations, together with the model error statistics. This generalised conditional pdf is expressed in the hierarchy

$$p(\mathbf{A}, \mathbf{Q}_{L:1}, \boldsymbol{x}_{L:0}|\boldsymbol{y}_{k:1}, \mathbf{R}_{K:0}) =$$
$$\frac{p(\boldsymbol{y}_{L:1}|\boldsymbol{x}_{L:0}, \mathbf{R}_{L:0})p(\boldsymbol{x}_{L:0}|\mathbf{A}, \mathbf{Q}_{L:1})p(\mathbf{A}, \mathbf{Q}_{L:1})}{p(\boldsymbol{y}_{L:1}, \mathbf{R}_{L:0})}, \qquad (3.115)$$

where the mutual independence of the observation and model error is used. Once again, this remarkably stresses how powerful and general the Bayesian framework can be.

The first term in the numerator of the right-hand side is the usual likelihood of the observations. The second term in the numerator is the prior on the trajectory, given a known model and known model error statistics. The

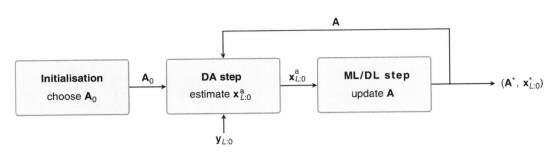

Figure 3.5 Estimation of both model and state trajectory using coordinate descent by alternately optimising on the state trajectory using DA and on the NN model parameters using ML/DL. The iterative loop stops when an accuracy criterion is met. Source: Farchi et al. (2021).

final term of the numerator is the joint prior of the model and the model error statistics, described as a *hyperprior*. The associated cost function is derived proportional to

$$\mathcal{J}_{\text{DA}-\text{ML}}(\mathbf{A}, \boldsymbol{x}_{L:0}, \mathbf{Q}_{L:1}) = -\log p(\mathbf{A}, \mathbf{Q}_{L:1}, \boldsymbol{x}_{L:0}|\boldsymbol{y}_{L:1}, \mathbf{R}_{L:1})$$
$$= \frac{1}{2}\sum_{k=1}^{L}\left\{ \|\boldsymbol{y}_k - \mathcal{H}_k(\boldsymbol{x}_k)\|^2_{\mathbf{R}_k} \right.$$
$$\left. + \log(|\mathbf{R}_k|) \right\}$$
$$+ \frac{1}{2}\sum_{k=1}^{L}\left\{ \|\boldsymbol{x}_k - \mathbf{F}^k_{\mathbf{A}}(\boldsymbol{x}_{k-1})\|^2_{\mathbf{Q}_k} \right.$$
$$\left. + \log(|\mathbf{Q}_k|) \right\} - \log p(\boldsymbol{x}_0, \mathbf{A}, \mathbf{Q}_{L:1}).$$

(3.116)

Note the resemblance of (3.116) with the weak-constraint 4D-VAR cost function of classical DA (Trémolet, 2006). Very importantly, this Bayesian formulation allows for a rigorous treatment of partial and noisy observations. The classical ML cost function that uses noiseless, complete observations of the physical system is derived from Eq. (3.116), assuming that $\mathbf{Q}_k$ is known, $\mathbf{H}_k \equiv \mathbf{I}_{N_x}$, and setting $\mathbf{R}_k$ to go to $\mathbf{0}$. Associating the initial data as $\boldsymbol{y}_0 \leftarrow \boldsymbol{x}_0$, $\mathcal{J}_{\text{DA}-\text{ML}}(\mathbf{A}, \boldsymbol{x}_{L:0}, \mathbf{Q}_{L:1})$ becomes, in its limit,

$$\mathcal{J}_{\text{ML}}(\mathbf{A}) = \frac{1}{2}\sum_{k=1}^{L} \|\boldsymbol{y}_k - \mathbf{F}^k_{\mathbf{A}}(\boldsymbol{y}_{k-1})\|^2_{\mathbf{Q}_k} - \log p(\boldsymbol{y}_0, \mathbf{A}, \mathbf{Q}_{L:1}).$$

(3.117)

Such connections between ML and DA were first highlighted by Hsieh and Tang (1998), Abarbanel et al. (2018), and Bocquet et al. (2019).

Solving the combined DA/ML problem, that is, minimising Eq. (3.116), leads to several key remarks. First, as mentioned earlier, this formalism allows one to learn a surrogate model of the true dynamics using partial, possibly sparse and noisy observations, as opposed to off-the-shelf ML techniques. This is obviously critical for geophysical systems. Second, in this framework, one fundamentally looks for a stochastically additive surrogate model of the form in Eq. (3.113) rather than a deterministic surrogate model, since $\boldsymbol{\eta}_k$ is drawn from the normal distribution of covariance matrices $\mathbf{Q}_k$. Third, there

are many ways to carry out this minimisation, as discussed by Bocquet et al. (2019) and Bocquet et al. (2020). Because the state vectors and the model parameters $\mathbf{A}$ are of fundamentally different nature, and yet are statistically interdependent, one idea is to minimise Eq. (3.116) through a coordinate descent (i.e. alternating minimisations on $\mathbf{A}$ and $\boldsymbol{x}_{L:0}$), as illustrated by Fig. 3.5. This was first suggested and successfully implemented by Brajard et al. (2020), by using an EnKF for the assimilation step (i.e. minimising on $\boldsymbol{x}_{L:0}$), and a *deep learning* (DL) optimiser for the ML step (i.e. minimising on $\mathbf{A}$). This work was extended to 4-D variational analysis using a 4D-VAR assimilation step with a DL optimiser for the ML step by Farchi et al. (2021).

3.5.3 Joint Estimation of the Model and the Error Statistics
A slightly different objective is to obtain a MAP estimate for the surrogate model, irrespective of any model state realisation, if, for example, one is interested in the MAP of the marginal conditional pdf

$$p(\mathbf{A}, \mathbf{Q}_{L:1}|\boldsymbol{y}_{L:1}, \mathbf{R}_{L:1}) = \int p(\mathbf{A}, \mathbf{Q}_{L:1}, \boldsymbol{x}_{L:0}|\boldsymbol{y}_{L:1}, \mathbf{R}_{L:1})\,\mathrm{d}\boldsymbol{x}_{L:0},$$

(3.118)

which is theoretically obtained by minimising

$$\mathcal{J}(\mathbf{A}, \mathbf{Q}_{L:1}) = -\log p(\mathbf{A}, \mathbf{Q}_{L:1}|\boldsymbol{y}_{L:1}, \mathbf{R}_{L:1}).$$ (3.119)

As pointed out by Bocquet et al. (2019), the marginal pdf Eg. (3.118) can be approximately related to the joint pdf through a Laplace approximation of the integral. Here, however, one is interested in the full solution to this problem. This can be solved numerically by using the *expectation-maximisation* (EM) statistical algorithm (Dempster et al., 1977) jointly with the variational numerical solution of Eq. (3.116). This was suggested in Ghahramani and Roweis (1999) and Nguyen et al. (2019) and implemented and validated by Bocquet et al. (2020).

3.6 Conclusions

Unifying techniques from statistical estimation, non-linear optimisation, and even machine learning, the Bayesian approach to DA provides a consistent treatment of a variety of topics in the classical DA problem, discussed throughout this chapter. Furthermore, the Bayesian approach can be reasonably extended to treat additional challenges not fully considered in this chapter, such as estimating process model parameters, handling significant modelling errors, and including the optimisation of various hyper-parameters of the observation-analysis-forecast cycle. Estimation may even include learning the dynamical process model itself, as in the emerging topic of DA/ML hybrid algorithms. In this context especially, a Bayesian analysis provides a coherent treatment of the problem where either DA or ML techniques themselves may be insufficient. This is particularly relevant where surrogate ML models are used to augment traditional physics-based dynamical models for, e.g. simulating unresolved dynamics at scales too fine to be dynamically represented. The hybrid approach in the Bayesian analysis provides a means to combine the dynamical and surrogate simulations with real-world observations, and to produce an analysis of the state and parameters for which subsequent simulations depend. This chapter thus presents one framework for interpreting the DA problem, both in its classical formulation and in the directions of the current state-of-the-art. In surveying a variety of widely used DA schemes, the key message of this chapter is how the Bayesian analysis provides a consistent framework for the estimation problem and how this allows one to formulate its solution in a variety of ways to exploit the operational challenges in the geosciences.

Acknowledgements

This chapter benefited from the opportunity to lecture on data assimilation at: (i) the CIMPA Research School on *Mathematics of Climate Science* in Kigali, Rwanda (28 June–9 July 2021); (ii) Utrecht University's Research school on *Data science and Beyond: Data Assimilation with Elements of Machine Learning* in Utrecht, Netherlands (23 August–27 August 2021); (iii) the Joint ICTP-IUGG Workshop on *Data Assimilation and Inverse Problems in Geophysical Sciences* in Trieste, Italy (18 October–29 October 2021); and (iv) at the University of Nevada–Reno, in Fall Semester 2021, where the combined lecture notes formed the basis of this manuscript. The authors would like to thank everyone, but especially the students, the named reviewer Polly Smith, and the editor Alik Ismail-Zadeh, who supported the development of these notes with their many and highly relevant suggestions. Finally, this chapter also benefited from many fruitful discussions with Alberto Carrassi, Alban Farchi, Patrick Raanes, Julien Brajard, and Laurent Bertino. CEREA is a member of Institut Pierre-Simon Laplace (IPSL).

References

Abarbanel, H. D. I., Rozdeba, P. J., and Shirman, S. (2018). Machine learning: Deepest learning as statistical data assimilation problems. *Neural Computation*, 30(8), 2025–55. https://doi.org/10.1162/neco_a_01094.

Ait-El-Fquih, B., and Hoteit, I. (2022). Filtering with one-step-ahead smoothing for efficient data assimilation. In S. K. Park and L. Xu eds., *Data Assimilation for Atmospheric, Oceanic and Hydrologic Applications*, vol. IV. Cham: Springer, pp. 69–96. https://doi.org/10.1007/978–3–030–77722-7_1.

Anderson, B. D. O., and Moore, J. B. (1979). *Optimal Filtering*. Englewood Cliffs, NJ: Prentice-Hall.

Asch, M., Bocquet, M., and Nodet, M. (2016). *Data Assimilation: Methods, Algorithms, and Applications*. Philadelphia, PA: Society for Industrial and Applied Mathematics. https://doi.org/10.1137/1.9781611974546.

Bannister, R. N. (2017). A review of operational methods of variational and ensemble-variational data assimilation. *Quarterly Journal of the Royal Meteorological Society*, 143, 607–33. https://doi.org/10.1002/qj.2982.

Baydin, A. G., Pearlmutter, B. A., Radul, A. A., and Siskind, J. M. (2018). Automatic differentiation in machine learning: A survey. *Journal of Machine Learning Research*, 18, 1–43. https://doi.org/0.5555/3122009.3242010.

Bishop, C. H., Etherton, B. J., and Majumdar, S. J. (2001). Adaptive sampling with the ensemble transform Kalman filter. Part I: Theoretical aspects. *Monthly Weather Review*, 129, 420–36. https://doi.org/10.1175/1520–0493

Bocquet, M. (2011). Ensemble Kalman filtering without the intrinsic need for inflation. *Nonlinear Processes in Geophysics*, 18, 735–50. https://doi.org/10.5194/npg-18–735–2011.

Bocquet, M., Brajard, J., Carrassi, A., and Bertino, L. (2019). Data assimilation as a learning tool to infer ordinary differential equation representations of dynamical models. *Nonlinear Processes in Geophysics*, 26, 143–62. https://doi.org/10.5194/npg-26-143-2019.

Bocquet, M., Brajard, J., Carrassi, A., and Bertino, L. (2020). Bayesian inference of chaotic dynamics by merging data assimilation, machine learning and expectation-maximization. *Foundations of Data Science*, 2, 55–80. https://doi.org/10.3934/fods.2020004.

Bocquet, M., Raanes, P. N., and Hannart, A. (2015). Expanding the validity of the ensemble Kalman filter without the intrinsic need for inflation. *Nonlinear Processes in Geophysics*, 22, 645–62. https://doi.org/10.5194/npg22–645-2015.

Bocquet, M., and Sakov, P. (2012). Combining inflation-free and iterative ensemble Kalman filters for strongly nonlinear systems. *Nonlinear Processes in Geophysics*, 19, 383–99. https://doi.org/10.5194/npg-19-383-2012.

Bocquet, M., and Sakov, P. (2013). Joint state and parameter estimation with an iterative ensemble Kalman smoother. *Nonlinear Processes in Geophysics*, 20, 803–18. https://doi.org/10.5194/npg-20-803-2013.

Bocquet, M., and Sakov, P. (2014). An iterative ensemble Kalman smoother. *Quarterly Journal of the Royal Meteorological Society*, 140, 1521–35. https://doi.org/10.1002/qj.2236.

Bonavita, M., Isaksen, L., and Hólm, E. (2012). On the use of EDA background error variances in the ECMWF 4D-VAR.

Quarterly Journal of the Royal Meteorological Society, 138, 1540–59. https://doi.org/10.1002/qj.1899.

Brajard, J., Carrassi, A., Bocquet, M., and Bertino, L. (2020). Combining data assimilation and machine learning to emulate a dynamical model from sparse and noisy observations: A case study with the Lorenz 96 model. *Journal of Computational Science*, 44, 101171. https://doi.org/10.1016/j.jocs.2020.101171.

Burgers, G., van Leeuwen, P. J., and Evensen, G. (1998). Analysis scheme in the ensemble Kalman filter. *Monthly Weather Review*, 126, 1719–24. https://doi.org/10.1175/1520-0493(1998)126<1719:ASITEK>2.0.CO;2.

Carrassi, A., Bocquet, M., Bertino, L., and Evensen, G. (2018). Data assimilation in the geosciences: An overview on methods, issues, and perspectives. *WIREs Climate Change*, 9, e535. https://onlinelibrary.wiley.com/doi/abs/10.1002/wcc.535.

Carrassi, A., Bocquet, M., Demaeyer, J., Gruzien, C., Raanes, P. N. and Vannitsem, S. (2022). Data assimilation for chaotic dynamics. In S. K. Park and L. Xu eds., *Data Assimilation for Atmospheric, Oceanic and Hydrologic Applications*, vol. IV. Cham: Springer, pp. 1–42. https://doi.org/10.1007/978-3-030-77722-7_1.

Carrassi, A., Bocquet, M., Hannart, A., and Ghil, M. (2017). Estimating model evidence using data assimilation. *Quarterly Journal of the Royal Meteorological Society*, 143, 866–80. https://doi.org/10.1002/qj.2972.

Chen, Y., and Oliver, D. S. (2012). Ensemble randomized maximum likelihood method as an iterative ensemble smoother. *Mathematical Geosciences*, 44, 1–26. https://doi.org/10.1007/s11004-011-9376-z.

Cohn, S. E., Sivakumaran, N. S., and Todling, R. (1994). A fixed-lag Kalman smoother for retrospective data assimilation. *Monthly Weather Review*, 122, 2838–67. https://doi.org/10.1175/1520-0493(1994)122<2838:AFLKSF>2.0.CO;2.

Corazza, M., Kalnay, E., Patil, D. J. et al. (2003). Use of the breeding technique to estimate the structure of the analysis errors of the day. *Nonlinear Processes in Geophysics*, 10, 233–43. https://doi.org/10.5194/npg-10-233-2003.

Cosme, E., Verron, J., Brasseur, P., Blum, J., and Auroux, D. (2012). Smoothing problems in a Bayesian framework and their linear Gaussian solutions. *Monthly Weather Review*, 140, 683–95. https://doi.org/10.1175/MWR-D-10-05025.1.

Courtier, P., Thépaut, J.-N., and Hollingsworth, A. (1994). A strategy for operational implementation of 4D-VAR, using an incremental approach. *Quarterly Journal of the Royal Meteorological Society*, 120, 1367–87. https://doi.org/10.1002/qj.49712051912.

Daley, R. (1991). *Atmospheric Data Analysis*. New York: Cambridge University Press.

Dempster, A. P., Laird, N. M., and Rubin, D. B. (1977). Maximum likelihood from incomplete data via the EM algorithm. *Journal of the Royal Statistical Society. Series B*, 1–38. https://doi.org/10.1111/j.2517-6161.1977.tb01600.x.

Desbouvries, F., Petetin, Y., and Ait-El-Fquih, B. (2011). Direct, prediction- and smoothing-based Kalman and particle filter algorithms. *Signal Processing*, 91, 2064–77. https://doi.org/10.1016/j.sigpro.2011.03.013.

Desroziers, G., Camino, J.-T., and Berre, L. (2014). 4DEnVar: link with 4D state formulation of variational assimilation and different possible implementations. *Quarterly Journal of the Royal Meteorological Society*, 140, 2097–110. https://doi.org/10.1002/qj.2325.

Evensen, G. (1994). Sequential data assimilation with a nonlinear quasi-geostrophic model using Monte Carlo methods to forecast error statistics. *Journal of Geophysical Research*, 99, 10143–62. https://doi.org/10.1029/94JC00572.

Evensen, G. (2003). The ensemble Kalman filter: Theoretical formulation and practical implementation. *Ocean Dynamics*, 53, 343–67. https://doi.org/10.1007/s10236-003-0036-9.

Evensen, G., and van Leeuwen, P. J. (2000). An ensemble Kalman smoother for nonlinear dynamics. *Monthly Weather Review*, 128, 1852–67. https://doi.org/10.1175/1520-0493(2000)128<1852:AEKSFN>2.0.CO;2.

Fablet, R., Chapron, B., Drumetz, L. et al. (2021). Learning variational data assimilation models and solvers. *Journal of Advances in Modeling Earth Systems*, 13, e2021MS002572. https://doi.org/10.1029/2021MS002572.

Farchi, A., and Bocquet, M. (2019). On the efficiency of covariance localisation of the ensemble Kalman filter using augmented ensembles. *Frontiers in Applied Mathematics and Statistics*, 5, 3. https://doi.org/10.3389/fams.2019.00003.

Farchi, A., Laloyaux, P., Bonavita, M., and Bocquet, M. (2021). Using machine learning to correct model error in data assimilation and forecast applications. *Quarterly Journal of the Royal Meteorological Society*, 147(739), 3067–84. https://doi.org/10.1002/qj.4116.

Fillion, A., Bocquet, M., Gratton, S., Gürol, S. and Sakov, P. (2020). An iterative ensemble Kalman smoother in presence of additive model error. *SIAM/ASA Journal of Uncertainty Quantification*, 8, 198–228. https://doi.org/10.1137/19M1244147.

Ghahramani, Z., and Roweis, S. T. (1999). Learning nonlinear dynamical systems using an EM algorithm. In M. Kearns, S. Solla, and D. Cohn, eds., *Advances in Neural Information Processing Systems*, vol. 11. Cambridge, MA: MIT Press, pp. 431–7. https://doi.org/0.5555/340534.340696.

Grewal, M. S., and Andrews, A. P. (2014). *Kalman Filtering: Theory and Practice with MATLAB*. Hoboken, NJ; John Wiley & Sons. https://doi.org/10.1002/9781118984987.

Griewank, A. (1989). On automatic differentiation. *Mathematical Programming: Recent Developments and Applications*, 6, 83–107.

Griewank, A., and Walther, A. (2003). Introduction to automatic differentiation. *Proceedings in Applied Mathematics and Mechanics*, 2(1), 45–9. https://doi.org/10.1002/pamm.200310012.

Grudzien, C., and Bocquet, M. (2022). A fast, single-iteration ensemble Kalman smoother for sequential data assimilation. *Geoscientific Model Development*, 15, 7641–7681. https://doi.org/10.5194/gmd-15-7641-2022.

Grudzien, C., Carrassi, A., and Bocquet, M. (2018). Chaotic dynamics and the role of covariance inflation for reduced rank Kalman filters with model error. *Nonlinear Processes in Geophysics*, 25, 633–48. https://doi.org/10.5194/npg-25-633-2018.

Gu, Y., and Oliver, D. S. (2007). An iterative ensemble Kalman filter for multiphase fluid flow data assimilation. *SPE Journal*, 12, 438–46. https://doi.org/10.2118/108438-PA.

Hamill, T. M., and Snyder, C. (2000). A hybrid ensemble Kalman filter-3D variational analysis scheme. *Monthly Weather Review*, 128, 2905–19. https://doi.org/10.1175/1520-0493 (2000)128<2905:AHEKFV>2.0.CO;2.

Härdle, W. K., Okhrin, O., and Okhrin, Y. (2017). *Basic Elements of Computational Statistics*. Cham: Springer. https://link .springer.com/book/10.1007/978-3-319-55336-8.

Hascoët, L. (2014). An introduction to inverse modelling and parameter estimation for atmosphere and ocean sciences. In É. Blayo, M. Bocquet, E. Cosme, and L. F. Cugliandolo, eds., *Advanced Data Assimilation for Geosciences*. Oxford: Oxford University Press, pp. 349–69.

Hsieh, W. W., and Tang, B. (1998). Applying neural network models to prediction and data analysis in meteorology and oceanography. *Bulletin of the American Meteorological Society*, 79, 1855–70. https://doi.org/10.1175/1520-0477 (1998)079<1855: ANNMTP>2.0.CO;2.

Hunt, B. R., Kostelich, E. J., and Szunyogh, I. (2007). Efficient data assimilation for spatiotemporal chaos: A local ensemble transform Kalman filter. *Physica D*, 230, 112–26. https://doi .org/10.1016/j.physd.2006.11.008.

Iglesias, M. A., Law, K. J. H., and Stuart, A. M. (2013). Ensemble Kalman methods for inverse problems. *Inverse Problems*, 29, 045001. https://doi.org/10.1088/0266-5611/ 29/4/045001.

Janjić, T., Bormann, N., Bocquet, M. et al. (2018). On the representation error in data assimilation. *Quarterly Journal of the Royal Meteorological Society*, 144, 1257–78. https://doi.org/ 10.1002/qj.3130.

Kalnay, E. (2003). *Atmospheric Modeling, Data Assimilation and Predictability*. Cambridge: Cambridge University Press. https://doi.org/10.1017/CBO9780511802270.

Kalnay, E., and Yang, S.-C. (2010). Accelerating the spin-up of ensemble Kalman filtering. *Quarterly Journal of the Royal Meteorological Society*, 136, 1644–51. https://doi.org/10 .1002/qj.652.

Kolmogorov, A. N. (2018). *Foundations of the Theory of Probability*, 2nd English ed. Mineola, NY: Dover Publications.

Kovachki, N. B., and Stuart, A. M. (2019). Ensemble Kalman inversion: a derivative-free technique for machine learning tasks. *Inverse Problems*, 35, 095005. https://doi.org/10.1088/ 1361-6420/ab1c3a.

Laloyaux, P., Bonavita, M., Dahoui, M., et al. (2020). Towards an unbiased stratospheric analysis. *Quarterly Journal of the Royal Meteorological Society*, 146, 2392–409. https://doi.org/ 10.1002/qj.3798.

Lawson, W. G., and Hansen, J. A. (2004). Implications of stochastic and deterministic filters as ensemble-based data assimilation methods in varying regimes of error growth. *Monthly Weather Review*, 132, 1966–81. https://doi.org/10.1175/1520-0493(2004)132<1966:IOSADF>2.0.CO;2.

Le Dimet, F.-X., and Talagrand, O. (1986). Variational algorithms for analysis and assimilation of meteorological observations: Theoretical aspects. *Tellus A*, 38, 97–110. https://doi.org/10 .3402/tellusa.v38i2.11706.√

Liu, C., Xiao, Q. and Wang, B. (2008). An ensemble-based four-dimensional variational data assimilation scheme. Part I: Technical formulation and preliminary test. *Monthly Weather Review*, 136, 3363–73. https://doi.org/10.1175/ 2008MWR2312.1.√

Lorenc, A. C. (1986). Analysis methods for numerical weather prediction. *Quarterly Journal of the Royal Meteorological Society*, 112, 1177–94. https://doi.org/10.1002/qj.49711247414.√

Lorenc, A. C. (2003). The potential of the ensemble Kalman filter for NWP: A comparison with 4D-Var. *Quarterly Journal of the Royal Meteorological Society*, 129, 3183–203. https://doi .org/10.1256/qj.02.132.√

Mahalanobis, P. C. (1936). On the generalized distance in statistics. *Proceedings of the National Institute of Science of India*, 2, 49–55.

Ménétrier, B., and Auligné, T. (2015). An overlooked issue of variational data assimilation. *Monthly Weather Review*, 143, 3925–30. https://doi.org/10.1175/MWRD-14-00404.1.

Meyer, C. D. (2000). *Matrix Analysis and Applied Linear Algebra*. Philadelphia, PA: Society for Industrial and Applied Mathematics.

Miller, R. N., Ghil, M., and Gauthiez, F. (1994). Advanced data assimilation in strongly nonlinear dynamical systems. *Journal of Atmospheric Sciences*, 51, 1037–56. https://doi.org/10.1175/1520-0469(1994)051<1037:ADAISN>2.0.CO;2.√

Nguyen, V. D., Ouala, S., Drumetz, L., and Fablet, R. (2019). EM-like learning chaotic dynamics from noisy and partial observations. *arXiv:1903.10335*. https://arxiv.org/abs/1903.10335.

Nocedal, J., and Wright, S. (2006). *Numerical Optimization*. New York: Springer. https://doi.org/10.1007/978-0-387-40065-5.

Pawitan, Y. (2001). *In All Likelihood: Statistical Modelling and Inference Using Likelihood*. Oxford: Oxford University Press.

Penny, S. G. (2017). Mathematical foundations of hybrid data assimilation from a synchronization perspective. *Chaos: An Interdisciplinary Journal of Nonlinear Science*, 27, 126801. https://doi.org/10.1063/1.5001819.√

Peyron, M., Fillion, A., Gürol, S., et al. (2021). Latent space data assimilation by using deep learning. *Quarterly Journal of the Royal Meteorological Society*, 147, 3759–77. https://doi.org/ 10.1002/qj.4153.√

Raanes, P. N. (2016). On the ensemble Rauch–Tung–Striebel smoother and its equivalence to the ensemble Kalman smoother. *Quarterly Journal of the Royal Meteorological Society*, 142, 1259–64. https://doi.org/10.1002/qj.2728.√

Raanes, P. N., Bocquet, M., and Carrassi, A. (2019a). Adaptive covariance inflation in the ensemble Kalman filter by Gaussian scale mixtures. *Quarterly Journal of the Royal Meteorological Society*, 145, 53–75. https://doi.org/10.1002/qj .3386.

Raanes, P. N., Stordal, A. S., and Evensen, G. (2019b). Revising the stochastic iterative ensemble smoother. *Nonlinear Processes in Geophysics*, 26, 325–38. https://doi.org/10.5194/ npg-26-325-2019.

Reich, S., and Cotter, C. (2015). *Probabilistic Forecasting and Bayesian Data Assimilation*. Cambridge: Cambridge University Press. https://doi.org/10.1017/CBO9781107706804.

Rojas, R. (2013). *Neural Networks: A Systematic Introduction*. New York: Springer. https://doi.org/10.1007/978-3-642-61068-4.

Ross, S. M. (2014). *Introduction to Probability Models, 11th edition*. San Diego, CA: Academic Press. https://doi.org/10.1016/C2012-0-03564-8.

Sakov, P., and Bertino, L. (2011). Relation between two common localisation methods for the EnKF. *Computational Geosciences*, 15, 225–37. https://doi.org/10.1007/s10596-010-9202-6.

Sakov, P., and Bocquet, M. (2018). Asynchronous data assimilation with the EnKF in presence of additive model error. *Tellus A*, 70, 1414545. https://doi.org/10.1080/16000870.2017.1414545.

Sakov, P., Haussaire, J.-M., and Bocquet, M. (2018). An iterative ensemble Kalman filter in presence of additive model error. *Quarterly Journal of the Royal Meteorological Society*, 144, 1297–309. https://doi.org/10.1002/qj.3213.

Sakov, P., and Oke, P. R. (2008). Implications of the form of the ensemble transformation in the ensemble square root filters. *Monthly Weather Review*, 136, 1042–53. https://doi.org/10.1175/2007MWR2021.1.

Sakov, P., Oliver, D. S., and Bertino, L. (2012). An iterative EnKF for strongly nonlinear systems. *Monthly Weather Review*, 140, 1988–2004. https://doi.org/10.1175/MWR-D-11-00176.1.

Schillings, C., and Stuart, A. M. (2018). Convergence analysis of ensemble Kalman inversion: The linear, noisy case. *Applicable Analysis*, 97, 107–23. https://doi.org/10.1080/00036811.2017.1386784.

Tabeart, J. M., Dance, S. L., Haben, S. A., et al. (2018). The conditioning of least-squares problems in variational data assimilation. *Numerical Linear Algebra with Applications*, 25, e2165. https://doi.org/10.1002/nla.2165.

Talagrand, O., and Courtier, P. (1987). Variational assimilation of meteorological observation with the adjoint vorticity equation. I: Theory. *Quarterly Journal of the Royal Meteorological Society*, 113, 1311–28. https://doi.org/10.1002/qj.49711347812.

Tandeo, P., Ailliot, P., Bocquet, M., et al. (2020). A review of innovation-based approaches to jointly estimate model and observation error covariance matrices in ensemble data assimilation. *Monthly Weather Review*, 148, 3973–94. https://doi.org/10.1175/MWRD-19-0240.1.

Taylor, M. E. (1996). *Partial Differential Equations. Vol. 1: Basic Theory*. New York: Springer. https://doi.org/10.1007/978-1-4419-7055-8.

Tippett, M. K., Anderson, J. L., Bishop, C. H., Hamill, T. M., and Whitaker, J. S. (2003). Ensemble square root filters. *Monthly Weather Review*, 131, 1485–90. https://doi.org/10.1175/1520-0493(2003)131<1485:ESRF>2.0.CO;2.

Tong, Y. L. (2012). *The Multivariate Normal Distribution*. New York: Springer. https://doi.org/10.1007/978-1-4613-9655-0.

Trémolet, Y. (2006). Accounting for an imperfect model in 4D-Var. *Quarterly Journal of the Royal Meteorological Society*, 132, 2483–504. https://doi.org/10.1256/qj.05.224.

Trevisan, A., and Uboldi, F. (2004). Assimilation of standard and targeted observations within the unstable subspace of the observation-analysis-forecast cycle. *Journal of the Atmospheric Sciences*, 61, 103–13. https://doi.org/10.1175/1520-0469(2004)061<0103:AOSATO>2.0.CO;2.

Whitaker, J. S., and Loughe, A. F. (1998). The relationship between ensemble spread and ensemble mean skill. *Monthly Weather Review*, 126, 3292–302. https://doi.org/10.1175/1520-0493(1998)126<3292:TRBESA>2.0.CO;2.

Yang, S.-C., Lin, K.-J., Miyoshi, T., and Kalnay, E. (2013). Improving the spinup of regional EnKF for typhoon assimilation and forecasting with Typhoon Sinlaku (2008). *Tellus A*, 65, 20804. https://doi.org/10.3402/tellusa.v65i0.20804.

Zupanski, M. (2005). Maximum likelihood ensemble filter: Theoretical aspects. *Monthly Weather Review*, 133, 1710–26. https://doi.org/10.1175/MWR2946.1.

Zupanski, M., Navon, I. M., and Zupanski, D. (2008). The maximum likelihood ensemble filter as a non-differentiable minimization algorithm. *Quarterly Journal of the Royal Meteorological Society*, 134, 1039–50. https://doi.org/10.1002/qj.251.

4

Third-Order Sensitivity Analysis, Uncertainty Quantification, Data Assimilation, Forward and Inverse Predictive Modelling for Large-Scale Systems

Dan Gabriel Cacuci

Abstract: This chapter presents a third-order predictive modelling methodology which aims at obtaining best-estimate results with reduced uncertainties (acronym: 3rd-BERRU-PM) for applications to large-scale models comprising many parameters. The building blocks of the 3rd-BERRU-PM methodology include quantification of third-order moments of the response distribution in the parameter space using third-order adjoint sensitivity analysis (which overcomes the curse of dimensionality), assimilation of experimental data, model calibration, and posterior prediction of best-estimate model responses and parameters with reduced best-estimate variances/ covariances for the predicted responses and parameters. Applications of these concepts to an inverse radiation transmission problem, to an oscillatory dynamical model, and to a large-scale computational model involving 21,976 uncertain parameters, respectively, are also presented, thus illustrating the actual computation and impacts of the first-, second-, and third-order response sensitivities to parameters on the expectation, variance, and skewness of the respective model responses.

4.1 Introduction: Basic Building Blocks of Third-Order BERRU Predictive Modelling

The results of measurements and computations are never perfectly accurate. On the one hand, results of measurements inevitably reflect the influence of experimental errors, imperfect instruments, or imperfectly known calibration standards. Around any reported experimental value, therefore, there always exists a range of values that may also be plausibly representative of the true but unknown value of the measured quantity. On the other hand, computations are afflicted by errors stemming from numerical procedures, uncertain model parameters, boundary and initial conditions, and/or imperfectly known physical processes or problem geometry. Therefore, knowing just the nominal values of the experimentally measured or computed quantities is insufficient for applications. The quantitative uncertainties accompanying measurements and computations are also needed, along with the respective nominal values. Determining the uncertainties in computed model responses (i.e. quantities of interest) is the scope of uncertainty quantification while determining changes in the responses that are induced by variations in the model parameters is the scope of sensitivity analysis. Extracting best-estimate values for model parameters and predicted results, together with best-estimate uncertainties for these parameters and results requires the combination of experimental and computational data and their uncertainties. This combination process is customarily called data assimilation. Such a combination process requires reasoning from incomplete, error-afflicted, and occasionally discrepant information. The combination of data from different sources involves a weighted propagation of parameter uncertainties, using the local functional derivatives of the system responses with respect to the input model parameters. These response derivatives are customarily called response sensitivities or simply sensitivities. At a fundamental level, the need for computing such sensitivities stems from the impossibility of measuring or computing any physical quantity with absolute precision, as has been discussed in the foregoing. Response sensitivities with respect to the model's parameters are also needed for the following purposes:

1. determining the effects of parameter variations on the system's behaviour;
2. understanding the system by highlighting important data;
3. eliminating unimportant data;
4. reducing over-design;

5. performing code verification, which addresses the question 'Are you solving the mathematical model correctly?';
6. prioritising introduction of data uncertainties;
7. quantification of response uncertainties due to parameter uncertainties;
8. performing forward predictive modelling, including data assimilation and model calibration, for the purpose of obtaining best-estimate predicted results with reduced predicted uncertainties; under ideal circumstances, predictive modelling aims at providing a probabilistic description of possible future outcomes based on all recognised errors and uncertainties.
9. performing inverse predictive modelling;
10. improving the system design, possibly reducing conservatism and redundancy;
11. prioritising possible improvements for the system under consideration;
12. designing and optimising the system (e.g. maximise availability/minimise maintenance);
13. performing code validation, which uses comparisons between computations and experiments to addresses the question 'Does the model represent reality?' Code verification and validation can be performed only by considering all of the sensitivities and uncertainties that affect the respective computations and experiments.

For a model comprising a total number TP of imprecisely known parameters denoted as $\alpha_1, ..., \alpha_{TP}$, a model response $R(\alpha_1, ..., \alpha_{TP})$ which is a suitably differentiable function of these model parameters, will admit TP first-order sensitivities $\partial R(\alpha_1, ..., \alpha_{TP})/\partial \alpha_{j_1}$, TP^2 second-order sensitivities $\partial^2 R(\alpha_1, ..., \alpha_{TP})/\partial \alpha_{j_2} \partial \alpha_{j_1}$, and TP^3 third-order sensitivities $\partial^3 R(\alpha_1, ..., \alpha_{TP})/\partial \alpha_{j_3} \partial \alpha_{j_2} \partial \alpha_{j_1}$, where $j_1, j_2, j_3 = 1, ..., TP$. However, because of the symmetry properties of the second- and third-order sensitivities, only $TP(TP+1)/2$ second-order sensitivities are distinct from each other, and only $TP(TP+1)(TP+2)/6$ third-order sensitivities are distinct from each other. The mathematical formalisms for uncertainty quantification and for data assimilation and inverse problems which will be presented in this chapter include sensitivities of first-, second-, and third-order. Therefore, these mathematical formalisms will be designated as third-order formalisms.

The material to be presented in this chapter is set in a general mathematical framework for describing generic physical (biological, engineering) systems involving imprecisely known parameters and/or processes, as outlined in Section 4.2. Generalising previous works, the generic physical systems considered in this chapter include not only uncertain parameters, initial and/or boundary conditions, but also include the physical systems' external and/or internal boundaries (interfaces) in the phase-space of the independent variables which are themselves considered to

be uncertain. This generalisation of previous concepts is important in engineering (where, e.g. it enables the consideration of uncertainties introduced by manufacturing tolerances) and is particularly important in the Earth sciences, where interfaces between various materials and the external boundaries of domains (e.g. reservoirs) are not known precisely.

Section 4.3 commences by presenting the mathematical framework of the Forward and Inverse 3rd-BERRU-PM methodology; the acronym BERRU-PM stands for Best-Estimate Results with Reduced Uncertainties Predictive Modelling. The *a priori* information considered by the 3rd-BERRU-PM methodology includes: (i) expected values and covariances of measured responses; (ii) *a priori* expectations and covariances of model parameters; (iii) *a priori* expectations and covariances of computed responses; (iv) first-, second-, and third-order response sensitivities with respect to the model's imprecisely known parameters, internal interfaces and external boundaries in the phase-space of independent variables. In addition to the information considered by the conventional data assimilation methodologies, the 3rd-BERRU-PM methodology also includes correlations between the model responses and parameters. This a priori information is 'assimilated' by using the maximum entropy principle (Jaynes, 1957) to produce a posterior distribution which enables: (i) model calibration, by yielding best-estimate predicted expectation values with reduced predicted uncertainties for the model responses, as well as best-estimate predicted parameter-response correlations; (ii) model extrapolation, by yielding best-estimate predicted expectation values with reduced predicted uncertainties for the model parameters; and (iii) estimation of the validation domain, by providing a chi-square consistency indicator which quantifies the mutual and joint consistency of the information available for data assimilation and model calibration. The 3rd-BERRU-PM methodology is best suited for forward and inverse problems involving large-scale models with many uncertain parameters. Section 4.3 also illustrates the superior prediction capabilities of the 3rd-BERRU-PM methodology for inverse problems by applying this methodology to an inverse deep-penetration problem of using uncertain measurements for predicting the dimensions of a material which is optically sufficiently thick to render useless the customary methods for inverse problems based on minimising user-defined generalised least-squares objective functions.

The sensitivities (i.e. functional derivatives) of model responses with respect to model parameters are notoriously difficult to determine *exactly* for large-scale models involving many parameters because the computation of even the first-order sensitivities by conventional methods requires at least as many computations as there are model parameters. Furthermore, the computation of higher-order sensitivities by conventional methods is subject to the 'curse of dimensionality' (Bellman, 1957), as the number of large-scale computations increases exponentially in the parameter phase-space. It is nowadays known that the adjoint method of sensitivity analysis is the most efficient method for exactly computing first-order sensitivities. As described by Práger and Kelemen (2014), the general adjoint sensitivity analysis methodology for non-linear systems was formulated by Cacuci (1981a,b), who has also introduced it to the environmental and Earth sciences. This adjoint sensitivity analysis methodology was extended by Cacuci (2015, 2018, 2019b) to enable efficiently the exact computation of second- and third-order response sensitivities for physical systems with imprecisely known boundaries.

In his original work, Cacuci (1981a,b) considered that the physical system's boundaries and interfaces in the phase-space of independent variables were perfectly well know. In subsequent works, Cacuci (2021a) generalised his original work to include the exact and efficient computations of first-order response sensitivities to imprecisely known boundaries/interfaces and parameters, calling this general methodology the *First-Order Comprehensive Adjoint Sensitivity Analysis Methodology for Non-linear Systems* (1st-CASAM-N), where the qualifier 'comprehensive' indicates that *all* model parameters, including the phase-space locations of internal and/or external boundaries, are considered to be imprecisely known (subject to uncertainties). The general mathematical framework of this comprehensive adjoint sensitivity analysis methodology for the computation of first-order sensitivities is summarised in Section 4.3, which also presents an illustrative application to perform a first-order sensitivity and uncertainty analysis of a typical oscillatory dynamical system which can undergo period-doubling bifurcations to transition from a stable state to an aperiodic, chaotic state. This illustrative model has been deliberately chosen since it is representative of similar systems of interest in the Earth sciences but cannot be analysed by conventional statistical methods.

The mathematical frameworks of the comprehensive adjoint sensitivity analysis methodologies for computing exactly and efficiently second- and third-order sensitivities are summarised in Section 4.4. Fundamentally, the second-order sensitivities will be derived by considering them to be first-order sensitivities of the first-order sensitivities, while the third-order sensitivities will be obtained by considering them to be first-order sensitivities of the second-order sensitivities. Thus, the 1st-CASAM-N is the starting point for developing the *Second-Order Comprehensive Adjoint Sensitivity Analysis Methodology for Non-linear Systems* (2nd-CASAM-N) and the 2nd-CASAM is the starting point for developing the *Third-Order Comprehensive Adjoint Sensitivity Analysis Methodology for Non-linear Systems* (3rd-CASAM-N). The mathematical frameworks of the 2nd-CASAM-N and 3rd-CASAM-N will be presented in Section 4.4. Along with these mathematical frameworks, Section 4.4 also presents an application of this high-order

comprehensive sensitivity analysis methodology to perform second- and third-order sensitivity and uncertainty analysis of a physical system (an OECD/NEA physics benchmark) which is representative of a large-scale system that involves many (21,976, in this illustrative example) parameters. The results presented in Section 4.4 illustrate the impact of the 21,976 first-order sensitivities, 482,944,576 second-order sensitivities (of which 241,483,276 are distinct from each other), and the largest 5,832,000 third-order sensitivities on the model response's expectation, variance, and skewness. This paradigm large-scale system cannot be analysed comprehensively by statistical methods. The concluding Section 4.5 summarises the significance of the new methodologies and paradigm results presented in this chapter and discusses further ongoing generalisations of these methodologies.

4.2 Mathematical Modelling of a Generic Non-linear Physical System Comprising Imprecisely Known Parameters and Boundaries

In general terms, the modelling of a physical system and/or the result of an indirect experimental measurement requires consideration of the following modelling components:

1. A mathematical model comprising independent variables (e.g. space, time), dependent variables (aka state functions; e.g. temperature, mass, momentum) and various parameters (appearing in correlations, coordinates of physical boundaries, etc.), which are all inter-related by equations (linear and/or non-linear in the state functions) that usually represent conservation laws.
2. Model parameters, which usually stem from processes that are external to the system under consideration and are seldom, if ever, known precisely. The known characteristics of the model parameters may include their nominal (expected/mean) values and, possibly, higher-order moments or cumulants (i.e. variance/covariances, skewness, kurtosis), which are usually determined from experimental data and/or processes external to the physical system under consideration. Occasionally, only inequality and/or equality constraints that delimit the ranges of the system's parameters are known.
3. One or several computational results, customarily called system responses (or objective functions, or indices of performance), which are computed using the mathematical model.
4. Experimentally measured values of the responses under consideration, which may be used to infer nominal (expected) values and uncertainties (variances, covariances, skewness, kurtosis, etc.) of the respective measured responses.

Without loss of generality, the imprecisely known model parameters can be considered to be real-valued scalar quantities. These model parameters will be denoted as $\alpha_1, \ldots, \alpha_{TP}$, where TP denotes the total number of imprecisely known parameters underlying the model under consideration. For subsequent developments, it is convenient to consider that these parameters are components of a vector of parameters denoted as $\boldsymbol{\alpha} \triangleq (\alpha_1, \ldots, \alpha_{TP})^\dagger \in \mathsf{E}_\alpha \in \mathbb{R}^{TP}$, where E_α is also a normed linear space and where $\mathbb{R}^{TP}$ denotes the TP-dimensional subset of the set of real scalars. The components of the TP-dimensional column vector $\boldsymbol{\alpha} \in \mathbb{R}^{TP}$ are considered to include imprecisely known geometrical parameters that characterise the physical system's boundaries in the phase-space of the model's independent variables. Matrices will be denoted using bold capital letters while vectors will be denoted using either capital or bold lower-case letters. The symbol $\triangleq$ will be used to denote 'is defined as' or 'is by definition equal to'. Transposition will be indicated by a dagger ($\dagger$) superscript.

The model is considered to comprise TI independent variables which will be denoted as x_i, $i = 1, \ldots, TI$, and which are considered to be components of a TI-dimensional column vector denoted as $\mathbf{x} \triangleq (x_1, \ldots, x_{TI})^\dagger \in \mathbb{R}^{TI}$, where the sub/superscript TI denotes the total (number of) independent variables. The vector $\mathbf{x} \in \mathbb{R}^{TI}$ of independent variables is considered to be defined on a phase-space domain which will be denoted as $\Omega(\boldsymbol{\alpha})$ and which is defined as follows: $\Omega(\boldsymbol{\alpha}) \triangleq \{-\infty \leq \lambda_i(\boldsymbol{\alpha}) \leq x_i \leq \omega_i(\boldsymbol{\alpha}) \leq \infty; \ i = 1, \ldots, TI\}$. The lower boundary-point of an independent variable is denoted as $\lambda_i(\boldsymbol{\alpha})$ (e.g. the inner radius of a sphere or cylinder, the lower range of an energy-variable), while the corresponding upper boundary-point is denoted as $\omega_i(\boldsymbol{\alpha})$ (e.g. the outer radius of a sphere or cylinder, the upper range of an energy-variable). A typical example of boundaries that depend on imprecisely known parameters is provided by the boundary conditions needed for models based on diffusion theory, in which the respective flux and/or current conditions for the boundaries facing vacuum are imposed on the extrapolated boundary of the respective spatial domain. The extrapolated boundary depends both on the imprecisely known physical dimensions of the problem's domain and also on the medium's properties, such as atomic number densities and microscopic transport cross sections. The boundary of $\Omega(\boldsymbol{\alpha})$, which will be denoted as $\partial\Omega(\boldsymbol{\alpha})$, comprises the set of all of the endpoints $\lambda_i(\boldsymbol{\alpha})$, $\omega_i(\boldsymbol{\alpha})$, $i = 1, \ldots, TI$, of the respective intervals on which the components of $\mathbf{x}$ are defined, such as $\partial\Omega(\boldsymbol{\alpha}) \triangleq \{\lambda_i(\boldsymbol{\alpha}) \cup \omega_i(\boldsymbol{\alpha}), \ i = 1, \ldots, TI\}$.

A *non-linear* physical system can be generally represented/modelled by means of coupled equations which can be represented in operator form thus:

$$\mathbf{N}[\mathbf{u}(\mathbf{x}), \boldsymbol{\alpha}] = \mathbf{Q}(\mathbf{x}, \boldsymbol{\alpha}), \ \mathbf{x} \in \Omega_x(\boldsymbol{\alpha}). \quad (4.1)$$

The quantities which appear in Eq. (4.1) are defined as follows:

1. $\mathbf{u}(\mathbf{x}) \triangleq [u_1(\mathbf{x}), \ldots, u_{TD}(\mathbf{x})]^\dagger$ is a TD-dimensional column vector of dependent variables; the abbreviation TD denotes total (number of) dependent variables. The functions $u_i(\mathbf{x})$, $i = 1, \ldots, TD$, denote the system's dependent variables (also called state functions); $\mathbf{u}(\mathbf{x}) \in \mathsf{E}_u$, where E_u is a normed linear space over the scalar field F of real numbers.

2. $\mathbf{N}[\mathbf{u}(\mathbf{x}), \boldsymbol{\alpha}] \triangleq [N_1(\mathbf{u}, \boldsymbol{\alpha}), \ldots, N_{TD}(\mathbf{u}, \boldsymbol{\alpha})]^\dagger$ denotes a TD-dimensional column vector The components $N_i(\mathbf{u}, \boldsymbol{\alpha})$, $i = 1, \ldots, TD$ are operators (including differential, difference, integral, distributions, and/or infinite matrices) acting (usually) non-linearly on the dependent variables $\mathbf{u}(\mathbf{x})$, the independent variables $\mathbf{x}$ and the model parameters $\boldsymbol{\alpha}$.

3. $\mathbf{Q}(\mathbf{x}, \boldsymbol{\alpha}) \triangleq [q_1(\mathbf{x}; \boldsymbol{\alpha}), \ldots, q_{TD}(\mathbf{x}; \boldsymbol{\alpha})]^\dagger$ is a TD-dimensional column vector which represents inhomogeneous source terms, which usually depend non-linearly on the uncertain parameters $\boldsymbol{\alpha}$; $\mathbf{Q} \in \mathsf{E}_Q$, where E_Q is also a normed linear space.

4. All of the equalities in this work are considered to hold in the weak (distributional) sense, since the right sides (sources) of Eq. (4.1) and of other various equations to be derived in this work may contain distributions (generalised functions/functionals), particularly Dirac-distributions and derivatives and/or integrals thereof.

In view of the definitions given above, $\mathbf{N}(\boldsymbol{\alpha}, \mathbf{u})$ represents the mapping $\mathbf{N}: \mathsf{D} \subset \mathsf{E} \to \mathsf{E}_Q$, where $\mathsf{D} = \mathsf{D}_u \oplus \mathsf{D}_\alpha$, $\mathsf{D}_u \subset \mathsf{E}_u$, $\mathsf{D}_\alpha \subset \mathsf{E}_\alpha$, and $\mathsf{E} = \mathsf{E}_u \oplus \mathsf{E}_\alpha$. Note that an arbitrary element $\mathbf{e} \in \mathsf{E}$ is of the form $\mathbf{e} = (\boldsymbol{\alpha}, \mathbf{u})$. When differential operators appear in Eq. (4.1), then a corresponding set of boundary and/or initial conditions, which are essential to define the domain of $\mathbf{N}(\mathbf{u}, \boldsymbol{\alpha})$, must also be given. These boundary and/or initial conditions can be represented in operator form as follows:

$$\mathbf{B}[\mathbf{u}(\mathbf{x}), \mathbf{x}; \boldsymbol{\alpha}] - \mathbf{C}(\mathbf{x}, \boldsymbol{\alpha}) = \mathbf{0}, \ \mathbf{x} \in \partial\Omega_x(\boldsymbol{\alpha}). \quad (4.2)$$

The components $B_i(\mathbf{u}; \boldsymbol{\alpha})$, $i = 1, \ldots, TD$ of the vector-valued operator $\mathbf{B}(\mathbf{u}, \boldsymbol{\alpha}) \triangleq [B_1(\mathbf{u}; \boldsymbol{\alpha}), \ldots, B_{TD}(\mathbf{u}; \boldsymbol{\alpha})]^\dagger$ in Eq. (4.2) are operators, defined on the boundary $\partial\Omega_x(\boldsymbol{\alpha})$ of the model's domain $\Omega_x(\boldsymbol{\alpha})$ in phase-space, which act non-linearly on $\mathbf{u}(\mathbf{x})$ and on $\boldsymbol{\alpha}$. The components $C_i(\mathbf{u}; \boldsymbol{\alpha})$, $i = 1, \ldots, TD$ of the column vector $\mathbf{C}(\mathbf{x}, \boldsymbol{\alpha}) \triangleq [C_1(\mathbf{u}; \boldsymbol{\alpha}), \ldots, C_{TD}(\mathbf{u}; \boldsymbol{\alpha})]^\dagger$ comprise inhomogeneous boundary sources which, in general, are non-linear functions of the model parameters $\boldsymbol{\alpha}$.

The nominal solution of Eqs. (4.1) and (4.2) is denoted as $\mathbf{u}^0(\mathbf{x})$, and is obtained by solving these equations at the nominal parameter values $\boldsymbol{\alpha}^0 \triangleq [\alpha_1^0, \ldots, \alpha_i^0, \ldots, \alpha_{TP}^0]^\dagger$. In other words, the vectors $\mathbf{u}^0(\mathbf{x})$ and $\boldsymbol{\alpha}^0$ satisfy the following equations:

$$\mathbf{N}[\mathbf{u}^0(\mathbf{x}), \boldsymbol{\alpha}^0] = \mathbf{Q}(\mathbf{x}, \boldsymbol{\alpha}^0), \ \mathbf{x} \in \Omega_x(\boldsymbol{\alpha}^0), \ \mathbf{B}[\mathbf{u}^0(\mathbf{x}), \mathbf{x}; \boldsymbol{\alpha}^0] - \mathbf{C}(\mathbf{x}, \boldsymbol{\alpha}^0) = \mathbf{0}, \ \mathbf{x} \in \partial\Omega_x(\boldsymbol{\alpha}^0). \quad (4.3)$$

Responses of particularly important interest are model representations/computations of physical measurements of the model's state functions $\mathbf{u}(\mathbf{x}_p)$ at a specific point, $\mathbf{x}_p$, in phase-space. Other responses of particular interest are averages over the phase-space domain (or segments thereof). Both point-measurements and average responses can be represented generically in the following integral form:

$$R[\mathbf{u}(\mathbf{x}), \boldsymbol{\alpha}; \mathbf{x}] \triangleq \prod_{i=1}^{TI} \int_{\lambda_i(\boldsymbol{\alpha})}^{\omega_i(\boldsymbol{\alpha})} S[\mathbf{u}(\mathbf{x}); \boldsymbol{\alpha}] dx_i, \; \prod_{i=1}^{TI} \int_{\lambda_i(\boldsymbol{\alpha})}^{\omega_i(\boldsymbol{\alpha})} [\;] dx_i \triangleq \int_{\lambda_1(\boldsymbol{\alpha})}^{\omega_1(\boldsymbol{\alpha})}$$

$$\dots \int_{\lambda_i(\boldsymbol{\alpha})}^{\omega_i(\boldsymbol{\alpha})} \dots \int_{\lambda_{TI}(\boldsymbol{\alpha})}^{\omega_{TI}(\boldsymbol{\alpha})} [\;] dx_1 \, dx_2 \dots dx_i \dots dx_{TI}, \qquad (4.4)$$

where $S[\mathbf{u}(\mathbf{x}); \boldsymbol{\alpha}]$ is suitably differentiable non-linear function of $\mathbf{u}(\mathbf{x})$ and of $\boldsymbol{\alpha}$. It is important to note that the components of $\boldsymbol{\alpha}$ are considered to *also* include parameters that may appear specifically just in the definition of the response under consideration, but which might not appear in Eqs. (4.1) and (4.2). For example, a measurement of a physical quantity can be represented as a response $R_p[\boldsymbol{\varphi}(\mathbf{x}_p), \%\psi(\mathbf{x}_p); \boldsymbol{\alpha}]$ located at a point, $\mathbf{x}_p$, in phase-space, which may itself be afflicted by uncertainties. Such a response can be represented mathematically in the form

$$R_p[\mathbf{u}(\mathbf{x}_p); \boldsymbol{\alpha}] \triangleq \prod_{i=1}^{TI} \int_{\lambda_i(\boldsymbol{\alpha})}^{\omega_i(\boldsymbol{\alpha})} R_p[\mathbf{u}(\mathbf{x}); \boldsymbol{\alpha}] \delta(\mathbf{x} - \mathbf{x}_p) dx_i, \text{ where}$$

$\delta(\mathbf{x} - \mathbf{x}_p)$ denotes the multidimensional Dirac-delta functional. The measurement point $\mathbf{x}_p$ appears only in the definition of the response but does not appear in Eqs. (4.1) and (4.2). Thus, the (physical) system defined in this work is considered to comprise both the system's computational model and the system's response.

In practice, the values of the parameters α_n are determined experimentally and are considered to be variates that obey an unknown multivariate probability distribution function, denoted as $p_\alpha(\boldsymbol{\alpha})$. Considering that the multivariate distribution $p_\alpha(\boldsymbol{\alpha})$ is formally defined on a domain D_α, the various moments (e.g. mean values, covariance and variances) of $p_\alpha(\boldsymbol{\alpha})$ can be defined in a standard manner by using the following notation:

$$\langle u(\boldsymbol{\alpha}) \rangle_\alpha \triangleq \int_{D_\alpha} u(\boldsymbol{\alpha}) p_\alpha(\boldsymbol{\alpha}) d\boldsymbol{\alpha}, \qquad (4.5)$$

where $u(\boldsymbol{\alpha})$ is a continuous function of the parameter $\boldsymbol{\alpha}$. Using the notation defined in Eq. (4.5), the expected (or mean) value of a model parameter α_i, denoted as α_i^0, is defined as follows:

$$\alpha_i^0 \triangleq \langle \alpha_i \rangle_\alpha \triangleq \int_{D_\alpha} \alpha_i p_\alpha(\boldsymbol{\alpha}) d\boldsymbol{\alpha}, \; i = 1, \dots, TP. \qquad (4.6)$$

Throughout this work, the superscript 0 will be used to denote nominal or expected values. The covariance, $\mathrm{cov}(\alpha_i, \alpha_j)$, of two parameters, α_i and α_j, is defined as follows:

$$\mu_2^{ij}(\boldsymbol{\alpha}) \triangleq \mathrm{cov}(\alpha_i, \alpha_j) \triangleq \langle (\alpha_i - \alpha_i^0)(\alpha_j - \alpha_j^0) \rangle_\alpha$$

$$\triangleq \rho_{ij} \sigma_i \sigma_j, \; i, j = 1, \dots, TP. \qquad (4.7)$$

The variance, $\mathrm{var}(\alpha_i)$, of a parameter α_i, is defined as follows: $\mathrm{var}(\alpha_i) \triangleq \langle (\alpha_i - \alpha_i^0)^2 \rangle_\alpha$, $i = 1, \dots, TP$. The standard deviation, σ_i, of α_i, is defined as follows: $\sigma_i \triangleq \sqrt{\mathrm{var}(\alpha_i)}$. The correlation, ρ_{ij}, between two parameters α_i and α_j, is defined as follows: $\rho_{ij} \triangleq \mathrm{cov}(\alpha_i, \alpha_j) / (\sigma_i \sigma_j)$; $i, j = 1, \dots, TP$. The third-order moment, μ_3^{ijk}, of the multivariate parameter distribution function $p(\boldsymbol{\alpha})$, and the third-order parameter correlation, t_{ijk}, respectively, are defined as follows:

$$\mu_3^{ijk}(\boldsymbol{\alpha}) \triangleq \int_{D_\alpha} (\alpha_i - \alpha_i^0)(\alpha_j - \alpha_j^0)(\alpha_k - \alpha_k^0) p(\boldsymbol{\alpha}) d\boldsymbol{\alpha}$$

$$\triangleq t_{ijk} \sigma_i \sigma_j \sigma_k, \; i, j, k = 1, \dots, TP. \qquad (4.8)$$

The fourth-order moment, μ_4^{ijkl}, of the multivariate parameter distribution function $p(\boldsymbol{\alpha})$, and the fourth-order parameter correlation, q_{ijkl}, respectively, are defined as follows:

$$\mu_4^{ijkl}(\boldsymbol{\alpha}) \triangleq \int_{D_\alpha} (\alpha_i - \alpha_i^0)(\alpha_j - \alpha_j^0)(\alpha_k - \alpha_k^0)(\alpha_l - \alpha_l^0) p(\boldsymbol{\alpha}) d\boldsymbol{\alpha}$$

$$\triangleq q_{ijkl} \sigma_i \sigma_j \sigma_k \sigma_l; \quad i, j, k, l = 1, \dots, TP. \qquad (4.9)$$

The uncertainties in the model parameters induce uncertainties in the distribution of the response in the phase-space of parameters, which can be obtained by using the propagation of errors formulas originally obtained by Tukey (1957) by formally expanding the computed response, denoted as $r_{i_1}^c(\boldsymbol{\alpha})$, in a Taylor-series around the nominal parameter values $\boldsymbol{\alpha}^0 \triangleq (\alpha_1^0, \dots, \alpha_{TP}^0)$. Up to third-order sensitivities, the Taylor-series of a computed response has the following form:

$$r_{i_1}^c(\boldsymbol{\alpha}) = r_{i_1}^c(\boldsymbol{\alpha}^0) + \sum_{i=1}^{TP} \left\{ \frac{\partial r_{i_1}^c(\boldsymbol{\alpha})}{\partial \alpha_i} \right\}_{\boldsymbol{\alpha}^0} (\alpha_i - \alpha_i^0)$$

$$+ \frac{1}{2} \sum_{i,j=1}^{TP} \left\{ \frac{\partial^2 r_{i_1}^c(\boldsymbol{\alpha})}{\partial \alpha_i \partial \alpha_j} \right\}_{\boldsymbol{\alpha}^0} (\alpha_i - \alpha_i^0)(\alpha_j - \alpha_j^0)$$

$$+ \frac{1}{6} \sum_{i,j,k=1}^{TP} \left\{ \frac{\partial^3 r_{i_1}^c(\boldsymbol{\alpha})}{\partial \alpha_i \partial \alpha_j \partial \alpha_k} \right\}_{\boldsymbol{\alpha}^0} (\alpha_i - \alpha_i^0)(\alpha_j - \alpha_j^0)(\alpha_k - \alpha_k^0)$$

$$+ \dots; \; i_1 = 1, \dots, TR. \qquad (4.10)$$

where $r_{i_1}^c(\boldsymbol{\alpha}^0)$ denotes the nominal value of the response computed at the nominal (mean) parameter values $\boldsymbol{\alpha}^0 \triangleq (\alpha_1^0, \ldots, \alpha_{TP}^0)$, where the superscript c denotes computed, and where the subscript $i_1 = 1, \ldots, TR$ denotes one of a total of TR responses that would be of interest. Using Eq. (4.10) yields the following expressions for the first three cumulants of the distribution of computed responses in the parameter phase-space:

1. The *expected* (mean) *value*, denoted as $E[r_{i_1}^c(\boldsymbol{\alpha})]$, of a response $r_{i_1}^c(\boldsymbol{\alpha})$:

$$
E[r_{i_1}^c(\boldsymbol{\alpha})] = r_{i_1}^c(\boldsymbol{\alpha}^0) + \frac{1}{2} \sum_{i,j=1}^{TP} \left\{ \frac{\partial^2 r_{i_1}^c(\boldsymbol{\alpha})}{\partial \alpha_i \partial \alpha_j} \right\}_{\boldsymbol{\alpha}^0} \rho_{ij}\, \sigma_i \sigma_j
$$

$$
+ \frac{1}{6} \sum_{i,j,k=1}^{TP} \left\{ \frac{\partial^3 r_{i_1}^c(\boldsymbol{\alpha})}{\partial \alpha_i \partial \alpha_j \partial \alpha_k} \right\}_{\boldsymbol{\alpha}^0} t_{ijk}\sigma_i\sigma_j\sigma_k;\ i_1 = 1, \ldots, TR.
$$

$$
(4.11)
$$

2. The *covariance*, denoted as $cov(r_{i_1}^c, r_{i_2}^c)$, of two responses, $r_{i_1}^c(\boldsymbol{\alpha})$ and $r_{i_2}^c(\boldsymbol{\alpha})$ for $i_1,\ i_2 = 1, \ldots, TR$:

$$
cov(r_{i_1}^c,\ r_{i_2}^c) = \sum_{i,j=1}^{TP} \left(\frac{\partial r_{i_1}^c}{\partial \alpha_i} \frac{\partial r_{i_2}^c}{\partial \alpha_j} \right) \rho_{ij}\sigma_i\sigma_j
$$

$$
+ \frac{1}{2} \sum_{i,j,\mu=1}^{TP} \left(\frac{\partial^2 r_{i_1}^c}{\partial \alpha_i \partial \alpha_j} \frac{\partial r_{i_2}^c}{\partial \alpha_\mu} + \frac{\partial r_{i_1}^c}{\partial \alpha_i} \frac{\partial^2 r_{i_2}^c}{\partial \alpha_j \partial \alpha_\mu} \right) t_{ij\mu}\sigma_i\sigma_j\sigma_\mu
$$

$$
+ \frac{1}{4} \sum_{i,j,\mu,\nu=1}^{TP} \left(\frac{\partial^2 r_{i_1}^c}{\partial \alpha_i \partial \alpha_j} \right) \left(\frac{\partial^2 r_{i_2}^c}{\partial \alpha_\mu \partial \alpha_\nu} \right) (q_{ij\mu\nu} - \rho_{ij}\rho_{\mu\nu})\sigma_i\sigma_j\sigma_\mu\sigma_\nu
$$

$$
+ \frac{1}{6} \sum_{i,j,\mu,\nu=1}^{TP} \left(\frac{\partial r_{i_1}^c}{\partial \alpha_i} \frac{\partial^3 r_{i_2}^c}{\partial \alpha_j \partial \alpha_\mu \partial \alpha_\nu} + \frac{\partial r_{i_2}^c}{\partial \alpha_i} \frac{\partial^3 r_{i_1}^c}{\partial \alpha_j \partial \alpha_\mu \partial \alpha_\nu} \right) q_{ij\mu\nu}\sigma_i\sigma_j\sigma_\mu\sigma_\nu.
$$

$$
(4.12)
$$

In particular, the variance of a response $r_{i_1}^c(\boldsymbol{\alpha})$ is obtained as by setting $i_1 = i_2$ in Eq. (4.12).

3. The *covariance of a response*, $r_{i_1}^c(\boldsymbol{\alpha})$ *and a parameter* α_ℓ, $i_1 = 1, \ldots, TR$ and $\ell = 1, \ldots, TP$, which is denoted as $cov(r_{i_1}^c,\ \alpha_\ell)$ and is given by the following expression:

$$
cov(r_{i_1}^c,\ \alpha_\ell) = \sum_{i=1}^{TP} \left\{ \frac{\partial r_{i_1}^c(\boldsymbol{\alpha})}{\partial \alpha_i} \right\}_{\boldsymbol{\alpha}^0} cov(\alpha_i, \alpha_\ell)
$$

$$
+ \frac{1}{2} \sum_{i,j=1}^{TP} \left\{ \frac{\partial^2 r_{i_1}^c(\boldsymbol{\alpha})}{\partial \alpha_i \partial \alpha_j} \right\}_{\boldsymbol{\alpha}^0} t_{ij\ell}\sigma_i\sigma_j\sigma_\ell
$$

$$
+ \frac{1}{6} \sum_{i,j,k=1}^{TP} \left\{ \frac{\partial^3 r_{i_1}^c(\boldsymbol{\alpha})}{\partial \alpha_i \partial \alpha_j \partial \alpha_k} \right\}_{\boldsymbol{\alpha}^0} q_{ij\mu\nu}\sigma_i\sigma_j\sigma_k\sigma_\ell. \quad (4.13)
$$

4. The *third-order cumulant*, $\mu_3(r_{i_1}^c, r_{i_2}^c, r_{i_3}^c)$, *for three responses*, $r_{i_1}^c(\boldsymbol{\alpha})$, $r_{i_2}^c(\boldsymbol{\alpha})$, and $r_{i_3}^c(\boldsymbol{\alpha})$, *for* $i_1, i_2, i_3 = 1, \ldots, TR$:

$$
\mu_3(r_{i_1}^c, r_{i_2}^c, r_{i_3}^c) = \sum_{i=1}^{TP}\sum_{j=1}^{TP}\sum_{\mu=1}^{TP} \frac{\partial r_{i_1}^c}{\partial \alpha_i} \frac{\partial r_{i_2}^c}{\partial \alpha_j} \frac{\partial r_{i_3}^c}{\partial \alpha_\mu} t_{ij\mu}\sigma_i\sigma_j\sigma_\mu
$$

$$
+ \frac{1}{2} \sum_{i,j,\mu,\nu=1}^{TP} \left\{ \frac{\partial r_{i_1}^c k}{\partial \alpha_i} \frac{\partial r_{i_2}^c}{\partial \alpha_j} \frac{\partial^2 r_{i_3}^c}{\partial \alpha_\mu \partial \alpha_\nu} (q_{ij\mu\nu} - \rho_{ij}\rho_{\mu\nu}) \right.
$$

$$
+ \frac{\partial r_{i_1}^c}{\partial \alpha_i} \frac{\partial^2 r_{i_2}^c}{\partial \alpha_j \partial \alpha_\mu} \frac{\partial r_{i_3}^c}{\partial \alpha_\nu} (q_{ij\mu\nu} - \rho_{i\nu}\rho_{j\mu})
$$

$$
+ \frac{\partial^2 r_{i_1}^c}{\partial \alpha_i \partial \alpha_j} \frac{\partial r_{i_2}^c}{\partial \alpha_\mu} \frac{\partial r_{i_3}^c}{\partial \alpha_\nu} (q_{ij\mu\nu} - \rho_{i\mu}\rho_{j\nu}) \right\}\sigma_i\sigma_j\sigma_\mu\sigma_\nu.
$$

$$
(4.14)
$$

It is important to note that the second-order sensitivities also contribute the leading correction terms to the response's expected value, causing it to differ from the response's computed value. The second-order sensitivities also contribute to the response variances and covariances. The skewness of a single response is customarily denoted as $\gamma_1(r_{i_1}^c)$, and is defined as follows: $\gamma_1(r_{i_1}^c) \triangleq \mu_3(r_{i_1}^c)/[\mathrm{var}(r_{i_1}^c)]^{3/2}$, where the expression of $\mu_3(r_{i_1}^c)$ is obtained by setting $r_{i_1}^c \equiv r_{i_2}^c \equiv r_{i_3}^c$ in Eq. (4.14). As is well-known, skewness indicates the direction and relative magnitude of a distribution's deviation from the normal distribution. In particular, if only first-order sensitivities are considered, the third-order moment of the response is always zero. Hence, a 'first-order sensitivity and uncertainty quantification' will always produce an erroneous third moment (and hence skewness) of the predicted response distribution, unless the unknown response distribution happens to be symmetrical. At least second-order sensitivities must be used in order to estimate the skewness of the response distribution. With pronounced skewness, standard statistical inference procedures such as constructing a confidence interval for the mean (expectation) of a computed/predicted model response will be not only incorrect, in the sense that the true coverage level will differ from the nominal (e.g. 95%) level, but the error probabilities will be unequal on each side of the predicted mean.

Although the formulas presented in Eqs. (4.11)–(4.14) were known since the work of Tukey (1957), they have not been used in practice beyond first-order in sensitivities because even the second-order sensitivities were already prohibitively expensive to compute for large-scale systems involving many uncertain parameters. In recent years, however, Cacuci (2015, 2018) has conceived the general-purpose second-order adjoint sensitivity analysis methodology, which was extended to fourth- and fifth-order by Cacuci (2022a,b). These methodologies enable the exact and efficient computation of high-order sensitivities of large-scale system responses, thereby enabling sensitivity analyses, uncertainty quantification, and forward and inverse data assimilation for large-scale physical systems, which would not otherwise be amenable for such analyses by any other methods, statistical or otherwise, as will be illustrated by the paradigm examples presented in the remainder of this chapter.

4.3 Forward and Inverse 3rd-BERRU-PM: Mathematical Framework and Illustrative Application

As will be shown in this section, the mathematical framework of the 3rd-BERRU-PM methodology subsumes the mathematical procedures for data assimilation and model calibration. The 3rd-BERRU-PM can be used both for forward problems, for predicting the outcome (response) of a model when all of the model's parameters are known, as well as for inverse problems, for predicting some unknown model parameters from independent measurements (or computations) of the model's response. The mathematical framework of the first-order BERRU-PM, which uses just *first-order sensitivities* of model response with respect to model parameters, has been presented in Cacuci (2019a), along with various applications to modelling of industrial processes (chemical processes, cooling towers, inverse radiation transport, etc.). The 3rd-BERRU-PM methodology presented in this section can incorporate high-order sensitivities and its mathematical framework will be presented in Section 4.3.1. An application to a simple inverse one-dimensional radiation problem, to determine the dimension of an 'optically thick' slab of material from uncertain radiation detector counts outside the slab, will be presented in Section 4.3.2. Conventional statistical methods will be shown to fail to predict the slab's thickness, because although apparently simple, this problem is a deep-penetration problem which has particularly important implications for similar inverse problems in the Earth sciences. The first-order version the 3rd-BERRU-PM methodology is available as a stand-alone software package (Cacuci, Fang, and Badea, 2018) for performing predictive modelling of coupled multi-physics systems (either time-dependent or time-independent).

4.3.1 3rd-BERRU-PM Mathematical Framework
This subsection presents the mathematical form of the *a priori* information used in the 3rd-BERRU-PM which will be combined/assimilated to obtain optimal, best-estimate predicted mean values for both the model

responses and model parameters, with reduced predicted uncertainties, in the *combined parameter-response phase space*. It is important to note that the information included in the 3rd-BERRU-PM methodology comprises not only first- and second-order sensitivities, but also third-order response sensitivities with respect to the model's imprecisely known parameters, internal interfaces, and external boundaries in the phase-space of independent variables. In contradistinction to the conventional data assimilation methodologies, the 3rd-BERRU-PM methodology also includes correlations between the model responses and parameters.

4.3.1.1 A Priori *Information Included in the 3rd-BERRU-PM*

Expected Values and Covariances of Measured Responses
Consider that TR model results (responses), which will be denoted as r_i^m, $i = 1, ..., TR$, have been experimentally measured. These measurements are used to determine the expectations and covariances or the measured responses. The responses r_i^m are considered to constitute the components of the TR-dimensional column vector $\mathbf{r}^m \triangleq (r_1^m, ..., r_{TR}^m)^\dagger$. The expected values, $E(r_i^m)$, of the measured responses will be considered to constitute the components of the column vector $\mathbf{E}(\mathbf{r}^m) \triangleq [E(r_1^m), ..., E(r_{TR}^m)]^\dagger$. The covariances, $cov(r_i^m, r_j^m)$, of two the measured responses are considered to be components of the $TR \times TR$-dimensional covariance matrix of measured responses $\mathbf{C}_m \triangleq \langle [\mathbf{r}^m - \mathbf{E}(\mathbf{r}^m)][\mathbf{r}^m - \mathbf{E}(\mathbf{r}^m)]^\dagger \rangle_r = [cov(r_i^m, r_j^m)]_{TR \times TR}$.

A Priori Expectations and Covariances of Model Parameters
The *a priori* information included in the 3rd-BERRU-PM methodology comprises the vector $\boldsymbol{\alpha}^0 \triangleq [\alpha_1^0, ..., \alpha_i^0, .., \alpha_{TP}^0]^\dagger$, comprising the parameter expectation values, and the $TP \times TP$-dimensional parameter covariance matrix $\mathbf{C}_\alpha \triangleq \langle (\boldsymbol{\alpha} - \boldsymbol{\alpha}^0)(\boldsymbol{\alpha} - \boldsymbol{\alpha}^0)^\dagger \rangle_\alpha \triangleq [cov(\alpha_i, \alpha_j)]_{TP \times TP}$ as defined in Eqs. (4.6) and (4.7), respectively.

A Priori Expectations and Covariances of Computed Responses:
The computed responses $r_k^c(\boldsymbol{\alpha})$, $k = 1, ..., TR$, are considered to be elements of an TR-dimensional vector $\mathbf{r}^c(\boldsymbol{\alpha}) \triangleq [r_1^c(\boldsymbol{\alpha}), ..., r_{N_r}^c(\boldsymbol{\alpha})]^\dagger$. The set of expectation values $E[r_k^c(\boldsymbol{\alpha})]$, $k = 1, ..., TR$, provided in Eq. (4.11), of the computed responses $r_k^c(\boldsymbol{\alpha})$, $k = 1, ..., TR$, are considered to be the components of the following vector of expected values of the computed responses $E[\mathbf{r}^c(\boldsymbol{\alpha})] \triangleq [E(r_1^c), ..., E(r_{N_r}^c)]^\dagger$. The response covariances defined in Eq. (4.12) are considered to be the components of a $(TR \times TR)$-dimensional covariance matrix $\mathbf{C}_r \triangleq \langle [\mathbf{r}^c - E(\mathbf{r}^c)][\mathbf{r}^c - E(\mathbf{r}^c)]^\dagger \rangle_\alpha$. The covariances between the computed responses and the model parameters defined in Eq. (4.13) are considered to be the components of an $(N_r \times N_\alpha)$-dimensional matrix $\mathbf{C}_{r\alpha} \triangleq \langle [\mathbf{r}^c - E(\mathbf{r}^c)](\boldsymbol{\alpha} - \boldsymbol{\alpha}^0)^\dagger \rangle_\alpha = \mathbf{C}_{\alpha r}^\dagger$.

4.3.1.2 Posterior Results: Best-Estimate Responses and Parameters with Reduced Uncertainties

The *a priori* information described in Section 4.3.1.1 is considered to stem from an unknown joint multivariate probability distribution of the form $p(\boldsymbol{\alpha}, \mathbf{r}) = p_r(\mathbf{r})p_\alpha(\boldsymbol{\alpha})$, which is obeyed in the joint phase-space of parameter and responses by the vector $\mathbf{z} \triangleq [\boldsymbol{\alpha}, \mathbf{r}^c(\boldsymbol{\alpha}), \mathbf{r}^m]^\dagger$. The least inform-ative (hence the most conservative) probability distribution for the vector $\mathbf{z} \triangleq [\boldsymbol{\alpha}, \mathbf{r}^c(\boldsymbol{\alpha}), \mathbf{r}^m]^\dagger$ that can be constructed by using the *a priori* information provided here can be con-structed, as shown by Cacuci (2019b), by applying the max-imum entropy principle (Jaynes, 1957) to obtain the posterior distribution. The expressions for the moments of this posterior distribution provide the predicted expect-ation, covariances, etc., for the model parameters and responses. They have been obtained – to a user-controlled degree of accuracy – by using the saddle-point method (also called Laplace approximation, or steepest descent method). To first-order of accuracy provided by the saddle-point approximation, Cacuci (2019a) has obtained the formal posterior expressions for the predicted best-estimate expect-ations and covariances for the parameters and responses, which are provided here.

Model Calibration:

a. Best-estimate predicted expectation value, denoted as $\mathbf{r}^{be}$, for the vector of model responses:

$$\mathbf{r}^{be} = \mathbf{r}^m + \mathbf{C}_m(\mathbf{C}_m + \mathbf{C}_r)^{-1}[E(\mathbf{r}^c) - E(\mathbf{r}^m)]. \qquad (4.15)$$

b. Best-estimate predicted covariance matrix, denoted as $\mathbf{C}_r^{be}$, for the best-estimate predicted responses:

$$\mathbf{C}_r^{be} = \mathbf{C}_m[\mathbf{I} - (\mathbf{C}_m + \mathbf{C}_r)^{-1}\mathbf{C}_m]. \qquad (4.16)$$

Model Extrapolation:

a. Best-estimate predicted expectation value, denoted as $\boldsymbol{\alpha}^{be}$, for the vector of model parameters, which can subsequently be used for model calibration:

$$\boldsymbol{\alpha}^{be} = \boldsymbol{\alpha}^0 - \mathbf{C}_{\alpha r}(\mathbf{C}_m + \mathbf{C}_r)^{-1}[E(\mathbf{r}^c) - E(\mathbf{r}^m)]. \qquad (4.17)$$

b. Best-estimate predicted covariance matrix, denoted as $\mathbf{C}_\alpha^{be}$, for the best-estimate predicted parameters:

$$\mathbf{C}_\alpha^{be} = \mathbf{C}_\alpha - \mathbf{C}_{\alpha r}(\mathbf{C}_m + \mathbf{C}_r)^{-1}\mathbf{C}_{r\alpha}. \qquad (4.18)$$

c. Best-estimate predicted parameter-response covariance matrix and its transpose, denoted as $\mathbf{C}_{\alpha r}$ and $\mathbf{C}_{r\alpha}^{be}$, respectively, for the best-estimate parameters $\boldsymbol{\alpha}^{be}$ and best-estimate responses $\mathbf{r}^{be}$:

$$\mathbf{C}_{\alpha r}^{be} = \mathbf{C}_{\alpha r}(\mathbf{C}_m + \mathbf{C}_r)^{-1}\mathbf{C}_m. \qquad (4.19)$$

$$\mathbf{C}_{r\alpha}^{be} = \mathbf{C}_m(\mathbf{C}_m + \mathbf{C}_r)^{-1}\mathbf{C}_{r\alpha} = (\mathbf{C}_{\alpha r}^{be})^\dagger. \qquad (4.20)$$

Estimation of the Validation Domain:
A chi-square consistency indicator – denoted as Q_{min} – which quantifies the mutual and joint consistency of the information

available for model calibration and is provided by the following expression:

$$Q_{min} = -\frac{1}{2}[E(\mathbf{r}^c) - E(\mathbf{r}^m)]^\dagger(\mathbf{C}_m + \mathbf{C}_r)^{-1}[E(\mathbf{r}^c) - E(\mathbf{r}^m)]. \qquad (4.21)$$

The quantity Q_{min} represents the square of the length of the vector $\mathbf{d} \triangleq [E(\mathbf{r}^c) - E(\mathbf{r}^m)]$, measuring (in the corres-ponding metric) the deviations between the experimental and nominally computed responses. The quantity Q_{min} obeys a chi-square distribution with TR degrees of freedom and can be evaluated directly from the given data (i.e. given parameters and responses, together with their original uncertainties) after having inverted the covariance matrix $(\mathbf{C}_m + \mathbf{C}_r)$. It is also important to note that Q_{min} is inde-pendent of calibrating (or adjusting) the original data. As the dimension of $[E(\mathbf{r}^c) - E(\mathbf{r}^m)]$ indicates, the number of degrees of freedom characteristic of the calibration under consideration is equal to the number TR of experimental responses. In the extreme case of absence of experimen-tal responses, no actual calibration takes place. An actual calibration (adjustment) occurs only when including at least one experimental response. The quan-tity Q_{min} can be used for estimating contours of con-stant uncertainty in the combined parameter and response (high-dimensional) phase-space, thereby quan-tifying the validation domain underlying the model under investigation.

It is important to note that the predicted best-estimate values $\mathbf{r}^{be}$ and $\boldsymbol{\alpha}^{be}$, together with their respective covariance matrices $\mathbf{C}_r^{be}$, $\mathbf{C}_\alpha^{be}$, and $\mathbf{C}_{\alpha r}$, which are presented in Eqs. (4.15)–(4.20) contain second-order and third-order sensi-tivities. Thus, the expressions presented in Eqs. (4.15)–(4.20) generalise all of the previous formulas of this type found in data assimilation procedures (Kalman filters, Bayesian linear statistics, etc.) published to date, which contain first-order and, occasionally, matrix-vector prod-ucts that include some second-order sensitivities. As indi-cated in Eq. (4.16), the *a priori* covariance matrix $\mathbf{C}_m$ is multiplied by the matrix $[\mathbf{I} - (\mathbf{C}_m + \mathbf{C}_r)^{-1}\mathbf{C}_m]$, which means that the variances contained on the diagonal of the best-estimate matrix $\mathbf{C}_r^{be}$ will be smaller than the experimentally measured variances contained in $\mathbf{C}_m$. Hence, the addition of new experimental information has reduced the predicted best-estimate response variances in $\mathbf{C}_r^{be}$ by comparison to the measured variances contained *a priori* in $\mathbf{C}_m$.

Both matrices $\mathbf{C}_\alpha$ and $\mathbf{C}_{\alpha r}(\mathbf{C}_m + \mathbf{C}_r)^{-1}\mathbf{C}_{r\alpha}$ are symmetric and positive definite. Therefore, the subtraction indicated in Eq. (4.18) implies that the components of the main diagonal of $\mathbf{C}_\alpha^{be}$ must have smaller values than the corresponding elements of the main diagonal of $\mathbf{C}_\alpha$. In this sense, the introduction of new computational and experimental infor-mation has reduced the best-estimate predicted parameter variances on the diagonal of $\mathbf{C}_\alpha^{be}$.

It is important to note from the results shown in Eqs. (4.15)–(4.20) that the computation of the best-estimate parameter and response values, together with their corresponding best-estimate covariance matrices, only requires the computation of $(\mathbf{C}_m + \mathbf{C}_r)^{-1}$, which entails the inversion of a matrix of size $TR \times TR$. This is computationally very advantageous, since the number of responses is much less than the number of model parameters (i.e. $TR \ll TP$) in the overwhelming majority of practical situations.

4.3.2 3rd-BERRU-PM Inverse Predictive Modelling: Determining a Material's Thickness from Detector Responses in the Presence of Counting Uncertainties

Consider a one-dimensional slab of homogeneous material extending from $z = 0$ to $z = a$ [cm], placed in air and characterised by a total interaction coefficient μ [cm^{-1}]. The slab contains a uniformly distributed source of strength Q [$photons/cm^3 sec$] emitting isotropically monoenergetic photons within the slab. It is assumed that there is no scattering into the energy lines. Under these conditions, the angular flux of photons within the slab is described by the Boltzmann transport equation without scattering and with 'vacuum' incoming boundary condition, that is,

$$\omega \frac{d\psi(z,\omega)}{dz} + \mu\psi(z,\omega) = \frac{Q}{2}, 0 < z \le a, \ \omega > 0, \ \psi(0,\omega) = 0,$$
(4.22)

where $\psi(z,\omega)$ denotes the neutron angular flux at position z and direction $\omega \triangleq \cos\theta$, and θ denotes the angle between the photon's direction and the z-axis. The solution of Eq. (4.22) is

$$\psi(z,\omega) = \frac{Q}{2\mu}[1 - \exp(\mu z/\omega)].$$
(4.23)

Consider further that the leakage flux of uncollided photons is measured by an 'infinite-plane detector placed in air at some location $z > a$ external to the slab. The detector's response function, denoted as Σ_d [cm^{-1}], is considered to be a perfectly well-known constant. If the detection process were a perfectly deterministic process, rather than a stochastic one, it would follow from Eq. (4.23) that the exact detector response, denoted as $r(\mu a)$, would be given by the following expression:

$$r(\mu a) \triangleq \Sigma_d \int_0^1 \psi(z,\omega)d\omega = \frac{Q\Sigma_d}{2\mu}[1 - E_2(\mu a)],$$

$$E_n(x) = \int_0^1 u^{n-2}e^{-x/u}du, n = 0, 1, 2, \ldots$$
(4.24)

In the absence of counting uncertainties, therefore, knowing the detector's response $r(\mu a)$ yields a unique slab thickness, by solving Eq. (4.24) to determine its unique real root, as shown by Cacuci (2017).

In the presence of detector counting uncertainties, the current state-of-the-art methods (see, e.g., Lewis et al., 2006) for solving inverse problems such as determining the optical dimension of a uniform homogeneous medium from K uncertain photon measurements, $r_{exp}^{(k)}$, $k = 1, \ldots, K$, external to the medium, rely on minimising a user-defined chi-square-type functional of the following form:

$$\chi^2 \triangleq \sum_{k=1}^K \left[\frac{\delta(k)}{std.dev\left(r_{exp}^{(k)}\right)}\right]^2; \ \delta(k) \triangleq r_{model}\left(\mu a^{(k)}\right) - r_{exp}^{(k)}.$$
(4.25)

The value $(\mu a)_{min}$ which yields the minimum value, χ^2_{min}, of χ^2, is considered to be the slab's optical thickness. For optically thin slabs, Cacuci (2017) has shown that the precision of measurements does not affect the location of the unique minimum of the quantity $\delta^2(k)$, and the actual thickness of the respective slab is determined sufficiently accurately, for practical purposes, by the unique location of this minimum.

On the other hand, the precision of the measurements decisively affects the results for optically thick slabs. If the measurements are inaccurate, then any minimisation of the expression in Eq. (4.25) will lead to erroneous physical results, in that the result delivered by any minimisation procedure will not be physically correct. Furthermore, the larger the optical thickness of the slab, the more unphysical will the result of the minimisation procedure likely be, since the very formulation of the χ^2-functional makes this functional extremely sensitive to the value of each measurement as the slab's optical thickness increases. This remains true of the χ^2-functional even when the measurements are precise. A typical result for a thick slab of actual optical thickness $\mu a = 10.0$ is presented in Fig. 4.1 for very precise measurements, assumed to be normally distributed with a mean equal to (the exact response) r_{model} and having a relative standard deviation of 0.001%. Cacuci (2017) has shown that if the measurements have a relative standard deviation of 10%, the traditional chi-square minimisation procedure can even fail to produce a real-valued minimum, hence failing to produce any inverse prediction of the slab's thickness. Even for very precise measurements, having a relative standard deviation of 0.001%, the customary chi-square minimisation procedure may fail to yield a real-valued minimum, as depicted in Fig. 4.1, which indicates that only 5 of the 10 measurements have yielded real-valued minima, with values $7 < \mu a_{model} < 9$, all of which underpredict the actual slab thickness of $\mu a = 10.0$. In contradistinction, the BERRU-PM (Cacuci, 2019a) predicts the slab's thickness to be $(\mu a)_{BERRU-PM} = 9.88$, as indicated by the results presented in Table 4.1, which is very close to the exact result $\mu a = 10.0$. The response sensitivities needed by the BERRU-PM procedure were obtained from Eq. (4.24).

Table 4.1 *Results predicted by the 3rd-BERRU-PM methodology for a slab of exact thickness* $\mu a = 10$ *after assimilating* K = 10 *experiments with* $\beta = 10^{-5}$, $\mu a = 10$

Experimental response mean value	Predicted response	Predicted response SD	Predicted parameter	Predicted parameter SD
4.999989×10^{-1}	4.999981×10^{-1}	2.7×10^{-6}	9.88	7.63×10^{-1}
	Exact Response	Exact Response SD	Exact Parameter	
	4.999981×10^{-1}	4.999981×10^{-6}	10.0	

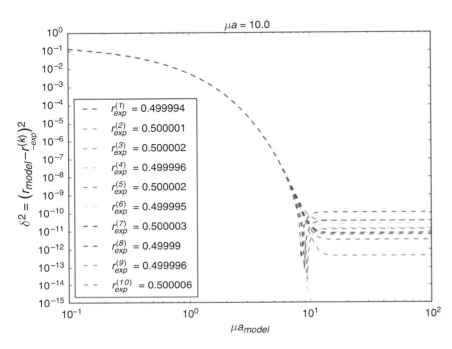

Figure 4.1 Variation of $\delta^2 \triangleq \left(r_{\text{model}} - r_{\text{exp}}^{(k)} \right)^2$ as a function of the model's optical thickness (μa_{model}) for a slab of actual optical thickness $\mu a = 10.0$, for very precise measurements with a relative standard deviation of 0.001%.

4.3 Illustrative Adjoint Sensitivity and Uncertainty Analysis of Oscillatory Dynamical Systems

Oscillatory dynamical systems are of considerable interest not only in the atmospheric and Earth sciences, where they were first studied by Lorenz (1963), but in all branches of the biological and physical sciences and engineering. The Lorenz (1963) model is relatively simple – comprising just three coupled first-order (in time) ordinary differential equations having quadratic non-linearities and just three positive parameters – and was therefore amenable to countless studies. Oscillatory dynamical systems that comprise many parameters have at best been analysed approximately, rather than exactly, because of the impractically large number of computations that would be required if using conventional methods for sensitivity and uncertainty analysis. Cacuci and Di Rocco (2020) have presented a pioneering adjoint sensitivity analysis of an oscillatory dynamical model which comprises sufficiently many parameters to render its analysis by

conventional statistical methods impractical. Since the mathematical tools used for this analysis are applicable to any large-scale dynamic model in the Earth and/or atmospheric sciences, they will be presented in Section 4.3.1. Representative sensitivity and uncertainty analysis results, which can serve as paradigm illustrations for the efficient and exact computation of the first-order sensitivities of any dynamic model's state/dependent variables with respect to all of the model's parameters in the stable and oscillatory regions (including the chaotic region) are presented in Section 4.3.2.

4.3.1 Mathematical Framework Underlying the First-Order Comprehensive Adjoint Sensitivity Analysis Methodology for Non-linear Systems (1st-CASAM-N)

The parameters that characterise a mathematical/computational model and its internal and external boundaries in phase-space are seldom known exactly: their true values (which are unknown) will differ from their nominal (or mean) values (which are known) by quantities denoted as $\delta\boldsymbol{\alpha} \triangleq (\delta\alpha_1, ..., \delta\alpha_{TP})$, where $\delta\alpha_i \triangleq \alpha_i - \alpha_i^0$. Evidently, the forward state functions $\mathbf{u}(\mathbf{x})$ are related to the model and boundary parameters $\boldsymbol{\alpha}$ through Eqs. (4.1) and (4.2). Hence, variations $\delta\boldsymbol{\alpha}$ in the model and boundary parameter will cause corresponding variations $\delta\mathbf{u}(\mathbf{x}) \triangleq [\delta u_1(\mathbf{x}), ..., \delta u_{TI}(\mathbf{x})]^\dagger$ around the nominal solution $\mathbf{u}^0(\mathbf{x})$ in the forward state functions. In turn, the variations $\delta\boldsymbol{\alpha}$ and $\delta\mathbf{u}(\mathbf{x})$ will induce variations in the system's response. For the mathematical derivations to follow, it is convenient to introduce the vector $\mathbf{e} \triangleq (\mathbf{u}, \boldsymbol{\alpha})$, which is defined on the product of the vector fields of the model parameters and state functions. The nominal value of $\mathbf{e} \triangleq (\mathbf{u}, \boldsymbol{\alpha})$ is denoted as $\mathbf{e}^0 \triangleq (\mathbf{u}^0, \boldsymbol{\alpha}^0)$. The most general definition of the first-order total sensitivity of the operator-valued model response $R(\mathbf{u}, \boldsymbol{\alpha})$ to variations $\delta\mathbf{u}(\mathbf{x})$ and $\delta\boldsymbol{\alpha}$ around the nominal values $\mathbf{u}^0$ and $\boldsymbol{\alpha}^0$, respectively, has been provided in the pioneering works of Cacuci (1981a, b) in terms of the first-order Gateaux-variation $\delta R(\mathbf{u}^0, \boldsymbol{\alpha}^0; \delta\mathbf{u}, \delta\boldsymbol{\alpha})$ of $R(\mathbf{u}, \boldsymbol{\alpha})$, which is defined as follows:

$$\delta R(\mathbf{e}^0; \mathbf{h}) \triangleq \left\{\frac{d}{d\varepsilon} \prod_{i=1}^{TI} \int_{\lambda_i(\boldsymbol{\alpha}^0+\varepsilon\delta\boldsymbol{\alpha})}^{\omega_i(\boldsymbol{\alpha}^0+\varepsilon\delta\boldsymbol{\alpha})} S(\mathbf{u}^0 + \varepsilon\delta\mathbf{u}, \boldsymbol{\alpha}^0 + \varepsilon\delta\boldsymbol{\alpha}; \mathbf{x})dx_i\right\}_{\varepsilon=0},$$
(4.26)

for a scalar quantity ε and arbitrary vector $\mathbf{h}(\mathbf{x}) \triangleq [\delta\mathbf{u}(\mathbf{x}); \delta\boldsymbol{\alpha}]^\dagger$ in a neighbourhood $(\mathbf{e}^0 + \varepsilon\mathbf{h})$ around $\mathbf{e}^0$. The G-differential $\delta R(\mathbf{e}^0; \mathbf{h})$ is an operator defined on the same domain as $R(\mathbf{e})$, has the same range as $R(\mathbf{e})$, and provides the total first-order total sensitivity of $R(\mathbf{e})$ with respect to variations in the model's parameters and state functions. The G-differential $\delta R(\mathbf{e}^0; \mathbf{h})$ satisfies the relation: $R(\mathbf{e}^0 + \varepsilon\mathbf{h}) - R(\mathbf{e}^0) = \delta R(\mathbf{e}^0; \mathbf{h}) + \Delta(\mathbf{h})$, with $\lim_{\varepsilon\to 0}[\Delta(\varepsilon\mathbf{h})]/\varepsilon = 0$. The existence of the G-variation $\delta R(\mathbf{e}^0; \mathbf{h})$ does not guarantee its numerical computability. Numerical methods most often require that $\delta R(\mathbf{e}^0; \mathbf{h})$ be linear in $\mathbf{h} \triangleq (\delta\mathbf{u}; \delta\boldsymbol{\alpha})$ in a neighbourhood $(\mathbf{e}^0 + \varepsilon\mathbf{h})$ around $\mathbf{e}^0$. Formally, the necessary and sufficient conditions for the G-variation $\delta R(\mathbf{e}^0; \mathbf{h})$ of a non-linear operator $R(\mathbf{e})$ to be linear and continuous in $\mathbf{h}$ in a neighbourhood $(\mathbf{e}^0 + \varepsilon\mathbf{h})$ around $\mathbf{e}^0 = (\boldsymbol{\alpha}^0, \mathbf{u}^0)$, and thus admit a total first-order G-derivative, are as follows:

(i) $R(\mathbf{e})$ must satisfy a weak Lipschitz condition at $\mathbf{e}^0$, that is,

$$\| R(\mathbf{e}^0 + \varepsilon\mathbf{h}) - R(\mathbf{e}^0) \| \leq k \|\varepsilon\mathbf{e}^0\|, \; k < \infty.$$
(4.27)

(ii) $R(\mathbf{e})$ must satisfy the following condition for two arbitrary vectors $\mathbf{h}_1$, $\mathbf{h}_2$ defined in the same vector space as $\mathbf{e}^0$:

$$R(\mathbf{e}^0 + \varepsilon\mathbf{h}_1 + \varepsilon\mathbf{h}_2) - R(\mathbf{e}^0 + \varepsilon\mathbf{h}_1) - R(\mathbf{e}^0 + \varepsilon\mathbf{h}_2) + R(\mathbf{e}^0)$$
$$= o(\varepsilon).$$
(4.28)

In practice, it is usually observed directly if the right-side of Eq. (4.26) is linear (or not) in $\delta\mathbf{u}(\mathbf{x})$. Numerical methods (e.g. Newton's method and variants thereof) for solving Eqs. (4.1) and (4.2) also require the existence of the first-order G-derivatives of original model equations, in which case the components of the operators which appear in these equations must also satisfy the conditions described in Eqs. (4.27) and (4.28). If these conditions are not satisfied, the system is unusually singular, and its solution cannot be obtained by usual numerical methods.

To proceed, the conditions described in Eqs. (4.27) and (4.28) will henceforth be considered to be satisfied by the operators underlying the physical system, in which case the *partial G-derivatives* of $R(\mathbf{e})$ at $\mathbf{e}^0$ with respect to $\mathbf{u}(\mathbf{x})$ and $\boldsymbol{\alpha}$ exist. It follows that that the first-order G-variation $\delta R(\mathbf{e}^0; \mathbf{h})$ defined in Eq. (4.26) can be written in the following form:

$$\delta R(\mathbf{e}^0; \mathbf{h}) = \{\delta R(\mathbf{e}^0; \mathbf{h})\}_{dir} + \{\delta R(\mathbf{e}^0; \mathbf{h})\}_{ind},$$
(4.29)

where the direct-effect term $\{\delta R(\mathbf{e}^0; \mathbf{h})\}_{dir}$ is defined as follows:

$$\{\delta R[\mathbf{u}(\mathbf{x}); \boldsymbol{\alpha}; \delta\boldsymbol{\alpha}]\}_{dir} \triangleq \sum_{j=1}^{TI} \prod_{\substack{i=1 \\ i\neq j}}^{TI}$$

$$\int_{\lambda_i(\boldsymbol{\alpha}^0)}^{\omega_i(\boldsymbol{\alpha}^0)} dx_i \left\{ S\left[\mathbf{u}\left(x_1, .., \omega_i(\boldsymbol{\alpha}), .., x_{N_x}\right); \boldsymbol{\alpha}\right] \frac{\partial \omega_i(\boldsymbol{\alpha})}{\partial\boldsymbol{\alpha}}\delta\boldsymbol{\alpha} \right\}_{(\mathbf{e}^0)}$$

$$-\sum_{j=1}^{TI} \prod_{\substack{i=1 \\ i\neq j}}^{TI} \int_{\lambda_i(\boldsymbol{\alpha}^0)}^{\omega_i(\boldsymbol{\alpha}^0)} dx_i \left\{ S\left[\mathbf{u}\left(x_1, .., \lambda_i(\boldsymbol{\alpha}), .., x_{N_x}\right); \boldsymbol{\alpha}\right] \frac{\partial \lambda_i(\boldsymbol{\alpha})}{\partial\boldsymbol{\alpha}}\delta\boldsymbol{\alpha} \right\}_{(\mathbf{e}^0)}$$

$$+\prod_{i=1}^{TI} \int_{\lambda_i(\boldsymbol{\alpha}^0)}^{\omega_i(\boldsymbol{\alpha}^0)} dx_i \left\{\frac{\partial S(\mathbf{e}; \boldsymbol{\alpha})}{\partial\boldsymbol{\alpha}}\right\}_{(\mathbf{e}^0)} \delta\boldsymbol{\alpha},$$
(4.30)

and where the indirect-effect term $\{\delta R(\mathbf{e}^0; \mathbf{h})\}_{ind}$ is defined as follows:

$$\{\delta R(\mathbf{e}^0; \mathbf{h})\}_{ind} \triangleq \prod_{i=1}^{TI} \int_{\lambda_i(\boldsymbol{\alpha}^0)}^{\omega_i(\boldsymbol{\alpha}^0)} dx_i \left\{\frac{\partial S(\mathbf{e}; \mathbf{x})}{\partial\mathbf{u}}\right\}_{(\mathbf{e}^0)} \delta\mathbf{u}(\mathbf{x}).$$
(4.31)

The quantity $\{\delta R(\mathbf{e}^0; \mathbf{h})\}_{dir}$ in Eq. (4.30) is called the *direct effect term* because it can be computed once the base-case values $\mathbf{e}^0$ are available. On the other hand, the *indirect effect term* $\{\delta R(\mathbf{e}^0; \mathbf{h})\}_{ind}$ defined in Eq. (4.31), can be quantified only after having determined the variations $\delta\mathbf{u}(\mathbf{x})$ in terms of the variations $\delta\boldsymbol{\alpha}$. The first-order relationship between the variations $\delta\mathbf{u}(\mathbf{x})$ and $\delta\boldsymbol{\alpha}$ is determined by taking the G-differentials of Eqs. (4.1) and (4.2), which yields the following equations:

$$\{\mathbf{V}^{(1)}(\mathbf{u}; \boldsymbol{\alpha})\mathbf{v}^{(1)}(\mathbf{x})\}_{\boldsymbol{\alpha}^0} = \{\mathbf{q}_V^{(1)}(\mathbf{u}; \boldsymbol{\alpha}; \delta\boldsymbol{\alpha})\}_{\boldsymbol{\alpha}^0}, \; \mathbf{x} \in \Omega_x,$$
(4.32)

$$\left\{\mathbf{b}_V^{(1)}\left(\mathbf{u};\boldsymbol{\alpha};\mathbf{v}^{(1)};\delta\boldsymbol{\alpha}\right)\right\}_{\boldsymbol{\alpha}^0} = \mathbf{0}, \quad \mathbf{x} \in \partial\Omega_x(\boldsymbol{\alpha}^0), \qquad (4.33)$$

where the superscript (1) indicates *first-level* and where the following definitions were used:

$$\mathbf{v}^{(1)}(\mathbf{x}) \triangleq [v_1^{(1)}(\mathbf{x}), \ldots, v_{TD}^{(1)}(\mathbf{x})]^{\dagger} \triangleq [\delta u_1(\mathbf{x}), \ldots, \delta u_{TD}(\mathbf{x})]^{\dagger};$$

$$\mathbf{q}_V^{(1)}(\mathbf{u};\boldsymbol{\alpha};\delta\boldsymbol{\alpha}) \triangleq \frac{\partial[\mathbf{Q}(\boldsymbol{\alpha}) - \mathbf{N}(\mathbf{u};\boldsymbol{\alpha})]}{\partial\boldsymbol{\alpha}}\delta\boldsymbol{\alpha} \triangleq \sum_{j_1=1}^{TP}\mathbf{s}_V^{(1)}(j_1;\mathbf{u};\boldsymbol{\alpha})\delta\alpha_{j_1};$$

$$\mathbf{V}^{(1)}(\mathbf{u};\boldsymbol{\alpha}) \triangleq \left\{\frac{\partial\mathbf{N}(\mathbf{u};\boldsymbol{\alpha})}{\partial\mathbf{u}}\right\}$$

$$\triangleq \begin{pmatrix} \frac{\partial N_1}{\partial u_1} & \cdots & \frac{\partial N_1}{\partial u_{TD}} \\ \vdots & \ddots & \vdots \\ \frac{\partial N_{TD}}{\partial u_1} & \cdots & \frac{\partial N_{TD}}{\partial u_{TD}} \end{pmatrix}; \frac{\partial\mathbf{N}(\boldsymbol{\alpha},\mathbf{u})}{\partial\boldsymbol{\alpha}}$$

$$\triangleq \begin{pmatrix} \frac{\partial N_1}{\partial \alpha_1} & \cdots & \frac{\partial N_1}{\partial \alpha_{TP}} \\ \vdots & \ddots & \vdots \\ \frac{\partial N_{TD}}{\partial \alpha_1} & \cdots & \frac{\partial N_{TD}}{\partial \alpha_{TP}} \end{pmatrix}. \qquad (4.34)$$

$$\mathbf{b}_V^{(1)}\left(\mathbf{u};\boldsymbol{\alpha};\mathbf{v}^{(1)};\delta\boldsymbol{\alpha}\right) \triangleq \frac{\partial\mathbf{B}(\mathbf{u};\boldsymbol{\alpha})}{\partial\mathbf{u}}\mathbf{v}^{(1)} + \frac{\partial[\mathbf{C}(\boldsymbol{\alpha}) - \mathbf{B}(\boldsymbol{\alpha},\mathbf{u})]}{\partial\boldsymbol{\alpha}}\delta\boldsymbol{\alpha}$$

$$+ \left\{\sum_{i=1}^{TI}\left[\frac{\partial\mathbf{B}[\mathbf{u}(\mathbf{x});\boldsymbol{\alpha}]}{\partial\omega_i}\frac{\partial\omega_i(\boldsymbol{\alpha})}{\partial\boldsymbol{\alpha}}\delta\boldsymbol{\alpha}\right]_{\mathbf{x}=\omega}\right.$$

$$\left. + \left[\frac{\partial\mathbf{B}[\mathbf{u}(\mathbf{x});\boldsymbol{\alpha}]}{\partial\lambda_i}\frac{\partial\lambda_i(\boldsymbol{\alpha})}{\partial\boldsymbol{\alpha}}\delta\boldsymbol{\alpha}\right]_{\mathbf{x}=\lambda}\right\}_{(e^0)}. \qquad (4.35)$$

The matrices which appear in Eq. (4.35) are defined in the same way as those appearing in Eq. (4.34). The system comprising Eqs. (4.32) and (4.33) is called the *First-Level Variational Sensitivity System* (first-LVSS). In order to determine the solutions of the 1st-LVSS that would correspond to every parameter variation $\delta\alpha_{j_1}$, $j_1 = 1, \ldots, TP$, the 1st-LVSS would need to be solved TP times, with distinct right-sides for each $\delta\alpha_{j_1}$, thus requiring TP large-scale computations.

Computing the (total) response sensitivity $\delta R(\mathbf{e}^0;\mathbf{h})$ by using the ($\delta\boldsymbol{\alpha}$-dependent) solution $\mathbf{v}^{(1)}(\mathbf{x})$ of the first-LVSS is called (Cacuci, 1981a,b) the Forward Sensitivity Analysis Procedure (FSAP). From the standpoint of computational costs and effort, the *FSAP* requires $O(TP)$ large-scale forward computations. Therefore, the *FSAP* is advantageous to employ only if, in the problem under consideration, the *total number of responses* of interest exceeds the number of system parameters and/or parameter variations of interest. In most practical situations, however, the number of model parameters significantly exceeds the number of scalar-

valued model responses of interest (i.e. $TP \gg TR$). In such cases, the *adjoint sensitivity analysis methodology* conceived and developed by Cacuci (1981a,b) is the most efficient method for computing exactly the first-order sensitivities since it requires only TR large-scale computations. The fundamental idea introduced by Cacuci (1981a,b) is to eliminate the appearance of the variation $\delta\mathbf{u}$ in the indirect-effect term defined in Eq. (4.31), by expressing the right-side of Eq. (4.31) in terms of adjoint functions that are the solutions of the *First-Level Adjoint Sensitivity System* (1st-LASS) which depends on the model's response but does not depend on parameter variations. In his original work, Cacuci (1981a,b) considered that the physical system's boundaries and interfaces in the phase-space of independent variables were perfectly well know. In subsequent works, Cacuci (2021a) has generalised his original work to include the exact and efficient computations of first-order response sensitivities to imprecisely known boundaries/interfaces and parameters, calling this general methodology *the First-Order Comprehensive Adjoint Sensitivity Analysis Methodology for Non-linear Systems* (1st-CASAM-N), where the qualifier 'comprehensive' indicates that *all* model parameters, including the phase-space locations of internal and/or external boundaries, are considered to be imprecisely known (subject to uncertainties).

As the name implies, the 'adjoint' sensitivity analysis methodology requires the introduction of adjoint operators, which can be defined in Banach spaces but are most useful in Hilbert spaces. The spaces E_u and E_Q are henceforth considered to be Hilbert spaces, denoted as $\mathsf{H}_u(\Omega_x)$ and $\mathsf{H}_Q(\Omega_x)$, respectively. The elements of $\mathsf{H}_u(\Omega_x)$ and $\mathsf{H}_Q(\Omega_x)$ are, as before, vector-valued functions defined on the open set $\Omega_x \subset \mathbb{R}^{\mathsf{J}_x}$, with smooth boundary $\partial\Omega_x$. In practice, $\mathsf{H}_u(\Omega_x)$ and $\mathsf{H}_Q(\Omega_x)$ are self-dual, so they can be represented by a Hilbert space denoted as $\mathsf{H}_1(\Omega_x)$, which can be considered to be real, without loss of generality, and which is endowed with an inner product of two vectors $\mathbf{u}^{(a)}(\mathbf{x}) \in \mathsf{H}_1$ and $\mathbf{u}^{(b)}(\mathbf{x}) \in \mathsf{H}_1$ denoted as $\langle\mathbf{u}^{(a)}, \mathbf{u}^{(b)}\rangle_1$ and defined as follows:

$$\langle\mathbf{u}^{(a)}, \mathbf{u}^{(b)}\rangle_1 \triangleq \left\{\int_{\lambda_1(\boldsymbol{\alpha})}^{\omega_1(\boldsymbol{\alpha})} \cdots \int_{\lambda_{TI}(\boldsymbol{\alpha})}^{\omega_{TI}(\boldsymbol{\alpha})} \left[\mathbf{u}^{(a)}(\mathbf{x}) \cdot \mathbf{u}^{(b)}(\mathbf{x})\right] dx_1 \ldots dx_{TI}\right\}_{\boldsymbol{\alpha}^0},$$
$$(4.36)$$

where the dot indicates the scalar product

$$\mathbf{u}^{(a)}(\mathbf{x}) \cdot \mathbf{u}^{(b)}(\mathbf{x}) \triangleq \sum_{i=1}^{TD} u_i^{(a)}(\mathbf{x})u_i^{(b)}(\mathbf{x}).$$

The construction of an alternative expression for the indirect-effect term defined in Eq. (4.31) proceeds by applying the principle outlined by Cacuci (1981a,b), which involve the following sequence of steps:

1. Using the inner product defined in Eq. (4.36), construct the inner product of Eq. (4.32) with a vector $\mathbf{a}^{(1)}(\mathbf{x}) \triangleq [a_1^{(1)}(\mathbf{x}), \ldots, a_{TD}^{(1)}]^{\dagger}$ to obtain the following relation:

$$\left\{\left\{\langle \mathbf{a}^{(1)},\ \mathbf{V}^{(1)}(\mathbf{u};\boldsymbol{\alpha})\mathbf{v}^{(1)}\rangle_1\right\}_{\boldsymbol{\alpha}^0}\right\}_{\boldsymbol{\alpha}^0} = \left\{\langle \mathbf{a}^{(1)},\ \mathbf{q}_V^{(1)}(\mathbf{u};\boldsymbol{\alpha};\delta\boldsymbol{\alpha})\rangle_1\right\}_{\boldsymbol{\alpha}^0},\ \mathbf{x}\in\Omega_x.$$
(4.37)

Using the definition of the adjoint operator in $H_1(\Omega_x)$, the left-side of Eq. (4.37) is transformed as follows:

$$\left\{\langle \mathbf{a}^{(1)},\ \mathbf{V}^{(1)}(\mathbf{u};\boldsymbol{\alpha})\mathbf{v}^{(1)}\rangle_1\right\}_{\boldsymbol{\alpha}^0} = \left\{\langle \mathbf{A}^{(1)}(\mathbf{u};\boldsymbol{\alpha})\,\mathbf{a}^{(1)},\ \mathbf{v}^{(1)}\rangle_1\right\}_{\boldsymbol{\alpha}^0}$$
$$+\left\{[P^{(1)}\big(\mathbf{u};\boldsymbol{\alpha};\ \mathbf{a}^{(1)};\mathbf{v}^{(1)}\big)]_{\partial\Omega_x}\right\}_{\boldsymbol{\alpha}^0},$$
(4.38)

where $\mathbf{A}^{(1)}(\mathbf{u};\boldsymbol{\alpha})$ is the operator adjoint to $\mathbf{V}^{(1)}(\mathbf{u};\boldsymbol{\alpha})$, i.e., $\mathbf{A}^{(1)}(\mathbf{u};\boldsymbol{\alpha})\triangleq[\mathbf{V}^{(1)}(\mathbf{u};\boldsymbol{\alpha})]^*$, and where $[P^{(1)}(\mathbf{u};\boldsymbol{\alpha};\ \mathbf{a}^{(1)};\mathbf{v}^{(1)})]_{\partial\Omega_x}$ denotes the associated bilinear concomitant evaluated on the space/time domain's boundary $\partial\Omega_x(\boldsymbol{\alpha}^0)$. The symbol $[\ ]^*$ is used in this work to indicate adjoint operator. In certain situations, it might be computationally advantageous to include certain boundary components of $[P^{(1)}\big(\mathbf{u};\boldsymbol{\alpha};\ \mathbf{a}^{(1)};\mathbf{v}^{(1)}\big)]_{\partial\Omega_x}$ into the components of $\mathbf{A}^{(1)}(\mathbf{u};\boldsymbol{\alpha})$.

2. The first term on the right-side of Eq. (4.38) is required to represent the indirect-effect term defined in Eq. (4.31) by imposing the following relationship:

$$\left\{\mathbf{A}^{(1)}(\mathbf{u};\boldsymbol{\alpha})\mathbf{a}^{(1)}(\mathbf{x})\right\}_{\boldsymbol{\alpha}^0} = \{\partial S(\mathbf{u};\boldsymbol{\alpha})/\partial\mathbf{u}\}_{\boldsymbol{\alpha}^0}\triangleq\mathbf{q}_A^{(1)}[\mathbf{u}(\mathbf{x});\boldsymbol{\alpha}],$$
$$\mathbf{x}\in\Omega_x.$$
(4.39)

3. The domain of $\mathbf{A}^{(1)}(\mathbf{u};\boldsymbol{\alpha})$ is determined by selecting appropriate adjoint boundary and/or initial conditions, which will be denoted in operator form as:

$$\left\{\mathbf{b}_A^{(1)}(\mathbf{u};\mathbf{a}^{(1)};\boldsymbol{\alpha})\right\}_{\boldsymbol{\alpha}^0} = \mathbf{0},\quad \mathbf{x}\in\partial\Omega_x(\boldsymbol{\alpha}^0).$$
(4.40)

These boundary conditions for $\mathbf{A}^{(1)}(\mathbf{u};\boldsymbol{\alpha})$ are usually inhomogeneous, that is, $\mathbf{b}_A^{(1)}(\mathbf{0};\mathbf{0};\boldsymbol{\alpha})\neq\mathbf{0}$, and are obtained requiring that: (i) they must be independent of unknown values of $\mathbf{v}^{(1)}(\mathbf{x})$ and $\delta\boldsymbol{\alpha}$; (ii) the substitution of the boundary and/or initial conditions represented by Eqs. (4.40) and (4.33) into the expression of $\left\{[P^{(1)}(\mathbf{u};\boldsymbol{\alpha};\ \mathbf{a}^{(1)};\mathbf{v}^{(1)})]_{\partial\Omega_x}\right\}_{\boldsymbol{\alpha}^0}$ must cause all terms containing unknown values of $\mathbf{v}^{(1)}(\mathbf{x})$ to vanish. Constructing the adjoint initial and/or boundary conditions for $\mathbf{A}^{(1)}(\mathbf{u};\boldsymbol{\alpha})$ reduces the bilinear concomitant $\left\{[P^{(1)}(\mathbf{u};\boldsymbol{\alpha};\ \mathbf{a}^{(1)};\mathbf{v}^{(1)})]_{\partial\Omega_x}\right\}_{\boldsymbol{\alpha}^0}$ to a quantity denoted as $\left\{[\hat{P}^{(1)}(\mathbf{u};\boldsymbol{\alpha};\ \mathbf{a}^{(1)};\delta\boldsymbol{\alpha})]_{\partial\Omega_x}\right\}_{\boldsymbol{\alpha}^0}$, which will contain boundary terms involving only known values of $\delta\boldsymbol{\alpha}$, $\boldsymbol{\alpha}^0$, $\mathbf{u}^0$, and $\boldsymbol{\psi}^{(1)}$ Since $\left\{[\hat{P}^{(1)}(\mathbf{u};\boldsymbol{\alpha};\ \mathbf{a}^{(1)};\delta\boldsymbol{\alpha})]_{\partial\Omega_x}\right\}_{\boldsymbol{\alpha}^0}$ is linear in $\delta\boldsymbol{\alpha}$, it can be expressed in the following form:

$$[\hat{P}^{(1)}(\mathbf{u};\boldsymbol{\alpha};\ \mathbf{a}^{(1)};\delta\boldsymbol{\alpha})]_{\partial\Omega_x} = \sum_{j_1=1}^{TP}[\partial\hat{P}^{(1)}(\mathbf{u};\boldsymbol{\alpha};\mathbf{a}^{(1)})/\partial\alpha_{j_1}]\delta\alpha_{j_1}.$$

4. The results obtained in Eqs. (4.39) and (4.38) are now replaced in Eq. (4.31) to obtain the following expression of the indirect-effect term as a function of $\mathbf{a}^{(1)}(\mathbf{x})$:

$$\left\{\delta R[\mathbf{u}(\mathbf{x});\boldsymbol{\alpha};\mathbf{v}^{(1)}(\mathbf{x})]\right\}_{ind} = \left\{\langle\mathbf{a}^{(1)},\ \mathbf{q}_V^{(1)}(\mathbf{u};\boldsymbol{\alpha};\delta\boldsymbol{\alpha})\rangle_1\right\}_{\boldsymbol{\alpha}^0}$$
$$-\left\{[\hat{P}^{(1)}(\mathbf{u};\boldsymbol{\alpha};\ \mathbf{a}^{(1)};\delta\boldsymbol{\alpha})]_{\partial\Omega_x}\right\}_{\boldsymbol{\alpha}^0}.$$
(4.41)

5. Replacing in Eq. (4.29) the result obtained in Eq. (4.41), together with the expression for the direct-effect term provided in Eq. (4.30), yields the following expression for the first G-differential of the response $R[\mathbf{u}(\mathbf{x});\boldsymbol{\alpha}]$:

$$\left\{\delta R[\mathbf{u}(\mathbf{x});\boldsymbol{\alpha};\mathbf{v}^{(1)}(\mathbf{x});\delta\boldsymbol{\alpha}]\right\}_{\boldsymbol{\alpha}^0} = \{\delta R[\mathbf{u}(\mathbf{x});\boldsymbol{\alpha};\delta\boldsymbol{\alpha}]\}_{dir}$$
$$+\left\{\langle\mathbf{a}^{(1)},\ \mathbf{q}_V^{(1)}(\mathbf{u};\boldsymbol{\alpha};\delta\boldsymbol{\alpha})\rangle_1\right\}_{\boldsymbol{\alpha}^0} - \left\{[\hat{P}^{(1)}\big(\mathbf{u};\boldsymbol{\alpha};\ \mathbf{a}^{(1)};\delta\boldsymbol{\alpha}\big)]_{\partial\Omega_x}\right\}_{\boldsymbol{\alpha}^0}$$
$$\triangleq \sum_{j_1=1}^{TP}\left\{R^{(1)}[j_1;\mathbf{u}(\mathbf{x});\mathbf{a}^{(1)}(\mathbf{x});\boldsymbol{\alpha}]\right\}_{\boldsymbol{\alpha}^0}\delta\alpha_{j_1},$$
(4.42)

where for each $j_1=1,\ldots,TP$, the quantity $R^{(1)}[j_1;\mathbf{u}(\mathbf{x});\mathbf{a}^{(1)}(\mathbf{x});\boldsymbol{\alpha}]$ denotes the first-order sensitivities of the response $R[\mathbf{u}(\mathbf{x});\boldsymbol{\alpha}]$ with respect to the model parameters α_{j_1} and has the following expression:

$$R^{(1)}[j_1;\mathbf{u}(\mathbf{x});\mathbf{a}^{(1)}(\mathbf{x});\boldsymbol{\alpha}]$$
$$= \int_{\lambda_1(\boldsymbol{\alpha})}^{\omega_1(\boldsymbol{\alpha})}\cdots\int_{\lambda_{TI}(\boldsymbol{\alpha})}^{\omega_{TI}(\boldsymbol{\alpha})}\mathbf{a}^{(1)}(\mathbf{x})\cdot\frac{\partial[\mathbf{Q}(\boldsymbol{\alpha})-\mathbf{N}(\mathbf{u};\boldsymbol{\alpha})]}{\partial\alpha_{j_1}}dx_1\ldots dx_{TI}$$
$$-\frac{\partial\hat{P}^{(1)}\big(\mathbf{u};\boldsymbol{\alpha};\mathbf{a}^{(1)}\big)}{\partial\alpha_{j_1}}+\int_{\lambda_1(\boldsymbol{\alpha})}^{\omega_1(\boldsymbol{\alpha})}\cdots\int_{\lambda_{TI}(\boldsymbol{\alpha})}^{\omega_{TI}(\boldsymbol{\alpha})}\frac{\partial S(\mathbf{u};\boldsymbol{\alpha};\boldsymbol{\alpha})}{\partial\alpha_{j_1}}dx_1\ldots dx_{TI}$$
$$+\sum_{j=1}^{TI}\left\{\int_{\lambda_1(\boldsymbol{\alpha})}^{\omega_1(\boldsymbol{\alpha})}\cdots\int_{\lambda_{j-1}(\boldsymbol{\alpha})}^{\omega_{j-1}(\boldsymbol{\alpha})}\int_{\lambda_{j+1}(\boldsymbol{\alpha})}^{\omega_{j+1}(\boldsymbol{\alpha})}\cdots\int_{\lambda_{TI}(\boldsymbol{\alpha})}^{\omega_{TI}(\boldsymbol{\alpha})}S\left[\mathbf{u}\big(x_1,..,\omega_j(\boldsymbol{\alpha}),..,x_{N_x}\big);\boldsymbol{\alpha}\right]\right.$$
$$\left.\frac{\partial\omega_j(\boldsymbol{\alpha})}{\partial\alpha_{j_1}}dx_1\ldots dx_{TI}\right\}_{\boldsymbol{\alpha}^0}$$
$$-\sum_{j=1}^{TI}\left\{\int_{\lambda_1(\boldsymbol{\alpha})}^{\omega_1(\boldsymbol{\alpha})}\cdots\int_{\lambda_{j-1}(\boldsymbol{\alpha})}^{\omega_{j-1}(\boldsymbol{\alpha})}\int_{\lambda_{j+1}(\boldsymbol{\alpha})}^{\omega_{j+1}(\boldsymbol{\alpha})}\cdots\int_{\lambda_{TI}(\boldsymbol{\alpha})}^{\omega_{TI}(\boldsymbol{\alpha})}S\left[\mathbf{u}\big(x_1,..,\lambda_j(\boldsymbol{\alpha}),..,x_{N_x}\big);\boldsymbol{\alpha}\right]\right.$$
$$\left.\frac{\partial\lambda_j(\boldsymbol{\alpha})}{\partial\alpha_{j_1}}dx_1\ldots dx_{TI}\right\}_{\boldsymbol{\alpha}^0}.$$
(4.43)

As indicated by Eq. (4.43), each of the first-order sensitivities $R^{(1)}[j_1;\mathbf{u}(\mathbf{x});\mathbf{a}^{(1)}(\mathbf{x});\boldsymbol{\alpha}]$ of the response $R[\mathbf{u}(\mathbf{x});\boldsymbol{\alpha}]$ with respect to the model parameters α_{j_1} (including boundary and initial conditions) can be computed inexpensively after having obtained the function $\mathbf{a}^{(1)}(\mathbf{x}) \in H_1$, using quadrature formulas to evaluate the various inner products involving $\mathbf{a}^{(1)}(\mathbf{x}) \in H_1$. The function $\mathbf{a}^{(1)}(\mathbf{x}) \in H_1$ is called the first-level adjoint function and is obtained by solving numerically the 1st-LASS which comprises Eqs. (4.39) and (4.40). It is very important to note that the 1st-LASS is independent of parameter variation $\delta\alpha_{j_1}, j_1 = 1, \ldots, TP$, and therefore needs to be solved only once, regardless of the number of model parameters under consideration. Thus, solving the 1st-LASS is the only large-scale computation needed for obtaining all of the first-order sensitivities. Since the 1st-LASS is linear in $\mathbf{a}^{(1)}(\mathbf{x})$, solving it requires less computational effort than solving the original non-linear system for $\mathbf{u}(\mathbf{x})$.

4.3.2 Illustrative Application of the 1st-CASAM-N to an Oscillatory Dynamical Model

March-Leuba, Cacuci, and Perez (1984) have conceived a reduced-order model for predicting the dynamics of boiling-water reactors (BWR). Remarkably, this reduced-order dynamic model predicted that a BWR could undergo large-amplitude period-doubling oscillations towards chaotic dynamics in certain unstable power/flow regions. On March 9, 1988, four years after the appearance of the BWR-model conceived by March-Leuba, Cacuci, and Perez (1984), the LaSalle County-2 BWR (in Seneca, Illinois, USA) underwent such predicted dynamic behaviour, leading to an automatic reactor shut-down (USNRC, 1988).

The BWR-model of March-Leuba, Cacuci, and Perez (1984) comprises five coupled, extremely stiff, first- and second-order differential-algebraic equations involving fifteen uncertain parameters and unsuitable for sensitivity and uncertainty analyses by conventional statistical methods because the use of such conventional methods would not only require an unrealistic amount of computational resources but would also produce results of unverifiable reliability. Cacuci and Di Rocco (2020) showed that the 1st-CASAM is free of such shortcomings and enables the efficient computation of the exact sensitivities of the BWR-model's state functions with respect to the model's uncertain parameters in the model's oscillatory regions, as will be illustrated in this section. The steps underlying the application of the 1st-CASAM to the BWR-model are typical for applying the 1st-CASAM-N to any model in the Earth sciences.

The differential equations underlying the reduced-order BWR-model developed by March-Leuba, Cacuci, and Perez (1984) for predicting the dynamic behaviour of a BWR when the reactor is perturbed from its originally critical steady-state condition describe the following time-dependent phenomena:

(i) a point reactor representation of neutron kinetics, including the neutron precursors:

$$\frac{dn(t)}{dt} = n(t)\frac{\rho(t) - \beta}{\Lambda} + \lambda c(t) + \frac{\rho(t)}{\Lambda}; \quad \frac{dc(t)}{dt} = n(t)\frac{\beta}{\Lambda} - \lambda c(t);$$

$$(4.44)$$

(ii) a lumped-parameter representation of the heat transfer process in the fuel:

$$\frac{dT(t)}{dt} = Q[n(t) + H(t)\Delta] - a_3 T(t); \qquad (4.45)$$

(iii) a lumped-parameter (two-nodes) representation of the channel thermal-hydraulics, accounting for the reactivity feedback:

$$\frac{d^2\gamma(t)}{dt^2} + a_2\frac{d\gamma(t)}{dt} + a_1\gamma(t) = mk_0 T(t); \rho(t) = \Gamma_1\gamma(t) + \Gamma_2 T(t).$$

$$(4.46)$$

The state variables in Eqs. (4.44)–(4.46) represent the following time-dependent quantities: (i) $n(t)$ denotes the excess neutron population; (ii) $c(t)$ denotes the excess population of delayed neutron precursors; (iii) $T(t)$ denotes the excess fuel temperature; (iv) $H(t)$ denotes the Heaviside functional; (v) $\gamma(t)$ denotes the relative excess coolant density; and (vi) $\rho(t)$ denotes the excess reactivity. At the initial time, $t = 0$, the reactor is assumed to be critical, operating in the steady-state condition. The qualifier 'excess' signifies that these quantities are defined as departures from the critical reactor configuration; consequently, all the state variables are therefore zero at $t = 0$. In addition, the excess neutron population, $n(t)$, and the excess precursors population, $c(t)$, are normalised with respect to the initial (critical) value and are therefore non-dimensional quantities.

The vector of parameters for this BWR-model is as follows: $\boldsymbol{\alpha} \triangleq (\alpha_1, \ldots, \alpha_{16})^\dagger \triangleq (k, \Delta, Q, \beta, \Lambda, \lambda, \Gamma_1, \Gamma_2, a_1, a_2, a_3, n_0, c_0, T_0, \gamma_{10}, \gamma_0)^\dagger$, where the quantities $n_0, c_0, T_0, \gamma_0, \gamma_{01}$ represent the initial conditions at $t = 0$ for Eqs. (4.44)–(4.46). The nominal values of the model parameters in Eqs. (4.44)–(4.46) are as follows: $\Gamma_1^0 = 0.15$, $\Gamma_2^0 = -2.61 \cdot 10^{-5} K^{-1}$, $\beta^0 = 0.0056$, $\Lambda^0 = 4.0 \cdot 10^{-5} s$, $\lambda^0 = 0.08 s^{-1}$, $Q^0 = 25.044 K/s$, $\Delta^0 = -0.1$, $a_1^0 = 6.8166 s^{-2}$, $a_2^0 = 2.2494 s^{-1}$, $a_3^0 = 0.2325 s^{-1}$, $k_0^0 = -0.01318 K^{-1} s^{-2}$. The nominal values of all initial conditions are zero. The reduced-order model defined by Eqs. (4.44)–(4.46) has two equilibrium points in phase-space: (i) a stable equilibrium, which corresponds to the critical reactor configuration at a point (denoted using the subscript A) in phase-space having the following coordinates: $n_A = -\Delta; c_A = -(\beta\Delta)/(\lambda\Lambda)$; $T_A = \gamma_A = \rho_A = 0$; and (ii) an unstable equilibrium at the point (denoted using the subscript B) in phase-space having the following coordinates: $n_B = -1; c_B = -\beta/(\lambda\Lambda)$; $T_B = Q(\Delta - 1)/a_3$; $\gamma_B = kQ(\Delta - 1)$; $\rho_B = (\Gamma_1 k/a_3 + \Gamma_2) Q(\Delta - 1)/a_3$.

Equations (4.44)–(4.46) are 'stiff' systems of non-linear ordinary differential equations. After intercomparing several solvers of stiff differential equations, as mentioned by Cacuci and Di Rocco (2020), the Rosenbrock solver (www.odeint.com, 2019) with a step-size varying between 0.001 and 0.01 seconds has been selected to solve Eqs. (4.44)–(4.46) since it provided the best performance in terms of computational time and accuracy. When the feedback gain m in Eq. (4.46) is increased, thereby increasing the heat transfer from the reactor core to the surrounding coolant, the equilibrium point A becomes unstable after the stability threshold is crossed. The phase-space trajectories of the BWR-model oscillate between the two unstable equilibrium points until a stable limit cycle sets in. Increasing the value of the feedback gain m beyond the stability of the first limit cycle will cause this cycle to bifurcate, through a period-doubling bifurcation, into a period-two cycle. Continuing to increase the feedback gain m leads to a cascade of period-doubling bifurcations which lead to aperiodic behaviour, as first observed by March-Leuba, Cacuci, and Perez (1984). For example, the bifurcation map for the excess temperature $T(t)$, as a function of the feedback gain m, is depicted in Fig. 4.2, which highlights the following regions:

- Region 1: Stable Region, before the first-order bifurcation, $m \cdot k_0 < k_{bif1} = 1.70898 \cdot k_0$;
- Region 2: Unstable Region between the first-order bifurcation and the second-order bifurcations, $k_{bif1} < m \cdot k_0 < k_{bif2} = 2.66657 \cdot k_0$;
- Region 3: Unstable Region between the second-order bifurcations and the third-order bifurcations, $k_{bif2} < m \cdot k_0 < k_{bif3} = 2.86991 \cdot k_0$;

- Region 4: Unstable Region between the third-order bifurcations and the fourth-order bifurcations, $k_{bif3} < m \cdot k_0 < k_{bif4} = 2.92027 \cdot k_0$;
- Region 5: Chaotic (Aperiodic) Region, arising from the cascade of period-doubling pitchfork bifurcations produced when the feedback gain m is increased past a critical value (called accumulation point), $m_c < m \cdot k_0$, where $m_c = 2.933995$.

The value attained by the excess temperature at a given instance in time, $t = t_d$, where the respective temperature could be measured by some detecting device, can be represented by the following particular form of Eq. (4.4):

$$T(t_d) \triangleq \int_0^{t_f} T(t)\delta(t - t_d)dt, \tag{4.47}$$

where t_f denotes the final time of interest; for this model, $t_f = 160s$, as will be discussed. Applying the principles of the 1st-CASAM-N detailed in Section 4.3.1 to the BWR-model described by Eqs. (4.44)–(4.46) yields (Cacuci and Di Rocco, 2020) the following final expression [which is a particular form of Eq. (4.43)] for the total first-order differential $\delta T(t_d)$ of $T(t_d)$ with respect to parameter variations:

$$\delta T(t_d) = \sum_{i=1}^{5} \int_0^{t_f} \varphi_i(t)q_i(t)dt + \varphi_1(0)\delta n_0 + \varphi_2(0)\delta c_0 \\ + \varphi_3(0)\delta T_0 + \varphi_4(0)\delta\gamma_{10} + \varphi_5(0)\delta\gamma_0. \tag{4.48}$$

The first-level adjoint sensitivity functions $\varphi_i(t)$ which appear in Eq. (4.48) are the solutions of the following 1st-LASS:

Figure 4.2 Bifurcation map of $T(t)$.

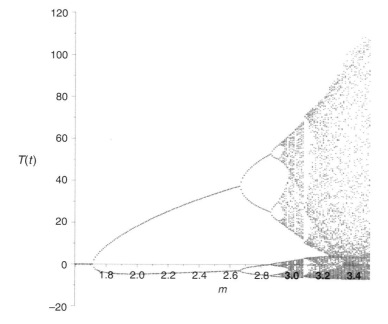

$$\left\{-\frac{d}{dt} - \left[\gamma(t)\frac{\Gamma_1}{\Lambda} + T(t)\frac{\Gamma_2}{\Lambda} - \frac{\beta}{\Lambda}\right]\right\}\varphi_1(t) - \frac{\beta}{\Lambda}\varphi_2(t)$$

$$-Q\varphi_3(t) = w_1\delta(t - t_d), \tag{4.49}$$

$$-\lambda\varphi_1(t) + \left(-\frac{d}{dt} + \lambda\right)\varphi_2(t) = w_2\delta(t - t_d) \tag{4.50}$$

$$\left[-n(t)\frac{\Gamma_2}{\Lambda} - \frac{\Gamma_2}{\Lambda}\right]\varphi_1(t) + \left(-\frac{d}{dt} + a_3\right)\varphi_3(t)$$

$$-k\varphi_4(t) = w_3\delta(t - t_d) \tag{4.51}$$

$$\left(-\frac{d}{dt} + a_2\right)\varphi_4(t) - \varphi_5(t) = w_4\delta(t - t_d) \tag{4.52}$$

$$\left[-n(t)\frac{\Gamma_1}{\Lambda} - \frac{\Gamma_1}{\Lambda}\right]\varphi_1(t) + a_1\varphi_4(t) - \frac{d\varphi_5(t)}{dt} = w_5\delta(t - t_d) \tag{4.53}$$

$$\varphi_i(t_f) = 0, \quad i = 1,...,5. \tag{4.54}$$

The graphs to be presented in this section extend in time up to $t_d = 150s$ from the onset of the perturbation in Δ effected in the initially steady-state critical reactor, which causes the BWR-model to embark on a dynamic evolution in a Region in phase-space that is governed by the magnitude of the feedback gain. This time-window – between 0 and 150 seconds – was optimised to enable the reproduction of the entire initial transitory phase of the curves followed by a representative segment of the subsequent periodic regime, which is sufficiently large for displaying several complete oscillation cycles but

not so large as to impair the readability of the entire graph. No fundamentally different phenomena occur after 150 seconds into the transient dynamics: in the stable Region 1, the state functions and their sensitivities practically reach their time-independent asymptotic values well before 150 seconds; in Regions 2 through 4, the respective periodic dynamics of the state functions and accompanying sensitivities also reach their respective asymptotic states before the 150 seconds final-time, while in the chaotic Region 5, the state functions and their sensitivities are not periodic. The asymptotic state of the sensitivities with respect to model parameters in Regions 2 through 4 itself oscillates with a small periodic amplitude even after 150 seconds, as discussed by Cacuci and Di Rocco (2020). However, solving the 1st-LASS is performed by commencing the numerical computations at $t_f = 160$ and proceeding towards $t = 0.0$. The reason for choosing $t_f = 160$ – that is, 10 seconds prior to the occurrence of the delta-impulse source at $t_d = 150s$ – as the starting point for the numerical computation of the first-level adjoint sensitivity functions $\varphi_i(t)$, $i = 1,...,5$, is to stabilise the Rosenbrock solver prior to the occurrence of the delta-impulse source.

In the stable Region 1, the largest sensitivity of $T(t)$ is with respect to the initial condition γ_0, which is depicted in Fig. 4.3 along with the sensitivities of $T(t)$ to the other initial conditions n_0, c_0 and T_0. All sensitivities presented in Fig. 4.3 display an initial oscillatory phase characterised by high-amplitude oscillations caused by the initial neutronic perturbation Δ. However, the amplitudes of these oscillations decrease to zero exponentially in time, thus confirming the expectation that all state functions $n(t)$, $c(t)$, $T(t)$, $\rho(t)$, and $\gamma(t)$ must become asymptotically independent of

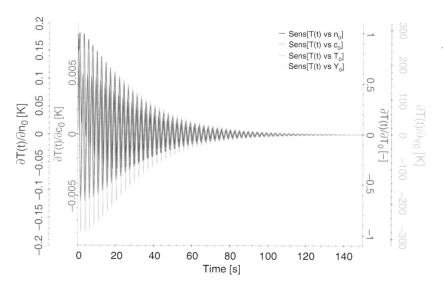

Figure 4.3 Time-evolutions of the sensitivities of $T(t)$ with respect to the initial conditions n_0, c_0, T_0, γ_0 in Region 1.

the values of the initial conditions as $t \to \infty$ in the stable Region 1.

The time-evolution of the nominal value of the state function $T(t)$ is depicted in dark grey in Fig. 4.4 along with the time evolution of the functions labelled $T(t) \pm \sigma$ and depicted in light grey, where σ denotes the total absolute standard deviation for $T(t)$ computed by using Eq. (4.12) in conjunction with the individual contributions $(S_i \sigma_i)^2$ from all 11 model parameters and 4 initial conditions. The standard deviation for $T(t)$ is very large immediately after the onset of the transient (by perturbing the initially critical reactor in steady state), reaching values that more than 10 times larger than the state function $T(t)$ itself. In time, the standard deviation decays exponentially, reaching values comparable to the state function $T(t)$ after about 140 seconds from the start of the transient. Very importantly, Fig. 4.4 indicates that predictions of the excess temperature $T(t)$ during the first 120 seconds after the BWR model is perturbed from its initial critical steady state are unreliable. Furthermore, it is possible that that the model would be 'pushed' in one of the oscillatory Regions (which will be investigated in the sections to follow here) by the large amplitudes reached by the standard deviation of $T(t)$ immediately after the onset of the perturbation that took the model out of its initial critical state. Similar conclusions were reached for $n(t)$ by Di Rocco and Cacuci (2020).

In the period-2 limit cycle Region 2, the time-evolutions of the excess fuel temperature $T(t)$ and of its sensitivities with respect to the initial conditions n_0, c_0, T_0, and γ_0, display an initial transitory phase which extends to about 70 seconds, after which all of these sensitivities settle into periodic oscillations of constant amplitudes, as depicted in Fig. 4.5. The fact that these oscillations continue with constant amplitudes indicates that perturbations in the initial conditions do not induce changes in the oscillation frequency of the system's state functions. In Fig. 4.6, the time-evolution of the nominal value of the state function $T(t)$ is depicted in dark grey in along with the time evolution of the functions labelled $T(t) \pm \sigma$, which are depicted in light grey, where σ denotes the total absolute standard deviation for $T(t)$ computed by using Eq. (4.12) in conjunction with the individual contributions $(S_i \sigma_i)^2$ from all 11 model parameters and 4 initial conditions. As indicated in Fig. 4.6, the standard deviation increases rapidly after the onset of the transient (by perturbing the initially critical reactor in steady state), oscillating with rapidly increasing amplitudes, reaching – in 40 seconds after the start of the transient – values that are 400 K larger than the amplitude of the excess temperature, $T(t)$, which oscillates around the initial value $T_0 = 0K$. It is evident from Fig. 4.5 that the standard deviation of $T(t)$ renders the prediction of the evolution of this state function highly unreliable after less than 10 seconds from the initiation of the transient.

In the chaotic Region 5, the sensitivities of $T(t)$ oscillate aperiodically with amplitudes that increase exponentially, reaching massive values. The standard deviation of $T(t)$ inherits the characteristics of the evolutions of the sensitivities of $T(t)$ with respect to the model parameters and initial conditions. As shown by Di Rocco and Cacuci (2020), considering a 5% standard deviation for all parameters and initial conditions in Region 5, the standard deviation of $T(t)$ increases from values of $O(10^8)$ after the first 25 seconds into the transient to values of $O(10^{43})$ towards 150 seconds into the transient, evidently defeating any attempt at predicting the behaviour of $T(t)$ in the chaotic Region 5.

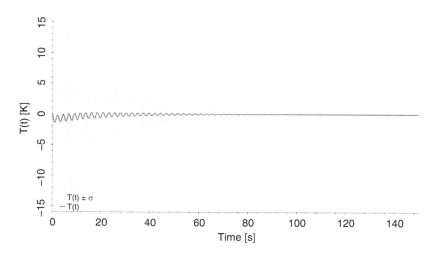

Figure 4.4 Time-evolution of $T(t)$ [dark grey graph] with uncertainty bands $T(t) \pm \sigma$ [light grey graphs], considering uniform 5% relative standard deviations for all parameters and 5% absolute standard deviations for the initial conditions in Region 1.

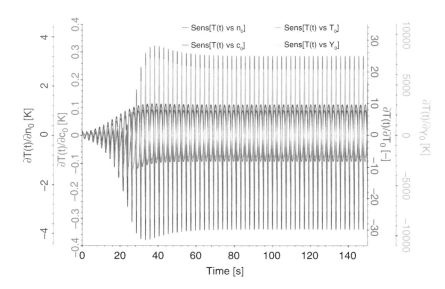

Figure 4.5 Time-evolutions of the sensitivities of $T(t)$ with respect to the initial conditions n_0, c_0, T_0, γ_0 in Region 2.

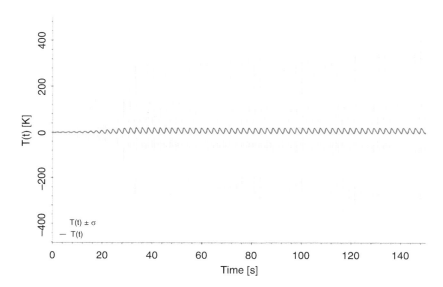

Figure 4.6 Time-evolution of $T(t)$ [dark grey graph] with uncertainty bands $T(t) \pm \sigma$ [light grey graphs], considering uniform 5% relative standard deviations for all parameters and 5% absolute standard deviations for the initial conditions in Region 2.

4.4 Illustrative Third-Order Uncertainty Analysis of a System Comprising Many Parameters: Application to the OECD/NEA Polyethylene-Reflected Plutonium Reactor Physics Benchmark

This section presents the mathematical framework of the *Third-Order Comprehensive Adjoint Sensitivity Analysis Methodology for Non-linear Systems* (3rd-CASAM-N) and an illustrative application of this methodology to a paradigm large-scale physical model to compute exactly its 21,976 first-order sensitivities of a measurable response to the model's parameters,

followed by the computation of the 482,944,576 second-order sensitivities (of which 241,483,276 are distinct from each other) and the largest 5,832,000 third-order sensitivities. The impacts of these sensitivities on the model response's expectation, variance and skewness are also quantified. This paradigm large-scale system cannot be analysed by statistical methods as exactly and as comprehensively as by using the 3rd-CASAM-N. The concepts and methodologies presented in this section can be applied equally well to any model in the Earth sciences.

4.4.1 Third-Order Comprehensive Adjoint Sensitivity Analysis Methodology for Non-linear Systems (3rd-CASAM-N): Mathematical Framework

Fundamentally, the second-order total G-differential of correspondingly differentiable function is defined inductively as 'the total first-order differential of the first-order total differential' of a function. Hence, the second-order sensitivities of the model response with respect to the model's parameters are defined as the first-order sensitivities of the first-order sensitivities. Thus, for each $j_1 = 1, \ldots, TP$, the G-variation $\{\delta R^{(1)}[j_1; \mathbf{u}(\mathbf{x}); \mathbf{a}^{(1)}(\mathbf{x}); \boldsymbol{\alpha}; \mathbf{v}^{(1)}(\mathbf{x}); \delta \mathbf{a}^{(1)}(\mathbf{x}); \delta \boldsymbol{\alpha}]\}_{\boldsymbol{\alpha}^0}$ of a first-order sensitivity $R^{(1)}[j_1; \mathbf{u}(\mathbf{x}); \mathbf{a}^{(1)}(\mathbf{x}); \boldsymbol{\alpha}]$ has the following expression:

$$\{\delta R^{(1)}[j_1; \mathbf{u}(\mathbf{x}); \mathbf{a}^{(1)}(\mathbf{x}); \boldsymbol{\alpha}; \mathbf{v}^{(1)}(\mathbf{x}); \delta \mathbf{a}^{(1)}(\mathbf{x}); \delta \boldsymbol{\alpha}]\}_{\boldsymbol{\alpha}^0}$$
$$= \{\delta R^{(1)}[j_1; \mathbf{u}(\mathbf{x}); \mathbf{a}^{(1)}(\mathbf{x}); \boldsymbol{\alpha}; \delta \boldsymbol{\alpha}]\}_{dir}$$
$$+ \{\delta R^{(1)}[j_1; \mathbf{u}(\mathbf{x}); \mathbf{a}^{(1)}(\mathbf{x}); \boldsymbol{\alpha}; \mathbf{v}^{(1)}(\mathbf{x}); \delta \mathbf{a}^{(1)}(\mathbf{x})]\}_{ind}. \quad (4.55)$$

In Eq. (4.55), the quantity $\{\delta R^{(1)}[j_1; \mathbf{u}(\mathbf{x}); \mathbf{a}^{(1)}(\mathbf{x}); \boldsymbol{\alpha}; \delta \boldsymbol{\alpha}]\}_{dir}$ denotes the direct-effect term, which comprises all dependencies on the vector $\delta \boldsymbol{\alpha}$ of parameter variations, and is defined as follows:

$$\{\delta R^{(1)}[j_1; \mathbf{u}(\mathbf{x}); \mathbf{a}^{(1)}(\mathbf{x}); \boldsymbol{\alpha}; \delta \boldsymbol{\alpha}]\}_{dir}$$
$$\triangleq \left\{ \frac{\partial R^{(1)}[j_1; \mathbf{u}(\mathbf{x}); \mathbf{a}^{(1)}(\mathbf{x}); \boldsymbol{\alpha}]}{\partial \boldsymbol{\alpha}} \delta \boldsymbol{\alpha} \right\}_{\boldsymbol{\alpha}^0}. \quad (4.56)$$

Also in Eq. (4.55), the indirect-effect term $\{\delta R^{(1)}[j_1; \mathbf{u}(\mathbf{x}); \mathbf{a}^{(1)}(\mathbf{x}); \boldsymbol{\alpha}; \mathbf{v}^{(1)}(\mathbf{x}); \delta \mathbf{a}^{(1)}(\mathbf{x})]\}_{ind}$ comprises all dependencies on the vectors $\mathbf{v}^{(1)}(\mathbf{x})$ and $\delta \mathbf{a}^{(1)}(\mathbf{x})$ of variations in the state functions $\mathbf{u}(\mathbf{x})$ and $\mathbf{a}^{(1)}(\mathbf{x})$, respectively, and is defined as follows:

$$\{\delta R^{(1)}[j_1; \mathbf{u}(\mathbf{x}); \mathbf{a}^{(1)}(\mathbf{x}); \boldsymbol{\alpha}; \mathbf{v}^{(1)}(\mathbf{x}); \delta \mathbf{a}^{(1)}(\mathbf{x})]\}_{ind}$$
$$\triangleq \{\partial R^{(1)}[j_1; \ldots; \boldsymbol{\alpha}]/\partial \mathbf{u}\}_{\boldsymbol{\alpha}^0} \mathbf{v}^{(1)}(\mathbf{x})$$
$$+ \{\partial R^{(1)}[j_1; \ldots; \boldsymbol{\alpha}]/\partial \mathbf{a}^{(1)}\}_{\boldsymbol{\alpha}^0} \delta \mathbf{a}^{(1)}(\mathbf{x}). \quad (4.57)$$

The functions $\mathbf{v}^{(1)}(\mathbf{x})$ and $\delta \mathbf{a}^{(1)}(\mathbf{x})$ could be obtained, in principle, by solving the *Second-Level Variational Sensitivity System* (2nd-LVSS), which is obtained by concatenating the 1st-LVSS with the system obtained by G-differentiating the 1st-LASS, and has the following form:

$$\{\mathbf{VM}^{(2)}[2 \times 2; \mathbf{U}^{(2)}(2; \mathbf{x}); \boldsymbol{\alpha}] \mathbf{V}^{(2)}(2; \mathbf{x})\}_{\boldsymbol{\alpha}^0}$$
$$= \{\mathbf{Q}_V^{(2)}[2; \mathbf{U}^{(2)}(2; \mathbf{x}); \boldsymbol{\alpha}; \delta \boldsymbol{\alpha}]\}_{\boldsymbol{\alpha}^0}, \quad \mathbf{x} \in \boldsymbol{\Omega}_x, \quad (4.58)$$

$$\{\mathbf{B}_V^{(2)}[2; \mathbf{U}^{(2)}(2; \mathbf{x}); \mathbf{V}^{(2)}(2; \mathbf{x}); \boldsymbol{\alpha}; \delta \boldsymbol{\alpha}]\}_{\boldsymbol{\alpha}^0}$$
$$= \mathbf{0}[2], \quad \mathbf{0}[2] \triangleq [\mathbf{0}, \mathbf{0}]^\dagger, \quad \mathbf{x} \in \partial \boldsymbol{\Omega}_x(\boldsymbol{\alpha}^0), \quad (4.59)$$

The argument '2' which appears in the list of arguments of the vector $\mathbf{U}^{(2)}(2; \mathbf{x})$ and the 'variational vector' $\mathbf{V}^{(2)}(2; \mathbf{x})$ in Eq. (4.58) indicates that each of these vectors is a 2-block

column vector (each block comprising a column-vector of dimension TD), defined as follows:

$$\mathbf{U}^{(2)}(2; \mathbf{x}) \triangleq \begin{pmatrix} \mathbf{u}^{(1)}(\mathbf{x}) \\ \mathbf{a}^{(1)}(\mathbf{x}) \end{pmatrix}; \quad \mathbf{V}^{(2)}(2; \mathbf{x}) \triangleq \delta \mathbf{U}^{(2)}(2; \mathbf{x})$$

$$\triangleq \begin{pmatrix} \mathbf{v}^{(2)}(1; \mathbf{x}) \\ \mathbf{v}^{(2)}(2; \mathbf{x}) \end{pmatrix} \triangleq \begin{pmatrix} \mathbf{v}^{(1)}(\mathbf{x}) \\ \delta \mathbf{a}^{(1)}(\mathbf{x}) \end{pmatrix}. \quad (4.60)$$

To distinguish block-vectors from block matrices, two bold capital letters have been used (and will henceforth be used) to denote block matrices, as in the case of the second-level variational matrix $\mathbf{VM}^{(2)}[2 \times 2; \mathbf{u}^{(2)}(\mathbf{x}); \boldsymbol{\alpha}]$. The second-level is indicated by the superscript (2). The argument 2×2, which appears in the list of arguments of $\mathbf{VM}^{(2)}[2 \times 2; \mathbf{u}^{(2)}(\mathbf{x}); \boldsymbol{\alpha}]$, indicates that this matrix is a 2×2-dimensional block-matrix comprising four matrices, each of dimensions $TD \times TD$, having the following structure:

$$\mathbf{VM}^{(2)}[2 \times 2; \mathbf{U}^{(2)}(2; \mathbf{x}); \boldsymbol{\alpha}] \triangleq \begin{pmatrix} \mathbf{V}^{(1)} & \mathbf{0} \\ \mathbf{V}_{21}^{(2)} & \mathbf{V}_{22}^{(2)} \end{pmatrix}. \quad (4.61)$$

The other quantities which appear in Eqs. (4.58) and (4.59) are two-block vectors having the same structure as $\mathbf{V}^{(2)}(2; \mathbf{x})$, and are defined as follows:

$$\mathbf{Q}_V^{(2)}[2; \mathbf{U}^{(2)}(2; \mathbf{x}); \boldsymbol{\alpha}; \delta \boldsymbol{\alpha}] \triangleq \begin{pmatrix} \mathbf{q}_V^{(2)}\left(1; \mathbf{U}^{(2)}(2; \mathbf{x}); \boldsymbol{\alpha}; \delta \boldsymbol{\alpha}\right) \\ \mathbf{q}_V^{(2)}\left(2; \mathbf{U}^{(2)}(2; \mathbf{x}); \boldsymbol{\alpha}; \delta \boldsymbol{\alpha}\right) \end{pmatrix}$$
$$\triangleq \begin{pmatrix} \mathbf{q}_V^{(1)}\left(\mathbf{u}; \boldsymbol{\alpha}; \delta \boldsymbol{\alpha}\right) \\ \mathbf{q}_2^{(2)}\left(\mathbf{u}; \mathbf{a}^{(1)}; \boldsymbol{\alpha}; \delta \boldsymbol{\alpha}\right) \end{pmatrix}; \quad (4.62)$$

$$\mathbf{q}_2^{(2)}\left(\mathbf{u}; \boldsymbol{\alpha}; \mathbf{a}^{(1)}; \delta \boldsymbol{\alpha}\right) \triangleq \frac{\partial \mathbf{q}_A^{(1)}[\mathbf{u}(\mathbf{x}); \boldsymbol{\alpha}]}{\partial \boldsymbol{\alpha}} \delta \boldsymbol{\alpha} - \frac{\partial [\mathbf{A}^{(1)}(\mathbf{u}; \boldsymbol{\alpha}) \mathbf{a}^{(1)}(\mathbf{x})]}{\partial \boldsymbol{\alpha}} \delta \boldsymbol{\alpha}$$
$$\triangleq \sum_{j_2=1}^{TP} \mathbf{s}_V^{(2)}\left[2; j_2; \mathbf{U}^{(2)}(2; \mathbf{x}); \boldsymbol{\alpha}\right] \delta \alpha_{j_2};$$

$$\mathbf{s}_V^{(2)}[2; j_2; \mathbf{U}^{(2)}(2; \mathbf{x}); \boldsymbol{\alpha}] \triangleq \frac{\partial \mathbf{q}_A^{(1)}[\mathbf{u}(\mathbf{x}); \boldsymbol{\alpha}]}{\partial \alpha_{j_2}} - \frac{\partial [\mathbf{A}^{(1)}(\mathbf{u}; \boldsymbol{\alpha}) \mathbf{a}^{(1)}(\mathbf{x})]}{\partial \alpha_{j_2}};$$
$$\quad (4.63)$$

$$\mathbf{B}_V^{(2)}[2; \mathbf{U}^{(2)}(2; \mathbf{x}); \mathbf{V}^{(2)}(2; \mathbf{x}); \boldsymbol{\alpha}; \delta \boldsymbol{\alpha}]$$
$$\triangleq \begin{pmatrix} \mathbf{b}_V^{(2)}[1; \mathbf{U}^{(2)}(2; \mathbf{x}); \mathbf{V}^{(2)}(2; \mathbf{x}); \boldsymbol{\alpha}; \delta \boldsymbol{\alpha}] \\ \mathbf{b}_V^{(2)}[2; \mathbf{U}^{(2)}(2; \mathbf{x}); \mathbf{V}^{(2)}(2; \mathbf{x}); \boldsymbol{\alpha}; \delta \boldsymbol{\alpha}] \end{pmatrix}$$
$$\triangleq \begin{pmatrix} \mathbf{b}_V^{(1)}\left(\mathbf{u}^{(1)}; \boldsymbol{\alpha}; \delta \mathbf{u}^{(1)}; \delta \boldsymbol{\alpha}\right) \\ \delta \mathbf{b}_A^{(1)}[\mathbf{U}^{(2)}(2; \mathbf{x}); \mathbf{V}^{(2)}(2; \mathbf{x}); \boldsymbol{\alpha}; \delta \boldsymbol{\alpha}] \end{pmatrix}; \quad (4.64)$$

$$\mathbf{V}_{21}^{(2)}\left(\mathbf{u}; \mathbf{a}^{(1)}; \boldsymbol{\alpha}\right) \triangleq \frac{\partial [\mathbf{A}^{(1)}(\mathbf{u}; \boldsymbol{\alpha}) \mathbf{a}^{(1)}]}{\partial \mathbf{u}} - \frac{\mathbf{q}_A^{(1)}[\mathbf{u}(\mathbf{x}); \boldsymbol{\alpha}]}{\partial \mathbf{u}};$$
$$\mathbf{V}_{22}^{(2)}(\mathbf{u}; \boldsymbol{\alpha}) \triangleq \mathbf{A}^{(1)}(\mathbf{u}; \boldsymbol{\alpha}); \quad (4.65)$$

$$\delta\mathbf{b}_A^{(1)}\left(\mathbf{u};\mathbf{a}^{(1)};\boldsymbol{\alpha}\right) \triangleq \frac{\partial\mathbf{b}_A^{(1)}}{\partial\mathbf{u}}\mathbf{v}^{(1)}(\mathbf{x}) + \frac{\partial\mathbf{b}_A^{(1)}}{\partial\mathbf{a}^{(1)}}\delta\mathbf{a}^{(1)}(\mathbf{x}) + \frac{\partial\mathbf{b}_A^{(1)}}{\partial\boldsymbol{\alpha}}\delta\boldsymbol{\alpha}.$$

$$(4.66)$$

The *Second-Order Comprehensive Adjoint Sensitivity Analysis Methodology for Non-linear Systems* (2nd-CASAM-N) aims at obtaining an alternative expression for the indirect-effect term defined by Eq. (4.57) which does not depend on the variational function $\mathbf{V}^{(2)}(2;\mathbf{x})$ but instead depends on a second-level adjoint function which is the solution of a *Second-Level Adjoint Sensitivity System* (2nd-LASS) constructed by using the 2nd-LVSS as the starting point and by applying the same principles as were applied in the foregoing to obtain the first-order sensitivities provided in Eq. (4.41). The 2nd-LASS is constructed in a Hilbert space, denoted as $\mathsf{H}_2(\Omega_x)$, which comprises as elements block-vectors of the same form as $\mathbf{V}^{(2)}(2;\mathbf{x})$. The inner product of two vectors $\boldsymbol{\Psi}^{(2)}(2;\mathbf{x}) \triangleq [\,\psi^{(2)}(1;\mathbf{x}),\ \psi^{(2)}(2;\mathbf{x})\,]^\dagger \in \mathsf{H}_2(\Omega_x)$ and $\boldsymbol{\Phi}^{(2)}(2;\mathbf{x}) \triangleq [\boldsymbol{\varphi}^{(2)}(1;\mathbf{x}),\boldsymbol{\varphi}^{(2)}(2;\mathbf{x})]^\dagger \in \mathsf{H}_2(\Omega_x)$ in the Hilbert space $\mathsf{H}_2(\Omega_x)$ will be denoted as $\langle\boldsymbol{\Psi}^{(2)}(2;\mathbf{x}),\boldsymbol{\Phi}^{(2)}(2;\mathbf{x})\rangle_2$ and defined as follows:

$$\langle\boldsymbol{\Psi}^{(2)}(2;\mathbf{x}),\boldsymbol{\Phi}^{(2)}(2;\mathbf{x})\rangle_2 \triangleq \sum_{i=1}^{2}\langle\psi^{(2)}(i;\mathbf{x}),\boldsymbol{\varphi}^{(2)}(i;\mathbf{x})\rangle_1.\quad(4.67)$$

Using the inner product defined in Eq. (4.67) and following the same steps as outlined for deriving the 1st-LASS yields the following 2nd-LASS for the second-level adjoint function $\mathbf{A}^{(2)}(2;j_1;\mathbf{x}) \triangleq [\mathbf{a}^{(1)}(1;j_1;\mathbf{x}),\mathbf{a}^{(2)}(2;j_1;\mathbf{x})]^\dagger \in \mathsf{H}_2(\Omega_x)$, for each $j_1 = 1,\ldots,TP$:

$$\{\mathbf{AM}^{(2)}[2\times2;\mathbf{U}^{(2)}(2;\mathbf{x});\boldsymbol{\alpha}]\mathbf{A}^{(2)}(2;j_1;\mathbf{x})\}_{\boldsymbol{\alpha}^0}$$

$$= \{\mathbf{Q}_A^{(2)}[2;j_1;\mathbf{U}^{(2)}(2;\mathbf{x});\boldsymbol{\alpha}]\}_{\boldsymbol{\alpha}^0},\ \mathbf{x}\in\Omega_x,\quad(4.68)$$

subject to boundary conditions represented as follows:

$$\{\mathbf{B}_A^{(2)}[2;\mathbf{U}^{(2)}(2;\mathbf{x});\mathbf{A}^{(2)}(2;j_1;\mathbf{x});\boldsymbol{\alpha}]\}_{\boldsymbol{\alpha}^0} = \mathbf{0}[2];\quad \mathbf{x}\in\partial\Omega_x(\boldsymbol{\alpha}^0),$$

$$(4.69)$$

where

$$\mathbf{Q}_A^{(2)}[2;j_1;\mathbf{U}^{(2)}(2;\mathbf{x});\boldsymbol{\alpha}] \triangleq \begin{pmatrix} \mathbf{q}_A^{(2)}\left(1;j_1;\mathbf{U}^{(2)};\boldsymbol{\alpha}\right) \\ \mathbf{q}_A^{(2)}\left(2;j_1;\mathbf{U}^{(2)};\boldsymbol{\alpha}\right) \end{pmatrix}$$

$$\triangleq \begin{pmatrix} \partial R^{(1)}[j_1;\mathbf{u}(\mathbf{x});\mathbf{a}^{(1)}(\mathbf{x});\boldsymbol{\alpha};\mathbf{v}^{(1)}(\mathbf{x})]/\partial\mathbf{u} \\ \partial R^{(1)}[j_1;\mathbf{u}(\mathbf{x});\mathbf{a}^{(1)}(\mathbf{x});\boldsymbol{\alpha};\mathbf{v}^{(1)}(\mathbf{x})]/\partial\mathbf{a}^{(1)} \end{pmatrix},\quad(4.70)$$

$$\mathbf{AM}^{(2)}[2\times2;\mathbf{U}^{(2)}(2;\mathbf{x});\boldsymbol{\alpha}] \triangleq [\mathbf{VM}^{(2)}(2\times2;\mathbf{U}^{(2)}(2;\mathbf{x});\boldsymbol{\alpha})]^*$$

$$= \begin{pmatrix} [\mathbf{V}^{(1)*}]^\dagger & [\mathbf{V}_{21}^{(2)*}]^\dagger \\ \mathbf{0} & [\mathbf{V}_{22}^{(2)*}]^\dagger \end{pmatrix}.\quad(4.71)$$

The matrix $\mathbf{AM}^{(2)}[2\times2;\mathbf{u}^{(2)}(\mathbf{x});\boldsymbol{\alpha}]$ comprises (2×2) block-matrices, each of dimensions TD^2, and is obtained from the following relation:

$$\{\langle\mathbf{A}^{(2)}(2;\mathbf{x}),\mathbf{VM}^{(2)}\mathbf{V}^{(2)}(2;\mathbf{x})\rangle_2\}_{\boldsymbol{\alpha}^0}$$

$$= \{[P^{(2)}(\mathbf{U}^{(2)};\mathbf{A}^{(2)};\mathbf{V}^{(2)};\boldsymbol{\alpha})]_{\partial\Omega_x}\}_{\boldsymbol{\alpha}^0}$$

$$+ \{\langle\mathbf{V}^{(2)}(2;\mathbf{x}),\mathbf{AM}^{(2)}[2\times2;\mathbf{U}^{(2)}(2;\mathbf{x});\boldsymbol{\alpha}]\mathbf{A}^{(2)}(2;\mathbf{x})\rangle_2\}_{\boldsymbol{\alpha}^0},$$

$$(4.72)$$

where the quantity $\{[P^{(2)}(\mathbf{U}^{(2)};\mathbf{A}^{(2)};\mathbf{V}^{(2)};\boldsymbol{\alpha})]_{\partial\Omega_x}\}_{\boldsymbol{\alpha}^0}$ denotes the corresponding bilinear concomitant on the domain's boundary, evaluated at the nominal values for the parameters and respective state functions. The second-level adjoint boundary/initial conditions represented by Eq. (4.59) are determined by requiring that: (i) they must be independent of unknown values of $\mathbf{V}^{(2)}(2;\mathbf{x})$; and (ii) the substitution of the boundary and/or initial conditions represented by Eqs. (4.59) and (4.69) into the expression of $\{[P^{(2)}(\mathbf{U}^{(2)};\mathbf{A}^{(2)};\mathbf{V}^{(2)};\boldsymbol{\alpha})]_{\partial\Omega_x}\}_{\boldsymbol{\alpha}^0}$ must cause all terms containing unknown values of $\mathbf{V}^{(2)}(2;\mathbf{x})$ to vanish.

The alternative expression of the total differential defined by Eq. (4.55) in terms of the second-level adjoint function $\mathbf{A}^{(2)}(2;j_1;\mathbf{x})$ is as follows:

$$\{\delta R^{(1)}[j_1;\mathbf{U}^{(2)}(2;\mathbf{x});\mathbf{A}^{(2)}(2;j_1;\mathbf{x});\boldsymbol{\alpha};\delta\boldsymbol{\alpha}]\}_{\boldsymbol{\alpha}^0}$$

$$= \sum_{j_2=1}^{TP}\{R^{(2)}[j_2;j_1;\mathbf{U}^{(2)}(2;\mathbf{x});\mathbf{A}^{(2)}(2;j_1;\mathbf{x});\boldsymbol{\alpha}]\}_{\boldsymbol{\alpha}^0}\delta\alpha_{j_2},$$

$$j_1 = 1,\ldots,TP,\quad(4.73)$$

where the quantity $R^{(2)}[j_2;j_1;\mathbf{U}^{(2)}(2;\mathbf{x});\mathbf{A}^{(2)}(2;j_1;\mathbf{x});\boldsymbol{\alpha}]$ denotes the second-order sensitivity of the generic scalar-valued response $R[\mathbf{u}(\mathbf{x});\boldsymbol{\alpha}]$ with respect to the parameters α_{j_1} and α_{j_2} computed at the nominal values of the parameters and respective state functions, and has the following expression:

$$For\ j_1,j_2 = 1,\ldots,TP : R^{(2)}[j_2;j_1;\mathbf{U}^{(2)}(2;\mathbf{x});\mathbf{A}^{(2)}(2;j_1;\mathbf{x});\boldsymbol{\alpha}]$$

$$\triangleq \frac{\partial R^{(1)}[j_1;\mathbf{u}(\mathbf{x});\mathbf{a}^{(1)}(\mathbf{x});\boldsymbol{\alpha}]}{\partial\alpha_{j_2}}$$

$$- \frac{\{\partial\hat{P}^{(2)}[\mathbf{U}^{(2)}(2;\mathbf{x});\mathbf{A}^{(2)}(2;j_1;\mathbf{x});\boldsymbol{\alpha}]\}_{\partial\Omega_x}}{\partial\alpha_{j_2}}$$

$$+ \sum_{i=1}^{2}\langle\mathbf{a}^{(2)}(i;j_1;\mathbf{x}),\mathbf{s}_V^{(2)}[i;j_2;\mathbf{U}^{(2)}(2;\mathbf{x});\boldsymbol{\alpha}]\rangle_1 \triangleq \frac{\partial^2 R[\mathbf{u}(\mathbf{x});\boldsymbol{\alpha}]}{\partial\alpha_{j_2}\partial\alpha_{j_1}}.$$

$$(4.74)$$

In Eq. (4.74), the quantity $\{[\hat{P}^{(2)}(\mathbf{U}^{(2)};\mathbf{A}^{(2)};\boldsymbol{\alpha};\delta\boldsymbol{\alpha})]_{\partial\Omega_x}\}_{\boldsymbol{\alpha}^0}$ denotes residual boundary terms which may not have vanished after having used the boundary and/or initial conditions represented by Eqs. (4.59) and (4.69) in Eq. (4.72). If the 2nd-LASS is solved TP-times, the second-order mixed sensitivities

$R^{(2)}[j_2; j_1; \mathbf{U}^{(2)}(2; \mathbf{x}); \mathbf{A}^{(2)}(2; j_1; \mathbf{x}); \boldsymbol{\alpha}] \equiv \partial^2 R / \partial \alpha_{j_2} \partial \alpha_{j_1}$ will be computed twice, in two different ways, in terms of two distinct second-level adjoint functions. Consequently, the symmetry property $\partial^2 R[\mathbf{u}(\mathbf{x}); \boldsymbol{\alpha}] / \partial \alpha_{j_2} \partial \alpha_{j_1} = \partial^2 R[\mathbf{u}(\mathbf{x}); \boldsymbol{\alpha}] / \partial \alpha_{j_1} \partial \alpha_{j_2}$ enjoyed by the second-order sensitivities provides an intrinsic (numerical) verification that the components of the second-level adjoint function $\mathbf{A}^{(2)}(2; j_1; \mathbf{x})$ and the first-level adjoint function $\mathbf{a}^{(1)}(\mathbf{x})$ are computed accurately.

The third-order sensitivities are obtained by considering them to be the first-order sensitivities of a second-order sensitivity. Thus, each of the second-order sensitivities $R^{(2)}[j_2; j_1; \mathbf{U}^{(2)}(2; \mathbf{x}); \mathbf{A}^{(2)}(2; j_1; \mathbf{x}); \boldsymbol{\alpha}] \equiv \partial^2 R / \partial \alpha_{j_2} \partial \alpha_{j_1}$ will be considered to be a model response which is assumed to satisfy the conditions stated in Eqs. (4.27) and (4.28) for each $j_1, j_2 = 1, \ldots, TP$, so that the first-order total G-differential of $R^{(2)}[j_2; j_1; \mathbf{U}^{(2)}(2; \mathbf{x}); \mathbf{A}^{(2)}(2; j_1; \mathbf{x}); \boldsymbol{\alpha}]$ will exist and will be linear in the variations $\mathbf{V}^{(2)}(2; \mathbf{x})$ and $\delta \mathbf{A}^{(2)}(2; j_1; \mathbf{x})$ in a neighbourhood around the nominal values of the parameters and the respective state functions. By definition, the first-order total G-differential of $R^{(2)}[j_2; j_1; \mathbf{U}^{(2)}(2; \mathbf{x}); \mathbf{A}^{(2)}(2; j_1; \mathbf{x}); \boldsymbol{\alpha}]$, which will be denoted as $\{\delta R^{(2)}[j_2; j_1; \mathbf{U}^{(2)}(2; \mathbf{x}); \mathbf{A}^{(2)}(2; j_1; \mathbf{x}); \boldsymbol{\alpha}; \mathbf{V}^{(2)}(2; \mathbf{x}); \delta \mathbf{A}^{(2)}(2; j_1; \mathbf{x}); \delta \boldsymbol{\alpha}]\}_{\boldsymbol{\alpha}^0}$, is given by the following expression:

$$
\begin{aligned}
&\{\delta R^{(2)}[j_2; j_1; \mathbf{U}^{(2)}(2; \mathbf{x}); \mathbf{A}^{(2)}(2; j_1; \mathbf{x}); \boldsymbol{\alpha}; \mathbf{V}^{(2)}(2; \mathbf{x}); \delta \mathbf{A}^{(2)} \\
&(2; j_1; \mathbf{x}); \delta \boldsymbol{\alpha}]\}_{\boldsymbol{\alpha}^0} \\
&\triangleq \left\{ \frac{\partial R^{(2)}[j_2; j_1; \mathbf{U}^{(2)}(2; \mathbf{x}); \mathbf{A}^{(2)}(2; j_1; \mathbf{x}); \boldsymbol{\alpha}]}{\partial \boldsymbol{\alpha}} \delta \boldsymbol{\alpha} \right\}_{\boldsymbol{\alpha}^0} \\
&+ \{\delta R^{(2)}[j_2; j_1; \mathbf{U}^{(2)}(2; \mathbf{x}); \mathbf{A}^{(2)}(2; j_1; \mathbf{x}); \boldsymbol{\alpha}; \mathbf{V}^{(2)}(2; \mathbf{x}); \delta \mathbf{A}^{(2)} \\
&(2; j_1; \mathbf{x})]\}_{ind},
\end{aligned}
$$ (4.75)

where

$$
\begin{aligned}
&\{\delta R^{(2)}[j_2; j_1; \mathbf{U}^{(2)}(2; \mathbf{x}); \mathbf{A}^{(2)}(2; j_1; \mathbf{x}); \boldsymbol{\alpha}; \mathbf{V}^{(2)}(2; \mathbf{x}); \\
&\delta \mathbf{A}^{(2)}(2; j_1; \mathbf{x})]\}_{ind} \\
&\triangleq \left\{ \frac{\partial R^{(2)}[j_2; j_1; \mathbf{U}^{(2)}; \mathbf{A}^{(2)}; \boldsymbol{\alpha}]}{\partial \mathbf{U}^{(2)}(2; \mathbf{x})} \right\}_{\boldsymbol{\alpha}^0} \mathbf{V}^{(2)}(2; \mathbf{x}) \\
&+ \left\{ \frac{\partial R^{(2)}[j_2; j_1; \mathbf{U}^{(2)}; \mathbf{A}^{(2)}; \boldsymbol{\alpha}]}{\partial \mathbf{A}^{(2)}(2; j_1; \mathbf{x})} \right\}_{\boldsymbol{\alpha}^0} \delta \mathbf{A}^{(2)}(2; j_1; \mathbf{x}) .
\end{aligned}
$$ (4.76)

The indirect-effect term $\{\delta R^{(2)}[j_2; j_1; \mathbf{U}^{(2)}(2; \mathbf{x}); \mathbf{A}^{(2)}(2; j_1; \mathbf{x}); \boldsymbol{\alpha}; \delta \boldsymbol{\alpha}]\}_{dir}$ can be computed after having determined the vectors $\mathbf{V}^{(2)}(2; \mathbf{x})$ and $\delta \mathbf{A}^{(2)}(2; j_1; \mathbf{x})$, which are the solutions of the following *Third-Level Variational Sensitivity System* (3rd-LVSS):

$$
\begin{aligned}
&\{\mathbf{VM}^{(3)}[4 \times 4; \mathbf{U}^{(3)}(4; \mathbf{x}); \boldsymbol{\alpha}] \mathbf{V}^{(3)}(4; \mathbf{x})\}_{\boldsymbol{\alpha}^0} \\
&= \{\mathbf{Q}_V^{(3)}[4; \mathbf{U}^{(3)}(4; \mathbf{x}); \boldsymbol{\alpha}; \delta \boldsymbol{\alpha}]\}_{\boldsymbol{\alpha}^0}, \quad \mathbf{x} \in \Omega_x,
\end{aligned}
$$ (4.77)

$$
\{\mathbf{B}_V^{(3)}[4; \mathbf{U}^{(3)}(4; \mathbf{x}); \mathbf{V}^{(3)}(4; \mathbf{x}); \boldsymbol{\alpha}; \delta \boldsymbol{\alpha}]\}_{\boldsymbol{\alpha}^0} = \mathbf{0}[4]; \quad \mathbf{x} \in \partial \Omega_x(\boldsymbol{\alpha}^0)
$$ (4.78)

where $\mathbf{0}[4] \triangleq [0, 0, 0, 0]^\dagger$ and where

$$
\mathbf{VM}^{(3)}\left(4 \times 4; \mathbf{U}^{(3)}; \boldsymbol{\alpha}\right) \triangleq \begin{pmatrix} \mathbf{VM}^{(2)}(2 \times 2) & \mathbf{0}[2 \times 2] \\ \mathbf{VM}_{21}^{(3)}(2 \times 2) & \mathbf{VM}_{22}^{(3)}(2 \times 2) \end{pmatrix};
$$ (4.79)

$$
\begin{aligned}
\mathbf{U}^{(3)}(4; \mathbf{x}) &\triangleq \begin{pmatrix} \mathbf{U}^{(2)}(2; \mathbf{x}) \\ \mathbf{A}^{(2)}(2; j_1; \mathbf{x}) \end{pmatrix}; \quad \mathbf{V}^{(3)}(4; \mathbf{x}) \triangleq \delta \mathbf{U}^{(3)}(4; \mathbf{x}) \\
&= \begin{pmatrix} \mathbf{V}^{(2)}(2; \mathbf{x}) \\ \delta \mathbf{A}^{(2)}(2; j_1; \mathbf{x}) \end{pmatrix};
\end{aligned}
$$ (4.80)

$$
\begin{aligned}
\mathbf{VM}_{21}^{(3)}(2 \times 2; \mathbf{x}) &\triangleq \frac{\partial \{\mathbf{AM}^{(2)}[2 \times 2; \mathbf{U}^{(2)}; \boldsymbol{\alpha}] \mathbf{A}^{(2)}(2; \mathbf{x})\}}{\partial \mathbf{U}^{(2)}(2; \mathbf{x})} \\
&- \frac{\partial \mathbf{Q}_A^{(2)}[2; \mathbf{u}^{(2)}(\mathbf{x}); \boldsymbol{\alpha}]}{\partial \mathbf{U}^{(2)}(2; \mathbf{x})};
\end{aligned}
$$ (4.81)

$$
\mathbf{VM}_{22}^{(3)}(2 \times 2; \mathbf{x}) \triangleq \mathbf{AM}^{(2)}[2 \times 2; \mathbf{U}^{(2)}; \boldsymbol{\alpha}]; \mathbf{0}[2 \times 2] \triangleq \begin{pmatrix} 0 & 0 \\ 0 & 0 \end{pmatrix};
$$ (4.82)

$$
\begin{aligned}
\mathbf{Q}_V^{(3)}[4; \mathbf{U}^{(3)}(4; \mathbf{x}); \boldsymbol{\alpha}; \delta \boldsymbol{\alpha}] &\triangleq \begin{pmatrix} \mathbf{Q}_V^{(2)}[2; \mathbf{U}^{(2)}(2; \mathbf{x}); \boldsymbol{\alpha}; \delta \boldsymbol{\alpha}] \\ \mathbf{Q}_2^{(3)}[2; \mathbf{U}^{(3)}(4; \mathbf{x}); \boldsymbol{\alpha}; \delta \boldsymbol{\alpha}] \end{pmatrix} \\
&\triangleq \begin{pmatrix} \mathbf{q}_V^{(3)}[1; \mathbf{U}^{(3)}(4; \mathbf{x}); \boldsymbol{\alpha}; \delta \boldsymbol{\alpha}] \\ \cdot \\ \mathbf{q}_V^{(3)}[4; \mathbf{U}^{(3)}(4; \mathbf{x}); \boldsymbol{\alpha}; \delta \boldsymbol{\alpha}] \end{pmatrix};
\end{aligned}
$$ (4.83)

$$
\begin{aligned}
&\mathbf{q}_V^{(3)}[i; \mathbf{U}^{(3)}(4; \mathbf{x}); \boldsymbol{\alpha}; \delta \boldsymbol{\alpha}] \equiv \sum_{j_3=1}^{TP} \mathbf{s}_V^{(3)}[i; j_3; \mathbf{U}^{(3)}(4; \mathbf{x}); \boldsymbol{\alpha}] \delta \alpha_{j_3}; \\
&i = 1, \ldots 4;
\end{aligned}
$$ (4.84)

$$
\begin{aligned}
\mathbf{Q}_2^{(3)}[2; \mathbf{U}^{(3)}(4; \mathbf{x}); \boldsymbol{\alpha}; \delta \boldsymbol{\alpha}] &\triangleq \frac{\partial \mathbf{Q}_A^{(2)}}{\partial \boldsymbol{\alpha}} \partial \boldsymbol{\alpha} \\
&- \frac{\partial \{\mathbf{AM}^{(2)}[2 \times 2; \mathbf{U}^{(2)}(2; \mathbf{x}); \boldsymbol{\alpha}] \mathbf{A}^{(2)}(2; j_1; \mathbf{x})\}}{\partial \boldsymbol{\alpha}} \partial \boldsymbol{\alpha};
\end{aligned}
$$ (4.85)

$$
\begin{aligned}
&\mathbf{B}_V^{(3)}[4; \mathbf{U}^{(3)}(4; \mathbf{x}); \mathbf{V}^{(3)}(4; \mathbf{x}); \boldsymbol{\alpha}; \delta \boldsymbol{\alpha}] \\
&\triangleq \begin{pmatrix} \mathbf{B}_V^{(2)}[2; \mathbf{U}^{(2)}(2; \mathbf{x}); \mathbf{V}^{(2)}(2; \mathbf{x}); \boldsymbol{\alpha}; \delta \boldsymbol{\alpha}] \\ \delta \mathbf{B}_A^{(2)}[2; \mathbf{U}^{(3)}(4; \mathbf{x}); \mathbf{V}^{(3)}(4; \mathbf{x}); \boldsymbol{\alpha}; \delta \boldsymbol{\alpha}] \end{pmatrix}.
\end{aligned}
$$ (4.86)

The right side of the 3rd-LVSS actually depends on the indices $j_1, j_2, j_3 = 1, \ldots, TP$, so the 3rd-LVSS would need to be solved TP^3 times to obtain each of the variational functions $\mathbf{V}^{(3)}(4; j_1, j_2, j_3; \mathbf{x})$. Thus, solving the 3rd-LVSS would require TP^3 large-scale computations, which is unrealistic for large-scale systems comprising many parameters. Since the 3rd-LVSS is never actually solved but is only used to construct the corresponding adjoint sensitivity system, the specific dependence of the 3rd-LVSS on the indices $j_1, j_2, j_3 = 1, ..., TP$ has been suppressed.

The 3rd-CASAM-N circumvents the need for solving the 3rd-LVSS by deriving an alternative expression for the indirect-effect term defined in Eq. (4.76), in which the function $\mathbf{V}^{(3)}(4; \mathbf{x})$ is replaced by a third-level adjoint function which is independent of parameter variations. This third-level adjoint function is the solution of a *Third-Level Adjoint Sensitivity System* (3rd-LASS) which is constructed by applying the same principles as those used for constructing the 1st-LASS and the 2nd-LASS. The Hilbert space appropriate for constructing the 3rd-LASS, denoted as $H_3(\Omega_x)$, comprises as elements block-vectors of the same form as $\mathbf{V}^{(3)}(4; \mathbf{x})$. Thus, a generic block-vector in $H_3(\Omega_x)$, denoted as $\mathbf{\Psi}^{(3)}(4; \mathbf{x}) \triangleq [\; \psi^{(3)}(1; \mathbf{x}), \psi^{(3)}(2; \mathbf{x}), \psi^{(3)}(3; \mathbf{x}), \quad \psi^{(3)}(4; \mathbf{x})]^\dagger \in H_3(\Omega_x)$, comprises four *TD*-dimensional vector-components of the form $\psi^{(3)}(i; \mathbf{x}) \triangleq [\psi_1^{(3)}(i; \mathbf{x}), ..., \psi_{TD}^{(3)}(i; \mathbf{x})]^\dagger \in H_1(\Omega_x)$, $i = 1, 2, 3, 4$, where each of these four components is a *TD*-dimensional column vector. The inner product of two vectors $\mathbf{\Psi}^{(3)}(4; \mathbf{x}) \in H_3(\Omega_x)$ and $\mathbf{\Phi}^{(3)}(4; \mathbf{x}) \in H_3(\Omega_x)$ in the Hilbert space $H_3(\Omega_x)$ will be denoted as $\langle \mathbf{\Psi}^{(3)}(4; \mathbf{x}), \mathbf{\Phi}^{(3)}(4; \mathbf{x}) \rangle_3$ and defined as follows:

$$\langle \mathbf{\Psi}^{(3)}(4; \mathbf{x}), \mathbf{\Phi}^{(3)}(4; \mathbf{x}) \rangle_3 \triangleq \sum_{i=1}^{4} \langle \psi^{(3)}(i; \mathbf{x}), \varphi^{(3)}(i; \mathbf{x}) \rangle_1.$$

(4.87)

The steps for constructing the 3rd-LASS are conceptually similar to those for constructing the 1st-LASS and 2nd-LASS, so they will be omitted for the sake of brevity. The final expressions for the third-order sensitivities are as follows:

$$\{\delta R^{(2)}[j_2; j_1; \mathbf{U}^{(3)}(4; \mathbf{x}); \mathbf{A}^{(3)}(4; j_2; j_1; \mathbf{x}); \boldsymbol{\alpha}; \delta\boldsymbol{\alpha}]\}_{\boldsymbol{\alpha}^0}$$

$$= \left\{ \frac{\partial R^{(2)}[j_2; j_1; \mathbf{U}^{(3)}; \boldsymbol{\alpha}]}{\partial \boldsymbol{\alpha}} \delta\boldsymbol{\alpha} \right\}_{\boldsymbol{\alpha}^0} - \{[\hat{P}^{(3)}\left(\mathbf{U}^{(3)}; \mathbf{A}^{(3)}; \delta\boldsymbol{\alpha}\right)]_{\partial\Omega_x}\}_{\boldsymbol{\alpha}^0}$$

$$+ \{\langle \mathbf{A}^{(3)}(4; j_2; j_1; \mathbf{x}), \mathbf{Q}_V^{(3)}[4; \mathbf{U}^{(3)}; \boldsymbol{\alpha}; \delta\boldsymbol{\alpha}] \rangle_3\}_{\boldsymbol{\alpha}^0},$$

(4.88)

where $\{[\hat{P}^{(3)}\left(\mathbf{U}^{(3)}; \mathbf{A}^{(3)}; \delta\boldsymbol{\alpha}\right)]_{\partial\Omega_x}\}_{\boldsymbol{\alpha}^0}$ denotes residual boundary terms which may have not vanished automatically, and where the third-level adjoint function $\mathbf{A}^{(3)}(4; \mathbf{x}) \triangleq [\; \mathbf{a}^{(3)}(1; \mathbf{x}), \mathbf{a}^{(3)}(2; \mathbf{x}), \mathbf{a}^{(3)}(3; \mathbf{x}), \mathbf{a}^{(3)}(4; \mathbf{x})]^\dagger \in H_3(\Omega_x)$ is the solution of the following 3rd-LASS, for $j_1 = 1, ..., TP; j_2 = 1, ..., j_1$:

$$\{\mathbf{AM}^{(3)}[4 \times 4; \mathbf{U}^{(3)}(4; \mathbf{x}); \boldsymbol{\alpha}]\mathbf{A}^{(3)}(4; j_2; j_1; \mathbf{x})\}_{\boldsymbol{\alpha}^0}$$

$$= \{\mathbf{Q}_A^{(3)}[4; j_2; j_1; \mathbf{U}^{(3)}(4; \mathbf{x}); \boldsymbol{\alpha}]\}_{\boldsymbol{\alpha}^0},$$

(4.89)

subject to boundary/interface/initial conditions represented in operator form as follows:

$$\{\mathbf{B}_A^{(3)}[4; \mathbf{U}^{(3)}(4; \mathbf{x}); \mathbf{A}^{(3)}(4; j_2; j_1; \mathbf{x}); \boldsymbol{\alpha}]\}_{\boldsymbol{\alpha}^0} = \mathbf{0}[4]; \quad \mathbf{x} \in \partial\Omega_x(\boldsymbol{\alpha}^0);$$

(4.90)

where

$$\mathbf{AM}^{(3)}\left[4 \times 4; \mathbf{U}^{(3)}(4; \mathbf{x}); \boldsymbol{\alpha}\right] \triangleq \left[\mathbf{VM}^{(3)}(4 \times 4; \mathbf{U}^{(3)}; \boldsymbol{\alpha})\right]^*;$$

(4.91)

$$\mathbf{Q}_A^{(3)}\left[4; j_2; j_1; \mathbf{U}^{(3)}(4; \mathbf{x}); \boldsymbol{\alpha}\right]$$

$$\triangleq \left[\mathbf{q}_A^{(3)}(1; j_2; j_1; \mathbf{U}^{(3)}; \boldsymbol{\alpha}), ..., \mathbf{q}_A^{(3)}\left(4; j_2; j_1; \mathbf{U}^{(3)}; \boldsymbol{\alpha}\right)\right];$$

(4.92)

$$\mathbf{q}_A^{(3)}\left(1; j_2; j_1; \mathbf{U}^{(3)}; \boldsymbol{\alpha}\right) \triangleq \partial R^{(2)}\left[j_2; j_1; \mathbf{u}^{(2)}; \mathbf{a}^{(2)}; \boldsymbol{\alpha}\right]/\partial \mathbf{u}^{(1)};$$

(4.93)

$$\mathbf{q}_A^{(3)}\left(2; j_2; j_1; \mathbf{U}^{(3)}; \boldsymbol{\alpha}\right) \triangleq \partial R^{(2)}\left[j_2; j_1; \mathbf{u}^{(2)}; \mathbf{a}^{(2)}; \boldsymbol{\alpha}\right]/\partial \mathbf{a}^{(1)};$$

(4.94)

$$\mathbf{q}_A^{(3)}\left(3; j_2; j_1; \mathbf{U}^{(3)}; \boldsymbol{\alpha}\right) \triangleq \partial R^{(2)}\left[j_2; j_1; \mathbf{u}^{(2)}; \mathbf{a}^{(2)}; \boldsymbol{\alpha}\right]/\partial \mathbf{a}^{(2)}(1; j_1; \mathbf{x});$$

(4.95)

$$\mathbf{q}_A^{(3)}\left(3; j_2; j_1; \mathbf{U}^{(3)}; \boldsymbol{\alpha}\right) \triangleq \partial R^{(2)}\left[j_2; j_1; \mathbf{u}^{(2)}; \mathbf{a}^{(2)}; \boldsymbol{\alpha}\right]/\partial \mathbf{a}^{(2)}(2; j_1; \mathbf{x}).$$

(4.96)

In component form, the total differential expressed by Eq. (4.75) has the following expression:

$$\{\delta R^{(2)}[j_2; j_1; \mathbf{U}^{(3)}(4; \mathbf{x}); \mathbf{A}^{(3)}(4; j_2; j_1; \mathbf{x}); \boldsymbol{\alpha}; \; \delta\boldsymbol{\alpha}]\}_{\boldsymbol{\alpha}^0}$$

$$= \sum_{j_3=1}^{TP} \{R^{(3)}[j_3; j_2; j_1; \mathbf{U}^{(3)}(4; \mathbf{x}); \mathbf{A}^{(3)}(4; j_2; j_1; \mathbf{x}); \boldsymbol{\alpha}]\}_{\boldsymbol{\alpha}^0} \delta\alpha_{j_3},$$

$$j_1; j_2 = 1, ..., TP,$$

(4.97)

where the quantity $R^{(3)}[j_3; j_2; j_1; \mathbf{U}^{(3)}(4; \mathbf{x}); \mathbf{A}^{(3)}(4; j_2; j_1; \mathbf{x}); \boldsymbol{\alpha}]$ denotes the third-order sensitivity of the generic scalar-valued response $R[\mathbf{u}(\mathbf{x}); \boldsymbol{\alpha}]$ with respect to any three model parameters α_{j_1}, α_{j_2}, α_{j_3}, and has the following expression, for $j_1, j_2, j_3 = 1, ..., TP$:

$$R^{(3)}[j_3; j_2; j_1; \mathbf{U}^{(3)}(4; \mathbf{x}); \mathbf{A}^{(3)}(4; j_2; j_1; \mathbf{x}); \boldsymbol{\alpha}]$$

$$\triangleq \frac{\partial R^{(2)}[j_2; j_1; \mathbf{U}^{(3)}(4; j_1; \mathbf{x}); \boldsymbol{\alpha}]}{\partial \alpha_{j_3}} - \frac{[\partial \hat{P}^{(3)}\left(\mathbf{U}^{(3)}; \mathbf{A}^{(3)}; \delta\boldsymbol{\alpha}\right)]_{\partial\Omega_x}}{\partial \alpha_{j_3}}$$

$$+ \sum_{i=1}^{4} \langle \mathbf{a}^{(3)}(i; j_2; j_1; \mathbf{x}), \mathbf{s}_V^{(3)}[i; j_3; j_1; \mathbf{U}^{(3)}(4; \mathbf{x}); \boldsymbol{\alpha}] \rangle_1$$

$$\triangleq \frac{\partial^3 R[\mathbf{u}(\mathbf{x}); \boldsymbol{\alpha}]}{\partial \alpha_{j_3} \partial \alpha_{j_2} \partial \alpha_{j_1}}.$$

(4.98)

The third-order sensitivities can be computed selectively, in the priority order pre-established by the user to solve the 3rd-LASS to obtain the requisite third-level adjoint sensitivity function $\mathbf{A}^{(3)}(4; j_2, j_1; \mathbf{x})$, $1 \le j_1, j_2 \le TP$. If the 3rd-LASS is solved $TP(TP+1)/2$ times, then each of the third-order mixed sensitivities $\partial^3 R[\mathbf{u}(\mathbf{x}); \boldsymbol{\alpha}]/\partial \alpha_{j_1} \partial \alpha_{j_3} \partial \alpha_{j_3}$ will be computed three times, each time using distinct adjoint functions. Thus, the symmetry property enjoyed by the third-order sensitivities provides an intrinsic verification of the numerical computation of the various adjoint sensitivity functions.

4.4.2 Uncertainty Analysis of a System Comprising Many Parameters: Application to the OECD/NEA Polyethylene-Reflected Plutonium Reactor Physics Benchmark

The *Third-Order Comprehensive Adjoint Sensitivity Analysis of Response-Coupled Linear Forward/Adjoint Systems* (3rd-CASAM-L) methodology developed by Cacuci (2019b) was applied to perform (Cacuci, Fang, and Favorite, 2019, 2020, and references therein) a comprehensive second-order sensitivity and uncertainty analysis of the leakage response of the OECD/NEA polyethylene-reflected plutonium (acronym: PERP) reactor physics benchmark (Valentine, 2006). The PERP benchmark is modelled by the linear Boltzmann neutron transport equation comprising 21,976 uncertain model parameters (180 group-averaged total microscopic cross sections, 21,600 group-averaged scattering microscopic cross sections, 120 parameters describing the fission process, 60 parameters describing the fission spectrum, 10 parameters describing the system's sources, and 6 isotopic number densities). Thus, the PERP benchmark comprises 21,976 first-order sensitivities of the leakage response with respect to the model parameters, and 482,944,576 second-order sensitivities (of which 241,483,276 are distinct from each other). These fundamental works (Cacuci, Fang, and Favorite, 2019, 2020, and references therein) have demonstrated that, for this benchmark, many second-order sensitivities were significantly (by over an order of magnitude) larger than the first-order sensitivities, and the cumulative effects of the second-order sensitivities on the predicted uncertainties of the PERP benchmark's response far exceeded the effects of the first-order sensitivities. In particular, the largest first-order relative sensitivity is $S^{(1)}(\sigma_{t,6}^{30}) = -9.366$ and the largest first-order relative sensitivity is $S^{(2)}(\sigma_{t,6}^{30}, \sigma_{t,6}^{30}) = 429.6$, both of which involve the microscopic total cross section in the lowest-energy group (i.e. the 30th-group) of hydrogen (isotope ^{1}H). Overall, the total microscopic cross section of isotopes ^{1}H and ^{239}Pu are the two most important parameters affecting the PERP benchmark's leakage response, since they are involved in all of the large first- and second-order sensitivities. Complete results for the 21,976 first-order sensitivities of the leakage response with respect to the model parameters, and the 241,483,276 distinct second-order sensitivities are provided by Cacuci, Fang, and Favorite (2020) and references therein. In particular, it was found that only 7,477 first-order sensitivities are non-zero.

The importance of the contributions to the response uncertainties stemming from the first-order and, respectively, second-order sensitivities are illustrated by the results presented in Table 4.2. The notations used in Table 4.1 are as follows:

(i) The superscript U denotes 'uncorrelated' while the superscript N denotes 'normally distributed'.

(ii) $L(\boldsymbol{\alpha}^0)$ denotes the PERP benchmark's leakage response computed at the nominal parameter values;

(iii) $[E(L)]_t^{(2,U)}$ denotes the contribution stemming from the second-order sensitivities to the total expectation, denoted as $[E(L)]_t^{(U)}$, of the leakage response;

(iv) $[\text{var}(L)]_t^{(1,U,N)}$ and $[\text{var}(L)]_t^{(2,U,N)}$ denote the contributions stemming from the first-order and, respectively, second-order sensitivities to the total variance, $[\text{var}(L)]_t^{(U,N)}$, of the leakage response;

(v) $[\mu_3(L)]_t^{(U,N)}$ denotes the third central moment of the leakage response for uncorrelated parameters;

(vi) $[\gamma_1(L)]_t^{(U,N)}$ denotes the skewness of the leakage response for uncorrelated parameters;

(vii) The subscript t denotes 'total cross sections', which are the model parameters considered for the results shown in Table 4.2.

As the expressions in Eqs. (4.11)–(4.13) indicate, only the unmixed second-order sensitivities contribute when the parameters are uncorrelated. On the other hand, all of the second-order sensitivities contribute when the parameters are correlated. The results presented in Table 4.2 show that the second-order sensitivities cause a larger deviation of the leakage response's expected value, $[E(L)]_t^{(U,N)}$, from its computed value, $L(\boldsymbol{\alpha}^0)$. For example, the results shown in the last column of Table 4.2 indicate that $[E(L)]_t^{(2,U,N)} \approx 260\% \times L(\boldsymbol{\alpha}^0) \approx 72\% \times [E(L)]_t^{(U,N)}$ for a 10% standard deviation in the model's cross sections. This result indicates that the term involving the second-order

Table 4.2 *Response moments for various relative standard deviations of total microscopic cross section*

Relative standard deviation	10%	5%	1%
$L(\boldsymbol{\alpha}^0)$	1.7648×10^6	1.7648×10^6	1.7648×10^6
$[E(L)]_t^{(2,FC)}$	2.9451×10^7	7.3627×10^6	2.9451×10^5
$[E(L)]_t^{(2,U)}$	4.598×10^4	1.149×10^6	4.598×10^6
$[E(L)]_t^{(FC)}$	3.1216×10^7	9.1275×10^6	2.0593×10^6
$[\text{var}(L)]_t^{(1,U,N)}$	3.419×10^{10}	8.549×10^{11}	3.419×10^{12}
$[\text{var}(L)]_t^{(1,FC,N)}$	4.7601×10^{13}	1.1900×10^{13}	4.7601×10^{11}
$[\text{var}(L)]_t^{(2,FC,N)}$	1.7347×10^{15}	1.0842×10^{14}	1.7347×10^{11}
$[\text{var}(L)]_t^{(FC,N)}$	1.7823×10^{15}	1.2302×10^{14}	6.4948×10^{11}
$[\mu_3(L)]_t^{(FC,N)}$	8.4113×10^{21}	5.2571×10^{20}	8.4113×10^{17}
$[\text{var}(L)]_t^{(2,U,N)}$	2.879×10^9	1.799×10^{12}	2.879×10^{13}
$[\gamma_1(L)]_t^{(U,N)}$	0.554	0.109	0.030
$[\gamma_1(L)]_t^{(FC,N)}$	0.1118	0.3983	1.6070

sensitivities is about 2.6 times larger than the computed leakage value $L(\boldsymbol{\alpha}^0)$, contributing 72% of the expected value $[E(L)]_t^{(U,N)}$ of the leakage response. As the results presented in Table 4.2 also indicate, the mixed second-order sensitivities play a very significant role in determining the moments of the leakage response distribution for correlated cross sections. The importance of the mixed second-order sensitivities increases as the relative standard deviations for the cross sections increase. For example, for fully correlated cross sections, neglecting the second-order sensitivities would cause an error as large as 2,000% in the expected value of the leakage response, and up to 6,000% in the variance of the leakage response. Furthermore, the effects of the mixed second-order sensitivities underscore the need for obtaining reliable data for the correlations that might exist among the total cross sections; such data is unavailable at this time.

Finally, the results shown in the last two rows of Table 4.2 highlight the impact of the second-order sensitivities on the response's skewness. If only first-order sensitivities are considered, the third-order moment of the response is always zero. Hence, a 'first-order sensitivity and uncertainty quantification' will always produce an erroneous third moment (and hence skewness) of the predicted response distribution, unless the unknown response distribution happens to be symmetrical. At least second-order sensitivities must be used in order to estimate the third-order moment (and hence the skewness) of the response distribution. With shifted expectation and pronounced skewness, standard statistical inference procedures such as constructing a confidence interval for the mean (expectation) of a computed/predicted model response will be not only incorrect, in the sense that the true coverage level will differ from the nominal (e.g. 95%) level, but the error probabilities will be unequal on each side of the predicted mean.

The results obtained by Cacuci, Fang, and Favorite (2020) indicate that the largest of the PERP benchmark's response sensitivities are with respect to the benchmark's 180 total cross sections. This finding has motivated the investigation of the third-order sensitivities of the PERP benchmark leakage response with respect to the 180 total cross sections, which were computed by Fang and Cacuci (2020) by applying the methodology summarised in Section 4.4.1. The results obtained by Fang and Cacuci (2020) for the 5,832,000 third-order sensitivities of the PERP response with respect to the PERP's 180 total cross sections are presented in Table 4.3, where $[\mathrm{var}(L)]_t^{(3,U,N)}$ denotes the contributions stemming from the third-order sensitivities to the total variance, denoted as $[\mathrm{var}(L)]_t^{(U,N)}$. The single largest third-order sensitivity is $S^{(3)}\left(\sigma_{t,1}^{g=30}, \sigma_{t,6}^{g'=30}, \sigma_{t,6}^{g''=30}\right) = -1.88 \times 10^5$, involving the microscopic total cross section for the 30th energy group of isotopes ^{239}Pu and ^{1}H (i.e. $\sigma_{t,1}^{g=30}$ and $\sigma_{t,6}^{30}$). Overall, the isotopes ^{1}H and ^{239}Pu are the two most important isotopes

Table 4.3 *Contribution of third-order sensitivities to the response variance for various relative standard deviations of the total microscopic cross section (all numbers denote neutrons/second).*

Relative standard deviation	1%	5%	10%
$[\mathrm{var}(L)]_t^{(3,U,N)}$	1.308×10^{10}	8.173×10^{12}	1.308×10^{14}
$[\mathrm{var}(L)]_t^{(U,N)}$ $= [\mathrm{var}(L)]_t^{(1,U,N)}$ $+[\mathrm{var}(L)]_t^{(2,U,N)}$ $+[\mathrm{var}(L)]_t^{(3,U,N)}$	5.015×10^{10}	1.083×10^{13}	1.630×10^{14}

affecting the PERP benchmark's leakage response, since their total microscopic cross sections are involved in all of the largest first-, second-, and third-order sensitivities.

Comparing the results shown in Table 4.3 to those shown in Table 4.2 indicates that the importance of the third-order sensitivities increases (compared to the importance of the first- and second-order sensitivities) as the parameters uncertainties increase. The contributions of the third-order sensitivities to the response's variance surpass the contributions of the first- and second-order sensitivities to the response's variance already for relatively small (ca. 5%) parameter standard deviations. These effects are rapidly amplified when the parameters uncertainties increase. In particular, for a uniform standard deviation of 10%, the third-order sensitivities contribute, through $[\mathrm{var}(L)]_t^{(3,U,N)}$, 80% of the response's total variance $[\mathrm{var}(L)]_t^{(U)}$, whereas the contribution $[\mathrm{var}(L)]_t^{(1,U,N)}$ stemming from the first-order sensitivities is only around 2%, while the contribution $[\mathrm{var}(L)]_t^{(2,U,N)}$ stemming from the second-order sensitivities is around 18%. Thus, neglecting the third-order contributions would cause a very large non-conservative error by under-reporting the response variance by a factor of 506%. Also, for a uniform standard deviation of 10%, the skewness $[\gamma_1(L)]_t^{(U,N)} = 0.030$ when the contributions from third-order sensitivities are considered, is much smaller than the corresponding skewness $[\gamma_1(L)]_t^{(U,N)} = 0.341$ when the contributions solely from the first- and second-order sensitivities are considered, which indicates that the effects of the third-order sensitivities are to reduce the skewness, thus reducing the asymmetry of the distribution of the leakage response around its expected value.

The computations of the response sensitivities were performed on a DELL desktop computer (AMD FX-8350) with an 8-core processor. Solving either the forward or the adjoint system using an angular expansion of order

ISN=256 (S_{256}) requires a CPU-time of 85 seconds. Furthermore, the typical CPU-time needed to perform the integration (quadrature) over the forward and/or adjoint functions, to compute one sensitivity (using S_{256}) is ca. 0.0012 seconds. Thus, the computation of the 21,976 first-order sensitivities using the adjoint methodology (1st-CASAM) required ca. 95 seconds CPU time. A two-point finite-difference scheme would have required ca. 4,000 hours CPU-time to compute *approximately* the first-order sensitivities. Using the 1st-CASAM, it was found that only 7,477 first-order response sensitivities were non-zero. Subsequently, the second-order sensitivities were computed only for the respective 7,477 parameters. Hence, the third-order sensitivities were computed only for the parameters that produced the largest second-order sensitivities (which turned out to be the total microscopic isotopic cross sections). Evidently, statistical methods would require orders-of-magnitude more CPU-time than finite-differences to obtain any resemblance of a 'response-surface' for the 21,976 *correlated* parameters and are therefore incapable of providing *reliable* results (e.g. including confidence intervals), for large-scale problems such as exemplified by the OECD/NEA reactor physics benchmark discussed in the foregoing.

4.5 Conclusions and Outlook

This chapter has presented the mathematical framework and building blocks of the 3rd-BERRU-PM methodology, which include, as prior information, measured expectations and covariances of responses along with the third-order computed moments of the response distribution in the parameter space. Following assimilation of experimental data, the 3rd-BERRU-PM methodology provides simultaneous model calibration and posterior prediction of best-estimate model responses and parameters, together with reduced best-estimate variances/covariances for the predicted responses and parameters. The 3rd-BERRU-PM methodology also provides a data consistency indicator which can be used to quantify contours of constant uncertainty in the combined phase-space of parameter and responses, thereby quantifying the validation domain underlying the model under investigation. The 3rd-BERRU-PM methodology can be used both for forward predictions of best-estimate results for calibrated model responses and parameters as well as for accurate and uniquely determined inverse predictions in the presence of uncertainties.

The model response sensitivities with respect to the model parameters are computed exactly and efficiently by applying the adjoint sensitivity methodology originally conceived by Cacuci (1981a,b) and subsequently extended comprehensively to higher-order by Cacuci (2015, 2018, 2019b, 2021a, 2021b, 2022a, 2022b, 2022c). The significant impacts of the higher-order sensitivities on the expected values and variances/covariances for the calculated and predicted model responses have also been highlighted. In particular, if only first-order sensitivities are considered, the third-order moment of the computed response vanishes. Hence, a 'first-order sensitivity and uncertainty quantification' will always produce an erroneous third moment (and hence skewness) of the predicted response distribution, unless the unknown response distribution happens to be symmetrical. At least second-order sensitivities must be used in order to estimate the third-order moment (and hence the skewness) of the response distribution. Skewness indicates the direction and relative magnitude of a distribution's deviation from the normal distribution. Since the second-order sensitivities impact decisively the expected values and the skewness of the calculated/predicted responses, they will also impact the computation of confidence intervals and tolerance limits for the predicted expectation of these responses. With pronounced skewness, standard statistical inference procedures such as constructing a confidence interval for the expectation of a computed/predicted model response will be not only incorrect, in the sense that the true coverage level will differ from the nominal (e.g. 95%) level, but the error probabilities will be unequal on each side of the predicted mean.

The paradigm illustrative applications presented in this chapter were deliberately chosen from other fields, for two main reasons: (i) these pioneering applications have direct correspondents in the Earth sciences, so they illustrate not only the steps involved in applying the high-order (3rd-CASAM-N) adjoint sensitivity analysis and the 3rd-BERRU-PM methodologies to the Earth sciences, but also indicate the breakthrough, fundamentally new, kind of results which these methodologies can produce; and (ii) the sensitivities obtained by the application of the 3rd-CASAM-N to the oscillatory dynamical system, on the one hand, and the large-scale OECD/NEA benchmark comprising 21,976 uncertain model parameters, on the other hand, cannot be obtained by using statistical methods.

Evidently, the number of sensitivities and, hence, the number of computations needed to determine them by using conventional (e.g. finite-differences, statistical) methods, increases exponentially with their order (the 'curse of dimensionality'). For small models involving few (less than a handful) parameters and responses that are trivial to obtain computationally, statistical methods – such as screening methods, sampling-based methods (including Monte Carlo), and variance-based methods (including Sobol's) – could be applied to obtain approximate quick-and-dirty values for the first-order sensitivities. These methods could also be used to do obtain similarly approximate values for variances of responses. None of these methods have been used to obtain higher-order sensitivities. A comprehensive review of all of these statistical methods has been provided by Saltelli, Chan, and Scott (2000). Of course, the user cannot *a priori* control the accuracy of values obtained by these statistical methods. The only way to verify the accuracy obtained by statistical methods is to compare them to the results produced by the adjoint methods of various orders (e.g. using the 5th-CASAM-N) (Cacuci, 2022b). Even finite-difference methods, involving recomputations using altered parameter values, yield erroneous results for second- and higher-order sensitivities, as has been discussed in this chapter. Furthermore, all of these statistical methods are inapplicable to oscillatory dynamical models, such as the one discussed in Section 4.3, and are also inapplicable to large systems with *many* (21,976) *correlated* parameters, as illustrated by the paradigm benchmark discussed in Section 4.4. Finally, statistical methods (screening, sampling-based, and/or variance-based methods) cannot be applied, by definition, to data assimilation and/or inverse problems, since these methods lack *ab initio* the mathematical capability of adding/assimilating data/information from sources (e.g. computational and experimental) outside the model being analysed; statistical methods provide no intrinsic mechanism to adjust/calibrate the model's parameters, etc., such as prided (e.g. by the 3rd-BERRU-PM) methodology. Performing the tasks within the scope of data assimilation and/or 'inverse problems' can only be achieved by using tools/methodologies beyond the possibilities of the aforementioned statistical methods.

The evident conclusion that can be drawn from this work is that the consideration of only the first-order sensitivities is insufficient for making credible predictions regarding the expected values and uncertainties (variances, covariances, skewness) of calculated and predicted/adjusted responses. At the very least, the second-order sensitivities must also be computed in order to enable the quantitative assessment of their impact on the predicted quantities. The high-order comprehensive adjoint sensitivity analysis methodology developed by Cacuci (2022) provides the tools for the exact and efficient computation of higher-order sensitivities while overcoming the curse of dimensionality that has hindered their computation thus far. Recently, Cacuci (2022a,b) has generalised the 3rd-CASAM-N to fifth-order (5th-CASAM-N), thereby enabling the efficient and exact computation of sensitivities up to fifth-order, thus overcoming the curse of dimensionality in the field of sensitivity analysis, which has significant impacts on all of the fields that use sensitivities (including optimisation, data assimilation, uncertainty quantification, model calibration and validation, etc.).

The question of when to stop computing progressively higher-order sensitivities has been addressed by Cacuci (2022c) in conjunction with the question of convergence of the Taylor-series expansion of the response in terms of the uncertain model parameters, since this Taylor-series expansion is the fundamental premise for the expressions provided by the 'propagation of errors' methodology for the cumulants of the model response distribution in the phase-space of model parameters. The convergence of this Taylor-series, which depends on both the response sensitivities to

parameters and the uncertainties associated with the parameter distribution, must be ensured. This can be done by ensuring that the combination of parameter uncertainties and response sensitivities are sufficiently small to fall inside the radius of convergence of the Taylor-series expansion. If the Taylor-series fails to converge, targeted experiments must be performed in order to reduce the largest sensitivities as well as the largest uncertainties (particularly standard deviations) that affect the most important parameters, by applying the principles of the 3rd-BERRU-PM to obtain best-estimate parameter values with reduced uncertainties (model calibration).

References

Bellman, R. E. (1957). *Dynamic Programming*. Princeton, NJ: Rand Corporation and Princeton University Press.

Cacuci, D. G. (1981a). Sensitivity theory for nonlinear systems: I. Nonlinear functional analysis approach, *Journal of Mathematical Physics*, 22, 2794–802.

Cacuci, D. G. (1981b). Sensitivity theory for nonlinear systems: II. Extensions to additional classes of responses, *Journal of Mathematical. Physics*, 22, 2803–12.

Cacuci, D. G. (2015). Second-order adjoint sensitivity analysis methodology (2nd-ASAM) for computing exactly and efficiently first- and second-order sensitivities in large-scale linear systems: I. Computational methodology. *Journal of Computational Physics.*, 284, 687–99. https://doi.org/10.1016/j.jcp.2014.12.042.

Cacuci, D. G. (2017). Inverse predictive modeling of radiation transport through optically thick media in the presence of counting uncertainties, *Nuclear Science and Engineering*, 186, 199–223. http://dx.doi.org/10.1080/00295639.2017.1305244.

Cacuci, D. G. (2018). *The Second-Order Adjoint Sensitivity Analysis Methodology*. Boca Raton, FL: CRC Press, Taylor & Francis Group.

Cacuci, D. G. (2019a). *BERRU Predictive Modeling: Best Estimate Results with Reduced Uncertainties*. New York: Springer Heidelberg.

Cacuci, D. G. (2019b). Towards overcoming the curse of dimensionality: The third-order adjoint method for sensitivity analysis of response-coupled linear forward/adjoint systems, with applications to uncertainty quantification and predictive modeling. *Energies*, 12, 4216. https://doi.org/10.3390/en12214216.

Cacuci, D. G. (2021a). First-order comprehensive adjoint sensitivity analysis methodology (1st-CASAM) for computing efficiently the exact response sensitivities for physical systems with imprecisely known boundaries and parameters: General theory and illustrative paradigm applications. *Annals of Nuclear Energy*, 151, 107913.

Cacuci, D. G. (2021b). On the need to determine accurately the impact of higher-order sensitivities on model sensitivity analysis, uncertainty quantification and best-estimate predictions, *Energies*, 14, 6318. https://doi.org/10.3390/en14196318.

Cacuci, D. G. (2022a). The fourth-order comprehensive adjoint sensitivity analysis methodology for nonlinear systems (4th-

CASAM-N): I. Mathematical framework. *Journal of Nuclear Engineering*, 3(1), 37–71. https://doi.org/10.3390/jne3010004jne-1568378.

Cacuci, D. G. (2022b). The fifth-order comprehensive adjoint sensitivity analysis methodology for nonlinear systems (5th-CASAM-N): I. Mathematical framework. *American Journal of Computational Mathematics*, 12, 44–78. https://doi.org/10.4236/ajcm.2022.121005.

Cacuci, D. G. (2022c). *The nth-Order Comprehensive Adjoint Sensitivity Analysis Methodology: Overcoming the Curse of Dimensionality. Vol. I: Linear Systems; Vol. II: Large-Scale Application; Vol. 3: Nonlinear Systems*. Cham: Springer Nature Switzerland.

Cacuci, D. G. and Di Rocco, F. (2020). Sensitivity and uncertainty analysis of a reduced-order model of nonlinear BWR dynamics sensitivity and uncertainty analysis of boiling water reactors nonlinear dynamics: II. Adjoint sensitivity analysis. *Annals of Nuclear Energy*, 148, 107748. https://doi.org/10.1016/j.anucene.2020.107748.

Cacuci, D. G. and Ionescu-Bujor, M. (2010), Sensitivity and uncertainty analysis, data assimilation and predictive best-estimate model calibration. In D. G. Cacuci, ed., *Handbook of Nuclear Engineering*, vol. 3. New York: Springer, pp. 1913–2051.

Cacuci, D. G., Fang, R., and Badea, M. C. (2018). MULTI-PRED: A software module for predictive modeling of coupled multi-physics systems, *Nuclear Science and Engineering*, 191, 187–202.

Cacuci, D. G., Fang, R., and Favorite, J. A. (2019). Comprehensive second-order adjoint sensitivity analysis methodology (2nd-CASAM) applied to a subcritical experimental reactor physics benchmark: I. Effects of imprecisely known microscopic total and capture cross sections. *Energies*, 12, 4219. https://doi.org/10.3390/en12214219.

Cacuci, D. G., Fang, R., and Favorite, J. A. (2020). Comprehensive second-order adjoint sensitivity analysis methodology (2nd-CASAM) Applied to a Subcritical Experimental Reactor Physics Benchmark: VI. Overall impact of 1st- and 2nd-order sensitivities on response uncertainties. *Energies*, 13, 1674. https://doi.org/10.3390/en13071674.

Cacuci, D. G., Navon, M. I., and Ionescu-Bujor, M. (2014). *Computational Methods for Data Evaluation and Assimilation*. Boca Raton, FL: Chapman & Hall/CRC.

Di Rocco, F. and Cacuci, D. G. (2020). Sensitivity and uncertainty analysis of a reduced-order model of nonlinear BWR dynamics. III. Uncertainty analysis results. *Annals of Nuclear Energy*, 148, 107749. https://doi.org/10.1016/j.anucene.2020.107749.

Fang, R. and Cacuci, D. G. (2020). Third order adjoint sensitivity and uncertainty Analysis of an OECD/NEA reactor physics benchmark: III. Response moments. *American Journal of Computational Mathematics*, 10, 559–70. https://doi.org/10.4236/ajcm.2020.104031.

Jaynes, E. T. (1957). Information theory and statistical mechanics. *Physics Review*, 106, 620–30. See also: Jaynes, E. T. *Probability Theory: The Logic of Science*, Cambridge: Cambridge University Press, 2003.

Lewis, J. M., Lakshmivarahan, S., and Dhal, S. K., *Dynamic Data Assimilation: A Least Squares Approach*. Cambridge: Cambridge University Press, 2006.

Lorenz, E. N. (1963). Deterministic nonperiodic flow. *Journal of Atmospheric Sciences*, 20 (2): 130–41.

March-Leuba, J., Cacuci, D. G., and Perez, R. B. (1984). Universality and Aperiodic Behavior of Nuclear Reactors. *Nuclear Science and Engineering*, 86, 401–4.

Práger T., and Kelemen, F. D. (2014). Adjoint methods and their application in earth sciences. In I., Faragó, Á. Havasi, and Z. Zlatev (eds.), *Advanced Numerical Methods for Complex Environmental Models: Needs and Availability*. Oak Park, IL: Bentham Science Publishers, pp. 203–75.

Saltelli, A., Chan. K., and Scott, E. M., eds. (2000). *Sensitivity Analysis*. Chichester: John Wiley & Sons.

Tukey, J. W. (1957). The propagation of errors. fluctuations and tolerances. Unpublished Technical Reports No. 10. 11. and 12. Princeton University.

USNRC (1988). United States Nuclear Regulatory Commission, Office of Nuclear Reactor Regulation, Washington, D.C. 20555, Information Notice No. 88-39: Lasalle Unit 2 Loss of Recirculation Pumps with Power Oscillation Event. www.nrc.gov/reading-rm/doc-collections/gen-comm/info-notices/1988/in88039.html.

Valentine, T. E. (2006). Polyethylene-reflected plutonium metal sphere subcritical noise measurements, SUB-PU-METMIXED-001. International Handbook of Evaluated Criticality Safety Benchmark Experiments, NEA/NSC/DOC(95)03/I-IX, Organization for Economic Co-Operation and Development, Nuclear Energy Agency, Paris, France. www.odeint.com, 2019.

.

PART II

**'Fluid' Earth Applications:
From the Surface to the Space**

5

Data Assimilation of Seasonal Snow

Manuela Girotto, Keith N. Musselman, and Richard L. H. Essery

Abstract: There is a fundamental need to understand and improve the errors and uncertainties associated with estimates of seasonal snow analysis and prediction. Over the past few decades, snow cover remote sensing techniques have increased in accuracy, but the retrieval of spatially and temporally continuous estimates of snow depth or snow water equivalent remains challenging tasks. Model-based snow estimates often bear significant uncertainties due to model structure and error-prone forcing data and parameter estimates. A potential method to overcome model and observational shortcomings is data assimilation. Data assimilation leverages the information content in both observations and models while minimising inherent limitations that result from uncertainty. This chapter reviews current snow models, snow remote sensing methods, and data assimilation techniques that can reduce uncertainties in the characterisation of seasonal snow.

5.1 Introduction and Motivations

Seasonal snow plays a major role in the water and energy budgets of many regions of the world. Seasonal snowmelt supplies provide freshwater to about 15% of the global population (Barrett, 2003; Viviroli et al., 2007). Additionally, snow-cover strongly influences weather and climate because of the highly reflective, emissive, and insulative properties of snow compared to other land surfaces. Seasonal snow also presents hazards such as flood and avalanche risks, disruption to transportation, and impacts on livestock, wildlife, and infrastructure (Nadim et al., 2006; Tachiiri et al., 2008; Berghuijs et al., 2016; Descamps et al., 2017; Croce et al., 2018; Musselman et al., 2018). Global warming is projected to critically impact snow water resources this century. Already, declines in snow-covered area and shifts to earlier snowmelt have been observed across the Northern Hemisphere in satellite and station records (Foster et al., 1999; Hammond et al., 2018; Musselman et al., 2021).

Accurate estimation of snow at fine spatial and temporal scales has historically been challenged by the complexity of land cover and terrain and the large global extent of snow-covered regions. Estimates of seasonal snow can be derived by both snow models and remote sensing observations (see reviews by Girotto et al., 2020; Largeron et al., 2020). Modelling of seasonal snow relies on accurate representation of snow physics and meteorological forcing data such as precipitation, temperature, humidity, radiation, and windspeed (Musselman et al., 2015; Raleigh et al., 2016). As in many Earth science disciplines, snow science is in an era of rapid advances as remote sensing products and models continue to gain granularity and physical fidelity (Lundquist et al., 2019). Satellite observations continue to revolutionise the way we monitor snow as new generations of sensors and platforms provide more extensive and global coverage of mountainous regions where seasonal snow accumulates (Schmugge et al., 2002; Frei et al., 2012). Despite the rapid growth in satellite technologies, a satellite mission dedicated to observing seasonal snow does not exist. Recently, international efforts such as NASA's SnowEx (Kim et al., 2017) and the Nordic Snow Radar Experiment (Lemmetyinen et al., 2018) have aimed to identify optimum multi-sensor synergies to map critical snowpack properties in future snow-focused satellite missions. Despite clear progress, the snow science community continues to face challenges related to the estimation of seasonal snow. Namely, advances in snow modelling remain limited by uncertainties in parameterisation schemes and input forcings, while remote sensing techniques have inherent limitations due to temporal, spatial, and technical constraints on the variables that can be observed.

A potential method to overcome such model and observational shortcomings is data assimilation (e.g. Houser et al., 1998; Andreadis and Lettenmaier, 2006; Girotto et al., 2014a; Girotto et al., 2014b) with the promise that, by combining information from remote sensing technologies with model estimates, data assimilation methods can produce optimal snow maps of snow properties including mass, also known as *snow water equivalent* (SWE) with sufficient global coverage and near real-time estimates. Data assimilation offers a way to integrate measurements from multiple sensors to improve snow model estimates (Girotto et al., 2020). Data assimilation has been used to improve modelled estimates of snow states, snow physics, model parameters, and sources of uncertainty (Helmert et al., 2018). There exists a wide variety of data assimilation techniques spanning degrees of complexity. Techniques vary from the simple direct insertion of observations into the model (e.g. Rodell and Houser, 2004; Li et al., 2019), to more mathematical Bayesian methods such as ensemble Kalman filter and particle filter (Evensen and Van Leeuwen, 2000) approaches. The latter are designed to account for the uncertainties of the model and observations using error statistics and an ensemble of possible model realisations (Section 5.4). This chapter aims to provide a summary of the techniques to assimilate seasonal snow in models, along with the key modelling frameworks and observation types that have been used to date. Detailed reviews can be found in Girotto et al. (2020) and Largeron et al. (2020).

5.2 Seasonal Snow Models

Much of Earth's seasonal snow is located in mountainous and other cold regions where the complex relationship among the effects of wind (e.g. Schmidt, 1982), topography, and/or vegetation (e.g. Golding and Swanson, 1978) continues to challenge the snow modelling scientific efforts (Jost et al., 2007). Despite these challenges, the need for accurate predictions of snow water resources has prompted the development of numerical snow models that are suitable for a range of applications including hydrological forecasting (e.g. Anderson and Burt, 1985), weather prediction (e.g. Niu et al., 2011), avalanche forecasting (e.g. Lehning et al., 1999), climate modelling (e.g. Bonan, 1998), and retrieval of snow characteristics by remote sensing (e.g. Wiesmann and Mätzler, 1999). The complexity of these models differs in their degree of process representation, and it ranges from the simplest temperature index models to multilayer energy balance models. Both model categories have been used for data assimilation (Section 5.5) applications.

Temperature index models use empirical relationships between local air temperature, precipitation, and snowmelt to estimate the evolution of seasonal snow (Ohmura, 2001). Because of their empirical nature, these models require careful calibration for the specific regions to which they are applied. Such heavily parameterised models have historically been favoured for the purposes of local to regional watershed runoff estimation, but have limited transferability to predict snow responses to climate change, or snowmelt beyond the spatial, temporal, and physiographic range of the conditions for which they are tuned (Kumar et al., 2013).

Energy balance snow models incorporate a detailed, sometimes multilayered, representation of all energy and mass fluxes into and out of a snowpack and are used to predict snowmelt and accumulation because of the computed net internal energy and mass balances. Besides snowmelt and accumulation, these models include representation of internal processes such as snow compaction, albedo

changes, temperature dynamics, and melt and refreeze processes. Modelling these variables is necessary for those data assimilation applications that leverage remotely sensed microstructure of snow. In fact, detailed knowledge of the internal snowpack structure is critical for radiative transfer applications in remote sensing (Wiesmann and Mätzler, 1999) and has utility in hydrological and climate change sensitivity applications (Bavay et al., 2009), presumably due to the correlation between snow material structure and surface–atmosphere interactions. Energy balance models differ in their representation of snowpack stratigraphy and vary from single layer (e.g. Schlosser et al., 1997; Essery et al., 1999), to three-layer (e.g. Shufen and Yongkang, 2001), to detailed multilayer (e.g. up to ~100 layers (Jordan, 1991; Brun et al., 1992; Lehning et al., 1999) snowpack representations.

Regardless of their degree of complexity, model estimates remain primarily hindered by two sources of uncertainty: (i) model inputs (e.g. meteorological conditions, (Raleigh et al. 2016) and (ii) fidelity of the equations used to represent the physical processes (structural uncertainty) (Slater et al., 2013; Lafaysse et al., 2017; Günther et al., 2019), and (iii) weaknesses in the user-specified parameter values (parameter uncertainty). Additional snow modelling sources of uncertainty can be introduced by the complex sub-grid landscape variability such as the slope, aspect, and vegetation that result in variable snow conditions, especially in mountainous terrain (e.g. Meromy et al., 2013; Todt et al., 2018). In the case of high uncertainty, simple snow models can be a viable alternative to physically based energy balance models because model complexity is not necessarily correlated with performance (Krinner et al., 2018). Physically based models could, on the other hand, offer more flexibility to benefit from the increasing availability and performance of satellite remote sensing techniques to validate prognostic model states that simpler models may not track (e.g. surface temperature, Hall et al., 2008).

5.3 Observations of Seasonal Snow

Direct ground measurements of snow provide the longest observational records, which now date back nearly a century or longer (see review by Kinar and Pomeroy, 2015). The most common observations are manual (i.e. snow courses, Church, 1933) and automated (i.e. snow pillows, Smith et al., 2017; Serreze et al., 1999) techniques. Most of the uncertainties associated with ground snow measurements are related to their spatial and temporal representativeness. Ground measurements are sparse, and some snow-covered areas are difficult to safely access. To overcome these limitations, recent decades have relied on remote sensing techniques as a more inclusive way, spatially and temporally, to observe seasonal snow. Existing efforts to estimate seasonal snowpack variability via remote sensing include microwave, visible and near infrared, radar, lidar, stereography, and gravimetric measurements. Snow can be distinguished from other land surfaces using airborne or satellite remote sensing techniques because of the nature of interactions among snow cover properties and electromagnetic radiation of different frequencies (Girotto et al., 2020). The spectral properties of snow depend upon several factors including grain size and shape, water content, impurity concentrations, temperature, and depth (e.g. Domine et al., 2006; Dietz et al., 2012; Skiles et al., 2018).

Snow remote sensing techniques have primarily focused on estimating two key variables of seasonal snow: (i) areal snow-cover extent and (ii) snow depth and SWE. The snow-cover extent is the surface area that is covered by snow, while depth and SWE provide estimates of snow volume and mass, respectively. Snow-cover extent is generally obtained reliably with high spatial and temporal resolution from visible and near infrared data (e.g. Hall et al., 2002; Painter et al., 2009; Riggs et al., 2017), but sensors capable of retrieving snow depth, such as the Advanced Topographic Laser Altimeter System (ATLAS) on ICESat-2 (Hagopian et al., 2016) are generally limited in spatial coverage. Comparatively, there is far less confidence in the measurement of SWE (Clifford, 2010; Kim et al., 2017).

Snow remote sensing techniques also fall into two general categories: active or passive sensing. Passive sensors are those that measure energy that is naturally available, such as the reflection of solar radiation or the emission of thermal infrared radiation. Active sensors use an onboard radiation (energy) source that is aimed at a target and the reflected wavelengths are detected and measured. Examples of active remote sensing technologies that have been used to estimate seasonal snow include active microwave (radar) and light detection and ranging (lidar) techniques. The most common passive remote sensing techniques for snow are visible (or optical), near infrared observations (e.g. Cline et al., 1998), and passive microwave detection (e.g. Foster et al., 1984; Li et al., 2012). Other snow remote sensing methods include thermal, gamma radiation (Cho et al., 2020), stereoscopic (Nolan et al., 2015; Deschamps-Berger et al., 2020), and gravimetry observations (Forman et al., 2012). For each of these techniques, the following sections summarise benefits and uncertainties that are leveraged by the data assimilation systems.

5.3.1 Passive Visible and Near Infrared Observations

Snow is highly reflective in the visible (optical) and less so in the near infrared (Vis/NIR) parts of the electromagnetic spectrum. Vis/NIR sensors are useful for providing observations of the surface area covered by snow and they have been used to detect snow cover since the mid-1960s. Vis/NIR observations can provide regional to global estimates of fractional snow-covered extent or area (Rosenthal and Dozier, 1996; Painter et al., 2009; Cortés et al., 2014) at spatial resolutions ranging from tens to hundreds of meters with varying temporal resolution (daily to every couple of weeks). These resolutions are generally considered acceptable for the mapping of snow patterns and changes, even in complex mountainous regions (Hammond et al., 2018). There are two main methods to derive snow cover from Vis/NIR. One method is based on normalised difference snow index (NDSI, Hall et al., 2002) that takes advantage of the striking difference of snow spectral reflectance in the Vis and NIR. Another method is based on the spectral unmixing (Painter et al., 2009) that uses linear optimisation methods to determine the fraction of the observed pixel that is covered by snow. These methods have been applied on satellite missions such as the advanced very high-resolution radiometer (e.g. Hüsler et al., 2014) the Landsat suites of satellite (e.g. Dozier, 1989), the moderate resolution imaging spectroradiometer (MODIS, Hall et al., 2002), and, more recently, the visible infrared imaging radiometer suite (VIIRS, Riggs et al., 2017) and Sentinel-2 (Gascoin et al., 2019).

The main source of error in snow cover estimation from Vis/NIR is the discrimination between clouds and snow due to their similar behaviour in the visible part of the spectrum (e.g. Miller et al., 2005). Uncertainties arise also because of the presence of factors influencing the spectral behaviour of snow and ice surfaces in the Vis/NIR spectrum. These are: snow grain size (Hall and Martinec, 1985; Rango, 1996), snow impurities (Aoki et al., 2007; Painter et al., 2012; Skiles and Painter, 2019), and snow temperature. Finally, snow-cover extent does not provide a direct estimate of SWE. Indirect methods that combine satellite-borne snow-cover information and physically based snow modelling, such as retrospective (or reconstruction) techniques (e.g. Molotch and Margulis, 2008; Rice et al., 2011; Girotto et al., 2014b; Raleigh et al., 2016; Jepsen et al., 2018), or empirical relationships, such as inversion of snow cover depletion curve (Toure et al., 2018) must be used to estimate SWE.

5.3.2 Passive Microwave Observations

An advantage of passive microwave sensors with respect to Vis/NIR sensors is that the microwave wavelengths are not influenced by cloud cover or sun angle, and, most importantly, can detect snow depth in addition to snow-cover extent (Pulliainen et al., 2020). Passive microwave sensors measure the brightness temperature, which is a signal of naturally emitted thermal radiance by the Earth surface. Passive microwave snow depth retrievals use a combination of microwave brightness temperature differences sensed at different frequencies, weighted by coefficients that are a function of the difference between vertical and horizontal polarisations. Examples of satellite-based passive microwave missions are the Scanning Multichannel

Microwave Radiometer (SMMR, e.g. Chang et al., 1987), the Special Sensor Microwave/Image (SSM/I, e.g. Tedesco et al., 2004) and the Advanced Microwave Scanning Radiometer (AMSR-E and AMSR2, e.g. Kelly, 2009).

The key limitation of passive microwave is its coarse spatial resolution, which is on the order of tens of kilometres (i.e. much coarser than Vis/NIR). This does not permit the representation of fine-scale surface processes, thus limiting the applicability of these observations for mountainous areas. Other uncertainties arise from the presence of liquid water in the snowpack (Kelly, 2009; Frei et al., 2012) and/or vegetation that alters the radiation emitted by the surface (Derksen, 2008). Finally, passive microwaves tend to saturate around 250 mm of SWE (Foster et al., 2005), and thus are of limited use to estimate deep snowpack typical of Earth's mountain water towers (Derksen, 2008; Viviroli et al., 2007).

5.3.3 Other Passive Remote Sensing Techniques

5.3.3.1 *Gamma Radiation*
Airborne gamma radiation measurements detect the natural terrestrial gamma radiation emitted from potassium, uranium, and thorium radioisotopes in the upper layer of soil. By measuring the difference in gamma radiation before and after the snow falls, these measurements can be used to estimate snowpack mass (Carroll, 1987; Carroll and Carroll, 1989). Aircraft-based gamma remote sensing efforts in the United States and Canada have produced operational high resolution SWE estimates for over 40 years (Cho et al., 2020). Among other limitations, soil moisture content and the mass of air between the sensor and the surface, both of which change at daily to seasonal rates, impact the accuracy of passive gamma-based SWE estimates. Similar to passive microwave remote sensing, the attenuation of the gamma signal increases with snow depth and mass reducing accuracy of the technique in regions where snow water resources are greatest.

5.3.3.2 *Gravimetry Observations*
Gravity data collected by the Gravity Recovery and Climate Experiment (GRACE) and GRACE Follow-On (GRACE-FO) satellites can be used to estimate changes in the mass of terrestrial water storage caused by snow and other hydrological factors such as soil moisture, groundwater, lakes, and rivers (Tapley et al., 2004). The spatial resolutions of such observations (~300 km at mid-latitudes) allow for continental and global studies but is a limitation to watershed-scale snow applications.

5.3.3.3 *Stereoscopic Observations*
Stereoscopic satellites (e.g. Pleiades 1a/1b and SPOT5/6/7) have provided a convenient method to estimate the spatial distribution of snow depth in high mountain catchments (Deschamps-Berger et al., 2020). A specific region is observed under different view angles at very high resolution (e.g. 50 cm to 2 m), which permits the creation of digital elevation models (DEMs) of the region. By differencing the snow-on and snow-off DEMs, snow depth maps are obtained. Vertical uncertainties are on the order of 50–80 cm at 3 m spatial resolution (Deschamps-Berger et al., 2020). Stereoscopic observations are available on demand at specific locations but provid estimates of snow depth and not SWE directly.

5.3.3.4 *Thermal Observations*
Recent work has demonstrated that MODIS can be used to estimate both snow and forest temperature in mixed pixels at 1 km resolution, with an accuracy of better than 1 K during the night (Lundquist et al., 2018). Thermal infrared satellites are useful to monitor snow surface temperature and are not restricted to daylight periods but cannot be used directly to estimate snow depth or mass.

5.3.4 Active Microwave Observations
Active microwave observations are not affected by cloud cover or sunlight conditions and have the advantage of permitting snow depth or SWE estimates from space at much higher resolution than passive methods. Active microwave sensors (i.e. radar) send their own energy and measure the total backscattered power from the snow-covered terrain. The total power received by the sensor can be expressed as the summation of backscatter from the air–snow boundary, the snow volume, and the snow–ground boundary, attenuated by a factor depending on the layered snowpack properties and incidence angle (Tedesco, 2014). Radar frequencies relevant for snow monitoring typically vary from 1 to 40 Ghz (<1 to 30 cm wavelengths, L to Ka bands (Largeron et al., 2020). While most active microwave studies have focused on the detection of snowmelt (Nagler et al., 2016), some early studies showed a very limited sensitivity of active microwave sensors to snow mass (Kendra et al., 1998; Strozzi and Matzler, 1998; Bernier et al., 1999; Shi and Dozier, 2000). A few recent studies have demonstrated the possibility of using active microwave data to estimate SWE (Lemmetyinen et al., 2018). Sentinel-1 or RADARSAT-2 are among the few Synthetic Aperture Radar (SAR) missions providing high-resolution backscatter measurements (at C-band; 5.4 GHz) with a revisit time of six days suitable for seasonal snow monitoring. Lievens et al. (2019) demonstrated the value of including cross-polarised backscatter measurements from C-band SAR to retrieve snow depth in mountainous areas at regional scales. Conde et al. (2019) used the SAR interferometry technique and Sentinel-1 C-band data to retrieve SWE estimates with sub-centimetre measurement accuracy and a 20 m spatial resolution.

The oblique viewing geometry of SAR systems enhances geometric distortions which make it particularly challenging to interpret in mountainous regions (Veyssière et al., 2019). Furthermore, retrieving unique snow mass solutions from active microwave is challenged by the dependencies on multiple snowpack characteristics (wetness, grain size, temperature, etc.) of backscattered signals. Despite these weaknesses, the high spatial and temporal resolution of these sensors makes their use potentially well-adapted for snow hydrological applications (Largeron et al., 2020). It is currently one of the most promising remote sensing techniques to monitor global snow-cover and mass at sufficient resolutions for a range of applications.

5.3.5 Active Optical Observations

Lidar technologies send a series of optical laser pulses and record the signals that reflect off the targeted surface and return to the platform. Lidar has provided high-resolution, high-accuracy surface elevation maps. Snow depth observations can be obtained from differencing two co-registered lidar images from snow surface ('snow-on') and bare-ground elevations before any snow accumulation ('snow-free' or 'snow-off'). Aircraft-based lidar systems provide high-resolution (~1 metre) accurate (~ few centimetre error estimates) snow depth estimates (Deems et al., 2013; Painter et al., 2016). Spatial coverage, typically at the watershed scale, requires substantial aircraft time that must be planned in advance. To estimate SWE from the resulting snow depth maps, models and/or complementary in situ observations must be used to estimate snow density (Smyth et al., 2019).

5.4 Data Assimilation

In general terms, data assimilation is a transdisciplinary tool that has been used in fields spanning Earth sciences and extending to medicine (Albers et al., 2017) and socio-economics (Houser, 2013). Figure 5.1 illustrates the concept. All estimates of a phenomenon or event (e.g. seasonal SWE) obtained either through modelling (Section 5.2) or observations (Section 5.3) have inherent uncertainties and limitations (Frei et al., 2012). As previously described, uncertainties in models are mainly associated with their parameterisation schemes or error-prone input forcings, while remote sensing techniques have inherent limitations due to temporal, spatial, and technical constraints on critical snow variables. To date, applications of snow data assimilation apply a wide range of techniques chosen on a case-by-case basis depending on the specific model, observation, level of complexity, etc. The simplest approach is the direct insertion method. The most common in snow research applications are the ensemble Kalman filter (EnKF, described in Section 5.4.2) and, more recently, the particle filter (PF, described in Section 5.4.3).

All assimilation methods require an 'observer' or 'observation operator' to translate variables from observation space to model space, and vice versa. The simplest observation operator is the identity matrix where the observed quantity is directly estimated by the model. That is the case when SWE observations are directly assimilated within a snow model where SWE is a prognostic variable. Sometimes, assimilating radiances can be more effective than SWE retrievals because difficulties arising from the

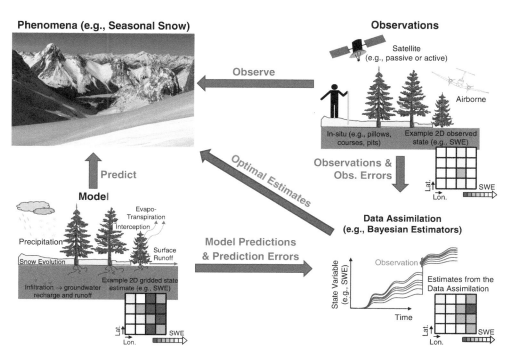

Figure 5.1 Estimates of an environmental variable (e.g. seasonal snow) can be obtained from model predictions or from observations (remote sensing or in situ). Neither are perfect and they contain errors and uncertainties. Data assimilation can be seen as a method that combines the strengths of modelled and observed estimates to obtain an optimised set of estimates for the environmental variable (with permission from Girotto et al., 2020).

non-unique and complex relationships linking the passive microwave signal to several snow properties are overcome (e.g. density, grain size/microstructure parameters, temperature, and wetness) (Helmert et al. 2018). In such cases, the observation operator can be a radiative transfer model (Picard et al., 2018) where the snowpack microstructure is converted into radiances. More recently, artificial intelligence methods (Forman and Reichle, 2014; Santi et al., 2022) have also been developed as an alternative to radiative transfer models.

5.4.1 Direct Insertion and Rule-Based Methods

Direct insertion or rule-based approaches are among the simplest data assimilation methods. These techniques inherently assume that observations are perfect and only models contain error, thus they directly replace the model variable values with observed ones. A rule-based method is described in Rodell and Houser (2004) and Toure et al. (2018). They use a rule-based approach that specifies whether to update the model with the measurements based on the difference between modelled and observed (from MODIS) snow cover extent. The rule-based approach improved both the snow cover extent and SWE estimates of their global land surface model simulations. Direct insertion methods are still widely used in snow applications. Some recent examples include the work by Hedrick et al. (2018) where lidar observations are directly inserted into a physically based model for estimating SWE in a domain in the Sierra Nevada, California. They conclude that the direct insertion of the lidar observations leads to improved SWE model estimates highlighting the potential benefits for managing water in the region. Li et al. (2019) directly insert a blended satellite- and model-based SWE product for initialising snow-dominated streamflow forecast models. They demonstrate that the direct insertion of the blended SWE product improves the efficiency of the streamflow model predictions compared to traditional SWE initialisation approaches. Lv and Pomeroy (2020) assimilate observations of vegetation snow interception within a physically based snow model. They compared direct insertion with more statistical assimilation methods and concluded that even the simplest of the assimilation methods (direct insertion) improved the simulation accuracy of snow interception amount and timing. Viallon-Galinier et al. (2020) use direct insertion to enter manually observed stratigraphy into a multilayer snow model. They found that the reinitialisation with observed profiles reduced modelling errors, especially during winter when snow layer heterogeneity is most pronounced.

While these examples highlight important model improvements obtained with a direct insertion approach, this simple approach has degraded performance compared to more statistical assimilation schemes such as ensemble Kalman or particle filters (e.g. Arsenault et al., 2013; Lv and Pomeroy, 2020). This degradation is often caused by the implicit assumption that errors come solely from the model and that observation errors are either acceptable or acceptably mitigated with rule-based insertion decisions. Filtering approaches such as the EnKF (Sect. 5.4.2) or PF (Sect. 5.4.3) should be more suited for estimating both the states and uncertainty of the observation-model systems (Largeron et al., 2020).

Nonetheless, several operational systems still use very simple assimilation methods (Helmert et al., 2018). For example, the GlobSnow product (Luojus et al., 2021) provides global gridded information on snow extent and SWE across the Northern Hemisphere by incorporating in situ station snow depth observations, microwave emission modelling, and spaceborne passive microwave observations using an iterative least squares minimisation scheme. Another widely used product is SNODAS, developed by the National Oceanic and Atmospheric Administration (NOAA) (Barrett, 2003). SNODAS incorporates in-situ and airborne observations with model estimates to provide daily SWE at 1 km resolution across the continental United States (Carroll, 2001). Its assimilation procedure is a simple nudging technique that calculates differences between estimated and observed SWE values and then spatially interpolates these differences to the model grid. Furthermore, the Canadian Meteorological Center Daily Snow Depth Analysis product (Brown and Brasnett, 2010) uses a simple statistical interpolation method to blend observations with model estimates of snow (Brown et al., 2003).

5.4.2 Ensemble Kalman Filter and Smoother Methods

The data assimilation approaches most used by the snow science community are the ensemble methods (e.g. the ensemble Kalman filter, EnKF), in which error statistics are determined from an ensemble of possible model realisations and model responses are assumed to be relatively linear. This assumption can lead to physical inconsistencies when EnKF methods are applied to complex non-linear snow models (Winstral et al., 2019). Through EnKF techniques (and variations), several works have assimilated SWE observations from in situ (e.g. Slater and Clark, 2006; Liu et al., 2013; Stigter et al., 2017) or satellite remote sensing (e.g. (De Lannoy et al., 2010; Dziubanski and Franz, 2016; Wang et al., 2021), or microwave radiance observations (e.g. Durand and Margulis, 2007; Dechant and Moradkhani, 2011; Xue and Forman, 2017). This chapter only reports key points for a few of these works. Some of the early applications of snow EnKF include the work by Slater and Clark (2006). They used an ensemble square root Kalman filter (EnSRF) to assimilate in situ observations of SWE into a snow model and showed improved SWE estimates were most evident during the early accumulation season and late melt period. However, due to the large temporal correlation inherent in the SWE of a seasonal snowpack, improvements were marginal. In the same year,

Durand and Margulis (2006) developed a point-scale radiometric data assimilation experiment to assimilate synthetic brightness temperature (i.e. passive microwave) observations. Because they assimilated brightness temperatures rather than directly assimilating SWE, they coupled a snow model with a radiative transfer model to assimilate and perform a synthetic assimilation experiment. They demonstrated that the EnKF was able to recover the 'true' snowpack states. By assimilating passive microwave radiation emitted from the land surface, as opposed to direct SWE retrievals, one can make inferences about snowpack properties, including SWE, depth, grain size, density, and liquid water content (Dechant and Moradkhani, 2011). In fact, a similar approach was taken by Dechant and Moradkhani (2011) in which EnKF assimilated remotely sensed brightness temperatures to improve SWE prediction and operational streamflow forecasts. Improved streamflow forecasts are also reported in Huang et al. (2017) where an EnKF was used to assimilate SWE retrieval in the Pacific Northwest, the Rocky Mountains, and the Sierra Nevada.

Some studies have also used EnKF to assimilate snow cover extent observations from a wide range of Vis/NIR satellite missions such as Landsat and/or MODIS. The assimilation of Vis/NIR snow cover extent (as opposed to microwave observations) leads, in general, to higher spatial resolutions of the resulting snow estimates. In the work by Andreadis and Lettenmaier (2006), Clark et al. (2006), and Su et al. (2008), an EnKF framework was used to assimilate satellite observed snow cover extent. Both studies conclude that the EnKF accurately simulated the seasonal variability and ensemble spread of snow cover extent. However, Andreadis and Lettenmaier (2006) indicated that the EnKF performance is modest for estimating ephemeral SWE and limited for deeper snowpack. This is because the EnKF leverages the instantaneous correlation between modelled snow cover extent and SWE. This correlation tends to diminish for larger values of SWE, as in when changes in SWE do not correspond to changes in snow cover extent (i.e. snow cover extent saturates at 100%). Thus, snow cover extent assimilation techniques are more effective during the beginning of the accumulation season and during the ablation season, when a change of snow cover area extent corresponds to changes in the SWE. To solve this issue, Durand et al. (2008) propose using a smoother version of the EnKF, the ensemble Kalman smoother (EnKS). In the EnKS, all snow cover extent observations within an assimilation window are assimilated, thus multiple strengths of the observed snow cover extent signal are leveraged, not only the instantaneous acquisition. The approach is also used in (Durand et al., 2008; Girotto et al., 2014a; Oaida et al., 2019) where a retrospective use of Vis/NIR satellite observations can provide accurate estimates of SWE. As an example, Fig. 5.2 illustrates how the use of an ensemble data assimilation method improves upon the model-only (i.e. without assimilation) when compared

to in situ SWE observations. The principle behind these Vis/NIR retrospective assimilation techniques is the same as deterministic reconstruction ones (e.g. Molotch and Margulis, 2008; Rice et al., 2011; Jepsen et al., 2018) where the maximum (or peak) SWE can be retrieved from a retrospective accumulation of spring-summer potential melt energy fluxes coupled with the disappearance date of snow as ascertained from visible and near infrared images.

The EnKF approach has also been used in a few studies focused of multi-spectral, multi-resolution, and multi-sensor data assimilation approaches. Durand and Margulis (2006) implemented the EnKF in a multiscale, multi-frequency radiometric data assimilation experiment using synthetic passive microwave radiance along with Vis/NIR snow cover extent observations. They state that the combined assimilation of passive microwave and Vis/NIR observations resulted in overall improved snow predictive skill because of the positive synergy due to the complementary nature of the two observation types. Liu et al. (2013) assimilated MODIS snow cover extent and AMSR-E snow depth products into the Noah land surface model and conclude that the assimilation of snow data consistently improves snow and streamflow predictions. De Lannoy et al. (2012) assimilated AMSR-E SWE retrievals and MODIS snow cover extent observations. Their joint SWE and snow cover extent EnKF assimilation significantly improves root-mean-square error and correlation values.

To conclude, the scientific community agrees that EnKF assimilation of in situ or satellite-based SWE, microwave, or Vis/NIR snow estimates leads to overall improved estimates of seasonal snow and related variables (e.g. streamflow, snow cover) with respect to observations or model estimates taken alone. However, because the EnKF performs best when applied to linear models and Gaussian distributed error statistics, it exhibits limited accuracy in cases of highly non-linear snow systems and has increased computation cost when large ensembles of computationally expensive models are needed (Largeron et al., 2020).

5.4.3 Particle Filter

Another ensemble-based method used in the snow assimilation community, arguably more sophisticated, includes particle filter (PF) techniques (e.g. Arulampalam et al., 2002; Van Leeuwen, 2009; Smyth et al., 2019). Several snow data assimilation studies directly compare the performances of the PF to the EnKF (Leisenring and Moradkhani, 2011; Margulis et al., 2016) and suggest that the PF is superior to the EnKF-based methods for predicting model states and parameters. These studies used both point-scale, in situ observations, and spatially distributed remote sensing observations.

Point scale experiments are summarised by Magnusson et al. (2014) and Smyth et al. (2019). Magnusson et al. (2014) directly compared the PF performances against a direct

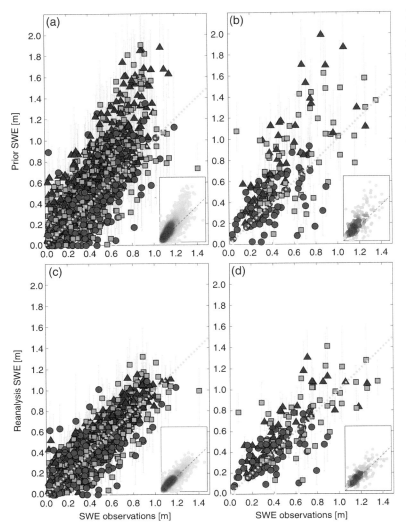

Figure 5.2 Observed SWE at the snow courses (left) and observed peak-SWE at the snow pillows (right) versus model-only (no assimilation) SWE (top) and reanalysis (i.e. with assimilation) SWE (bottom). Grey error bars represent the ensemble interquartile range of the reanalysis SWE. Brown (triangle), dark green (circle), and light green (square) colours (marker type) indicate low-, mid-, and high-elevation observation locations (observation elevation ranges over 2300–3500 m asl were organised in three bands of 400 m each). Inset figures show the densities for each scatter plot around the 1:1 line; high-to-low density is represented by dark-to-light blue colours (with permission from Girotto et al., 2014b).

insertion method for a multilayer energy-balance snow model. They found that both schemes reduced errors in SWE, snowpack runoff, and soil temperature. While the direct insertion is likely to produce inconsistencies between modelled variables, the particle filter avoids such limitations without loss of performance. Smyth et al. (2019) tested whether the assimilation of snow depth using a PF can lead to improved estimates of snow density, and consequently SWE. They conclude that the particle filter reduced density and SWE root-mean-square error relative to open loop simulations.

The use of a PF in a spatially distributed context was presented by Leisenring and Moradkhani (2011) where SWE was assimilated within a National Weather Service model. Similarly, Liu et al. (2021) and Margulis et al. (2015) derived an ensemble PF approach to estimate SWE from the assimilation of snow cover extent. Using the same approach, Margulis et al. (2019) assimilated infrequent (i.e. a couple of observations per year) lidar

snow depth observations within a land surface model. They found that data assimilation provides a useful framework for leveraging infrequent remotely sensed snow depth observations to derive continuous (spatially and temporally) accurate estimates of unobserved variables such as SWE and snowmelt, even at times when observations are unavailable. Another successful application is reported in Han et al. (2021) where the PF assimilates a high-spatial-resolution remotely sensed snow depth data set within a snow model. One key benefit of improved spatial distribution of snow cover and SWE estimates is the resulting improvement in runoff and streamflow predictions (Thirel et al., 2013; Li et al., 2019).

To conclude, the current literature on snow data assimilation suggests that PF applications resolve some of the limitations reported in the EnKF studies for two reasons: (i) the PF can use non-linear whereas the EnKF perform best for linear systems; and (ii) the PF does not involve assumptions regarding the distribution of the errors. However, the computational demand of the PF remains a key limitation in its application for snow studies as the PF typically requires larger ensembles to characterise the full probability distribution of state variables and consequently their uncertainties by resampling sets of state variables.

5.4.4 Spatial Correlations in Snow Data Assimilation Systems

Spatial distribution updates are essential in operational analyses of in situ snow depth measurements. Most snow data assimilation research efforts have been one-dimensional approaches, where one satellite observation type (i.e. SWE, snow depth, or snow cover extent) is used to update co-located modelled snow estimates. That is, snow updates can only be performed at the locations where an observation is available. One-dimensional techniques disregard spatial correlation across observations and model errors. In a few exceptions, De Lannoy et al. (2010) and Cantet et al. (2019) tested the effect of introducing spatial error correlation into snow data assimilation updates. De Lannoy et al. (2010) assimilated coarse-scale (25 km) SWE observations into fine-scale (1 km) land model simulations and tested the effect of different spatial aggregation and correlation methods. Their results indicate that assimilating disaggregated fine-scale observations independently is less efficient than assimilating a collection of neighbouring correlated coarser scale observations. Cantet et al. (2019) assimilated SWE data from a sparse network of in situ snow observation stations using a PF. Their PF formulation included error spatial correlations to allow for snow states to be updated at locations where observations were not directly available. These few studies indicate that underlying spatial error correlations should be exploited to improve spatial estimates of seasonal snow.

5.5 Summary and Conclusions

This chapter reviews current data assimilation techniques used to estimate seasonal snow and elucidates key benefits and remaining challenges associated with each system. To date, modelling efforts have provided the most spatially and temporally complete estimates of local, regional, and global snow properties; however, the accuracy of snow model estimates remains hindered by uncertain forcing and parameters, and error-prone model structures and process representations. Satellite and airborne remote sensing allow for extensive and global coverage of seasonal snow estimates even in remote, complex mountainous regions. While snow cover extent and related surface properties are generally obtained reliably with high spatial and temporal resolutions from visible and near infrared data, we critically lack similar robust estimates of snow mass relevant to water resource

applications, especially in mountainous regions (Clifford, 2010). Compared to Vis/NIR data, microwave measurements are more directly related to the mass of snow. While active microwave data have recently been found suitable for providing the temporal and spatial resolutions required to monitor seasonal snow for a range of applications, passive microwave techniques are not useful for estimating deep or wet snow at an acceptable spatial resolution capable of resolving global snow processes inclusive of mountain water towers. Airborne lidar systems are, to date, the most accurate methods to retrieve seasonal snow, but they only observe snow depth (not SWE) and are limited to targeted regions and for specific, infrequent times.

Data assimilation is a viable way to converge different temporal and spatial resolutions of in situ and remotely sensed observations and a useful technology to bridge gaps in scale and accuracy between these observations and models. In fact, the assimilation of satellite and airborne observations lead, in general, to overall improved estimates of seasonal snow and related variables. Some remaining challenges in the snow data assimilation field include research into the effects of underlying spatial error correlations in data assimilation to improve the spatial estimates of SWE, and possibly merging multiple observations to improve snow model accuracy. Finally, even if the research field in snow data assimilation has evolved significantly, operational and weather forecasting systems use methods (if any) that are much simpler than the state of the art. The inclusion of a broader range of observations is an active and emergent research field as multi-sensor, multi-resolution snow observations become available. These efforts promise to provide robust and diverse information to improve our ability to map, model, and project past, current, and future seasonal snow characteristics, and the effects of snow on the Earth system.

References

Albers, D. J., Levine, M., Gluckman, B. et al. (2017). Personalized glucose forecasting for type 2 diabetes using data assimilation. *PLoS Computational Biology*, 13(4), e1005232.

Anderson, M. G., and Burt, T. P., eds. (1985). *Hydrological forecasting*. Chichester: Wiley.

Andreadis, K. M., and Lettenmaier, D. P. (2006). Assimilating remotely sensed snow observations into a macroscale hydrology model. *Advances in Water Resources*, 29(6), 872–86.

Aoki, T., Motoyoshi, H., Kodama, Y., Yasunari, T. J., and Sugiura, K. (2007). Variations of the snow physical parameters and their effects on albedo in Sapporo, Japan. *Annals of Glaciology*, 46, 375–81.

Arsenault, K. R., Houser, P. R., De Lannoy, G. J., and Dirmeyer, P. A. (2013). Impacts of snow cover fraction data assimilation on modeled energy and moisture budgets. *Journal of Geophysical Research: Atmospheres*, 118(14), 7489–504.

Arulampalam, M. S., Maskell, S., Gordon, N., and Clapp, T. (2002). A tutorial on particle filters for online nonlinear/non-Gaussian Bayesian tracking. *IEEE Transactions on Signal Processing*, 50(2), 174–88.

Barrett, A. P. (2003). *National Operational Hydrologic Remote Sensing Center SNOw Data Assimilation System (SNODAS) Products at NSIDC*. NSIDC Special Report 11. Boulder, CO: National Snow and Ice Data Center. https://nsidc.org/sites/default/files/nsidc_special_report_11.pdf.

Bavay, M., Lehning, M., Jonas, T., and Löwe, H. (2009). Simulations of future snow cover and discharge in Alpine headwater catchments. *Hydrological Processes: An International Journal*, 23(1), 95–108.

Berghuijs, W. R., Woods, R. A., Hutton, C. J., and Sivapalan, M. (2016). Dominant flood generating mechanisms across the United States. *Geophysical Research Letters*, 43(9), 4382–90.

Bernier, M., Fortin, J.-P., Gauthier, Y. et al. (1999). Determination of snow water equivalent using RADARSAT SAR data in eastern Canada. *Hydrological Processes*, 13(18), 3041–51.

Bonan, G. B. (1998). The land surface climatology of the NCAR Land Surface Model coupled to the NCAR Community Climate Model. *Journal of Climate*, 11(6), 1307–26.

Brown, R. D., and Brasnett, B. (2010). Canadian Meteorological Centre (CMC) daily snow depth analysis data. Version 1 [Data Set]. Boulder, CO: National Snow and Ice Data Center. https://doi.org/10.5067/W9FOYWH0EQZ3.

Brown, R. D., Brasnett, B., and Robinson, D. (2003). Gridded North American monthly snow depth and snow water equivalent for GCM evaluation. *Atmosphere-Ocean*, 41(1), 1–14.

Brun, E., David, P., Sudul, M., and Brunot, G. (1992). A numerical model to simulate snow-cover stratigraphy for operational avalanche forecasting. *Journal of Glaciology*, 38(128), 13–22.

Cantet, P., Boucher, M. A., Lachance-Coutier, S., Turcotte, R., and Fortin, V. (2019). Using a particle filter to estimate the spatial distribution of the snowpack water equivalent. *Journal of Hydrometeorology*, 20(4), 577–94.

Carroll, S. S., and Carroll, T. R. (1989). Effect of uneven snow cover on airborne snow water equivalent estimates obtained by measuring terrestrial gamma radiation. *Water Resources Research*, 25(7), 1505–10.

Carroll, T. (1987). Operational airborne measurements of snow water equivalent and soil moisture using terrestrial gamma radiation in the United States. *IAHS-AISH Publication*, 166, 213–23.

Carroll, T. (2001). *Airborne gamma radiation snow survey program: A user's guide*, version 5.0. National Operational Hydrologic Remote Sensing Center (NOHRSC), Chanhassen, 14.

Chang, A. T., Foster, J. L., and Hall, D. K. (1987). Nimbus-7 SMMR derived global snow cover parameters. *Annals of Glaciology*, 9, 39–44.

Cho, E., Jacobs, J. M., and Vuyovich, C. M. (2020). The value of long-term (40 years) airborne gamma radiation SWE record for evaluating three observation-based gridded SWE data sets by seasonal snow and land cover classifications. *Water Resources Research*, 56(1). https://doi.org/10.1029/2019WR025813

Church, J. E. (1933). Snow surveying: Its principles and possibilities. *Geographical Review*, 23(4), 529–63.

Clark, M. P., Slater, A. G., Barrett, A. P. et al. (2006). Assimilation of snow covered area information into hydrologic and land-surface models. *Advances in Water Resources*, 29(8), 1209–21.

Clifford, D. (2010). Global estimates of snow water equivalent from passive microwave instruments: History, challenges and future developments. *International Journal of Remote Sensing*, 31(14), 3707–26.

Cline, D. W., Bales, R. C., and Dozier, J. (1998). Estimating the spatial distribution of snow in mountain basins using remote sensing and energy balance modeling. *Water Resources Research*, 34(5), 1275–85.

Conde, V., Nico, G., Mateus, P. et al. (2019). On the estimation of temporal changes of snow water equivalent by spaceborne SAR interferometry: A new application for the Sentinel-1 mission. *Journal of Hydrology and Hydromechanics*, 67(1), 93–100.

Cortés, G., Girotto, M., and Margulis, S. A. (2014). Analysis of sub-pixel snow and ice extent over the extratropical Andes using spectral unmixing of historical Landsat imagery. *Remote Sensing of Environment*, 141, 64–78.

Croce, P., Formichi, P., Landi, F. et al. (2018). The snow load in Europe and the climate change. *Climate Risk Management*, 20, 138–54.

De Lannoy, G. J., Reichle, R. H., Arsenault, K. R. et al. (2012). Multiscale assimilation of Advanced Microwave Scanning Radiometer–EOS snow water equivalent and Moderate Resolution Imaging Spectroradiometer snow cover fraction observations in northern Colorado. *Water Resources Research*, 48(1). https://doi.org/10.1029/2011WR010588.

De Lannoy, G. J., Reichle, R. H., Houser, P. R. et al. (2010). Satellite-scale snow water equivalent assimilation into a high-resolution land surface model. *Journal of Hydrometeorology*, 11(2), 352–69.

Dechant, C., and Moradkhani, H. (2011). Radiance data assimilation for operational snow and streamflow forecasting. *Advances in Water Resources*, 34(3), 351–64.

Deems, J. S., Painter, T. H., and Finnegan, D. C. (2013). Lidar measurement of snow depth: A review. *Journal of Glaciology*, 59(215), 467–79.

Derksen, C. (2008). The contribution of AMSR-E 18.7 and 10.7 GHz measurements to improved boreal forest snow water equivalent retrievals. *Remote Sensing of Environment*, 112(5), 2701–10.

Descamps, S., Aars, J., Fuglei, E. et al. (2017). Climate change impacts on wildlife in a High Arctic archipelago – Svalbard, Norway. *Global Change Biology*, 23(2), 490–502.

Deschamps-Berger, C., Gascoin, S., Berthier, E. et al. (2020). Snow depth mapping from stereo satellite imagery in mountainous terrain: Evaluation using airborne laser-scanning data. *The Cryosphere*, 14(9), 2925–40.

Dietz, A. J., Kuenzer, C., Gessner, U., and Dech, S. (2012). Remote sensing of snow: A review of available methods. *International Journal of Remote Sensing*, 33(13), 4094–134.

Domine, F., Salvatori, R., Legagneux, L. et al. (2006). Correlation between the specific surface area and the short wave infrared (SWIR) reflectance of snow. *Cold Regions Science and Technology*, 46(1), 60–8.

Dozier, J. (1989). Spectral signature of alpine snow cover from the Landsat Thematic Mapper. *Remote Sensing of Environment*, 28, 9–22.

Durand, M., and Margulis, S. A. (2006). Feasibility test of multi-frequency radiometric data assimilation to estimate snow water equivalent. *Journal of Hydrometeorology*, 7(3), 443–57.

Durand, M., and Margulis, S. A. (2007). Correcting first-order errors in snow water equivalent estimates using a multifrequency, multiscale radiometric data assimilation scheme. *Journal of Geophysical Research: Atmospheres*, 112 (D13). https://doi.org/10.1029/2006JD008067.

Durand, M., Molotch, N. P., and Margulis, S. A. (2008). A Bayesian approach to snow water equivalent reconstruction. *Journal of Geophysical Research: Atmospheres*, 113(D20). https://doi.org/10.1029/2008JD009894.

Dziubanski, D. J., and Franz, K. J. (2016). Assimilation of AMSR-E snow water equivalent data in a spatially-lumped snow model. *Journal of Hydrology*, 540, 26–39.

Essery, R., Martin, E., Douville, H., Fernandez, A., and Brun, E. (1999). A comparison of four snow models using observations from an alpine site. *Climate Dynamics*, 15(8), 583–93.

Evensen, G., and Van Leeuwen, P. J. (2000). An ensemble Kalman smoother for nonlinear dynamics. *Monthly Weather Review*, 128(6), 1852–67.

Forman, B. A., and Reichle, R. H. (2014). Using a support vector machine and a land surface model to estimate large-scale passive microwave brightness temperatures over snow-covered land in North America. *IEEE Journal of Selected Topics in Applied Earth Observations and Remote Sensing*, 8(9), 4431–41.

Forman, B. A., Reichle, R. H., and Rodell, M. (2012). Assimilation of terrestrial water storage from GRACE in a snow-dominated basin. *Water Resources Research*, 48(1).

Foster, J. L., Hall, D. K., Chang, A. T. C., and Rango, A. (1984). An overview of passive microwave snow research and results. *Reviews of Geophysics*, 22(2), 195–208.

Foster, J. L., Hall, D. K., Chang, A. T. et al. (1999). Effects of snow crystal shape on the scattering of passive microwave radiation. *IEEE Transactions on Geoscience and Remote Sensing*, 37(2), 1165–8.

Foster, J. L., Sun, C., Walker, J. P. et al. (2005). Quantifying the uncertainty in passive microwave snow water equivalent observations. *Remote Sensing of Environment*, 94(2), 187–203.

Frei, A., Tedesco, M., Lee, S. et al. (2012). A review of global satellite-derived snow products. *Advances in Space Research*, 50(8), 1007–29.

Gascoin, S., Grizonnet, M., Bouchet, M., Salgues, G., and Hagolle, O. (2019). Theia Snow collection: High-resolution operational snow cover maps from Sentinel-2 and Landsat-8 data. *Earth System Science Data*, 11(2), 493–514.

Girotto, M., Cortés, G., Margulis, S. A., and Durand, M. (2014a). Examining spatial and temporal variability in snow water equivalent using a 27 year reanalysis: Kern River watershed, Sierra Nevada. *Water Resources Research*, 50(8), 6713–34.

Girotto, M., Margulis, S. A., and Durand, M. (2014b). Probabilistic SWE reanalysis as a generalization of deterministic SWE reconstruction techniques. *Hydrological Processes*, 28(12), 3875–95.

Girotto, M., Musselman, K. N., and Essery, R. L. (2020). Data assimilation improves estimates of climate-sensitive seasonal snow. *Current Climate Change Reports*, 6, 81–94.

Golding, D. L., and Swanson, R. H. (1978). Snow accumulation and melt in small forest openings in Alberta. *Canadian Journal of Forest Research*, 8(4), 380–8.

Günther, D., Marke, T., Essery, R., and Strasser, U. (2019). Uncertainties in snowpack simulations: Assessing the impact of model structure, parameter choice, and forcing data error on point-scale energy balance snow model performance. *Water Resources Research*, 55(4), 2779–800.

Hagopian, J., Bolcar, M., Chambers, J. et al. (2016). Advanced topographic laser altimeter system (ATLAS) receiver telescope assembly (RTA) and transmitter alignment and test. Proc. SPIE 9972, *Earth Observing Systems*, XXI, 997207. https://doi.org/10.1117/12.2240241.

Hall, D. K., Box, J. E., Casey, K. A. et al. (2008). Comparison of satellite-derived and in-situ observations of ice and snow surface temperatures over Greenland. *Remote Sensing of Environment*, 112(10), 3739–49.

Hall, D. K., and Martinec, J. (1985). *Remote Sensing of Ice and Snow*. London: Chapman & Hall.

Hall, D. K., Riggs, G. A., Salomonson, V. V., DiGirolamo, N. E., and Bayr, K. J. (2002). MODIS snow-cover products. *Remote Sensing of Environment*, 83(1–2), 181–94.

Hammond, J. C., Saavedra, F. A., and Kampf, S. K. (2018). Global snow zone maps and trends in snow persistence 2001–2016. *International Journal of Climatology*, 38(12), 4369–83.

Han, P., Long, D., Li, X. et al. (2021). A dual state-parameter updating scheme using the particle filter and high-spatial-resolution remotely sensed snow depths to improve snow simulation. *Journal of Hydrology*, 594, 125979.

Hedrick, A. R., Marks, D., Havens, S. et al. (2018). Direct insertion of NASA Airborne Snow Observatory-derived snow depth time series into the iSnobal energy balance snow model. *Water Resources Research*, 54(10), 8045–63.

Helmert, J., Şensoy Şorman, A., Alvarado Montero, R. et al. (2018). Review of snow data assimilation methods for hydrological, land surface, meteorological and climate models: Results from a COST HarmoSnow survey. *Geosciences*, 8(12), 489.

Houser, P. R. (2013). Improved disaster management using data assimilation. In J. Tiefenbacher, ed., *Approaches to Disaster Management: Examining the Implications of Hazards, Emergencies and Disasters*. London: IntechOpen, 83–103. https://doi.org/10.5772/3355.

Houser, P. R., Shuttleworth, W. J., Famiglietti, J. S. et al. (1998). Integration of soil moisture remote sensing and hydrologic modeling using data assimilation. *Water Resources Research*, 34(12), 3405–20.

Huang, C., Newman, A. J., Clark, M. P., Wood, A. W., and Zheng, X. (2017). Evaluation of snow data assimilation using the ensemble Kalman filter for seasonal streamflow prediction in the western United States. *Hydrology and Earth System Sciences*, 21(1), 635–50.

Hüsler, F., Jonas, T., Riffler, M., Musial, J. P., and Wunderle, S. (2014). A satellite-based snow cover climatology (1985–2011) for the European Alps derived from AVHRR data. *The Cryosphere*, 8(1), 73–90.

Jepsen, S. M., Harmon, T. C., Ficklin, D. L., Molotch, N. P., and Guan, B. (2018). Evapotranspiration sensitivity to air temperature across a snow-influenced watershed: Space-for-time substitution versus integrated watershed modeling. *Journal of Hydrology*, 556, 645–59.

Jordan, R. E. (1991). *A One-Dimensional Temperature Model for a Snow Cover: Technical Documentation for SNTHERM. 89.* (No. CRREL-SR-91-16). Hanover, NH: Cold Regions Research and Engineering Lab,

Jost, G., Weiler, M., Gluns, D. R., and Alila, Y. (2007). The influence of forest and topography on snow accumulation and melt at the watershed-scale. *Journal of Hydrology*, 347(1–2), 101–15.

Kelly, R. (2009). The AMSR-E snow depth algorithm: Description and initial results. *Journal of the Remote Sensing Society of Japan*, 29(1), 307–17.

Kendra, J. R., Sarabandi, K., and Ulaby, F. T. (1998). Radar measurements of snow: Experiment and analysis. *IEEE Transactions on Geoscience and Remote Sensing*, 36(3), 864–79.

Kim, E., Gatebe, C., Hall, D., et al. (2017). NASA's SnowEx campaign: Observing seasonal snow in a forested environment. 2017 IEEE International Geoscience and Remote Sensing Symposium (IGARSS), Fort Worth, TX, USA, July 23–28, 2017, pp. 1388–90, https://doi.org/10.1109/IGARSS.2017 .8127222.

Kinar, N. J., and Pomeroy, J. W. (2015). Measurement of the physical properties of the snowpack. *Reviews of Geophysics*, 53(2), 481–544.

Krinner, G., Derksen, C., Essery, R. et al. (2018). ESM-SnowMIP: Assessing snow models and quantifying snow-related climate feedbacks. *Geoscientific Model Development*, 11(12), 5027–49.

Kumar, M., Marks, D., Dozier, J., Reba, M., and Winstral, A. (2013). Evaluation of distributed hydrologic impacts of temperature-index and energy-based snow models. *Advances in Water Resources*, 56, 77–89.

Lafaysse, M., Cluzet, B., Dumont, M. et al. (2017). A multiphysical ensemble system of numerical snow modelling. *The Cryosphere*, 11(3), 1173–98.

Largeron, C., Dumont, M., Morin, S. et al. (2020). Toward snow cover estimation in mountainous areas using modern data assimilation methods: A review. *Frontiers in Earth Science*, 8, 325. https://doi.org/10.3389/feart.2020.00325.

Lehning, M., Bartelt, P., Brown, B. et al. (1999). SNOWPACK model calculations for avalanche warning based upon a new network of weather and snow stations. *Cold Regions Science and Technology*, 30(1–3), 145–57.

Leisenring, M., and Moradkhani, H. (2011). Snow water equivalent prediction using Bayesian data assimilation methods. *Stochastic Environmental Research and Risk Assessment*, 25(2), 253–70.

Lemmetyinen, J., Derksen, C., Rott, H. et al. (2018). Retrieval of effective correlation length and snow water equivalent from radar and passive microwave measurements. *Remote Sensing*, 10(2), 170. https://doi.org/10.3390/rs10020170.

Li, D., Durand, M., and Margulis, S. A. (2012). Potential for hydrologic characterization of deep mountain snowpack via passive microwave remote sensing in the Kern River basin, Sierra Nevada, USA. *Remote Sensing of Environment*, 125, 34–48.

Li, D., Lettenmaier, D. P., Margulis, S. A., and Andreadis, K. (2019). The value of accurate high-resolution and spatially

continuous snow information to streamflow forecasts. *Journal of Hydrometeorology*, 20(4), 731–49.

Lievens, H., Demuzere, M., Marshall, H.-P. et al. (2019). Snow depth variability in the Northern Hemisphere mountains observed from space. *Nature Communications*, 10(1), 1–12.

Liu, Y., Fang, Y., and Margulis, S. A. (2021). Spatiotemporal distribution of seasonal snow water equivalent in High Mountain Asia from an 18-year Landsat-MODIS era snow reanalysis dataset. *The Cryosphere*, 15, 5261–80.

Liu, Y., Peters-Lidard, C. D., Kumar, S. et al. (2013). Assimilating satellite-based snow depth and snow cover products for improving snow predictions in Alaska. *Advances in Water Resources*, 54, 208–27.

Lundquist, J. D., Chickadel, C., Cristea, N. et al. (2018). Separating snow and forest temperatures with thermal infrared remote sensing. *Remote Sensing of Environment*, 209, 764–79.

Lundquist, J., Hughes, M., Gutmann, E., and Kapnick, S. (2019). Our skill in modeling mountain rain and snow is bypassing the skill of our observational networks. *Bulletin of the American Meteorological Society*, 100(12), 2473–90.

Luojus, K., Pulliainen, J., Takala, M. et al. (2021). GlobSnow v3. 0 Northern Hemisphere snow water equivalent dataset. *Scientific Data*, 8(1), 1–16.

Lv, Z., and Pomeroy, J. W. (2020). Assimilating snow observations to snow interception process simulations. *Hydrological Processes*, 34(10), 2229–46.

Magnusson, J., Gustafsson, D., Hüsler, F., and Jonas, T. (2014). Assimilation of point SWE data into a distributed snow cover model comparing two contrasting methods. *Water Resources Research*, 50(10), 7816–35.

Margulis, S. A., Cortés, G., Girotto, M., and Durand, M. (2016). A Landsat-era Sierra Nevada snow reanalysis (1985–2015). *Journal of Hydrometeorology*, 17(4), 1203–21.

Margulis, S. A., Fang, Y., Li, D., Lettenmaier, D. P., and Andreadis, K. (2019). The utility of infrequent snow depth images for deriving continuous space-time estimates of seasonal snow water equivalent. *Geophysical Research Letters*, 46(10), 5331–40.

Margulis, S. A., Girotto, M., Cortés, G., and Durand, M. (2015). A particle batch smoother approach to snow water equivalent estimation. *Journal of Hydrometeorology*, 16(4), 1752–72.

Meromy, L., Molotch, N. P., Link, T. E., Fassnacht, S. R., and Rice, R. (2013). Subgrid variability of snow water equivalent at operational snow stations in the western USA. *Hydrological Processes*, 27(17), 2383–400.

Miller, S. D., Lee, T. F., and Fennimore, R. L. (2005). Satellite-based imagery techniques for daytime cloud/snow delineation from MODIS. *Journal of Applied Meteorology*, 44(7), 987–97.

Molotch, N. P., and Margulis, S. A. (2008). Estimating the distribution of snow water equivalent using remotely sensed snow cover data and a spatially distributed snowmelt model: A multi-resolution, multi-sensor comparison. *Advances in Water Resources*, 31(11), 1503–14.

Musselman, K. N., Addor, N., Vano, J. A., and Molotch, N. P. (2021). Winter melt trends portend widespread declines in snow water resources. *Nature Climate Change*, 11(5), 418–24.

Musselman, K. N., Lehner, F., Ikeda, K. et al. (2018). Projected increases and shifts in rain-on-snow flood risk over western North America. *Nature Climate Change*, 8(9), 808–12.

Musselman, K. N., Pomeroy, J. W., Essery, R. L., and Leroux, N. (2015). Impact of windflow calculations on simulations of alpine snow accumulation, redistribution and ablation. *Hydrological Processes*, 29(18), 3983–99.

Nadim, F., Kjekstad, O., Peduzzi, P., Herold, C., and Jaedicke, C. (2006). Global landslide and avalanche hotspots. *Landslides*, 3(2), 159–73.

Nagler, T., Rott, H., Ripper, E., Bippus, G., and Hetzenecker, M. (2016). Advancements for snowmelt monitoring by means of Sentinel-1 SAR. *Remote Sensing*, 8(4), 348. https://doi.org/10.3390/rs8040348.

Niu, G.-Y., Yang, Z.-L., Mitchell, K. E. et al. (2011). The community Noah land surface model with multiparameterization options (Noah-MP): 1. Model description and evaluation with local-scale measurements. *Journal of Geophysical Research: Atmospheres*, 116(D12).

Nolan, M., Larsen, C., and Sturm, M. (2015). Mapping snow depth from manned aircraft on landscape scales at centimeter resolution using structure-from-motion photogrammetry. *The Cryosphere*, 9(4), 1445–63.

Oaida, C. M., Reager, J. T., Andreadis, K. M. et al. (2019). A high-resolution data assimilation framework for snow water equivalent estimation across the western United States and validation with the airborne snow observatory. *Journal of Hydrometeorology*, 20(3), 357–78.

Ohmura, A. (2001). Physical basis for the temperature-based melt-index method. *Journal of Applied Meteorology*, 40(4), 753–61.

Painter, T. H., Berisford, D. F., Boardman, J. W. et al. (2016). The Airborne Snow Observatory: Fusion of scanning lidar, imaging spectrometer, and physically-based modeling for mapping snow water equivalent and snow albedo. *Remote Sensing of Environment*, 184, 139–52.

Painter, T. H., Bryant, A. C., and Skiles, S. M. (2012). Radiative forcing by light absorbing impurities in snow from MODIS surface reflectance data. *Geophysical Research Letters*, 39(17).

Painter, T. H., Rittger, K., McKenzie, C. et al. (2009). Retrieval of subpixel snow covered area, grain size, and albedo from MODIS. *Remote Sensing of Environment*, 113(4), 868–79.

Picard, G., Sandells, M., and Löwe, H. (2018). SMRT: An active–passive microwave radiative transfer model for snow with multiple microstructure and scattering formulations (v1. 0). *Geoscientific Model Development*, 11(7), 2763–88.

Pulliainen, J., Luojus, K., Derksen, C. et al. (2020). Patterns and trends of Northern Hemisphere snow mass from 1980 to 2018. *Nature*, 581(7808), 294–8.

Raleigh, M. S., Livneh, B., Lapo, K., and Lundquist, J. D. (2016). How does availability of meteorological forcing data impact physically based snowpack simulations? *Journal of Hydrometeorology*, 17(1), 99–120.

Rango, A. (1996). Spaceborne remote sensing for snow hydrology applications. *Hydrological Sciences Journal*, 41(4), 477–94.

Rice, R., Bales, R. C., Painter, T. H., and Dozier, J. (2011). Snow water equivalent along elevation gradients in the Merced and Tuolumne River basins of the Sierra Nevada. *Water Resources Research*, 47(8).

Riggs, G. A., Hall, D. K., and Román, M. O. (2017). Overview of NASA's MODIS and visible infrared imaging radiometer suite (VIIRS) snow-cover earth system data records. *Earth System Science Data*, 9(2), 765–77.

Rodell, M., and Houser, P. R. (2004). Updating a land surface model with MODIS-derived snow cover. *Journal of Hydrometeorology*, 5(6), 1064–75.

Rosenthal, W., and Dozier, J. (1996). Automated mapping of montane snow cover at subpixel resolution from the Landsat Thematic Mapper. *Water Resources Research*, 32(1), 115–30.

Santi, E., Brogioni, M., Leduc-Leballeur, M. et al. (2022). Exploiting the ANN Potential in Estimating Snow Depth and Snow Water Equivalent From the Airborne SnowSAR Data at X-and Ku-Bands. *IEEE Transactions on Geoscience and Remote Sensing*, 60, 1–16. https://do.org/10.1109/TGRS.2021.3086893.

Schlosser, C. A., Robock, A., Vinnikov, K. Y., Speranskaya, N. A., and Xue, Y. (1997). 18-year land-surface hydrology model simulations for a midlatitude grassland catchment in Valdai, Russia. *Monthly Weather Review*, 125(12), 3279–96.

Schmidt, R. A. (1982). Properties of blowing snow. *Reviews of Geophysics*, 20(1), 39–44.

Schmugge, T. J., Kustas, W. P., Ritchie, J. C., Jackson, T. J., and Rango, A. (2002). Remote sensing in hydrology. *Advances in Water Resources*, 25(8–12), 1367–85.

Serreze, M. C., Clark, M. P., Armstrong, R. L., McGinnis, D. A., and Pulwarty, R. S. (1999). Characteristics of the western United States snowpack from snowpack telemetry (SNOTEL) data. *Water Resources Research*, 35(7), 2145–60.

Shi, J., and Dozier, J. (2000). Estimation of snow water equivalence using SIR-C/X-SAR. I. Inferring snow density and subsurface properties. *IEEE Transactions on Geoscience and Remote Sensing*, 38(6), 2465–74.

Shufen, S., and Yongkang, X. (2001). Implementing a new snow scheme in simplified simple biosphere model. *Advances in Atmospheric Sciences*, 18(3), 335–54.

Skiles, S. M., Flanner, M., Cook, J. M., Dumont, M., and Painter, T. H. (2018). Radiative forcing by light-absorbing particles in snow. *Nature Climate Change*, 8(11), 964–71.

Skiles, S. M., and Painter, T. H. (2019). Toward understanding direct absorption and grain size feedbacks by dust radiative forcing in snow with coupled snow physical and radiative transfer modeling. *Water Resources Research*, 55(8), 7362–78.

Slater, A. G., Barrett, A. P., Clark, M. P., Lundquist, J. D., and Raleigh, M. S. (2013). Uncertainty in seasonal snow reconstruction: Relative impacts of model forcing and image availability. *Advances in Water Resources*, 55, 165–77.

Slater, A. G., and Clark, M. P. (2006). Snow data assimilation via an ensemble Kalman filter. *Journal of Hydrometeorology*, 7(3), 478–93.

Smith, C. D., Kontu, A., Laffin, R., and Pomeroy, J. W. (2017). An assessment of two automated snow water equivalent instruments during the WMO Solid Precipitation Intercomparison Experiment. *The Cryosphere*, 11(1), 101–16.

Smyth, E. J., Raleigh, M. S., and Small, E. E. (2019). Particle filter data assimilation of monthly snow depth observations improves estimation of snow density and SWE. *Water Resources Research*, 55(2), 1296–311.

Stigter, E. E., Wanders, N., Saloranta, T. M. et al. (2017). Assimilation of snow cover and snow depth into a snow model to estimate snow water equivalent and snowmelt runoff in a Himalayan catchment. *The Cryosphere*, 11(4), 1647–64.

Strozzi, T., and Matzler, C. (1998). Backscattering measurements of alpine snowcovers at 5.3 and 35 GHz. *IEEE Transactions on Geoscience and Remote Sensing*, 36(3), 838–48.

Su, H., Yang, Z.-L., Niu, G.-Y., and Dickinson, R. E. (2008). Enhancing the estimation of continental-scale snow water equivalent by assimilating MODIS snow cover with the ensemble Kalman filter. *Journal of Geophysical Research: Atmospheres*, 113(D8).

Tachiiri, K., Shinoda, M., Klinkenberg, B., and Morinaga, Y. (2008). Assessing Mongolian snow disaster risk using livestock and satellite data. *Journal of Arid Environments*, 72(12), 2251–63.

Tapley, B. D., Bettadpur, S., Ries, J. C., Thompson, P. F., and Watkins, M. M. (2004). GRACE measurements of mass variability in the Earth system. *Science*, 305(5683), 503–5.

Tedesco, M. (2014). *Remote Sensing of the Cryosphere*. Chichester: John Wiley & Sons.

Tedesco, M., Pulliainen, J., Takala, M., Hallikainen, M., and Pampaloni, P. (2004). Artificial neural network-based techniques for the retrieval of SWE and snow depth from SSM/I data. *Remote Sensing of Environment*, 90(1), 76–85.

Thirel, G., Salamon, P., Burek, P., and Kalas, M. (2013). Assimilation of MODIS snow cover area data in a distributed hydrological model using the particle filter. *Remote Sensing*, 5(11), 5825–50.

Todt, M., Rutter, N., Fletcher, C. G. et al. (2018). Simulation of longwave enhancement in boreal and montane forests. *Journal of Geophysical Research: Atmospheres*, 123(24), 13731–47.

Toure, A. M., Reichle, R. H., Forman, B. A., Getirana, A., and De Lannoy, G. J. (2018). Assimilation of MODIS snow cover fraction observations into the NASA catchment land surface model. *Remote Sensing*, 10(2), 316.

Van Leeuwen, P. J. (2009). Particle filtering in geophysical systems. *Monthly Weather Review*, 137(12), 4089–114.

Veyssière, G., Karbou, F., Morin, S., Lafaysse, M., and Vionnet, V. (2019). Evaluation of sub-kilometric numerical simulations of c-band radar backscatter over the French alps against sentinel-1 observations. *Remote Sensing*, 11(1), 8. https://doi.org/10.3390/rs11010008.

Viallon-Galinier, L., Hagenmuller, P., and Lafaysse, M. (2020). Forcing and evaluating detailed snow cover models with stratigraphy observations. *Cold Regions Science and Technology*, 180, 103163.

Viviroli, D., Dürr, H. H., Messerli, B., Meybeck, M., and Weingartner, R. (2007). Mountains of the world, water towers for humanity: Typology, mapping, and global significance. *Water Resources Research*, 43(7).

Wang, J., Forman, B. A., Girotto, M., and Reichle, R. H. (2021). Estimating terrestrial snow mass via multi-sensor assimilation of synthetic AMSR-E brightness temperature spectral differences and synthetic GRACE terrestrial water storage retrievals. *Water Resources Research*, 57 (9), e2021WR029880.

Wiesmann, A., and Mätzler, C. (1999). Microwave emission model of layered snowpacks. *Remote Sensing of Environment*, 70(3), 307–316.

Winstral, A., Magnusson, J., Schirmer, M., and Jonas, T. (2019). The bias-detecting ensemble: A new and efficient technique for dynamically incorporating observations into physics-based, multilayer snow models. *Water Resources Research*, 55(1), 613–31.

Xue, Y., and Forman, B. A. (2017). Integration of satellite-based passive microwave brightness temperature observations and an ensemble-based land data assimilation framework to improve snow estimation in forested regions. *2017 IEEE International Geoscience and Remote Sensing Symposium (IGARSS)*, 311–14.

6
Data Assimilation in Glaciology

Mathieu Morlighem and Daniel Goldberg

Abstract: Data assimilation has always been a particularly active area of research in glaciology. While many properties at the surface of glaciers and ice sheets can be directly measured from remote sensing or in situ observations (surface velocity, surface elevation, thinning rates, etc.), many important characteristics, such as englacial and basal properties, as well as past climate conditions, remain difficult or impossible to observe. Data assimilation has been used for decades in glaciology in order to infer unknown properties and boundary conditions that have important impact on numerical models and their projections. The basic idea is to use observed properties, in conjunction with ice flow models, to infer these poorly known ice properties or boundary conditions. There is, however, a great deal of variability among approaches. Constraining data can be of a snapshot in time, or can represent evolution over time. The complexity of the flow model can vary, from simple descriptions of lubrication flow or mass continuity to complex, continent-wide Stokes flow models encompassing multiple flow regimes. Methods can be deterministic, where only a best fit is sought, or probabilistic in nature. We present in this chapter some of the most common applications of data assimilation in glaciology, and some of the new directions that are currently being developed.

6.1 Background and definitions

6.1.1 Observations of Ice Sheets and Glaciers

A *glacier* can be defined as a persistent body of ice that flows and deforms under its own weight. There are approximately 160,000 glaciers on Earth ranging in size from ~1 to thousands of square kilometres. An *ice sheet* is a body of glacial ice over 50,000 km^2 in area. There are only two surviving ice sheets today: Antarctica and Greenland.

For over half a century, scientists have recorded comprehensive quantitative measurements of the shape and speed of glaciers and ice sheets – to the point where currently satellite measurements are made of the elevation and surface velocity of glaciers and ice sheets with high accuracy and resolution. These data complement lower-quality satellite observations made in previous decades. Geophysical and airborne observing platforms are additionally able to capture the shape of internal reflecting layers, which encapsulate ice-sheet history, as well as the ice thickness.

Many of the methods discussed in this chapter rely on data which are spatially (and in some cases temporally) extensive, and current glaciological remote sensing methods provide a wealth of such information.

6.1.2 Modelling of Ice Sheets and Glaciers

Assimilation of the vast glaciological data collected requires physically based models. The flow of glaciers and ice sheets is modelled by solving conservation laws: conservation of momentum (stress balance), conservation of mass (continuity equation), and conservation of energy. For simplicity, we only describe here the first two conservation laws because the thermal regime of the ice is rarely directly involved in data assimilation methods.

Stress Balance

The stress balance of ice is generally modelled as an incompressible Stokes flow:

$$\begin{cases} \nabla \cdot \boldsymbol{\sigma}' - \nabla P + \rho \mathbf{g} = \mathbf{0} \\ \nabla \cdot \mathbf{v} = 0, \end{cases} \tag{6.1}$$

where $\boldsymbol{\sigma}'$ is the deviatoric stress tensor, P is the ice pressure, and $\mathbf{g}$ is the acceleration due to gravity. While ice may behave elastically over short time scales (e.g. tidal cycle), its long-term behaviour is described by a viscous constitutive law:

$$\boldsymbol{\sigma}' = 2\mu \dot{\boldsymbol{\varepsilon}}, \tag{6.2}$$

where the ice viscosity, μ, follows a Norton–Hoff viscoplastic law referred to as Glen's flow law (Glen, 1955):

$$\mu = \frac{B}{2\,\dot{\varepsilon}_e^{(n-1)/n}}, \tag{6.3}$$

and where B is the ice viscosity factor ($A = B^{-n}$, the ice rate factor, is sometimes used instead of B), n is Glen's exponent, generally taken as 3, and $\dot{\varepsilon}_e$ is the effective strain rate.

In terms of boundary conditions, it is generally assumed that the ice–air interface is stress free: $\sigma \cdot \mathbf{n} = \mathbf{0}$, and hydrostatic water pressure is applied at the ice–ocean interface:

$$\sigma \cdot \mathbf{n} = \rho_w \, g \, z \, \mathbf{n}, \tag{6.4}$$

where $\mathbf{n}$ is the outward pointing unit vector, ρ_w is the density of seawater, and z is the vertical coordinate (zero at sea level). Finally, at the ice–bed interface, we have a non-penetration condition ($\mathbf{v} \cdot \mathbf{n} = 0$, if there is no basal melt) and basal friction is applied along the tangential plane. To remain general, we write this friction condition as:

$$\tau_b = \sigma \cdot \mathbf{n} - (\mathbf{n} \cdot \sigma \cdot \mathbf{n})\mathbf{n} = -C \, f_\tau(|\mathbf{v}_b|, N) \, \mathbf{v}_b, \tag{6.5}$$

where $C \geq 0$ is a friction coefficient and f_τ is a function of the basal velocity, $\mathbf{v}_b$, and basal effective pressure, N.

Since solving Stokes flow at large scale can be computationally expensive, approximations are sometimes used. The simplest is the *Shallow-Ice Approximation*, which assumes that ice motion is determined solely by vertical shearing stress, and is appropriate for slow-moving glaciers and ice-sheet interiors (Greve and Blatter, 2009). The *Shallow-Shelf Approximation* (SSA; sometimes *Shelfy-Stream Approximation*) considers flow to be vertically uniform and controlled by membrane stresses (MacAyeal, 1989), and applies to fast-flowing ice streams and ice shelves. The *First-Order Approximation* includes vertical shearing as well as membrane stresses but neglects 'bridging' (non-hydrostatic) stresses (Blatter, 1995; Pattyn, 2003). The majority of continental-scale ice-sheet modelling studies implement one or a combination of these approximations.

Mass Transport

Depth integrating the continuity equation (the second equation of Eq. (6.1)) and applying the necessary boundary conditions at the base and surface of the ice sheet leads to the two-dimensional equation of mass transport, which governs the change in geometry of the ice sheet:

$$\frac{\partial H}{\partial t} = -\nabla \cdot H\bar{\mathbf{v}} + \dot{m}_s - \dot{m}_b, \tag{6.6}$$

where H is the ice thickness, $\bar{\mathbf{v}}$ is the depth-averaged velocity, $\dot{m}_s$ is the surface mass balance (positive when snow accumulation), and $\dot{m}_b$ is the melt rate at the base of the ice (positive when melting).

6.1.3 Model-Data Assimilation

We begin by defining what is meant by ice-sheet data assimilation. Ice-sheet data assimilation combines models of ice-sheet and glacier physics with observations in order to infer unresolved or unobservable physical properties. The aim of ice-sheet data assimilation varies by implementation: in some cases, the properties

themselves are of direct interest, whereas in other cases the assimilation is a calibration procedure for model projections. Ice-sheet data assimilation is an extremely diverse topic, with the types of data assimilated ranging from internal radar layers to satellite altimetry, and with models ranging from simple mass-balance models to sophisticated Stokes flow. In order to establish a common language for discussing these different types of assimilation, we make some formal definitions. Broadly, ice-sheet data assimilation can be described by the following optimisation problem: find α such that:

$$\alpha = \underset{\mathbf{x}=f(\alpha)}{\arg \min} \, \mathcal{J}(\alpha), \tag{6.7}$$

$$\mathcal{J}(\alpha) = \frac{1}{2} \int_\Omega \frac{\left(y^{obs} - \mathcal{H}(\mathbf{x})\right)^2}{\sigma_y^2} d\Omega, \tag{6.8}$$

where Ω is the computational domain of the function f. The functional $\mathcal{J}$ is referred to as a *cost function* (or sometimes an *objective function*), and quantifies the misfit between the model and available observations. The model *state* is $\mathbf{x}$, the solution of the *forward model*, f, for a given α. The minimisation is carried out over all possible values of α – which is a parameter, or set of parameters, for the model. For instance, α might be the surface mass balance over a finite historical period, or it might be the set of all values describing the basal friction field under an ice sheet. These parameters are often referred to as the *controls* of the assimilation. Here, y^{obs} describes a spatial field of an ice-sheet *observable* (e.g. surface elevation), and Ω is the model domain. Here σ_y represents the uncertainty in y^{obs}, and $\mathcal{H}$ is the *observation operator* that maps the model state, $\mathbf{x}$, to the space of observables.

Eqs. (6.7) and (6.8) are meant as representative of the general form of the assimilation, but can vary depending on the approach. Here y^{obs} is written as a scalar-valued function, but the observable can be a vector-valued function (e.g. ice velocity). The integral in Eq. (6.8) is a spatial integral but integration can be over time as well. The parameters α are in general spatially (and temporally) varying but by convention are represented by a discrete vector. While many assimilation problems use the integral formulation in Eq. (6.8), $\mathcal{J}$ is sometimes constructed as a discrete sum over spatiotemporal observation points i:

$$\mathcal{J}(\alpha) = \frac{1}{2} \sum_i \frac{\left(y_i^{obs} - \mathcal{H}_i(\mathbf{x})\right)^2}{\sigma_i^2}. \tag{6.9}$$

As shown in Section 6.3.3, the distinction between Eqs. (6.8) and (6.9) is critical to Bayesian interpretation of uncertainties.

Strictly, Eq. (6.7) describes *variational* data assimilation (VDA), in which optimal fit to observations is found subject to model physics. VDA is the approach most widely used in glaciological modelling, although other types are discussed in this chapter as well. The form of Eq. (6.7) is intentionally simplistic and, in practice, cost functions can differ in various ways, as discussed in this and later sections.

6.1.4 Regularisation

In many cases, the observables in an assimilation framework are insensitive to small variations in the controls. The inverse problem as defined by Eq. (6.7) is typically *ill posed*: there may be many solutions α to the minimisation problem (Hadamard, 1902), or the problem may be ill conditioned (i.e. sensitive to small uncertainties in the observational data). A common strategy to address this problem is to add a *regularisation* term to the cost function, which penalises non-physical behaviours. For instance, a common approach in estimating ice-sheet basal friction is to use Tikhonov regularisation (Morlighem et al., 2010), which penalises the square gradient of the control field:

$$\mathcal{J}(\boldsymbol{\alpha}) = \underbrace{\frac{1}{2} \int_{\Omega} \frac{\left(y^{obs} - \mathcal{H}(\mathbf{x})\right)^2}{\sigma_y^2} d\Omega}_{\text{misfit, } \mathcal{J}_0} + \underbrace{\frac{1}{2} \int_{\Omega} \gamma |\nabla \alpha|^2 d\Omega}_{\text{regularisation}}, \qquad (6.10)$$

where γ is a parameter that sets the level of regularisation. The first term, $\mathcal{J}_0$, is often referred to as the *misfit* or *data* norm, while the integral in the second term is sometimes referred to as the *model* norm (Waddington et al., 2007). The minimisation of the cost function $\mathcal{J}$ is thus a trade-off between minimising the two. Careful selection of the regularisation parameter γ is important. An overly strong regularisation can lead to poor model-data fit, while overly weak regularisation can lead to non-physical features in the optimal controls, or over-fitting. A commonly used approach to identify optimal parameters is the *L*-curve analysis (Hansen, 2000; Gillet-Chaulet et al., 2012), which examines the relationship between misfit and model norm as the regularisation parameter is varied. In approaches with more than one control field (or more than one regularisation parameter) an *L*-'surface' may be used (Furst et al., 2015). There are alternative methods of finding the appropriate regularisation, however, such as the discrepancy principle (Habermann et al., 2013) or the Lagrange multiplier method of Waddington et al. (2007).

The regularisation term is sometimes referred to as a 'prior' (Barnes et al., 2021), which lends itself to a Bayesian interpretation of assimilation (see Section 6.3.3). In this sense, the regularisation is imposing *a priori* knowledge regarding the controls of the assimilation. Care should be taken, however, not to impose knowledge that one does not possess as this could influence quantifications of uncertainty (Arthern, 2015).

6.1.5 Cost Function Minimisation

Equation (6.7) represents, in most glaciological assimilations, a very large-scale non-linear optimisation (or rather, minimisation) problem. Methods to solve this minimisation problem have included ad-hoc approaches (Chandler et al., 2006; Price et al., 2011) or explicit reduction of the control dimension to create a more manageable problem (Payne et al., 2004) among other methods. However, by far the most common approach taken is an iterative one in which search directions are updated based on cost function gradients (MacAyeal, 1993). Such approaches include simple gradient-descent methods (e.g. Joughin et al., 2004) and quasi-Newton methods (e.g. Gillet-Chaulet et al., 2012; Barnes et al., 2021). Quasi-Newton methods, such as the Limited-memory Broyden–Fletcher–Goldfarb–Shanno algorithm (L-BFGS, Gilbert and Lemaréchal, 1989; Nocedal and Wright, 2006), construct a low-rank approximation to the *Hessian* (or second derivative matrix) of the cost function through consecutive gradient calculations, and generally converge faster than simple gradient descent. 'True' Newton methods, which use the exact Hessian of the cost function, are expected to converge faster than quasi-Newton methods, but to date such methods have not been used for large-scale glaciological data assimilation, predominantly because of the computational cost in finding the inverse of the Hessian (see Section 6.3). Recently however, a study has made use of a more sophisticated Hessian approximation (Shapero et al. (2021); see also Section 6.3.6), which converges faster than L-BFGS in experiments carried out.

Methods for calculating forward model gradients vary greatly across various assimilation frameworks. A common approach is the *adjoint* method, which solves the mathematical adjoint of the linearised model (see Section 6.2.2). Such methods are advantageous as they require a single adjoint solution, as opposed to individually perturbing potentially tens of thousands of parameters (MacAyeal, 1992a). While most approaches to date derive the continuous adjoint model, which is then discretised, there are an increasing number of studies making use of an *algorithmic differentiation* (AD; Heimbach and Bugnion, 2009). AD software tools generate a discretised form of the adjoint model directly from the discretised forward model through repeated application of the chain rule at the level of the numerical code (Griewank and Walther, 2008). In the following sections, different applications of these approaches are examined.

6.2 Inferring Ice Dynamical Properties and Boundary Conditions

In this section, we describe some of the popular data assimilation techniques that have been employed to constrain poorly known boundary conditions and ice material properties from surface observations. We first describe ad hoc methods that are easy to implement but require running the model over long time scales and assume that the system is in steady state. We then present the time-independent (or snapshot) variational approaches that have been particularly popular in glaciology, and then the time-dependent approaches that are also based on variational methods.

6.2.1   Ad Hoc Methods

Inference from Surface Speed
Described in Price et al. (2011), this method is used to infer basal drag from surface speed. The approach starts with a very large value for the basal drag (i.e. no sliding), as described in Eq. (6.5). The model is then run forward while keeping the geometry fixed, until the ice temperature, velocity, and viscosity reach a steady state. Once a steady state is reached, the basal stress $|\boldsymbol{\tau}_b|$ is evaluated. The modelled surface speed is then subtracted from the observed surface speed to estimate the sliding speed, $\mathbf{v}_b$, that was previously assumed to be $\mathbf{0}$. These sliding speeds, together with the inferred basal stress, are used to determine the sliding coefficient by solving:

$$C = \frac{|\boldsymbol{\tau}_b|}{f_\tau(|\mathbf{v}_b|, N)|\mathbf{v}_b|}. \tag{6.11}$$

Using this new distribution of C, the model is run again to steady state. The evaluation of C by determining the basal drag, $|\boldsymbol{\tau}_b|$, and adding the misfit between modelled and observed surface velocities to the sliding velocities, is repeated until convergence is reached.

To our knowledge, this method has not been used outside this study, but is easy to implement and relies on the simple concept that decreasing the friction coefficient has a direct, positive impact on the surface velocities.

Inference from Surface Height
Pollard and DeConto (2012) proposed a similar method but using surface height, s, as a target instead of surface velocities. The idea is to run a model forward in time and adjust the friction coefficient, C, so that the surface heights match observed elevations, assuming that the ice sheet is in steady state. In regions where the modelled surface height is too low, friction is increased, which allows for the ice to thicken, whereas in regions where the ice is too high, friction is decreased to allow for more sliding and thinning.

In practice, the friction is updated every 5,000 years as follows:

$$C_{\text{new}} = C_{\text{old}} \times 10^{\Delta z}, \tag{6.12}$$

where:

$$\Delta z = \max(-1.5, \min(1.5, (s_{\text{obs}} - s)/s_0)), \tag{6.13}$$

where $s_0 = 500$ m is a scaling parameter. This method allows for the modelled height to reach an average of 50 m of misfit between the model and observation within 200,000 to 400,000 years. While simple, it still requires running the model for a long period of time and is difficult to apply to more sophisticated ice sheet models that typically only run for a few centuries.

6.2.2   Time-Independent/Snapshot Variational Approaches
One of the limitations of these ad hoc methods is that they require the model to be in steady state, which is not ideal when studying rapidly changing systems. A popular alternative approach that removes this assumption was introduced by MacAyeal (1992b, 1993) in the 1990s, and has been applied almost exclusively, until relatively recently, to the two-dimensional SSA. Known as a 'control method', it is a direct application of the optimal control theory to glaciology (Bryson and Ho, 2018). A type of VDA, the method consists of minimising the cost function through a gradient-descent approach. We first describe VDA in a general context, for both discrete and continuous model equations, and then show how it is applied to the SSA.

Discrete Adjoint Method
In the discrete form, the model state is a vector $\mathbf{x} \in \mathbb{R}^m$, and the control is $\boldsymbol{\alpha} \in \mathbb{R}^n$. We write $J(\mathbf{x}, \boldsymbol{\alpha}) \in \mathbb{R}_+$ the cost function that we want to minimise, and define $\mathcal{J}(\boldsymbol{\alpha}) = J(\mathbf{x}(\boldsymbol{\alpha}), \boldsymbol{\alpha})$ such that $\mathbf{x}(\boldsymbol{\alpha})$ satisfies the model equation $\mathbf{g}(\mathbf{x}, \boldsymbol{\alpha}) = \mathbf{0}$, which is an m-dimensional zero vector. For instance, if one uses the finite-element method, we would have $\mathbf{g}(\mathbf{x}, \boldsymbol{\alpha}) = \mathbf{K}\mathbf{x} - \mathbf{F}$, where $\mathbf{K}$ is the stiffness matrix and $\mathbf{F}$ the load vector (where $\mathbf{K}$ and $\mathbf{F}$ in general depend on the control $\boldsymbol{\alpha}$). We would like to find the gradient of the cost function in order to minimise it, as described in Section 6.1.5.

The most intuitive approach to compute this gradient would be to apply a small perturbation to each of the components of the control, $\boldsymbol{\alpha}$ using a finite-difference method to construct the gradient of the cost function. This method, however, requires to solve n times the model equations, $\mathbf{g}(\mathbf{x}, \boldsymbol{\alpha}) = \mathbf{0}$. While simple, this approach becomes very inefficient and computationally prohibitive when $n \gtrsim 10$. Variational data assimilation allows to compute this gradient by solving only two sets of equations. The gradient of the cost function is:

$$\nabla \mathcal{J} = \frac{\partial J}{\partial \boldsymbol{\alpha}} + \frac{\partial J}{\partial \mathbf{x}} \frac{\partial \mathbf{x}}{\partial \boldsymbol{\alpha}}. \tag{6.14}$$

Since J has an explicit expression, the first two vectors are easy to compute. The Jacobian matrix, $\partial \mathbf{x}/\partial \boldsymbol{\alpha} \in \mathbb{R}^{n \times m}$, however, is difficult to evaluate, especially if n and m are large. If we take the derivative of the forward model with respect to the control, $\boldsymbol{\alpha}$, we find:

$$\frac{d\mathbf{g}}{d\alpha} = \frac{\partial \mathbf{g}}{\partial \alpha} + \frac{\partial \mathbf{g}}{\partial \mathbf{x}}\frac{\partial \mathbf{x}}{\partial \alpha} = \mathbf{0}, \tag{6.15}$$

which is $\mathbf{0}$ since the forward model must be satisfied. Using this equation, we can rewrite the problematic term as:

$$\frac{\partial \mathbf{x}}{\partial \alpha} = -\left[\frac{\partial \mathbf{g}}{\partial \mathbf{x}}\right]^{-1}\frac{d\mathbf{g}}{d\alpha}. \tag{6.16}$$

Substituting this expression in Eq. (6.14) gives:

$$\nabla \mathcal{J} = \frac{\partial J}{\partial \alpha} - \frac{\partial J}{\partial \mathbf{x}}\left[\frac{\partial \mathbf{g}}{\partial \mathbf{x}}\right]^{-1}\frac{d\mathbf{g}}{d\alpha}. \tag{6.17}$$

We define the *adjoint state*, $\lambda \in \mathbb{R}^m$, as:

$$\lambda^T = -\frac{\partial J}{\partial \mathbf{x}}\left[\frac{\partial \mathbf{g}}{\partial \mathbf{x}}\right]^{-1}, \tag{6.18}$$

or, written differently:

$$\left[\frac{\partial \mathbf{g}}{\partial \mathbf{x}}\right]^T \lambda = -\left[\frac{\partial J}{\partial \mathbf{x}}\right]^T, \tag{6.19}$$

which is generally referred to as the *adjoint model*. The calculation of the adjoint state requires solving a system of equations which has the same size as the forward model. Once the adjoint state is calculated, we can compute the gradient of the cost function as:

$$\nabla \mathcal{J} = \frac{\partial J}{\partial \alpha} + \lambda^T \frac{d\mathbf{g}}{d\alpha}. \tag{6.20}$$

It can be shown that the adjoint state is equivalent to the vector of Lagrange multipliers introduced to enforce that the initial model equation, or forward model, is satisfied.

Continuous Adjoint Method

If we consider a set of continuous model equations, we can follow the same steps as in the previous section, but this time we introduce a Lagrangian to enforce the constraint that the model equations need to be satisfied. The model state, $\mathbf{x}$, and the control, α, are now fields in the appropriate Banach spaces (note that, for simplicity and brevity, we will assume that all variables are defined in the appropriate space). The objective function is $J(\mathbf{x}, \alpha) \in \mathbb{R}_+$ and, again, we define $\mathcal{J}(\alpha) = J(\mathbf{x}(\alpha), \alpha)$ such that $\mathbf{x}(\alpha)$ satisfies the model equation $\mathbf{g}(\mathbf{x}, \alpha) = \mathbf{0}$.

We want to find α such that it minimises the cost function. The idea of optimal control is to introduce a Lagrangian for this optimisation problem:

$$\mathcal{L}(\mathbf{x}, \lambda, \alpha) = J(\mathbf{x}, \alpha) + \int_\Omega \lambda \cdot \mathbf{g}(\mathbf{x}, \alpha) d\Omega. \tag{6.21}$$

Similarly, the new variable, λ, is the Lagrange multiplier (or adjoint state) for the constraint $\mathbf{g}(\mathbf{x}, \alpha) = \mathbf{0}$. If $\mathbf{x}$ is solution of the forward model, then

$$\mathcal{J}(\alpha) = \mathcal{L}(\mathbf{x}, \lambda, \alpha) \tag{6.22}$$

since, in this case, $\mathbf{g}(\mathbf{x}, \alpha) = \mathbf{0}$. If we take the directional derivative of this expression with respect to the control along the direction β, and apply the chain rule, this yields:

$$\langle \nabla \mathcal{J}, \beta \rangle = \left\langle \frac{\partial \mathcal{L}}{\partial \mathbf{x}}, \mathbf{x}'(\beta) \right\rangle + \left\langle \frac{\partial \mathcal{L}}{\partial \lambda}, \lambda'(\beta) \right\rangle + \left\langle \frac{\partial \mathcal{L}}{\partial \alpha}, \beta \right\rangle. \tag{6.23}$$

Because we want the forward model to be satisfied, the second term on the right-hand side is zero:

$$\forall \mu, \quad \left\langle \frac{\partial \mathcal{L}}{\partial \lambda}, \mu \right\rangle = \langle \mu, \mathbf{g}(\mathbf{x}, \alpha) \rangle = 0. \tag{6.24}$$

If we choose the adjoint state such that the Lagrangian is also stationary along $\mathbf{x}$, the first term is zero:

$$\forall \mathbf{w}, \quad \left\langle \frac{\partial \mathcal{L}}{\partial \mathbf{x}}, \mathbf{w} \right\rangle = \left\langle \frac{\partial J(\mathbf{x}, \alpha)}{\partial \mathbf{x}}, \mathbf{w} \right\rangle + \left\langle \lambda, \frac{\partial \mathbf{g}}{\partial \mathbf{x}}\{\mathbf{w}\} \right\rangle = 0, \tag{6.25}$$

where the $\{\cdot\}$ indicates the action of the differential operator in a given direction. Using the properties of inner products, this can be rearranged:

$$\left\langle \left(\frac{\partial \mathbf{g}}{\partial \mathbf{x}}\right)^T \{\lambda\}, \mathbf{w} \right\rangle = -\left\langle \frac{\partial J(\mathbf{x}, \alpha)}{\partial \mathbf{x}}, \mathbf{w} \right\rangle, \tag{6.26}$$

yielding the *adjoint* equations to be solved for λ. This then provides a simple expression for the gradient of the cost function with respect to the control:

$$\langle \nabla \mathcal{J}, \beta \rangle = \left\langle \frac{\partial \mathcal{L}}{\partial \alpha}, \beta \right\rangle = \left\langle \frac{\partial J}{\partial \alpha}, \beta \right\rangle + \left\langle \lambda \cdot \frac{\partial \mathbf{g}}{\partial \alpha}(\mathbf{x}, \alpha), \beta \right\rangle \tag{6.27}$$

and we identify the gradient as:

$$\nabla \mathcal{J} = \frac{\partial J}{\partial \alpha} + \lambda \cdot \frac{\partial \mathbf{g}}{\partial \alpha}(\mathbf{x}, \alpha). \tag{6.28}$$

Application to SSA

Here we apply this approach to the Shallow-Shelf Approximation. The equations of SSA are derived from the full-Stokes equations (Eq. (6.1)), and rely on two assumptions. The first one is that there is no bridging effect, which means that the vertical normal stress is purely lithostatic, or equal to the weight of ice vertically above. The second assumption is that vertical shear is negligible, so that velocities are assumed depth-independent. These two assumptions lead to the following depth-integrated model:

$$\begin{cases} \frac{\partial}{\partial x}\left(4H\mu\frac{\partial v_x}{\partial x} + 2H\mu\frac{\partial v_y}{\partial y}\right) + \frac{\partial}{\partial y}\left(H\mu\frac{\partial v_x}{\partial y} + H\mu\frac{\partial v_y}{\partial x}\right) = \rho g H\frac{\partial s}{\partial x} + Cv_x \\ \frac{\partial}{\partial x}\left(H\mu\frac{\partial v_x}{\partial y} + H\mu\frac{\partial v_y}{\partial x}\right) + \frac{\partial}{\partial y}\left(4H\mu\frac{\partial v_y}{\partial y} + 2H\mu\frac{\partial v_x}{\partial x}\right) = \rho g H\frac{\partial s}{\partial y} + Cv_y \end{cases}, \tag{6.29}$$

where we assume here, for simplicity, that the friction law is $\tau_b = -C\,\mathbf{v}$.

MacAyeal (1992b) applied an adjoint-based inversion to infer the basal friction under Ice Stream E, Antarctica. Joughin et al. (2004) modified the method to determine the friction under Ross Ice Stream assuming a weak plastic bed. Rommelaere and MacAyeal (1997) extended this approach to infer the ice rheological parameter, B, of the Ross ice shelf, Antarctica, with the same Shallow-Shelf model, from surface velocities. Larour (2005) applied a similar method to infer B of the Ronne ice shelf. This method was then extended to higher-order and full-Stokes models by Morlighem et al. (2010) and applied to the entire Antarctic ice sheet (Morlighem et al., 2013b). It is now used by many glaciologists and available in several ice sheet models (Elmer, ISSM, STREAMICE, Uʹa, BISICLES, MALI, etc).

As an illustration, we describe here how to derive the gradient of a classical cost function:

$$J(\mathbf{v}) = \frac{1}{2}\int_\Omega (v_x - v_x^{obs})^2 + (v_y - v_y^{obs})^2 d\Omega. \quad (6.30)$$

In this case, the observation operator is the identity function since we observe the model state itself, $\mathbf{x} = (v_x, v_y)^T$. Let's assume that we are using the SSA equations with homogeneous Dirichlet conditions along the boundary and we are inferring the basal friction coefficient, $\alpha = C$, from observed surface velocities. Let's further ignore the dependency of the ice viscosity, μ, on the ice velocity in the computation of the adjoint state (which is a common assumption in glaciology; see, for example, MacAyeal, 1992a). Using the continuous approach (Section 6.2.2), the adjoint equations (Eq. (6.26)) are: $\forall \mathbf{w}$

$$\left\langle \frac{\partial J(\mathbf{v},\alpha)}{\partial \mathbf{v}}, \mathbf{w} \right\rangle + \left\langle \lambda, \frac{\partial \mathbf{g}}{\partial \mathbf{v}}\{\mathbf{w}\} \right\rangle = \int_\Omega (v_x - v_x^{obs})w_x$$
$$+ (v_y - v_y^{obs})w_y d\Omega \quad (6.31)$$
$$+ \int_\Omega \lambda_x \frac{\partial}{\partial x}\left(4H\mu\frac{\partial w_x}{\partial x} + 2H\mu\frac{\partial w_y}{\partial y}\right)$$
$$+ \lambda_x \frac{\partial}{\partial y}\left(H\mu\frac{\partial w_x}{\partial y} + H\mu\frac{\partial w_y}{\partial x}\right) - Cw_x\lambda_x d\Omega$$
$$+ \int_\Omega \lambda_y \frac{\partial}{\partial x}\left(H\mu\frac{\partial w_x}{\partial y} + H\mu\frac{\partial w_y}{\partial x}\right)$$
$$+ \lambda_y \frac{\partial}{\partial y}\left(4H\mu\frac{\partial w_y}{\partial y} + 2H\mu\frac{\partial w_x}{\partial x}\right) - Cw_y\lambda_y d\Omega = 0.$$

If we integrate the two last lines by parts twice, and assume homogeneous boundary conditions, this equation becomes:

$$\int_\Omega (v_x - v_x^{obs})\,w_x + (v_y - v_y^{obs})w_y d\Omega$$
$$+ \int_\Omega w_x \frac{\partial}{\partial x}\left(4H\mu\frac{\partial \lambda_x}{\partial x} + 2H\mu\frac{\partial \lambda_y}{\partial y}\right)$$
$$+ w_x \frac{\partial}{\partial y}\left(H\mu\frac{\partial \lambda_x}{\partial y} + H\mu\frac{\partial \lambda_y}{\partial x}\right) - Cw_x\lambda_x d\Omega$$
$$+ \int_\Omega w_y \frac{\partial}{\partial y}\left(4H\mu\frac{\partial \lambda_y}{\partial y} + 2H\mu\frac{\partial \lambda_x}{\partial x}\right)$$
$$+ w_y \frac{\partial}{\partial x}\left(H\mu\frac{\partial \lambda_x}{\partial y} + H\mu\frac{\partial \lambda_y}{\partial x}\right) - Cw_y\lambda_y d\Omega = 0, \quad (6.32)$$

and since this is true for all $\mathbf{w}$, the adjoint equations are:

$$\begin{cases} \frac{\partial}{\partial x}\left(4H\mu\frac{\partial \lambda_x}{\partial x} + 2H\mu\frac{\partial \lambda_y}{\partial y}\right) + \frac{\partial}{\partial y}\left(H\mu\frac{\partial \lambda_x}{\partial y} + H\mu\frac{\partial \lambda_y}{\partial x}\right) - C\lambda_x = -(v_x - v_x^{obs}) \\ \frac{\partial}{\partial y}\left(4H\mu\frac{\partial \lambda_y}{\partial y} + 2H\mu\frac{\partial \lambda_x}{\partial x}\right) + \frac{\partial}{\partial x}\left(H\mu\frac{\partial \lambda_x}{\partial y} + H\mu\frac{\partial \lambda_y}{\partial x}\right) - C\lambda_y = -(v_y - v_y^{obs}). \end{cases}$$
$$(6.33)$$

The adjoint state can therefore be solved using very similar equations as the forward model but with a different right-hand side. Once the adjoint state is calculated, the gradient of the cost function is simply:

$$\nabla \mathcal{J} = -\lambda \cdot \mathbf{v}. \quad (6.34)$$

A similar approach can be applied to the discretised equations, leading to a similar gradient expression. We must however pay attention to the fact that the discretised gradient provides the sensitivity of the cost function with respect to the value of the control at each node or cell, which is not the same as the continuous gradient. In other words, the discretised gradient for node i is related to the continuous gradient as follows:

$$\nabla \mathcal{J}_i = \langle \nabla \mathcal{J}, \varphi_i \rangle = \int_\Omega -\lambda \cdot \mathbf{v}\, \varphi_i\, d\Omega, \quad (6.35)$$

where φ_i is the shape function associated with C_i, the friction coefficient at node i. The discrete and continuous gradients can therefore be converted from one form to the other by using the mass matrix (e.g. Li et al., 2017).

Simultaneous Inference of Multiple Parameter Fields

For reasons of simplicity, the preceding example considers only a single control field, C. However, in many cases, other key spatially varying properties must be inferred, such as the ice viscosity factor, B, which depends on ice temperature (Cuffey and Paterson, 2010) but also on the degree of damage (Borstad et al., 2013). The approach for the Shallow-Shelf model can be easily generalised to account for two parameter fields, as has been done in numerous studies (e.g. Lee et al., 2015; Barnes et al., 2021). Concerns arise in such cases regarding potential compensating effects of the parameters, which can lead to multiple solutions (Ranganathan et al., 2020), but to date there is not a consensus on whether such 'dual inversions' lead to non-uniqueness.

Alternative Approaches

Arthern and Gudmundsson (2010) proposed another method, which is based on reformulating the optimisation as an inverse Robin problem. The forward model (called here the Neumann problem as a Neumann boundary condition is applied at the surface) is modified to solve a Dirichlet problem, where the surface velocity is now imposed as being equal to observed surface speeds. If the model is perfectly calibrated, these two problems should have the same final state. By introducing a specific cost function proposed initially by Kohn and Vogelius (1984):

$$\mathcal{J}(\boldsymbol{\alpha}) = \int_{\Gamma_s} (\mathbf{v}^N - \mathbf{v}^D) \cdot (\boldsymbol{\sigma}^N - \boldsymbol{\sigma}^D) \cdot \mathbf{n} \, d\Gamma, \quad (6.36)$$

where Γ_s is the ice surface where surface speeds are imposed in the Dirichlet problem, and the superscript N and D denote the velocity and stress tensors from the Neumann and Dirichlet problem solutions. Using this approach, if basal friction is assumed to be $\tau_b = -C\mathbf{v}_b$, the gradient of this cost function with respect C is simply

$$\nabla \mathcal{J} = |\mathbf{v}^D|^2 - |\mathbf{v}^N|^2. \quad (6.37)$$

This method can also be employed to find the rheology factor, B. This approach is simple in the sense that it is easy to implement, but is not flexible in terms of choice of cost function.

6.2.3 Time-Dependent Approaches

Using a time-dependent model (i.e. the stress balance coupled with mass transport) provides stronger constraints for assimilating time-dependent data, and expands the types of data that can be used for constraints in the cost function, such as time-evolving altimetry. It also removes the assumption that all observational data are contemporaneous, which must be made for time-independent assimilation and could lead to inaccuracies for highly dynamic systems.

As in time-independent approaches, minimisation of the cost function involves calculating its derivatives with respect to the assimilation controls – but with a time-dependent model the complexity is increased. A continuous adjoint method (Section 6.2.2) can be applied, as is done in Michel et al. (2014), which assimilates transient surface elevation data using a Shallow-Ice model to estimate ice thickness. With non-local stress balances such as Stokes and Shallow-Shelf, however, model complexity makes such approaches challenging. As such, studies which focus on time-

dependent adjoint-based assimilation have started to adopt algorithmic differentiation (AD).

Algorithmic Differentiation

Algorithmic differentiation calculates gradient information for a given cost function and forward model without the need for explicit derivation of the adjoint equations. To illustrate, we define $\mathbf{x}^k$ as the discretised state vector for a given model at time step k. $\mathbf{x}^k$ includes velocity, ice thickness, and possibly additional prognostic variables (such as ice temperature). A model time step can be expressed as

$$\mathbf{x}^{k+1} = F(\mathbf{x}^k, \boldsymbol{\alpha}^k), \quad (6.38)$$

where $F(\cdot, \cdot)$ represents the calculation of velocities and update of thickness (and other prognostic variables) over a single time step. The parameters, $\boldsymbol{\alpha}$, can, in general, vary per time step (as indicated by the superscript) but can also vary on much longer intervals, or not at all (Goldberg et al., 2015). In the special case where the cost function $\mathcal{J}$ depends only on the final state and the parameters are time-constant (i.e. $\boldsymbol{\alpha}^k \equiv \boldsymbol{\alpha}$), the gradient of the cost function is

$$\begin{aligned}
\nabla \mathcal{J} &= \frac{\partial \mathcal{J}}{\partial \mathbf{x}^k} \frac{d\mathbf{x}^k}{d\alpha} \\
&= \frac{\partial \mathcal{J}}{\partial \mathbf{x}^k} \left(\frac{\partial F}{\partial \alpha} + \frac{\partial F}{\partial \mathbf{x}^{k-1}} \frac{d\mathbf{x}^{k-1}}{d\alpha} \right) \\
&= \frac{\partial \mathcal{J}}{\partial \mathbf{x}^k} \left(\frac{\partial F}{\partial \alpha} + \frac{\partial F}{\partial \mathbf{x}^{k-1}} \left(\frac{\partial F}{\partial \alpha} + \frac{\partial F}{\partial \mathbf{x}^{k-2}} \frac{d\mathbf{x}^{k-2}}{d\alpha} \right) \right) \\
&= \cdots
\end{aligned} \quad (6.39)$$

The expression can be continually expanded using the chain rule for calculus. *Reverse mode* AD implements Eq. (6.39) through repeated left-multiplication of the linear operators. The chain rule is applied not only at the level of a time step but to individual numerical operations within functions and subroutines. It can be shown (Heimbach and Bugnion, 2009) that this computation is mathematically equivalent to solving the adjoint model backward in time. Note that the gradients of F are evaluated at the state $\mathbf{x}^i$, where $i = k, k-1, \ldots$ so solving the adjoint requires either storage or recomputation of the state at each backward time step.

To date, AD has been applied to a number of ice-sheet models for the purpose of time-dependent assimilation. Larour et al. (2014) applied ADOL-C, an AD tool for C++, to a Shallow-Shelf version of the finite-element Ice Sheet System Model (ISSM). The resulting adjoint was used to assimilate surface elevation data derived from ICESat altimetry for the Northeast Greenland Ice Stream over the 2003–9 period. Two assimilations were attempted, one using surface mass balance as a control, and one using basal friction. In each assimilation, the data consisted of annual elevation change in each year.

Both OpenAD, an open-source AD tool for Fortan (Utke et al., 2008), and the commercial tool TAF (Giering et al., 2005) have been used to differentiate STREAMICE, the ice-sheet component of the MITgcm (Goldberg and Heimbach, 2013; Goldberg et al., 2016), a hybrid Shallow-Shelf, Shallow-Ice model. Using the adjoint model, Goldberg et al. (2015) assimilated a dataset consisting of 10 years of annual surface elevation (with velocities available for 4 of these years) for Smith Glacier, a small but fast-thinning marine glacier in West Antarctica. Surface elevation was not available for ice shelves, and the control set consisted of basal friction parameters and the stress tensor along the 1996 grounding line (the latter representing buttressing from the shelves). In contrast with Larour et al. (2014), the controls were either constant in time, or varied linearly in time over the assimilation period. It was found that time-constant controls were sufficient to reproduce the majority of grounding-line variability over the period, and that allowing the controls to vary linearly in time led to marginal reduction of the cost function. The results suggested that observed 2001–11 retreat was due to the state of the glacier in 2001 rather than ice-shelf melting.

Larour et al. (2014) used controls that varied on the model time step. In other words, the number of degrees of freedom to describe the control was $N_\Omega \times N_t$, where N_Ω is the number of degrees of freedom of a finite-element function on the model grid and N_t is number of (monthly) time steps – giving a control dimension far larger than that of the observations. While this demonstrates the power of the adjoint in efficient derivative calculation, the findings of Goldberg et al. (2015) suggest that a low number of temporal degrees of freedom should be investigated first to avoid over-fitting.

Assimilation of Internal Layers

Time-dependent assimilation has also been implemented in the context of inferring past climate and ice-sheet behaviour from internal ice layers. While distinct from the applications of Larour et al. (2014) and Goldberg et al. (2015) in terms of investigative questions and types of data utilised, the computational approach is similar.

Internal reflecting horizons (IRHs) in ice sheets are caused by buried impurities deposited on the surface in past epochs, and can often be traced via airborne ice-penetrating radar over large expanses (Karlsson et al., 2014). As material surfaces, they reflect past deformation, as well as surface and basal mass balance (Bell et al., 2011), and, coupled with borehole data, can provide age constraints, and potentially knowledge of past flow, climate, and basal conditions (Bodart et al., 2021). Age-depth information has been used to infer Holocene surface accumulation rates (e.g. Bodart et al., 2021) as well as records of historic changes in flow direction and speed (e.g. Conway et al., 2002; Siegert et al., 2013). These and similar studies, however, used methods that are either qualitative in nature or did not take heterogeneity in ice-sheet flow into account (Dansgaard et al., 1969).

Waddington et al. (2007), by contrast, minimised the model–data misfit to infer accumulation rate patterns, layer ages, and basal conditions. The forward model of Waddington et al. (2007) is a steady-state thermomechanical Shallow-Ice model in a flow-band ($x - z$) domain, which provides surface elevation and surface velocity as observables. In addition, the model implements a Lagrangian particle-tracking method for particle position (x, z) at time t:

$$\begin{bmatrix} x(t) \\ z(t) \end{bmatrix} = \begin{bmatrix} x_0 \\ s(x_0) \end{bmatrix} + \int_0^t \mathbf{v}(x,z,t')dt', \qquad (6.40)$$

where $\mathbf{v} = (v_x, v_z)^T$ is the ice velocity, and s is the steady surface elevation profile. Equation (6.40) is used to find profiles of layers of a given age. The authors used this model to minimise the misfit to observed surface elevation and velocity as well as layer geometry, using surface mass balance and layer age as controls – providing a time-constant estimate of accumulation. The method was further developed to allow for a time-evolving model (Koutnik and Waddington, 2012), allowing for assimilation of ice-core and radar layer data in the West Antarctic Ice Sheet divide (Koutnik et al., 2016).

Very often, ages are not available for radar layers, making it challenging to infer past ice-sheet conditions. The model of Hindmarsh et al. (2009) uses three-dimensional velocities from a steady Shallow-Ice approximation model to solve the age-depth equation:

$$\frac{D\mathcal{A}}{Dt} = \frac{\partial \mathcal{A}}{\partial t} + \mathbf{v} \cdot \nabla \mathcal{A} = 1, \qquad (6.41)$$

where $\mathcal{A}$ is age (and $\mathcal{A}$ is set to zero at the surface $z = s$) and $D\mathcal{A}/Dt$ is its material derivative. A sophisticated least-square fitting algorithm is then used to identify isolines of modelled $\mathcal{A}$, which best fit observed layers. Karlsson et al. (2014) used this model to evaluate the fit of potential Holocene mass balance scenarios using three-dimensional layer data from Pine Island Glacier.

To date, a formal assimilation of englacial radar data with a continental-scale, three-dimensional dynamic ice flow model has not been carried out. Such assimilation requires cost function gradient information in order to be tractable. In Waddington et al. (2007), the simplicity and limited spatial scale of the model allowed the gradients to be calculated simply by perturbing individual controls to assess cost function partial derivatives, but this finite-difference approach would become intractable for a large-scale study. However, Heimbach and Bugnion (2009) and Logan et al. (2020) successfully applied OpenAD to the ice-sheet model SICOPOLIS (Greve et al., 2011), a thermomechanical ice-sheet model used for studies over paleo-time scales, which implements layer-tracing algorithms. Combined with an increasingly comprehensive set of dated IRH's in the Antarctic and Greenland Ice Sheets (Ashmore et al., 2020;

MacGregor et al., 2015), there is potential for assimilation of ice-sheet histories to now take place.

6.3 Ensemble and Probabilistic Approaches

The methods discussed in Section 6.2 were developed with the aim of minimising a cost function (i.e. of finding an optimal set of parameters to obtain model–data fit). In general, they do not provide formal uncertainties for these parameters – uncertainties which are important to know, especially if the parameters, or the optimised models, are used for analysis or projection. Obtaining uncertainties for inverted parameter sets of large dimension is a challenging problem, and methodologies are only beginning to be developed for glaciological assimilation. In this section, the leading methods are briefly discussed.

6.3.1 Bayes' Theorem
Most probabilistic approaches can be understood as an implementation of Bayes' theorem, which describes how new information changes the likelihood of hidden parameters. Bayes' theorem states that, for a parameter α, and data $\mathbf{y}^{obs}$,

$$p(\alpha|\mathbf{y}^{obs}) = \frac{p(\mathbf{y}^{obs}|\alpha)p(\alpha)}{p(\mathbf{y}^{obs})}. \qquad (6.42)$$

With this formula, the *posterior* probability distribution $p(\alpha|\mathbf{y}^{obs})$ – that is, the likelihood of the parameters conditioned on what is observed – can be quantified. Different probabilistic approaches to assimilation make use of Eq. (6.42) in different ways. In the case of VDA, this leads to a continuous function to be optimised for maximum likelihoods (see Section 6.3.3); while in Monte Carlo analysis (see Sections 6.3.4 and 6.4.3), it is used to directly sample the posterior, and in ensemble methods it is used to assign 'likelihoods' to members of large ensembles. These methods are explored in the following.

6.3.2 Bayesian Calibration via Ensemble-Based Emulation and Dimensional Reduction
As mentioned, Bayesian calibration can be applied to large ensembles of model output, generated by varying key parameters. In such approaches, an ensemble $\{\alpha_1, \ldots, \alpha_N\}$ is generated, where α_i represents a specific realisation of model parameters and its associated model output. Likelihoods are assigned based on a probability distribution of observables. For instance, if a model output or set of outputs yields unlikely results, the likelihoods of the corresponding parameter realisations are scaled by this small probability. It is generally assumed that the parameter realisations are uniformly distributed (i.e. that the prior

distribution $p(\alpha)$ is uniform). Ritz et al. (2015) generated a 1,000-member ensemble of Antarctic-wide simulations from 2000–2200 with GRISLI, a three-dimensional thermomechanical Shallow-Ice model (Quiquet et al., 2018) with a set of 13 parameters governing parameterised marine ice retreat. Likelihoods were assessed based on observed mass loss values from a number of Antarctic catchments and their uncertainties, in order to generate a posterior distribution for Antarctic ice loss over the next two centuries.

The ensemble approach can be computationally expensive when using regional or continental scale models, particularly if the models implement higher-order stress balances or use a high spatial resolution to capture phenomena such as marine ice retreat. Moreover, an ensemble generated *a priori* may not have sufficient resolution in parameter space in the neighbourhood of the point of maximum likelihood (sometimes referred to as the *maximum a posteriori*, or MAP point). For this reason, statistical *emulators* are sometimes used (Edwards et al., 2021). An emulator takes $\widetilde{\mathbf{Y}}$, an *m*-by-*n* matrix in which the i_{th} column contains model outputs (of dimension *m*) from ensemble member *i* (i.e. for the given set of parameters α_i), and carries out a singular value decomposition:

$$\widetilde{\mathbf{Y}} = \mathbf{U}\mathbf{\Sigma}\mathbf{V}^T, \qquad (6.43)$$

where $\mathbf{U}$ and $\mathbf{V}$ are unitary matrices, and $\mathbf{\Sigma}$ is a diagonal matrix containing the *principal values* of the ensemble. All but the largest values of $\mathbf{\Sigma}$ are set to zero, resulting in a mapping between parameter values α_i and a low-dimensional representation of model output. (In Wernecke et al., 2020), five principal values are retained.) This discrete mapping is then statistically interpolated (e.g. with a Gaussian process model) to provide a continuous mapping, which can be queried in a Bayesian calibration based on observations. Rather than using aggregated data as in Ritz et al. (2015), Wernecke et al. (2020) project observational data into principal spatial components for comparison with emulator output.

6.3.3 Bayesian Inference with Laplacian Approximation
Ensemble methods are somewhat limited by the small number of parameters, $(\mathcal{O}(\sim 10))$, that can be considered. As discussed in Section 6.2, the parameters that determine ice-sheet dynamics are spatial fields, requiring, in general, a very large (i.e. $10^4 - 10^6$) number of degrees of freedom to describe them. While these parameter fields likely exhibit spatial correlations, effectively reducing their dimensionality, their correlation scales are unknown, preventing *a priori* dimension reduction. However, there is a connection between the variational data assimilation described in Section 6.2 and Bayesian methods, which can be exploited. Assume a model, which calculates ice-sheet surface velocities, $\mathbf{v}$, based on parameters C, and there are observations, $\mathbf{v}^{obs}$,

for which error is normally distributed. In the following, observations are treated as measurements at discrete locations and described by a vector (although of different dimension than the control parameters C).

The likelihood operator (cf. Eq. (6.42)) can be expressed as

$$p(\mathbf{v}^{obs}|C) \propto e^{-\frac{1}{2}|\mathbf{v}^{obs} - \mathcal{H}(\mathbf{v}(C))|^2_{\Gamma_{obs}^{-1}}}, \qquad (6.44)$$

where $\mathcal{H}$ is the observation operator (introduced in Section 6.1.3). Here the constant of proportionality is simply a normalisation constant which does not bear on the analysis, and which we subsequently ignore. We can equivalently consider the negative log-likelihood:

$$-\log\left(p(\mathbf{v}^{obs}|C)\right) = \frac{1}{2}|\mathbf{v}^{obs} - \mathcal{H}(\mathbf{v}(C))|^2_{\Gamma_{obs}^{-1}}. \qquad (6.45)$$

Here, Γ_{obs} is the *observational covariance matrix*, and the norm $|\mathbf{x}|_{\Gamma_{obs}^{-1}}$ is defined as $\sqrt{x^T\Gamma_{obs}^{-1}x}$. If the prior distribution $p(C)$ is also characterised as a multivariate Gaussian with covariance Γ_{prior} and mean C_0, the negative log of the posterior distribution can be written

$$-\log\left(p(C|\mathbf{v}^{obs})\right) = \frac{1}{2}|\mathbf{v}^{obs} - \mathcal{H}(\mathbf{v}(C))|^2_{\Gamma_{obs}^{-1}} + \frac{1}{2}|C - C_0|^2_{\Gamma_{prior}^{-1}}. \qquad (6.46)$$

Note that both Γ_{prior} and Γ_{obs} are symmetric positive-definite matrices.

It can be shown (e.g. Isaac et al., 2015) that finding the point in parameter space with maximum posterior likelihood, C_{MAP}, is equivalent to variational data assimilation with the appropriate data and model norms. If Γ_{obs} is diagonal, then the first term on the right-hand side of Eq. (6.46) is equivalent to the misfit norm (Eq. (6.8)).

Additionally, the second term on the right-hand side of Eq. (6.46) corresponds to the discretised form of a wide range of commonly used regularisation cost terms, or model norms. For instance, if a finite-element model is used and C_0 is spatially invariant, the regularisation cost term in Eq. (6.10) can be written

$$\gamma \int_\Omega |\nabla C|^2 dx = (C - C_0)^T \mathbf{A}(C - C_0), \qquad (6.47)$$

where C is a vector representing the coefficient vector of the finite-element function C and a_{ij}, the entries of $\mathbf{A}$, are $\int_\Omega \gamma \nabla\varphi_i \cdot \nabla\varphi_j \, d\Omega$, with $\varphi_{i,j}$ the finite-element basis functions. Note that $\mathbf{A}$ as defined here is not (the inverse of) a true covariance matrix, as it is not positive definite. Barnes et al. (2021) define the regularisation term for the U´ a ice model as

$$\int_\Omega \gamma|\nabla(C - C_0)|^2 + \delta(C - C_0)^2 dx. \qquad (6.48)$$

When discretised, this term expresses a proper prior distribution of C. In such a prior, δ expresses our

confidence in the value of C at a location, while γ expresses our confidence in its persistence over a certain length scale. This regularisation term uses a gradient operator, but such an operator leads to infinite pointwise variance (Bui-Thanh et al., 2013), and so methods which quantify posterior uncertainty with Eq. (6.46) use Laplacian-based regularisation (Isaac et al., 2015).

The equivalence between cost function and negative-log posterior density (with the caveat that the regularisation term must be a proper probability distribution) suggests that minimising a deterministic function with respect to the control parameters is equivalent to finding the point of maximum likelihood, or *mode*, of the posterior distribution. This alone does not aid in quantifying the uncertainty of C, however: the probability space is of very high dimension (tens to hundreds of thousands in realistic applications) and querying the posterior density requires solving a large-scale non-linear model (Martin et al., 2012). A *local* approximation of the posterior density can be found through a second-order approximation of Eq. (6.46) in the neighbourhood of C_{MAP}. It can be shown that the covariance of this local approximation is given by the inverse of the *Hessian* of the cost function at C_{MAP}:

$$\Gamma_{post}^{-1} = \mathbf{H}(C_{MAP}) = \frac{\partial^2 \mathcal{J}}{\partial C^2}\Big|_{C_{MAP}} = \mathbf{H}_{mis}(C_{MAP}) + \Gamma_{prior}^{-1}, \qquad (6.49)$$

where $\mathbf{H}_{mis}(C_{MAP})$ is the Hessian of the data norm only. Essentially, the approach approximates $p(C|\mathbf{v}^{obs})$ by $\mathcal{N}(C_{MAP}, \mathbf{H}(C_{MAP})^{-1})$, that is, a Gaussian distribution with mean C_{MAP} and covariance $\mathbf{H}(C_{MAP})^{-1}$. This approximation is sometimes referred to as a *Laplacian* approximation.

Calculating the complete Hessian is challenging as it is a large, dense matrix requiring extensive computation, and inverting directly is even more challenging as it is often singular or near-singular. Previous authors (Bui-Thanh et al., 2013; Petra et al., 2014; Isaac et al., 2015) find a low-rank approximation to the Hessian that is easy to invert (see also Kalmikov and Heimbach, 2014). Details can be found in these and other references, but the low-rank approximation is generated via a singular value decomposition of the *prior-preconditioned* misfit Hessian, $\Gamma_{prior}\mathbf{H}_{mis}$. The space of C is decomposed into principal components – with the significance that the leading principal components, or directions, are those most strongly constrained by the assimilation (Koziol et al., 2021).

Together with an adjoint of the forward model, the parameter uncertainty Γ_{post} can be projected to estimate uncertainties related to various scalar Quantities of Interest (*QoIs*), such as future sea level rise contribution. Using properties of joint Gaussian distributions,

$$\sigma^2_{QoI} = \left(\frac{\partial QoI}{\partial C}\right)^T \Gamma_{post} \left(\frac{\partial QoI}{\partial C}\right),\tag{6.50}$$

where σ^2_{QoI} is the variance of the Quantity of Interest. (σ_{QoI}, the standard deviation, is then a proxy for uncertainty). Isaac et al. (2015) used this framework to estimate posterior uncertainty of sliding parameters in a full-Stokes model of Antarctica, and the resulting uncertainty in ice loss. In this work, their parameter dimension is approximately 1.2 million, but only ~1,000 principal components are retained in their low-rank approximation. The framework represents a powerful method to deal with uncertainty with a parameter space of very high dimension, but it is computationally challenging for a number of reasons, which we discuss in Section 6.3.6. The discussion here emphasises the connection between cost functions and posterior densities. However, it is important that the observational covariance be taken into account correctly. Equation (6.45) is an inner product involving modelled and observed values at discrete locations, and these locations do not depend on model resolution. On the other hand, Eq. (6.30) is a functional which considers observations to be a continuous function, and many inversions use integral-based cost functions. Such an approach introduces mesh-dependent terms into the posterior covariance (Koziol et al., 2021). Essentially, treating such a cost function as a likelihood function implies a very high density of independent observations, effectively adding information that does not exist and therefore underestimating parameter uncertainty.

6.3.4 Stochastic Newton MCMC Methods

The posterior distribution of parameter uncertainty, Eq. (6.46), is non-Gaussian (even if the observational and prior distributions are Gaussian). As such, the Hessian-based uncertainty quantification (UQ) framework described in Section 6.3.3 is an approximation, and potentially a poor one in the case of highly skewed distributions. Markov-Chain Monte Carlo (MCMC) methods provide a means for sampling the posterior density, but the high-dimensional probability spaces and computational expense of the forward model make this challenging. Stochastic Newton MCMC methods build on the Hessian-based methods discussed in Section 6.3.3 to make this more tractable.

Markov-Chain Monte Carlo methods generate new samples of a probability space by selecting based on a *proposal density* centred at the previous sample. For a given sample α_t, Stochastic Newton MCMC uses as its proposal density the local approximation to the posterior given by

$$\mathcal{N}(\alpha_t, \mathbf{H}(\alpha_t)^{-1}).\tag{6.51}$$

Importantly, the evaluation of the posterior density is not approximated – only the basis on which new samples are

taken. The low-rank approximation discussed in Section 6.3.3, which retains the directions in probability space which are most constrained by the assimilation, ensures these directions are explored.

Stochastic Newton MCMC has been shown to accelerate 'standard' MCMC schemes by two orders of magnitudes in seismic inversion problems (Martin et al., 2012). However, it requires a new, Hessian-based approximation to the posterior for each new sample, which is computationally costly. The approach can be modified, however, by computing the Hessian only at the MAP point, and using the corresponding density as a proposal distribution, rather than recomputing the proposal density with every sample. It was found that using the MAP-based Hessian led to convergence at least as rapidly as the original Stochastic Newton MCMC method.

The problems considered in Martin et al. (2012) and Petra et al. (2014) were of large dimension ($\mathcal{O}(\sim100)$) relative to the ensemble-based methods discussed in Section 6.3, but quite small compared to adjoint-based inversions in glaciology (e.g. Cornford et al., 2015). Thus, more work is needed to determine whether such methods are tractable for continental ice-sheet models.

6.3.5 Ensemble Kalman Filter Approach

The time-dependent variational approach in Section 6.2.3 has similarities to a variational assimilation approach developed in numerical weather prediction (NWP), 4DVAR. Another method used in NWP is the EnKF (Lorenc, 2003), essentially a Monte Carlo analysis in which an ensemble of model states are run forward in time. The states are updated at discrete time intervals based on observational data.

Gillet-Chaulet (2020) applied EnKF techniques to glaciological assimilation. Only the basic idea of the algorithm is given here; for further details, please see Gillet-Chaulet (2020). The forward model used in this study is a Shallow-Shelf Approximation. The model 'state' considered is composed of (time-dependent) ice surface elevation and (time-independent) bed elevation and basal friction, and observations are of surface elevation and surface velocity, available at times t_k. An ensemble of state vectors $\{x\}$ is generated, where x_i represents the discretised fields comprising the state for realisation i. The filter then consists of a sequence of forecast-analysis steps. Each member of $\{x\}$ is evolved from time t_{k-1} to time t_k, yielding the *forecast* $\{x\}^{f,k}$, the *forecast* ensemble at time t_k. An update is then applied to generate the *analysis* ensemble:

$$\mathbf{S}^{a,k} = (\mathbf{B}^T \Gamma_{obs} \mathbf{B} + \mathbf{S}^{f,k})^{-1},\tag{6.52}$$

$$\bar{\mathbf{x}}^{a,k} = \bar{\mathbf{x}}^{f,k} + \mathbf{S}^{a,k} \mathbf{B}^T (\Gamma^k_{obs})^{-1} (\mathbf{y}^k_{obs} - \mathcal{H}(\mathbf{x}^{f,k})).\tag{6.53}$$

Here $\bar{\mathbf{x}}^{f,k}$ and $\mathbf{S}^{f,k}$ are the mean and covariance matrix of the forecast ensemble, $\bar{\mathbf{x}}^{a,k}$ and $\mathbf{S}^{a,k}$ similarly for the analysis, and (Γ^k_{obs}) is the observational covariance at t_k. The parameter-to-observable operator is $\mathcal{H}$, and its linearised form is **B**. The *analysis* ensemble is then evolved in time to t_{k+1}. Under the assumption of Gaussianity, Eqs. (6.52) and (6.53) represent a *Bayesian update* of the state.

Gillet-Chaulet (2020) implements a type of EnKF called the error subspace ensemble transform Kalman filter (ESTKF). In applications such as ice-sheet assimilation, the size of the ensemble is much smaller than the state dimension, meaning the forecast covariance matrix is a low-rank approximation of the full covariance matrix. The ESTKF reformulates the analysis update as a linear combination of the forecast ensemble, such that Eqs. (6.52) and (6.53) are satisfied. To address spurious correlations within the covariance matrix due to the small ensemble size, covariances over long-length scales are set to zero in an ad hoc manner (known as localisation).

Gillet-Chaulet (2020) apply their ESTKF to the finite-element model Elmer/Ice in an idealised, one-dimensional configuration and use synthetic observations taken over a 35-year period.

The model dimension is $\sim 10^4$ and ensemble sizes of $\sim 10^2$ are used. By testing forecasts from the ESTKF against those of the 'true' model state, it is seen that the fidelity of forecasts is greatly improved by assimilating over time periods of rapid change. However, the method has yet to be tested in larger-scale or more realistic settings.

Figure 6.1 provides a schematic diagram showing the key differences between the two approaches to assimilation of time-varying data discussed: the time-dependent variational approach discussed in Section 6.2.3; and the EnKF approach.

6.3.6 Comparison of Probabilistic Methods

The preceding sections discuss a number of diverse approaches, which aim to quantify uncertainty in ice-sheet assimilation for the purpose of estimating forecast uncertainty. The approaches attempt, in various ways, to resolve trade-offs between large model dimension, large parameter dimension, full representation of probability spaces, and computational complexity. Hessian-based uncertainty quantification and ensemble Kalman filters approximate probability densities as Gaussian, potentially biasing estimates of uncertainty. Stochastic Newton MCMC makes a Gaussian approximation in its proposal density, while it queries the true posterior distribution.

Meanwhile, ensemble methods make no assumptions regarding the form of the posterior and have potential to reveal multimodal distributions; however, such methods are limited to parameter dimensions on the order of ~ 10, while Hessian-based and EnKF methods allow parameter dimensions on the order of $\sim 10^4 - 10^6$ (Note that Stochastic Newton MCMC can theoretically accommodate large parameter dimensions as well, but such studies have not yet been implemented in glaciology). Additionally, where higher-order models are used to generate ensembles, each ensemble member may involve a deterministic inversion involving tens of thousands of parameters (Nias et al., 2016; Wernecke et al., 2020) – and the uncertainty inherent in these inversions is not accounted for. An exception is the study of Brinkerhoff et al. (2021), who use emulation to constrain parameter uncertainty in a coupled model of

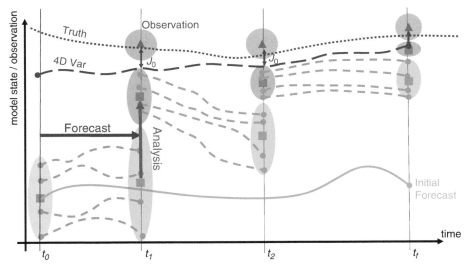

Figure 6.1 A schematic diagram (adapted from Gillet-Chaulet, 2020) describing the two leading approaches to assimilating time-dependent glaciological data into ice-sheet models, VDA, or 4DVar (Section 6.2.3), and ensemble Kalman filtering (EnKF; Section 6.3.5). The true time-evolving 'state' of the ice sheet between t_0 and t_f is represented by the blue dotted line, with error-prone observations at intermediate times t_1 and t_2. The 4DVar approach, beginning from the initial forecast (cyan line) generated using *a priori* internal parameters, iteratively adjusts its parameters to reduce the model data misfit, $\mathcal{J}_0$ (as indicated by the dashed violet line). EnKF generates multiple forecasts over each analysis window ($[t_0, t_1]$, $[t_1, t_2]$, etc.). Importantly, 4DVar adjusts its parameters *globally* in each iteration, meaning the final result is a continuous, physically consistent trajectory. EnKF introduces 'jumps' in each analysis step, but also provides estimates of uncertainty, while 4DVar generates only the optimal state.

subglacial hydrology and higher-order glacier physics – and therefore directly model basal slipperiness rather than representing it with a high-dimensional parameter space.

The methods all consider low-rank approximations in order to make their computations tractable. Ensemble and EnKF methods carry out principal component decompositions on computed ensembles, meaning the methods ultimately depend on the information contained within the ensembles, which must be generated prior to the analysis. By contrast, Hessian-based UQ and Stochastic Newton MCMC prioritise the directions of parameter space that are most strongly determined by the data. Thus, the components analysed are, by design, the most important to consider.

Ensemble and EnKF methods do not require extensive modifications to the ice-sheet models themselves. Emulation-based ensemble studies require only the output of a model ensemble, and in EnKF methods the analysis computations can be run in an 'offline' mode, where the forecast state is processed externally to provide the analysis state. Hessian-based UQ and Stochastic Newton MCMC for non-linear models require second-derivative information; such information requires either higher-order AD (Kalmikov and Heimbach, 2014; Maddison et al., 2019) or the solution of differential equations not solved by most ice-sheet models (Petra et al., 2012; Isaac et al., 2015). Applying AD remains a challenge for established ice-sheet models, with successes primarily only where the model code adheres to a certain structure (Goldberg and Heimbach, 2013; Goldberg et al., 2016; Logan et al., 2020). There is growing use of automated code generation finite-element libraries such as FEniCS and Firedrake in glaciological model development (Kyrke-Smith et al., 2017; Brinkerhoff et al., 2021), and features of these libraries can easily be exploited to find higher-order functional derivatives (Maddison et al., 2019; Koziol et al., 2021). It also may be sufficient in certain cases to avoid high-order derivative calculation by using the Gauss–Newton approximation to the Hessian (Koziol et al., 2021; Shapero et al., 2021):

$$\mathbf{H}_{mis} \approx \mathbf{B}^T \Gamma_{obs}^{-1} \mathbf{B}, \tag{6.54}$$

where $\mathbf{B}$ is the gradient of the observation operator (cf. Section 6.3.3). Essentially, AD and derivative calculation for probabilistic ice-sheet data assimilation is not as great a barrier as it once was.

All of the methods discussed here have strengths and weaknesses and the ideal method will depend on the problem being investigated. The way forward may lie not in choosing a method, but in combining different methods. For instance, elements of Hessian-based UQ and Stochastic Newton MCMC may aid in generating ensembles for EnKF which efficiently capture variability near the mode of the posterior and enable smaller ensembles.

6.3.7 Alternative Approaches

Many of the issues mentioned in Section 6.3.6 (non-Gaussianity, computational complexity) arise because of the non-linearity of the glaciological flow equations. A completely different approach has been proposed by Raymond and Gudmundsson (2009), which makes use of a linear theory developed by Gudmundsson (2003) to estimate basal properties of glaciers using a probabilistic Bayesian approach. In this approach, basal topography and slipperiness are inferred from surface elevation and velocity. However, due to the linear approximations made, the method is limited to small perturbations around a known reference state. The framework is not suitable for constraining projection uncertainty but may be effective in characterising basal environments. To date, the framework has been applied to a flowline on Rutford Ice Stream (Pralong and Gudmundsson, 2011).

6.3.8 Model Uncertainty

The likelihood operator as written (Eq. (6.45)), implicitly assumes that the model $v(\alpha)$ is *perfect*, and that any uncertainty is due to imperfect observations. In truth, there is a great degree of *model uncertainty* (i.e. error in mathematically and numerically representing the physical system). Model uncertainty is extremely difficult to characterise (Chatfield, 1995), and various studies take different approaches. The idealised studies of Gillet-Chaulet (2020) and Koziol et al. (2021) ignore model uncertainty completely; Wernecke et al. (2020) consider model uncertainty as uncorrelated Gaussian noise based on the variance inherent in their ensemble; and Werder et al. (2020) (see Section 6.4.3) treats per-glacier model uncertainty as additional parameters in the Bayesian estimation.

Babaniyi et al. (2021) applied a Bayesian Approximation Error (BAE) approach to estimate ice-sheet basal properties in the presence of rheological uncertainty. Ice-sheet rheology is governed by parameters which are themselves uncertain (such as the Glen's law exponent, n). The BAE step involves generating an ensemble of model output by sampling from prior distribution parameters, and using the ensemble statistics to produce a modified likelihood covariance matrix which accounts for the impacts of rheological uncertainty. The BAE framework could, in theory, be applied to a range of structural uncertainties in ice-sheet models, leading to a more accurate accounting of model uncertainty – but this, as well as application to realistic problems, still needs to be done.

6.4 Reconstructing of Bed Topography and Ice Thickness

Data assimilation methods have also been employed to map the beds of glaciers and ice sheets. Ice thickness and bed topography are most efficiently measured by ice-penetrating radar (either ground-based or airborne; Evans and Robin, 1966), but the vast majority of mountain glaciers, and many parts of the ice sheets remain unmapped. Data assimilation offers a cost-effective way to construct estimates where no data are available. Several approaches have been proposed (see, e.g., the Ice Thickness Models Intercomparison eXperiment projects; Farinotti et al., 2017, 2021), and we will only focus on a few popular ones. Most approaches are based on the conservation of mass (Eq. (6.6)), the conservation of momentum (Eq. (6.1)) or a combination of the two.

6.4.1 Two-Step Methods

The majority of bed/thickness mapping methods rely on a two-step approach, which consists of using the conservation of mass first, followed by the conservation of momentum.

Flux Conservation

In the first step, we rely on the mass transport equation to determine the depth-integrated flux $\mathbf{q} = q \, \mathbf{n} = H\bar{\mathbf{v}}$:

$$\nabla \cdot q \, \mathbf{n} = \dot{m}_s - \dot{m}_b - \frac{\partial H}{\partial t} = \dot{a}, \qquad (6.55)$$

where $\mathbf{n}$ is a unit vector describing the two-dimensional direction of the depth-integrated flux, q, and $\dot{a}$ is the right-hand side, sometimes referred to as *apparent mass balance*. If the right-hand side and the flux directions are known, the distribution of the flux can be determined by using a routing algorithm (e.g. Budd and Warner, 1996) or by solving this transport equation using standard approaches such as finite-elements or finite-differences.

To estimate the flux directions, one can use surface gradients, as dictated by the Shallow-Ice Approximation. This approximation will only be valid at a length scale of a few ice thicknesses and some smoothing of the surface topography must be applied. Another possible approach is to use observed surface velocities instead, but many mountain glaciers remain poorly mapped.

For the right-hand side, the basal melt term is generally negligible under grounded ice compared to the other terms. The surface mass balance can come from regional climate models but is, most of the time, simply parameterised as a function of the elevation with respect to the Equilibrium-Line Altitude (ELA), the presence of debris and other parameters (e.g. Farinotti et al., 2009). The thinning rate is generally ignored, but can be estimated from surface altimetry.

Momentum Balance

The flux conservation step yields a complete map of the depth-integrated flux $q = H|\bar{\mathbf{v}}|$. In order to reconstruct the ice thickness and bed topography, an estimate of the depth-averaged velocity $\bar{v} = |\bar{\mathbf{v}}|$ is needed. The vast majority of bed

mapping methods for mountain glaciers relies on the Shallow-Ice Approximation (SIA; Hutter, 1983), which is a zeroth-order approximation of ice sheet flow and only accounts for vertical shear stresses. Under the SIA assumptions, the ice velocity reads:

$$\mathbf{v}(z) = \mathbf{v}_b - 2(\rho g)^n |\nabla s|^{n-1} \nabla s \int_b^z (s - z)^n A \, dz, \qquad (6.56)$$

where $\mathbf{v}_b$ is the basal sliding speed, and s is the surface elevation. If it is assumed that ice is isothermal (i.e. A is uniform with depth), and that the ice basal velocity is negligible ($\mathbf{v}_b = \mathbf{0}$), the ice thickness can be reconstructed as:

$$H = \left(\frac{(n+2)q}{2A(c\rho g)^n |\nabla s|^n} \right)^{1/(n+2)}, \qquad (6.57)$$

where c is a parameter introduced to account for potential sliding. This method, or similar approaches, have been used in many studies (e.g. Fastook et al., 1995; Warner and Budd, 2000; Farinotti et al., 2009; Huss and Farinotti, 2012) and led to the first global mapping of mountain glacier volumes (Huss and Farinotti, 2012).

Fürst et al. (2017) proposed a more advanced approach, where the surface mass balance is first used as a control in Eq. (6.55) in order to have a flux that is positive and smooth, and use the SIA to infer the bed topography. The flow rate factor is calibrated with radar-derived thickness measurements. Then, in a second step, a second optimisation is performed in regions where surface velocities are available. In these regions, another cost function is minimised that measures the misfit between the calculated ice thickness and radar measurements (together with other terms to ensure that the inferred ice thickness is physically consistent). This second optimisation is only performed in regions of fast flow, where ice flow is primarily controlled by basal sliding, and the controls are the surface mass balance and ice velocities. This approach was first applied to Svalbard (Fürst et al., 2018).

6.4.2 Mass Conservation with Observed Velocities

There is now complete coverage of surface velocities for both ice sheets and several large ice caps from satellite interferometry (Rignot et al., 2011; Rignot and Mouginot, 2012; Joughin et al., 2018). In this case, one can rely directly on these observations instead of the SIA. However, assumptions are needed regarding how to convert these surface velocities to depth-averaged velocities. In some studies (Rasmussen, 1988; Morlighem et al., 2010), it is assumed that surface velocities are good approximations for depth-averaged velocities. In the worst-case scenario, when there is no basal sliding, Eq. (6.56) shows that depth-averaged velocities can be 20% smaller than surface velocities ($\bar{v} = (n+1)/(n+2) \, v(s)$). In regions of fast flow, far from the ice divide and along ice streams, this is an excellent approximation that avoids having to introduce basal conditions and rheological factors. In this case, the ice thickness is estimated by solving for ice flux (Eq. (6.55)), where the ice thickness is the unknown (i.e. the velocity and velocity directions are given). This hyperbolic equation requires measurements of ice thickness at the inflow boundary, Γ_-, of the domain in order to properly constrain the system:

$$\begin{cases} \nabla \cdot H\overline{\mathbf{v}} = \dot{a} & \text{in } \Omega \\ H = H_{obs} & \text{on } \Gamma_- \end{cases}. \tag{6.58}$$

Morlighem et al. (2011, 2013a) further refined this method by assimilating radar-derived ice thickness measurements in order to constrain the calculation so that the final bed map is as close as possible to observations where they are available. To account for all measurements of ice thickness H_{obs}, along flight tracks T, that lie within the model domain Ω, the following cost function is minimised:

$$\mathcal{J}(H) = \int_T \frac{1}{2}(H - H_{obs})^2 dT + \int_\Omega \gamma_\parallel (\nabla H \cdot \mathbf{n}_\parallel)^2 d\Omega$$
$$+ \int_\Omega \gamma_\perp (\nabla H \cdot \mathbf{n}_\perp)^2 d\Omega, \tag{6.59}$$

where $\mathbf{n}_\parallel$ and $\mathbf{n}_\perp$ are unit vectors parallel/perpendicular to ice velocity, and $\gamma_\parallel$ and $\gamma_\perp$ are constant regularisation parameters. The controls are the observed velocity and the apparent mass balance. These input fields are allowed to change within their respective error margins. This method led to the development of BedMachine Greenland (Morlighem et al., 2014, 2017) and BedMachine Antarctica (Morlighem et al., 2020).

6.4.3 Bayesian Method

Werder et al. (2020) proposed to combine the two-step approach from Huss and Farinotti (2012) with a Bayesian inversion scheme. In Werder et al. (2020), the parameters α (cf. Eq. (6.42)) are apparent mass balance (parameterised in Huss and Farinotti, 2012), ice temperature, basal sliding factor, and extrapolation parameters for thickness and velocity. The last two parameters are considered because the method of Huss and Farinotti (2012) collapses each glacier into a flowline, and then extends to a thickness map through an ad-hoc extrapolation. The observations considered are surface velocities (which the model of Huss and Farinotti, 2012 was augmented to calculate) and (where available) thickness measurements. A Markov-Chain Monte Carlo (MCMC) method is used, allowing for efficient investigation of the posterior probability density. The method is very powerful as it does not presume the form of the posterior distribution (as opposed to methods which make the Laplacian approximation) and allows a means to incorporate the growing repository of glacier velocities. However, the parameters are not spatially resolved within a glacier due to computational expense, even with the use of MCMC.

6.4.4 Simultaneous Mass and Momentum Method

Perego et al. (2014) introduced a variational assimilation method, which infers both the basal sliding parameter C (c. f. Section 6.2) and ice-sheet topography. The method uses an approach similar to that described in Section 6.2.2 – with key differences being that the cost function includes a term penalising any imbalance in the mass-transport equation, and also that ice thickness (or equivalently ice bed) is included as a control parameter. The method was developed primarily to initialise climate-forced ice-sheet models without large non-physical transients in thickness evolution, and was successfully applied to a model of the Greenland Ice Sheet.

As with other 'dual' inversions for multiple parameters (Section 6.2.2), there is risk of parameter mixing and non-uniqueness of solution (Goldberg and Heimbach, 2013). For this reason, it may be a preferable strategy to adopt an approach for bed inversion which does not depend on the momentum balance (such as mass continuity) and then use the resulting topography to infer ice-dynamical parameters.

6.4.5 Artificial Intelligence and Machine Learning

Neural networks are being used more and more frequently in environmental science to improve the computational efficiency of certain processes in numerical model (Krasnopolsky and Schiller, 2003), or to infer unknown quantities from a set of input data.

One of the first applications of neural networks in glaciology is from Clarke et al. (2009), who used a neural network to infer the subglacial topography of mountain glaciers from surface topography with some success.

More recently, Leong and Horgan (2020) proposed a new technique that is based on adapted architecture of the enhanced super-resolution Generative Adversarial Network (GAN) to reconstruct basal roughness in bed topography maps from ice surface elevation, velocity, and snow accumulation. The neural network is trained in a few regions of Antarctica where high-resolution (250 m) bed elevation maps are available, and applied over the entire ice sheet to generate high resolution bed topography. Meanwhile Jouvet et al. (2022) and Jouvet (2023) introduce an approach, which a deep learning-based ice dynamics emulator acts as the forward model in the assimilation of observations. The emulator is trained on the output of an ensemble of runs carried out with a physically-based glacier model. These sorts of new approaches have a lot of potential as the amount of remote sensing data collected in polar regions increases.

6.5 Perspectives and Outlook

While data assimilation has been an important and popular research topic since the 1990s, glaciology has been rapidly shifting from a data-poor state to a data-rich one. In the 1990s, modellers only had partial coverage of ice velocity and the thinning rates were mostly unknown. Data assimilation was a powerful tool to add knowledge to the model, by inferring boundary or initial conditions. Today, we have time series of complete velocity maps and thinning rate maps at an ever-increasing spatial and temporal resolution. Modellers are not yet equipped to take advantage of this data revolution, as the vast majority of glaciological assimilation platforms cannot assimilate this breadth of information. There may additionally be data redundancies (e.g. ice velocities may be of such high resolution that more resolved velocities do not add information); however, without sophisticated uncertainty quantification, these will go undetected. In order that both existing and newly collected data are used to their full potential in improving projections of sea level, transformative data science and machine learning solutions may be needed.

References

Arthern, R. J. (2015). Exploring the use of transformation group priors and the method of maximum relative entropy for Bayesian glaciological inversions. *Journal of Glaciology*, 61(229), 947–62.

Arthern, R. J., and Gudmundsson, G. H. (2010). Initialization of ice-sheet forecasts viewed as an inverse Robin Problem, *Journal of Glaciology*, 56(197), 527–33.

Ashmore, D. W., Bingham, R. G., Ross, N. et al. (2020). Englacial architecture and age-depth constraints across the West Antarctic ice sheet. *Geophysical Research Letters*, 47(6), e2019GL086663. https://doi.org/10.1029/2019GL086663.

Babaniyi, O., Nicholson, R., Villa, U., and Petra, N. (2021). Inferring the basal sliding coefficient field for the stokes ice sheet model under rheological uncertainty. *The Cryosphere*, 15(4), 1731–50.

Barnes, J. M., dos Santos, T. D., Goldberg, D., Gudmundsson, G. H., Morlighem, M., and De Rydt, J. (2021). The transferability of adjoint inversion products between different ice flow models. *The Cryosphere*, 15(4), 1975–2000. https://doi.org/10.5194/tc-15-1975-2021.

Bell, R. E., et al. (2011). Widespread persistent thickening of the east Antarctic Ice Sheet by freezing from the base. *Science*, 331(6024), 1592–5. https://doi.org/10.1126/science.1200109.

Blatter, H. (1995). Velocity and stress-fields in grounded glaciers: A simple algorithm for including deviatoric stress gradients. *Journal of Glaciology*, 41(138), 333–44.

Bodart, J. A., Bingham, R. G., Ashmore, D. et al. (2021). Age-depth stratigraphy of Pine Island Glacier inferred from airborne radar and ice-core chronology. *Journal of Geophysical Research: Earth Surface*, 126(4), e2020JF005,927.

Borstad, C. P., Rignot, E., Mouginot, J., and Schodlok, M. P. (2013). Creep deformation and buttressing capacity of damaged ice shelves: Theory and application to Larsen C ice shelf. *The Cryosphere*, 7, 1931–47. https://doi.org/10.5194/tc-7-1931-2013.

Brinkerhoff, D., Aschwanden, A., and Fahnestock, M. (2021). Constraining subglacial processes from surface velocity observations using surrogate-based Bayesian inference. *Journal of Glaciology*, 67(263), 385–403.

Bryson, A. E., and Ho, Y.-C. (2018). *Applied Optimal Control: Optimization, Estimation, and Control*. Boca Raton, FL:CRC Press.

Budd, W. F., and Warner, R. C. (1996). A computer scheme for rapid calculations of balance-flux distributions. *Annals of Glaciology*, 23, 21–27.

Bui-Thanh, T., Ghattas, O., Martin, J., and G. Stadler. (2013). A computational framework for infinite-dimensional Bayesian inverse problems Part I: The linearized case, with application to global seismic inversion. *SIAM Journal on Scientific Computing*, 35(6). https://doi.org/10.1137/12089586X.

Chandler, D. M., Hubbard, A. L., Hubbard, B., and Nienow, P. (2006). A Monte Carlo error analysis for basal sliding velocity calculations. *Journal of Geophysical Research: Earth Surface*, 111(F4). https://doi.org/10.1029/2006JF000476.

Chatfield, C. (1995). Model uncertainty, data mining and statistical inference. *Journal of the Royal Statistical Society: Series A (Statistics in Society)*, 158(3), 419–44.

Clarke, G. K. C., Berthier, E., Schoof, C. G., and Jarosch, A. H. (2009). Neural networks applied to estimating subglacial topography and glacier volume. *Journal of Climate*, 22(8), 2146–60. https://doi.org/10.1175/2008JCLI2572.1.

Conway, H., Catania, G., Raymond, C. F., Gades, A. M., Scambos, T. A., and Engelhardt, H. (2002). Switch of flow direction in an Antarctic ice stream. *Nature*, 419(6906), 465–67. https://doi.org/10.1038/nature01081.

Cornford, S. L., et al. (2015). Century-scale simulations of the response of the West Antarctic Ice Sheet to a warming climate. *The Cryosphere*, 9, 1579–600. https://doi.org/10.5194/tc-9-1579-2015.

Cuffey, K. M., and Paterson, W. S. B. (2010). *The Physics of Glaciers*, 4th ed. Oxford: Elsevier.

Dansgaard, W., Johnsen, S. J., Møller, J., and Langway, C. C. (1969). One thousand centuries of climatic record from camp century on the Greenland ice sheet. *Science*, 166(3903), 377–80.

Edwards, T., et al. (2021). Quantifying uncertainties in the land ice contribution to sea level rise this century. *Nature*, 593. https://doi.org/10.1038/s41586-021-03302-y.

Evans, S., and Robin, G. d. Q. (1966). Glacier depth-sounding from air. *Nature*, 210(5039), 883–5. https://doi.org/10.1038/210883a0.

Farinotti, D., Huss, M., Bauder, A., Funk, M., and Truffer, M. (2009). A method to estimate the ice volume and ice-thickness distribution of alpine glaciers. *Journal of Glaciology*, 55(191), 422–30.

Farinotti, D., et al. (2017). How accurate are estimates of glacier ice thickness? Results from ITMIX, the Ice Thickness Models Intercomparison eXperiment. *The Cryosphere*, 11(2), 949–70. https://doi.org/10. 5194/tc-11-949-2017.

Farinotti, D., et al. (2021). Results from the Ice Thickness Models Intercomparison eXperiment Phase 2 (ITMIX2). *Frontiers in Earth Science*, 8, 484. https://doi.org/10.3389/feart.2020.571923.

Fastook, J. L., Brecher, H. H., and Hughes, T. J. (1995). Derived bedrock elevations, strain rates and stresses from measured surface elevations and velocities: Jakobshavns Isbrae, Greenland. *Journal of Glaciology*, 41(137), 161–73.

Furst, J. J., Durand, G., Gillet-Chaulet, F. et al. (2015). Assimilation of Antarctic velocity observations provides evidence for uncharted pinning points. *The Cryosphere*, 9, 1427–43. https://doi.org/10.5194/tc-9-1427-2015.

Fürst, J. J., et al. (2017). Application of a two-step approach for mapping ice thickness to various glacier types on Svalbard. *The Cryosphere*, 11(5), 2003–32. https://doi.org/10.5194/tc-11-2003-2017.

Fürst, J. J., et al. (2018). The ice-free topography of Svalbard. *Geophysical Research Letters*, 45(21), 11,760–9. https://doi.org/10.1029/2018GL079734.

Giering, R., Kaminski, T., and Slawig, T. (2005). Generating efficient derivative code with TAF. *Future Generation Computer Systems*, 21(8), 1345–55. https://doi.org/10.1016/j.future.2004.11.003.

Gilbert, J. C., and Lemaréchal, C. (1989). Some numerical experiments with variable-storage quasi-Newton algorithms. *Mathematical Programming*, 45(1–3), 407–35. https://doi.org/10.1007/BF01589113.

Gillet-Chaulet, F. (2020). Assimilation of surface observations in a transient marine ice sheet model using an ensemble Kalman filter. *The Cryosphere*, 14, 811–32. https://doi.org/10.5194/tc-14-811-2020.

Gillet-Chaulet, F., Gagliardini, O., Seddik, H. et al. (2012). Greenland Ice Sheet contribution to sea-level rise from a new-generation ice-sheet model. *The Cryosphere*, 6, 1561–76. https://doi.org/10.5194/tc-6-1561-2012.

Glen, J. W. (1955). The creep of polycrystalline ice. *Proceedings of the Royal Society A*, 228(1175), 519–38.

Goldberg, D. N., and Heimbach, P. (2013). Parameter and state estimation with a time-dependent adjoint marine ice sheet model. *The Cryosphere*, 17, 1659–78.

Goldberg, D. N., Heimbach, P., Joughin, I., and Smith, B. (2015). Committed retreat of Smith, Pope, and Kohler Glaciers over the next 30 years inferred by transient model calibration. *The Cryosphere*, 9(6), 2429–46. https://doi.org/10.5194/tc-9-2429-2015.

Goldberg, D. N., Narayanan, S. H. K., Hascoet, L., and Utke, J. (2016). An optimized treatment for algorithmic differentiation of an important glaciological fixed-point problem. *Geoscientific Model Development*, 9(5), 1891–904.

Greve, R., and Blatter, H. (2009). *Dynamics of Ice Sheets and Glaciers*. Berlin: Springer Science & Business Media.

Greve, R., Saito, F., and Abe-Ouchi, A. (2011). Initial results of the SeaRISE numerical experiments with the models SICOPOLIS and IcIES for the Greenland ice sheet. *Annals of Glaciology*, 52(58), 23–30.

Griewank, A., and Walther, A. (2008). *Evaluating Derivatives: Principles and Techniques of Algorithmic Differentiation*, vol. 19, 2nd ed. Philadelphia, PA: SIAM Frontiers in Applied Mathematics.

Gudmundsson, G. H. (2003). Transmission of basal variability to a glacier surface. *Journal of Geophysical Research: Solid Earth*, 108(B5), 1–19. https://doi.org/10.1029/2002JB002107.

Habermann, M., Truffer, M., and Maxwell, D. (2013). Changing basal conditions during the speed-up of Jakobshavn Isbræ, Greenland. *The Cryosphere*, 7(6), 1679–92. https://doi.org/10.5194/tc-7-1679-2013.

Hadamard, J. (1902). Sur les probl`emes aux Dérivées partielles et leur signification physique, *Princeton University Bulletin*, 13, 49–52.

Hansen, P. C. (2000). The L-curve and its use in the numerical treatment of inverse problems. In P. Johnston, ed., *Computational Inverse Problems in Electrocardiology: Advances in Computational Bioengineering*. Southampton: WIT Press, pp. 119–42.

Heimbach, P., and Bugnion, V. (2009). Greenland ice-sheet volume sensitivity to basal, surface and initial conditions derived from an adjoint model. *Annals of Glaciology*, 50(52), 67–80.

Hindmarsh, R. C., Leysinger-Vieli, G. J.-M., and Parrenin, F. (2009). A large-scale numerical model for computing isochrone geometry. *Annals of Glaciology*, 50(51), 130–40.

Huss, M., and Farinotti, D. (2012). Distributed ice thickness and volume of all glaciers around the globe. *Journal of Geophysical Research: Earth Surface*, 117(F4). https://doi.org/10.1029/2012JF002523.

Hutter, K. (1983). *Theoretical Glaciology: Material Science of Ice and the Mechanics of Glaciers and Ice Sheets*. Dordrecht: D. Reidel Publishing.

Isaac, T., Petra, N., Stadler, G., and Ghattas, O. (2015). Scalable and efficient algorithms for the propagation of uncertainty from data through inference to prediction for large-scale problems, with application to flow of the Antarctic ice sheet. *Journal of Computational Physics*, 296, 348–38.

Joughin, I., MacAyeal, D. R., and Tulaczyk, S. (2004). Basal shear stress of the Ross ice streams from control method inversions. *Journal of Geophysical Research: Solid Earth*, 109(B9), 1–62. https://doi.org/10.1029/2003JB002960.

Joughin, I., Smith, B. E., and Howat, I. M. (2018). A complete map of Greenland ice velocity derived from satellite data collected over 20 years. *Journal of Glaciology*, 64(243), 1–11. https://doi.org/10.1017/jog.2017.73.

Jouvet, G. (2023). Inversion of a Stokes glacier flow model emulated by deep learning. *Journal of Glaciology*, 69(273), 13–26.

Jouvet, G., Cordonnier, G., Kim, B., Lüthi, M., Vieli, A. and Aschwanden, A. (2022). Deep learning speeds up ice flow modelling by several orders of magnitude. *Journal of Glaciology*, 68(270), 651–64.

Kalmikov, A. G., and Heimbach, P. (2014). A hessian-based method for uncertainty quantification in global ocean state estimation. *SIAM Journal on Scientific Computing*, 36(5), S267–S295.

Karlsson, N. B., Bingham, R. G., Rippin, D. M. et al. (2014). Constraining past accumulation in the central Pine Island Glacier basin, West Antarctica, using radio-echo sounding. *Journal of Glaciology*, 60(221), 553–62.

Kohn, R., and Vogelius, M. (1984). Determining conductivity by boundary measurements. *Communications on Pure and Applied Mathematics*, 37(3), 289–98. https://doi.org/10.1002/cpa.3160370302.

Koutnik, M. R., and Waddington, E. D. (2012). Well-posed boundary conditions for limited-domain models of transient ice flow near an ice divide. *Journal of Glaciology*, 58(211), 1008–20.

Koutnik, M. R., Fudge, T., Conway, H. et al. (2016). Holocene accumulation and ice flow near the West Antarctic ice sheet divide ice core site. *Journal of Geophysical Research: Earth Surface*, 121(5), 907–924.

Koziol, C. P., Todd, J. A., Goldberg, D. N., and Maddison, J. R. (2021). fenics ice 1.0: A framework for quantifying initialization uncertainty for time-dependent ice sheet models. *Geoscientific Model Development*, 14, 5843–61. https://doi.org/10.5194/gmd-14-5843-2021.

Krasnopolsky, V. M., and Schiller, H. (2003). Some neural network applications in environmental sciences. Part I: Forward and inverse problems in geophysical remote measurements. *Neural Networks*, 16(3–4), 321–34.

Kyrke-Smith, T. M., Gudmundsson, G. H., and Farrell, P. E. (2017). Can seismic observations of bed conditions on ice streams help constrain parameters in ice flow models? *Journal of Geophysical Research: Earth Surface*, 122(11), 2269–82.

Larour, E. (2005). Modélisation numérique du comportement des banquises flottantes, validée par imagerie satellitaire, Ph.D. thesis, Ecole Centrale Paris.

Larour, E., Utke, J., Csatho, B. et al. (2014). Inferred basal friction and surface mass balance of the Northeast Greenland Ice Stream using data assimilation of ICESat (Ice Cloud and land Elevation Satellite) surface altimetry and ISSM (Ice Sheet System Model). *The Cryosphere*, 8(6), 2335–51. https://doi.org/10.5194/tc-8–2335-2014.

Lee, V., Cornford, S. L., and Payne, A. J. (2015). Initialization of an ice-sheet model for present-day Greenland. *Annals of Glaciology*, 56(70), 129–40. https://doi.org/10.3189/2015AoG70A121.

Leong, W. J., and Horgan, H. J. (2020). DeepBedMap: A deep neural network for resolving the bed topography of Antarctica. *The Cryosphere*, 14(11), 3687–705. https://doi.org/10.5194/tc-14-3687-2020.

Li, D., Gurnis, M., and Stadler, G. (2017). Towards adjoint-based inversion of time-dependent mantle convection with nonlinear viscosity. *Geophysical Journal International*, 209(1), 86–105. https://doi.org/10.1093/gji/ggw493.

Logan, L. C., Narayanan, S. H. K., Greve, R., and Heimbach, P. (2020). Sicopolis-ad v1: an open-source adjoint modeling framework for ice sheet simulation enabled by the algorithmic differentiation tool OpenAD. *Geoscientific Model Development*, 13(4), 1845–64.

Lorenc, A. C. (2003). The potential of the ensemble Kalman filter for NWP: A comparison with 4d-var. *Quarterly Journal of the Royal Meteorological Society: A Journal of the Atmospheric Sciences, Applied Meteorology and Physical Oceanography*, 129(595), 3183–203.

MacAyeal, D. R. (1989). Large-scale ice flow over a viscous basal sediment: Theory and application to Ice Stream B, Antarctica. *Journal of Geophysical Research Atmospheres*, 94(B4), 4071–87.

MacAyeal, D. R. (1992a). Irregular oscillations of the West Antarctic ice sheet. *Nature*, 359(6390), 29–32.

MacAyeal, D. R. (1992b). The basal stress distribution of Ice Stream E, Antarctica, inferred by control methods. *Journal of Geophysical Research: Solid Earth*, 97(B1), 595–603.

MacAyeal, D. R. (1993). A tutorial on the use of control methods in ice-sheet modelling. *Journal of Glaciology*, 39(131), 91–8.

MacGregor, J. A. et al. (2015). Radiostratigraphy and age structure of the Greenland Ice Sheet. *Journal of Geophysical Research: Earth Surface*, 120(2), 212–41. https://doi.org/10.1002/2014JF003215.

Maddison, J. R., Goldberg, D. N., and Goddard, B. D. (2019). Automated calculation of higher order partial differential equation constrained derivative information. *SIAM Journal on Scientific Computing*, 41(5), C417–C445.

Martin, J., Wilcox, L. C., Burstedde, C., and Ghattas, O. (2012). A stochastic Newton MCMC method for large-scale statistical inverse problems with application to seismic inversion. *SIAM Journal on Scientific Computing*, 34(3), A1460–A1487.

Michel, L., Picasso, M., Farinotti, D., Funk, M., and Blatter, H. (2014). Estimating the ice thickness of shallow glaciers from surface topography and mass-balance data with a shape optimization algorithm. *Computers & Geosciences*, 66, 182–99.

Morlighem, M., Rignot, E., Seroussi, H. et al. (2010). Spatial patterns of basal drag inferred using control methods from a full-Stokes and simpler models for Pine Island Glacier, West Antarctica. *Geophysical Research Letters*, 37(L14502), 1–6. https://doi.org/10.1029/2010GL043853.

Morlighem, M., Rignot, E., Seroussi, H. et al. (2011). A mass conservation approach for mapping glacier ice thickness. *Geophysical Research Letters*, 38(L19503), 1–6. https://doi.org/10.1029/2011GL048659.

Morlighem, M., Rignot, E., Mouginot, J. et al. (2013a). High-resolution bed topography mapping of Russell Glacier, Greenland, inferred from Operation IceBridge data. *Journal of Glaciology*, 59(218), 1015–23. https://doi.org/10.3189/2013JoG12J235.

Morlighem, M., Seroussi, H., Larour, E., and Rignot, E. (2013b). Inversion of basal friction in Antarctica using exact and incomplete adjoints of a higher-order model. *Journal of Geophysical Research: Earth Surface*, 118(3), 1746–53. https://doi.org/10.1002/jgrf.20125.

Morlighem, M., Rignot, E., Mouginot, J. (2014). Deeply incised submarine glacial valleys beneath the Greenland Ice Sheet. *Nature Geoscience*, 7(6), 418–22. https://doi.org/10.1038/ngeo2167.

Morlighem, M., et al. (2017). BedMachine v3: Complete bed topography and ocean bathymetry mapping of Greenland from multi-beam echo sounding combined with mass conservation. *Geophysical Research Letters*, 44(21), 11,051–61. https://doi.org/10.1002/2017GL074954,2017GL074954.

Morlighem, M., et al. (2020). Deep glacial troughs and stabilizing ridges unveiled beneath the margins of the Antarctic ice sheet. *Nature Geoscience*, 13(2), 132–7. https://doi.org/10.1038/s41561-019-0510-8.

Nias, I. J., Cornford, S. L., and Payne, A. J. (2016). Contrasting the modelled sensitivity of the Amundsen Sea Embayment ice streams. *Journal of Glaciology*, 62(233), 552–62. https://doi.org/10.1017/jog.2016.40.

Nocedal, J., and Wright, S. (2006). *Numerical Optimization*. New York: Springer.

Pattyn, F. (2003). A new three-dimensional higher-order thermomechanical ice sheet model: Basic sensitivity, ice stream development, and ice flow across subglacial lakes. *Journal of Geophysical Research: Solid Earth*, 108(B8), 1–15. https://doi.org/10.1029/2002JB002329.

Payne, A. J., Vieli, A., Shepherd, A. P., Wingham, D. J., and Rignot, E. (2004). Recent dramatic thinning of largest West

Antarctic ice stream triggered by oceans. *Geophysical Research Letters*, 31(23), 1–4. https://doi.org/10.1029/2004GL021284.

Perego, M., Price, S., and Stadler, G. (2014). Optimal initial conditions for coupling ice sheet models to Earth system models. *Journal of Geophysical Research: Earth Surface*, 119, 1–24. https://doi.org/10.1002/2014JF003181.

Petra, N., Zhu, H., Stadler, G., Hughes, T. J. R., and Ghattas, O. (2012). An inexact Gauss-Newton method for inversion of basal sliding and rheology parameters in a nonlinear Stokes ice sheet model. *Journal of Glaciology*, 58(211), 889–903. https://doi.org/10.3189/2012JoG11J182.

Petra, N., Martin, J., Stadler, G., and Ghattas, O. (2014). A computational framework for infinitedimensional Bayesian inverse problems, Part II: Stochastic Newton MCMC with application to ice sheet flow inverse problems. *SIAM Journal on Scientific Computing*, 36(4), A1525–A1555. https://doi.org/10.1137/130934805.

Pollard, D., and DeConto, R. M. (2012). A simple inverse method for the distribution of basal sliding coefficients under ice sheets, applied to Antarctica. *The Cryosphere*, 6(5), 953–71. https://doi.org/10.5194/tc-6–953–2012.

Pralong, M. R., and Gudmundsson, G. H. (2011). Bayesian estimation of basal conditions on Rutford Ice Stream, West Antarctica, from surface data. *Journal of Glaciology*, 57(202), 315–24.

Price, S. F., Payne, A. J., Howat, I. M., and Smith, B. E. (2011). Committed sea-level rise for the next century from Greenland ice sheet dynamics during the past decade. *Proceedings of the National Academy of Sciences, USA*, 108(22), 8978–83.

Quiquet, A., Dumas, C., Ritz, C., Peyaud, V., and Roche, D. M. (2018). The GRISLI ice sheet model (version 2.0): Calibration and validation for multi-millennial changes of the Antarctic ice sheet. *Geoscientific Model Development*, 11, 5003–25. https://doi.org/10.5194/gmd-11-5003-2018.

Ranganathan, M., Minchew, B. Meyer, C. R., and Gudmundsson, G. H. (2020). A new approach to inferring basal drag and ice rheology in ice streams, with applications to West Antarctic Ice Streams. *Journal of Glaciology*, 67(262), 229–42. https://doi.org/10.1017/jog.2020.95.

Rasmussen, L. A. (1988). Bed topography and mass-balance distribution of Columbia Glacier, Alaska, USA, determined from sequential aerial-photography. *Journal of Glaciology*, 34(117), 208–16.

Raymond, M. J., and Gudmundsson, G. H. (2009). Estimating basal properties of glaciers from surface measurements: a non-linear Bayesian inversion approach. *The Cryosphere*, 3(1), 181–222.

Rignot, E., and Mouginot, J. (2012). Ice flow in Greenland for the International Polar Year 2008–2009. *Geophysical Research Letters*, 39(11). https://doi.org/10.1029/2012GL051634.

Rignot, E., Mouginot, J., and Scheuchl, B. (2011). Ice Flow of the Antarctic Ice Sheet. *Science*, 333(6048), 1427–1430. https://doi.org/10.1126/science.1208336.

Ritz, C., Edwards, T. L. Durand, G. et al. (2015). Potential sea-level rise from Antarctic ice-sheet instability constrained by observations. *Nature*, 528(7580), 115–18. https://doi.org/10.1038/nature16147.

Rommelaere, V., and MacAyeal, D. R. (1997). Large-scale rheology of the Ross Ice Shelf, Antarctica, computed by a control method. *Annals of Glaciology*, 24, 43–8.

Shapero, D., Badgeley, J., Hoffmann, A., and Joughin, I. (2021). icepack: A new glacier flow modeling package in Python, version 1.0. *Geoscientific Model Development*, 14, 4593–616. https://doi.org/10.5194/gmd-14-4593-2021.

Siegert, M., Ross, N., Corr, H., Kingslake, J., and Hindmarsh, R. (2013). Late Holocene ice-flow reconfiguration in the Weddell Sea sector of West Antarctica. *Quaternary Science Reviews*, 78, 98–107.

Utke, J., Naumann, U., Fagan, M. (2008). OpenAD/F: A modular open-source tool for automatic differentiation of Fortran codes. *ACM Transactions on Mathematical Software*, 34(4) 1–36. https://doi.org/10.1145/1377596.1377598.

Waddington, E. D., Neumann, T. A., Koutnik, M. R., Marshall, H.-P., and Morse, D. L. (2007). Inference of accumulation-rate patterns from deep layers in glaciers and ice sheets. *Journal of Glaciology*, 53(183), 694–712.

Warner, R., and Budd, W. (2000). Derivation of ice thickness and bedrock topography in data-gap regions over Antarctica, *Annals of Glaciology*, 31, 191–7.

Werder, M. A., Huss, M., Paul, F., Dehecq, A., and Farinotti, D. (2020). A Bayesian ice thickness estimation model for large-scale applications, *Journal of Glaciology*, 66(255), 137–52. https://doi.org/10.1017/jog. 2019.93.

Wernecke, A., Edwards, T. L., Nias, I. J., Holden, P. B. and N. R. Edwards. (2020). Spatial probabilistic calibration of a high-resolution Amundsen Sea embayment ice sheet model with satellite altimeter data. *The Cryosphere*, 14(5), 1459–74.

7

Data Assimilation in Hydrological Sciences

Fabio Castelli

Abstract: Hydrological sciences cover a wide variety of water-driven processes at the Earth's surface, above, and below it. Data assimilation techniques in hydrology have developed over the years along many quite independent paths, following not only different data availabilities but also a plethora of problem-specific model structures. Most hydrologic problems that are addressed through data assimilation, however, share some distinct peculiarities: scarce or indirect observation of most important state variables (soil moisture, river discharge, groundwater level, to name a few), incomplete or conceptual modelling, extreme spatial heterogeneity, and uncertainty of controlling physical parameters. On the other side, adoption of simplified and scale-specific models allows for substantial problem reduction that partially compensates these difficulties, opening the path to the assimilation of very indirect observations (e.g. from satellite remote sensing) and efficient model inversion for parameter estimation. This chapter illustrates the peculiarities of data assimilation for state estimation and model inversion in hydrology, with reference to a number of representative applications. Sequential ensemble filters and variational methods are recognised to be the most common choices in hydrologic data assimilation, and the motivations for these choices are also discussed, with several examples.

7.1 Introduction

Hydrological sciences cover a vast variety of water-driven processes at the Earth's surface, above, and below it. By a consensus definition, 'Hydrology is the science that encompasses the occurrence, distribution, movement and properties of the waters of the earth and their relationship with the environment within each phase of the hydrologic cycle' (Fabryka-Martin et al., 1983). On more practical grounds, disciplines other than hydrology deal with the parts of the hydrologic cycle regarding the open atmosphere and the oceans, so that the term 'land components of the water cycle' may provide a more accurate term for what most of the hydrologic scientific community deals with. Even with this restriction, land hydrology covers an exceptionally large variety of processes on spatial scales ranging from nanometres (e.g. adsorption and transport of contaminants in porous media) to thousands of kilometres (e.g. continental land–atmosphere interactions). Over the last hundred years, hydrology has progressively evolved from the simple 'conceptual surface hydrology' driven by engineering design problems to a rich ensemble of ramifications into sub-disciplines and new specialisations such as hydrometeorology, geohydrology, ecohydrology, sociohydrology, and so forth. Main drivers for this continuing evolution are the physical connectivity of water through many environmental and social sciences and, for what is central for the subject of this chapter, by the capacity of improving the physical base of the hydrologic modelling through the assimilation of an increasing amount and variety of data. On the other side, the growth of many – and often largely independent – sub-disciplines has not favoured the development of main leading or guiding principles, beyond simple mass and energy balances, which would remain robust and usable across the many relevant spatial and temporal scales. When compared to atmospheric sciences, as an example, the route to a holistic modelling approach is still in its infancy (e.g. Frey et al., 2021). Therefore, data assimilation techniques in hydrology have developed over the years along many quite independent paths, following not only different data availability but also a plethora of problem-specific model structures (Liu et al., 2012). Even when dealing with a specific hydrologic process (e.g. river flow), operational flood prediction models may differ from hydrologic balance models used in water resources management.

Most hydrologic problems that are addressed through data assimilation, however, share some distinct peculiarities that contrast with data assimilation in nearby disciplines that deal with fluid flows, such as meteorology.

(a) Most of the complexity resides not in the flow itself, but in the environment that contains/constraints the flow, and in the forcing. As an example, what may diminish the predictability of the propagation of a flood wave along a river reach, and of the possible floodplain inundation, is not the turbulent nature of the flow, but the uncertainty in the representation of a complex river geometry (its vegetation, levees, transverse structure, movable bed, etc.; Mudashiru et al., 2021). Groundwater flow is laminar in most of the cases, but the porous media that contains the flow is characterised, at all relevant scales, by huge variabilities that are minimally resolved by available direct survey techniques (Maliva, 2016). Evapotranspiration reflects the complexity of both the turbulent atmospheric boundary layer and the plant structure and physiology (Wang and Dickinson, 2012); variations of vegetation canopy structure and soil matric potential are again enormous, both in time and space. Even though precipitation is routinely measured at the ground and from space, its intermittent nature still leaves a significant amount of error in the reconstruction of the precipitation fields at the spatial detail that is required by many hydrologic prediction models, especially during intense events over small- (Ochoa-Rodriguez et al., 2015) and medium-size watersheds (Maggioni et al., 2013). Model structural uncertainties are strictly bound to parameter-estimation uncertainties (Moges et al., 2021). Process conceptualisation or oversimplification in many commonly used hydrological models is often the consequence of the limitation in calibrating the parameters of a more physically based, complex model, so that the response from quite different models may turn out to have similar likelihoods with respect to some validation dataset (Beven and Binley, 2014).

(b) Strictly connected with the previous point, the most important governing equations in hydrology are instead *convergent* in nature: for a given environment (e.g. a given set of model parameters), the mathematical model will converge in time towards the same solution even when starting from quite different initial conditions. A vertical column of soil will tend to dry out, towards a certain equilibrium soil moisture profile, until a new precipitation event arrives (Zeng and Decker, 2009). The initial soil moisture content will control only the timing to approach equilibrium, not the shape of the equilibrium profile, which is instead determined by other environmental parameters (the soil structure, the density and the osmotic potential of the plant roots, the depth to the water table ...). The convergent, non-chaotic nature of hydrologic equations is a clear advantage in terms of general system predictability (no butterfly effect in most hydrologic processes!), but this theoretical advantage can be fully used only in case a perfect model is available with precisely known forcing. For imperfect models and forcing, the convergent nature affects the efficiency of many DA techniques. As an example, in sequential ensemble approaches it may be problematic to explore the model space with standard filter formulations (Bauser et al., 2021), requiring ad-hoc techniques such as bias correction or ensemble inflation. Errors in model parameters and/or forcing will drive all the ensemble members towards a biased solution. Regardless of perturbed model states, new observations following at too large a distance from the ensemble of the model predictions may cause a rapid filter degeneracy.

(c) Some key state variables are routinely observed on a very sparse set of discrete points (e.g. river discharge, groundwater level) or inferred quite imprecisely from very indirect measurements. Among these last variables, soil moisture is the most notable one. On one side, it provides the key control on substantially all the surface hydrologic fluxes so that its estimation is a central issue in most hydrologic problems (Entekhabi et al., 1996; Seneviratne et al., 2010; Berghuijs et al., 2016). On the other side, while station density in ground measurement networks is too low to resolve soil moisture spatial variability, current satellite remote sensing techniques provide soil moisture retrievals that are characterised by accuracy levels that, in many cases, are just a few tens of percent better than open-loop model simulations with reliable forcing (Dong et al., 2020). Also, accuracy of satellite remote sensing retrievals of soil moisture dramatically drops below the very few top centimetres of the soil and over regions characterised by high variability of terrain slope or by dense vegetation cover (Chan et al., 2017; Babaeian et al., 2019).

(d) Scarce or indirect observation of most important state variables, together with extreme spatial heterogeneity and uncertainty of controlling physical parameters, has traditionally pushed towards the development and operational use of incomplete or conceptual modelling. It is then still pending the enormous (and too-often neglected) problem of dealing with large and statistically unknown model structural errors. On the other side, adoption of simplified and scale-specific models allows for a substantial problem reduction that partially compensate these difficulties, opening the path to the assimilation of very indirect observations and efficient model inversion for parameter estimation.

In the remaining sections, we will illustrate, with reference to a selection from a large literature and a few specific studies, how the techniques of data assimilation in hydrology have been shaped by these peculiarities. We will distinguish in the first case among state estimation and geophysical inversion problems, with broad distinction in the first case between large-scale state estimation for land–atmosphere interaction and basin-scale hydrologic prediction, and in the second case, between static and dynamic inversion. Data assimilation in snow hydrology is intentionally omitted in this review, being the topic of a dedicated chapter in this same book. A short last section addresses some emergent topics, fuelled by the unprecedented availability of new types of data and by the spreading of the joint use of physically based and artificial intelligence (AI)-based modelling.

7.2 State Estimation

7.2.1 Soil Moisture, Global Earth Observation, and Large-Scale Land Assimilation Systems

State estimation (in the sense of the DA terminology) in land surface hydrology has been developed with reference to a number of different variables, but soil moisture is by far the one that has received the most attention. This is because: (a) soil moisture is the key variable controlling and coupling the water cycle, energy cycle, and carbon cycle at the land surface (Seneviratne et al., 2010); it also controls the occurrence of hydrologic extremes such as droughts (Bolten et al., 2010) and floods (Berthet et al. 2009); (b) soil moisture measurements at the ground are way too sparse and discontinuous to provide enough data for most state estimation problems, while satellite remote sensing provides indirect and partial measurements limited to the very few top centimetres of the soil layer (Babaeian et al., 2019); moreover, only very recent missions provide soil moisture estimates at kilometre-scale resolution useful for small- and mid-size watershed studies, to the cost of quite long revisit times (Bauer-Marschallinger et al., 2019). Aside from empirical techniques based on short-wave infrared remote sensing (Verstraeten et al., 2006; Zhang et al., 2014), retrieval of soil moisture with active and/or passive microwave remote sensing is itself a geophysical inversion (data assimilation) problem (Entekhabi et al., 1994; Kerr et al., 2012; Das et al., 2018). Solutions of such inversion problems are however generally provided to the hydrologic users' community as remote sensing products. They are mainly based on single-imagery processing with eventual ancillary information on atmospheric state and land use, with no actual use of a true hydrologic model. More advanced inversion techniques are recently used to provide enhanced products (in terms of spatial resolution) by post-processing the overlapping of consecutive measurement footprints or by merging the observations from different satellite platforms (Hajj et al., 2017; Lievens et al., 2017). As an example of the first case, the Backus–Gilbert interpolator, that is an optimal solution to ill-posed inverse linear problems, is used to obtain a soil moisture product at 9 km resolution starting from the native resolution of about 36 km of the passive radiometer of the Soil Moisture Active Passive (SMAP) mission (Chan et al., 2017). Still provided to the users as a SMAP soil moisture product (precisely, a Level-4 product) is the result of the sequential assimilation of the SMAP radiometer brightness temperature observations into a physically based hydrologic model (Reichle et al., 2019); the assimilation algorithm is based on the NASA Catchment land surface model with

a spatially distributed ensemble Kalman filter (EnKF). Several important advantages are obtained through this extensive data assimilation approach in a model capable of resolving the joint water and energy balance in the soil-vegetation layers: production of a continuous 9-km, 3-hour product, covering also the regions where high vegetation or terrain slope variability are known to decorrelate the surface soil moisture from the sensor signal (Chan et al., 2017); estimation of the soil moisture state and profile down to about 100 cm, a depth considerably larger than the very few top centimetres whose moisture directly influences the surface brightness temperature measured by the radiometer, hence covering most of what is considered as the 'hydrologically active' soil layer (Castillo et al., 2015); use of a very informative signal for the soil moisture dynamics, that is precipitation, with data from a variety of sources (Balsamo et al., 2018). On the flip side, there is the potential bias introduced by the model structural error and poorly known model parameters, in particular the soil hydraulic properties, aggravated by the typical convergent behaviour mentioned in the introduction; a delicate pre-calibration of the hydrologic model is then needed, that may be periodically improved as new calibration datasets become available, but still resulting in quite nonhomogeneous product accuracy over different regions (Reichle et al., 2017).

Together with land surface temperature, soil moisture is the central state variable in the large-scale Land Data Assimilation Systems (LDAS; Balsamo et al., 2018), that have been developed as hydrological companions to operational weather forecasting systems to improve their estimation and prediction of near-surface temperature and moisture. Advanced Scatterometer (ASCAT) soil wetness observations (Wagner et al., 2013) are assimilated together with standard ground meteorological data in the UK MetOffice LDAS using an extended Kalman filter (Gomez et al., 2020). The North American hydrologic community model 'Noah' is used at the US NCEP to assimilate daily soil moisture fields from the NOAA-NESDIS Soil Moisture Operational Product System (SMOPS), based on both ASCAT and Soil Moisture and Ocean Salinity (SMOS) data, with an EnKF (Zheng et al., 2018). The more complete global Land Surface Model (LMS) of this kind is probably the CHTESSEL used at ECMWF, also representing the surface–atmosphere carbon exchanges together with the water and energy ones (Boussetta et al., 2013). A large revision effort is underway for such an LDAS, with the aim of better exploiting the use of different data sources to improve the hydrologic components, such as the vegetation dynamics or the fluxes from snow cover and large water bodies, and also to directly connect the land analysis to downstream operational applications such as flood prediction (Mason et al., 2020; Boussetta et al., 2021).

Most of these LSMs here have been initially designed with the main purpose of providing lower-boundary conditions to the atmosphere by describing the vertical fluxes of energy and water (recently carbon too), between the land surface and the atmosphere; much less attention was originally devoted on horizontal fluxes such as runoff (Massari et al., 2015; Zsoter et al., 2019). This allowed the development of quite simplified modelling and data assimilation frameworks, with the main goal of improving the accuracy of the weather forecast but not necessarily the hydrologic one too. Up to quite recently, LSMs have suffered from significant limitations in the representation of important hydrologic fluxes and states, also at the surface; not to mention groundwater, the generation and propagation of runoff is the most important one (Wang et al., 2016). From the data assimilation point of view, with most techniques based on sequential filters, the increments of the surface moisture state, while improving the lower boundary condition for the weather prediction model, usually add or remove water independently from each computational cell. This may have a potentially strong negative impact on the correct representation of the hydrologic cycle by opening the water budget, causing potential problems when the focus switches to hydrologic runoff forecasting such as in the Global Flood Awareness System (GloFAS), based on the ECMWF-HTESSEL LDAS (Harrigan et al., 2020). As an example, Zsoter et al. (2019) again point out that while this version of the LDAS system largely improves the accuracy of the estimation of some key variable for the vertical land–atmosphere exchanges, such as the 2-m temperature and snow depth, it is, at the same time, deteriorating the prediction of peak river flow in high latitude snowmelt-dominated areas. Localised improvements in flood prediction are also detected, but, in general, the overall water budget suffers from adding or removing water during the data assimilation, introducing possible inconsistency between upstream and downstream flood predictions in the same watershed.

To overcome these drawbacks, global LMSs have been coupled with river routing models, constructing Global Hydrologic Models. The main purpose is to quantify freshwater flows and storage changes at regional and continental scales, hence preserving the water budget of world major watersheds (Sood and Smakhtin, 2015; Iscitsuka et al., 2021). As a further indication of the widespread hydrologic approach where model structure (represented or neglected processes, resolved scales, etc.) is strongly bound to the possibility of assimilating specific data sources, these type of modelling are considered to be the ideal framework for assimilating data from upcoming satellite missions specifically designed to monitor global surface waters, such as the SWOT (Surface Water and Ocean Topography; Biancamaria et al., 2016), launched in December 2022. In a set of OSSEs experiments, Emery et al. (2020) propose an asynchronous EnKF assimilation scheme to solve the problem of very long and asynchronous SWOT revisit times (21 days) over the modelled domain. However, this family of Global Hydrologic Models still uses quite coarse spatial resolutions (typically a 0.5 x 0.5 cell grid). In each model grid cell, the surface reservoir is a unique river channel that may gather multiple real river branches, not fully resolving yet the full details of river-waters extent (river widths larger than about 100 m) that SWOT will provide (Altenau et al., 2021).

7.2.2 State Estimation in Distributed Hydrologic and Hydraulic Models for Runoff and Flood Prediction

When the main focus of state estimation (and eventual prediction) is channel runoff, river flow (stage or discharge) observations are the most commonly assimilated data in watershed distributed hydrologic (Clark et al., 2008; Xie et al., 2014; Ercolani and Castelli, 2017) and river hydraulic models (Neal et al., 2007). Assimilation of river flow data faces a number of specific problems: (a) discharge data, or translation of stage data into discharge estimates, are characterised by a non-trivial error structure, with relative error increasing in magnitude for extreme flow conditions (Domeneghetti et al., 2012); and (b) river gauging stations are very sparse and usually cover a very limited number of main river branches also in quite large watersheds; small size (e.g. areas of the order of 100 km2 or less) watersheds remain substantially ungauged and the global number of functioning river gauges is progressively decreasing (Allen et al., 2020).

As an alternative to relying on sparse ground observations, in particular for prediction or estimation of flood inundation dynamics, several recent studies explored the use of satellite altimetry data over open water for improving the accuracy of hydrologic and hydraulic modelling (Grimaldi et al., 2016). Most common choice, dictated by the need of resolving meter-scale river and flood plain features (Annis et al., 2019), is the use of SAR products (Hostache et al., 2018, Cooper et al., 2019). High-resolution SAR imagery comes to the detriment of the revisit time frequency, so that the available data for assimilation in the flood inundation model may be considered as sporadic: it is quite commonly the case that only one inundation map (or a very few in the largest basins) falls within the time-span of the flood wave (Dasgupta et al., 2021). Also, quality of SAR observation of flood extent may vary dramatically from case to case, depending on the geometrical characteristics of the inundated environment (Pierdicca et al., 2009), and this needs to be taken explicitly into account in the design of the assimilation algorithm (Waller et al., 2018). As previously discussed, mathematical models of river flooding are typical convergent systems, where the dynamics is strongly constrained by boundary conditions (e.g. hydrologic runoff converging into the river network) and model parameters (e.g. geometry and roughness of the river and floodplain). For such a system, assimilation of observations that are sporadic in time is affected by a serious problem of the short persistence of the improvements of the model performances. Regardless of the assimilation technique, the model predictions after the state update rapidly converge, in a few hours or even minutes depending on the watershed size, to the same open-loop prediction without the SAR altimetry assimilation (Andreadis and Schumann, 2014). This problem intersects with the one, common to many other assimilation problems over large multidimensional domains tackled with ensemble techniques, of spurious error growth due to under-sampling. Two-dimensional flood inundation models assimilating SAR water altimetry data would require, if a global filtering approach were to be adopted, a number of ensemble members larger than the dimension of the state vector. This is usually prohibitive from a computational perspective, and unavoidable under-sampling may bring inconsistent updates with spurious, unphysical spatial correlations over distances that are too large. On the other hand, surface water flow in rivers and floodplains is again very strongly constrained by the channel and terrain geometries. At spatial resolutions of tens of meters commonly used by such models, river flow direction is given; in the floodplains, where the depth-averaged flow may follow more complex 2D patterns, terrain slope, and land singularities (road embankments, buildings, etc.) strongly reduce its degrees of freedom. This prior constraint to the modelled dynamics opens the door to the possibility of tackling the under-sampling problem through localised filters (Anderson, 2007), where the localisation of the covariance may be based on physical grounds. As an example, Garcia-Pintado et al. (2015) use a spatial metric, to localise the covariance, defined as the maximum among the Euclidean and the 'along-the-channels-network' distance. The main

assumption is that the physical connectivity of flows along predefined paths, that is the channel networks, would strongly influence the development of the forecast error covariance.

A promising path to improve the estimation and predictability of flood inundations is the joint assimilation of traditional river gauge data and satellite imagery. Sinergy among the two different data sources may potentially solve both the issue of spatial sparsity of river gauge data and the sporadicity of high-resolution satellite flood mapping. Along this path, Annis et al. (2022) use an EnKF to jointly assimilate ground observations from a set of river gauging stations and inundation extent imagery from a Landsat-7 overpass. Among other advantages, the quite precise level measurements at the river gauging stations balance the lower informative content of the optical imagery (i.e. flooding extent only), with respect to SAR altimetry; as in many other possible cases, especially for fast flooding dynamics, SAR imagery was not available for the studied event. As shown in Fig. 7.1, the estimation uncertainty of the flood extent is sensitive to the data assimilation configuration. If only river gauge data were available, one river stage time series only would reduce the uncertainty by a substantial amount; almost an order of magnitude reduction is attained assimilating the Landsat water extent, but such a reduction is limited to a very few hours if Landsat is assimilated alone; river gauge data at four different locations are necessary in this case study to reach the same level of uncertainty of the Landsat assimilation, with further improvement for a few hours in case of joint assimilation. Localisation of the update needs to be applied for the assimilation of point river stage observations too. The along channel upstream and downstream

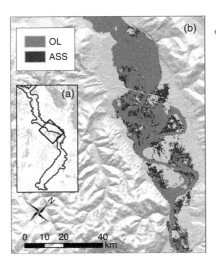
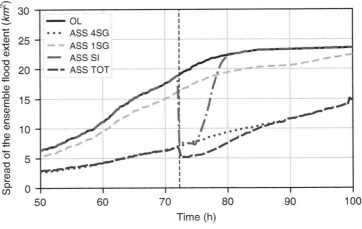

Figure 7.1 Flood extent estimation with joint assimilation of river gauge data and Landsat imagery in a 2D hydrologic-hydraulic model (adapted from Annis et al., 2022). Left panel: flood extent at time of Landsat overpass estimated in open loop (black) and with data assimilation (light grey). Right panel: estimation uncertainty of total flooded area in open loop (dark continuous line), assimilation of river gauge data only (dots – 4 gauges; light dash – 1 gauge), Landsat only (dash-dot) and both river gauges and Landsat (dark dash).

water level correction is performed applying a distance-based gain function in this case, then propagated to the nearest floodplain cells. Furthermore, to coherently assimilate more than one simultaneous stage observation, updates propagated to the same computational cell from different gauge locations are weighted with the inverse of the distance.

Sequential filters such as the EnKF are quite straightforward to implement in their standard formulation, also for multidimensional problems, and this is one of the main reasons for the wide spread of its use. When filter modifications need to be introduced, such as the covariance or update localisation as described here, the issue is not the more complex implementation but rather the introduction of quite arbitrary choices in the form and parameters of the filter modifiers. Not to be forgotten is the issue of convergent behaviour of many hydrologic systems, including the flood modelling. Recognising that the problem of the rapid attraction of the model trajectories towards a biased solution can be efficiently addressed by reducing the model structural error, the most adopted filter modification is the filter augmentation to estimate the most uncertain model parameters and/or model inputs and/or boundary conditions on top of model states (Moradkhani et al., 2005; Zhang et al., 2017; Liu et al., 2021). As an example, in the case of SAR imagery assimilation for flood estimation, Garcia-Pintado et al. (2015) augment the filtered bathymetry state with channel Manning roughness parameter, downstream river level, and lateral hydrologic input flows to channels.

As mentioned, river gauging sparsity may be such that a given watershed of interest may result totally ungauged. State augmentation with model parameters has also been introduced to address this issue. Xie et al. (2014) propose an EnKF assimilation strategy for improving the runoff prediction of a distributed hydrologic model in nested ungauged subbasins (i.e. ungauged subbasins that have gauged upstream or downstream neighbours); state augmentation with model parameters for both ungauged and gauged subbasin, with prescribed correlation among them, is demonstrated to successfully propagate state updates from gauged to ungauged locations.

Variational assimilation techniques, and variational assimilation with an adjoint in particular, may be more challenging to implement, with respect to ensemble-based sequential filters, as they require the capability of implementing the task of coding the adjoint model. This may be an impossible task in cases where the numerical details of the chosen prediction hydrologic model are not available. However, when the adjoint model can be implemented, the variational assimilation approach has the undoubted merit of a physically based representation of the sensitivities among the model states and between the model states and model parameters (Margulis and Entekhabi, 2001). This physically based sensitivity estimate may turn out to be

crucial in case of sparse observations, and it may then be a valid alternative to the empirical definition of the localisation structures in EnKF techniques. Ercolani and Castelli (2017) take advantage of the predefined paths and directions of the flow routing along the channel network to reduce the assimilation of sparse river discharge data, with a fully distributed (2D) hydrologic model for flood prediction, into a multivariate 1D variational assimilation scheme (see Appendix for a general formulation of such a scheme). The scheme allows, without the need of predefined correlations, the estimation of non-observed model states and parameters. In a first hindcast experiment on a set of real high-flow events in a mid-size (approx. 8,000 Km2) Mediterranean watershed, simultaneous assimilation of half-hourly discharge data at multiple locations was demonstrated to improve the accuracy of the flood peak prediction along the entire channel network, provided that assimilated data at least cover part of the rising limb of the flood wave. As largely expected, accuracy of prediction at downstream locations further increases when flood peak is observed upstream (Fig. 7.2). However, as a further demonstration of the convergent nature of the problem, full advantage of the flow data assimilation is reached by updating not only the flow state along the channel network, but also the hydrologic input to the channel (i.e. the runoff in the direct contributing hillslope cells).

The possibility of updating the prior estimate of the event-antecedent soil moisture condition by assimilating the river flow data is also explored in the same study, based on the frequently verified hypothesis that the runoff response to precipitation events is strongly sensitive to such a condition (DeChant and Moradkhani, 2012). To overcome the difficulty of computing the Jacobian around threshold-switching soil moisture state, a mixed Monte Carlo–Variational approach is proposed, addressing at the same time the need of mass conservation that would be violated by updating runoff only (Fig. 7.3). At each iteration of the variational algorithm, an iteration of a particle filter is also run for the component of the model generating the hill-slope runoff, where main stochastic perturbation is inserted in the structure of the spatial interpolation of rain gauge data and in the initial soil moisture condition. The update of the lateral runoff input to the channel network is then used to estimate the posterior likelihood of the runoff fields resulting by the various particles.

Direct assimilation of various soil moisture products in hydrologic flood prediction models has been also demonstrated to improve the prediction accuracy in several cases (Cenci et al., 2017; Massari et al., 2018; Azimi et al., 2020; Baugh et al., 2020). The main target of the prediction is now the formation of the flood-wave and its propagation along the main rivers, while eventual floodplain inundations are resolved in downstream, loosely coupled hydraulic models. In most of such studies, the EnKF, the particle filter, and their variants are the preferred choices given the direct

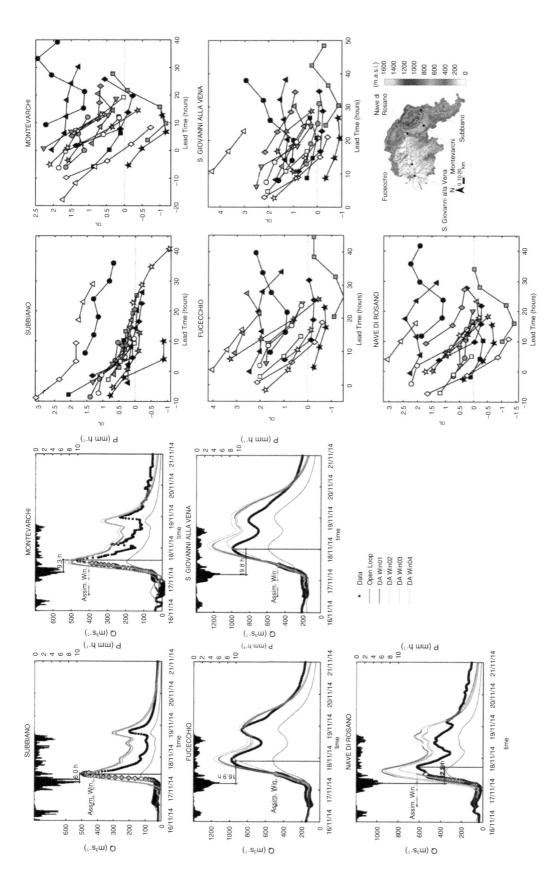

Figure 7.2 Flood prediction performance in a set of hindcast experiments with variational assimilation of multiple river flow data in a distributed hydrologic model (adapted from Ercolani and Castelli, 2017). Left panels: predicted hydrographs of a chosen event at gauged locations for different assimilation windows; rainfall occurred after the end of the assimilation window is assumed to be perfectly predictable. Right panels: improvement (positive values) of accuracy of flood prediction, measured by the log of the Nash–Sutcliffe efficiency index, for varying prediction lead times with respect to flood peak (different symbols represent different flood events).

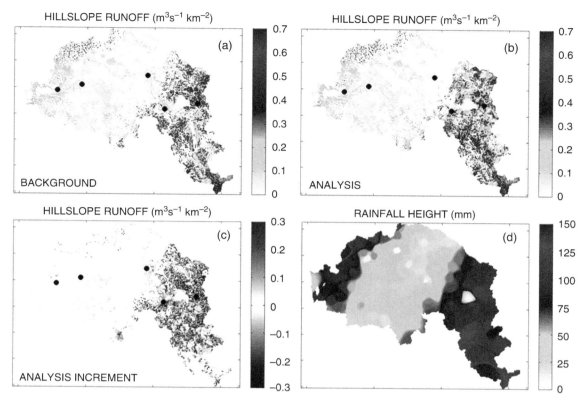

Figure 7.3 Antecedent soil moisture analysis prior to a flood event obtained with a mixed particle filter and variational assimilation of point river flow data (adapted from Eercolani and Castelli, 2017).

spatial observation of the main state variable. However, accuracy of EnKF depends on the ensemble size. The ensemble size may remain quite small for lumped or semi-distributed models (Alvarez-Garreton et al., 2015). For operational spatially extensive or high resolution fully distributed models, CPU and storage availability may limit the ensemble size to number much smaller than the one needed for a potentially reachable accuracy. Alternative paths have also been recently proposed in the hydrologic literature to overcome this limitation, using Polynomial Chaos Expansion to implement what is termed Probabilistic (or Stochastic) Collocation Method as a substitute of the Monte Carlo simulations that are at the base of the classical EnKF (Fan et al., 2015). If the computation of the PCE polynomial coefficients is still challenging for high-dimensional problems (the number of coefficients to be estimated scales quadratically with the number of unknowns), such a framework allows for a more efficient implementation of the multi-fidelity decomposition of the filter. As an example, Man et al. (2020) use an analysis of variance (ANOVA type) to decompose the low-fidelity and the high-fidelity-residual model responses in a limited set of members. A quite subjective choice is however left in the multi-fidelity approach: which simplification the low-fidelity model operates needs to be specified by choices of terms to

be cancelled from the high-fidelity one. It results in an experience-based compromise between the need to cut off the most computationally demanding terms in the hydrologic model and the danger of an excessive structural error in the low-fidelity model (Ng and Willcox, 2014).

7.3 Geophysical Inversions

7.3.1 Inversion for Static Parameters

A classical inversion problem in hydrological sciences, dating back long before the era of widespread availability of remote sensing data, has been formulated in the field of hydrogeology, where the main challenge to be faced was the estimation of spatially variable hydraulic properties of aquifers using very sparse wells data (Carrera and Neuman, 1986). What was termed the Bayesian geostatistical approach in the earlier groundwater data assimilation literature (Kitanidis, 1996) essentially relied on the formulation, with the use of Bayes' theorem, of the joint posterior probability density function (pdf) of the unknown model parameters and their prior expectations. The pdf of unknown parameters is itself parameterised through a generalised covariance function. Optimal parameters are obtained by seeking the minimum of the negative log likelihood of the pdf through an iterative Gaussian–Newton method. Geophysical inversion for groundwater modelling leads quite often to ill-posed problems, which can be tackled by prescribing a probabilistic model for the unknown priors as in the method here. An alternative, when a prior pdf cannot be conveniently prescribed, is the addition of one or more regularisation terms to the function to be minimised (Steklova and Haber, 2017). These terms also need some sort of 'prescription', which can be formulated on the base of preliminary geological studies or Kriging-like geostatistical analysis.

Continuous advances in hydrogeological sensing technology now provide increasing volumes of hydrogeophysical and geochemical data that can be assimilated to estimate high resolution fields of aquifer properties (Harvey and Gorelick, 1995; Hochstetler et al., 2016; Rajabi et al., 2018). Two main implementation bottlenecks arise, in consideration also that very few hydrogeological studies are supported by such large projects to gain access to high-performance computing resources: storage and matrix operations (e.g. inversion) on large and dense covariances of the order of the square of the model size; the computation of the variations of the forward map (i.e. the sensitivity of the groundwater hydrogeochemical model with respect to the unknown parameters) from the model parameters to the available measurements on model states. At each iteration, a number of model runs of the order of min (M,N) is required to compute a numerical approximation to the Jacobian, M being the number of unknown parameters and N the number of states. The use of the adjoint technique (Talagrand and Courtier, 1987), when the model structure allows the development of an adjoint model (see again Appendix for a general formulation of a variational assimilation/inversion problem with an adjoint in a multivariate dynamics), increases the accuracy of the estimation of the model sensitivity to parameters also for very non-linear problems, hence increasing the convergence speed of the iterative procedure, but the number of model runs needed in each iteration remains substantially the same. Recent study efforts have been then focused on the goal of decreasing the computational burden of the original Bayesian Geophysical Inversion (Ghorbanidehno et al., 2020), introducing such techniques as the eigenspectrum-based compression of covariances, the use of principal components to reduce the number of needed model runs (Lee et al., 2016), the use of proper orthogonal decomposition to reduce the adjoint model space in order to save both computational and coding costs (Altaf et al., 2013), or of meta-models such as polynomial chaos expansion (Laloy et al., 2013).

As in other fields, the geophysical inversion for aquifers characterisation also proceeds in the direction of exploring the usability of observations on state variables describing also processes that are not the main traditional hydrologic ones but that are physically linked to them.

The conceptual path is, as in other applications recalled further down, having the use of new data driving the increase in complexity of the modelling, and not vice versa. As an example, Ceccatelli et al. (2021) coupled a surface hydrologic model with a groundwater model including aquifer compaction dynamics to study the subsidence induced by excessive groundwater abstraction. Unknown aquifer parameters (hydraulic conductivity, elastic and inelastic storage coefficients) have been estimated by assimilating three years of dense Sentinel-1 ground displacement data and sparse (in space and time) wells data (Fig. 7.4), with aquifer recharge provided by the coupled surface hydrologic model whose main parameters were calibrated through river discharge data. Given the ill-posedness of the problem, a variational data assimilation scheme with non-linear Tikhonov regularisation (Hou and Jin, 1997) has been used in this case.

Finally, it can be considered as a prototypal form of geophysical inversion the use of states observations for the more traditional calibration of distinct types of (hydrologic) model parameters, both conceptual and physically based ones. The review of the main approaches to hydrologic model calibration would require an entire chapter; we just mention here a few recent examples, such as Yang et al. (2016) and Pinnington et al. (2021), where the calibration is more formally framed as a data assimilation problem.

7.3.2 Inversion for Dynamic Parameters

Geophysical inversion may be dynamic too, that is, finalised at the estimation of time-varying model parameters. From a formal point of view, a variable that would be considered as a state in a more complex model may be considered a time-varying parameter in a simpler model; a prediction equation for the variable is not provided in the first case, so that the variable can be mathematically treated as a model parameter. This may be a convenient alternative to direct/sequential state estimation when the variable to be estimated is not observed (or, more precisely, an invertible observation operator cannot be explicitly provided for that variable).

A typical example is the use of simplified surface energy balance models (Chen and Liu, 2020) to infer the partitioning of residual solar energy into latent (i.e. evapotranspiration) and sensible heat flux by assimilation of Land Surface Temperature data from geostationary (Xu et al., 2014) and polar orbiting (Xu et al., 2019) satellites. Eventually, these methods can simultaneously assess the turbulent fluxes and the control of soil moisture on their partitioning, through the definition of ad hoc soil moisture indices or soil-moisture related dynamic parameters such as the evaporative fraction (Caparrini et al., 2003, 2004; Bateni et al., 2013, 2014). Parsimony of the model, beyond obvious computational aspects, is chosen in these cases for a variety of reasons: the possibility of easily formulating and coding an adjoint model, reducing the problem from a full 4D to a lower

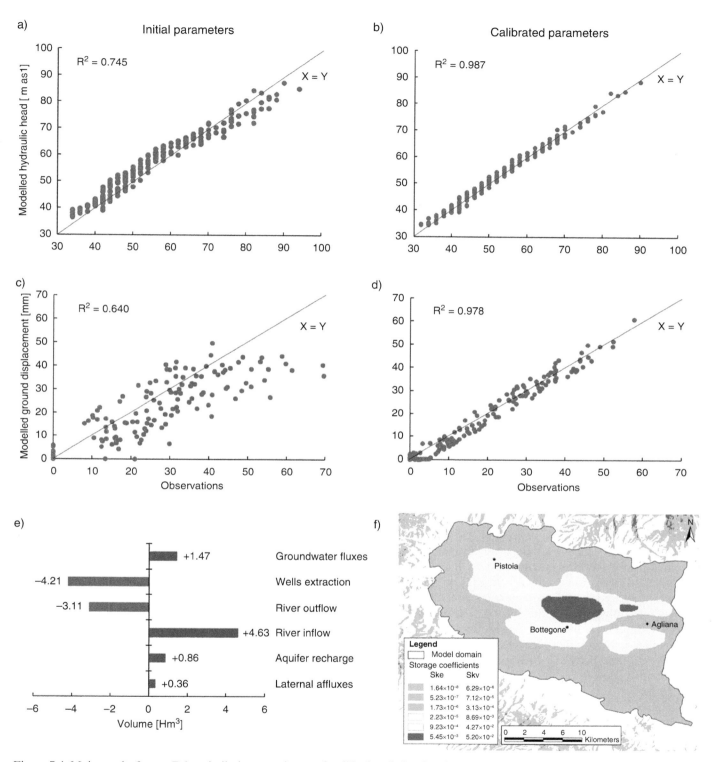

Figure 7.4 Main results from a DA assimilation experiment of well hydraulic head and ground displacements for geophysical inversion of a phreatic aquifer in Central Italy (adapted from Ceccatelli et al., 2021). Top and mid panels: scatter plots of modelled vs. observed hydraulic heads and displacements at selected verification locations (not used in DA) prior (left) and after (right) parameters optimisation. Bottom left: resulting overall aquifer balance, highlighting strong interactions with surface hydrology. Bottom right: one of the spatial fields of aquifer properties estimated through geophysical inversion.

dimensional domain; minimising the further noise that may be introduced by model components whose parameters or forcing are poorly constrained or observed at the scale of interest. Specifically, the use of the simple surface energy balance, eventually separating the contribution of soil and vegetation as sources of flux with distinct dynamics (Kustas and Norman, 1999), is chosen to the very purpose of not using a full soil–water mass balance. At scales from a few hundred meters up, horizontal redistribution of energy at the surface can be neglected and the problem treated as multivariate rather than multidimensional in the horizontal directions; model spatial resolution is immediately adapted to the characteristics of available land surface temperature (LST) satellite observations; no systematically biased soil moisture estimates are forced by inadequate representation of soil hydraulic properties or rainfall input; even more important for the development of a variational assimilation techniques, the complexity of the variational treatment of the threshold behaviour of soil moisture dynamics and infiltration and switching boundary conditions is avoided.

With reference to the mathematical formulation of the 1-D variational assimilation provided in the Appendix, in the simpler formulation of Caparrini et al. (2003, 2004) the energy balance is represented as an ordinary differential equation in time for LST, in the so-called force–restore method (Noilhan and Planton, 1989) with weighted contribution from vegetation canopy and bare soil, solved independently for each 'pixel' of LST observations. Inverted dynamical parameters are the evaporative fraction for soil and vegetation, implicitly controlled by soil moisture in water-limited hydrologic regimes, and the bulk neutral heat transfer coefficient, depending on vegetation type and phenological state. To accommodate a possible underdetermination of the inversion problem ($M = 2N$, in the notation of the Appendix), the inverted parameters are forced to remain constant on different time scales while assimilating hourly LST observations: daily for the evaporative fraction and half-monthly for the bulk heat transfer coefficient. The assimilation, fully resolving the diurnal cycle of land energy balance components including the ground heat flux, proceeds with a daily assimilation cycle nested in an outer half-monthly one. Simplicity of the formulation allowed to easily merge hourly, lower resolution, LST observations from geostationary satellites (GOES, METEOSAT) with higher resolution, daily observations from instruments on polar orbiting platforms (AVHRR, MODIS). More recent improvements of this quite simple scheme followed what can be termed a 'data-driven physically based' approach, where further physically based model equations are added, limited to the need of assimilating the observations of other variables to resolve some parameter indeterminacy in the main reference model equation. As an example, Bateni et al. (2014) added a predictive equation for the Leaf Area Index to the surface energy balance in order to assimilate SEVIRI estimates of Fraction of Photosynthetically Active Radiation absorbed by vegetation (FPAR) and hence to further constrain the inverted values of the vegetation heat transfer coefficients; even more, adding a new assimilated observation allowed a more robust inversion for the evaporation fraction with distinct values for vegetation and bare soil. A similarly very parsimonious bulk water balance model has been finally added in Abdolghafoorian and Farhadi (2020) to assimilate also soil moisture retrieval from the SMAP mission.

7.4 Some Emerging Topics

Progress in the development and testing of new data assimilation systems for hydrologic applications is in continuous acceleration, with the number of scientific publications doubling about every decade. A search on Scopus of the terms 'data AND assimilation AND hydrology' (quite restrictive, but still indicative) on 29 September 2021, returned 12 items for year 2000, 31 for 2010, and 62 for 2020. Important drivers for such growth have been the large availability of new data sources, mainly from satellite, such as the development of efficient data distribution platforms in the early 2000s or the new satellite missions for soil moisture monitoring about a decade later. On top of the increase of available computational power, which has always been and remains to be a crucial limiting factor, fuel for some important recent advancements is coming from two sides: the availability of new sources of data and the inclusion of AI algorithms in the data assimilation systems.

A notable example of new types of observations, well representative of the 'big-data era', that just started to enter in formal assimilation processes are the so-called crowdsourced data produced by a variety of activities mainly (but not only) in the realm of 'citizen science' (Nardi et al., 2021). These observations may constitute a valuable complement of standard hydrological monitoring, useful to cover the large number of still ungauged or poorly gauged small- and mid-size watersheds. Typical examples are images and video taken from cellular phones during flooding events (Le Coz et al., 2016), simplified river stage measurements submitted by volunteers via cellular phone text messages (Weeser et al., 2018), or hydrometeorological data from a variety of opportunistic sensors (e.g. smartphones, personal weather stations, commercial microwave links; de Vos et al., 2020). Main challenges in the assimilation of such data are related to their being characterised by random accuracies, larger errors than standard professional observations, and irregular temporal frequencies. Avellaneda et al. (2020) used an augmented EnKF to calibrate the parameters of a distributed hydrologic model for daily river flow simulation using stream stage and stream temperature volunteer data, assessing the sensitivity of the model accuracy with respect to the temporal sampling characteristics. The effects of varying space-time coverage from social sensors are analysed, with reference to the flood prediction of two real events, in Mazzoleni et al. (2021).

There is a much longer measurement history and application in various hydrological studies of water-stable isotopes, which have been demonstrated to provide new insights into the underlying physics of many hydrological processes, such as the partitioning of the soil vadose zone water into evapotranspiration, percolation and discharge (Good et al., 2015). However, despite the rich literature on the use of stable water isotopes data in regional and watershed studies (see Sprenger et al., 2016a, for a quite extensive review), the formal data assimilation of such data is still at its infancy and has been so far substantially limited to model calibration exercises (Birkel et al., 2014; Sprenger et al., 2016b). Only very recently has the use of stable water isotopes been framed in state estimation frameworks, such as the Bayesian update of prior isotopic signatures to assess the altitude dependence of groundwater recharge (Arellano et al., 2020), while a first true data assimilation experiment in a physically based prediction model just appeared in the atmospheric sciences literature (Tada et al., 2021), with the use of a local ensemble transform Kalman filter to assimilate water vapor isotopes in a general circulation model.

About the second point of attention, that is, the use of AI-based algorithms, the novelty of interest here is the transition from the mere use of machine learning models as black-box substitutes of the 'physically based' hydrological prediction models (Abrahart et al., 2012) to the joint use of both type of models to improve the efficiency of data assimilation systems. The ability of AI algorithms to learn non-linear relationships between different sets of variables is being proved to overcome certain DA difficulties, with applications on hydrologic prediction and simulation too. In describing a DA framework as a two-component system (i.e. a prediction model for the sates dynamics and a measurement operator relating measurements to model states), machine learning has been proposed to replace a physically based measurement operator that may be biased with respect to model states or even unknown, while leaving untouched the states dynamics model. Rodriguez-Fernandez et al. (2019) use a neural network to directly assimilate SMOS brightness temperature in the H-TESSEL LSM for updating the soil moisture rather than assimilating, with a standard linear operator, the SMOS soil moisture product; the neural network is trained against the H-TESSEL simulations to guarantee the same climatology for the SMOS-neural network and the H-TESSEL soil moisture. A similar approach was also previously followed by Kolassa et al. (2017) with SMAP data. While these two studies propose an alternative to existing measurement operators (or measurement models) upon which the standard soil moisture products are based,

Boucher et al. (2020) even push forward the concept of using a machine learning algorithm in the place of a classical measurement operator. They explore the case where this measurement operator could not be defined on the base of the prediction model structure. In particular, they use two distinct types of neural networks to assimilate, with an EnKF, river discharge and temperature at different time lags in a lumped hydrologic model where the state variables to be updated are the levels in conceptual reservoirs (i.e. conceptual state variables that cannot be measured, neither directly nor indirectly). With classical techniques, such a problem could have been solved by using a variational assimilation with an adjoint such as in the Appendix; the use of the AI component allows to use the EnKF instead, avoiding the difficulty of having to develop and code the adjoint model. As previously discussed, hydrological models are highly likely to have quite large structural errors with systematic biases. In this respect, AI techniques are also proposed to postprocess the detectable bias of model ensembles in Monte Carlo type of assimilation with respect to measurements. As an example, King et al. (2020) demonstrate how a random forest algorithm outperforms other classical linear postprocessing algorithms in analysing the non-linear and seasonal behaviour of the bias in a regional snow data assimilation system.

7.5 Conclusions

A broad review has been presented in this chapter about how the main data assimilation techniques span the various fields and key problems in such a wide discipline as hydrology. Main ingredients of data assimilation (i.e. observations and models), vary dramatically across hydrologic applications in terms of space-time density (the first), complexity and computational demand (the second), and accuracy and error structure (both), so that dominant or preferred pathways cannot be identified. On the other side, two main drivers distinguish DA in hydrology from other broad geophysical disciplines, namely the large uncertainty in many environmental parameters (e.g. soil hydrologic properties) that control the model dynamics, and the very uncertain observation of key state variables (soil moisture being the most notable case).

The first driver was initially employed in data assimilation to solve inverse problems, both static and time-dependent. Variational techniques have been often preferred over sequential (filter-type) techniques in this case, mainly due to the possibility of more easily and directly exploring the sensitivity of model response with respect to model parameters even when the observable model states and unknown parameters to be inverted for were quite 'conceptually distant' (e.g. the discussed case of inverting for aquifer properties while observing ground surface subsidence). Among different variational techniques, the use of adjoint models had quite a success due to the possibility, in specific application-oriented problems, to use quite parsimonious models' structure; a parsimony that brought the double advantage of reducing the dimension of model states and parameters and strongly facilitating the coding of the adjoint model. Among the various aspects that still require scientific advancement in hydrologic inversion problems, the most challenging remains the solution of many underdetermined problems, where the number of available observations is smaller than the number of unknown parameters and, on top of that, such observations are affected by large measurement errors. The use of new types of data, in particular the ones coming from citizen science initiatives, is opening newer challenges along this general pathway.

Treatment of large measurement errors, even in the presence of frequent and/or dense observations such as from remote sensing, is again the main challenge along the path of state estimation problems. Various types of non-linear sequential filters (the EnKF being the most widely used) coupled to more and more complex models has been the main answer so far to this challenge: in the face of large measurement errors, it is attempted to reduce the model errors as much as possible by increasing the number of represented processes (i.e. the model's 'completeness'); sequential filtering, especially the ensemble one, allow for a more precise and consistent treatment of both error types and the objective definition of their optimal weighting. Computational demand and algorithmic efficiency remain among the key issues, despite the continuous increase in available computational power, when this approach is applied to large data assimilation systems; merging AI algorithms with 'process-based' data assimilation techniques is already established as the most promising direction to tackle these issues.

7.6 Appendix: Multivariate 1D-VAR Assimilation with an Adjoint

Let the hydrologic model be described by a set of coupled ordinary differential equations for the set of different state variables $\mathbf{X}_1, ..., \mathbf{X}_N$ (e.g. soil moisture, surface storage, river discharge), possibly distributed in space, controlled by the set of model parameters and forcing inputs $\boldsymbol{\theta}_1, ..., \boldsymbol{\theta}_M$ (e.g. soil hydraulic conductivity, channel roughness, but also precipitation, solar radiation, etc.), eventually distributed in space too:

$$\frac{d\mathbf{X}_i}{dt} = \mathbf{F}_i(\mathbf{X}_1, ..., \mathbf{X}_N | \boldsymbol{\theta}_1, ..., \boldsymbol{\theta}_M) \quad ; \quad i = 1, ..., N. \quad \text{A7.1}$$

For the sake of conciseness, both stationary model parameters and time-varying forcing inputs are included in the same set of variables $\boldsymbol{\theta}$, whose main difference from the states $\mathbf{X}$ is the lack of a physical constraint for the dynamics (i.e. a predicting equation).

Let $\mathbf{Y}_1, ..., \mathbf{Y}_N$ be the set of observations, related to the state variables by measurement operators $\mathbf{H}_1, ..., \mathbf{H}_N$. Without much loss of generality, we may assume that the measurement operators are linear and mutually independent, i.e.:

$$\mathbf{Y}_i = \mathbf{H}_i \mathbf{X}_i \quad ; \quad i = 1, ..., N. \quad \text{A7.2}$$

A variational data assimilation algorithm may be then constructed as an optimisation problem by minimising the following penalty functional, defined over a time assimilation window $[t_0, t_1]$ (e.g. Le Dimet and Talagrand, 1986):

$$\begin{aligned} J = &\sum_{i=1}^{N} \frac{1}{t_1 - t_0} \int_{t_0}^{t_1} (\mathbf{H}_i\mathbf{X}_i - \mathbf{Y}_i)^T \mathbf{G}_{Y_i}^{-1}(\mathbf{H}_i\mathbf{X}_i - \mathbf{Y}_i)dt \\ &+ \sum_{i=1}^{N} (\mathbf{H}_i\mathbf{X}_i - \mathbf{Y}_i)^T \mathbf{G}_{Y_i}^{-1}(\mathbf{H}_i\mathbf{X}_i - \mathbf{Y}_i)\big|_{t_0} \\ &+ \sum_{j=1}^{M} \frac{1}{t_1 - t_0} \int_{t_0}^{t_1} (\boldsymbol{\theta}_j - \hat{\boldsymbol{\theta}}_j)^T \mathbf{G}_{\theta_j}^{-1}(\boldsymbol{\theta}_j - \hat{\boldsymbol{\theta}}_j)dt \\ &+ \sum_{i=1}^{N} \int_{t_0}^{t_1} \boldsymbol{\lambda}_i^T \left[\frac{d\mathbf{X}_i}{dt} - \mathbf{F}_i(\mathbf{X}_1, ..., \mathbf{X}_N | \boldsymbol{\theta}_1, ..., \boldsymbol{\theta}_M) \right]dt. \quad \text{A7.3} \end{aligned}$$

The first term represents the misfit of the estimated states with respect to observations. Different misfit components are weighted by the inverse of a covariance matrix $\mathbf{G}_{Y_i}$. This inverse may be simply replaced by the null matrix at any time when a state variable cannot be observed. The second term in Eq. A7.3 is similar to the first one, but it is evaluated at the start time of the assimilation window. It is introduced separately from the first one in case the assimilation goal is the estimation of some initial condition (e.g. the soil saturation prior to some flooding event). The third term is the update of the model parameters with respect to some prior $\hat{\boldsymbol{\theta}}$, with the corresponding covariance $\mathbf{G}_{\theta_j}$. The last term is the 'physical constraint' provided by the hydrologic prediction model. As a constraint to the minimisation problem, the prediction equations are added to the penalty functional through the Lagrange multipliers $\boldsymbol{\lambda}_1, ..., \boldsymbol{\lambda}_N$. The physical constraint brings implicitly in the optimisation problem all the correlations (sensitivities) among model states and parameters. Based on this consideration, the entire states–parameters covariance matrix is simplified as a block-diagonal one, composed by the blocks $\mathbf{G}_{Y_i}$ and $\mathbf{G}_{\theta_j}$. These covariance blocks need to be prescribed, and their magnitude plays a significant role in the convergence of the iterative procedure described in the next paragraph. To this end, prescription criteria are not that different from the one used for prior covariance in recursive state estimation approaches. In particular, while $\mathbf{G}_{Y_i}$ may be easily inferred from available information on the error structure of the used measurement technique, $\mathbf{G}_{\theta_j}$ requires a more subjective evaluation on the (co)-variabilities on unknown parameters, which may be based on existing relevant literature of the case study.

In the adjoint technique, the solution to the minimisation problem is sought with an iterative procedure based on the Euler–Lagrange equations. These are obtained by first taking the full derivative of the functional and then integrating by parts the Lagrangian term. After some algebra, equating to zero all independent variations of J with respect to $\mathbf{X}_i$ and $\boldsymbol{\theta}_j$ gives the following system of equations for $i = 1, ..., N$ and $j = 1, ..., M$:

$$\frac{d\boldsymbol{\lambda}_i}{dt} = -\boldsymbol{\lambda}_i^T \frac{\partial \mathbf{F}_i}{\partial \mathbf{X}_i} + (\mathbf{H}_i \mathbf{X}_i - \mathbf{Y}_i)^T \mathbf{G}_{Y_i}^{-1} \mathbf{H}_i, \qquad \text{A7.4}$$

$$\boldsymbol{\lambda}_i|_{t_1} = 0, \qquad \text{A7.5}$$

$$(\mathbf{H}_i \mathbf{X}_i - \mathbf{Y}_i)^T \mathbf{G}_{Y_i}^{-1} \mathbf{H}_i|_{t_0} = \boldsymbol{\lambda}_i|_{t_0}, \qquad \text{A7.6}$$

$$(\boldsymbol{\theta}_j - \hat{\boldsymbol{\theta}}_j)^T \mathbf{G}_{\theta_j}^{-1} = \sum_{i=1}^{N} \boldsymbol{\lambda}_i^T \frac{\partial \mathbf{F}_i}{\partial \boldsymbol{\theta}_j}. \qquad \text{A7.7}$$

For those model parameters that are to be considered as constant in time, Eq. A7.7 may be replaced by:

$$(\boldsymbol{\theta}_j - \hat{\boldsymbol{\theta}}_j)^T \mathbf{G}_{\theta_j}^{-1} = \frac{1}{t_1 - t_0} \sum_{i=1}^{N} \int_{t_0}^{t_1} \boldsymbol{\lambda}_i^T \frac{\partial \mathbf{F}_i}{\partial \boldsymbol{\theta}_j} dt. \qquad \text{A7.8}$$

The iterative procedure starts by integrating forward the hydrologic model (Eq. A7.1) with the prior model parameters $\hat{\boldsymbol{\theta}}_j$ and a first guess initial condition $\mathbf{X}_i|_{t_0}$. Then the adjoint model (Eq. A7.4) is integrated backward with homogenous 'end condition' (Eq. A7.5). Note that if the forward hydrologic model has a convergent behaviour, the adjoint is also convergent when integrated backward. In particular, the adjoint model is convergent towards homogeneous null values for vanishing forcing term (the last term in Eq. A7.4), that is, when model states and observations ideally match. Non-homogeneous solutions for the Lagrange multipliers $\boldsymbol{\lambda}_i$ are then used to update the initial condition (Eq. A7.6) and the model parameters (Eq. A7.7 or A7.8). The cycle is repeated until some converge criterion is reached, such as a negligible value for the norm of the Lagrange multipliers.

Contrary to what is commonly believed, the variational method with an adjoint can also provide a quantitative measure of the uncertainty in the estimation. The Hessian of the cost function J can be used to this purpose. To a good level of approximation, in the vicinity of its minimum, the cost function has a quadratic shape. With the further assumption of Gaussian distribution of the parameters to be estimated (more precisely, of the parameters update with respect to a prior near the minimum of the cost function), the Hessian provides a good approximation to the inverse of

the parameters' covariance (Thacker, 1989). However, iterations for seeking the minimum of the cost function are terminated in the usual practice at some non-negligible distance from the 'true' minimum. A correct computation of the Hessian should take care of all the residual cross-correlations among the cost function components represented by both the forward and the adjoint model. An accurate method for computing the Hessian is based on considering the gradient of the cost function as a model, with each component of the gradient corresponding to a new cost function for each of the parameters, to be minimised with new Lagrange multipliers $\mathbf{q}_j$ and $\mathbf{p}_j$ (Burger et al., 1992):

$$\boldsymbol{\Phi}_j = \frac{\partial J}{\partial \boldsymbol{\theta}_j} + \sum_{i=1}^{N} \int_{t_0}^{t_1} \mathbf{q}_j^T \left[\frac{d\mathbf{X}_i}{dt} - \mathbf{F}_i(\mathbf{X}_1, ..., \mathbf{X}_N | \boldsymbol{\theta}_1, ..., \boldsymbol{\theta}_M) \right] dt$$
$$+ \sum_{i=1}^{N} \int_{t_0}^{t_1} \mathbf{p}_j^T \left[\frac{d\boldsymbol{\lambda}_i}{dt} + \boldsymbol{\lambda}_i^T \frac{\partial \mathbf{F}_i}{\partial \mathbf{X}_i} - (\mathbf{H}_i \mathbf{X}_i - \mathbf{Y}_i)^T \mathbf{G}_{Y_i}^{-1} \mathbf{H}_i \right] dt. \qquad \text{A7.9}$$

Taking again the first variations, a new set of Euler-Lagrange equations is obtained:

$$\frac{d\mathbf{p}_j}{dt} = -\mathbf{p}_j^T \sum_{1=1}^{N} \frac{\partial \mathbf{F}_i}{\partial \mathbf{X}_i} - \sum_{1=1}^{N} \frac{\partial \mathbf{F}_i}{\partial \boldsymbol{\theta}_j}, \qquad \text{A7.10}$$

$$\frac{d\mathbf{q}_j}{dt} = \mathbf{q}_j^T \sum_{1=1}^{N} \frac{\partial \mathbf{F}_i}{\partial \mathbf{X}_i} - \mathbf{p}_j^T \sum_{1=1}^{N} \mathbf{G}_{Y_i}^{-1}, \qquad \text{A7.11}$$

with conditions:

$$\mathbf{p}_j|_{t_0} = 0, \qquad \text{A7.12}$$

$$\mathbf{q}_j|_{t_1} = 0. \qquad \text{A7.13}$$

Finally, the condition of minimum for the new penalty functions provides the following formula for the diagonal and off-diagonal components of the Hessian matrix:

$$H_{\theta_j \theta_j} = \sum_{i=1}^{N} \left[\mathbf{p}_j^T \frac{\partial^2 \mathbf{F}_i}{\partial \boldsymbol{\theta}_j \partial \mathbf{X}_i} \boldsymbol{\lambda}_i - \mathbf{q}_j^T \frac{\partial \mathbf{F}_i}{\partial \boldsymbol{\theta}_j} \right] + \mathbf{G}_{\theta_j}^{-1}. \qquad \text{A7.14}$$

$$H_{\theta_j \theta_k} = \sum_{i=1}^{N} \left[\mathbf{p}_j^T \frac{\partial^2 \mathbf{F}_i}{\partial \boldsymbol{\theta}_k \partial \mathbf{X}_i} \boldsymbol{\lambda}_i - \mathbf{q}_j^T \frac{\partial \mathbf{F}_i}{\partial \boldsymbol{\theta}_k} \right], \qquad \text{A7.15}$$

or, for time constant parameters:

$$H_{\theta_j \theta_j} = \frac{1}{t_1 - t_0} \sum_{i=1}^{N} \int_{t_0}^{t_1} \left[\mathbf{p}_j^T \frac{\partial^2 \mathbf{F}_i}{\partial \boldsymbol{\theta}_j \partial \mathbf{X}_i} \boldsymbol{\lambda}_i - \mathbf{q}_j^T \frac{\partial \mathbf{F}_i}{\partial \boldsymbol{\theta}_j} \right] dt + \mathbf{G}_{\theta_j}^{-1}, \qquad \text{A7.16}$$

$$H_{\theta_j \theta_k} = \frac{1}{t_1 - t_0} \sum_{i=1}^{N} \int_{t_0}^{t_1} \left[\mathbf{p}_j^T \frac{\partial^2 \mathbf{F}_i}{\partial \boldsymbol{\theta}_k \partial \mathbf{X}_i} \boldsymbol{\lambda}_i - \mathbf{q}_j^T \frac{\partial \mathbf{F}_i}{\partial \boldsymbol{\theta}_k} \right] dt. \qquad \text{A7.17}$$

References

Abdolghafoorian, A., and Farhadi, L. (2020). LIDA: A Land Integrated Data Assimilation framework for mapping land surface heat and evaporative fluxes by assimilating space-borne soil moisture and land surface temperature. *Water Resources Research*, 56(8), e2020WR027183.

Abrahart, R. J., Anctil, F., Coulibaly, P. et al. (2012). Two decades of anarchy? Emerging themes and outstanding challenges for neural network river forecasting. *Progress in Physical Geography*, 36(4), 480–513.

Allen, G. H., Olden, J. D., Krabbenhoft, C. et al. (2020). Is our finger on the pulse? Assessing placement bias of the global river gauge network. *AGU Fall Meeting Abstracts*, H010-0016.

Altaf, M. U., El Gharamti, M., Heemink, A. W., and Hoteit, I. (2013). A reduced adjoint approach to variational data assimilation. *Computer Methods in Applied Mechanics and Engineering*, 254, 1–13.

Altenau, E. H., Pavelsky, T. M., Durand, M. T. et al. (2021). The Surface Water and Ocean Topography (SWOT) Mission River Database (SWORD): A global river network for satellite data products. *Water Resources Research*, 57(7), e2021WR030054.

Alvarez-Garreton, C., Ryu, D., Western, A. W. et al. (2015). Improving operational flood ensemble prediction by the assimilation of satellite soil moisture: Comparison between lumped and semi-distributed schemes. *Hydrology and Earth System Sciences*, 19(4), 1659–76.

Anderson, J. L. (2007). Exploring the need for localization in ensemble data assimilation using a hierarchical ensemble filter. *Physica D: Nonlinear Phenomena*, 230(1–2), 99–111.

Andreadis, K. M., and Schumann, G. J. (2014). Estimating the impact of satellite observations on the predictability of large-scale hydraulic models. *Advances in Water Resources*, 73, 44–54.

Annis, A., Nardi, F., and Castelli, F. (2022). Simultaneous assimilation of water levels from river gauges and satellite flood maps for near-real time flood mapping. *Hydrology and Earth System Sciences*, 26, 1019–41.

Annis, A., Nardi, F., Morrison, R. R., and Castelli, F. (2019). Investigating hydrogeomorphic floodplain mapping performance with varying DTM resolution and stream order. *Hydrological Sciences Journal*, 64(5), 525–38.

Arellano, L. N., Good, S. P., Sanchez-Murillo, R. et al. (2020). Bayesian estimates of the mean recharge elevations of water sources in the Central America region using stable water isotopes. *Journal of Hydrology: Regional Studies*, 32, 100739.

Avellaneda, P. M., Ficklin, D. L., Lowry, C. S., Knouft, J. H., and Hall, D. M. (2020). Improving hydrological models with the assimilation of crowdsourced data. *Water Resources Research*, 56, e2019WR026325.

Azimi, S., Dariane, A. B., Modanesi, S. et al. (2020). Assimilation of Sentinel 1 and SMAP – based satellite soil moisture retrievals into SWAT hydrological model: The impact of satellite revisit time and product spatial resolution on flood simulations in small basins. *Journal of Hydrology*, 581, 124367.

Babaeian, E., Sadeghi, M., Jones, S. B. et al. (2019). Ground, proximal, and satellite remote sensing of soil moisture. *Review of Geophysics*, 57(2), 530–616.

Balsamo, G., Agusti-Panareda, A., Albergel, C. et al. (2018). Satellite and in situ observations for advancing global earth surface modelling: A review. *Remote Sensing*, 10, 38.

Bateni, S. M., Entekhabi, D., and Castelli, F. (2013). Mapping evaporation and estimation of surface control of evaporation using remotely sensed land surface temperature from a constellation of satellites. *Water Resources Research*, 49, 950–68.

Bateni, S., Entekhabi, D., Margulis, S., Castelli, F., and Kergoat, L. (2014). Coupled estimation of surface heat fluxes and vegetation dynamics from remotely sensed land surface temperature and fraction of photosynthetically active radiation. *Water Resources Research*, 50, 8420–40.

Bauser, H.H., Berg, D., and Roth, K. (2021). Technical Note: Sequential ensemble data assimilation in convergent and divergent systems. *Hydrology and Earth System Sciences*, 25, 3319–29.

Bauer-Marschallinger, B., Freeman, V., Cao, S. et al. (2019). Toward global soil moisture monitoring with Sentinel-1: Harnessing assets and overcoming obstacles. *IEEE Transactions on Geoscience and Remote Sensing*, 57(1), 520–39.

Baugh, C., de Rosnay, P., Lawrence, H. et al. (2020). The impact of SMOA soil moisture data assimilation within the operational global flood awareness system (GloFAS). *Remote. Sensing*, 12(9), 1490, https://doi.org/10.3390/rs12091490.

Berghuijs, W. R., Woods, R. A., Hutton, C. J., and Sivapalan, M. (2016). Dominant flood generating mechanisms across the United States. *Geophysical Research Letters*, 43(9), 4382–90.

Berthet, L., Andréassian, V., Perrin, C., and Javelle, P. (2009). How crucial is it to account for the antecedent moisture conditions in flood forecasting? Comparison of event-based and continuous approaches on 178 catchments. *Hydrology and Earth System Sciences*, 13(6), 819–31.

Beven, K., and Binley, A. (2014). GLUE: 20 years on. *Hydrological Processes*, 28(24), 5897–918.

Biancamaria, S., Lettenmaier, D. P., and Pavelsky, T. M. (2016). The SWOT mission and its capabilities for land hydrology. In A. Cazenave, N. Champollion, J. Benveniste, and J. Chen, eds., *Remote Sensing and Water Resources*. Cham: Springer, pp. 117–47.

Birkel, C., Soulsby, C., and Tetzlaff, D. (2014), Developing a consistent process-based conceptualization of catchment functioning using measurements of internal state variables. *Water Resources Research*, 50, 3481–501.

Bolten, J. D., Crow, W. T., Jackson, T. J., Zhan, X., and Reynolds, C. A. (2010). Evaluating the utility of remotely sensed soil moisture retrievals for operational agricultural drought monitoring. *IEEE Journal of Selected Topics in Applied Earth Observations and Remote Sensing*, 3(1), 57–66.

Boucher, M.-A., Quilty, J., and Adamowski, J. (2020). Data assimilation for streamflow forecasting using extreme learning machines and multilayer perceptrons. *Water Resources Research*, 56, e2019WR026226. https://doi.org/10.1029/2019WR026226.

Boussetta, S., Balsamo, G., Beljaars, A. et al. (2013). Natural land carbon dioxide exchanges in the ECMWF Integrated Forecasting System: Implementation and offline validation, *Journal of Geophysical Research: Atmospheres*, 118, 5923–46.

Boussetta, S., Balsamo, G., Arduini, G. et al. (2021). ECLand: The ECMWF land surface modelling system. *Atmosphere*, 12, 723.

Burger, J., Le Brizaut, J. S., and Pogu, M. (1992). Comparison of two methods for the calculation of the gradient and of the Hessian of cost functions associated with differential systems. *Mathematics and Computers in Simulation*, 34, 551–62.

Caparrini, F., Castelli, F., and Entekhabi, D. (2003). Mapping of land-atmosphere heat fluxes and surface parameters with remote sensing data. *Boundary-Layer Meteorology*, 107, 605–33.

Caparrini, F., Castelli, F., and Entekhabi, D. (2004). Variational estimation of soil and vegetation turbulent transfer and heat flux parameters from sequences of multisensor imagery. *Water Resources Research*, 40, W12515.

Carrera, J., and Neuman, S.P. (1986). Estimation of aquifer parameters under transient and steady state conditions: 1. Maximum likelihood method incorporating prior information. *Water Resources Research*, 22(2), 199–210.

Castillo, A., Castelli, F., and Entekhabi, D. (2015). Gravitational and capillary soil moisture dynamics for distributed hydrologic models. *Hydrology and Earth System Sciences*, 19(4), 1857–69.

Ceccatelli, M., Del Soldato, M., Solari, L. et al. (2021). Numerical modelling of land subsidence related to groundwater withdrawal in the Firenze-Prato-Pistoia basin (central Italy). *Hydrogeology Journal*, 29, 629–49.

Cenci, L., Pulvirenti, L., Boni, G. et al. (2017). An evaluation of the potential of Sentinel 1 for improving flash flood predictions via soil moisture-data assimilation. *Advances in Geosciences*, 44, 89–100.

Chan, S. K., Bindlihs, R., O'Neill, P. et al. (2016). Assessment of the SMAP passive soil moisture product. *IEEE Transactions on Geoscience and Remote Sensing*, 54(8), 4994–5007.

Chan, S. K., Bindlihs, R., O'Neill, P. et al. (2017). Development and assessment of the SMAP enhanced passive soil moisture product. *Remote Sensing of Environment*, 204, 931–41.

Chen, J. M., and Liu, J. (2020). Evolution of evapotranspiration models using thermal and shortwave remote sensing data. *Remote Sensing of Environment.*, 237, 111594.

Clark, M. P., Rupp, D. E., Woods, R. A. et al. (2008). Hydrological data assimilation with the ensemble Kalman filter: Use of streamflow observations to update states in a distributed hydrological model. *Advances in Water Resources*, 31(10), 1309–24.

Cooper, E. S., Dance, S. L., Garcia-Pintado, J., Nichols, N. K., and Smith, P. J. (2019). Observation operators for assimilation of satellite observations in fluvial inundation forecasting. *Hydrology and Earth System Sciences*, 23(6), 2541–59.

Das, N. N., Entekhabi, D., Dunbar, R. S. et al. (2018). The SMAP mission combined active-passive soil moisture product at 9 km and 3 km spatial resolutions. *Remote Sensing of Environment*, 211, 204–17.

Dasgupta, A., Hostache, R., Ramsankaran, R. et al. (2021). On the impacts of observation location, timing and frequency on flood extent assimilation performance. *Water Resources Research*, 57, e2020WR028238

DeChant, C. M., and Moradkhani, H. (2012). Examining the effectiveness and robustness of sequential data assimilation methods for quantification of uncertainty in hydrologic forecasting. *Water Resources Research*, 48, W04518.

Domeneghetti, A., Castellarin, A., and Brath, A. (2012). Assessing rating-curve uncertainty and its effects on hydraulic model calibration. *Hydrology and Earth System Sciences*, 16(4), 1191–202.

Dong, J., Crow, W. T., Tobin, K. J. et al. (2020). Comparison of microwave remote sensing and land surface modeling for surface soil moisture climatology estimation. *Remote Sensing of Environment.*, 242, 111756.

Emery, C. M., Biancamaria, S., Boone, A. et al. (2020). Assimilation of wide-swath altimetry water elevation anomalies to correct large-scale river routing model parameters. *Hydrology and Earth System Sciences*, 24, 2207–33.

Entekhabi, D., Nakamura, H., and Njoku, E.G. (1994). Solving the inverse problem for soil moisture and temperature profiles by sequential assimilation of multifrequency remotely sensed observations. *IEEE Transactions on Geoscience and Remote Sensing*, 32(2), 438–48.

Entekhabi, D., Rodriguez-Iturbe, I., and Castelli, F. (1996). Mutual interaction of soil moisture state and atmospheric processes. *Journal of Hydrology*, 184(1–2), 3–17.

Ercolani, G., and Castelli, F. (2017). Variational assimilation of streamflow data in distributed flood forecasting. *Water Resources Research*, 53(1), 158–83.

Fabryka-Martin, J., Merz, J., and Universities Council on Water Resources (1983). *Hydrology: The Study of Water and Water Problems: A Challenge for Today and Tomorrow*. Carbondale, IL: Universities Council on Water Resources.

Fan, Y. R., Huang, W. W., Li, Y. P., Huang, G. H., and Huang, K. (2015). A coupled ensemble filtering and probabilistic collocation approach for uncertainty quantification of hydrological models. *Journal of Hydrology*, 530, 255–72.

Frey, S. K., Miller, K., Khader, O. et al. (2021). Evaluating landscape influences on hydrologic behavior with a fully-integrated groundwater–surface water model. *Journal of Hydrology*, 602, 126758.

Garcia-Pintado, J., Mason, D. C., Dance, S. L. et al. (2015). Satellite-supported flood forecasting in river networks: A real case study. *Journal of Hydrology*, 523, 706–24.

Ghorbanidehno, H., Kokkinaki, A., Lee, J., and Darve, E. (2020). Recent developments in fast and scalable inverse modeling and data assimilation methods in hydrology. *Journal of Hydrology*, 591, 125266.

Gomez, B., Charlton-Perez, C. L., Lewis, H., and Candy, B. (2020). The Met Office operational soil moisture analysis system. *Remote*, 12, 3691.

Good, S. P., Noone, D., and Bowen, G. (2015), Hydrologic connectivity constrains partitioning of global terrestrial water fluxes. *Science*, 349(6244), 175–7.

Grimaldi, S., Li, Y., Pauwels, V. R. N., and Walker, J. P. (2016). Remote sensing-derived water extent and level to constrain hydraulic flood forecasting models: Opportunities and challenges. *Surveys in Geophysics*, 37(5), 977–1034.

Hajj, M. E., Baghdadi, N., Zribi, M., and Bazzi, H. (2017). Synergic use of Sentinel-1 and Sentinel-2 images for operational soil moisture mapping at high spatial resolution over agricultural areas. *Remote Sensing*, 9(12), 1292. https://doi .org/10.3390/rs9121292.

Harrigan, S., Zsoter, E., Alfieri, L. et al. (2020). GloFAS-ERA5 operational global river discharge reanalysis 1979–present. *Earth System Science Data*, 12, 2043–60

Harvey, C. F., and Gorelick, S. M. (1995). Mapping hydraulic conductivity: Sequential conditioning with measurements of solute arrival time, hydraulic head, and local conductivity. *Water Resources Research*, 31(7), 1615–26.

Hochstetler, D. L., Barrash, W., Leven, C. et al. (2016). Hydraulic tomography: Continuity and discontinuity of high-K and low-K zones. *Groundwater*, 54(2), 171–85.

Hostache, R., Chini, M., Giustarini, L. et al. (2018). Near-real-time assimilation of SAR-derived flood maps for improving flood forecasts. *Water Resources Research*, 54(8), 5516–35.

Hou, Z.-Y., and Jin, Q.-N. (1997). Tikhonov regularization for nonlinear ill-posed problem. *Nonlinear Analysis: Theory, Methods & Applications*, 28(11), 1799–809.

Ishitsuka, Y., Gleason, C. J., Hagemann, M. W. et al. (2021). Combining optical remote sensing, McFLI discharge estimation, global hydrologic modeling, and data assimilation to improve daily discharge estimates across an entire large watershed. *Water Resources Research*, 56, e2020WR027794.

Kerr, Y. H., Waldteufel, P., Richaume, P. et al. (2012). The SMOS soil moisture retrieval algorithm. *IEEE Transactions on Geoscience and Remote Sensing*, 50(5), 1384–403.

King, F., Erler, A. R., Frey, S. K., and Flechter, C. G. (2020). Application of machine learning techniques for regional bias correction of snow water equivalent estimates in Ontario, Canada. *Hydrology and Earth System Sciences*, 24(10), 4887–902.

Kitanidis, P. K. (1996). On the geostatistical approach to the inverse problem. *Advances in Water Resources*. 19(6), 333–42.

Kolassa, J., Reichle, R. H., Liu, Q. et al. (2017). Data assimilation to extract soil moisture information from SMAP observations. *Remote Sensing*, 9, 1179.

Kustas, W. P., and Norman, J. M. (1999). Evaluation of soil and vegetation heat flux predictions using a simple two-source model with radiometric temperatures for partial canopy cover. *Agriculture an Forest Meteorology*, 94(1), 13–29.

Laloy, E., Rogiers, B., Vrugt, J. A., Mallants, D., and Jacques, D. (2013). Efficient posterior exploration of a high-dimensional groundwater model from two-stage Markov chain Monte Carlo simulation and polynomial chaos expansion. *Water Resources Research*, 49(5), 2664–82.

Le Coz, J., Patalano, A., Collins, D. et al. (2016). Crowdsourced data for flood hydrology: Feedback from recent citizen science projects in Argentina, France and New Zealand. *Journal of Hydrology*, 541, 766–77.

Le Dimet, F. X., and Talagrand, O. (1986). Variational algorithms for analysis and assimilation of meteorological observations: theoretical aspects. *Tellus A*, 38:2, 97–110.

Lee, J., Yoon, H., Kitanidis, P. K., Werth, C. J., and Valocchi, A. J. (2016). Scalable subsurface inverse modeling of huge data sets with an application to tracer concentration breakthrough data from magnetic resonance imaging, *Water Resources Research*, 52, 5213–31.

Lievens, H. et al. (2017). Joint Sentinel-1 and SMAP data assimilation to improve soil moisture estimates. *Geophysical Research Letters*, 44(12), 6145–53.

Liu, Y., Weerts, AH., Clark, M. et al. (2012). Advancing data assimilation in operational hydrologic forecasting: Progresses, challenges, and emerging opportunities. *Hydrology and Earth System Sciences*, 16(10), 3863–87.

Liu, K., Huang, G., Simunek, J. et al. (2021). Comparison of ensemble data assimilation methods for the estimation of time-varying soil hydraulic parameters. *Journal of Hydrology*, 594, 125729.

Maggioni, V., Vergara, H. J., Anagnostou, E. N. et al. (2013). Investigating the applicability of error correction ensembles of satellite rainfall products in river flow simulations. *Journal of Hydrometeorology*, 14(4), 1194–211.

Maliva, G. (2016). *Aquifer Characterization Techniques*. Cham Springer.

Man, J., Zheng, Q., Wu, L., and Zeng, L. (2020). Adaptive multi-fidelity probabilistic collocation-based Kalman filter for subsurface flow data assimilation: Numerical modeling and real-world experiment. *Stochastic Environmental Research and Risk Assessment*, 34, 1135–46. https://doi.org/10.1007/s00477-020-01815-y

Margulis, S. A., and Entekhabi, D. (2001). A coupled land surface-boundary layer model and its adjoint. *Journal of Hydrometeorology*, 2(3), 274–96.

Mason, D., Garcia-Pintado, J., Cloke, H. L., Dance, S. L., and Munoz-Sabatier, J. (2020) Assimilating high resolution remotely sensed data into a distributed hydrological model to improve run off prediction. *ECMWF Tech Memo, 867*, European Centre for Medium-Range Weather Forecasts.

Massari, C., Brocca, L., Tarpanelli, A., and Moramarco, T. (2015). Data assimilation of satellite soil moisture into rainfall-runoff modelling: A complex recipe? *Remote Sensing*, 7(9), 11403–33.

Massari, C., Camici, S., Ciabatta, L., and Brocca, L. (2018). Exploiting satellite-based surface soil moisture for flood forecasting in the Mediterranean area: State update versus rainfall correction. *Remote Sensing*, 10(2), 292. https://doi.org/10.3390/rs10020292.

Mazzoleni, M., Alfonso, L., and Solomatine, D.P. (2021). Exploring assimilation of crowdsourcing observations into flood models. In A. Scozzari, S. Mounce, D. Han, F. Soldovier, and D. Solomatine, eds., *ICT for Smart Water Systems: Measurements and Data Science. Handbook of Environmental Chemistry*, vol. 102. Cham: Springer, pp. 209–34.

Moges, E., Demissie, Y., Larsen, L., and Yassin, F. (2021). Review: Sources of hydrological model uncertainties and advances in their analysis. *Water*, 13(1), 28, https://doi.org/10.3390/w13010028.

Moradkhani, H., Hsu, K.-L., Gupta, H., and Sorooshian, S. (2005). Uncertainty assessment of hydrologic model states and parameters: Sequential data assimilation using the particle filter. *Water Resources Research*, 41, W05012.

Mudashiru, R. B., Sabtu, N., Abustan, I., and Balogun, W. (2021). Flood hazard mapping methods: A review. *Journal of Hydrology*, 603, 126846.

Nardi, F., Cudennec, C., Abrate, T. et al. (2021). Citizens AND HYdrology (CANDHY): conceptualizing a transdisciplinary framework for citizen science addressing hydrological challenges. *Hydrological Sciences Journal*. https://doi.org/10.1080/02626667.2020.1849707.

Neal, J. C., Atkinson, P. M., and Hutton, C. W. (2007). Flood inundation model updating using an ensemble Kalman filter and spatially distributed measurements. *Journal of Hydrology*, 336, 401–15.

Ng, L. W.-T., and Willcox, K. E. (2014). Multifidelity approaches for optimization under uncertainty. *International Journal for Numerical Methods in Engineering*, 100(10), 746–72.

Noilhan, J., and Planton, S. (1989). A simple parameterization of land-surface processes for meteorological models. *Monthly Weather Review*, 117, 536–50.

Ochoa-Rodriguez, S., Wang, L.-P., Gires, A. et al. (2015). Impact of spatial and temporal resolution of rainfall inputs on urban hydrodynamic modelling outputs: A multi-catchment investigation. *Journal of Hydrology*, 531(2), 389–407.

Pierdicca, N., Chini, M., Pulvirenti, L. et al. (2009). Using COSMO-SkyMed data for flood mapping: Some case-studies. *2009 IEEE International Geoscience and Remote Sensing Symposium*, 2, II-933–6.

Pinnington, E., Amezcua, J., Cooper, E. et al. (2021). Improving soil moisture prediction of a high-resolution land surface model by parameterising pedotransfer functions through assimilation of SMAP satellite data. *Hydrology and Earth System Sciences*, 25, 1617–41.

Rajabi, M. M., Ataie-Ashtiani, B., and Simmons, C. T. (2018). Model-data interaction in groundwater studies: Review of methods, applications and future directions. *Journal of Hydrology*, 567, 457–77.

Reichle, R. H., de Lannoy, G. J. M., Liu, Q. et al. (2017). Assessment of the SMAP Level-4 surface and root-zone soil moisture product using in situ measurements. *Journal of Hydrometeorology*, 18(10), 2621–45.

Reichle, R. H, Liu, Q., Koster, R. D. et al. (2019). Version 4 of the SMAP Level-4 soil moisture algorithm and data product. *Journal of Advances in Modeling Earth Systems*, 11(10), 3106–30.

Rodriguez-Fernandez, N., de Rosnay, P., Albergel, C. et al. (2019). SMOS neural network soil moisture data assimilation in a land surface model and atmospheric impact. *Remote Sensing*, 11, 1334.

Seneviratne, S. I., Corti, T., Davin, E. L. et al. (2010). Investigating soil moisture–climate interactions in a changing climate: A review. *Earth-Science Reviews*, 99(3–4), 125–61.

Sood, A., and Smakhtin, V. (2015). Global hydrological models: A review. *Hydrological Sciences Journal*, 60, 549–65.

Sprenger, M., Leistert, H., Gimbel, K., and Weiler, M. (2016a). Illuminating hydrological processes at the soil-vegetation-atmosphere interface with water stable isotopes. *Reviews of Geophysics*, 54(3), 674–704.

Sprenger, M., Seeger, S., Blume, T., and Weiler, M. (2016b). Travel times in the vadose zone: Variability in space and time. *Water Resources Research*, 52(8), 5727–54.

Steklova, K., and Haber, E. (2017). Joint hydrogeophysical inversion: State estimation for seawater intrusion models in 3D. *Computational Geosciences*, 21, 75–94.

Tada, M., Yoshimura, K., and Toride, K. (2021). Improving weather forecasting by assimilation of water vapor isotopes. *Scientific Reports*, 11(1), 18067.

Talagrand, O., and Courtier, P. (1987). Variational assimilation of meteorological observations with the adjoint vorticity equation. I: Theory. *Quarterly Journal of the Royal Meteorological Society*, 113(478), 1311–28.

Thacker, W. C. (1989). The role of the Hessian matrix in fitting models to measurements. *Journal of Geophysical Research: Oceans*, 94(C5), 6177–96.

Verstraeten, W. W., Veroustraete, F., van der Sande, C. J., Grootaers, I., and Feyen, J. (2006). Soil moisture retrieval using thermal inertia, determined with visible and thermal spaceborne data, validated for European forests. *Remote Sensing of Environment*, 101(3), 299–314.

de Vos, L. W., Droste, A .M., Zander, M. J. et al. (2020). Hydrometeorological monitoring using opportunistic sensing networks in the Amsterdam metropolitan area. *Bulletin of the American Meteorological Society*, 101(2), E167–E185.

Wagner, W., Hahn, S., Kidd, R. et al. (2013). The ASCAT soil moisture product: A review of its specifications, validation results, and emerging applications. *Meteorologische Zeitschrift*, 22(1), 5–33.

Waller, J. A., Garcia-Pintado, J., Mason, D. C., Dance, S. L., and Nichols, N. K. (2018) Technical note: Assessment of observation quality for data assimilation in flood models. *Hydrology and Earth System Sciences*, 22(7), 3983–92.

Wang, K., and Dickinson, R. E. (2012). A review of global terrestrial evapotranspiration: Observation, modeling, climatology, and climatic variability. *Reviews of Geophysics*, 50(2), RG2005.

Wang, L. L., Chen, D. H., and Bao, H. J. (2016). The improved Noah land surface model based on storage capacity curve and Muskingum method and application in GRAPES model. *Atmospheric Science Letters*, 17, 190–8.

Weeser, B., Stenfert Kroese, J., Jacobs, S. R. et al. (2018). Citizen science pioneers in Kenya: A crowdsourced approach for hydrological monitoring. *Science of the Total Environment*, 631–632, 1590–9.

Xie, X., Meng, S., Liang, S., and Yao, Y. (2014). Improving streamflow predictions at ungauged locations with real-time updating: Application of an EnKF-based state-parameter estimation strategy. *Hydrology and Earth System Sciences*, 18(10), 3923–36.

Xu, T., Bateni, S. M., Liang, S., Entekhabi, D., and Mao, K. (2014). Estimation of surface turbulent heat fluxes via variational assimilation of sequences of land surface temperatures from Geostationary Operational Environmental Satellites. *Journal of Geophysical Research: Atmospheres*, 119(18), 10780–98.

Xu, T., He, X., Bateni, S. M. et al. (2019). Mapping regional turbulent heat fluxes via variational assimilation of land surface temperature data from polar orbiting satellites. *Remote Sensing of Environment*, 221, 444–61.

Yang, K., Zhu, L., Chen, Y. et al. (2016). Land surface model calibration through microwave data assimilation for improving soil moisture simulations. *Journal of Hydrology*, 533, 266–76.

Zeng, X., and Decker, M. (2009). Improving the numerical solution of soil moisture–based Richards equation for land models with a deep or shallow water table. *Journal of Hydrometeorology.*, 1081, 308–19.

Zhang, H., Franssen, H.-J. H., Han, X., Vrugt, J. A., and Vereecken, H. (2017). State and parameter estimation of two land surface models using the ensemble Kalman filter and the particle filter. *Hydrology and Earth System Sciences*, 21(9), 4927–58.

Zhang, F., Zhang, L. W., Shi, J. J., and Huang, J. F. (2014). Soil moisture monitoring based on land surface temperature-vegetation index space derived from MODIS data. *Pedosphere*, 24(4), 450–60.

Zheng, W., Zhan, X., Liu, J. and Ek, M. (2018). A preliminary assessment of the impact of assimilating satellite soil moisture data products on NCEP Global Forecast System. *Advances in Meteorology*, 1–12, 7363194.

Zsoter, E., Cloke, H., Stephens, E. et al. (2019). How well do operational numerical weather prediction configurations represent hydrology? *Journal of Hydrometeorology*, 20(8), 1533–52.

8

Data Assimilation and Inverse Modelling of Atmospheric Trace Constituents

Dylan Jones

Abstract: During the past two decades, there have been significant efforts to better quantify emissions of environmentally important trace gases along with their trends. In particular, there has been a clear need for robust estimates of emissions on policy-relevant scales of trace gases that impact air quality and climate. This need has driven the expansion of the observing network to better monitor the changing composition of the atmosphere. This chapter will discuss the use of various data assimilation and inverse modelling approaches to quantify these emissions, with a focus on the use of satellite observations. It will discuss the inverse problem of retrieving the atmospheric trace gas information from the satellite measurements, and the subsequent use of these satellite data for quantifying sources and sinks of the trace gases.

8.1 Introduction

Human activity has produced dramatic changes in the composition of the atmosphere, with profound implications for climate and air quality. Air pollution is now a leading cause of human mortality globally (GBD 2019 Risk Factors Collaborators, 2020). In North America and Europe, air quality regulation has led to dramatic improvements in air quality in the past two decades. However, there are still many people in North America and Europe exposed to high levels of air pollution. In the United States, for example, in 2020, there were 79 million people living in regions where ozone (O_3) levels exceeded the air quality standard.[1] Anthropogenic emissions of nitrogen oxides (NOx = NO + NO_2), which are key ozone precursors, have declined in North America, Europe, and China, but there is uncertainty in the trend of these emissions (Jiang et al., 2022).

Anthropogenic emissions of carbon have led to unprecedented high levels of atmospheric CO_2 and CH_4. Developing effective carbon emission policies to limit future atmospheric increases in these greenhouse gases (GHGs) will require robust estimates of their sources and sinks on policy-relevant scales. For example, the atmospheric growth rate of CH_4 began decreasing in the early 1990s, stabilised between 2000 and 2007, and has begun increasing again. The processes that led to the stabilisation and subsequent recovery of the growth rate are uncertain (Rigby et al., 2017; Turner et al., 2019). Emissions of CO_2 from fossil fuel combustion are about 9–10 Pg C/yr (Le Quéré et al., 2018), but 50–60% of this is taken up, almost equally, by the oceans and the terrestrial biosphere. However, the terrestrial biospheric sink is highly variable and there are large uncertainties in regional estimates of these fluxes (e.g. Sitch et al., 2015). Inverse modelling using atmospheric trace gas observations has emerged as a widely used approach to better quantify these fluxes. This chapter will focus on these inverse modelling efforts for both GHGs and air quality applications.

In the past, a significant challenge for data assimilation and inverse modelling of atmospheric composition measurements in the troposphere has been that the observing network was too sparse. This limitation led to a major expansion of the observing system, both ground-based and space-based. In particular, space agencies around the world have invested significant resources during the past three decades to better monitor the changing composition of the atmosphere. In situ observations are typically more accurate and precise than space-based atmospheric composition data, but, as a result of the global nature of the human influence on the carbon cycle, as well as the impact of intercontinental transport of air pollution on local air quality, space-based observations have become a key component of the expanding global observing system.

Continuous measurements of atmospheric composition from space have been made since the 1970s with the launch of instruments such as the Solar Backscatter Ultraviolet (SBUV) instrument and the Total Ozone Mapping Spectrometer (TOMS) on the NIMBUS-7

---

[1] www.epa.gov/air-trends/air-quality-national-summary

spacecraft in 1978. These instruments measured reflected solar radiation in the nadir from which atmospheric ozone abundances were retrieved. Data from SBUV and TOMS, and the follow-on SBUV/2 that was launched in 1984, were critical to understanding the halogen-catalysed ozone loss that led to the stratospheric ozone hole. By the 1990s, these data were increasingly assimilated into models to constrain stratospheric ozone. The Stratospheric Aerosol and Gas Experiment II (SAGE II), launched in 1984, provided information on the vertical distribution of ozone, water vapour, NO_2, and aerosol optical depth using solar occultation. The Microwave Limb Sounder (MLS), launched on the Upper Atmospheric Research Satellite (UARS) in 1991, provided vertical profile information on a much larger suite of trace gases in the stratosphere. Data from SAGE II and MLS were widely used in the emerging data assimilation systems in the 1990s. For example, Levelt et al. (1998) assimilated MLS ozone data into a chemical transport model (CTM) using an optimal interpolation approach. They found that the assimilation reduced the mean bias and the root-mean-square errors in the modelled ozone. Khattatov et al. (1999) used a variational approach and an extended Kalman filter to assimilate data from MLS and the Cryogenic Limb Array Etalon Spectrometer (CLAES) into a photochemical box model to improve the short-lived chemical constituents in the model. At the NASA Global Modeling and Assimilation Office, continuous assimilation of stratospheric ozone data began with the Goddard Earth Observing System Data Assimilation System (GEOS-DAS), which assimilated data from TOMS and SBUV/2 (Riishøjgaard et al., 2000).

Atmospheric composition measurements from limb viewing satellite instruments such as SAGE II and MLS offer valuable information on the vertical distribution of trace gases with greater vertical resolution than is possible from nadir viewing instruments. But these profiles are confined mainly to the stratosphere since limb viewing instruments are unable to probe the lower atmosphere because of the long paths associated with the limb view and the greater likelihood of interference of clouds in the troposphere. It was the availability of nadir measurements from the growing suite of satellite instruments during the past two decades that drove the increasing focus on tropospheric chemical data assimilation. This chapter will focus on the use of these nadir measurements for data assimilation and inverse modelling of trace constituents in the troposphere. It will begin with a discussion of the inverse problem associated with retrieving the tropospheric abundance of trace gases from space-based measurement, and then examine the use of these retrievals for data assimilation and inverse modelling of the sources and sinks of the gases.

8.2 Remote Sounding Retrievals of Atmospheric Composition

The Bayesian inversion approach is one of the most widely used methods for inferring atmospheric composition information from remote sounding measurements, as well as for inverse modelling of surface sources and sinks of atmospheric trace gases. In this context, these inversion analyses typically employ the maximum *a posteriori* (MAP) estimator, following the approach described by Rodgers (2000). In the MAP framework, we assume a linear relationship of the following form between the observations $\mathbf{y} \in \mathbb{R}^m$, which are the radiances measured by the satellite, and the atmospheric state $x \in \mathbb{R}^n$, discretised on n levels:

$$\mathbf{y} = \mathbf{Hx} + \epsilon, \tag{8.1}$$

with observation errors ϵ. If the errors are Gaussian in their distribution, then the estimate of the state can be obtained by minimising the cost function

$$J(\mathbf{x}) = \frac{1}{2}(\mathbf{y} - \mathbf{Hx})^T \mathbf{R}^{-1}(\mathbf{y} - \mathbf{Hx}) + \frac{1}{2}(\mathbf{x} - \mathbf{x}^b)^T \mathbf{B}^{-1}(\mathbf{x} - \mathbf{x}^b), \tag{8.2}$$

where $\mathbf{H} \in \mathbb{R}^{m \times n}$ is the observation operator (or forward model) that maps the state to the observation space, and ϵ are the observation errors, with observation error covariance matrix $\mathbf{R} \in \mathbb{R}^{m \times m}$. Here $\mathbf{x}^b \in \mathbb{R}^n$ is the *a priori* (or background) estimate of the state, with *a priori* error covariance matrix $\mathbf{B} \in \mathbb{R}^{n \times n}$. The optimised estimate of the state is given by

$$\mathbf{x}^a = \mathbf{x}^b + (\mathbf{H}^T \mathbf{R}^{-1} \mathbf{H} + \mathbf{B}^{-1})^{-1} \mathbf{H}^T \mathbf{R}^{-1}(\mathbf{y} - \mathbf{Hx}^b). \tag{8.3}$$

This form of the solution is used when $n \ll m$, which is typically the case for satellite retrievals. For the case in which $m \ll n$, Eq. (8.3) can be expressed as

$$\mathbf{x}^a = \mathbf{x}^b + \mathbf{BH}^T(\mathbf{HBH}^T + \mathbf{R})^{-1}(\mathbf{y} - \mathbf{Hx}^b), \tag{8.4}$$

as it would be more efficient computationally to deal with the matrix $(\mathbf{HBH}^T + \mathbf{R})$ rather than $(\mathbf{H}^T \mathbf{R}^{-1} \mathbf{H} + \mathbf{B}^{-1})$.

Atmospheric trace gases in the troposphere are retrieved from remote sounding measurements in the UV-visible (UV/VIS), shortwave infrared (SWIR), or thermal infrared (TIR) regions of the spectrum. The different spectral regions can provide different information on the vertical structure of the trace gases. Nadir measurements in the UV/VIS region capture the absorption signature of the trace gas of interest in backscattered solar radiation. As a consequence, these measurements offer little information on the vertical distribution of the trace gas. In contrast, nadir measurements of thermal emission can provide more vertical profile information. In either case, it is important to recognise that the vertical information is limited.

The Measurement of Pollution in the Troposphere (MOPITT) satellite instrument, which was launched in 1999 to measure air pollution in the lower atmosphere, measures carbon monoxide (CO) in the SWIR and TIR regions of the spectrum, at 2.3 μm and 4.6 μm, respectively. The initial MOPITT retrievals were conducted on a 7-element state vector, corresponding to seven pressure levels from the surface to 150 hPa. Despite the 7-element state vector, the retrieved CO profile information is highly correlated, with less than two independent pieces of information, or degrees of freedom for signal (DOFS). Starting with version 5 of the MOPITT product (Deeter et al., 2012), the retrievals have been conducted on a 10-element vertical grid, but the TIR retrievals still provide less than two DOFS. For the SWIR retrievals, the DOFS is ≤ 1. However, combined TIR and SWIR retrievals (Worden et al., 2010) often have DOFS > 2, but only over limited land scenes. The Tropospheric Emission Spectrometer (TES), which was launched on the Aura spacecraft in 2004, measures thermal emission between 3.3 μm and 15.4 μm, from which a wide range of trace gases are retrieved. The initial TES retrievals were conducted on a 65-element state vector, but for the ozone retrievals, for example, the whole atmospheric profile has 2–4 DOFS, and the tropospheric portion of the profile has DOFS < 2.

The limited vertical information in the retrieved vertical profile of the trace gases must be properly accounted for when using the data. This smoothing of the vertical profile is captured by the averaging kernel matrix ($\mathbf{A} \in \mathbb{R}^{n \times n}$), which is retained as a diagnostic product of the retrieval. Using Eq. (8.1), we can express the retrieval, Eq. (8.3) as

$$\mathbf{x}^a = \mathbf{x}^b + \mathbf{K}(\mathbf{y} - \mathbf{H}\mathbf{x}^b) + \mathbf{K}\epsilon, \tag{8.5}$$

where $\mathbf{K} = \mathbf{B}\mathbf{H}^T(\mathbf{H}\mathbf{B}\mathbf{H}^T + \mathbf{R})^{-1}$ is the gain matrix. If we let

$$\mathbf{A} = \mathbf{K}\mathbf{H}, \tag{8.6}$$

we obtain

$$\mathbf{x}^a = \mathbf{x}^b + \mathbf{A}(\mathbf{x} - \mathbf{x}^b) + \mathbf{K}\epsilon, \tag{8.7}$$

$$= (\mathbf{I} - \mathbf{A})\mathbf{x}^b + \mathbf{A}x + \mathbf{K}\epsilon, \tag{8.8}$$

which expresses the retrieval as a linear representation of the true atmospheric state ($\mathbf{x}$), with $\mathbf{A} = \partial\mathbf{x}^a/\partial\mathbf{x}$, corresponding to the sensitivity of the retrieval to the true state. Equation (8.8) shows that the retrieval is a combination of the *a priori* information $(\mathbf{I} - \mathbf{A})\mathbf{x}^b$ and a smooth representation of the true state $\mathbf{A}\mathbf{x}$. The DOFS is given by the trace of the averaging kernel matrix,

$$\text{DOFS} = \text{tr}(\mathbf{A}). \tag{8.9}$$

The ideal averaging kernel would be the identity matrix ($\mathbf{I}$), but in reality $\mathbf{A}$ differs significantly from $\mathbf{I}$. In the case where the retrieval sensitivity is low, the retrieval will reflect mainly the *a priori* (background) information ($\mathbf{x}^b$).

An example of the averaging kernel for the ozone retrieval from TES is shown in Fig. 8.1. The individual lines in the figure correspond to the rows of $\mathbf{A}$ (the n retrieval levels). As

can be seen, the levels in the lower troposphere (between the surface and 500 hPa) have similar sensitivity, that is low near the surface and peaks around 700 hPa. This suggests that ozone retrieved on these levels will be a weighted combination of the actual ozone on these levels. At the surface, the sensitivity is low since TIR measurements cannot capture variations in the trace gases near the surface unless there is a large thermal contrast between the ground and the overlying atmosphere. The need to account for the smoothing influence was described by Rodgers and Connor (2003) and demonstrated by Crawford et al. (2004), who compared MOPITT retrievals with coincident in situ aircraft measurements of CO over the North Pacific. They found that the MOPITT CO profiles differed significantly from the aircraft profiles and did not capture the localised pollution plume seen by the aircraft. However, there was much greater agreement between MOPITT and the aircraft profile after transforming the aircraft profile with the MOPITT averaging kernels and *a priori* profile as follows:

$$\hat{\mathbf{x}}_{ac} = \mathbf{x}^b + \mathbf{A}(\mathbf{x}_{ac} - \mathbf{x}^b), \tag{8.10}$$

where $\mathbf{x}_{ac}$ is the in situ aircraft profile interpolated onto the retrieval grid and $\hat{\mathbf{x}}_{ac}$ is the transformed aircraft profile. With this transformation, the in situ profile will appropriately reflect the contribution of the MOPITT *a priori* information and the smoothing influence of the retrieval. It is critical to perform this transformation when comparing any in situ or modelled profile with a satellite retrieval.

Since trace retrievals using UV/VIS or SWIR measurements offer limited information on the vertical distribution of the gases, these retrievals often report the vertically integrated column abundance of the trace gas (e.g. in molecules cm^{-2}). For these retrievals, Eq. (8.8) can be expressed as

$$c^a = \mathbf{h}^T\mathbf{x}^b + \mathbf{h}^T\mathbf{A}(\mathbf{x} - \mathbf{x}^b) + \mathbf{h}^T\mathbf{K}\epsilon \tag{8.11}$$

$$= c^b + \mathbf{a}^T(\mathbf{x} - \mathbf{x}^b) + \epsilon_c, \tag{8.12}$$

where c^a and c^b are the retrieved and background column abundances, respectively, $\mathbf{h}$ is the vertical integration operator (which can take the form of a pressure weighting function), $\mathbf{a}$ is the column averaging kernel, and ϵ_c is the error on the retrieved column. The column averaging kernels typically are unity throughout the troposphere and decrease sharply in the stratosphere, suggesting a uniform vertical weighting of the trace gas in the troposphere. An example of this is shown in Yoshida et al. (2011) for the column averaging kernels for CO_2 and CH_4 from the Greenhouse gases Observing SATellite (GOSAT). For long-lived trace gases, which are well-mixed in the atmosphere, variations in mass of the atmospheric column associated with surface topography can contribute significantly to spatial variations in the retrieved column abundance of the gases, and can confound analyses using the data to estimate sources and sinks of the

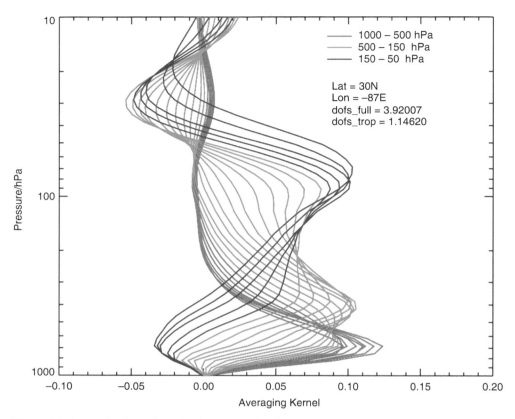

Figure 8.1 Averaging kernels for TES ozone retrievals at 30°N and 87°W, on 15 August 2007. The averaging kernels for the retrieval levels in the lower troposphere, the upper troposphere, and the stratosphere are shown in red, green, and blue, respectively. Source: Parrington et al., 2008.

gases. Consequently, the retrieved column abundances for these gases are normalised by the column abundance of dry air, producing a column-averaged dry-air mole fraction of the trace gas, designated XG for trace gas G. It is this column-averaged dry-air mole fraction that is used in data assimilation and inverse modelling analyses to estimate sources and sinks of these long-lived trace gases.

8.3 Surface Source and Sink Estimation

As with the remote sounding inversions, the Bayesian approach has emerged during the past two decades as one of the most widely used frameworks for inverse modelling of sources and sinks of environmentally important trace gases. For source and sink estimation, the inverse problem is similar to the satellite retrieval problem except that the observations are the atmospheric trace gas measurements, that are distributed in time, and the inversion state vector consists of the surface sources or sinks that are lagged in time compared to the observations. To relate the sources or sinks to the atmospheric abundance of a given trace gas requires the use of an atmospheric model that simulates the evolution of the atmospheric distribution of the trace gas. This relationship can be characterised as

$$\mathbf{x}_i = M(\mathbf{x}_{i-1}, \mathbf{p}_{i-1}), \tag{8.13}$$

where M is the atmospheric model, $\mathbf{x}_i$ is the model state (the atmospheric distribution of the trace gas) at the time step i, and $\mathbf{p}$ are the surface fluxes of the trace gas (or other model parameters, such as chemical reactions). The observation model (Eq. (8.1)) takes the form

$$\mathbf{y} = H(M(\mathbf{x}, \mathbf{p})) + \epsilon, \tag{8.14}$$

with observation operator H. In the case of satellite observations, H would account for any spatial interpolation as well as the smoothing influence of the retrieval, as described in Eq. (8.10). It should be noted that the errors here in Eq. (8.14) are different from those in Eq. (8.1) as they represent the errors in the model simulation of the trace gas retrievals that would arise even if the true sources and sinks were known. Assuming Gaussian errors, the corresponding cost function is given by

$$J(\mathbf{p}) = \tfrac{1}{2}\,[\mathbf{y} - H(M(\mathbf{x}, \mathbf{p}))]^T \mathbf{R}^{-1} [\mathbf{y} - H(M(\mathbf{x}, \mathbf{p}))]$$
$$+ \tfrac{1}{2}\,(\mathbf{p} - \mathbf{p}^b)^T \mathbf{B}_p^{-1} (\mathbf{p} - \mathbf{p}^b), \tag{8.15}$$

where $\mathbf{p}^b$ is the *a priori* emission estimate and $\mathbf{B}_p$ is the *a priori* error covariance matrix. The MAP solution is given by

$$\mathbf{p}^a = \mathbf{p}^b + (\mathbf{H}^T \mathbf{R}^{-1} \mathbf{H} + \mathbf{B}_p^{-1})^{-1} \mathbf{H}^T \mathbf{R}^{-1} (\mathbf{y} - \mathbf{H} \mathbf{p}^b) \,, \qquad (8.16)$$

with $\mathbf{H} = \partial H(M(\mathbf{x}, \mathbf{p}))/\partial \mathbf{p}$. In the case of a linear problem, Eq. (8.15) is perfectly quadratic and Eq. (8.16) is easily obtained. For moderately non-linear problems, a numerical scheme that iteratively solves for $\mathbf{p}^a$ would be needed. To simplify the problem, atmospheric inversion analyses typically employ CTMs, in which the model meteorology is prescribed by meteorological reanalyses. Use of meteorological reanalyses to drive the models ensures that transport of the trace constituents will be reliable without the need to also assimilate meteorological data in the inversions to constraint the model transport.

8.3.1 Bayesian Synthesis Inversions

Many of the early inverse modelling studies used a Bayesian synthesis approach, which employed a coarse discretisation of the state vector and constructed the $\mathbf{H}$ matrix in Eq. (8.16) by individually perturbing each element in the state vector and sampling the resulting field at the observation locations and time. This approach yields an 'influence function' that captures the sensitivity of the atmospheric abundance of the trace gas to a unit change in the surface fluxes represented by the perturbed state vector element. For each element of the state vector, the influence function corresponds to a column of $\mathbf{H}$. This approach was widely used throughout the 1990s and early 2000s for inverse modelling of observations of trace gases such as atmospheric CO_2, CO, nitrous oxide (N_2O), CH_4, and some halocarbons. A key benefit of this approach is that the Bayesian framework provides a means of obtaining an estimate of the *a posteriori* uncertainty of the inferred sources or sinks for the trace gas of interest. For example, using observations of atmospheric CO_2 and $^{13}CO_2$, Enting et al. (1995) estimated an uncertainty of ± 1.2 Gt C y^{-1} for the net exchange of CO_2 between the ocean and the atmosphere, and found that their estimated fluxes were sensitive to the observations ingested. When the inversion ingested ship data from the South Pacific, it suggested that the Southern Ocean was a source of CO_2, whereas without the ship data the region was estimated to be a weak sink. Limited observational coverage from the surface observing network was a major challenge for these early inverse modelling studies and, as mentioned in the Introduction, this limitation was one of the factors that drove the development of satellite instruments to measure the global distribution of CO_2, CO, and other trace gases.

The MOPITT instrument has provided the longest continuous record of CO in the lower atmosphere, and during the past two decades there have been numerous studies focused on quantifying emissions of CO. Atmospheric CO is produced as a by-product of incomplete combustion and from the oxidation of CH_4 and volatile organic compounds (VOCs). The combustion sources comprise fossil fuel combustion as well as biomass burning from natural wildfires and agricultural-related vegetation fires. It is removed from the atmosphere by reaction with the hydroxyl radical (OH), the main atmospheric oxidant. In fact, CO is the dominant sink for OH, and, consequently, changes in the abundance of CO will drive changes in the abundance of OH, which will impact the lifetime of other gases, such as CH_4. It was because of its influence on the oxidative capacity of the atmosphere that interest emerged in the 1970s and 1980s in understanding how human activity was driving changes in atmospheric CO.

Pétron et al. (2004), conducted the first time-dependent Bayesian synthesis inversion of MOPITT CO data to estimate monthly mean emissions of CO. The inversion was set up such that each month of observations was used to update emissions from the previous two months, with the *a posteriori* emissions for each month based on the last update. They linearised the CO chemistry by using prescribed OH fields from a full-chemistry version of the model with detailed tropospheric chemistry. But to account for the impact of the non-linearity in the chemistry on the inversion, they conducted three iterations of the inversion in which they ran the full-chemistry version of the model after each iteration with the optimised CO emissions to obtained new OH fields that were then used in the subsequent iteration of the inversion. They found that the inferred CO emissions from all regions were higher in the first iteration compared to the two subsequent iterations, illustrating the importance of capturing the feedback of the optimised CO emissions on the OH distribution. Their inversion analysis suggested higher wintertime emissions from fossil and biofuel fuel combustion, which were 30% and 100% larger, respectively, in winter than in summer. Pétron et al. (2004) also estimated significantly greater emissions from biomass burning in Africa than their *a priori*.

Arellano et al. (2006) also conducted a time-dependent Bayesian synthesis inversion of MOPITT CO data and estimated significantly greater CO emissions from biomass burning compared to their *a priori*. They estimated higher biomass burning emissions from northern Africa, Oceania, and Indonesia, and from Europe and Russia. For some source regions, the estimated emissions from Pétron et al. (2004) and Arellano et al. (2006) were inconsistent. For example, inferred biomass burning emissions from South America and northern Africa were much higher in Arellano et al. (2006) than in Pétron et al. (2004). The estimated biomass burning emissions in Arellano et al. (2006) for Europe and Russia were also greater than those in Pétron et al. (2004) and peaked in May, whereas the Pétron et al. (2004) emissions peaked in July and August.

The differences in the regional CO emission estimates between Pétron et al. (2004) and Arellano et al. (2006) are emblematic of the discrepancies found between different inverse modelling emission estimates of CO in the literature (e.g. Heald et al., 2004; Stavrakou and Müller, 2006; Jones et al., 2009; Kopacz et al., 2010; Hooghiemstra et al., 2012; Jiang et al., 2015, 2017; Zheng et al., 2019; Miyazaki et al., 2020). The CO inversion analyses using satellite observations yield consistent source estimates on global and hemispheric scales, but can produce large differences on regional scales, reflecting differences in *a priori* emissions, differences in the information content of the data sets assimilated, and discrepancies in the chemistry and transport in the atmospheric models. A particular issue with the Bayesian synthesis approach is the implicit assumption that the spatial distribution of the sources or sinks are known at the scales at which the state vector is discretised, and that only the strength of the sources or sinks is uncertain. However, errors in the spatial distribution of the sources or sinks can result in aggregation errors. Ideally, the inversion should be conducted at the highest resolution possible, and then aggregated to the scales that are actually constrained by the observations, but high-resolution Bayesian synthesis inversions can be computationally expensive.

As a result of the increasing availability of massively parallel computing systems, it is becoming more feasible to conduct Bayesian synthesis inversions at the native resolution of global CTMs, which are run at lower resolution than weather forecast models. For example, Maasakkers et al. (2019) used the GEOS-Chem model at a resolution of 4° × 5° to estimate mean 2010–15 emissions of CH_4. The **H** matrix in their inversion was constructed by perturbing the individual 4° × 5° emission grid boxes in their 1,009-element state vector. They also solved for the 2010–15 trend in the emissions on the same 4° × 5° grid. A benefit of the Bayesian synthesis approach is that it provides a complete representation of the **H**, **B**, and **R** matrices, which offers the means of constructing an averaging kernel matrix (Eq. (8.6)) for the inversion, from which the DOFS (Eq. (8.9)) can be estimated.

Maasakkers et al. (2019) used space-based CH_4 measurements from GOSAT and although their inversion consisted of 1,009 emission elements in the state vector, they estimated that the inversion provided only 128 DOFS. Shown in Fig. 8.2 are the diagonal elements of their averaging kernel matrix. The inversion exhibited strong sensitivity to CH_4 emissions in central Africa, South America, East Asia, and South Asia, and weak sensitivity to emissions in North America and Europe. For the emission trends, Maasakkers et al. (2019) obtained only 7 DOFS, but suggested that emissions from wetlands, livestock, and oil and gas accounted for 43%, 16%, and 11%, respectively, of the observed increase in atmospheric CH_4 after 2007. The Maasakkers et al. (2019) analysis illustrates well the utility of the Bayesian synthesis inversion approach, when conducted at the resolution of the model, and the value of averaging kernels for characterising the information content of the inversion.

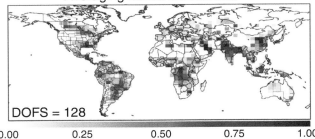

Figure 8.2 Diagonal of the averaging kernel matrix for a GOSAT CH_4 inversion analysis to estimate mean CH_4 emissions for 2010–14. Source: Maasakkers et al., 2019.

8.3.2 Variational Approaches

Adjoint methods offer a computationally efficient means of conducting inversions at the resolution of the atmospheric model. In particular, the four-dimensional variational (4D-Var) scheme is an adjoint-based approach that is widely used for inverse modelling of emissions of environmentally important trace gases. The 4D-Var was initially used in atmospheric science for numerical weather prediction, and was first used operationally at the European Centre for Medium Range Weather Forecasts (ECMWF) (Rabier et al., 2000). As typically used, the 4D-Var scheme seeks to optimise the initial model state to best match a set of observations distributed in time (over an assimilation window). The model trajectory, Eq. (8.13), is used as a strong constraint in the optimisation. An early application of the 4D-Var scheme for regional air quality forecasting over Europe was presented by Elbern and Schmidt (2001), who showed that optimisation of the initial ozone state resulted in improved short-term (6–12 hours) ozone forecasts. For the application of the 4D-Var scheme for source and sink estimation, the cost function takes the form

$$J(\mathbf{p}) = \frac{1}{2}\sum_{i=0}^{N}[\mathbf{y}_i - \mathbf{H}_i(\mathbf{x}_i)]^T\mathbf{R}_i^{-1}[\mathbf{y}_i - \mathbf{H}_i(\mathbf{x}_i)]$$
$$+ \frac{1}{2}(\mathbf{p} - \mathbf{p}^b)^T\mathbf{B}_p^{-1}(\mathbf{p} - \mathbf{p}^b), \quad (8.17)$$

where $\mathbf{y}_i$ are the observations distributed in time, and the other variables are as defined in Eq. (8.15). The cost function is minimised by constructing the Lagrangian function

$$L(\mathbf{p}, \mathbf{x}_i, \lambda_i) = \frac{1}{2}(\mathbf{p} - \mathbf{p}^b)^T\mathbf{B}_p^{-1}(\mathbf{p} - \mathbf{p}^b) + \sum_{i=0}^{N}\frac{1}{2}[\mathbf{y}_i - \mathbf{H}_i(\mathbf{x}_i)]^T$$
$$\mathbf{R}_i^{-1}[\mathbf{y}_i - \mathbf{H}_i(\mathbf{x}_i)] + \sum_{i=1}^{N}\lambda_i^T[\mathbf{x}_i - M(\mathbf{x}_{i-1}, \mathbf{p})], \quad (8.18)$$

where λ_i are the Lagrange multipliers. The gradients of L with respect to $\mathbf{x}_i$ and $\mathbf{p}$ are given by

$$\frac{\partial L}{\partial \mathbf{x}_i} = -\mathbf{H}_i^T \mathbf{R}_i^{-1} [\mathbf{y}_i - \mathbf{H}_i(\mathbf{x}_i)] - \lambda_i + \left(\frac{\partial M}{\partial \mathbf{x}_i}\right)^T \lambda_{i+1}, \qquad (8.19)$$

$$\frac{\partial L}{\partial \mathbf{x}_N} = -\mathbf{H}_i^T \mathbf{R}_i^{-1} [\mathbf{y}_N - \mathbf{H}_i(\mathbf{x}_N)] - \lambda_N, \qquad (8.20)$$

$$\frac{\partial L}{\partial \mathbf{p}} = \mathbf{B}_p^{-1}(\mathbf{p} - \mathbf{p}^b) - \sum_{i=1}^{N} \left(\frac{\partial M}{\partial \mathbf{p}}(\mathbf{x}_{i-1}, \mathbf{p})\right)^T \lambda_i, \qquad (8.21)$$

where $\mathbf{M}^T = \left(\frac{\partial M}{\partial \mathbf{x}_i}\right)^T$ is the adjoint of the tangent linear model $\mathbf{M}$. At the minimum, the gradients are equal to zero, and the adjoint model equations are given by

$$\lambda_N = -\mathbf{H}_i^T \mathbf{R}_i^{-1} [\mathbf{y}_N - \mathbf{H}_i \mathbf{x}_N],$$

$$\lambda_i = \left(\frac{\partial M}{\partial \mathbf{x}_i}\right)^T \lambda_{i+1} - \mathbf{H}_i^T \mathbf{R}_i^{-1} [\mathbf{y}_i - \mathbf{H}_i \mathbf{x}_i]. \qquad (8.22)$$

A useful feature of the 4D-Var scheme is that it is a smoother, which means it can use observations in the future to inform emission estimates in the past. This is particularly useful when the spatio-temporal coverage of the observations is sparse, as the inversion can use a long assimilation window to ingest sufficient data to obtain constraints on the sources or sinks. However, the length of the window is constrained by the assumption of linearity in the adjoint. For gases such as CO_2, for which the inversion problem is linear (if the winds are prescribed and variations in CO_2 do not feed back on the meteorology, as is the case in CTMs), the inversion can employ exceedingly long windows. For example, Basu et al. (2013) estimated monthly mean fluxes of CO_2 using a 22-month assimilation window, and Liu et al. (2014) and Deng et al. (2016) estimated monthly CO_2 fluxes using a 12-month window. For shorter-lived gases such as CO, which has a global mean lifetime of about two months, 4D-Var inversions to estimate monthly mean CO emissions have typically used an assimilation window of about one month (e.g. Jiang et al., 2015, 2017; Kopacz et al., 2010), although some studies such as Hooghiemstra et al. (2012) used a three-month window. In a 4D-Var multiconstituent assimilation, Zhang et al. (2019) assimilated space-based observations of CO, NO_2, and O_3 to quantify emissions of CO and NOx using a two-week assimilation window. In their analysis, they also optimised the O_3 state, and because of the short lifetime of O_3 and NO_2 in the lower troposphere, and the non-linearity in the chemistry, it was necessary to use a short assimilation window. They argued that the constraints on the chemistry provided by the O_3 and NO_2 observations enabled them to use a short window for the estimation of the CO sources in contrast to a CO-only inversion analysis. They had selected a two-week window because of the 16-day repeat cycle of the orbit of TES, which provided the O_3 observations used in the assimilation.

However, they conducted a sensitivity analysis and found that using a window as short as two days would not have had a significant impact on the CO and NOx emission estimates.

The value of jointly assimilating CO and NO_2 observations to quantify emissions of CO was first shown by Müller and Stavrakou (2005), who used an adjoint-based approach to assimilate surface observations of CO with NO_2 data from the Global Ozone Monitoring Experiment (GOME) satellite instrument. Müller and Stavrakou (2005) found that the assimilated NO_2 data impacted the unobserved constituents in the model chemistry, which corrected discrepancies in the chemical sink for CO in the model, and resulted in *a posteriori* CO fields that were in better agreement with independent observations than when only CO data were assimilated.

The joint assimilation of multiple chemical constituents that have similar emission types also offers a means of quantifying emissions from different fossil fuel sectors. Qu et al. (2022) used a 4D-Var scheme to assimilate space-based observations of CO, NOx, and SO_2 to quantify emissions of these gases in China from biomass burning and the following six fossil fuel sectors: transportation, industry, residential, aviation, shipping, and energy. Because of the computational cost of the assimilation, they only estimated emissions for January of each year between 2005 and 2012 to determine trends in the emissions. Their analysis suggested that NOx emissions increased in China until 2011 due to emissions from transportation, energy, and industry. In contrast, CO emissions decreased during the same period due to changes in residential and industrial emissions. Qu et al. (2022) also found that SO_2 emissions peaked in 2007 and were associated with changes in emissions from the energy, residential, and industry sectors.

Another well-established application of the 4D-Var scheme in an air quality context is for the assimilation of space-based observations to quantify emissions of isoprene. Tropospheric ozone is produced by the oxidation of CH_4 and VOCs in the presence of NOx, and emissions of isoprene represent the dominant biogenic source of VOCs. During the past two decades, there has been much effort (e.g. Palmer et al., 2003; Millet et al., 2006; Marais et al., 2012; Barkley et al., 2013) to better quantify isoprene emissions using space-based observations of formaldehyde (HCHO), which is a by-product of isoprene oxidation. Bauwens et al. (2016) used an adjoint-based scheme to assimilate HCHO observations from the Ozone Monitoring Instrument (OMI) to quantify isoprene emissions between 2005 and 2013. Their analysis suggested large reductions (30–40%) in tropical emissions, mainly in the Amazon and northern Africa, relative to their *a priori* estimates. They also inferred greater emissions in Eurasia, with a positive trend in emissions in Siberia that they attributed to increasing temperatures and forest expansion in the region.

A powerful feature of the adjoint-based inversions is that the adjoint model offers a computationally efficient means of assessing the sensitivity of a given observing network to the sources and sinks to be optimised. Liu et al. (2015) examined the source–receptor relationship for XCO_2 observations from GOSAT and found that because of the atmospheric transport pathways, XCO_2 data over North America have strong sensitivity to CO_2 fluxes over northern South America and central Africa. Similarly, they found that XCO_2 over Europe exhibited strong sensitivity to North American fluxes. In contrast, they found that in the tropics, the XCO_2 data were sensitive mainly to local fluxes. In a similar analysis, Byrne et al. (2017) assessed the sensitivity of observations from the in situ surface network, the Total Carbon Column Observing Network (TCCON), GOSAT, and the Orbiting Carbon Observatory-2 (OCO-2) to surface fluxes of CO_2. These four observing networks exhibit large differences in observational coverage, from the relatively sparse TCCON observations to the much more dense OCO-2 data. Byrne et al. (2017) used an adjoint model to calculate the sensitivity of the modelled XCO_2 at the

et al. (2017). The in situ network, which has good spatio-temporal coverage over North America, provides strong sensitivity to North American CO_2 fluxes during all seasons, but offers limited sensitivity to fluxes in the tropics and southern hemisphere. The OCO-2 observational coverage, in contrast, provides the greatest sensitivity to fluxes in the tropics and subtropics, and to fluxes in the extratropical northern hemisphere during boreal summer. The Byrne et al. (2017) results highlighted the complementarity between the ground-based and space-based observing systems for CO_2 and demonstrated the utility of the type of sensitivity studies that are possible with adjoint-based analyses.

Despite the utility of the 4D-Var scheme, a major limitation of the scheme is that it does not provide a full description of the *a posteriori* error covariance matrix. Consequently, it is difficult to calculate the DOFS in the inversion and identify which components of the *a posteriori* estimate of the state vector reflect information from the observations rather than the *a priori*. For example, the Maasakkers et al. (2019) Bayesian synthesis inversion using GOSAT XCH_4 data obtained only 128 DOFS for the 1,009 emission grid boxes

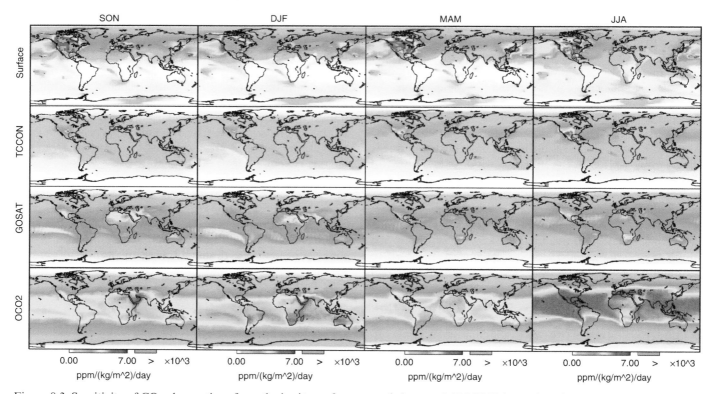

Figure 8.3 Sensitivity of CO_2 observations from the in situ surface network (top row), TCCON (second row), GOSAT (third row), and OCO-2 (bottom row) to surface fluxes of CO_2 for September–November (first column), December–February (second column), March–May (third column), and June–August (fourth column). Source: Byrne et al., 2017.

locations and times of the observations to the CO_2 sources and sinks. Fig. 8.3 shows the spatial distribution of the sensitivity, on seasonal time scales, estimated by Byrne

optimised in the inversion. This means that different 4D-Var inversions using different *a priori* CH_4 emissions would produce different *a posteriori* CH_4 emission estimates, and it would be challenging to determine where the differences are due mainly to the different *a priori* and where they reflect actual information from the observations. The inability to obtain a full representation of the posteriori error covariance matrix is a major limitation of the 4D-Var scheme that makes it difficult to assess the degree to which the inversion reduces uncertainty in the estimated sources and sinks.

8.3.3 Ensemble Kalman Filters

In a Bayesian inversion context, the ensemble Kalman filter (EnKF) (e.g. Anderson, 2001) is a powerful alternative to the 4D-Var scheme as it uses an ensemble approach to characterise and propagate the error covariance. This alleviates the need for an adjoint model and it can easily take advantage of massively parallel computing systems to efficiently conduct long inversions. As a Bayesian scheme, the fundamental formulation of the inverse problem is similar to that described in the previous sections, with the notable difference being that the forecast error covariance ($\mathbf{P}$) at a given time step, used in constructing the gain matrix $\mathbf{K} = \mathbf{P}\mathbf{H}^T(\mathbf{H}\mathbf{P}\mathbf{H}^T + \mathbf{R})^{-1}$, is given by

$$\mathbf{P} = \mathbf{X}\mathbf{X}^T, \tag{8.23}$$

where $\mathbf{X} \in \mathbb{R}^{n \times p}$ is a matrix of the deviations with respect to the ensemble mean ($\bar{\mathbf{x}}$) for the p ensemble members

$$\mathbf{X} = \frac{1}{\sqrt{p-1}}(\mathbf{x}_1 - \bar{\mathbf{x}}, \mathbf{x}_2 - \bar{\mathbf{x}}, \ldots, \mathbf{x}_p - \bar{\mathbf{x}}). \tag{8.24}$$

For the inverse problem, the state here is augmented to include the distributions of the trace gases of interest as well as their surface sources and sinks. The EnKF approach described here is known as the square-root scheme. An alternative approach also perturbs the observations to generate an ensemble of observations and is referred to as the stochastic scheme. Another form of the EnKF is the local ensemble transform Kalman filter (LETKF) (Hunt et al., 2007), in which the analysis is conducted locally at the model grid box using only observations in a limited domain around the grid box. This offers significant computational benefits as it enables the assimilation to more effectively exploit massively parallel computing architectures.

A challenge that arises with using the EnKF for estimating sources and sinks is that the inversion typically lacks a dynamical model for the sources and sinks. As a result, ad hoc approaches are often used to represent the temporal evolution of the sources and sinks. The simplest of these approaches is to assume persistence, but that does not allow for error growth in the sources and sinks. Another well-known issue with the EnKF is that the ensemble size is typically much smaller than that of the state (i.e. $p \ll n$), which results in a rank-deficient covariance matrix. As

a result, there will be spurious correlations in $\mathbf{P}$, which can be mitigated using localisation (Hamill et al., 2001), but the localisation must be tuned for the inversion.

The CarbonTracker model from the National Oceanic and Atmospheric Administration (NOAA) is a well-known ensemble-based CO_2 flux inversion system.[2] The assimilation system is based on a square-root EnKF scheme and assimilates ground-based, shipboard, and aircraft in situ measurements of CO_2 to obtain weekly estimates of CO_2 fluxes. CarbonTracker does not optimise the model state; only the surface fluxes of CO_2 are included in the state vector. To limit aggregation error while reducing the computational cost to run the assimilation, the surface fluxes are discretised into a maximum of 239 elements in the state vector, consisting of 30 ocean regions and 11 large land regions, with each land region further disaggregated into 19 ecoregions based on vegetation type. The assimilation solves for scaling factors for the fluxes in each of these ecoregions and ocean regions. The most recent version of CarbonTracker, version CT2019B (Jacobson et al., 2020), provides estimates of weekly CO_2 fluxes from 2000–18. CarbonTracker has also been used to assimilate space-based observations of CO_2 as part of a model intercomparison project (MIP) in support of the Orbiting Carbon Observatory-2 (OCO-2) (Crowell et al., 2019).

Another example of the use of an ensemble-based approach for CO_2 flux estimation is Feng et al. (2011, 2016), who used an LETKF scheme for inverse modelling of space-based CO_2 data, and also contributed to the OCO-2 MIP. Using the same LETKF approach, Palmer et al. (2019) assimilated data from GOSAT and OCO-2 to estimate CO_2 fluxes for 2015 and 2016. They found that the satellite data suggested an unexpected large source of CO_2 from tropical Africa, estimated to be 1.48 Pg C in 2015 and 1.65 Pg C in 2016, with the largest emissions in western tropical Africa and western Ethiopia. The processes responsible for this large putative source are unknown, reflecting current uncertainty about the tropical carbon cycle.

The EnKF has also been used for assimilation of space-based measurements of CO together with meteorological observations (Arellano et al., 2007; Barré et al., 2015; Gaubert et al., 2020). The Gaubert et al. (2020) analysis was conducted for the Korea–United States Air Quality (KORUS-AQ) experiment in May–June 2016. Their inversion suggested that the *a priori* CO emissions significantly underestimated CO emissions, particularly in northern China where the underestimate was as large as 80%. Their *a posteriori* CO emissions also led to improvements in the model simulation of other trace gases such as O_3, HO_2, OH, and long-lived VOCs.

The utility of the ensemble-based approach for chemical data assimilation is illustrated by Miyazaki et al. (2012,

---

[2] https://gml.noaa.gov/ccgg/carbontracker/.

2017, 2020), who assimilated space-based observations of a suite of trace constituents to produce a chemical reanalysis for tropospheric chemistry. In Miyazaki et al. (2012) they assimilated observations of O_3, NO_2, CO, and HNO_3 to constrain the concentrations of all of the predicted constituents in the chemical mechanism in the model, together with emissions of CO and NOx (from the surface and from lightning). Because of the correlations in the chemistry, which is captured by the covariance matrices, the assimilation is able to provide constraints on the unobserved constituents. As shown in Miyazaki et al. (2012) (see their figure 3), there were strong correlations between many of the constituents in their analysis, reflecting the chemical coupling in the model. In particular, they found that in the middle troposphere the correlations were particularly strong between the various hydrocarbons and their oxidation products. In the most recent version of their tropospheric chemistry reanalysis (TCR-2) (Miyazaki et al., 2020), they extended the reanalysis from 2005–18 at a spatial resolution of 1.1° globally. The availability of long-term chemical reanalyses raises the possibility of using these products to better quantify trends in atmospheric composition as well as in the emissions of gases such as CO and NOx.

8.4 Summary

Despite the considerable expansion of the observing system during the past three decades, it is still a challenge for inverse modelling analyses to provide reliable emission estimates on policy-relevant scales. Comparisons of emission estimates in the literature reveal large discrepancies between the inferred emissions. For example, Elguindi et al. (2020) compared various emission estimates for CO, NOx, SO_2, non-methane volatile organic compounds (NMVOCs), black carbon (BC), and organic carbon (OC) and found a large spread in the reported inventories. They found that inferred emissions of CO in Europe and western Africa in 2010 varied between 26.8–40.9 and 16.8–63.5 Tg CO, respectively. One possible contributing factor to these discrepancies is that there may be insufficient observational coverage from the previous generation of satellite instruments to constrain the emission estimates on the desired scales. For example, the MOPITT instrument has a footprint of 22×22 km^2 and achieves global coverage every three days, with significant data loss due to cloud cover. In contrast, the TROPOspheric Monitoring Instrument (TROPOMI), which was launched in 2017, has a footprint of 7×3.5 km^2 and achieves global coverage daily. Next-generation instruments such as TROPOMI offer significantly greater observational coverage to better quantify local emissions. This observational capability will be further enhanced by the availability of atmospheric composition measurements from geostationary satellites, which will provide observations at much greater spatio-temporal density than is possible from low-Earth orbiting satellites. The Geostationary Environment Monitoring Spectrometer (GEMS) was launched in 2020 to monitor air quality in Asia, the Tropospheric Emissions: Monitoring of Pollution (TEMPO) instrument will be launched in 2023 to provide geostationary observations of North American air quality, and the Sentinel-4 satellite will be launched by the European Space Agency in 2024 to monitor air quality in northern Africa and Europe.

Another factor that may contribute to discrepancies between inferred emission estimates is the implicit assumption that the atmospheric models employed in the inversions are unbiased. Biases in the models due to discrepancies in the chemical mechanisms in the models or the representation of atmospheric transport will get projected onto the inferred emission estimates. However, characterising and mitigating these model biases is not straightforward. Stanevich et al. (2021) suggested the use of a weak constraint 4D-Var scheme (Derber, 1989; Trémolet, 2006, 2007), which relaxes the use of the model trajectory as a strong constraint in the 4D-Var optimisation, as a possible means of characterising and mitigating the impact of systematic model errors in chemical data assimilation. There is a clear need for new approaches for mitigating biases online in the context of the assimilation to produce more robust emission estimates.

A significant challenge in chemical data assimilation is the large range of spatial and temporal scales involved. It is important to capture the impact of local emissions on the distribution of the tracers, accounting for the influence of atmospheric transport from local to synoptic scales and larger, for tracers with lifetimes that differ by orders of magnitude, from hours for a gas like NO_2 to months for CO. One potential solution to this problem might be approaches that combine machine learning with traditional data assimilation schemes, with the machine learning used to represent the chemistry and transport on the small scales not well represented in global models. Such a hybrid data assimilation approach could help bridges the scales.

References

Anderson, J. L. (2001). An ensemble adjustment Kalman filter for data assimilation. *Monthly Weather Review*, 129, 2884–903.

Arellano, A. F., Jr., Kasibhatla, P. S., Giglio, L. et al. (2006). Time dependent inversion estimates of global biomass-burning CO emissions using measurement of pollution in the troposphere (MOPITT) measurements. *Journal of Geophysical Research*, 111. https://doi.org/10.1029/2005JD006613.

Arellano, A. F., Jr., Raeder, K., Anderson, J. L. et al. (2007). Evaluating model performance of an ensemble-based chemical data assimilation system during INTEX-B field mission. *Atmospheric Chemistry and Physics*, 7, 5695–710.

Barkley, M. P., Smedt, I. D., Roozendael, M. V. et al. (2013). Top-down isoprene emissions over tropical South America inferred from SCIAMACHY and OMI formaldehyde columns. *Journal of Geophysical Research: Atmospheres*, 118(12), 6849–68. https://doi.org/10.1002/jgrd.50552.

Barré, J., Gaubert, B., Arellano, A. F. J. et al. (2015). Assessing the impacts of assimilating IASI and MOPITT CO retrievals using CESMCAM-chem and DART. *Journal of Geophysical Research*, 120, 10501–29. https://doi.org/10.1002/2015JD023467.

Basu, S., Guerlet, S., Butz, A. et al. (2013). Global CO_2 fluxes estimated from GOSAT retrievals of total column CO_2. *Atmospheric Chemistry and Physics*, 13, 8695–717. https://doi.org/10.5194/acp-13-8695-2013.

Bauwens, M., Stavrakou, T., Muller, J.-F. et al. (2016). Nine years of global hydrocarbon emissions based on source inversion of OMI formaldehyde observations. *Atmospheric Chemistry and Physics*, 16, 10133–58. https://doi.org/10.5194/acp-16-10133-2016.

Byrne, B., Jones, D. B. A., Strong, K. et al. (2017). Sensitivity of CO_2 surface flux constraints to observational coverage. *Journal of Geophysical Research*, 122, 6672–94. https://doi.org/10.1002/2016JD026164.

Crawford, J. H., Heald, C. L., Fuelberg, H. E. et al. (2004). Relationship between Measurements of Pollution in the Troposphere (MOPITT) and in situ observations of CO based on a large-scale feature sampled during TRACE-P. *Journal of Geophysical Research*, 109, D15S04. https://doi.org/10.1029/2003JD004308.

Crowell, S., Baker, D., Schuh, A. et al. (2019). The 2015–2016 carbon cycle as seen from OCO-2 and the global in situ network. *Atmospheric Chemistry and Physics*, 19, 9797–831. https://doi.org/10.5194/acp-19-9797-2019.

Deeter, M. N., Worden, H. M., Edwards, D. P., Gille, J. C., and Andrews, A. E. (2012). Evaluation of MOPITT retrievals of lower-tropospheric carbon monoxide over the United States. *Journal of Geophysical Research*, 117 (D13306). https://doi.org/10.1029/2012JD017553.

Deng, F., Jones, D., O'Dell, C. W., Nassar, R., and Parazoo, N. C. (2016). Combining GOSAT XCO_2 observations over land and ocean to improve regional CO_2 flux estimates. *Journal of Geophysical Research*, 121, 1896–913. https://doi.org/10.1002/2015JD024157.

Derber, J. C. (1989). A variational continuous assimilation technique. *Monthly Weather Review*, 117, 2437–46. https://doi.org/10.1175/1520-0493(1989)117<2437:AVCAT>2.0.CO;2.

Elbern, H., and Schmidt, H. (2001). Ozone episode analysis by four-dimensional variational chemistry data assimilation. *Journal of Geophysical Research*, 106 (D4), 3569–90.

Elguindi, N., Granier, C., Stavrakou, T. et al. (2020). Intercomparison of magnitudes and trends in anthropogenic surface emissions from bottom-up inventories, top-down estimates, and emission scenarios. *Earth's Future*, 8 (e2020EF001520). https://doi.org/10.1029/2020EF001520.

Enting, I. G., Trudinger, C. M., and Francey, R. J. (1995). A synthesis inversion of the concentration and $\delta^{13}C$ of atmospheric CO_2. *Tellus*, 47B, 35–52.

Feng, L., Palmer, P. I., Parker, R. J. et al. (2016). Estimates of European uptake of CO_2 inferred from GOSAT XCO_2 retrievals: Sensitivity to measurement bias inside and outside

Europe. *Atmospheric Chemistry and Physics*, 16, 1289–302. https://doi.org/10.5194/acp-16-1289-2016.

Feng, L., Palmer, P. I., Yang, Y. et al. (2011). Evaluating a 3-D transport model of atmospheric CO_2 using ground-based, aircraft, and space-borne data. *Atmospheric Chemistry and Physics*, 11, 2789–803. https://doi.org/10.5194/acp-11-2789-.

Gaubert, B., Emmons, L. K., Raeder, K. et al. (2020). Correcting model biases of CO in East Asia: Impact on oxidant distributions during KORUS-AQ. *Atmospheric Chemistry and Physics*, 20, 14617–47. https://doi.org/10.5194/acp20-.

GBD 2019 Risk Factors Collaborators. (2020). Global burden of 87 risk factors in 204 countries and territories, 1990–2019: A systematic analysis for the global burden of disease study 2019. *Lancet*, 396, 1223–49. https://doi.org/10.1016/S0140-6736(20)30752-2.

Hamill, T. M., Whitaker, J. S., and Snyder, C. (2001). Distance-dependent filtering of background error covariance estimates in an ensemble Kalman filter. *Monthly Weather Review*, 129, 2776–90.

Heald, C. L., Jacob, D. J., Jones, D. B. A. et al. (2004). Comparative inverse analysis of satellite (MOPITT) and aircraft (TRACE-P) observations to estimate Asian sources of carbon monoxide. *Journal of Geophysical Research*, 109 (D23306). https://doi.org/10.1029/2004JD005185.

Hooghiemstra, P. B., Krol, M. C., Bergamaschi, P. et al. (2012). Comparing optimized CO emission estimates using MOPITT or NOAA surface network observations. *Journal of Geophysical Research*, 117 (D06309). https://doi.org/https://doi.org/10.1029/2011JD017043.

Hunt, B. R., Kostelich, E. J., and Szunyogh, I. (2007). Efficient data assimilation for spatiotemporal chaos: A local ensemble transform Kalman filter. *Physica D*, 230, 112–26.

Jacobson, A. R., Schuldt, K. N., Miller, J. B. et al. (2020). CarbonTracker CT2019B. https://doi.org/10.25925/20201008.

Jiang, Z., Jones, D. B. A., Worden, H. M., and Henze, D. K. (2015). Sensitivity of top-down CO source estimates to the modeled vertical structure in atmospheric CO. *Atmospheric Chemistry and Physics*, 15, 1521–37. https://doi.org/10.5194/acp-15-1521-2015.

Jiang, Z., Worden, J. R., Worden, H. et al. (2017). A 15-year record of CO emissions constrained by MOPITT CO observations. *Atmospheric Chemistry and Physics*, 17, 4565–83. https://doi.org/10.5194/acp-17-4565-2017.

Jiang, Z., Zhu, R., Miyazaki, K. et al. (2022). Decadal variabilities in tropospheric nitrogen oxides over United States, Europe, and China. *Journal of Geophysical Research: Atmospheres*, 127. https://doi.org/10.1029/2021JD035872.

Jones, D. B. A., Bowman, K. W., Logan, J. A. et al. (2009). The zonal structure of tropical O_3 and CO as observed by the Tropospheric Emission Spectrometer in 2004. Part 1: Inverse modeling of CO emissions. *Atmospheric Chemistry and Physics*, 9, 3547–62.

Khattatov, B. V., Gille, J. C., Lyjak, L. V. et al. (1999). Assimilation of photochemically active species and a case analysis of UARS data. *Journal of Geophysical Research* (D15), 18715–37.

Kopacz, M., Jacob, D. J., Fisher, J. A. et al. (2010). Global estimates of CO sources with high resolution by adjoint inversion of multiple satellite datasets (MOPITT, AIRS, SCIAMACHY, TES). *Atmospheric Chemistry and Physics*, 10, 855–76.

Le Quéré, C., Andrew, R. M., Friedlingstein, P. et al. (2018). Global Carbon Budget 2017. *Earth System Science Data*, 405–48. https://doi.org/10.5194/essd10–405–2018.

Levelt, P. F., Khattatov, B. V., Gille, J. C. et al. (1998). Assimilation of MLS ozone measurements in the global three-dimensional chemistry transport model ROSE. *Geophysical Research Letters*, 25(24), 4493–96.

Liu, J., Bowman, K., Lee, M. et al. (2014). Carbon monitoring system flux estimation and attribution: Impact of ACOS-GOSAT XCO_2 sampling on the inference of terrestrial biospheric sources and sinks. *Tellus B*, 66, 22486. https://doi.org/10.3402/tellusb.v66.22486.

Liu, J., Bowman, K. W., and Henze, D. K. (2015). Source-receptor relationships of column average CO_2 and implications for the impact of observations on flux inversions. *Journal of Geophysical Research*, 120, 5214–36. https://doi.org/10.1002/2014JD022914.

Maasakkers, J. D., Jacob, D. J., Sulprizio, M. P. et al. (2019). Global distribution of methane emissions, emission trends, and OH concentrations and trends inferred from an inversion of GOSAT satellite data for 2010–2015. *Atmospheric Chemistry and Physics*, 19, 7859–81. https://doi.org/10.5194/acp-19-7859-2019.

Marais, E. A., Jacob, D. J., Kurosu, T. P. et al. (2012). Isoprene emissions in Africa inferred from OMI observations of formaldehyde columns. *Atmospheric Chemistry and Physics*, 12(14), 6219–35. https://doi.org/10.5194/acp-12–6219–2012

Millet, D. B., Jacob, D. J., Turquety, S. et al. (2006). Formaldehyde distribution over North America: Implications for satellite retrievals of formaldehyde columns and isoprene emission. *Journal of Geophysical Research: Atmospheres*, 111(D24). https://doi.org/10.1029/2005JD006853.

Miyazaki, K., Bowman, K., Sekiya, T. et al. (2020). Updated tropospheric chemistry reanalysis and emission estimates, TCR-2, for 2005–2018. *Earth System Science Data*, 12, 2223–59. https://doi.org/10.5194/essd-12-2223-2020.

Miyazaki, K., Eskes, H., Sudo, K. et al. (2017). Decadal changes in global surface NOx emissions from multi-constituent satellite data assimilation. *Atmospheric Chemistry and Physics*, 17, 807–37. https://doi.org/10.5194/acp-17-807-2017.

Miyazaki, K., Eskes, H. J., Sudo, K. (2012). Simultaneous assimilation of satellite NO_2, O_3, CO, and HNO_3 data for the analysis of tropospheric chemical composition and emissions. *Atmospheric Chemistry and Physics*, 12(20), 9545–79. https://doi.org/10.5194/acp-12-9545-2012.

Müller, J.-F., and Stavrakou, T. (2005). Inversion of co and NOX emissions using the adjoint of the images model. *Atmospheric Chemistry and Physics*, 5(5), 1157–86. https://doi.org/10.5194/acp-5-1157-2005.

Palmer, P. I., Feng, L., Baker, D. et al. (2019). Net carbon emissions from African biosphere dominate pan-tropical atmospheric CO2 signal. *Nature Communications*, 10, 3344. https://doi.org/10.1038/s41467-019-11097-w.

Palmer, P. I., Jacob, D. J., Fiore, A. M. eet al. (2003). Mapping isoprene emissions over North America using formaldehyde column observations from space. *Journal of Geophysical Research: Atmospheres*, 108 (D6). https://doi.org/10.1029/2002JD002153.

Parrington, M., Jones, D. B. A. Bowman, K. W. et al. (2008). Estimating the summertime tropospheric ozone distribution over North America through assimilation of observations from the tropospheric emission spectrometer. *Journal of Geophysical Research*, 113 (D18307). https://doi.org/10.1029/2007JD009341.

Pétron, G., Granier, C., Khattotov, B. et al. (2004). Monthly CO surface sources inventory based on the 2000–2001 MOPITT satellite data. *Geophysical Research Letters*, 31 (L21107). https://doi.org/10.1029/2004GL020560.

Qu, Z., Henze, D. K., Worden, H. M. et al. (2022). Sector-based top-down estimates of NOx, SO_2, and CO emissions in East Asia. *Geophysical Research Letters*, 49 (e2021GL096009). https://doi.org/10.1029/2021GL096009.

Rabier, F., J¨arvinen, H., Klinker, E., Mahfouf, J.-F., and Simmons, A. (2000). The ECMWF operational implementation of four-dimensional variational assimilation. I: Experimental results with simplified physics. *Quarterly Journal of the Royal Meteorological Society*, 126, 1143–70.

Rigby, M., Montzka, S. A., Prine, R. G. et al. (2017). Role of atmospheric oxidation in recent methane growth. *Proceedings of the National Academy of Sciences*, 114(21), 5373–7. https://doi.org/10.1073/pnas.1616426114.

Riishøjgaard, L. P., Štajner, I., and Lou, G.-P. (2000). The GEOS ozone data assimilation system. *Advances in Space Research*, 25, 1063–72. https://doi.org/10.1016/S0273-1177(99)00443-.

Rodgers, C. D. (2000). *Inverse Methods for Atmospheric Sounding: Theory and Practice*. Singapore: World Scientific Publishing.

Rodgers, C. D., and Connor, B. J. (2003). Intercomparison of remote sounding instruments. *Journal of Geophysical Research*, 108 (D3), 4116. https://doi.org/10.1029/2002jd002299.

Sitch, S., Friedlingstein, P., Gruber, N. et al. (2015). Recent trends and drivers of regional sources and sinks of carbon dioxide. *Biogeosciences*, 12, 653–79. https://doi.org/10.5194/bg-12-653-2015.

Stanevich, I., Jones, D. B. A., Strong, K. et al. (2021). Characterizing model errors in chemical transport modeling of methane: Using GOSAT XCH_4 data with weak-constraint four-dimensional variational data assimilation. *Atmospheric Chemistry and Physics*, 21, 9545–72. https://doi.org/10.5194/acp-21-9545-2021.

Stavrakou, T., and Müller, J.-F. (2006). Grid-based versus big region approach for inverting CO emissions using measurement of pollution in the troposphere (MOPITT) data. *Journal of Geophysical Research*, 111(D15), 304. https://doi.org/10.1029/2005JD006896.

Trémolet, Y. (2006). Accounting for an imperfect model in 4d-var. *Quarterly Journal of the Royal Meteorological Society*, 132, 2483–504. https://doi.org/10.1256/qj.05.224.

Trémolet, Y. (2007). Model-error estimation in 4d-var. *Quarterly Journal of the Royal Meteorological Society*, 133, 1267–80. https://doi.org/10.1002/qj.94.

Turner, A. J., Frankenberg, C., and Kort, E. A. (2019). Interpreting contemporary trends in atmospheric methane. *Proceedings of the National Academy of Sciences*, 116(8), 2805–13. https://doi.org/10.1073/pnas.1814297116.

Worden, H. M., Deeter, M. N., Edwards, D. P. et al. (2010). Observations of near-surface carbon monoxide from space using MOPITT multispectral retrievals. *Journal of Geophysical Research*, 115,(D18), 314. https://doi.org/10.1029/2010JD014242.

Yoshida, Y., Ota, Y., Eguchi, N. et al. (2011). Retrieval algorithm for CO_2 and CH_4 column abundances from short-wavelength infrared spectral observations by the greenhouse gases observing satellite. *Atmospheric Measurement Techniques*, 4, 717–34. https://doi.org/10.5194/amt-4-717-2011.

Zhang, X., Jones, D. B. A., Keller, M. et al. (2019). Quantifying emissions of co and nox using observations from MOPITT, OMI, TES, and OSIRIS. *Journal of Geophysical Research*, 124 (1029). https://doi.org/11701193/2018JD028670.

Zheng, B., Chevallier, F., Yin, Y. et al. (2019). Global atmospheric carbon monoxide budget 2000–2017 inferred from multi-species atmospheric inversions. *Earth System Science Data*, 11, 1411–36. https://doi.org/10.5194/essd-11-1411-2019.

9

Data Assimilation of Volcanic Clouds: Recent Advances and Implications on Operational Forecasts

Arnau Folch and Leonardo Mingari

Abstract: Operational forecasts of volcanic clouds are a key decision-making component for civil protection agencies and aviation authorities during the occurrence of volcanic crises. Quantitative operational forecasts are challenging due to the large uncertainties that typically exist on characterising volcanic emissions in real time. Data assimilation, including source term inversion, has long been recognised by the scientific community as a mechanism to reduce quantitative forecast errors. In terms of research, substantial progress has occurred during the last decade following the recommendations from the ash dispersal forecast workshops organised by the International Union of Geodesy and Geophysics (IUGG) and the World Meteorological Organization (WMO). The meetings held in Geneva in 2010–11 in the aftermath of the 2010 Eyjafjallajökull eruption identified data assimilation as a research priority. This Chapter reviews the scientific progress and its transfer into operations, which is leveraging a new generation of operational forecast products.

9.1 Introduction

Explosive volcanic eruptions can inject large quantities of particles and aerosols into the atmosphere that, after entraining and mixing with the ambient air, develop sustained buoyant plumes (eruption columns) able to rise up to stratospheric levels (Sparks et al., 1997). The particles that result from magma fragmentation and its subsequent quenching can span in size from volcanic bombs (diameter $d > 64$mm), which easily decouple from the ascending mixture and settle to the ground following ballistic trajectories, to coarse ($2\,mm > d > 64\,\mu m$) and fine ($d < 64\,\mu m$) ash, which can be carried upwards efficiently to form volcanic ash clouds that disperse downwind from the volcano. On the other hand, volcanic emissions can also include a substantial component of SO_2 that can lead to the formation of sulphate aerosols and be co-located with the emissions of ash or injected at different atmospheric layers (SO_2 is not always an ash proxy). The long-range dispersal of both types of clouds, at scales varying from regional to global, jeopardises aerial navigation (Miller and Casadevall, 2000). High concentrations of ash particles, angular in shape and highly abrasive, can damage turbine blades, airplane windscreens, and fuselage, disrupt navigation instruments and, in the worst scenario, can even melt in the combustion chamber resulting in clogging of cooling passages and potential engine stall (Dunn and Wade, 1994). Diluted clouds may not pose an immediate threat to safety but, nonetheless, aircraft operations under harsh aerosol environments degrade overall engine performance, yielding to a lower time on wing and increasing the engine maintenance costs (Clarkson and Simpson, 2017). The likelihood of aircraft encounters is enhanced because volcanic clouds are often injected at or near the tropopause and, in many latitudes of the planet, this happens to coincide with the jet streams and the airplane cruise levels. However, larger eruptions have the potential for stratospheric injection of ultra-fine particles and sulfate aerosols that affect the atmospheric radiative budget and cause measurable atmospheric alterations persisting for months to years (Robock and Oppenheimer, 2003). Considering these aspects, the importance of early warning systems and operational forecasts of volcanic ash/SO_2 clouds becomes obvious.

The modelling and forecasting of volcanic clouds aims at obtaining their location in the atmosphere and time evolution of concentration, and involves three different components (Folch, 2012), namely: (i) a meteorological or numerical weather prediction model, which describes the 4-D state of the atmosphere (wind field, air density, temperature, moisture, precipitation rate, etc.), (ii) a particle (ash)/SO_2 dispersal model, which accounts for transport (advection by wind, turbulent diffusion, particle sedimentation), removal (dry and wet deposition mechanisms, eventual particle aggregation), and chemical reactions and/or phase changes and, finally, (iii) an emission model, which defines the source term (i.e. the eruptive column) in time and space and that, typically, is embedded in the dispersal model. Clearly, uncertainties exist in all these components, either in the physics of the models (e.g. model parameterisations) or in their inputs and underlying meteorological drivers. However, the characterisation and quantification of the emission term is, in most cases, the most challenging aspect and constitutes a first-order factor limiting the accuracy of volcanic cloud forecasts. These volcanic ash/SO_2 emission terms are characterised by the so-called Eruption Source Parameters (ESPs) that include the starting and the end times of each eruptive phase, its maximum injection height, the vertical distribution of mass released along the eruption column, and the emission rate (source strength or total mass emitted per unit time). Moreover, in the case of volcanic ash, the ESPs also include the granulometric characteristics of the particles, needed by dispersal models to compute sedimentation velocities (i.e. size, density, and shape factor). The in situ quantification of the ESPs is very difficult or simply impossible in many occasions (e.g. remote and/or unmonitored volcanoes, obscured observation conditions). As a result, large epistemic uncertainties can exist from uncertain model inputs, yielding to model error propagation and amplification. For example, variations (uncertainties) in the cloud top height can result in progressively larger cloud shape and location mismatches with observations under wind shear scenarios. Similarly, uncertainties in the total ash/SO_2 emission rates translate into concentration errors downwind and, consequently, into poorer quantitative forecasts.

Until a decade ago, all operational forecast systems in place worldwide only delivered qualitative ash cloud forecast products, consisting of 'ash/no ash' delineation zones. This was largely motivated by the conservative 'zero ash tolerance' regulatory Convention on International Civil Aviation which, in practice, made the source strength quantification unnecessary in operational model setups. However, the eruptions of the Eyjafjallajökull (Iceland, 2010) and Cordón Caulle (Chile, 2011) volcanoes dramatically revealed how such a precautionary criterion for flight banning and airspace closure could yield to overreaction and billions of US$ economic losses to the aviation sector, particularly when volcanic clouds pass through congested air traffic regions. The adoption of new guidelines based on quantitative ash concentration thresholds (e.g. the 4, 2, and 0.2 mg/m^3 defined at that time in Europe) had important implications for operational systems, which had to face the question of how to better constrain the source parameters and their uncertainties. In parallel, the aviation sector has experienced a growing demand from engine lessors and maintenance, repair, and overhaul (MRO) stakeholders to quantify engine ingestions and, thereby, optimise the costs of engine maintenance cycles. All these aspects boosted research progress on monitoring and modelling of volcanic clouds, leveraged also by community efforts such as the 2010/2011 IUGG-WMO workshops on Ash Dispersal Forecast and Civil Aviation (Bonadonna et al., 2012), which delineated a roadmap to implement forecasting strategies to better

deal with uncertainties in model inputs and ultimately transfer developments into operations. In particular, data assimilation (DA) was soon identified as a priority to reduce quantitative forecast errors, where the term 'data assimilation' is understood here in a very broad sense, from a simple manual update of model inputs by forecasters to truly automated variational or sequential data assimilation methodologies. In the field of DA and related observations, advances have occurred both in terms of in situ ground-based monitoring techniques and in terms of fusing models with distal cloud observations from satellites, distal ground-based networks, or even from instruments on board aircrafts. Note that the first strategy makes use of in situ ground-based observations, typically from volcano observatories (VOs), and aims at direct measurement of the ESPs. In contrast, the second approach builds on *distal* observations of clouds and, consequently, is more tailored to long-range dispersal forecasts. Given the practical impossibility of exhaustively monitoring in situ the 1500+ active volcanoes that exist worldwide and, at the same time, the aviation requirement for global-coverage forecasts, the second approach has naturally been adopted by almost all operational settings. These include the nine Volcanic Ash Advisory Centers (VAACs) but also some national-level institutions and VOs with operational mandates.

This chapter reviews the scientific advances that have occurred in DA of volcanic clouds during the last decade and discusses how its implementation and transfer into operations can contribute to a new generation of forecast products. Section 9.2 summarises the different observation platforms and detection/retrieval mechanisms, including their pros and cons. With this background in mind, Section 9.3 presents the recent advances in DA for volcanic clouds, which span from simpler data insertion or source term inversion mechanisms to more sophisticated variational and sequential data assimilation techniques. Finally, Section 9.4 discusses the implications of transferring research findings into operations and the emerging perspectives.

9.2 Volcanic Cloud Observation

Volcanic clouds can be observed using a myriad of instruments including active or passive satellite-based sensors, ground-based monitoring, or even in situ particle sampling from research aircrafts. Ground-based instrumentation networks, for example, laser remote sensors such as ceilometers or multi-wavelength polarisation LIDARs, are used for measuring properties and distribution of tropospheric aerosols, vertical structure of meteorological clouds, or height of the atmospheric boundary layer. These instruments are deployed for atmospheric research and monitoring purposes but, nonetheless, they have been successfully used to detect and characterise volcanic components. For example, during the passage of the Eyjafjallajökull ash clouds over Europe, data from the European Aerosol Research LIDAR NETwork (EARLINET) was available in near real-time (e.g. Balis et al., 2016). However, the limited network densities, the data process latency, and the lack of oceanic and global coverage, are limiting factors for considering ground-based instrumentation in operational forecast systems. In contrast, space-based sensors can furnish high-resolution temporally resolved global observations, something advantageous for punctuated, sparse, sporadic, and short-lived events like volcanic eruptions. In fact, DA of volcanic clouds relies almost exclusively on satellite observations, leaving other data sources for model validation purposes. This section gives a succinct review of satellite detection and retrieval of volcanic ash/SO_2. The reader is referred to the vast literature existing on this topic including, for example, the recent review by Prata and Lynch (2019).

The absorption of electromagnetic radiation by particles varies with the wavelength. In the long-wavelength infrared (IR) region of the spectrum (8–15 μm), silicate-rich particles like volcanic ash are more absorbent at shorter wavelengths whereas ice and water droplets in meteorological clouds show an opposite behaviour. This 'reverse absorption' characteristic of volcanic ash motivated the introduction of dual-band IR measurements to discriminate between ash and meteorological clouds from space by taking the Brightness Temperature Difference (BTD) at two different channels, for example, at 10.8 and 12 μm (Prata, 1989), with negative-value pixels indicative of the presence of ash. This simple idea underpins all real-time automated ash detection algorithms developed since then (e.g. Pavolonis et al., 2006), which nowadays include combinations of multiple IR channels and other detection corrections. It should be mentioned that, despite their enormous success and popularity, passive IR detection methods present limitations, including that overlying meteorological clouds can obscure the ash cloud, that absorption signals can be masked by the presence of ice,

or that the BTD method fails in detecting too optically thick or thin clouds. Table 9.1 lists the main passive IR sensors and platforms currently used to detect (and retrieve) volcanic ash and SO_2 worldwide. This includes the latest generation of geostationary satellites that provide high temporal (10–15 min) and spatial (1–2 km) resolution observations. Ash detection algorithms give a binary answer (yes/no) on the presence of ash or, at most, an ash presence probability based on the level of detection confidence (e.g. Francis et al., 2012). However, detection algorithms do not provide any quantitative result and, consequently, are of more limited utility in terms of DA. In contrast, retrieval algorithms go further and invert the raw observation signal with a microphysical model of the ash particles and a radiative transfer model to derive an 'effective' particle size, cloud opacity (optical depth), and column mass loading, all with constrained retrieval uncertainties (30–50% on average). These are more meaningful quantities for models and assimilation of 'observations', and underpin most DA strategies discussed next. It is important to note that passive-based retrievals can only give a top (vertically integrated) view of the cloud and an estimation of its height; vertically resolved quantification typically requires active remote sensing (e.g. LIDARs). Unfortunately, only polar-orbiting satellites carry active sensors (e.g. CALIOP) and, consequently, the narrow field-of-view of the instrument and the frequency of satellite overpasses limit the coverage and the quantity of vertically resolved observations.

On the other hand, the detection of SO_2 clouds from satellite observations is less challenging than volcanic ash although this depends on the wavelength the sensor is measuring, for example, IR versus ultraviolet (UV) (Carn et al., 2009; Clarisse et al., 2012). First, in the long-wavelength IR region, the SO_2 retrieval for imager satellites such as Himawari-8/AHI is based on the strongest IR absorption band for SO_2 (i.e. the 7.3 μm band) (e.g. Muser et al., 2020, Prata et al., 2021). Second, the UV region (280–340 nm) is

Table 9.1 *Summary of the main passive IR sensors and platforms used for detection/retrieval of volcanic ash and SO_2. (*) p: polar-orbiting, g: geostationary (image period in parentheses). Modified from Prata (2016) and Prata and Lynch (2019).*

Sensor	Acronym	Platform	Platform type (*)	Resolution (km)
Advanced Very High Resolution Radiometer	AVHRR-2/3	NOAA-POES	p	1
High-Resolution Infrared Radiation Sounder	HIRS-2/3	MET-Op	p	26
Moderate-Resolution Imaging Spectroradiometer	MODIS	Aqua / Terra	p	1
Visible Infrared Imaging Radiometer Suite	VIIRS	Suomi-NPP	p	1
Atmospheric Infrared Sounder	AIRS	Aqua	p	13
Cross-Track Infrared Sounder	CrIS	Suomi-NPP	p	14
Infrared Atmospheric Sounding Interferometer	IASI	METOP-A	p	12
Advanced Himawari Imager	AHI	Himawari-8/9	g (10 min)	2
Spinning Enhanced Visible and InfraRed Imager	SEVIRI	MSG	g (15 min)	1–2
Advanced Baseline Imager	ABI	GOES-R	g (15 min)	2

Table 9.2 *Summary of the main nadir-viewing spectrometers and platforms used for detection/retrieval of volcanic SO$_2$ using UV (and shortwave IR)*

Sensor	Acronym	Platform	Platform type	Resolution (km)
Ozone Monitoring Instrument	OMI	Aura	p	13 x 25
Global Ozone Monitoring Experiment	GOME-2	ERS-2, MetOp	p	40 x 320
Ozone Mapping and Profiler Suite	OMPS	JPSS	p	50
TROPOspheric Monitoring Instrument	TROPOMI	Sentinel-5P	p	3 x 5

also very suitable for measurements of volcanic SO$_2$. Several hyperspectral spectrometers (see Table 9.2) give observations of global SO$_2$, including retrieval algorithms from backscattered radiance measurements. The sensitivity of SO$_2$ retrievals depends on *a priori* assumption of the SO$_2$ cloud height, and the retrievals often provide different products depending on this assumption. Among the hyperspectral spectrometers, it is worth mentioning the success of TROPOMI with its very high detection sensitivity (~1 DU; Theys et al., 2019).

9.3 Data Assimilation Advances

9.3.1 Data Insertion

Data insertion is a simple DA strategy and, essentially, consists of initialising a dispersal model with an effective 'virtual source' inserted away from the volcano. The obvious advantage of such a forecast initialisation is that no prior knowledge on the uncertain ESPs is needed. Moreover, if multiple observations are available at different time steps, successive data insertions can be used to restart a forecast and to halt the forward propagation in time of other model uncertainties. A possible forecast strategy built upon data insertion is to generate an ensemble of runs, for example, with each ensemble member initialised from observations at different times. The ensemble mean or any other weighted combination of the ensemble members can then furnish a deterministic forecast, but with the additional advantage that some observation snapshots may reveal parts of the cloud obscured during other initialisation steps (e.g. due to overlaying of meteorological clouds). In contrast, the main disadvantage of the data insertion mechanism is that the virtual sources are normally obtained from passive column load satellite retrievals, something that may require additional assumptions on the cloud top height, thickness, and vertical distribution of concentration in order to reconstruct the three-dimensional structure of the source. This issue can be partially addressed by co-locating the column load retrievals with some additional vertically resolved observation, for example, profiles from the CALIOP LIDAR aboard the CALIPSO polar-orbiting satellite (if a coincident overpass exists). On the other hand,

data insertion does not consider the mass emitted since the last retrieval and, for this reason, it is particularly tailored to cases where the cloud is already detached from the volcano at the time of initialising the forecast model. Finally, another important aspect of data insertion to consider is that retrieval errors are actually ingested into the model and propagated forward.

Wilkins et al. (2014) used SEVIRI retrievals of ash column load and cloud top height from the 13–14 May 2010 Eyjafjallajökull ash cloud to initialise four different Numerical Atmospheric-dispersion Modelling Environment (NAME) dispersal model simulations driven by the global version of the Met Office Unified Model (MetUM). With these, the authors proposed a 'conservative' composite forecast considering the greatest ash column load values at each grid cell. This pioneering work on data insertion showed the potential of the multiple retrieval analysis strategy, which improved the forecast skills of the reference run with no data insertion. Clearly, this method is likely to be more effective under little cloud cover and to worsen whenever parts of the cloud are consistently undetected by all retrievals. This idea of a multiple data insertion composite forecasts was further expanded by Wilkins et al. (2016) including a more in-depth forecast validation using the structure, amplitude, and location (SAL) and the Figure Metric of Space (FMS) metrics for the 8 May 2010 Eyjafjallajökull and the 24 May 2011 Grímsvötn ash clouds. In this case, up to six IR retrievals in a 35-hour time window were used, and the results showed that all single data insertion forecasts performed similarly to the NAME best guess forecast, which was initialised from the volcanic source alone and included measurements of the ESPs (i.e. with fewer uncertainties than in operational settings). Not surprisingly, the six-member composite forecast using the maximum loads scored worse than any single member, essentially because this conservative estimate was actually designed to ensure that most observable ash is captured. Wilkins et al. (2016) concluded that, for scenarios in which a lot of ash is obscured, the data insertion method alone is likely to be insufficient. Crawford et al. (2016) initialised the HYSPLIT dispersal model with Moderate Resolution Imaging Spectroradiometer (MODIS) mass and cloud top retrievals complemented with particle effective radius estimations. The 2008 Kasatochi eruption was used here as the test case, for which a high-resolution and long-lasting observation dataset with little presence of meteorological clouds was available. These ideal observation conditions were complemented by co-locating CALIOP data from several polar-orbiting passes to derive further information on the vertical structure of the ash cloud. The data insertion option was compared against two operational source model settings, the uniform line source and the cylindrical source, using the Critical Success Index (CSI) metrics to measure the overlap of the two datasets. Their conclusion was also that the data insertion option performed as well or better than initialising the model at the vent, particularly when using the early retrievals (i.e. when the cloud is optically thick and has a simpler structure). On the other hand, Folch et al. (2020) implemented a simple ash/SO_2 data insertion mechanism in the version 8.0 of the FALL3D model. In the model initialisation step, an option was added to interpolate gridded satellite data from latest generation geostationary satellites into the model grid imposing conservation of mass, that is, ensuring that the resulting column mass in the model (computed concentration multiplied by cloud thickness) equals that of satellite data over the same cell area. As test cases, Prata et al. (2021) used the FALL3D-8.0 data insertion mechanism with fine ash and SO_2 for the June 2011 Cordón Caulle (Meteosat-9 SEVIRI data) and the June 2019 Raikoke (Himawari-8 AHI data) eruptions, respectively. Time series of SAL and FMS metrics were used to validate simulations with and without data insertion initialisations, showing that simulations initialised with data insertion consistently outperformed their counterparts. For illustrative purposes, Fig. 9.1 shows an example of data insertion and its benefits on the forecast. However, Prata et al. (2021) also pointed out that retrievals could be affected by several factors (e.g. cloud interference, high water vapour burdens, chosen detection thresholds) meaning that the ash/SO_2 detection schemes may miss some legitimate ash or SO_2 that the model is otherwise predicting. These aspects can partially explain the reluctance in implementing data insertion mechanisms in operational forecast model settings. In fact, in terms of operational implementation, data insertion strategies are of limited use. To our knowledge, only the Tokyo VAAC includes an initial particle distribution consistent with the observed cloud boundaries (Eliasson and Yoshitani, 2015) and the Darwin VAAC fuses data insertion with source characterisation to filter members during the analysis stage (see next sections).

9.3.2 Source Term Inversion

Source term inversion is a modelling strategy that, essentially, aims at finding the combination of key eruption source parameters (ESPs) that bring the model into best

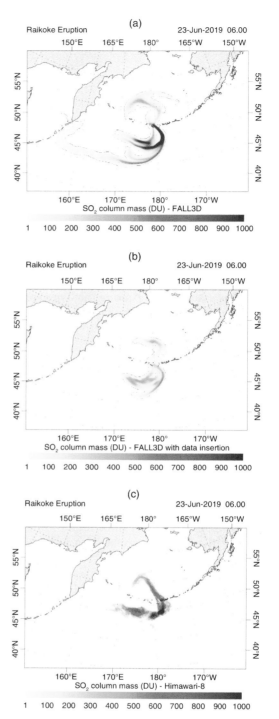

Figure 9.1 Example of a data insertion forecast using FALL3D for the 2019 Raikoke SO_2 cloud. (a) Cloud column mass of the control run forecast (no DA) at 23 June 06:00 UTC, 36h after the eruption start. (b) Forecast with data insertion at the same instant (insertion at 22 June 18:00 UTC, 12h before the plot). (c) Column mass retrieval from AHI Himawari-8 observations. All plots in Dobson Units (DU); same colour scale.

agreement with observations, typically consisting on column mass retrievals. Like in the simpler data insertion mechanisms, the different source term inversion strategies also provide an initial condition to the forecast but with two important advantages. On the one hand, the source inversion explicitly resolves the vertical structure of the cloud so that, as opposed to data insertion, no additional hypotheses are needed on cloud top height, thickness, and mass concentration distribution. On the other hand, the emission profiles resulting from the inversion procedure can be used to produce a complete forecast, a desirable feature when the cloud is not yet detached from the vent.

Eckhardt et al. (2008) did the first implementation of an inverse modelling procedure for volcanic clouds using the FLEXPART Lagrangian dispersion model and considering SO_2 observations from various satellite-borne instruments (AIRS, OMI, and SEVIRI). Their inversion method, built upon the original approach of Seibert (2000), uses an iterative algorithm to minimise a cost function using an *a priori* for the unknown sources and a Bayesian formulation involving uncertainties for both the prior and the observations. In this strategy, the prior source profiles are first decomposed in a number of elementary points above the volcano and the dispersal model is solved forward in time up to the 'assimilation' instant(s), when the inversion mechanism computes corrections for each elementary component. The cost function in Eckhardt et al. (2008) results from adding three different contributions, namely, model-observation misfits at receptors, deviation from the *a priori* values, and deviation from smoothness. This results in a linear system of equations to be solved for the elementary source corrections (i.e. giving the increments with respect to the prior) and providing the 'most likely' solution in the statistical sense. It is important to note that this strategy is very optimal for Lagrangian model formulations because the contribution to the receptors from each elementary point source is always available (Lagrangian 'puffs' can be tagged trivially). As a result, only one forward run of the dispersal model with all the prior elementary sources is, in theory, needed. This is not true in Eulerian frameworks because, in this case, model grid points do not know the elementary source contributions (only the total aggregated concentration/load exists), preventing the construction of a source–receptor matrix. Nonetheless, this elementary-based method can be extended to Eulerian models, but this comes at the cost of running an ensemble of n single-point prior simulations, where n is the number of elementary sources in the emission profile. Considering that n is typically of the order or larger than 100, this can impose computational constraints that make Eulerian models less suitable for this particular inversion strategy in operational settings.

The elementary Bayesian inversion strategy was first applied by Eckhardt et al. (2008) to reconstruct the 2007 Jebel at Tair (Yemen) SO_2 cloud to obtain emission profiles and altitudes, allowing further simulation of sulphur transport across Asia and over the Pacific Ocean for about a week after the inversion time instant. Later on, Kristiansen et al. (2010) did a second application example to derive the vertical SO_2 emission height profile for the 7–8 August 2008 Kasatochi eruption. The simulated Kasatochi SO_2 cloud was compared against independent satellite data up to six days after the eruption, with a good overall agreement. The same inversion method existing in FLEXPART for SO_2 was later extended to time-dependent volcanic ash emissions by Stohl et al. (2011), in this case considering coincident SEVIRI and IASI measurements of the 14 April to 24 May 2010 Eyjafjallajökull clouds, yielding to a vertically and temporally resolved *a posteriori* emission quantification. It is worth mentioning that the inversion reduced the root-mean-square (RMS) errors in FLEXPART by 28% when comparing *a priori* and *a posteriori* source inversion results. Kristiansen et al. (2012) ported this time-resolved source term option to the Lagrangian NAME model, in this case driven by the MetUM meteorological model. For both FLEXPART and NAME Lagrangian dispersal models, the *a posteriori* Eyjafjallajökull source terms differed significantly from the *a priori* guess in that the inverted emissions were more pulse-like, resulting in ash being released near the top of the eruptive column. In addition, and thanks to the exceptionally good set of observations available for this well-studied eruption, the inverted simulations could be compared against ground-based and research aircraft measurements, resulting in a factor of 2 uncertainty reduction and improved overall forecast skill scores. More applications can be found in Steensen et al. (2017) and Moxnes et al. (2014), who did the first ash/SO_2 joint inversion applying the fully resolved approach in Stohl et al. (2011) to the May 2011 Grímsvötn clouds and considering IASI data up to four days after the start of the eruption. At operational level, a variant of the elementary Bayesian inversion method was implemented in the London VAAC to assist with the provision of forecast guidance. In this case, the UK Met. Office inversion system (INTEM) was extended to estimate volcanic ash source parameters and furnish the NAME dispersal model with vertical emission profiles in an automated way (Pelley et al., 2015). For INTEM, the *a priori* ash profiles are first set with observations of the eruptive column height (with associated uncertainty estimations) and then the embedded INTEM system runs to obtain the refined (inverted) source term profile given all the SEVIRI satellite retrievals available from the start of the eruption up to the inversion (analysis) time. This procedure can run cyclically (e.g. every six hours) to give automated updates to the ESPs as more retrievals become available. Similar Bayesian-based approaches have been implemented in other models, basically differing from the original formulation of Eckhardt et al. (2008) on the definition of the cost function. For example, Chai et al. (2017) implemented a source inversion system in the HYSPLIT Lagrangian model to estimate optimal ash profiles emission rates from MODIS mass loadings and ash cloud top heights, and tested it with the 2008 Kasatochi eruption. In this case the cost function only integrates the differences between the model predictions and the observations and the deviations of the (optimal) final solution from the first guess. This study found that multiple-retrieval assimilations at different times produced a better hindcast than only assimilating the latest observation available.

On the other hand, and in parallel, another inversion mechanism for SO$_2$ clouds was proposed by Boichu et al. (2014) using the CHIMERE regional Eulerian chemistry-transport model driven by the Global Forecasting System (GFS) reanalysis. The method considers a feedback loop between the model and the polar-orbiting Infrared Atmospheric Sounding Interferometer (IASI) retrievals. The inverse problem is solved here by determining the time history of the SO$_2$ flux that minimises (in the least squares sense) the misfit between observed and modelled spatial and temporal distributions of SO$_2$ within one or several images and, as opposed to the Bayesian approach, no *a priori* knowledge on mass flux is required. More recently, Zidikheri and Potts (2015) introduced a completely different source inversion approach, initially only to invert SO$_2$ emissions but later on extended to volcanic ash by Zidikheri et al. (2016). As opposed to the previous Bayesian approaches, this method does not decompose the prior emission profiles into elementary source terms but considers instead that the emission profiles are characterised by known functions depending on a small number of parameters that need to be optimally determined by the inversion. For example, the vertical distribution of mass could be given as a two-parameter Suzuki parameterisation (Suzuki, 1983), the total source strength can depend on the fraction of fine ash and eruption column height through a Mastin-like relationship (Mastin et al., 2009), and so on. In this way, a set of prior dispersal model simulations is run with each simulation characterised by a single combination of parameter values sampled within a constrained range. The novel idea introduced by Zidikheri and Potts (2015) was to find the optimal combination(s) of values using cloud mass loading pattern correlations as a measure of the model–observations agreement. To this purpose, a pattern correlation is computed for each model run considering the scalar product of two normalised vectors in an *n*-dimensional space, where *n* is the number of points with observations. The first vector contains local deviations of observations from the spatial (over all points) and temporal means, whereas the second vector is analogous but for model forecasts. The inversion consists of finding the combination of source parameters with maximum pattern correlation, which mathematically implies a maximum alignment of these two vectors. In fact, an optimal pattern correlation of 1 implies that the model

and the observations have, at every receptor point and time slab, the same relative deviations with respect to mean quantities. It is clear that, compared with the elementary Bayesian inversion, this method requires running a number of prior simulations depending on the dimension of the parameter space and its discretisation. However, it presents multiple advantages. First of all, no assumptions are strictly needed on model/observation uncertainties. Second, the resulting emissions are, by construction, positive-definite and no extra procedure is required to filter out spurious non-physical negative solutions as occurs with the previous functional forms. Third, the method is independent of the modelling framework so that Eulerian models pay no particular penalty. Last but not least, during the inversion process, the prior simulations can be ranked according to their correlation pattern and this can be used to define the members for an ensemble forecast, for example, by selecting (filtering) the subset of runs in the analysis with most 'optimal' parameter values.

Zidikheri et al. (2017a, 2017b) tested the pattern-based inversion approach using the HYSPLIT model driven by the ACCESS-R regional meteorological model on various recent eruptions affecting the Darwin VAAC domain – Kelut (February 2014), Sangeang Api (May 2014), Rabaul (August 2014), Manam (July 2015), and Soputan (January 2016) – using mass load retrievals from the Japanese Meteorological Agency's geostationary satellite MTSAT and, for the most recent events after 2015, also from Himawari-8. In this case, the parametric emission profiles were defined in terms of vertical mass distribution (a blend of top-Gaussian and vertically constant forms), time-dependent column height, and particle grain size (log-normal distribution). In addition to showing how pattern-based inversion can improve forecasts, a main finding was that tuning the vertical mass distribution is normally the most important aspect to match observed cloud mass load patterns.

9.3.3 Variational Data Assimilation

Variational DA methods aim at estimating the best model pattern that is consistent with the observations within a given observing window (Carrassi et al., 2018). Variational methods seek an optimal analysis solution by minimising a cost function depending on the difference between model outputs and observations in order to estimate the control variables of a model, which can include both state variables and model parameters (Lu et al., 2016a). Variational methods are widely employed in many fields (e.g. meteorology) but their application to volcanic clouds is limited to a few studies. Flemming and Inness (2013) proposed a technique based on the 4D-Var method to estimate the total SO$_2$ emission rate and plume height, without determining the vertical emission distribution. This limitation was corrected by Lu et al. (2016a), who conducted several twin experiments to determine the vertical profile of emission rate by assimilating synthetic observations of volcanic ash mass loading. Their conclusion was that the standard 4D-Var method was unable to correctly estimate the vertical structure of the emission source term due to the lack of vertical information in the observational dataset. For this reason, a modified 4D-Var method was proposed and evaluated using twin experiments. This alternative sought an optimal linear combination of model realisations by minimising a reformulated 4D-Var cost function in order to estimate the vertical distribution of the emission source term (assumed to be a constant input parameter) over an assimilation time window. According the authors, this approach showed better performance than the traditional 4D-Var method and has the potential to be applied to real cases using satellite retrievals. Lu et al. (2016b) did a further step by integrating observations of the ash column height and the mass loading in the data assimilating system. Twin experiments were conducted to evaluate the performance of this technique and it was concluded that it is possible to reconstruct the vertical profile if large assimilation windows (>6 h) are considered. Moreover, it was suggested that the assimilation system is unable to produce reliable forecasts by assimilating only two-dimensional mass loading and therefore vertical information must be integrated to remove spurious vertical correlations. In fact, Inness et al. (2022) used the Raikoke eruption in June 2019 to show how adding TROPOMI SO_2 height data to the total SO_2 column products improved the Copernicus Atmosphere Monitoring Service (CAMS) operational forecasts, which make no use of prior knowledge of the plume height. On the other hand, Vira et al. (2017) explored the well-known relationship between the 4D-Var data assimilation and the source term inversion to estimate temporal and vertically resolved SO_2 emission fluxes using the SILAM atmospheric chemistry model during the 2010 eruption of Eyjafjallajökull. The authors found that assimilating the plume height retrievals reduced the overestimation of injection height during individual periods of one to three days.

9.3.4 Ensemble-Based Sequential Data Assimilation
Sequential DA methods have been extensively used to study and forecast geophysical systems (Carrassi et al., 2018). This approach is based on the estimation theory and can be applied to a broad range of operational and research scenarios. In sequential schemes, the assimilation process is characterised by a sequence of steps involving a forecast step and an analysis step in which model variables are corrected whenever observations are available. During the analysis step, the *a posteriori* estimate is computed from the *a priori* forecast state and the observational dataset to be assimilated.

9.3.4.1 Particle Filter Methods
Ensemble-based forecasting has long been recognised and used as a proper way to deal with model uncertainties, providing an adequate framework for full data assimilation strategies. However, before reviewing the more classical Kalman-based sequential DA schemes, it is worth mentioning a strategy that makes use of data insertion and/or source term inversion to define the ensemble members and their spread. Zidikheri and Lucas (2021a, 2021b) used a large number of ensemble runs generated from perturbations of the volcano source term, data insertion, or a combination of both. The novel contribution from these recent studies was to filter the trial runs based on their degree of agreement with observations within the analysis time window, resulting in a kind of 'particle filter' strategy. Moreover, Zidikheri and Lucas (2021a, 2021b) considered not only quantitative mass load retrievals but also a mix of several observation fields including also ash detection, particle size, or cloud top height. With these, a multiple-step filtering process is used to determine the optimal subset of analysis ensemble members that is retained for the successive forecast stage.

9.3.4.2 Kalman-Filter Methods
Sequential DA techniques are mostly based on the Kalman filter (KF; Kalman, 1960), which represents the optimal sequential technique for linear dynamics. Given a model state vector and an observation vector, the KF gives the optimal solution for the dynamical and observational models (both assumed to be linear and with Gaussian noise). However, it was soon recognised that the KF is not feasible for realistic geophysical problems involving high-dimensional numerical models and, for this reason, several algorithms based on the original KF have been proposed to reduce the computational requirements and handle the non-linearities characteristic of real physical problems. The ensemble-based KFs are a popular family of methods in which the probability distributions are approximated by an ensemble of system states and the error covariance matrix of the original KF formulation is replaced by a sampled covariance matrix computed from the ensemble. One of the most important practical advantages of ensemble-based techniques is the independence of the filter algorithm on the specific forward model. In addition, ensemble-based KFs are relatively easy to implement (no adjoint operators are required) and can benefit from massively parallel computer architectures due to the large degree of parallelism that can be achieved. In each assimilation step, a forecast is used as a first guess to generate an ensemble of improved model states or posterior ensemble which is compatible with available observations. The analysis is an estimate of the three-dimensional model state and can be used to initialise the next forecast step. In a sequential approach, this procedure can be repeated multiple times to continuously correct the dispersal model state by assimilating a sequence of observations recorded at

regular time intervals. Most ensemble-based formulations can be divided into two major categories depending on how the analysis is generated: *stochastic* and *deterministic* approaches (Houtekamer and Zhang, 2016). The ensemble Kalman filter (EnKF) introduced by Evensen (1994) is one of the most popular methods based on a stochastic Monte Carlo approach, and has been widely used in oceanography and meteorology applications (Evensen, 2003). Subsequently, Burgers et al. (1998) showed that the observations must be treated as random variables to arrive at a consistent analysis scheme and proposed the perturbed observations-based EnKF formulation. In contrast, the deterministic approaches do not perturb observations randomly. Instead, deterministic filters use algorithms to generate the analysis ensemble through an explicit transformation of the state ensemble. A remarkable example is the Ensemble Transform Kalman Filter (ETKF, Bishop et al., 2001), a popular square-root filter formulation (Nerger et al., 2012) which is suitable for high-dimensional systems, relatively easy to implement, and is computationally efficient. The application of ensemble filters in geophysical systems can lead to spurious correlations and underestimations of the ensemble spread due to a limited size of the ensemble, sampling errors, and model errors (Anderson and Anderson, 1999). The problem of variance underestimation (filter collapse) is usually addressed by using inflation methods, whereas localisation is adopted to suppress spurious correlations. The localised version of the ETKF (i.e. LETKF) proposed by Hunt et al. (2007) is a popular method for data assimilation with localisation suitable for high-dimensional systems and computationally efficient. Both ETKF and LETKF methods have previously been used to assimilate volcanic observations.

Most work on ensemble assimilation dealing with volcanic ash/SO_2 has focused on ensemble KF techniques. Numerical experiments are based on both global (Fu et al., 2015, 2017; Osores et al., 2020) and localised (Fu et al., 2016; Pardini et al., 2020; Mingari et al., 2022) filters (i.e. on applying the filter over the whole computational domain or over a part of it). For example, Osores et al. (2020) proposed an ensemble-based data assimilation system for volcanic ash using the global ETKF. Alternatively, the localised version of the ETKF (i.e. LETKF) was implemented in the assimilation systems used by Pardini et al. (2020) and Mingari et al. (2022) in order to improve volcanic ash/SO_2 forecasts. In the numerical studies conducted by Fu et al. (2015, 2016, 2017), the LOTOS-EUROS model was combined with an ensemble square-root filter to reconstruct the optimal model state and improve quantitative forecasts of the 2010 Eyjafjallajökull volcanic eruption.

Typically, the ensemble model states are constructed by perturbing uncertain model parameters such as eruption source parameters (ESPs) and meteorological fields (assumed to be an input parameter when the meteorological and dispersal models are coupled according to an off-line approach) with typical ensemble sizes ranging between 32

and 128 members. In fact, uncertainties in ESPs are known to be first-order contributors to model errors (Costa et al., 2016) and the eruption column height is recognised as one of the most relevant source parameters, perturbed in all published studies. Other perturbed EPS include mass eruption rate, vertical mass distribution, and eruption start time and duration. In addition, horizontal wind components have also been perturbed. Some preliminary work is based on twin experiments (e.g. Fu et al., 2015; Osores et al., 2020; Mingari et al., 2022), where observations are generated from a control model simulation by adding random errors. In contrast, other numerical experiments are based on real observations corresponding to recent volcanic eruptions, including the 2010 Eyjafjallajökull (Fu et al., 2016, 2017), 2018 Etna (Pardini et al., 2020), and 2019 Raikoke (Mingari et al., 2022) eruptions. Fig. 9.2 illustrates how this sequential DA strategy improves the forecast skills for the latter case.

In the filter-based methods, the observation operator that translates the model state into the observation space must be defined depending on the specific observational dataset to be assimilated. Typically, this entails satellite-retrieved or synthetic two-dimensional column mass loading for assimilation purposes (Osores et al., 2020; Pardini et al., 2020; Mingari et al., 2022). In this case, the observation operator involves a vertical integration of concentration, a sum over different species (if multi-species observations are being assimilated) and, finally, the interpolation to the observation coordinates. Alternatively, the full concentration field can also be assimilated as in Fu et al. (2017), where the authors defined a pre-processing operator to translate ash mass loading and cloud top height satellite retrievals into three-dimensional concentrations. This procedure avoids the problem of the artificial vertical correlations that could potentially be introduced by the assimilation of a two-dimensional dataset of observations. Note that the observational operator in this case is trivial as it only translates concentrations at model grid into observation locations. In addition, the ash cloud was assumed uniformly distributed within a layer with a given vertical thickness sampled from a range of plausible values obtained from a literature review. However, as this thickness was assumed to be uniform and constant, artificial vertical correlations were not completely removed. Note that the definition of a linear observation operator is required to assimilate mass loading and mass concentration, while other observables (e.g. optical depth), would lead to a non-linear observation operator. So far, all published work has focused on linear operators.

9.4 Conclusions and Emerging Perspectives

Until a decade ago, all operational forecast systems in place worldwide only delivered qualitative ash cloud forecast products, consisting of 'ash/no ash' delineation zones. Today, the operational Volcanic Ash Advisory

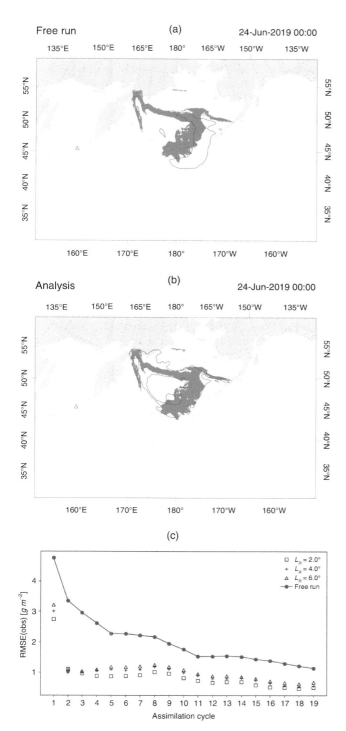

Figure 9.2 Comparison between SO_2 cloud observations and forecasts for the 2019 Raikoke eruption using the FALL3D model. (a) Comparison between 1 gm^{-2} mass loading contour for observations (green shaded area) and model (solid red line) at 23 June 00:00 UTC. Model results correspond to the free run without DA. (b) Same for the analysis stage using LETKF to assimilate satellite retrievals. (c) Root-mean-square error in the observation space for each assimilation cycle (19 cycles, assimilation every 3h). The LETKF simulations consider three different

Centres (VAACs) are under a complex transition process that includes updating of products with quantitative forecasts, use of the System Wide Information Management (SWIM) of the International Civil Aviation Organization (ICAO) or integration with the WMO information system (WIS) global infrastructure. At a scientific level, substantial progress has occurred during the last decade on implementation, verification, and validation of DA strategies for volcanic clouds. However, technical challenges still exist for transfer of DA into operations, including the definition of complex workflows able to access data from distributed sources, manage data streams, and fuse data and models in a robust and automated way.

A range of possible DA methodologies exists depending on the available observations. Satellite-based observations are clearly the preferred choice, although column mass retrievals have no vertical resolution and this can yield to spurious correlations in some DA methodologies. This limitation can be circumvented if mass observations are complemented with column height retrievals, more difficult to obtain at operational level. An alternative could be to assimilate radiance products but, to this purpose, dispersal models need to embed a radiative transfer model.

In terms of porting DA strategies into operations, source inversion is the simplest promising option, particularly for Lagrangian models. Unlike data insertion, inversion needs no additional observations on cloud top height, thickness, and vertical distribution of concentration in order to reconstruct the three-dimensional structure of the source. More sophisticated options like sequential DA (Kalman filters) have shown to be a promising alternative for assimilation of volcanic ash/SO_2. According to several evaluation metrics, this approach results in a significant improvement of quantitative forecast. However, some limitations of this method leading to suboptimal filter performance should be highlighted. Specifically, the assimilation of ash/SO_2 challenges the Gaussian assumption of model and observation errors that underpins KF methods. It has been shown that the skewness of probability distributions is a significant issue that violates the Gaussian assumption in ensemble forecasting, resulting in suboptimal behaviour of the ensemble KF. As a result, the analysis step can yield an unrealistic posterior estimate that is not consistent with the non-Gaussian Bayes' theorem, introducing artificial negative concentrations. Even if the previous results reported in the literature are encouraging, further work needs to be carried out to establish more appropriate methodologies for positively skewed,

Figure 9.2 (Cont.)

localisation radii ($L_R = 2°, 4°, 6°$), all clearly outperforming the free run. Multiple assimilation cycles were required to keep the evaluation metrics below the reference values given by the free run. Overall, the analysis errors were decreased by more than 50% relative to the free run errors, showing the importance of DA to improve the predictive capability of dispersal models.

non-Gaussian prior distributions. Different approaches have been proposed to deal with non-Gaussianity, including variable transformations (e.g. Zhou et al., 2011; Amezcua and Van Leeuwen, 2014), Bayesian approaches, such as particle filter (Van Leeuwen and Ades, 2013), and ensemble Kalman filtering methods for highly skewed non-negative uncertainty distributions (Bishop, 2016).

References

Amezcua, J., and Van Leeuwen, P. J. (2014). Gaussian anamorphosis in the analysis step of the EnKF: A joint state-variable/observation approach. *Tellus A: Dynamic Meteorology and Oceanography*, 6(1), 23493. http://doi.org/10.3402/tellusa.v66.23493.

Anderson, J. L., and Anderson, S. L. (1999). A Monte Carlo implementation of the nonlinear filtering problem to produce ensemble assimilations and forecasts. *Monthly Weather Review*, 127(12), 2741–58.

Balis, D., Koukouli, M. E., Siomos, N. et al. (2016). Validation of ash optical depth and layer height retrieved from passive satellite sensors using EARLINET and airborne lidar data: The case of the Eyjafjallajokull eruption. *Atmospheric Chemistry and Physics*, 16, 5705–20.

Bishop, C. H., Brian, J. E., and Sharanya, J. M. (2001). Adaptive sampling with the Ensemble Transform Kalman Filter. Part 1: Theoretical aspects. *Monthly Weather Review*, 129(3), 420–36.

Bishop, C. H. (2016). The GIGG-EnKF: Ensemble Kalman filtering for highly skewed non-negative uncertainty distributions. *Quarterly Journal of the Royal Meteorological Society*, 142 (696), 1395–412.

Boichu, M., Clarisse, L., Khvorostyanov, D., and Clerbaux, C. (2014). Improving volcanic sulfur dioxide cloud dispersal forecasts by progressive assimilation of satellite observations, Geophysical Research Letters. *American Geophysical Union*, 41, 2637–43.

Bonadonna, C., Folch, A., Loughlin, S., and Puempel, H. (2012). Future developments in modelling and monitoring of volcanic ash clouds: Outcomes from the first IAVCEI-WMO workshop on ash dispersal forecast and civil aviation. *Bulletin of Volcanology*, 74, 1–10.

Burgers, G., Leeuwen, P. J. van, and Evensen, G. (1998). Analysis scheme in the ensemble Kalman filter. *Monthly Weather Review*, 126(6), 1719–24.

Carrassi, A., Bocquet, M., Bertino, L., and Evensen, G. (2018). Data assimilation in the geosciences: An overview of methods, issues, and perspectives. *WIREs Climate Change*, 9(e535), 1–50.

Carn, S. A., Krueger, A. J., Krotkov, N. A., Yang, K., and Evans, K. (2009). Tracking volcanic sulfur dioxide clouds for aviation hazard mitigation. *Natural Hazards*, 51, 325–43.

Chai, T., Crawford, A., Stunder, B. et al. (2017). Improving volcanic ash predictions with the HYSPLIT dispersion model by assimilating MODIS satellite retrievals. *Atmospheric Chemistry and Physics*, 17, 2865–79.

Clarisse, L., Hurtmans, D., Clerbaux, C. et al. (2012). Retrieval of sulphur dioxide from the infrared atmospheric sounding interferometer (IASI). *Atmospheric Measurement Techniques*, 5, 581–94.

Clarkson, R., and Simpson, H. (2017). Maximising airspace use during volcanic eruptions: Matching engine durability against ash cloud occurrence. In *Proceedings of the NATO STO AVT-272 Specialists Meeting on Impact of Volcanic Ash Clouds on Military Operations*, vol. 1, pp. 17-1–17-19. www.sto.nato.int/publications/STO%20Meeting%20Proceedings/STO-MP-AVT-272/MP-AVT-272-17.pdf.

Costa, A., Suzuki, Y., Cerminara, M. et al. (2016). Results of the eruptive column model inter-comparison study, *Journal of Volcanology and Geothermal Research*, 326, 2–25.

Crawford, A., Stunder, B., Ngan, F., and Pavalonis, M. (2016). Initializing HYSPLIT with satellite observations of volcanic ash: A case study of the 2008 Kasatochi eruption. *Journal of Geophysical Research: Atmospheres*, 121(10), 786–10.

Dunn, M. G., and Wade, D. P. (1994). Influence of volcanic ash clouds on gas turbine engines. In T. J. Casadevall, ed., *Volcanic ash and aviation safety; Proceedings of the First International Symposium on Volcanic Ash and Aviation Safety held in Seattle, Washington, in July 1991*. Reston, VA: US Geological Survey Bulletin 2047, pp. 107–17. https://doi.org/10.3133/b2047.

Eckhardt, S., Prata, A. J., Seibert, P., Stebel, K., and Stohl, A. (2008). Estimation of the vertical profile of sulfur dioxide injection into the atmosphere by a volcanic eruption using satellite column measurements and inverse transport modeling. *Atmospheric Chemistry and Physics*, 8, 3881–97.

Eliasson, J., and Yoshitani, J. (2015). Airborne measurements of volcanic ash and current state of ash cloud prediction, *Disaster Prevention Research Institute Annuals*, 58B, 35–41. www.dpri.kyoto-u.ac.jp/nenpo/no58/ronbunB/a58b0p03.pdf.

Evensen, G. (1994). Sequential data assimilation with a nonlinear quasi-geostrophic model using Monte Carlo methods to forecast error statistics. *Journal of Geophysical Research: Oceans*, 99(C5), 10143–62.

Evensen, G. (2003). The ensemble Kalman filter: Theoretical formulation and practical implementation. *Ocean Dynamics*, 53 (4), 343–67.

Flemming, J., and Inness, A. (2013). Volcanic sulfur dioxide plume forecasts based on UV satellite retrievals for the 2011 Grímsvötn and the 2010 Eyjafjallajökull eruption. *Journal of Geophysical Research: Atmospheres*, 118(17), 10172–89.

Folch, A. (2012). A review of tephra transport and dispersal models: Evolution, current status, and future perspectives. *Journal of Volcanology and Geothermal Research*, 235–236, 96–115.

Folch, A., Mingari, L., Gutierrez, N. et al. (2020). FALL3D-8.0: a computational model for atmospheric transport and deposition of particles, aerosols and radionuclides. Part 1: Model physics and numerics. *Geoscientific Model Development*, 13, 1431–58.

Francis, P. N., Cooke, M. C., and Saunders, R.W. (2012). Retrieval of physical properties of volcanic ash using Meteosat: A case study from the 2010 Eyjafjallajökull eruption. *Journal of Geophysical Research: Atmospheres*, 117, D00U09. https://doi.org/10.1029/2011JD016788

Fu, G., Lin, H.X., Heemink, A.W. et al. (2015). Assimilating aircraft-based measurements to improve forecast accuracy of volcanic ash transport. *Atmospheric Environment*, 115, 170–84.

Fu, G., Heemink, A., Lu, S. et al. (2016). Model-based aviation advice on distal volcanic ash clouds by assimilating aircraft in situ measurements. *Atmospheric Chemistry and Physics*, 16 (14), 9189–200.

Fu, G., Prata, F., Lin, H. X. et al. (2017). Data assimilation for volcanic ash plumes using a satellite observational operator: A case study on the 2010 Eyjafjallajökull volcanic eruption. *Atmospheric Chemistry and Physics*, 17(2), 1187–205.

Houtekamer, P. L., and Zhang, F. (2016). Review of the ensemble Kalman filter for atmospheric data assimilation. *Monthly Weather Review*, 144(12), 4489–532.

Hunt, B. R., Kostelich, E. J., and Szunyogh, I. (2007). Efficient data assimilation for spatiotemporal chaos: A local ensemble transform Kalman filter. *Physica D*, 230(1), 112–26.

Inness, A., Ades, M., Balis, D. et al. (2022). The CAMS volcanic forecasting system utilizing near-real time data assimilation of S5P/TROPOMI SO_2 retrievals. *Geoscientific Model Development*, 15, 971–94.

Kalman, R. E. (1960). A new approach to linear filtering and prediction problems. *Journal of Basic Engineering*, 82, 35–45.

Kristiansen, N. I., Stohl, A., Prata, A. J. et al. (2010). Remote sensing and inverse transport modeling of the Kasatochi eruption sulfur dioxide cloud. *Journal of Geophysical Research: Atmospheres*, 115, 1–18.

Kristiansen, N., Stohl, A., Prata, A. et al. (2012). Performance assessment of a volcanic ash transport model mini-ensemble used for inverse modeling of the 2010 Eyjafjallajökull eruption. *Journal of Geophysical Research: Atmospheres*, 117, 1–25.

Lu, S., Lin, H. X., Heemink, A. W., Fu, G., and Segers, A. J. (2016a). Estimation of volcanic ash emissions using trajectory-based 4D-Var data assimilation. *Monthly Weather Review*, 144(2), 575–89.

Lu, S., Lin, H. X., Heemink, A., Segers, A., and Fu, G. (2016b). Estimation of volcanic ash emissions through assimilating satellite data and ground-based observations. *Journal of Geophysical Research: Atmospheres*, 121(18), 10971–94.

Mastin, L. G., Guffanti, M. C., and Servranckx, R. et al. (2009). A multidisciplinary effort to assign realistic source parameters to models of volcanic ash-cloud transport and dispersion during eruptions. *Journal of Volcanology and Geothermal Research*, 186(1–2), 10–21.

Miller, T. P., and Casadevall, T. J. (2000). Volcanic ash hazards to aviation. In H. Sigurdsson, ed., *Encyclopedia of Volcanoes*. San Diego, CA: Academic Press, pp. 915–30.

Mingari, L., Folch, A., Prata, A. T. et al. (2022). Data assimilation of volcanic aerosols using FALL3D+PDAF. *Atmospheric Chemistry and Physics*, 22, 1773–92.

Moxnes, E. D., Kristiansen, N. I., Stohl, A. et al. (2014). Separation of ash and sulfur dioxide during the 2011 Grímsvötn eruption. *Journal of Geophysical Research: Atmospheres*, 119, 7477–01.

Muser, L. O., Hoshyaripour, G. A., Bruckert, J. et al. (2020). Particle aging and aerosol–radiation interaction affect volcanic plume dispersion: Evidence from the Raikoke 2019 eruption. *Atmospheric Chemistry and Physics*, 20, 15015–36.

Nerger, L., Janjić, T., Schröter, J., and Hiller, W. (2012). A unification of ensemble square root Kalman filters. *Monthly Weather Review*, 140(7), 2335–45.

Osores, S., Ruiz, J., Folch, A., and Collini, E. (2020). Volcanic ash forecast using ensemble-based data assimilation: An ensemble transform Kalman filter coupled with the Fall3d-7.2 Model (ETKF–Fall3d Version 1.0). *Geoscientific Model Development*, 13(1), 1–22.

Pardini, F., Corradini, S., Costa, A. et al. (2020). Ensemble-based data assimilation of volcanic ash clouds from satellite observations: Application to the 24 December 2018 Mt. Etna Explosive Eruption. *Atmosphere*, 11(4), 359.

Pavolonis, M. J., Feltz, W. F., Heidinger, A. K., and Gallina, G. M. (2006). A daytime complement to the reverse absorption technique for improved automated detection of volcanic ash. *Journal of Atmospheric and Oceanic Technology*, 23, 1422–44.

Pelley, R., Cooke, M., Manning, A. et al. (2015). Initial implementation of an inversion technique for estimating volcanic ash source parameters in near real time using satellite retrievals: Forecasting Research Technical Report, vol. 604. Met Exeter, UK: Met Office. https://library.metoffice.gov.uk/Portal/Default/en-GB /DownloadImageFile.ashx?objectId=415&ownerType=0 &ownerId=212804.

Prata, A. J. (1989). Observations of volcanic ash clouds in the 10– 12-micron window using AVHRR/2 Data. *International Journal of Remote Sensing*, 10, 751–61.

Prata, A. T. (2016). Remote sensing of volcanic eruptions. In J. C. Duarte and W. P. Schellart, eds., *Plate Boundaries and Natural Hazards*, American Geophysical Union (AGU). Hoboken, NJ: Wiley and Sons, pp. 289–322.

Prata, F., and Lynch, M. (2019). Passive Earth observations of volcanic clouds in the atmosphere. *Atmosphere*, 10, 199.

Prata, A. T., Mingari, L., Folch, A., Macedonio, G., and Costa, A. (2021). FALL3D-8.0: A computational model for atmospheric transport and deposition of particles, aerosols and radionuclides. Part 2: Model validation. *Geoscientific Model Development*, 14, 409–36.

Robock, A., and Oppenheimer, C. (2003). Volcanism and the Earth's atmosphere: American Geophysical Union, *Geophysical Monograph*, 139, 360 pp.

Seibert, P. (2000). Inverse modelling of sulfur emissions in Europe based on trajectories. In P. Kasibhatla, M. Heimann, P. Rayner, N. Mahowald, R. G. Prinn, and D. E. Hartley, eds., *Inverse Methods in Global Biogeochemical Cycles*, Geophysical Monograph 114. Washington, DC: American Geophysical Union, pp. 147–54,

Sparks, R. S. J., Bursik, M. I., Carey, S. N. et al. (1997). *Volcanic Plumes*. Chichester: John Wiley & Sons.

Steensen, B. M., Kylling, A., Kristiansen, N. I., and Schulz, M. (2017). Uncertainty assessment and applicability of an inversion method for volcanic ash forecasting. *Atmospheric Chemistry and Physics*, 17, 9205–22.

Stohl, A., Prata, A., Eckhardt, S. et al. (2011). Determination of time- and height-resolved volcanic ash emissions and their use for quantitative ash dispersion modeling: The 2010 Eyjafjallajokull eruption. *Atmospheric Chemistry and Physics*, 11, 4333–51.

Suzuki, T. (1983). A theoretical model for dispersion of tephra. In D. Shimozuru and I. Yokoyama, eds., *Volcanism: Physics and Tectonics*. Tokyo: Arc, pp. 95–113.

Theys, N., Hedelt, P., De Smedt, I. et al. (2019). Global monitoring of volcanic SO$_2$ degassing with unprecedented resolution from TROPOMI onboard Sentinel-5 Precursor. *Scientific Reports*, 9, 2643.

Van Leeuwen, P. J., and Ades, M. (2013). Efficient fully nonlinear data assimilation for geophysical fluid dynamics. *Computers & Geosciences*, 55, 16–27.

Vira, J., Carboni, E., Grainger, R. G., and Sofiev, M. (2017). Variational assimilation of IASI SO2 plume height and total column retrievals in the 2010 eruption of Eyjafjallajökull using the SILAM v5.3 chemistry transport model. *Geoscientific Model Development*, 10, 1985–2008.

Wilkins, K. L., Mackie, S., Watson, M. et al. (2014). *Data insertion in volcanic ash cloud forecasting. Annals of Geophysics, Fast Track*, 2. https://doi.org/10.4401/ag-6624.

Wilkins, K. L., Watson, I. M., Kristiansen, N. I. et al. (2016). Using data insertion with the NAME model to simulate the 8 May 2010 Eyjafjallajökull volcanic ash cloud. *Journal of Geophysical Research: Atmospheres*, 121, 306–23. https://doi.org/10.1002/2015JD023895.

Zhou, H., Gómez-Hernández, J., Hendricks Franssen, H. J., and Li, L. (2011). An approach to handling non-Gaussianity of parameters and state variables in ensemble Kalman filtering. *Advances in Water Resources*, 34(7), 844–64.

Zidikheri, M. J., and Potts, R. J. (2015). A simple inversion method for determining optimal dispersion model parameters from satellite detections of volcanic sulfur dioxide. *Journal of Geophysical Research: Atmospheres*, 120, 9702–17.

Zidikheri, M. J., and Lucas, C. (2021a). A computationally efficient ensemble filtering scheme for quantitative volcanic ash forecasts. *Journal of Geophysical Research: Atmospheres*, 126, e2020JD033094.

Zidikheri, M. J., and Lucas, C. (2021b). Improving ensemble volcanic ash forecasts by direct insertion of satellite data and ensemble filtering. *Atmosphere*, 12, 1215.

Zidikheri, M., Potts, R. J., and Lucas, C. (2016). A probabilistic inverse method for volcanic ash dispersion modelling. *The ANZIAM Journal*, 56, 194–209.

Zidikheri, M., Lucas, C., and Potts, R. (2017a), Estimation of optimal dispersion model source parameters using satellite detections of volcanic ash. *Journal of Geophysical Research: Atmospheres*, 122, 8207–32.

Zidikheri, M., Lucas, C., and Potts, R. (2017b). Toward quantitative forecasts of volcanic ash dispersal: Using satellite retrievals for optimal estimation of source terms. *Journal of Geophysical Research: Atmospheres*, 122, 8187–206.

10

Data Assimilation in the Near-Earth Electron Radiation Environment

Yuri Y. Shprits, Angelica M. Castillo, Nikita Aseev,
Sebastian Cervantes, Ingo Michaelis, Irina Zhelavskaya,
Artem Smirnov, and Dedong Wang

Abstract: Energetic charged particles trapped by the Earth's magnetic field present a significant hazard for Earth-orbiting satellites and humans in space. Application of the data assimilation tools allows us to reconstruct the global state of the radiation particle environment from sparse single-point observations. The measurements from different satellites with different observational errors can be blended in an optimal way with physics-based models. The mathematical formulation on the diffusion and diffusion-advection equations for the Earth's Van Allen radiation belts and ring current is described. We further describe several recent studies that successfully applied the data assimilation tools to the near-Earth space radiation environment. The applications to the reanalysis of the radiation belts and ring current, real-time predictions, and analysis of the missing physical processes are described and motivation for these studies is provided. We further discuss various assimilation techniques and potential topics for future research.

10.1 Introduction: Societal Impact of Space Weather and Current State of Understanding of the Physical Processes

Energetic particles in the Van Allen radiation belts, commonly referred to as trapped radiation, pose a significant risk to Earth-orbiting satellites and humans in space. During solar storms, the radiation in the near-Earth space can significantly increase. During disturbed geomagnetic conditions numerous anomalies have been reported by satellite operators (Robinson Jr, 1989; Baker, 2000). Currently, there are over 4000 satellites orbiting the Earth (UCS, 2021). There are a number of communication satellites that assist in navigation, weather predictions, telecommunications, and national defence.

The near-Earth space environment is a central focus of many ongoing international missions, including the ESA Cluster mission (Escoubet et al., 1997), NASA's

Time History of Events and Macroscale Interactions during Substorms (THEMIS) mission (Sibeck and Angelopoulos, 2008), NOAA's Polar Orbiting Environmental Satellites (POES), and the Japanese Exploration of energization and Radiation in Geospace 'ARASE' (ERG) project (Miyoshi et al., 2018) (Fig. 10.1). Engineering analysis of anomalies in space and determination of the cause of anomalies require accurate knowledge of fluencies (flux of particles integrated over a period of time) along the satellite orbit. The understanding of satellite observations from different spacecraft is complicated by the fact that measurements are given at various locations in space, have different instrumental errors, and often vary by orders of magnitude. Fluencies can only be obtained either when particle detectors are installed on a spacecraft or when the global reconstruction of the fluxes is available at all times, radial distances, and geomagnetic latitudes and longitudes. Models and observations can be combined by means of data assimilation tools to reconstruct the global evolution of the radiation environment which allows us to calculate fluencies for any given satellite orbit.

The Van Allen radiation belts are two donut-shaped regions of the near-Earth space, where the magnetic field of the Earth traps high energy particles. The highly energetic particles in the Van Allen belts can produce deep dielectric charging in satellites and damage satellite electronics (Rosen, 1976; Baker et al., 1987). Another population of particles in the near-Earth space which is hazardous for satellites is usually referred to as 'ring current'. The name originates from the current that is generated by these particles and the magnetic field that is produced by this current. The hazardous effects of the ring current include degradation of satellite solar panels, charging of satellite surfaces that can damage them, and induction of currents in power grids that can damage transformers or can cause voltage instabilities (Lanzerotti, 2001; Baker, 2002, 2005).

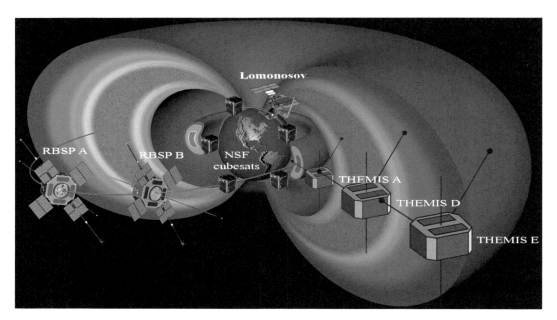

Figure 10.1 Radiation belts and orbits of satellites including Van Allen Probes (formerly named Radiation Belt Storm Probes (RBSP A and RBSP B)), THEMIS, NSF-supported Lomonosov mission (launched on April 28, 2016), and multiple new CubeSat missions on Low-Earth Orbit (LEO). Modified figure; original from NASA's Goddard Space Flight Center.

10.2 Modelling the Near-Earth Radiation Environment

10.2.1 Earth's Van Allen Radiation Belts

Electrons in the radiation belts (kinetic energy above several hundred keV) exhibit a two-zone structure. The inner electron belt is typically located between 1.2 and 2.0 Earth radii (R_E), while the outer zone extends from 4 to 8 R_E. Relativistic electrons in the Earth's radiation belts can penetrate through satellite protective shielding, get deposited in dielectric materials, and produce discharges that can damage miniature electronics (Rosen, 1976; Baker et al., 1987).

The dynamics of energetic particles are still poorly quantified. Reeves et al. (2003) found that geomagnetic storms can increase, significantly decrease, or not substantially change the fluxes of relativistic electrons during storms. The variability in the responses of the radiation belts to geomagnetic disturbances has been attributed to the complex competing nature of acceleration and loss (see review by Friedel et al. (2002); Shprits et al. (2008a); Shprits et al. (2008b)). As radiation belt electrons undergo azimuthal drift in the Earth's magnetosphere, they encounter several distinct classes of plasma waves that can exchange energy with them, produce acceleration or deceleration, and scatter electrons into the atmosphere where they will be lost from the system. Electrons can also be lost to the interplanetary medium when they hit the boundary of the magnetosphere (magnetopause) (Shprits et al., 2006; Turner et al., 2012).

The motion of relativistic electrons in the Earth's radiation belts can be simplified and described in terms of the three basic periodic motions: gyro-motion around the field line, the bounce motion in the Earth's magnetic mirror field along the field line, and the azimuthal drift around the Earth due to magnetic gradients and curvature (Fig. 10.2).

Charged particles in the magnetosphere undergo three types of periodic motion: (1) gyration about a geomagnetic field line, (2) bounce motion along the field line between mirror points, and (3) drift around the Earth. Each type of periodic motion is associated with an adiabatic invariant, which stays approximately constant under slow variations of parameters of the system, compared to the period of the corresponding motion. In order to simplify the six-dimensional description of particles in space, one can consider the description of the phase space density in terms of adiabatic invariants and ignore the variations in phases of invariants. The calculation of invariants is a challenging task as invariants depend on the time-varying three-dimensional dynamic changes of the magnetic field.

The first adiabatic invariant μ, which corresponds to the cyclotron motion of a particle, can be expressed as follows:

$$\mu = \frac{p_\perp^2}{2m_0 B}, \text{ where } p_\perp = p \sin \alpha, \tag{10.1}$$

$p_\perp$ is the component of the particle momentum that is perpendicular to the magnetic field line, m_0 is the rest mass

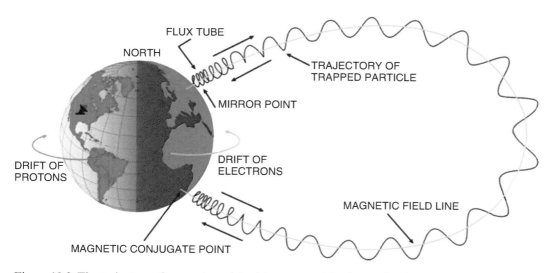

Figure 10.2 The trajectory of trapped particles (electrons and ions) experiencing gyro-motion, magnetic mirroring between the mirror points, and gradient and curvature drifts in the geomagnetic field. Each periodic motion can be associated with an adiabatic invariant. Modified figure; original from Cohen et al. (2005).

of the particle, and B is the magnitude of the magnetic field at the location of the particle. It is convenient to consider $p_\perp$ as a function of pitch angle α (the angle between magnetic field and the total momentum of the particle p, when the particle passes the equator). If there is no electric field parallel to the magnetic field line at which the particle resides, p is conserved, and $B = B_m$ is also conserved, as follows from the conservation of μ.

The second adiabatic invariant J is associated with the particle's bounce motion between the mirror points. It can be written as:

$$J = 2\frac{p}{\sqrt{B_m}} \int_{s'}^{s''} \sqrt{B_m - B(s)}\, ds, \qquad (10.2)$$

where p is the momentum of the particle, and the integral is taken along the magnetic field line between mirror points s' and s''. In practice, it is convenient to use the modified second adiabatic invariant K instead of J:

$$K = \int_{s'}^{s''} \sqrt{B_m - B(s)}\, ds. \qquad (10.3)$$

Unlike the adiabatic invariant J, K does not depend on particle energy and can be considered as a purely field-geometric quantity. Note that calculation of K requires knowledge of the magnetic field which is changing with solar activity and can be compressed, twisted, and stretched. Subbotin and Shprits (2012) suggested a new adiabatic invariant $V = \mu(K + 0.5)^2$ that is convenient for radiation belt modelling. The factor 0.5 has been chosen empirically in order to simplify the interpolation, but a different value could also be used. A grid in V and K coordinates facilitates

implementation of numerical schemes and improves accuracy, stability, and performance of radiation belt codes.

The third adiabatic invariant Φ is the magnetic flux enclosed by the drift path (within surface S) of a charged particle. This invariant is defined only for particles on closed drift orbits and is important for the most energetic part of the particle population. A more intuitive form of the third adiabatic invariant is the L^* parameter (Roederer, 2012):

$$L^* = \frac{2\pi B_E R_E^2}{\Phi}, \text{ where } \Phi = \int_S B \cdot dS, \qquad (10.4)$$

which is the radial distance from the centre of the Earth to the field line in the equatorial plane, if the field changes slowly to dipolar (note that L^* is by definition unitless); R_E is the Earth's radius, and B_E is the magnetic field magnitude at the Earth's surface at the geomagnetic equator in a dipole field approximation. The calculation of invariants K and L^* (Eqs. (10.3) and (10.4)) is particularly computationally expensive and requires tracing particles in realistic empirical magnetic field models.

By ignoring diffusion in terms of the phases of adiabatic invariants, the Fokker–Planck equation for diffusive evolution of the phase-averaged six-dimensional PSD of the relativistic electrons may be simplified by specifying the phase averaged PSD in terms of only three adiabatic invariants and written as follows:

$$\frac{\partial f}{\partial t} = \sum_{i,\, j=1,2,3} \frac{\partial}{\partial J_i} D_{ij} \frac{\partial f}{\partial J_j}, \qquad (10.5)$$

where D_{ij} are diffusion coefficients, and J_i are adiabatic invariants. Note that the modified Fokker–Planck equation (Eq. (10.5)) does not contain advective terms as, in collisionless plasmas, second-order coefficients can be related to the first-order coefficients. Although some diffusive processes may violate all three adiabatic invariants, it is customary and convenient to separately consider radial diffusion, which violates the third adiabatic invariant, pitch angle and energy diffusion, which violate the first and second invariants (Schulz and Lanzerotti, 1974). This separation of processes allows us to rewrite the 3-D time-dependent Fokker–Planck equation for PSD evolution as a superposition of radial diffusion and local processes in terms of L^*, the pitch angle α, and relativistic momentum p (Shprits et al., 2008b; Subbotin et al., 2011):

$$
\frac{\partial f}{\partial t} = L^{*2} \left.\frac{\partial}{\partial L^*}\right|_{\mu, J} \left(D_{L^*L^*} L^{*-2} \left.\frac{\partial f}{\partial L^*}\right|_{\mu, J} \right)
$$

$$
+ \frac{1}{p^2} \left.\frac{\partial}{\partial p}\right|_{\alpha, L} \left(p^2 \langle D_{pp}(\alpha, p) \rangle \left.\frac{\partial f}{\partial p}\right|_{\alpha, L} + \langle D_{p\alpha}(\alpha, p) \rangle \left.\frac{\partial f}{\partial \alpha}\right|_{p, L} \right)
$$

$$(10.6)$$

$$
+ \frac{1}{T(\alpha)\sin(2\alpha)} \left.\frac{\partial}{\partial \alpha}\right|_{p, L} T(\alpha)\sin(2\alpha) \left(\langle D_{\alpha\alpha}(\alpha, p) \rangle \left.\frac{\partial f}{\partial \alpha}\right|_{p, L} \right.
$$

$$
\left. + \langle D_{\alpha p}(\alpha, p) \rangle \left.\frac{\partial f}{\partial p}\right|_{\alpha, L} \right)
$$

$$
- \frac{f}{\tau} + S,
$$

where $\langle D_{pp} \rangle$, $\langle D_{p\alpha} \rangle = \langle D_{\alpha p} \rangle$, and $\langle D_{\alpha\alpha} \rangle$ are bounce and drift averaged scattering rates (or momentum, mixed, and pitch-angle diffusion coefficients) due to resonant wave-particle interactions with ULF and VLF electromagnetic waves

inside the magnetosphere. In Eq. (10.6), α is assumed to be the equatorial pitch angle, which is generally denoted with the subscript 'eq' as in the following equations; $T(\alpha)$ is a function related to the pitch-angle dependence of the bounce frequency; S represents the convective source of particles; and τ is a characteristic lifetime parameter accounting for the loss to the atmosphere which is assumed to be infinite outside the loss cone (cone of pitch angles at which particles' mirror points lie below the top of ionosphere). The estimation of the averages over the bounce orbit (Eqs. (10.17) to (10.19)) is discussed in the following paragraphs.

The three-dimensional Versatile Radiation Belt code (VERB-3D code) (Shprits et al., 2009) models the violation of adiabatic invariants by solving the modified 3-D Fokker–Planck diffusion equation (Eq. (10.6)) that incorporates energy diffusion, pitch-angle scattering, mixed diffusion, and radial diffusion into the drift and bounce-averaged particle PSD.

The VERB-3D code has been validated against 100 days of phase space density reanalysis and 100 days of flux data (Subbotin et al., 2011), a year (Drozdov et al., 2015; Wang and Shprits, 2019), individual storms (NSF Geospace Environmental Modeling (GEM) workshop events; Kim et al., 2012; Wang et al., 2020), and super storms (Shprits et al., 2011). Results from the GEM challenge have been made publicly available (Kim et al., 2012; see example in Fig. 10.3). The VERB-3D can resolve small pitch angles near the edge of the loss cone, can model fluxes at LEO (e.g. Shprits et al., 2011), and thus can be used to link equatorial (such as THEMIS, Van Allen Probes) and low latitude, non-equatorial, particle measurements (e.g. POES, SAMPEX, Lomonosov). The VERB-3D code uses a non-uniform pitch-angle grid, which also allows for accurate

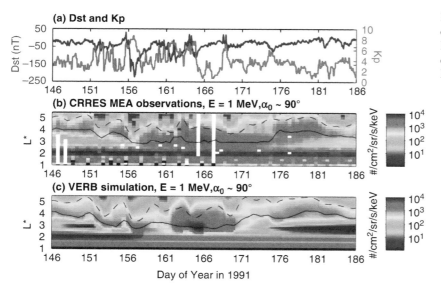

Figure 10.3 The VERB-3D model is able to reproduce key features of Combined Release and Radiation Effects Satellite (CRRES) radiation belt observations during storms. Modified figure, from Kim et al. (2012).

calculation of precipitating fluxes. The code has also been recently used to model 87 years of electron dynamics (Saikin et al., 2021). Simulations with the VERB code require calculation of the diffusion coefficients. Radial diffusion is usually parameterised as a function of the geomagnetic index Kp and local diffusion coefficients need to be calculated from wave models by means of quasilinear diffusion equations.

According to quasilinear theory (e.g. Kennel and Engelmann, 1966; Lyons et al., 1972; Albert, 2005; Glauert and Horne, 2005) the diffusion coefficients can be given by:

$$D_{\alpha\alpha} = \sum_{n=n_l}^{n_h} \int_{X_{\min}}^{X_{\max}} X dX D_{\alpha\alpha}^{nX} \tag{10.7}$$

$$D_{\alpha p} = D_{p\alpha} = \sum_{n=n_l}^{n_h} \int_{X_{\min}}^{X_{\max}} X dX D_{\alpha p}^{nX} \tag{10.8}$$

$$D_{pp} = \sum_{n=n_l}^{n_h} \int_{X_{\min}}^{X_{\max}} X dX D_{pp}^{nX}, \tag{10.9}$$

where the summation is over harmonics (n) of the cyclotron frequency, $X = \tan(\psi)$, where ψ is the wave normal angle, α is the local pitch angle, and

$$D_{\alpha\alpha}^{nX} = \sum_i \frac{q_\sigma^2 \omega_i^2}{4\pi(1+X^2)N(\omega_i)} \left[\frac{n\Omega_\sigma/(\gamma\omega) - \sin^2\alpha}{\cos\alpha} \right]^2$$

$$\cdot B^2(\omega_i) g(X) \frac{|\Phi_{n,k}|^2}{\left| v_\parallel - \frac{\partial\omega}{\partial k_\parallel} \right|} \Bigg|_{k_{\parallel i}}, \tag{10.10}$$

$$D_{\alpha p}^{nX} = D_{\alpha\alpha}^{nX} \left[\frac{\sin\alpha\cos\alpha}{n\Omega_\sigma/(\gamma\omega) - \sin^2\alpha} \right]_{k_{\parallel i}}, \tag{10.11}$$

$$D_{pp}^{nX} = D_{\alpha\alpha}^{nX} \left[\frac{\sin\alpha\cos\alpha}{n\Omega_\sigma/(\gamma\omega) - \sin^2\alpha} \right]_{k_{\parallel i}}^2, \tag{10.12}$$

where q_σ is the charge of the particle species σ, Ω_σ is the particle gyrofrequency including the sign of its charge, and γ is the relativistic correction factor. $B(\omega)$ is the intensity of the magnetic field squared per unit frequency and $N(\omega)$ is a normalisation factor ensuring that the wave energy per unit frequency is given by $B^2(\omega)$. The dependence of wave magnetic field energy with wave normal angle is given by $g(X)$:

$$g(X) = \begin{cases} \exp\left(-\left(\frac{X-X_m}{X_w}\right)^2\right) & X_{\min} \leq X \leq X_{\max}, \\ 0 & \text{otherwise}, \end{cases} \tag{10.13}$$

where X_w is the angular width and X_m is the peak. The term $|\Phi_{nk}|^2$ depends on the wave refractive index $\mu = c|k|/\omega$ for the particular wave mode:

$$|\Phi_{nk}|^2 = \left[\left(\frac{\mu^2 - L}{\mu^2 - S} J_{n+1} + \frac{\mu^2 - R}{\mu^2 - S} J_{n-1} \right) \right.$$

$$\left. \cdot \left(\frac{\mu^2 \sin^2\psi - P}{2\mu^2} + \cot\alpha \sin\psi \cos\psi J_n \right) \right]^2$$

$$\cdot \left[\left(\frac{R-L}{2(\mu^2-S)} \right)^2 \left(\frac{P-\mu^2\sin^2\psi}{\mu^2} \right)^2 + \left(\frac{P\cos\psi}{\mu^2} \right)^2 \right]^{-1}, \tag{10.14}$$

where J_n is the n-th order Bessel function of first kind with argument $k_\perp p_\perp/(m_\sigma \Omega_\sigma)$

It should be noted that $D_{\alpha\alpha}$ and D_{pp} are always positive, while $D_{\alpha p}$ can become negative. The integrands in Eqs. (10.7)–(10.9) are evaluated at the resonant parallel wave number $k_{\parallel,i}$ and the resonant frequency ω_i, which satisfies the resonance condition, given by:

$$\omega - k_\parallel v_\parallel = n\Omega_\sigma/\gamma, \tag{10.15}$$

and the dispersion relation $D(k, \omega, X) = 0$, which for a cold magnetised plasma is given by

$$D(\omega, k, X) = (SX^2 + P)\mu^4 - (RLX^2 + PS(2 + X^2))\mu^2 + PRL(1 + X^2) = 0, \tag{10.16}$$

where R, L, S, and P are the usual Stix parameters (Stix, 1962). For a given wave of normal angle, cyclotron resonance n, and energy, there may be several resonant frequencies for a particular wave mode.

As a particle moves along the field line between the mirror points, it will experience different magnetic field intensities, plasma density, and ion composition, and it will change its pitch angle. Since, the local diffusion coefficients will be changing along the bounce orbit, the local diffusion coefficients should be bounce-averaged. The bounce-averaged diffusion coefficients, $\langle D_{\alpha_{eq}\alpha_{eq}} \rangle$, $\langle D_{\alpha_{eq}p} \rangle$, and $\langle D_{pp} \rangle$ are given by

$$\langle D_{\alpha_{eq}\alpha_{eq}} \rangle = \frac{1}{\tau_B} \int_0^{\tau_B} D_{\alpha\alpha} \left(\frac{\partial\alpha_{eq}}{\partial\alpha} \right)^2 dt, \tag{10.17}$$

$$\langle D_{\alpha_{eq}p} \rangle = \frac{1}{\tau_B} \int_0^{\tau_B} D_{\alpha p} \left(\frac{\partial\alpha_{eq}}{\partial\alpha} \right) dt, \tag{10.18}$$

$$\langle D_{pp} \rangle = \frac{1}{\tau_B} \int_0^{\tau_B} D_{pp} dt, \tag{10.19}$$

where τ_B is the period of bouncing between the mirror points. Changing the integration from time to magnetic latitude and assuming a dipole magnetic field gives

$$\langle D_{\alpha_{eq}\alpha_{eq}} \rangle = \frac{1}{T} \int_0^{\lambda_m} D_{\alpha\alpha} \frac{\cos\alpha}{\cos^2\alpha_{eq}} \cos^7\lambda d\lambda, \qquad (10.20)$$

$$\langle D_{\alpha_{eq}p} \rangle = \frac{1}{T} \int_0^{\lambda_m} D_{\alpha p} \frac{(1+3\sin\lambda)^{1/4}}{\cos\alpha} \cos^4\lambda d\lambda, \qquad (10.21)$$

$$\langle D_{pp} \rangle = \frac{1}{T} \int_0^{\lambda_m} D_{pp} \frac{(1+3\sin\lambda)^{1/2}}{\cos\alpha} \cos\lambda d\lambda, \qquad (10.22)$$

where λ_m is the latitude at which the particle mirrors and $T(\alpha_{eq})$ is a function related to the variation of τ_B with α_{eq}. In a dipole magnetic field, $T(\alpha_{eq})$ can be approximated following Hamlin et al. (1961) by

$$T(\alpha_{eq}) = 1.30 - 0.56\sin\alpha_{eq}. \qquad (10.23)$$

The mirror latitude is found by solving the polynomial

$$C_l^6 + 3C_l\sin^4\alpha_{eq} - 4\sin^4\alpha_{eq} = 0 \qquad (10.24)$$

for $C_l = \cos^2\lambda_m$. In practice, one assumes distributions of waves $B(Kp, X, MLT, L, \lambda)$ where MLT is the magnetic local time, $X(\lambda, MLT)$, and distributions in frequency of chorus and hiss plasma waves that are inferred from the statistics of satellite measurements. Those distributions are statistically derived from wave observations.

10.2.2 Electron Ring Current

Lower energy electrons (kinetic energy < 100 keV), often referred to as the magnetospheric ring current, are capable of producing surface charging and can cause satellite anomalies. The ring current, unlike radiation belts, is asymmetric around the Earth and closes in the ionosphere through so-called region 1 and region 2. The dynamics of the ring current electron population are modelled using the modified Fokker–Planck equation with additional advective terms, that are often referred to as 'convective' terms (Shprits et al., 2015):

$$\frac{\partial f}{\partial t} = -v_\varphi \frac{\partial f}{\partial \varphi} - v_{R_0} \frac{\partial f}{\partial R_0} + \frac{1}{G_{(V,K,L^*)}} \frac{\partial}{\partial L^*} G_{(V,K,L^*)} D_{L^*L^*} \frac{\partial f}{\partial L^*}$$
$$+ \frac{1}{G_{(V,K,L^*)}} \frac{\partial}{\partial V} G_{(V,K,L^*)} \left(D_{VV} \frac{\partial f}{\partial V} + D_{VK} \frac{\partial f}{\partial K} \right)$$
$$+ \frac{1}{G_{(V,K,L^*)}} \frac{\partial}{\partial K} G_{(V,K,L^*)} \left(D_{KV} \frac{\partial f}{\partial V} + D_{KK} \frac{\partial f}{\partial K} \right) - \frac{f}{\tau}, \qquad (10.25)$$

where L^*, K, and V are adiabatic invariants, $f(\varphi, R_0, V, K)$ is the phase space density, t represents time, φ is MLT, R_0 is the radial distance from the centre of the Earth at the geomagnetic equator, τ is the electron lifetime related to scattering into the loss cone and magnetopause shadowing, v_φ and v_{R_0} are bounce-averaged drift velocities, and $D_{L^*L^*}$, D_{VV} D_{VK}, D_{KV} and D_{KK} are bounce-averaged diffusion

coefficients. Here $G_{(V,K,L^*)}$ is the Jacobian of the coordinate transformation from (μ, J, Φ) to (V, K, L^*) coordinates defined as:

$$G_{(V,K,L^*)} = -2\pi B_E R_E^2 \sqrt{8m_0 V}/(K+0.5)^3/L^{*2} \qquad (10.26)$$

The first and second terms on the right-hand side of Eq. (10.25) describe advection dynamics, driven by the $E \times B$, gradient, and curvature drifts; the third term describes radial diffusion; terms 4 and 5 represent local diffusion; and sources and losses are estimated by the sixth term.

Bounce-average drift velocities (Eq. (10.27)) (Volland, 1973; Maynard and Chen, 1975; Stern, 1975).

$$\langle v_0 \rangle = \frac{E_0 \times B_E}{B_E^2} + \frac{1}{q\tau B_E^2} \nabla J \times B_E \qquad (10.27)$$

where E_0 and B_E are the electric and magnetic fields at the equator, respectively, and are generally calculated using an empirical magnetic field model (e.g. T89 (Tsyganenko, 1989)) and the Volland–Stern Kp-dependent empirical electric field model. In order to prepare measurements for the ring current simulations, invariants L^*, K, and V have to be estimated along the satellite orbit for each measurement.

10.3 Data Assimilation

10.3.1 Kalman Filter (KF)

The Kalman filter (KF) is a celebrated data assimilation method that allows for optimal combination of model results and sparse data from various sources contaminated by noise (Kalman, 1960). Using a physics-based model and available satellite observations, we can estimate the optimal state of the electron PSD in the radiation belts (i.e. f at time k, denoted in Eqs. (10.28) to (10.30) as $\mathbf{z}_k^a$) and the uncertainty of the state estimate (described by the error covariance matrix $\mathbf{P}_k^a$) associated with errors in the model and the data. The Kalman filter allows us to determine estimates of the state and covariance analytically by defining an initial state vector $\mathbf{z}_0^a$ (estimated as the steady state solution of the radial diffusion equation) and initial covariance $\mathbf{P}_0^a$. Iteration over two elementary steps is then performed:

(1)   The *forecast step*: the time evolution of the state vector $\mathbf{z}$ is assumed to be governed by the numerically discretised linear partial differential operator $\mathbf{M}_k$ (Eq. 10.6):

$$\mathbf{z}_k^f = \mathbf{M}_k \mathbf{z}_{k-1}^a, \qquad (10.28)$$

where $\mathbf{z}_k^f$ is the PSD state vector in the 3-D phase space volume advanced by the model $\mathbf{M}_k$ in time, and superscripts 'f' indicate forecasted state. Deviations of the forecast state estimate from the true state of the system are defined by the forecast error covariance matrix $\mathbf{P}_k^f$ which can be calculated from a previous analysis step as

$$\mathbf{P}_k^f = \mathbf{M}_k \mathbf{P}_{k-1}^a \mathbf{M}_k^T + \mathbf{Q}_k. \qquad (10.29)$$

The model errors are commonly assumed to be a sequence of uncorrelated white noise with zero mean. In our field of application, the corresponding model error covariance $\mathbf{Q}_k$ is usually a diagonal matrix defined as a fraction of the squared forecast state, $\mathbf{Q}_k = \eta(\mathbf{z}_k^f)^2$, and is therefore time dependent. The factor η modulates the statistical impact of the model on the estimated analysis state. In radiation belt studies, the value of η ranges between 0.2 and 0.7 (Naehr and Toffoletto, 2005; Kondrashov et al., 2007, 2011; Shprits et al., 2007; Daae et al., 2011; Schiller et al., 2012; Shprits et al., 2012) depending on the focus of the study and on the particular parameters of the model (e.g. diffusion coefficients, boundary and initial conditions). Such a scaling of model errors is important as the state can change very fast by up to several orders of magnitude and errors need to be adjusted accordingly.

(2) The *analysis step or update step*: the observations of the system $\mathbf{y}_k^{\text{obs}}$ are assumed to have uncertainties described by uncorrelated white noise with zero mean and observation error covariance $\mathbf{R}_k$. This error covariance is chosen to be a diagonal matrix and errors are proportional to the state. If at time step k, a new data point is available, the covariance is defined as $\mathbf{R}_k = \beta(\mathbf{y}_k^{\text{obs}})^2$. In this case, the coefficient β controls the statistical influence of the observation on the estimated analysis state. The value of β commonly ranges between 0.3 and 0.5. Combining the forecast error covariance matrix $\mathbf{P}_k^f$ with the uncertainty of the data $\mathbf{R}_k$, the KF finds optimal weights (defined in the Kalman gain $\mathbf{K}_k$) that minimise the error covariance $\mathbf{P}_k^a$ of the optimal state estimate $\mathbf{z}_k^a$ at time k,

$$\begin{aligned} \mathbf{K}_k &= \mathbf{P}_k^f \mathbf{H}_k^T (\mathbf{R}_k + \mathbf{H}_k \mathbf{P}_k^f \mathbf{H}_k^T)^{-1}, \\ \mathbf{z}_k^a &= \mathbf{z}_k^f + \mathbf{K}_k(\mathbf{y}_k^{\text{obs}} - \mathbf{H}_k \mathbf{z}_k^f), \\ \mathbf{P}_k^a &= (\mathbf{I} - \mathbf{K}_k \mathbf{H}_k) \mathbf{P}_k^f, \end{aligned} \qquad (10.30)$$

Where the observation operator $\mathbf{H}_k$ maps the model space onto the observation space at times when new observations become available, and accounts for differences in dimensionality between data and model, due to the sparsity of the observations.

10.3.2 Ensemble Kalman Filter (EnKF)

The ensemble Kalman filter (EnKF) (Evensen, 2003) can be seen as a Monte Carlo approximation of the KF. In this case, the optimal state of the system $\mathbf{z}_k^a$ at time k is approximated by the mean $\overline{Z}_k^a$ of an ensemble of samples $\{\mathbf{z}_{i,k}^a\}$, where $i = 1, .., N_{\text{ens}}$ and $N_{\text{ens}} =$ number of ensemble members:

$$\mathbf{z}_k^a \approx \overline{\mathbf{z}}_k^a = \frac{1}{N_{\text{ens}}} \sum_{i=1}^{N_{\text{ens}}} \mathbf{z}_{i,k}^a \qquad (10.31)$$

The ensemble error covariance ($\mathbf{P}_e$) gives the spread of the ensemble distribution and can be linked to the error covariance matrices ($\mathbf{P}_k^f$ and $\mathbf{P}_k^a$) of the optimal state estimate by the following empirical approximations:

$$\begin{aligned} \mathbf{P}_k^f \approx \mathbf{P}_e^f &= \frac{1}{N_{\text{ens}} - 1} \sum_{i=1}^{N_{\text{ens}}} \left(\mathbf{z}_{i,k}^f - \overline{\mathbf{z}}_k^f\right)\left(\mathbf{z}_{i,k}^f - \overline{\mathbf{z}}_k^f\right)^T \\ \mathbf{P}_k^a \approx \mathbf{P}_e^a &= \frac{1}{N_{\text{ens}} - 1} \sum_{i=1}^{N_{\text{ens}}} \left(\mathbf{z}_{i,k}^a - \overline{\mathbf{z}}_k^a\right)\left(\mathbf{z}_{i,k}^a - \overline{\mathbf{z}}_k^a\right)^T \end{aligned} \qquad (10.32)$$

Perturbations ($\epsilon_{i,k}$) are drawn from a Gaussian distribution with mean equal to the observed value and covariance $\mathbf{R}_k$ are used to generate an ensemble of observations, when a new measurement $\mathbf{y}_k^{\text{obs}}$ is available:

$$\mathbf{y}_{i,k}^{\text{obs}} = \mathbf{y}_k^{\text{obs}} + \epsilon_{i,k} \qquad (10.33)$$

where $i = 1, \ldots, N_{\text{ens}}$. Every state in the ensemble is propagated in the update step, as follows:

$$\mathbf{z}_{i,k}^a = \mathbf{z}_{i,k}^f + \mathbf{K}_k \left(\mathbf{y}_{i,k}^{\text{obs}} - \mathbf{H}_k \mathbf{z}_{i,k}^f\right) \qquad (10.34)$$

where the Kalman gain ($\mathbf{K}_k$) with the optimal weighting factors is calculated as in Eq. (10.30). A number of studies have used this approach (e.g. Koller et al., 2005; Bourdarie and Maget, 2012; Castillo et al., 2021)

10.4 What Can Be Done with Data Assimilation in the Van Allen Radiation Belts?

10.4.1 Global Analysis of the State of the Radiation Belts Using Data Assimilation

In the past, most of the radiation belt and ring current research concentrated on the analysis of data from individual spacecraft. Such an analysis does not allow us to infer the global evolution of the radiation environment. While measurements in space are sparse, satellite data of pitch-angle distributions, energy distribution, and fluxes at different energies can add up to terabytes. In situ satellite observations are often restricted to a limited range of radial distances and energies, and also have different observational errors. Additionally, observations at different radial distances are taken at different times. These unavoidable data acquisition limitations complicate the analysis of the radial profiles of the PSD, which is essential for understanding the relative contribution of the local acceleration and radial transport on the evolution of the radiation belt electrons. Data assimilation allows us to fill in the temporal and spatial gaps left by sparse in situ measurements and combine measurements and the model according to the underlying error structure of different observations and the model.

Equation (10.6) can be abstracted to:

$$\frac{\partial f}{\partial t} = \mathbf{M}(D)f, \tag{10.35}$$

where M represents the Fokker–Planck evolution operator (linearised Eq. (10.6)) for the PSD f, and D represents various model parameters (e.g. diffusion coefficients). This equation requires knowledge of the initial condition of f and boundary conditions to make predictions into the future. Data assimilation methods use partial observations of f within a time period (observation period) to estimate the initial condition using an inverse problem. Then the reconstructed initial condition is used to issue predictions for PSD beyond the observation period. This process can be repeated in a sequential manner in time.

A variety of initial studies using simple one-dimensional radial models show the efficiency of Kalman filtering applied to the field of radiation belts. Friedel et al. (2003) performed direct insertion of geosynchronous and GPS data into the Salammbô code (Beutier and Boscher, 1995). Naehr and Toffoletto (2005) used a log-transform of the PSD and an extended Kalman filter (EKF) for a performance study using synthetic data. Shprits et al. (2007) used a KF with a simple one-dimensional radial diffusion model to reconstruct radiation belt PSD for a period of 50 days. Figure 10.4 shows the sparse Combined Release and Radiation Effects Satellite (CRRES) spacecraft data (top), and the result of the reanalysis with a one-dimensional radial diffusion model (bottom). Using an EnKF, Koller et al. (2007) analysed a storm in 2002 and showed that data assimilation can be used not only to reconstruct the state of the system, but also to infer missing physical processes. Kondrashov et al. (2007) and Schiller et al. (2012) augmented the KF to perform parameter estimation studies.

Additionally, Ni et al. (2009b) tested four different empirical external magnetic field models and found that combined

reanalyses have rather negligible response to the choice of magnetic field model. The authors also showed that the errors of PSD obtained by assimilating multi-satellite measurements at different locations can be reduced in comparison with the errors of individual satellite assimilation. The sensitivity of the reanalysis of electron radiation belt PSD to various outer boundary conditions and loss models was studied by Daae et al. (2011), who showed that the KF has a remarkable performance when enough satellite data is available at all considered locations. Ni et al. (2009a) assimilated Akebono electron flux measurements into the VERB-3D code and validated their results against CRRES data. Using the same methodology, Shprits et al. (2012) and Ni et al. (2013) performed a long-term multi-spacecraft reanalysis and analysed the location of the peak of the PSD, finding good correlation with the plasmapause location. The authors were also able to find a link between PSD dropouts and solar wind dynamic pressure increases.

Two methods have been recently tested for specific events to assimilate data in 3-D. First, Bourdarie and Maget (2012) applied EnKF (Evensen, 2003) to the Salammbô code to reconstruct the 3-D state of the radiation environment. While EnKF may be convenient in its implementation, as it uses the physics-based codes as a black box, it may present a significant computational challenge when the number of ensemble members is high. Another known complication associated with the EnKF is that underestimation of the forecast covariance can result in a 'filter divergence'. The divergence is a state of the filter in which observations cannot influence the solution. Another approach was implemented by Shprits et al. (2013) who applied 3-D data assimilation using CRRES data and the VERB-3D model using a suboptimal operator-splitting method. They applied the standard formulation of a Kalman filter, but only for the 1-D diffusion operators in radial distance, energy, and pitch angle sequentially. For each of the 1-D diffusion operators,

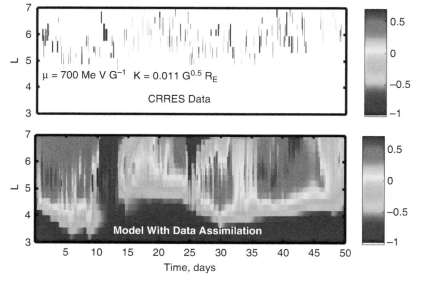

Figure 10.4 Example of assimilation of sparse satellite data with a simple 1-D radial diffusion model and Kalman Filter. (Top) Hourly averaged PSD inferred from CRRES MEA observations for $K = 0.11\,G^{0.5}\,RE$ and $\mu = 700$ MeV/G for a 50-day period starting on 18 August 1990. (Bottom) Results of the data assimilation with a radial diffusion model. Modified figure, from Shprits et al. (2007).

they used only data points that lie along the direction of the one-dimensional diffusion.

10.4.2 Real-Time Predictions

Data assimilation allows us to blend data from very different sources with state-of-the-art models to globally reconstruct the evolution of energetic and relativistic electrons in the near-Earth environment. Sequential Kalman filtering has been recently implemented for real-time data assimilation and data assimilative predictions. This stable framework assimilates data from different sources and provides real-time data assimilative nowcast and predictions. The data assimilative predictions using the split-operator KF (Shprits et al., 2013) are now broadcasted (Fig. 10.5) at the website of the German Research Center for Geosciences (GFZ-RB-forecast, 2021) and have been made available for satellite operators. Further, the resulting reanalysis, models, and data assimilation framework is made available to the community through the GFZ website (GFZ-data, 2021).

We are currently using a very stable version of the VERB-3D code that does not include mixed terms. That significantly simplifies the operations of the code and speeds up the code execution. Neglect of the mixed terms also allows us to use a relatively coarse grid of $25 \times 25 \times 25$ in pitch angle, energy, and L^*. Numerical experiments presented by Subbotin and Shprits (2009), and Aseev et al. (2016) showed that errors associated with such a grid

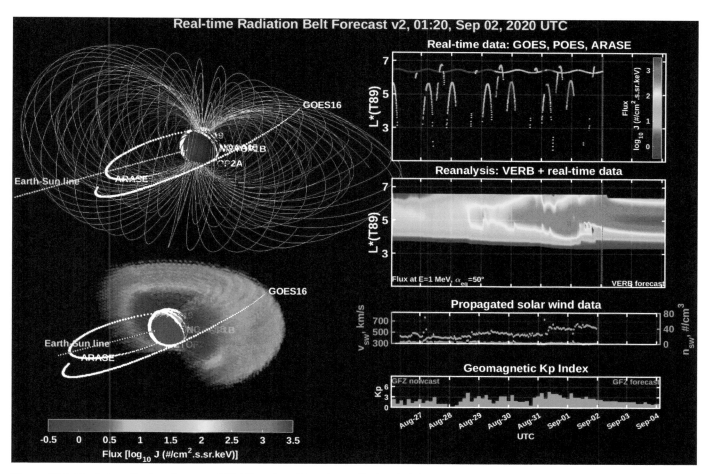

Figure 10.5 Real-time display shown at GFZ-Potsdam. Two-day radiation belt forecast of 1 MeV electrons using the data-assimilative VERB code, real-time ARASE, ACE, POES and GOES data. (Top left) Real-time satellite trajectories and geometry of the magnetic field lines used for calculation of the transformation into the coordinate system of adiabatic invariants. (Bottom left) 3-D visualisation of the radiation belts. (Right panels) Top panel depicts real-time flux measurements. Second panel shows the slice in energy and equatorial pitch angles interpolated at 1 MeV kinetic energy and 50° equatorial pitch angle. Third panel shows the resulting reanalysis of the radiation belts and a prediction with a horizon of two days. The dashed red line demarcates the historical reanalysis from the forecast for 2 days ahead of real-time. The bottom panels show propagated solar wind parameters and the geomagnetic index Kp index of geomagnetic activity, respectively. Source: GFZ-RB-forecast (2021).

resolution are small in comparison with the uncertainties of the model, while stability and speed of the real-time algorithm are maintained. Higher grid resolution and inclusion of mixed terms may be needed when data assimilation is used to infer physical processes from the innovation vector (as recently done in Cervantes et al., 2020a). The prediction framework first uploads the available data in real time. Then the measurements of fluxes at the satellite orbit are remapped to the equatorial pitch angles. After that, we calculate the PSD as a function of the adiabatic invariants for assimilation into the VERB code that provides the model matrix for the data assimilation. The outer boundary condition in L^* is assumed to be a zero derivative in the radial gradient at $L^* = 6.6$. If the L^* value calculated with T89 model magnetopause is inside $L^* = 6.6$, we assume that the outer boundary condition is a zero value. That condition allows the code to adjust to real-time data at all L-shells and also allows the algorithm to promptly respond to the magnetopause loss and to outward radiation diffusion caused by it (Shprits et al., 2006) during the main phase of storms. The Kp forecast is provided by GFZ-Potsdam and is utilising machine learning (ML) (Shprits et al., 2019). This predictive system has previously worked with NASA's Van Allen Probes measurements and is now operating with GOES, Arase, and POES data that is provided in real-time.

10.4.3 Radiation Belt Reanalysis

A statistical study of a data assimilation reanalysis of the relativistic electron PSD for a 200-day period was presented by Shprits et al. (2012). The authors used the innovation vector to locate the peaks in PSD and sources of local acceleration. Additionally, correlation of PSD dropouts with sudden jumps of solar wind dynamic pressure is shown in their study. In a recent study, Cervantes et al. (2020a) successfully reconstructed the long-term evolution of the radiation belt fluxes, which now provides a continuous evolution of the 3-D cube of the radiation belt fluxes (Fig. 10.6) for four consecutive years. In this study, the VERB-3D code including the mixed diffusion was used. Such reconstruction, or reanalysis as it is referred to in data assimilation, will help future radiation belt research and may help develop more accurate empirical models. Reanalysis allows the operators of spacecraft to fly a virtual spacecraft through the 4-D (three spatial dimensions and one temporal dimension) cube of the radiation belt to infer the radiation environment at the spacecraft orbit and perform anomaly resolution analysis.

Uncertainty quantification is a key element in the development of data assimilation techniques as proper implementation of the Kalman filter algorithm is linked to prior knowledge of the statistical characteristics of uncertainties in the model and the satellite data. Inaccurate parameterisations of physical processes may arise from

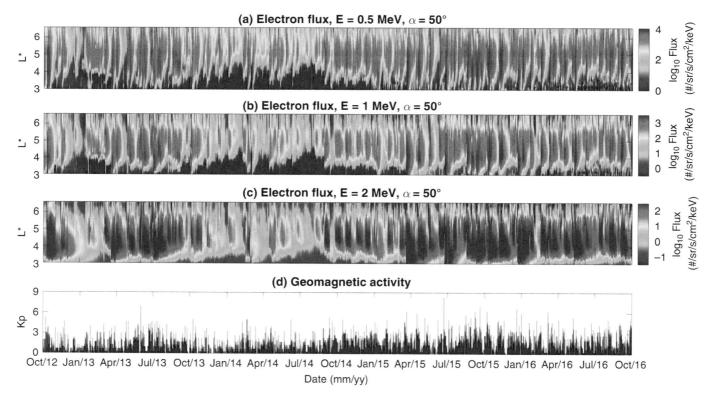

Figure 10.6 The assimilative results of the combined reanalysis of electron PSD in this figure account for 3-D diffusion, mixed pitch angle energy diffusion, scattering by EMIC waves, and magnetopause shadowing. (Top) Evolution of electron flux as a function of L^* and time from 1 October 2012 to 1 October 2016 for pitch angle $\alpha = 50°$, and: (a) $E = 0.5°$ MeV, (b) $E = 1$ MeV, and (c) $E = 2$ MeV; and (d) geomagnetic activity Kp index, inferred by GFZ from ground observations of the magnetic field fluctuations. Modified figure from Cervantes et al. (2020b).

assumptions of the quasilinear theory that simplify the diffusion equation. Since the environmental conditions of the plasma in the radiation belts cannot be reproduced in experimental laboratory set-ups on Earth, the true state of the system remains unknown to us, and therefore quantification of model error statistics and its properties is a rather difficult task in practice. On the other hand, uncertainty could be further increased due to the presence of observation errors. Measurement errors are introduced by the limitations of the satellite instruments measuring electron fluxes in space, by the interaction of the instruments with electronics on board the satellite itself, by background noise in the near-Earth environment, and by strong geomagnetic conditions that can cause bias measurements and degeneration of the detectors over the years (Galand and Evans, 2000; Asikainen and Mursula, 2011, 2013). Such uncertainties are difficult to estimate due to unknown conditions in space and limited access to the instruments after the launch of the satellite (McFadden et al., 2007). Significant errors also arise from mapping from the observational space of fluxes as a function of pitch angle and energy to modelling space of phase space density as a function of adiabatic invariants (e.g. Green and Kivelson, 2004).

Podladchikova et al. (2014b) proposed an approach to the identify the model errors of a 1-D radial diffusion model by estimating the unknown bias and the model error covariance matrix from the sparse CRRES observations over a period of 441 days. Both the identified bias and the covariance matrix of model errors appear to depend on the L-shell. Sensitivity of the PSD reanalysis to model error statistics showed that neglecting the bias can cause significant errors in the state estimate when satellite data is not available. An identification technique to estimate the observation error statistics was presented by Podladchikova et al. (2014a). The authors estimated the residuals describing the mismatch between the satellite observations and independent pseudo measurement vectors. The expectation and covariance matrix of the residuals deliver coefficients of proportionality characterising the dependence of the observation error on the measurement itself for every L-shell. Further work on uncertainty estimation for filtering applications in the radiation belts is necessary and will be the focus of future studies.

Reanalysis in the inner magnetosphere will soon become a standard for data analysis in space physics and will change the way data are analysed. The impact of the data assimilation tools will be comparable to the impact that NCEP/NCAR reanalysis (Kalnay, 2003) had on meteorological climate studies. In the future, the results of reanalysis will be used to develop next generation specification models. Reanalysis will be used by engineers to obtain the average fluencies during a mission and understand potential limits of the variability.

10.4.4 Search for Missing Physics by Means of Data Assimilation Tools

Applications of data assimilation may also help to reveal the underlying physical processes in the near-Earth space that will help us to better understand the space environment. Figure 10.7 illustrates how missing physical processes can be determined from the innovation vector. The innovation vector $\mathbf{D}_k$ is generally defined as $\mathbf{D}_k = \mathbf{K}_k(\mathbf{y}_k^{obs} - \mathbf{H}_k \mathbf{z}_k^f)$ and gives a notion of the difference between the observations and the forecast state. A number of studies have defined the innovation vector as $\mathbf{d}_k = \mathbf{y}_k^{obs} - \mathbf{H}_k \mathbf{z}_k^f$. The figure shows the magnitude of the innovation vector of radiation belt PSD for 50 days of simulations, using a simple one-dimensional radial diffusion model of radiation belts (Shprits et al., 2007). The innovation vector peaks in the heart of the outer radiation belt, suggesting a missing acceleration process that the one-dimensional radial diffusion model is not able to reproduce. A number of radiation belt studies used reanalysis or innovation to quantify and understand the radiation belt dynamics (Kondrashov et al., 2007; Daae et al., 2011; Schiller et al., 2012; Shprits et al., 2012; Ni et al., 2013; Cervantes et al., 2020a,b).

In a recent study by Cervantes et al. (2020b), various acceleration and loss mechanisms in the radiation belts were quantified by means of analysing the innovation in the data assimilation framework. Cervantes et al. (2020a) developed a framework for data assimilation of the 3-D code including mixed diffusion. Previous efforts simply ignored numerically difficult mixed terms, which complicated the physical interpretation of the results. Inclusion of the mixed terms that facilitate the loss of particles allowed the authors to analyse the innovation and to make conclusions on the underlying physical process, that may be missing from the simulations.

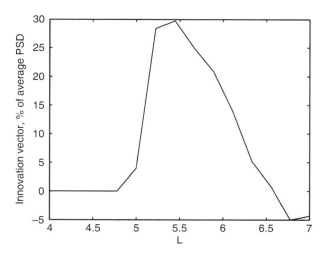

Figure 10.7 A 50-day averaged innovation vector, after Shprits et al. (2007).

10.4.5 Data Assimilation in Electron Ring Current

While there are a number of recently developed ring current physics-based models (e.g. RAM (Jordanova et al., 1997; Jordanova and Miyoshi, 2005); RBE model (Fok et al., 2008, 2011), IMPTAM (Ganushkina et al., 2013, 2014), VERB-4D code (Shprits et al., 2012; Aseev et al., 2016)) and a number of satellites providing measurements of ring current fluxes, there have only been a few attempts to assimilate the measurements in the physics-based model (Nakano et al., 2008; Godinez et al., 2016; Aseev and Shprits, 2019) and there is currently no model that is using data assimilation for real-time forecasting or nowcasting.

In general, for the ring current (in the most general case), we need to specify the four-dimensional distribution function, as the ring current has an asymmetry in the local time (fourth variable). Another complication of modelling the ring current is that the evolution of phase space density is described by the advection-diffusion equation, which is more difficult to solve numerically than just a diffusion equation. The ring current is also much more dynamic than the radiation belts and it is difficult to distinguish between the spatial and temporal evolution.

Numerical solutions of Eq. (10.25) in the full four-dimensional space are computationally expensive. In a recent study by Aseev and Shprits (2019), preliminary synthetic data simulations and sensitivity tests with Kalman filtering of the log of electron fluxes were performed by solving for convection (advective term in Eq. 10.25) and assimilating data in MLT and radial distance for fixed first and second adiabatic invariants. Diffusion terms were substituted by exponential decay rates (parameterised lifetimes τ) and radial diffusion (third term in (10.25)) was neglected. A comparison of the model with and without data assimilation is shown in Fig. 10.8. These initial results indicate that data assimilation can be successfully applied for the ring current electrons and the sparse data is capable of providing enough information to change the entire state of the system.

The innovation vector may help us to identify the impact of changes related to the electric field, magnetic field, or inappropriate lifetimes parameterisation in the ring current system. Magnetic and electric fields play an important role for the ring current electron transport, but it is difficult to observe them in space, because available satellite measurements are very sparse, and the fields' strength varies dramatically in time. For these reasons, global models of electric and magnetic fields inevitably contain significant uncertainties that lead, in turn, to errors in the model forecast. Model errors and what causes them can be estimated from the innovation vector. On the basis of the sign and magnitude of the innovation vector and telltale signatures of different processes, it is possible to identify if the errors are due to inaccuracies of electric or magnetic field models, missing particle injections, or other processes.

10.5 Discussion and Future Directions of the Research

Both the KF and the EnKF have their own advantages. The EnKF does not require linearisation of the physics-based model and the physics-based model can be used as a black box. However, unlike the suboptimal EnKF, the KF provides an exact solution for optimal blending of model and observations. While the exact KF requires handling of large covariance matrices, the EnKF requires multiple ensemble runs of the code. Implementation of the EnKF may also suffer from occasional filter divergence and may require implementation of inflation and possibly localisation. However, the EnKF may be easily adopted to be used with a non-liner observation operator. Therefore, the choice of the filter should depend on the application.

In a recent work, Castillo et al. (2021) performed a detailed comparison of the split-operator KF with a stochastic split-operator EnKF and studied how these simulations should be set up to provide the most accurate reconstruction of the radiation belt fluxes in 3-D. This study has provided a detailed description of how to properly set up the code for the EnKF and demonstrated how EnKF can be used and extended for future applications for ring current and for parameter estimation (Castillo et al., 2021). It is possible that similar methods can be used for the ring current electron population. Nevertheless, there is currently no operational model that is using data assimilation for the predictions of ring current.

Generally, particle detectors aboard satellites measure particle fluxes at specified energies and pitch-angle ranges along the satellite orbits. However, since the theoretical description of plasma dynamics is stated in terms of PSD, observations need to be converted to PSD and remapped into the invariant space. Measurements from some satellites (e.g. NOAA POES), do not provide pitch-angle resolved measurements, but only single pitch angles. The development of parametrisations of the pitch-angle distributions can help to extrapolate data to large pitch angles and to assimilate data. Such measurements can also be used for the validation. Additionally, dependence on MLT has to be taken into account for ring current particles, which is not necessarily the case for radiation belts. Once all measurements have been transformed into invariant coordinates, we can convert particle fluxes (either differential or integral fluxes, depending on the instrument) to PSD. As data assimilation routinely estimates the innovation vector, the systematic difference between the model and data may be eliminated as a part of the assimilation cycle. Similarly, the bias in data may be also corrected as a part of assimilation cycle simplifying the introduction of new data into ring current and radiation belt models. Intercalibration of the processed electron data for ring current from various satellites, may be performed using satellite conjunctions and using data assimilation tools.

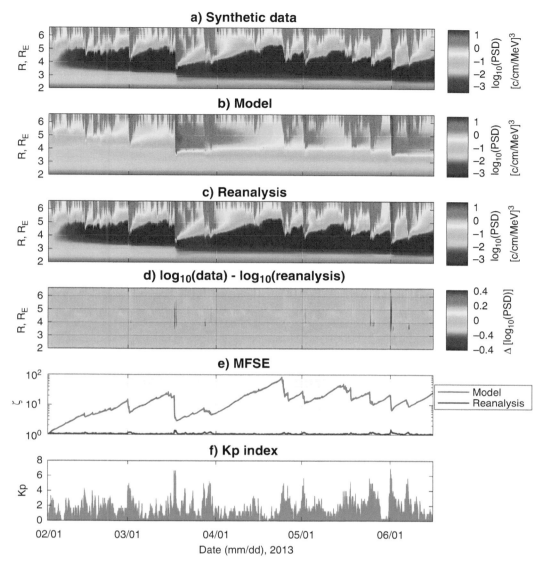

Figure 10.8 (a) MLT-averaged synthetic 'true' state obtained by running the VERB-CS code = Convection Simplified Versatile Electron Radiation Belt code for $\mu = 2.3$ MeV/G and $K = 0.3$ $G1/2$ RE electrons. (b) MLT-averaged model results when the electron lifetimes are 10 times higher than in the model used to calculate the true state. (c) MLT-averaged reanalysis results. (d) Logarithmic difference between the MLT-averaged synthetic true state and reanalysis. (e) Evolution of the MFSE. (f) The Kp index. PSD = phase space density; MFSE = mean fractional symmetric error; MLT = magnetic local time. Source: Aseev and Shprits (2019).

References

Albert, J. (2005). Evaluation of quasi-linear diffusion coefficients for whistler mode waves in a plasma with arbitrary density ratio. *Journal of Geophysical Research: Space Physics*, 110 (A3).

Aseev, N., and Shprits, Y. (2019). Reanalysis of ring current electron phase space densities using Van Allen Probe observations, convection model, and log-normal Kalman Filter. *Space Weather*, 17(4), 619–38.

Aseev, N. A., Shprits, Y. Y., Drozdov, A. Y., and Kellerman, A. C. (2016). Numerical applications of the advective-diffusive

codes for the inner magnetosphere. *Space Weather*, 14(11), 993–1010.

Asikainen, T., and Mursula, K. (2011). Recalibration of the long-term NOAA/MEPED energetic proton measurements. *Journal of Atmospheric and Solar-Terrestrial Physics*, 73(2), 335–47.

Asikainen, T., and Mursula, K. (2013). Correcting the NOAA/ MEPED energetic electron fluxes for detector efficiency and proton contamination. *Journal of Geophysical Research: Space Physics*, 118(10), 6500–10.

Baker, D., Belian, R., Higbie, P., Klebesadel, R., and Blake, J. (1987). Deep dielectric charging effects due to high-energy electrons in Earth's outer magnetosphere. *Journal of Electrostatics*, 20(1), 3–19.

Baker, D. N. (2000). The occurrence of operational anomalies in spacecraft and their relationship to space weather. *IEEE Transactions on Plasma Science*, 28(6), 2007–16.

Baker, D. N. (2002). How to cope with space weather. *Science*, 297 (5586), 1486–7.

Baker, D. N. (2005). Specifying and forecasting space weather threats to human technology. In I. A. Daglis, ed., *Effects of Space Weather on Technology Infrastructure, NATO Science Series II: Mathematics, Physics and Chemistry*, vol. 176. Dordrecht: Springer, pp. 1–25.

Beutier, T., and Boscher, D. (1995). A three-dimensional analysis of the electron radiation belt by the Salammbô code. *Journal of Geophysical Research: Space Physics*, 100 (A8), 14853–61.

Bourdarie, S., and Maget, V. (2012). Electron radiation belt data assimilation with an ensemble Kalman filter relying on the Salammbô code. *Annales Geophysicae*, 30, 929–43.

Castillo, A. M., de Wiljes, J., Shprits, Y. Y., and Aseev, N. A. (2021). Reconstructing the dynamics of the outer electron radiation belt by means of the standard and ensemble Kalman filter with the VERB-3D code. *Space Weather*, 19 (10), e2020SW002672.

Cervantes, S., Shprits, Y. Y., Aseev, N. et al. (2020a). Identifying radiation belt electron source and loss processes by assimilating spacecraft data in a three-dimensional diffusion model. *Journal of Geophysical Research: Space Physics*, 125(1), 1–16.

Cervantes, S., Shprits, Y. Y., Aseev, N. A., and Allison, H. J. (2020b). Quantifying the effects of EMIC wave scattering and magnetopause shadowing in the outer electron radiation belt by means of data assimilation. *Journal of Geophysical Research: Space Physics*, 125(8).

Cohen, D., Spanjers, G., Winter, J. et al. (2005). Design and Systems Engineering of AFRL's Demonstration and Sciences Experiment. *Proceedings of the GATech Space Systems Engineering Conference*, 8–10 November 2005. Paper No. GT-SSEC.D.1, pp. 17. http://hdl.handle.net/1853/8037.

Daae, M., Shprits, Y. Y., Ni, B. et al. (2011). Reanalysis of radiation belt electron phase space density using various boundary conditions and loss models. *Advances in Space Research*, 48(8), 1327–34.

Drozdov, A. Y., Shprits, Y. Y., Orlova, K. G. et al. (2015). Energetic, relativistic, and ultrarelativistic electrons: Comparison of long-term VERB code simulations with Van Allen Probes measurements. *Journal of Geophysical Research: Space Physics*, 120(5), 3574–87.

Escoubet, C., Schmidt, R., and Goldstein, M. eds. (1997). Cluster-science and mission overview. In C. P. Escoubet, C. T. Russell, and R. Schmidt, eds., *The Cluster and Phoenix Missions*. Dordrecht: Springer, pp. 11–32.

Evensen, G. (2003). The ensemble Kalman filter: Theoretical formulation and practical implementation. *Ocean Dynamics*, 53 (4), 343–67.

Fok, M.-C., Glocer, A., Zheng, Q. et al. (2011). Recent developments in the radiation belt environment model. *Journal of Atmospheric and Solar-Terrestrial Physics*, 73(11–12), 1435–43.

Fok, M.-C., Horne, R. B., Meredith, N. P., and Glauert, S. A. (2008). Radiation belt environment model: Application to

space weather nowcasting. *Journal of Geophysical Research: Space Physics*, 113(A3).

Friedel, R., Bourdarie, S., Fennell, J., Kanekal, S., and Cayton, T. (2003). 'Nudging' the Salammbo Code: First results of seeding a diffusive radiation belt code with in situ data: GPS, GEO, HEO and POLAR. *Eos Transactions of the American Geophysical Union*, 84(46), Fall Meeting, 8–12 December 2003, San Francisco, 476 Suppl., abstract, SM11D–06.

Friedel, R., Reeves, G., and Obara, T. (2002). Relativistic electron dynamics in the inner magnetosphere: A review. *Journal of Atmospheric and Solar-Terrestrial Physics*, 64, 265–82.

Galand, M., and Evans, D. S. (2000). Radiation damage of the proton MEPED detector on POES (TIROS/NOAA) satellites. *NOAA Technical Memorandum*. Boulder, CO. OAR 456-SEC 42.

Ganushkina, N. Y., Amariutei, O., Shprits, Y., and Liemohn, M. (2013). Transport of the plasma sheet electrons to the geostationary distances. *Journal of Geophysical Research: Space Physics*, 118(1), 82–98.

Ganushkina, N. Y., Liemohn, M., Amariutei, O., and Pitchford, D. (2014). Low-energy electrons (5–50 kev) in the inner magnetosphere. *Journal of Geophysical Research: Space Physics*, 119(1), 246–59.

GFZ-data (2021). Radiation belts data. www.gfzpotsdam.de/sektion/magnetosphaerenphysik/daten-produkte-dienste/.

GFZ-RB-forecast (2021). Radiation belts forecast. https://isdc.gfz-potsdam.de/data-assimilative-radiation-belt-forecast/.

Glauert, S. A., and Horne, R. B. (2005). Calculation of pitch angle and energy diffusion coefficients with the PADIE code. *Journal of Geophysical Research: Space Physics*, 110(A4).

Godinez, H. C., Yu, Y., Lawrence, E. et al. (2016). Ring current pressure estimation with RAM-SCB using data assimilation and Van Allen Probe flux data. *Geophysical Research Letters*, 43(23), 11948–56.

Green, J. C., and Kivelson, M. G. (2004). Relativistic electrons in the outer radiation belt: Differentiating between acceleration mechanisms. *Journal of Geophysical Research: Space Physics*, 109(A3).

Hamlin, D. A., Karplus, R., Vik, R. C., and Watson, K. M. (1961). Mirror and azimuthal drift frequencies for geomagnetically trapped particles. *Journal of Geophysical Research (1896–1977)*, 66(1), 1–4.

Jordanova, V., Kozyra, J., Nagy, A., and Khazanov, G. (1997). Kinetic model of the ring current–atmosphere interactions. *Journal of Geophysical Research: Space Physics*, 102(A7), 14279–91.

Jordanova, V., and Miyoshi, Y. (2005). Relativistic model of ring current and radiation belt ions and electrons: Initial results. *Geophysical Research Letters*, 32(14).

Kalman, R. (1960). A new approach to linear filtering and prediction problems. *Trans. ASME Journal of Basic Engineering*, 82 (1), 35–45.

Kalnay, E. (2003). *Atmospheric Modeling, Data Assimilation and Predictability*. Cambridge: Cambridge University Press.

Kennel, C., and Engelmann, F. (1966). Velocity space diffusion from weak plasma turbulence in a magnetic field. *The Physics of Fluids*, 9(12), 2377–88.

Kim, K., Shprits, Y., Subbotin, D., and Ni, B. (2012). Relativistic radiation belt electron responses to GEM magnetic storms:

Comparison of CRRES observations with 3-D VERB simulations. *Journal of Geophysical Research: Space Physics*, 117(A8).

Koller, J., Chen, Y., Reeves, G. D. et al. (2007). Identifying the radiation belt source region by data assimilation. *Journal of Geophysical Research: Space Physics*, 112(A6).

Koller, J., Friedel, R., and Reeves, G. (2005). Radiation belt data assimilation and parameter estimation. *LANL Reports*, LA-UR-05-6700. http://library.lanl.gov/cgi-bin/getfile?LA-UR-05-6700.pdf.

Kondrashov, D., Ghil, M., and Shprits, Y. Y. (2011). Lognormal Kalman filter for assimilating phase space density data in the radiation belts. *Space Weather*, 9(11).

Kondrashov, D., Shprits, Y. Y., Ghil, M., and Thorne, R. (2007). A Kalman filter technique to estimate relativistic electron lifetimes in the outer radiation belt. *Journal of Geophysical Research: Space Physics*, 112(A10).

Lanzerotti, L. J. (2001). Space weather effects on technologies. *Washington DC American Geophysical Union Geophysical Monograph Series*, 125, 11–22.

Lyons, L. R., Thorne, R. M., and Kennel, C. F. (1972). Pitch-angle diffusion of radiation belt electrons within the plasmasphere. *Journal of Geophysical Research*, 77(19), 3455–74.

Maynard, N. C., and Chen, A. J. (1975). Isolated cold plasma regions: Observations and their relation to possible production mechanisms. *Journal of Geophysical Research (1896–1977)*, 80(7), 1009–13.

McFadden, J., Evans, D., Kasprzak, W. et al. (2007). In-flight instrument calibration and performance verification. In M. Wüest, D. S. Evansvon and R. Steiger, eds., *Calibration of Particle Instruments in Space Physics*, vol. SR-007. Bern: International Space Science Institute, pp. 277–385.

Miyoshi, Y., Shinohara, I., Takashima, T. et al. (2018). Geospace exploration project ERG. *Earth, Planets and Space*, 70(1), 1–13.

Naehr, S. M., and Toffoletto, F. R. (2005). Radiation belt data assimilation with an extended Kalman filter. *Space Weather*, 3(6).

Nakano, S., Ueno, G., Ebihara, Y. et al. (2008). A method for estimating the ring current structure and the electric potential distribution using energetic neutral atom data assimilation. *Journal of Geophysical Research: Space Physics*, 113(A5).

Ni, B., Shprits, Y., Nagai, T. et al. (2009a). Reanalyses of the radiation belt electron phase space density using nearly equatorial CRRES and polar-orbiting Akebono satellite observations. *Journal of Geophysical Research: Space Physics*, 114(A5).

Ni, B., Shprits, Y., Thorne, R., Friedel, R., and Nagai, T. (2009b). Reanalysis of relativistic radiation belt electron phase space density using multisatellite observations: Sensitivity to empirical magnetic field models. *Journal of Geophysical Research: Space Physics*, 114(A12).

Ni, B., Shprits, Y. Y., Friedel, R. H. et al. (2013). Responses of Earth's radiation belts to solar wind dynamic pressure variations in 2002 analyzed using multisatellite data and Kalman filtering. *Journal of Geophysical Research: Space Physics*, 118(7), 4400–14.

Podladchikova, T. V., Shprits, Y. Y., Kellerman, A. C., and Kondrashov, D. (2014a). Noise statistics identification for

Kalman filtering of the electron radiation belt observations: 2. Filtration and smoothing. *Journal of Geophysical Research: Space Physics*, 119(7), 5725–43.

Podladchikova, T. V., Shprits, Y. Y., Kondrashov, D., and Kellerman, A. C. (2014b). Noise statistics identification for Kalman filtering of the electron radiation belt observations: 1. Model errors. *Journal of Geophysical Research: Space Physics*, 119(7), 5700–24.

Reeves, G. D., McAdams, K. L., Friedel, R. H. W., and O'Brien, T. P. (2003). Acceleration and loss of relativistic electrons during geomagnetic storms. *Geophysical Research Letters*, 30(10).

Robinson Jr., P. A. (1989). Spacecraft environmental anomalies handbook. Technical report, Jet Propulsion Lab, Pasadena, CA.

Roederer, J. G. (2012). *Dynamics of Geomagnetically Trapped radiation*, vol. 2. Berlin: Springer Science & Business Media.

Rosen, A. (1976). Spacecraft charging by magnetospheric plasmas. *IEEE Transactions on Nuclear Science*, 23(6), 1762–68.

Saikin, A. A., Shprits, Y. Y., Drozdov, A. Y. et al. (2021). Reconstruction of the radiation belts for solar cycles 17–24 (1933–2017). *Space Weather*, 19(3), e2020SW002524.

Schiller, Q., Li, X., Koller, J., Godinez, H., and Turner, D. L. (2012). A parametric study of the source rate for outer radiation belt electrons using a Kalman filter. *Journal of Geophysical Research: Space Physics*, 117(A9).

Schulz, M., and Lanzerotti, L. (1974). *Particle Diffusion in the Radiation Belts*. New York: Springer-Verlag.

Shprits, Y., Daae, M., and Ni, B. (2012). Statistical analysis of phase space density buildups and dropouts. *Journal of Geophysical Research: Space Physics*, 117(A1).

Shprits, Y., Elkington, S., Meredith, N., and Subbotin, D. (2008a). Review of modeling of losses and sources of relativistic electrons in the outer radiation belts: I. Radial transport. *Journal of Atmospheric and Solar-Terrestrial Physics*, 70(14), 1679–93.

Shprits, Y. Y., Kellerman, A. C., Drozdov, A. Y. et al. (2015). Combined convective and diffusive simulations: VERB-4D comparison with 17 March 2013 Van Allen Probes observations. *Geophysical Research Letters*, 42(22), 9600–8.

Shprits, Y., Kellerman, A., Kondrashov, D., and Subbotin, D. (2013). Application of a new data operator-splitting data assimilation technique to the 3-D VERB diffusion code and CRRES measurements. *Geophysical Research Letters*, 40(19), 4998–5002.

Shprits, Y., Kondrashov, D., Chen, Y. et al. (2007). Reanalysis of relativistic radiation belt electron fluxes using CRRES satellite data, a radial diffusion model, and a Kalman filter. *Journal of Geophysical Research: Space Physics*, 112(A12216).

Shprits, Y., Subbotin, D., Meredith, N., and Elkington, S. (2008b). Review of modeling of losses and sources of relativistic electrons in the outer radiation belts: II. Local acceleration and loss. *Journal of Atmospheric and Solar-Terrestrial Physics*, 70(14), 1694–713.

Shprits, Y., Subbotin, D., and Ni, B. (2009). Evolution of electron fluxes in the outer radiation belt computed with the VERB code *Journal of Geophysical Research: Space Physics*, 114(A11).

Shprits, Y., Subbotin, D., Ni, B. et al. (2011). Profound change of the near-Earth radiation environment caused by solar superstorms. *Space Weather*, 9(8).

Shprits, Y., Thorne, R., Friedel, R. et al. (2006). Outward radial diffusion driven by losses at magnetopause. *Journal of Geophysical Research: Space Physics*, 111(A11).

Shprits, Y. Y., Vasile, R., and Zhelavskaya, I. S. (2019). Nowcasting and predicting the Kp index using historical values and real-time observations. *Space Weather*, 17(8), 1219–29.

Sibeck, D., and Angelopoulos, V. (2008). THEMIS science objectives and mission phases. *Space Science Reviews*, 141(1), 35–59.

Stern, D. P. (1975). The motion of a proton in the equatorial magnetosphere. *Journal of Geophysical Research (1896–1977)*, 80(4), 595–9.

Stix, T. H. (1962). *The Theory of Plasma Waves*. New York: McGraw-Hill.

Subbotin, D. A. and Shprits, Y. Y. (2009). Three-dimensional modeling of the radiation belts using the Versatile Electron Radiation Belt (VERB) code. *Space Weather*, 7(10).

Subbotin, D. A. and Shprits, Y. Y. (2012). Three-dimensional radiation belt simulations in terms of adiabatic invariants using a single numerical grid. *Journal of Geophysical Research: Space Physics*, 117(A05205).

Subbotin, D., Shprits, Y., and Ni, B. (2011). Long-term radiation belt simulation with the VERB 3-D code: Comparison with CRRES observations. *Journal of Geophysical Research: Space Physics*, 116(A12).

Tsyganenko, N. (1989). A magnetospheric magnetic field model with a warped tail current sheet. *Planetary and Space Science*, 37(1), 5–20.

Turner, D., Shprits, Y., Hartinger, M., and Angelopoulos, V. (2012). Explaining sudden losses of outer radiation belt electrons during geomagnetic storms. *Nature Physics*, 8(3), 208–12.

UCS (2021). UCS Satellite Database, Union of Concerned Scientists. https://ucsusa.org/resources/satellite-database/.

Volland, H. (1973). A semiempirical model of large-scale magnetospheric electric fields. *Journal of Geophysical Research (1896–1977)*, 78(1), 171–80.

Wang, D. and Shprits, Y. Y. (2019). On how high-latitude chorus waves tip the balance between acceleration and loss of relativistic electrons. *Geophysical Research Letters*, 46(14), 7945–54.

Wang, D., Shprits, Y. Y., Zhelavskaya, I. S. et al. (2020). The effect of plasma boundaries on the dynamic evolution of relativistic radiation belt electrons. *Journal of Geophysical Research: Space Physics*, 125(5), e2019JA027422.

PART III

'Solid' Earth Applications: From the Surface to the Core

11

Trans-Dimensional Markov Chain Monte Carlo Methods Applied to Geochronology and Thermochronology

Kerry Gallagher

Abstract: Trans-dimensional Markov chain Monte Carlo (MCMC) treats the number of model parameters as an unknown, and provides a natural approach to assess models of variable complexity. We demonstrate the application of these methods to geochronology and thermochronology. The first is mixture modelling, physically a finite dimension problem, which aims to extract the number and characteristics of component age distributions from an overall distribution of radiometric age data. We demonstrate the MCMC method with Gaussian and skew-t component distributions, the latter containing the former as a special case, applied to a suit of U-Pb zircon data from a sediment in northern France. When considering the posterior distributions obtained from the MCMC samplers, the asymmetrical skew distribution models imply fewer components than the symmetrical Gaussian distribution models. We present some heuristic criteria based on different ways to look the results and aid in model choice in the mixture modelling problem. The second application is a thermal history model, physically a continuous time-temperature function but here parametrised in terms of a finite number of time temperature nodes. We consider a suite of synthetic data from a vertical profile to demonstrate the variable resolution in models constrained from single and multiple samples. Provided the implicit assumptions made when grouping multiple samples are valid, the multi-sample approach is preferable as we exploit the variable information on the model (thermal history) contained in different samples.

11.1 Introduction

This chapter presents some example applications of inverse methods, specifically trans-dimensional Markov chain Monte Carlo (MCMC), to geochronological and thermochronological data. We first give some background to geo/thermochronology, then present applications from two different contexts: mixture modelling applied to detrital geochrononological data, and thermal history modelling applied to thermochronological data. The purpose is not to present all the details of the inverse method, but more

the nature of the applied problems and how the results of the inverse modelling process can be presented to assess the quality of the results.

Geochronology is the science of dating rocks and minerals, based on the radioactive decay of one or more parent isotopes to produce atoms of a daughter isotope of a different element. The simplest case is that of a single parent isotope, and the decay of one parent atom producing one daughter isotope in a single decay step. Then the ratio of the number of daughter atoms ($\mathbf{D}$) to the number of parent atoms ($\mathbf{P}$), combined with the decay constant (λ) for the parent, allows us to calculate a date ($t_{\mathbf{date}}$) for the host rock or mineral, according to

$$t_{date} = \frac{1}{\lambda} Ln\left(\frac{D}{P} + 1\right). \tag{11.1}$$

Here, we note that geologists often differentiate between the date as defined in Eq. (11.1), calculated directly from measured data, and an age. An age is effectively an interpretation of a calculated date (or series of dates) in terms of some kind of geological event, or activity (e.g. Schoene, 2014). One reason for this distinction is related to the concept of closed and open systems in geochronology. We can define a geochronological system in terms of a given radioactive decay scheme (the combination of parent and daughter isotopes in a given mineral, such as fission track analysis in apatite, or U-Pb dating of zircon). In brief, a closed system is one in which both the parent and daughter isotopes only change over time due to the radioactive decay/production process. Then Eq. (11.1) can provide a geologically valid age (e.g. the time of formation of a rock or mineral). In contrast, an open system is one in which either or both the parent and daughter can escape over time, for example, due to thermally activated chemical diffusion. This typically leads to loss of the daughter rather than the parent as the former is generally smaller and often a gas, so diffuses more easily than the latter, for example. Then, in the case of an open system, the measured number of atoms for the daughter is not necessarily truly representative of the decay/production over the lifetime of the mineral. Consequently, the calculated date will be younger than that expected for a closed system. Then the date may not represent the age of a geological event, and certainly not the time of formation of the mineral.

A rigorous mathematical formulation for the loss of the daughter isotope due to thermally activated diffusion was developed in detail by Dodson (1973). He considered diffusion/production equation given below

$$\frac{\partial C}{\partial t} = \frac{1}{x^n}\frac{\partial}{\partial x}\left(x^n D(T)\frac{\partial C}{\partial x}\right) + P, \qquad (11.2)$$

where C can be number of atoms, or concentration, of the daughter isotope, t is time (s), x is a spatial coordinate, for example, radius in a sphere or cylinder (m), n is an integer that describes the geometry (n = 0,1,2 represent respectively an infinite slab, an infinite cylinder, and a sphere), and P is the rate of production of the daughter isotope by radioactive decay of the parent (C s^{-1} with C the appropriate units for the daughter isotope). Then D(T) is the temperature dependent chemical diffusion coefficient, which is defined as

$$D(T) = D_o e^{-E/RT}, \qquad (11.3)$$

where T is absolute temperature (K), R is the universal gas constant ($Jmol^{-1}K^{-1}$), E is activation energy ($Jmol^{-1}$), and D_o (m^2s^{-1}) is a constant, equivalent to the diffusivity at infinite temperature. The last two parameters, known as the kinetic parameters, are specific to a given geochronological (parent-daughter-mineral) system and can be estimated from diffusion experiments in the laboratory (e.g. McDougall and Harrison, 1988; Farley, 2000).

Dodson's major contribution was to provide an expression for the *closure temperature* as a function of the diffusivity parameters (E, D_o, R), grain size (r) and geometry (A), both of which define the diffusion domain, and the cooling rate over geological time ($\frac{dT}{dt}$). The concept of the closure temperature is illustrated in Fig. 11.1 and the expression derived by Dodson (1973) is given as

$$T_c - \frac{E}{RLn\left(\frac{AD_oRT_c^2}{r^2 E\frac{dT}{dt}}\right)} = 0. \qquad (11.4)$$

The closure temperature is then taken to be the temperature of the mineral at the time given by the calculated date, or the calculated date represents the time at which the mineral was at the closure temperature. Figure 11.2 summarises estimates of closure temperatures for different systems over a range of cooling rates typical of geological processes and we can see there is a wide range of temperature sensitivity, depending on the method and the geological question under consideration. For example, questions concerning the formation of the continental crust typically exploit high temperature systems such as U-Pb in zircon, while questions

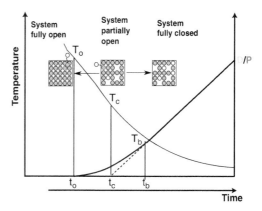

Figure 11.1 The concept of closure temperature (modified from Braun et al., 2009) which reflects the rate of thermally activated diffusion decreasing with decreasing temperature. The system is represented by the box, with the green and orange circles representing the parent (P – dark grey) and daughter (D – light grey) elements. We consider a continuous cooling history (thin black curve) starting at a temperature (> T_o) high enough such that the system is fully open (the daughter product is lost by thermally activated diffusion faster than it is produced by radioactive decay). Once the temperature drops below T_o (at t_o) D is retained, the ratio D/P (thicker black curve) increases, tending to linear growth over time. Once the temperature drops below the blocking temperature (T_b at time t_b), diffusive loss is negligible, and the D/P ratio increases linearly to the present day value. The closure temperature (T_c) corresponds to the temperature at the time equivalent to the calculated age (t_c), (based on the assumption of linear growth, the dashed line, and no diffusive loss).

concerning erosion and landscape evolution will use the lower temperature systems such as U-Th-He in apatite, or even cosmogenic isotopes produced in minerals in the top few metres from the Earth's surface. The recognition of the temperature dependence of dating systems and the complexities introduced due to that dependency led to the definition thermochronology as a sub-discipline of geochronology. Thermochronology exploits the temperature dependence of different dating systems (in particular systems sensitive to temperatures < 250–300°C) to understand the rates and magnitudes of near surface geological processes such as tectonics, erosion, and transport to sedimentary basins.

Useful though the closure temperature concept is, it has some limitations. For example, as presented here, it assumes monotonic cooling at a constant rate and also that the temperature range over which diffusive loss of the daughter occurs is a small proportion of the closure temperature itself. Since the 1990s, thermochronology has advanced through laboratory experiments aimed specifically at improving understanding of the temperature dependence of some thermochronological methods. These methods involve fission track analysis (primarily in the mineral apatite), (U-Th)/He dating (in apatite, zircon, and more recently Fe-oxides) and [40]Ar/[39]Ar dating (e.g. feldspars and mica).

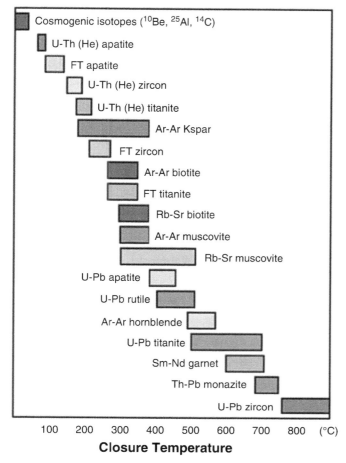

Closure Temperature

Figure 11.2 Estimated closure temperature ranges for different geochronological systems with a system being defined by a parent-daughter-mineral combination (based on a colour original from Pollard et al., 2003, New Departures in Structural Geology and Tectonics NSF White Paper, www.pangea.stanford.edu/~dpollard/NSF/).The range represents different cooling rates, grain size and geometries, and the difference in the temperatures when the system is fully open or fully blocked for diffusive loss of the daughter (T$_o$ or T$_b$ and see Fig. 11.1) for each system. For lower temperature systems the range is often referred to as the partial retention or annealing zone (PRZ, PAZ).

In the following, we will present some applications of trans-dimensional inverse methods to geo- and thermochronological data to identify the number of potential source components for detrital material in sedimentary basins and to extract temperature histories of rocks over geological time. Although the forward models are generally computationally simple and fast, the relationships between the observed data and these processes are often complex and non-linear. Here we are interested in applications of inverse methods to understand and interpret geochronological/thermochronological data and so we will not go into details of the analytical methods. However, useful overviews of those and the underlying principles are available in

Reiners and Ehlers (2005), Reiners et al. (2018), and Malusa and Fitzgerald (2019).

11.2 Application of Inverse Methods Applied to Geothermochronology

11.2.1 Bayesian Mixture Modelling to Extract Age Component Distribution

Identifying underlying component distributions contained in a general distribution with mixture modelling is a classic problem in statistics (e.g. Everitt and Hand, 1981; Titterington et al., 1985). In general the problem is specified in terms of

(i) the number of component distributions,
(ii) the form of the component distributions (typically parametric, e.g. Gaussian),
(iii) the parameters of the component distributions (e.g. mean, variance),
(iv) the proportion of each component distribution (with the constraint that these sum to 1).

The inverse problem is then, given the observed discrete data, what are the underlying component distributions? We note that this problem is naturally discrete in that there is a finite number of parameters (number of component distribution and their parameters). In the geochronological context, probably the first application was that of Galbraith and Green (1990) applied to fission track age data, based on using a Poissonian statistical model appropriate to discrete fission track count data that leads to a binomial probability mass function, the parameter of which is related to the component age. They estimated the relevant parameters (parameter for the binomial distribution, and proportions) using maximum likelihood methods (see also chapters 5 and 6 of Galbraith (2005) for more details of this approach). Sambridge and Compston (1994) presented a similar maximum likelihood approach, but formulated the problem in terms of Gaussian and double exponential probability density functions, applied to U-Pb zircon age data. Both of these approaches specified the number of component distributions in advance, estimated the means and proportions using maximum likelihood, and assessed uncertainty in terms of covariance matrices. The choice of the number of components was made primarily by visual inspection of either radial plots (Galbraith and Green, 1990) or relative misfit as a function of the number of components (Sambridge and Compston, 1994) or a test statistic (sum of squares of residuals, Kolmogorov–Smirnoff) comparing the predicted and observed distributions.

Identifying the number of components is obviously a key aspect of any mixture modelling method. Motivated by the limitations in previous approaches mentioned here, Jasra et al. (2006) presented a Bayesian mixture modelling method, following Richardson and Green (1997) and

Richardson et al. (2002), implemented with reversible jump (RJ) or trans-dimensional MCMC. In this case, the number of component distributions is a parameter to estimate, and in the context of the Bayesian approach, we obtain a distribution on that parameter. To make comparisons with previous approaches, Jasra et al. (2006) considered normal distributions for the components, each defined by the mean (or location parameter), μ, and the variance (or scale parameter), σ^2. They also presented a more general form of component distribution, a skew-t distribution. This distribution is unimodal and defined in terms of its location parameter (μ), inverse scale parameter $w = \frac{1}{\sigma^2}$, and also two skew parameters (v, ζ) as in the following:

$$f(x; \mu, \omega, v, \zeta) = C_{v,\zeta} \left(1 + \frac{\omega(x - \mu)}{\left(v + \zeta + (\omega(x - \mu))^2 \right)^{\frac{1}{2}}} \right)^{v + \frac{1}{2}}$$

$$\left(1 - \frac{\omega(x - \mu)}{\left(v + \zeta + (\omega(x - \mu))^2 \right)^{\frac{1}{2}}} \right)^{\zeta + \frac{1}{2}}, \quad (11.5a)$$

with the normalisation constant defined as

$$C_{v,\zeta} = \frac{\omega}{2^{v + \zeta + 1} B(v, \zeta)(v + \zeta)^{\frac{1}{2}}}, \quad (11.5b)$$

where $B(.,.)$ is the Beta function. The location of the mode of this distribution is given as

$$x_{mode} = \mu + \frac{(v - \zeta)}{\omega} \left(\frac{(v + \zeta)}{(2v + 1)(2\zeta + 1)} \right)^{\frac{1}{2}}. \quad (11.5c)$$

The two skew parameters v, ζ control the amount of skewness, leading to left (negative) when $v > \zeta$ and right (positive) skew when $v < \zeta$, and the skewness is more pronounced for smaller values of these parameters. Finally, if $v = \zeta$, the distribution is symmetric with heavier tails when the values are small and approaches a Gaussian when the values are large, similar to the behaviour of standard t distribution (as a function of the degrees of freedom).

The inverse problem was set up as follows: firstly, we allow for uncertainty in the observed data ($\mathbf{d_{obs}}$), assuming that each datum is sampled from a Gaussian distribution centred on the unknown true datum values ($\mathbf{d_{true}}$) with a standard deviation equal to the reported measurement error (σ as a standard deviation), that is,

$$\mathbf{p}(\mathbf{d_{obs}}|\mathbf{d_{true}}; \sigma^2) \sim N(\mathbf{d_{true}}, \sigma^2), \quad (11.6)$$

then the mixture model is formulated in terms of the unknown true, or measurement error free, values. The justification for this is that we should not try to fit the mixture model parameters to the noisy observations, but prefer to sample the true, but unknown, values. In doing this, we explicitly allow for the uncertainty (measurement error) on

the observations. The distribution of these values for a mixture model is then defined as

$$\mathbf{p}(\mathbf{d_{true}}|k, \pi, \theta) \sim \sum_{i=1}^{k} \pi_{\mathbf{i}} \mathbf{f}(\mathbf{d_{true}}; \theta_{\mathbf{i}}), \quad (11.7)$$

where k indicates the number of components, $\pi_{\mathbf{i}}$ is the proportion of the i-th component with distribution $\mathbf{f}(.)$, and finally $\theta_{\mathbf{i}}$ represents the distribution parameters (e.g. μ, σ for a Gaussian, or μ, ω, v, ζ for a skew). The aim is to use MCMC sampling to approximate the posterior distribution, which can be written (up to a proportionality constant) as

$$\mathbf{p}(\mathbf{d_{true}}, k, \pi, \theta|\mathbf{d_{obs}}; \sigma^2) \, \alpha \prod_{i=1}^{N} \mathbf{p}(\mathbf{d_{obs, i}}|\mathbf{d_{true, i}}; \sigma^2)$$

$$\mathbf{p}(\mathbf{d_{true, i}}|k, \pi, \theta) \times \mathbf{p}(k)\mathbf{p}(\pi|k) \prod_{j=1}^{k} \mathbf{p}(\mu_j)\mathbf{p}(\omega_j)\mathbf{p}(v_j, \zeta_j), \quad (11.8)$$

where N is the number of data and k the number of components. The first two terms on the right effectively represent the data likelihood and the term after the x contains the prior terms for the parameters to be estimated.

To solve the problem, it is necessary to define priors for all parameters k, μ, ω, v, ζ, π. A range of possible priors (and associated hyper-parameter priors) was presented in Jasra et al. (2006) together with the appropriate form of the posterior for different hyper-parameters. We refer the reader to that publication for the mathematical details. In general, they present one prior for each parameter type, except for number of components and the skew parameters. Some priors use hyper-parameters (parameters defining the prior distributions themselves to be treated as unknowns with a prior to be resampled as part of the MCMC process). For the results we present here, we used a uniform prior on the number of components ($\mathbf{p}(k) = 1/k_{max}$) with k_{max} the maximum allowed number of components.

For the prior on the skew parameters, we present results based on both of the two priors described in Jasra et al. (2006). The first is a weakly informative prior (prior 1), which allows equally for symmetrical, positive, or negative distributions, but tends to avoid distributions with heavy tails and high skew. The second (prior 2) favours heavy tails and more highly skewed distributions. Thus, as symmetrical distributions will tend to imply more components than an asymmetric one, we expect prior 2 to suggest fewer components than prior 1 (and also a Gaussian distribution component model).

The details of the reversible jump, or trans-dimensional sampler, are also given in Jasra et al. (2006). The approach sweeps over the different parameters (as in Eq. 11.8), or groups of parameters, in succession with an acceptance condition assessed at each stage. As usual with MCMC sampling, we obtain an ensemble of estimates for all parameters that approximates the full posterior distribution. The output from a trans-dimensional model also allows us to examine the posterior distribution for the number of components, assess what number is the most probable, and then examine the models associated with that chosen number of components. We can also monitor the maximum likelihood (i.e. equivalent of the best data-fit) solution which tends to be more complex (more components) than the maximum posterior model.

Finally, once the posterior distribution has been estimated, we can use that to examine the expected model calculated under different assumptions. The expected model is a form of weighted average, with the weighting being the posterior distribution. As the MCMC samples are selected (accepted) according to the posterior distribution, the expected model is then just the average of the MCMC samples. With the usual definition of the expected model as an integral, we can write

$$E(\mathbf{m}) = \int \mathbf{m} p(\mathbf{m}) \, d\mathbf{m} = \frac{1}{M} \sum_{i=1}^{M} \mathbf{m}_i, \qquad (11.9a)$$

where M is the number of MCMC samples of the parameter vector $\mathbf{m}$. Similarly, the expected value of a function of the model parameters, $f(\mathbf{m})$ can be written as

$$E(f(\mathbf{m})) = \int f(\mathbf{m}) \, p(\mathbf{m}) \, d\mathbf{m} = \frac{1}{M} \sum_{i=1}^{M} f(\mathbf{m})_i. \qquad (11.9b)$$

In the results presented in Section 11.2.1.1, we will consider the expected distribution calculated from the distributions calculated for each accepted set of model parameters $E(f(\mathbf{m}))$ and also the distribution calculated with the expected values of the component distribution parameters $E(f(\mathbf{m}))$. We note we expect $E(f(\mathbf{m}))$ to be equal to $f(E(\mathbf{m}))$ if the function f is linear (as expectation is linear).

11.2.1.1 Mixture Modelling Example
We choose a set of U-Pb ages measured on zircons separated from a Devonian sedimentary sample (SA5) from the Ancenis Basin in northern France (Ducassou, 2009). There are 40 ages ranging from 401 to 2,343 Myr, and errors (1σ) from 12 to 55 Myr, with the younger ages tending to have the larger error (lower precision). The measured data are summarised in Fig. 11.3 in histogram form, which does not take into account the errors, and as a radial plot, which does. The radial plot (Galbraith, 1988) is a graphical summary of data with different precisions. The data are transformed such that x co-ordinate = 1/error is a measure of relative precision and the y-axis = measured value/error is an error-normalised value. Often the y-axis is centred on the mean (or better the variance weighted mean) to give a z-score (measured value − weighted mean)/error). In this form, each point on the plot has the same standard error (on the y-scale), more precise data plot further from the origin, and the slope of a line from the data point to the origin is equivalent to the measured value (or measured value − weighted mean). The radial plot is a convenient graphical representation to assess heterogeneity in a data set and in the

case of mixture modelling, to assess visually the possibility of having different components in the data. We can see that both visual representations of the data imply at least two components.

All the model results presented here are based on 10^6 burn-in and 10^6 post-burn-in iterations, and a thinning factor of 50. We first ran the mixture modelling using Gaussian distributions for the component distribution and the results are summarised in Fig. 11.4. Figure 11.4a shows the likelihood, posterior, and number of components as a function of iteration. This demonstrates the wide range (2–15) in the number of components sampled, and also the natural parsimony of the Bayesian approach. This is manifested as the best data fitting (higher likelihood) models with a relatively large number of components having lower posterior probabilities (e.g. where the blue and green curves touch, around 1.35×10^6 and just before 1.6×10^6 iterations). The preferred number of components from the posterior distribution (Fig. 11.4b) is three or four (having about 75% of the posterior probability of three components), with up to eight with p(k) > 0.05. Figure 11.4c presents the expected model conditional on three components. As mentioned, we consider two models, the expected model formed by summing the predicted distributions for all models with three components and the predicted distribution using the expected values for the parameters (μ, σ, π), equivalent to $E(f(\mathbf{m}))$ and $f(E(\mathbf{m}))$, respectively. We can see in the case for three components, the two expected models are not too different, and neither are the individual component distributions (Fig 11.4d), particularly for the two well-defined components, defining the youngest and oldest components. Figures 11.4c and 11.4d also show the 95% credible interval on the expected location parameters (i.e. the range that is defined between the 2.5% lowest values and the 2.5% highest values for a given parameter). We see that these are pretty much symmetrical about the expected values for the youngest and oldest components, but a little skewed to younger values for the central component. The inference of three components with a Gaussian distribution component model is predictable with this particular data set, given the two well-defined peaks at younger and older ages and the smaller central peak (ages around 1,200 Myr ago [Ma]).

Figures 11.5a and 11.5b show the results for four Gaussian distribution components and we can see more obvious differences in the two expected models mentioned previously. The total distribution based on the expected values of the parameter estimates does not capture the form of the data histogram as the mean values reflect that bimodal sampling of the two central components of the four to try and capture what is the well-resolved central component in the three-component model. This suggests that consideration of the two different forms of expected models can be an additional criterion for model selection, given that the posterior probability of having four components is not that different to having three components (see Fig. 11.4a).

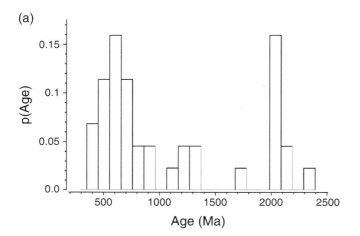

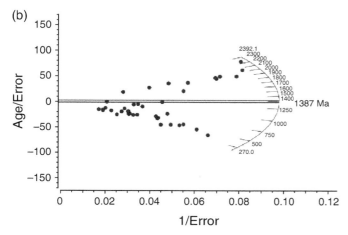

Figure 11.3 (a) Histogram of U-Pb-zircon age data (sample SA5) from a Devonian aged sediment (St Anne Grit) in the Ancenis Basin in northern France (Ducassou, 2009). (b) Radial plot of the data in (a). This representation (Galbraith, 1988) uses age/error as the *y*-ordinate and 1/error as the *x*-ordinate. The slope of a line joining a data point to the origin is the age of that data point. All data have the same relative error (e.g. ±2 on the *y*-axis as shown by the two red lines bounding the mean age (with slope = 0). The radial plot provides a rapid visual assessment of age components, and also explicitly incorporates the error for each observation such that more precise data plot further to the right.

The results for the skew model with the two prior distributions are shown in Figs. 11.6 (prior 1, preference for lighter skew) and 11.7 (prior 2, preference for heavier skew), respectively. As expected, prior 1 tends to infer more components than prior 2. However, relative to the Gaussian distribution results, prior 1 implies a higher probability for models with two components, reflecting the ability to introduce some asymmetry in the component distributions. Those models tend to imply a skew to older ages for the younger of the two age components, to accommodate the ages around 1,200 Ma. Prior 2 clearly prefers two inferred components, reflecting the preference for higher skew models in that prior choice. Although not shown here, we

note that the maximum likelihood models for the two skew priors (1 and 2) contain 12 and 4 components respectively.

Figures 11.6c and 11.7c show another aspect of the mixture modelling process, that is the probabilistic classification of the data in terms of the different components identified in a given model. These are shown for the observed data and the expected value of the resampled data. For these examples, the resampled data tend to be less distinctly classified than the observed values, but the preferred classifications do not change. Figures 11.6d and 11.7d show the relationship between the observed and resampled data (with the error bars on the former being the input errors and those on the latter being the 95% credible range for the resampled values over all accepted models). Here we can see that this resampling process does not drastically change the values used in the modelling, as the input errors are fairly small (at least relative to the actual ages).

The results of the three different model formulations (Gaussian v. skew-t, and two different priors for the skew-t models) on the same data illustrate the effect of these model assumptions on the results. Visual inspection, coupled perhaps with additional geological information/constraints, allows us to assess which model may be preferable. While we do not address this aspect here, it is possible to make a more quantitatively based model choice by calculating the evidence from Bayes' rule (e.g. Sambridge et al., 2006; Trotta, 2008; Friel and Wyse, 2012). In words, this is the probability of having the observed data for a given hypothesis. In doing so, the data are fixed (i.e. the observed data), and this allows us to assess the probability of different hypotheses. The hypothesis that has the maximum evidence (or maximum support from the data) is then the most likely. The evidence is defined as

$$\mathbf{p}(\mathbf{d}|\mathcal{M}) = \int \mathbf{p}(\mathbf{d}|\mathbf{m}, \mathcal{M}) \, \mathbf{p}(\mathbf{m}|\mathcal{M}) \, \mathbf{dm},$$

where $\mathcal{M}$ is a given model formulation and we can see that the evidence is the expected likelihood (sometimes called the marginal likelihood) for that model formulation. From the evidence, we can calculate the posterior probability of the model formulation, given the data, again from Bayes' rule, that is,

$$\mathbf{p}(\mathcal{M}|\mathbf{d}) \; \alpha \; \mathbf{p}(\mathbf{d}|\mathcal{M})\mathbf{p}(\mathcal{M}),$$

Where $\mathbf{p}(\mathcal{M})$ is the prior probability of the model formulation. Typically, the priors are chosen to be equal for different models. Then that leads to the ratio of posterior probabilities for two models, a and b, being equal to the ratio of the evidence (also known as the Bayes factor, B_{ab}):

$$B_{ab} = \frac{\mathbf{p}(\mathbf{d}|\mathcal{M}_{\mathbf{a}})}{\mathbf{p}(\mathbf{d}|\mathcal{M}_{\mathbf{b}})}.$$

There are now various approaches have been proposed that provide practical recipes to do the appropriate calculations for the evidence and the Bayes factor in

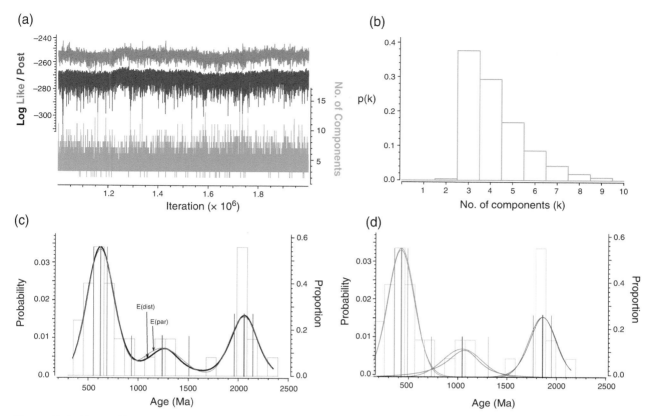

Figure 11.4 (a) Log likelihood (red), log posterior (blue), and number of components (green) of accepted models for the post-burn-in iterations for a mixture model for the data in Fig. 11.3, based on Gaussian component distributions. (b) The inferred posterior distribution on the number of components (the green curve in (a)), and most probable number of components is three. (c) The total distribution $(E(\mathbf{par}) = f(E(\mathbf{m})))$ for the three-component model based on the expected values of the mean and standard deviation for each component (Eq. 11.9a) and the expected distributionmathnormal $(E(\mathbf{dist}) = \mathbf{E}(f(\mathbf{m}))$, (Eq. 11.9b). The coloured vertical lines show the expected values for the location parameter (mean) of each Gaussian component with the dotted lines defining the 95% credible range. The height of each line indicates the expected value of the proportion (right-hand scale) for each component. (d) The individual component distributions for the three-component model. The dotted line distributions are those calculated using the expected values of the mean and standard deviation for each component. The solid line distributions are the expected distributions (as in Eq. 11.9b) for each component. The vertical lines are as for (c). The solid line distribution for the central component is slightly displaced relative to dotted line equivalent, but otherwise the two different estimates are similar.

a reasonable time (Liu et al., 2016; Pooley and Marion, 2018; Higson et al., 2019). One of the more reliable methods, thermodynamic integration (e.g. Lartillot and Philippe, 2006), can be implemented using an approach that also deals elegantly with the problem of multimodal likelihood functions mentioned earlier. This approach is known as parallel tempering (e.g. Sambridge, 2014) and requires running many MCMC simulations in parallel. Each MCMC simulation uses a slightly modified measure of how well we fit the data and allows the sampling to move between different modes in the likelihood function. Subsequently, all the sampling chains can be used to calculate the evidence (support provided by the data) for a given model hypothesis and so allow quantitative comparisons between different hypotheses.

11.2.2 Inverse Thermal History Modelling in Thermochronology

As outlined in the Introduction (Section 11.1), thermochronology takes advantage of the temperature sensitivity of different dating systems to infer rate and magnitudes of geological processes that involve changing temperature over time. The aim of inverse modelling is to use measured thermochronological data to reconstruct thermal history of a rock or mineral. In practice, this has been most widely considered in the context of low-temperature thermochronology (loosely defined as systems sensitive to temperatures < 300°, which include the fission track, (U-Th)/He, and ^{40}Ar/^{39}Ar methods (see Reiners and Ehlers, 2005; Braun et al., 2009). These systems are sensitive to temperatures from near surface values (~30–40°C) to a maximum around 300°C, or a little higher. The range

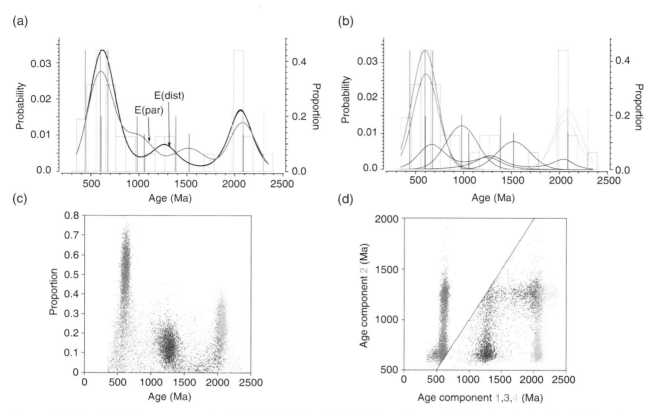

Figure 11.5 Results for four components for the data in Fig. 11.3. (a) shows the total distributions (as in Fig. 11.4c) and we see the distribution based on the expected values of the parameters, E(par), does not really capture the form of the data. The vertical bars represent the expected values for the location (mean) parameter, the dashed lines the 95% credible range, and the height of each line is the expected value of the proportion for each component. The lower limit for the credible range on the green component pretty much corresponds with the expected value for the red (youngest) component, and the upper limit for the blue component corresponds to the expected value for the cyan (oldest) component. This suggests that the four-component model is not particularly well resolved. (b) The individual component distributions (as in Fig. 11.4d). The expected distributions for the two of the central component are both bimodal reflecting a trade-off between the different components and also implies that the four-component model is not particularly well resolved. (c) Proportion v. location parameter (means for each component Gaussian distribution). As the second (green) component mean reduces, its proportion tends increase and in a form of compensation or trade-off, the first component age (red) also tends to decrease as does its proportion. (d) The mean of component distributions 1, 3, and 4 as a function of the mean age for component distribution 2 (the components have the same colour code as (a) and (b)). The black diagonal line is a 1:1 relation for the two axes. Components 1 (red) and 3 (blue) tend to trade off with component 2 such as if the 2 becomes younger, 1 also becomes younger and as 2 becomes older, 3 becomes older (and also 4 becomes older as 3 becomes older). This trade-off behaviour suggests that the four-component model is not particularly well resolved.

in temperature sensitivity is often referred to as the partial annealing zone (PAZ) for fission track analysis and the partial retention zone (PRZ) for noble gas diffusion (He, Ar). In these low-temperature systems, the range of temperature sensitivity which leads to loss of the daughter product is manifested as a progressive decrease in measured age depending on how long a given rock sample has resided in, or has taken to cool across, the partial retention/annealing zone. In this case, a measured age will not necessarily reflect any particular discrete geological or thermal event, but the data can still be used to try and recover information on the thermal history.

To undertake (inverse) thermal history modelling requires understanding, often empirical, of the controls

on the temperature sensitivity for different thermochronometric systems. The most important physical process is some form of thermally activated chemical diffusion (e.g. Reiners and Ehlers, 2005), which, in its most simple form, leads to the loss of the daughter product (e.g. the isotope ^{4}He in the case of (U-Th)/He dating) from the host mineral. Due to this loss, the age of the mineral will be reduced, and will be zero if all the daughter product has been lost. Partial loss, the extent of which depends on the thermal history, will lead to a reduced age (reduced relative to that expected with no loss of the daughter product).

Although not well understood, some kind of diffusion type process is also likely to be the main control on the

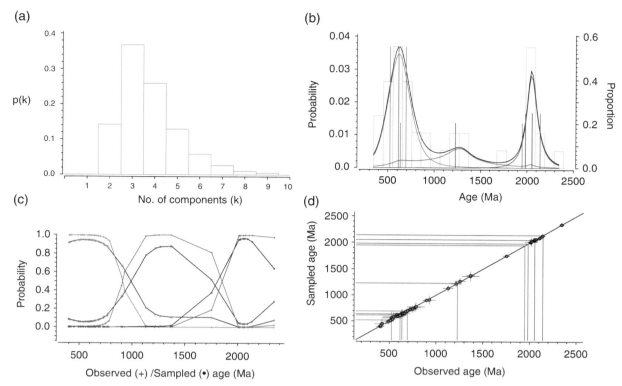

Figure 11.6 Mixture modelling results based on skew-t component distributions, with a prior oriented to low degrees of skew. (a) The inferred posterior distribution on the number of components and most probable number of components is three, similar to the Gaussian component distribution model (Fig. 11.4) but the probability of two components is higher. (b) The total and individual component distributions based on the expected distributions (Eq. 11.9b). The coloured vertical lines show the expected values for the location parameter (mean) of each component with the dotted lines defining the 95% credible range. The height of each line indicates the expected value of the proportion (right-hand scale) for each component. Note the central component is slightly trimodal, with minor peaks corresponding to the expected values of the minimum and maximum age components. The 95% credible intervals for the expected value of this central component also extend close to the expected values for two components. This behaviour suggests that the three-component model is not that well resolved (support by the higher probability of having two components for this skew-t model). (c) The classification probabilities (the probability that a given value is assignable to a given component) for the observed (+ plotted on the coloured lines) and resampled (solid circle plotted on black lines) ages. (d) The relationship between the observed and resampled ages, together with the expected values for the location parameter (mean) of each component shown by the solid lines and the dotted lines define the 95% credible range. The diagonal line is 1:1.

shortening, or annealing, of fission tracks (Li et al., 2011). A track, meta-stable physical damage to the crystal lattice due to the passage of positive charged ions, is considered as the relevant daughter product of the spontaneous fission of ^{238}U. New fission tracks are formed continuously over time and the final length of each individual track will depend on the temperature history it has experienced. The fission track age is a function of the distribution of lengths (for details, see Galbraith, 2005). Analogous with ^{4}He loss, track shortening (annealing) leads to a reduction in the fission track age. If all tracks are totally annealed, there is then no daughter product, and the age will be zero.

In terms of an inverse problem, we want to reconstruct temperature as a function of time (i.e. the thermal history). While some form of the heat transfer equation can be used to predict the thermal history (e.g. Braun et al., 2009; Licciardi et al., 2020), this requires specification of boundary conditions (heat flow, surface temperature) over time and also the relevant thermophysical properties (e.g. thermal conductivity, heat capacity, density, heat production). It is more common to try and extract the thermal history directly without a physically based model for heat transfer. While this implies a continuous function, the forward problem is generally formulated to represent the thermal history as a series of piecewise continuous linear segments, and the model parameters for a single sample are then defined as a finite number (N) of time-temperature points or nodes:

$$\mathbf{m} = \{t_i, T_i\}, i = 1, N.$$

In some cases, we deal with several samples (joint modelling) at the same time, under the assumption that they have all experienced the same form of thermal history (e.g. the same heating and cooling episodes) but not necessarily the same

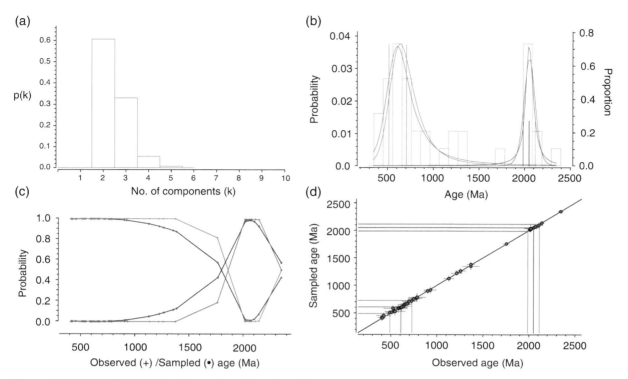

Figure 11.7 As for Fig. 11.6, but for a prior oriented to high degrees of skew. In this case, the preference for high skew leads to a preferred model with two components (see the caption of Fig. 11.6 for caption details). The classification (c and d) changes with the ages assigned to the central component in Fig. 11.6 (low skew model) tend to be taken up by the high skew of the younger component.

temperature values. Given that temperature generally increases with depth in the Earth's crust, it is often valid to assume that samples that are deeper in a borehole (or perhaps currently at lower elevations if exposed on the Earth's surface in a recently cut valley for example) were hotter than those at shallower depth (or higher elevations). This is referred to as vertical profile modelling. In this case, the model parameters are augmented by the temperature difference, or offset, between the top and bottom samples in the vertical profile, to give

$$\mathbf{m} = \{t_i, T_i, O_i\}, i = 1, N.$$

We choose either the top or the bottom sample as a reference point, and the offset then allows us to define a thermal history for each sample as a function of the depth or elevation difference between the reference sample and the sample of interest. The first MCMC approach proposed to solve this problem was presented by Gallagher et al. (2005), in which the number of nodes, N, was specified in advance. Gallagher (2012) later presented a trans-dimensional MCMC approach with N treated as an unknown.

Stephenson et al. (2006) presented an extension of the Gallagher et al. (2005) approach to allow joint modelling of spatially distributed samples, which uses partition modelling to group or cluster samples with a similar (but unknown) thermal history. The locations of the partitions are defined by the centres of Voronoi cells. This approach

has the advantage of allowing discontinuous spatial changes in the thermal histories, and therefore can readily deal with the presence of faults, for example. The inference of the partition structure was treated as a trans-dimensional problem (e.g. Denison et al., 2002), such that the number the partitions is allowed to vary, while the thermal history modelling was fixed dimensional (as in Gallagher et al., 2005). A fully trans-dimensional approach incorporating the RJ-MCMC thermal history modelling method (Gallagher, 2012) together with the partition model is currently under development. This is motivated in part by recent exchanges (Herman et al., 2013; Schildgen et al., 2018; Willett et al., 2021) related to spatial modelling of thermochronological data and the possibility of artefacts introduced in model results when samples are grouped together. If the grouping is based on distance criteria (closer is more likely to be similar) but may lead to incompatible thermal histories, due to faulting, for example.

Here we will look at the approaches for modelling single and vertical profile samples based on synthetic data, the aim being to illustrate how well we can recover a known thermal history model. The details of MCMC algorithms are presented in Gallagher (2012) and Licciardi et al. (2020) and we will not repeat them here. As described here, a thermal history for a vertical profile is defined as $\{t_i, T_i, O_i\}, i = 1, N$, and in a trans-dimensional approach, N, the number of nodes, is treated as an unknown. The prior is uniform, $N \sim U(2, N_{max})$, such that the thermal history will have a minimum of two nodes (the present-day temperature and one time-temperature point in the geological past) and up to N_{max}. The priors on the temperature, time, and offset are also uniform with the upper and lower limit determined by the nature of the data (different thermochronological systems having different temperature sensitivities). For example, a range from 0 to 140°C is appropriate for fission track analysis and (U-Th)/He dating on the mineral apatite, providing a slightly large range than the PAZ/PRZ defined earlier, to allow some flexibility to ensure that the sample can experience total annealing/diffusive loss (open system at higher temperature) and to be totally retentive (closed system at low temperature).

A characteristic of thermochronometric systems is that the data reflect a cumulative or time-integrated signal to variable degrees. The ability to recover the thermal history is often dominated by the maximum temperature experienced. In principle, thermal histories that involve multiple heating/cooling episodes can be recovered to variable degrees depending on the magnitude of the reheating event (maximum temperature experienced at the peak of heating). One consideration is that a reheating event will tend to overprint (or even totally erase) any existing thermal history information contained in the data if the maximum temperature exceeds the maximum temperature from any previous heating events. McDannell and Issler (2021) have demonstrated that some reheating events can be recovered (to variable extents), depending on the nature of the data available. Often, however, the data themselves do not necessarily require reheating events (even if the true thermal history contains them). In some cases, well-defined geological constraints are available. These can be imposed on the thermal models (as a form of prior information) to force all thermal histories to respect these constraints. Typical constraints can be the age of formation of an igneous rock (either intrusive or extrusive), or inferred ages of unconformities, implying that a particular sample was close to or at the surface at a given time. These constraints are generally defined in terms of a range for both temperature and time, the magnitudes of which should depend on the reliability of/confidence in the geological information available.

11.2.2.1 Predictive Models in Thermochronology
Diffusion Models For diffusion problems, we need to solve a differential equation (11.2). For (U-Th-Sm)/He dating, the production term defined in Eq. (11.2) is given as

$$P = 8^{238}\lambda[^{238}U] + 7^{235}\lambda[^{235}U] + 6^{232}\lambda[^{232}Th] + {}^{147}\lambda[^{147}Sm],$$
(11.10a)

where [] is concentration, or number of atoms, and λ is the decay constant for a given isotope. This reflects the production of He by decay of one atom of each the different parent isotopes

^{238}U (eight atoms of He), ^{235}U (seven atoms of He), ^{232}Th (six atoms of He) and ^{147}Sm (one atom of He). In the absence of diffusion, the concentration, or number of atoms, of the daughter (in this case He) produced after time t is given as

$$[^{4}He] = 8[^{238}U](e^{238_{\lambda t}} - 1) + 7[^{235}U](e^{235_{\lambda t}} - 1)$$
$$+ 6[^{232}TH](e^{232_{\lambda t}} - 1) + [^{147}Sm](e^{147_{\lambda t}} - 1).$$
(11.10b)

Of course, the interest in thermochronology comes from the temperature dependent diffusion aspect and so we need so solve the diffusion equation (Eq. 11.2).

Usually, we adopt some numerical approach to deal with the diffusion equation (e.g. Meesters and Dunai, 2002a; Braun et al., 2009) and we solve for the time-dependent distribution of the daughter isotope in a volume, or diffusion domain. The diffusion domain may represent a complete grain (as is the assumed for He in apatite, Farley, 2000), or an effective domain, not necessarily representing a physically identifiable region in a grain, as is the case for multi-domain diffusion models (e.g. Ar in feldspar, Lovera et al. (1989); He in calcite, Cros et al. (2014); or He in haematite, Farley and McKeon (2015)). Consequently, the size and geometry of the grain (or diffusion domain) are important inputs for the modelling procedure. In many cases, we use an effective grain size (e.g. the radius of a spherical grain with the same volume/surface area ratio as the actual grain) and solve a 1D diffusion problem rather than for a full 3-D geometry. This speeds up the forward model and has little effect on the final solution (Meesters and Dunai, 2002a). The predicted age is given as the ratio of the daughter to the production rate, integrated over the grain/domain volume, such as

$$Age = \int_{V} \frac{[D(\mathbf{r})]}{P(\mathbf{r})} \, d\mathbf{r}.$$
(11.11)

For He dating, we also need also allow for the fact that alpha particles are ejected some distance (10–30 µm; Farley et al., 1996) from the parent nucleus. The ejection distance depends on the nature of both the parent nucleus and the host grain lattice. If the distance from the parent atom to the edge of the grain is less than the ejection distance, then there is a finite probability that the alpha particle will be ejected out of the grain, leading to a lower than expected He age. However, once ejection distances are known, it is straightforward to allow for this effect (Meesters and Dunai, 2002b) at the same time as modelling diffusive loss of He.

In some cases (e.g. apatite and zircon He dating in cratonic domains), we may need to allow for the dependence of the diffusivity on radiation damage (apatite: Flowers et al., 2009, Gautheron et al., 2009, Recanati et al., 2017; zircon: Willet et al., 2017, Guenthner et al., 2013, Ginster et al., 2019). During radioactive decay of U and Th isotopes, an alpha particle is ejected out of the parent nucleus. In response, the parent nucleus recoils (like the kick of a rifle). This recoil creates damage to the surrounding grain lattice (Shuster et al., 2006). Such damage zones can then trap diffusing He which then requires an anomalously high activation energy (or higher effective diffusivity) to escape the damaged zones, or traps. As the amount of radiation damage increases, the effective diffusivity increases up to a point where the damage zones may become connected, and then the effective diffusivity may decrease (Ketcham et al., 2013). This effect is more pronounced in zircon than apatite due to the higher concentrations of U and Th. Given the low contribution to the total He budget by decay of Sm and its low value of decay constant, we can ignore this contribution. When considering this radiation damage effect, models are typically calibrated in terms of effective uranium, $eU = U + 0.235Th$. The value of eU is effectively the equivalent concentration of U, allowing for the different concentrations and decay rates of U and Th.

Annealing Models The approach used for modelling fission track shortening, or annealing, is somewhat different, although again based on calibrating parameters from laboratory experiments. In a series of landmark papers in the mid 1980s, researchers at the University of Melbourne presented the results of laboratory experiments calibrating the temperature and time dependence of fission track length in apatite (Green et al., 1986; Laslett et al., 1987; Duddy et al., 1988, 1989). They heated grains of Durango apatite to remove any existing natural fission tracks, and produced new fission tracks by irradiating the grains with thermal neutrons in a nuclear reactor. Subgroups of the grains when exposed to different thermal conditions (i.e. a constant time at a constant temperature), similar in a way to the diffusion experiments described earlier. The lengths (l) of fission tracks under the different time temperature conditions were measured to provide calibration data for the annealing model. Subsequently, data from additional annealing experiments were presented on apatite (Carlson et al., 1999; Donelick et al., 1999) and zircon (Tagami et al., 1998; Yamada et al., 2007) and similarly used to calibrate annealing models.

The initial length (l_0) of a newly formed fission track is more or less constant (around 16–17 m in apatite and 12–13 m in zircon. The predictive models are parametrised in terms of the fractional reduction ($r = l/l_0$)) in track length (i.e. annealing), as a function of temperature and time. A typical mathematical formulation for an annealing model (e.g. Ketcham et al., 1999, 2007) with a time duration of t at temperature T, is given as

$$\frac{\left[\frac{(1-r^\beta)}{\beta}\right]^\alpha - 1}{\alpha} = c_{0+}c_1 \left[\frac{ln(\Delta t) - c_2}{\left(\frac{1}{T}\right) - c_3}\right], \qquad (11.12)$$

where α, β, and c_{1-4} are all parameters estimated from the laboratory data. Note this model is essentially empirical and is not based on any profound understanding of the annealing process at the atomic level. In practice, the models can be modified to allow for additional factors such as chemical

composition (e.g. Cl-rich apatites are more resistant to annealing than F-rich apatites), anisotropic annealing (tracks parallel to the c crystallographic axis of apatite tend to anneal more slowly than tracks perpendicular), and segmentation of tracks at high degrees of annealing (Green, 1988; Ketcham et al., 1999, 2007). Calibration of models based on different apatite composition leads to different values of the parameters in Eq. (11.12). Thus, the basic formulation remains the same but the predictions (track length) will differ depending on the apatite compositional parameters.

The formulation in Eq. (11.12) involves a constant temperature over a given time interval. To apply these models to a time-varying thermal history, the principle of equivalent time (Goswami et al., 1984) is used to set up a time-stepping approach. This is developed as follows: we calculate the fractional track length (r_1) for the first time step, t_1, at the first (constant) temperature (e.g. the average temperature over the time step, given as T_1). For the next time step, t_2 with a different temperature, T_2, we first calculate how long we would need to anneal a new track at temperature, T_2, to produce same the fractional track length (r_1) as the first step. This done by writing Eq. (11.12) as

$$\frac{\left[\frac{(1-r_1^\beta)}{\beta}\right]^\alpha - 1}{\alpha} = c_{0+}c_1 \left[\frac{ln(\Delta t_{eq}) - c_2}{\left(\frac{1}{T_2}\right) - c_3}\right], \qquad (11.13)$$

and solving for t_{eq}, the equivalent time. Then we use Eq. (11.12), setting $t = t_2 + t_{eq}$, and $T = T_2$ to calculate the fractional track length (r_2) at the end of the second time step. We simulate the formation of fission tracks at a continuous rate over time. In practice, this involves simulating new track formation at constant time intervals over the duration of the thermal history (t_{TOTAL}). The equivalent time approach process is repeated for each time step over the duration of thermal history relevant for each simulated track (which depends on its time of formation).

The fission track age can be calculated in different ways, but is effectively based on the mean fractional track length of the simulated tracks, converted into an equivalent fractional track density using a formulation similar to

$$\rho = a + br, \qquad (11.14a)$$

where a and b are constants estimated from laboratory data (e.g. Green, 1988; Ketcham, 2005). The predicted age can be estimated using a function such as

$$Age_{FT} = \frac{\rho}{\rho_0} t_{TOTAL}. \qquad (11.14b)$$

The term ρ_0 is a correction factor based on the observation that the longest natural (spontaneous) tracks are shorter than the longest artificial (induced) tracks. The predicted track length distribution is given by first converting the fractional track length to an actual length by choosing a value for the initial length, l_0, which also depends on the composition of apatite. Then taking each of these lengths in turn, we use a Gaussian distribution kernel, using the length as the mean, and a standard deviation (which is inversely proportional to the length, and calibrated from laboratory experiments). These individual distributions are then summed to produce the final predicted track length distribution.

11.2.2.2 Data and Likelihood

The data used to constrain the thermal histories are primarily ages but in the case of fission track analysis, we also have the track length distribution. When considering (U-Th/He) dating, the data set typically contains several ages often made on grains of different sizes and or eU contents. For fission track analysis, the typical data set is a series (20–30) of ages measured on single grains, and the track length distributions (typically based on 50–100 individual length measurements). In simple terms, we can think of the oldest ages as providing constraints on the overall duration of the thermal history we can recover. Detail of the temperature variations is contained in the variability of ages, reflecting the temperature sensitivity of diffusion as a function of grain size and composition (the eU effect discussed earlier), and for fission track analysis, in the form of the track length distribution.

To implement MCMC, we need to define the likelihood function and the form of this depends on the data type. For example, when using (U-Th)/He age data, we use a Gaussian likelihood,

$$\mathbf{p}(\mathbf{d}|\mathbf{m}) = \prod_{i=1}^{N} \frac{1}{\sigma_i\sqrt{2\pi}} e^{-\frac{1}{2}\left(\frac{d_i^{obs}-d_i^{pred}}{\sigma_i}\right)^2}, \quad (11.15)$$

where N is the number of observations (ages), d_i^{obs} is the i-th observation (age), d_i^{pred} is the equivalent predicted value for a particular thermal history model (m), and σ_i is the measurement error or uncertainty associated with d_i^{obs}.

Gallagher (1995) presented a likelihood function appropriate for fission track count (age) and length data for the external detector method (EDM). The count data represent the number of natural (spontaneous) and induced counts (N_s and N_i). The spontaneous tracks represent the daughter product of natural fission of ^{238}U, and the induced tracks represent the daughter product of reactor-induced fission of ^{235}U. The induced tracks are used to estimate the present day concentration of ^{238}U, and the calculated age with the EDM approach uses the ratio N_s/N_i and the daughter parent ratio in a slightly modified version of the age equation (Eq. 11.1, and see Galbraith, 2005, for more details).

The likelihood for the count data is based on the assumption that these spontaneous and induced counts (N_s and N_i) follow independent Poisson distributions for measured single grain age. It is possible to derive a distribution for the spontaneous counts, conditional on the total number of tracks counted, and this is a binomial distribution given as

$$\mathbf{p}(N_s|N_s+N_i) = \frac{(N_s+N_i)!}{N_s!N_i!}\theta^{N_s}(1-\theta)^{N_i}, \quad (11.16)$$

where $\theta = \frac{\rho_s}{\rho_s+\rho_i}$ and $\frac{\rho_s}{\rho_i}$ is the predicted ratio of spontaneous to induced track densities, calculated from the predicted age (*Age*), for a given thermal history model, as

$$\frac{\rho_s}{\rho_i} \approx (e^{\lambda Age}-1)\beta r, \quad (11.17)$$

where λ is the alpha decay constant for ^{238}U, r is the predicted fractional length $r = l/l_0$, and β is a constant determined from the observed age value (see Gallagher 1995).

The likelihood for observed track lengths uses the predicted length distribution, $f(l)$, as a probability distribution, and the likelihood of having a measured length, l_k, is obtained by integrating this distribution over a small range centred on the measured length, for example,

$$\mathbf{p}(l_k) = \int_{l_{k-0.1}}^{l_{k+0.1}} f(l)dl. \quad (11.18)$$

Taking all fission track (FT) count and length data, the combined likelihood is just the product of all the individual likelihoods for N_{age} single grain ages and N_{length} track length measurements

$$\mathbf{p}(FT\text{-}data) = \prod_{j=1}^{N_{age}} \mathbf{p}(N_s^j|N_s^j+N_i^j) \prod_{k=1}^{N_{length}} \mathbf{p}(l_k). \quad (11.19)$$

To avoid numerical problems in the implementation of the MCMC approach, the log of the likelihood is used in preference to the actual likelihood functions defined in this section.

11.2.2.3 Examples of MCMC Inversion with Thermochronological Data

Here we will present an example of inverse modelling based on a known thermal history and synthetic data produced from thermal history. The thermal history starts at temperatures above the effective closure temperature for all samples and subsequently contains two heating/cooling events in the geological past that produce a range of maximum temperatures across the partial annealing/retention zones described earlier. The idea is to explore how we much of the thermal history information we can recover in a relatively standard inversion approach with relatively standard data. The thermal history incorporates a constant temperature gradient of 25°C/km (or temperature offset of 50°C) over time until the present day. The present day temperature gradient was 7°C/km (typical of the atmospheric temperature lapse rate) to represent the samples ending up on the Earth's surface (e.g. exposed at different elevations in a valley). The prescribed thermal history and associated predicted data are shown in Fig. 11.8.

We produce synthetic data for nine samples from a vertical profile covering an elevation range of 2,000 m. For each sample, we have apatite fission track data (with 30 single grain ages, and 100 track length measurements), and 5 (U-Th)/He ages, with grain size of 30, 40, 50, 60, and to 70 m. We add noise to the He data (sampled from a Gaussian distribution with mean = 0 and standard deviation of 5% the predicted age). For both types of data, we do not incorporate any variation in chemical composition, nor the radiation damage effect for the (U-Th)/He data. The fission track data (counts and lengths) are randomly sampled from predicted distributions so have some inherent uncertainty but probably less that we might expect for most real data sets. Given this, both types of data would be considered high quality.

Single Sample Modelling We first consider just one sample for the inverse modelling to demonstrate some of the characteristics associated with recovering the original thermal history in that case. We chose the upper and lower samples (whose thermal histories correspond to the upper and lower temperature paths shown in Fig. 11.8) and will examine the results for each in turn. The priors for the thermal history are all uniform with a maximum of 50 time-temperature points. The time prior range was defined as 75 ± 75 Myr and the temperature prior range is $70 \pm 70°C$, except for the present day temperature which has a prior of $10 \pm 10°C$. The MCMC sampling chain was run for 10^5 burn-in iterations and 10^5 post-burn-in iterations.

Figure 11.9 summarises the results for the upper sample. Figure 11.9a shows the log likelihood (red), log posterior

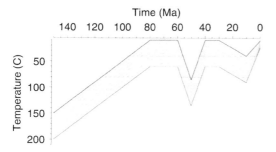

Figure 11.8 The specified thermal history used to produce synthetic thermochronological data. The upper and lower lines are the thermal histories for the highest and lowest elevation samples respectively and the thinner grey lines indicate the thermal histories for the samples at intervening elevations. The light horizontal lines denote the approximate partial annealing zone (PAZ) for fission track annealing and the darker horizontal lines the partial retention zone (PRZ) for He diffusion in apatite. If a thermal history exceeds the upper limit of the PAZ/PRZ, the relevant thermochronological system is effectively reset to zero. Then the data do not contain any information on the thermal history prior to that time.

(blue), and number of time-temperature points (green) of accepted models for the post-burn-in thermal history models. While the likelihood remains fairly flat, the posterior shows a tendency to have an inverse correlation with the complexity (number of time-temperature points) in the thermal history, illustrating the natural parsimony of the Bayesian approach. Figure 11.9b summarises a range of different thermal history models for this one sample. The coloured pixel image represents what is known more formally as the marginal posterior distribution of temperature. We could take a slice from this image at a given time (e.g. 100 Ma), and produce a histogram of temperature at 100 Ma. This is a conditional probability distribution, that is, the probability of temperature at a given time conditional on the variation of accepted temperature values at all other times. Note this is not the same as the probability of the temperature at 100 Ma (due to the conditional dependence on the other parts of the thermal history). We can see that the green-yellow colours (higher probability) highlight two segments of effectively linear cooling (from about 150 to 100 Ma and 60–70 Ma to the present day, 0 Ma) with little obvious indication of reheating.

Figure 11.9b also presents some individual thermal histories models: the maximum likelihood (best data flitting equivalent to the highest value of the likelihood in Fig. 11.9a), maximum posterior (highest value of the posterior in Fig. 11.9a), and also the expected model, with the 95% credible interval range also shown. The expected model (averaged over all accepted thermal histories irrespective of the number of time-temperature points), is generally relatively smooth as it is a form of (weighted) average. The highest values of any peaks of the marginal distribution as a function of time define the maximum mode distribution. When the marginal distribution is highly asymmetrical at any given time, the expected model does not lie along the peaks (modes), but deviates towards the direction of maximum skew. For example, in comparison to the max. likelihood/max.posterior (ML/MP) models, we can see in Fig. 11.9b that the expected model underestimates the maximum temperature around 70–60 Ma. This timing is a little earlier than the true thermal history (50 Ma). However, we can see that the credible intervals become narrower around 50 Ma, and also support the possibility of a reheating. For example, we can see that the lower limit of the credible interval prior to 50 Ma approaches the lower limit of the temperature prior while the upper limit tends to be lower than the value around 50 Ma. This implies that the magnitude of the reheating after the initial cooling ending around 100 Ma is not well resolved, but the data perhaps inform us better about the timing of the most recent reheating. We also note that the maximum likelihood model is relatively complex, with some structure that is not present in the true model (e.g. prior to 120 Ma). This is often the case as variations can lead to just a slight improvement, or even no change, in the likelihood, but these models can still be accepted.

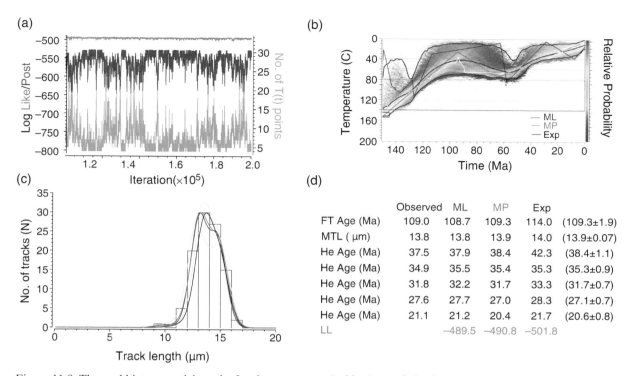

Figure 11.9 Thermal history model results for the uppermost (coldest) sample in the vertical profile. (a) Log Likelihood/posterior and number of time-temperature points as a function of the postburn-iteration. (b) Thermal history models. The maximum likelihood (ML – dark red), maximum posterior (MP – dark green) and expected (EXP – black) thermal history models are shown, with the 95% credible range shown as the black lines either side of the expected model. The colour coding shows marginal posterior thermal history, indicating relative probability (red = higher probability, blue = lower probability) for the thermal history paths and the white thermal history is constructed by joining the peaks in the marginal distribution at 1-Myr time intervals. The red box indicates the prior range for time and temperature. The small black box centred on 150 Ma and 150°C is introduced as a constraint on the thermal history to force the thermal history models to start at a temperature above the closure temperatures. (c) Predicted track length distributions for the maximum likelihood, maximum posterior, and expected thermal history models (using the same colour coding as (b)) and the grey lines indicate the 95% credible intervals on the predicted distributions. (d) Summary of the predicted ages (and mean track length, MTL) for the for the maximum likelihood, maximum posterior, and expected thermal history models. The log likelihood value for each is also given (LL, in red). The values in parentheses are the expected (mean) values for the predictions for all accepted thermal history models with the ± being two standard deviation range.

Relative to the ML and MP models, the predictions from expected model are not close to the observed values (as can be seen in Figs. 11.9c and d) with lower log likelihood values. Also we can see that the predicted ages, and also the track length distributions, are a little older/longer. This is a relatively commonly feature and is due to the lower maximum temperatures in the reheating episode around 50–60 Ma that do not produce enough diffusive loss/annealing. Of course, the expected model is not an individual model sampled during the MCMC iterations so is not directly constrained through fitting the data. Figure 11.10 illustrates a selection of 1,000 models selected from the accepted post-burn-in models, colour coded in terms of the relative likelihood/posterior. The individual models tend to have a form of cooling from the initial constraint at 150 Ma at 150°C, a reheating starting anywhere between 120 and 60 Ma to a maximum temperature of up to about between 70 and 50 Ma, 90°C and a final cooling to near surface temperatures at the present day. When classified

according to the likelihood (Fig. 11.10a), we see higher likelihood models (e.g. those coloured red/pink) can be relatively complex (e.g. some structure early in the thermal history, multiple minor reheating/cooling events after 50 Ma, a wide range on timing associated with the initiation of the main reheating event and the maximum temperature of that reheating relatively loosely constrained). We see a collection of less complex models when classified by the posterior value (Fig. 11.10b). This is as expected given that the Bayesian approach tends to penalise complexity and more complex models tend to have lower posterior values. The models with higher posterior (again the red/pink models in Fig 11.10b) capture the form of the true thermal history fairly well, with some dispersion in the initial timing of the reheating event, but less than that implied based on the likelihood classification.

Overall, the reconstructed thermal histories using the uppermost sample seem to identify the timing of the peak temperature for the reheating event around 50 Ma. The data

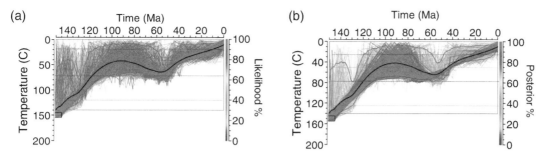

Figure 11.10 (a) The expected thermal history, as shown in Fig. 11.9b, overlain by a sample of 1,000 accepted thermal histories from the MCMC chain, colour coded by the relative likelihood, such that 100% is the maximum and 0% is the minimum of all the accepted model likelihoods. (b) The expected thermal history, as in (a), overlain by a sample of 1,000 accepted thermal histories from the MCMC chain, colour coded by the relative posterior probabilities, such that 100% is the maximum and 0% is the minimum of all the accepted model posterior probabilities.

provide little information on the rates of heating/cooling either side of this event and do not seem to require a significant structure in the thermal histories after 50 Ma. This is expected due to the fact that the temperatures for the uppermost sample never exceed 40°C after 40 Ma. This is outside the PRZ/PAZ range and so the data are not sensitive to any temperature variations in the true thermal history.

We now consider the results based on the lowest (hottest) sample in the vertical profile and use the same prior as the upper sample. Given the nature of the data, the PRZ/PAZ ranges, we do not expect the data to be sensitive to temperatures > 130–140°C (i.e. the range of the prior on temperature) so the fact the prior does not span the temperature range of the true model for this sample is not a problem. We just want to compare how much of the thermal history we can recover for this single sample. Given the true thermal history (Fig. 11.8), we expect that the resolution of the thermal history prior to 50 Ma will be poor. This is because the maximum temperature of the lowest sample exceeds the upper limit or the PRZ/PAZ and so all information on the earlier thermal history is lost. The results are given in Fig. 11.11 in the same format as Fig. 11.9. Again, we see the inverse correlation between the posterior and the model complexity (Fig. 11.11a) with the maximum posterior model having the lowest number of time-temperature of all accepted post-burn-in models. We see the anticipated lack of resolution of the thermal history prior to 50 Ma in terms of the credible intervals effectively span the prior range and the expected model temperature over that time is the mean of the prior. Between 50 and 10 Ma, the resolution of the thermal history is still relatively poor. However, in contrast to the uppermost sample, this sample does constrain the recent cooling event starting at 10 Ma and also constraints the upper range of the marginal posterior to be less than the maximum temperature at 10 Ma (i.e. < 90°C). Again, we see the expected model predictions are not as good as those from the maximum likelihood/posterior models and the likelihood values are significantly lower. This is because this sample primarily constrains the last cooling event, and due to the

nature of smoothing implicit in the expected mode, the maximum temperature at that time is underestimated relative to the individual models. Then the observed length distribution is not well reproduced (Fig. 11.11c) and the ages are overpredicted (Fig. 11.11d).

Having considered two (upper and lower) samples independently, we have seen how they can inform us differently about the temperature history with variable resolution depending on the temperatures relative to the PRZ/PAZ (i.e. the temperature sensitivity of the thermochronological systems under consideration). Therefore, an obvious approach is to model all samples jointly, to take advantage of any common information and exploit information particular to individual samples.

Multi-sample (Vertical Profile) Modelling For the multi-sample (vertical profile) inverse model, we use just five of the nine samples, but make predictions of the data for the other four samples to highlight the fact that vertical profiles samples contain common information that can improve the resolution of the combined thermal history. In this case, we define the prior on time and temperature for the uppermost sample in the profile and use the same priors as for the single sample models. When modelling a vertical profile, we add the extra temperature offset (or gradient) parameters and we use a temperature gradient prior range that was 30 ± 15°C/km (with 10 ± 10°C/km for the present day). The MCMC sampling chain was run for 120,000 burn-in iterations, then 50,000 post-burn-in iterations.

The likelihood, posterior, and number of time-temperature points for the accepted models in the post-burn-in stage are shown in Fig. 11.12. As in the single sample model runs, we can see that while the likelihood is relatively constant, the number of time-temperature points (representing in part the complexity of a continuous function) varies from 5 to 27, but generally stays at relatively low values (< 15). Again, there is a clear visual anti-correlation between the number of time-temperature points and the posterior value, reflecting the

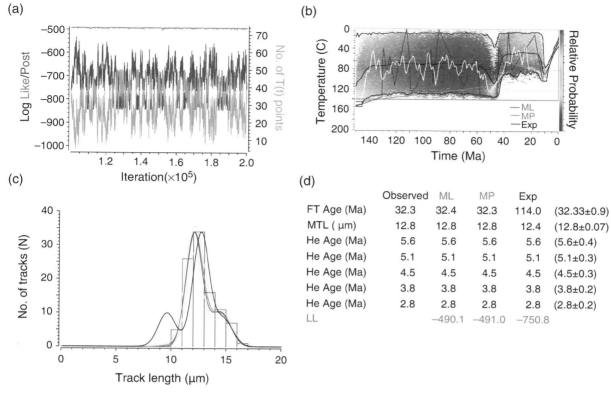

Figure 11.11 As for Fig. 11.9, but for the lowermost (hottest) sample in the vertical profile.

natural parsimony of the trans-dimensional MCMC approach (i.e. the preference for simpler solutions). However, this does not preclude sometimes accepting more complex solutions but reduces the tendency of accepting them, unless the likelihood improves significantly.

The recovered thermal histories are shown in Fig. 11.12 with the expected model (Fig. 11.12a) and the maximum posterior model (Fig. 11.12b) – the maximum likelihood model is not significantly different to the posterior. We can see the advantages of the multi-sample relative to the single-sample modelling. As we expect, the lower temperature parts of the thermal history, prior to a reheating, are not well resolved. However the timing and maximum temperatures of the two reheating events (50 and 10 Ma) are well resolved (as indicated by the narrow credible intervals at those times in Fig. 11.12b). The data are well reproduced (Fig. 11.12d), including for the samples that were not used for the inversion (marked by the * on the error bars). The expected model again slightly, but systematically, over-predicts the ages, but less dramatically than for the single samples. This is again attributable to the smoothing of the thermal history, and in particular, a reduction in the peak temperatures. In practice, each sample has the possibility to constrain different parts of the thermal history better than others. Effectively, this is the part of the thermal history where a given sample is in the appropriately sensitive temperature range (e.g. the PRZ/PAZ

temperature ranges for apatite). As was discussed for the single sample models, the upper sample spends a significant part of the total thermal history (from about 140 Ma to the present day) in that range, while the lower sample cools into that range around 50 Ma. Clearly, if the vertical profile has remained structurally intact, the samples in between will record the thermal history over a range of durations between those two end members.

Finally, Fig. 11.13 shows a selection of individual thermal history models colour coded by the log likelihood (Fig. 11.13a) and the log posterior (Fig. 11.13b). In contrast to the same representation for single sample models, the two are fairly similar, except for that part of the thermal history poorly or not at all constrained by the data (prior to around 120 Ma). Again, this demonstrates the advantages of jointly modelling the data such that the true signal is reinforced, and random noise tends to destructively interfere (i.e. cancels out). Furthermore, the unwarranted complexity is reduced as the combined data does not allow such variation in the thermal histories to be accepted, as the combined likelihood is more sensitive than that for individual samples.

11.3 Summary

We have considered two applications of Bayesian trans-dimensional MCMC modelling in the context of

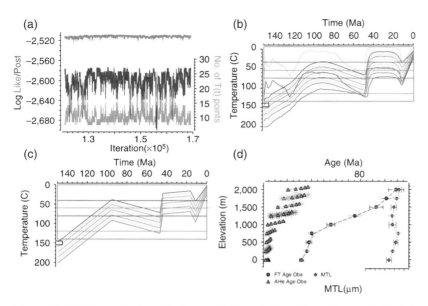

Figure 11.12 The vertical profile thermal history inferred from the synthetic data produced from the thermal history in Fig. 11.8. (a) Log-likelihood/posterior and number of time-temperature points as a function of post-burn iteration. (b) The expected model with the 95% credible ranges on the upper (blue) and lower (red) thermal histories (cyan and magenta lines respectively). The grey lines are the thermal histories for samples used in the inversion and the yellow lines are the thermal histories for samples not used in the inversion. The large red box is the prior for time and temperature for the upper sample thermal history. The small black box centred on 150 Ma and 150°C is introduced as a constraint on the upper thermal history to force the thermal histories to start at a temperature above the closure temperatures. (c) The maximum posterior thermal history model (and see the caption for (a) for details). (d) The observed (input synthetic) data (symbols) and predictions (lines) for the maximum posterior model (Fig. 11.10b). The predicted FT ages for the expected model (Fig 11.10a) are also shown (the cyan line) which are a little older than those for the maximum posterior model (a similar tendency to slightly older ages is seen for the AHe ages and the mean track length (MTL) are little longer than those for the posterior model, but not shown here). This is due to the inherent smoothing in the expected model leading to slightly lower maximum temperature and/or less time at higher temperature, and so less annealing/diffusion (loss of the daughter product).

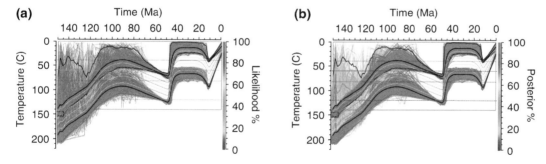

Figure 11.13 (a) The expected thermal history, as in Fig. 11.11b, overlain by a sample of 1,000 accepted thermal histories from the MCMC chain, colour coded by the relative likelihood, such that 100% is the maximum and 0% is the minimum of all the accepted model likelihoods. (b) The expected thermal history, as in (a), overlain by a sample of 1,000 accepted thermal histories from the MCMC chain, colour coded by the relative posterior probabilities, such that 100% is the maximum and 0% is the minimum of all the accepted model posterior probabilities. The posterior probabilities of complex models that do not change the likelihood significantly are low (e.g. those with temperatures less than 80°C prior to around 120 Ma).

geothermochronology. The advantage of this approach is that the model complexity, typically in terms of the number of model parameters, adapts to the information contained in the data. The first application, mixture modelling, has a truly finite number of parameters and the model complexity is reflected in the number of components, and the number of

parameters associated with the component distributions. We presented results with a real data example, and two different component distribution models, Gaussian (with two parameters) and skew-t (with four parameters, and contains the Gaussian as a special case). The skew-t approach was implemented with two different priors, one preferring light skew and the other preferring heavier skew. As we expect, the results show that the heavier skew models imply fewer components than more symmetrical models. In terms of choosing a preferred model, the posterior probability distribution for the number of components is a useful guide. However, it was also shown that some models that seem reasonable (e.g. the second most probable number of components) can reflect a form of trade-off between the components. Examination of the expected distributions (total or the individual components), calculated under different assumptions (based on the distributions themselves or the parameters of the distributions) can reveal this effect, and aid in model choice for a given set of model assumptions (e.g. Gaussian or skew-t distributions). The second application, extracting thermal histories from thermochronological data. Physically, the model is a continuous function (temperature over time) but the inverse model parameterisation is typically in terms of a finite number of parameters (nodes of the time-temperature function, and linear interpolation between the nodes). Again, the model complexity is reflected in the number of time-temperature nodes that define the form of the thermal history. We considered examples based on synthetic data in a vertical profile for which we know the true thermal history. The aim was to illustrate how different data inform us about different parts of the thermal history. Modelling constrained by just one sample illustrates this aspect. The upper (coldest) sample in the vertical profile contains the earlier part of the thermal history, but loses detail once the temperature is below the lower limit of temperature sensitivity of the data types used. In contrast, the lower (hottest) sample spends most of the time above the upper limit of temperature sensitivity, and we only recover the most recent cooling event during which the sample rapidly crosses the full range of temperature sensitivity. Jointly modelling multiple samples exploits this aspect of different information available from different samples to define a common form of thermal history experience by all samples. The actual thermal history for a given sample (at a specific elevation or depth today) is defined relative to a reference thermal history (e.g. that for the upper sample) using a temperature offset/gradient parameter that may or may not be constant over time. The results of the joint modelling demonstrate improved resolution of the thermal history relative to that inferred from single samples. Work is ongoing to extend the multi-sample approach to samples distributed over a region (rather than just a local vertical profile) based on trans-dimensional partition models (for spatial clustering, allowing for local discontinuities) coupled with the trans-dimensional thermal history models as presented here.

References

Braun, J., Van der Beek, P., and Batt, G. (2009). *Quantitative Thermochronology*. Cambridge: Cambridge University Press.

Carlson, W. D., Donelick, R. A., and Ketcham, R. A. (1999). Variability of apatite fission-track annealing kinetics: I. Experimental Results. *American Mineralogist*, 84, 1213–23

Cros, A., Gautheron, C., Pagel, M. et al. (2014). ^{4}He behavior in calcite filling viewed by (U-Th)/He dating, ^{4}He diffusion and grainlographic studies. *Geochimica et Cosmochimica Acta*, 125, 414–32.

Denison, D. G. T., Holmes, C. C., Mallick, B. K., and Smith, A. F. M. (2002). *Bayesian Methods for Non-linear Classification and Regression*. Chichester: John Wiley and Sons.

Dodson, M. H. (1973). Closure Temperature in Cooling Geochronological and Petrological Systems. *Contributions to Mineralogy and Petrology*, 40, 259–74.

Donelick, R. A., Ketcham, R. A., and Carlson, W. D. (1999). Variability of apatite fission-track annealing kinetics II: Grainlographic orientation effects. *American Mineralogist*, 84, 1224–34.

Ducassou, C. (2009). Age et origine des premiers reliefs de la chaîne hercynienne: le dévonocarbonifère du bassin d'Ancenis. Ph.D. thesis (in French), Université Rennes 1.

Duddy, I. R., Green, P. F., and Laslett, G. M. (1988). Thermal annealing of fission tracks in apatite 3. Variable temperature annealing. *Chemical Geology: Isotope Geoscience Section*, 73, 25–38.

Duddy, I. R., Green, P. F., Laslett, G. M. et al. (1989). Thermal annealing of fission tracks in apatite 4. Quantitative modelling techniques and extension to geological timescales. *Chemical Geology: Isotope Geoscience Section*, 79, 155–82.

Everitt, B.S., and Hand, D.J. (1981). *Finite Mixture Distributions*. New York: Chapman and Hall.

Farley, K. A. (2000). Helium diffusion from apatite: General behavior as illustrated by Durango fluorapatite. *Journal of Geophysical Research*, 105(B2), 2903.

Farley, K. A., and McKeon, R. (2015). Radiometric dating and temperature history of banded iron formation–associated hematite, Gogebic iron range, Michigan, USA. *Geology* 43(12), 1083–6. https://doi.org/10.1130/G37190.1.

Farley, K. A., Wolf, R. A., and Silver, L. T. (1996). The effects of long alpha-stopping distances on (U-Th)/He ages. *Geochimica et Cosmochimica Acta*, 60, 4223–9.

Flowers, R. M., Ketcham, R. A., Shuster, D. L., and Farley, K. A. (2009). Apatite (U-Th)/He thermochronometry using a radiation damage accumulation and annealing model. *Geochimica et Cosmochimica Acta*, 73, 2347–65

Friel, N., and Wyse, J. (2012). Estimating the evidence: A review. *Statistica Neerlandica*, 66, 288–308. https://doi.org/10.1111/j.1467-9574.2011.00515.x.

Galbraith, R. (1988). Graphical display of estimates having differing standard errors. *Technometrics*, 30(3), 271–81. https://doi.org/10.2307/1270081.

Galbraith, R.F. (2005). *Statistics for Fission-Track Analysis*. New York: Chapman and Hall.

Galbraith, R. F., and Green, P. F. (1990). Estimating the component ages in a finite mixture. *Nuclear Tracks and Radiation Measurements*, 17(3), 197–206.

Gallagher, K. (1995). Evolving thermal histories from fission track data. *Earth Planetary Science Letters*, 136, 421–35.

Gallagher, K. (2012). Transdimensional inverse thermal history modeling for quantitative thermochronology. *Journal of*

Geophysical Research: Solid Earth, 117, B02408. https://doi.org/10.1029/2011JB00882.

Gallagher, K., Stephenson, J., Brown R., Holmes, C., and Fitzgerald, P. (2005). Low temperature thermochronology and modelling strategies for multiple samples 1: Vertical profiles. *Earth Planetary Science Letters*, 237, 193–208.

Gautheron, C., Tassan-Got, L., Barbarand, J., and Pagel, M. (2009). Effect of alpha-damage annealing on apatite (U-Th)/He thermochronology. *Chemical Geology*, 266, 157–70.

Ginster, U., Reiners, P. W., Nasdala, L., and Chutimun Chanmuangg, N. (2019). Annealing kinetics of radiation damage in zircon. *Geochimica et Cosmochimica Acta*, 249, 235–46.

Goswami, J. N., Jha, R., and Lal, D. (1984). Quantitative treatment of annealing of charged particle tracks in common rock minerals. *Earth Planetary Science Letters*, 71, 120–8.

Green, P. F. (1988). The relationship between track shortening and fission track age reduction in apatite: Combined influences of inherent instability, annealing anisotropy, length bias and system calibration. *Earth Planetary Science Letters*, 89, 335–52.

Green, P. F., Duddy, I. R., Gleadow, A. J. W., Tingate, P. R., and Laslett, G. M. (1986). Thermal annealing of fission tracks in apatite: 1. A qualitative description. *Chemical Geology: Isotope Geoscience Section*, 59, 237–53

Guenthner, W. R., Reiners, P. W., Ketcham, R. A., Nasdala, L., and Giester, G. (2013). Helium diffusion in natural zircon: Radiation damage, anisotropy, and the interpretation of zircon (U- Th)/He thermochronology. *American Journal of Science*, 313, 145–98.

Herman F., Seward, D., Valla, P. G. et al. (2013). Worldwide acceleration of mountain erosion under a cooling climate. *Nature*, 504(7480), 423–6. https://doi.org/10.1038/nature12877.

Higson, E., Hadnley, W., Hobson, M, and Lasenby, A. (2019). Dynamic nested sampling: An improved algorithm for parameter estimation and evidence calculation. *Statistics and Computing*, 29, 891–913

Jasra, A., Stephens, D. A., Gallagher, K., and Holmes, C. C. (2006). Analysis of geochronological data with measurement error using Bayesian mixtures, *Mathematical Geology*, 38(3), 269-300.

Ketcham, R. A. (2005). Forward and inverse modeling of low-temperature thermochronometry data. In P. W. Reiners and T. A. Ehlers, eds., *Reviews in Mineralogy and Geochemistry. Vol 58: Low-Temperature Thermochronology: Techniques, Interpretations and Applications*. Chantilly, VA: Mineralogical Society of America, pp. 275–314.

Ketcham, R. A., Carter, A., Donelick, R. A., Barbarand, J., and Hurford, A. J. (2007). Improved modeling of fission-track annealing in apatite. *American Mineralogist*, 92, 799–810.

Ketcham, R. A., Donelick, R. A., and Carlson, W. D. (1999). Variability of apatite fission-track annealing kinetics. III. Extrapolation to geological timescales. *American Mineralogist*, 84, 1235–55.

Ketcham, R. A., Guenthner, W. R., and Reiners, P. W. (2013). Geometric analysis of radiation damage connectivity in zircon, and its implications for helium diffusion. *American Mineralogist*, 98, 350–60

Lartillot, N., and Philippe, H. (2006). Computing Bayes factors using thermodynamic integration. *Systematic Biology*, 55, 195–207.

Laslett, G. M., Green, P. F., Duddy, I. R., and Gleadow, A. J. W. (1987). Thermal annealing of fission tracks, 2. A quantitative analysis. *Chemical Geology: Isotope Geoscience Section*, 65, 1–13.

Li, W., Wang, L., Lang, M., Trautmann, C., and Ewing, R. C. (2011). Thermal annealing mechanisms of latent fission tracks: Apatite vs. zircon. *Earth and Planetary Science Letters*, 203, 227–35.

Licciardi, A., Gallagher, K., and Clark, S. A. (2020). A Bayesian approach for thermal history reconstruction in basin modelling, *Journal of Geophysical Research: Solid Earth*, 125 (7), e2020JB019384. https://doi.org/10.1029/2020JB019384.

Liu, P., Elshall, A. S., Ye, M. et al. (2016). Evaluating marginal likelihood with thermodynamic integration method and comparison with several other numerical methods. *Water Resources Research*, 52(2), 734–58. https://doi.org/10.1002/2014WR016718.

Lovera, O. M., Richter, F. M, and Harrison, T. M. (1989). 40Ar/39Ar geothermometry for slowly cooled samples having a distribution of domain sizes. *Journal of Geophysical Research*, 94, 17917–35.

Malusa, M., and Fitzgerald, P. F. (2019). *Fission-Track Thermochronology and its Application to Geology, Springer Textbooks in Earth Sciences*. Cham: Springer International Publishing.

McDannell, K. T., and Issler, D. R. (2021). Simulating sedimentary burial cycles. Part 1: Investigating the role of apatite fission track annealing kinetics using synthetic data. *Geochronology*, 3, 321–35. https://doi.org/10.5194/gchron-3-321-2021.

McDougall, I., and Harrison, T. M. (1988). *Geochronology and Thermochronology by the 40Ar/39Ar method*, 2nd ed. New York: Oxford University Press.

Meesters, A. G., and Dunai, T. J (2002a). Solving the production-diffusion equation for finite diffusion domains of various shapes. Part I: Implications for low-temperature (U-Th)/He-thermochronology. *Chemical Geology*, 186, 333–44

Meesters, A. G., and Dunai, T. J. (2002b). Solving the production-diffusion equation for finite diffusion domains of various shapes. Part II: Application to cases with -ejection and non-homogenous distribution of the source. *Chemical Geology*, 186, 347–63

Pooley, C. M., and Marion, G. (2018). Bayesian model evidence as a practical alternative to deviance information criterion. *Royal Society Open Science*, 5(3), 171519. http://dx.doi.org/10.1098/rsos.171519.

Recanati, A., Gautheron, C., Barbarand, J. et al. (2017). Helium trapping in apatite damage: Insights from (U-Th-Sm)/He dating of different granitoid lithologies. *Chemical Geology* 470, 116–31

Reiners, P. W., Carlson, R.W., Renne, P.R. et al. (2018). *Geochronology and Thermochronology*. Hoboken, NJ: Wiley.

Reiners, P. W., and Ehlers, T. A., eds. (2005). *Reviews in Mineralogy and Geochemistry. Vol. 58: Low-Temperature Thermochronology: Techniques, Interpretations and Applications*. Chantilly, VA: Mineralogical Society of America.

Richardson, S., and Green, P. J. (1997). On Bayesian analysis of mixture models with an unknown number of components. *Journal of the Royal Statistical Society Series B*, 59(4), 731–92.

Richardson, S., Leblond, L., Jaussent, I., and Green, P. J. (2002). Mixture models in measurement error problems, with reference to epidemiological studies. *Journal of the Royal Statistical Society Series A*, 165(3), 549–66.

Sambridge, M. (2014). A parallel tempering algorithm for probabilistic sampling and multimodal optimization. *Geophysical Journal International*, 196, 357–74. https://doi.org/10.1093/gji/ggt342.

Sambridge, M. S., and Compston, W. (1994). Mixture modelling of multi-component data sets with application to ion-probe zircon ages. *Earth Planetary Science Letters*, 128, 373–90.

Sambridge, M., Gallagher, K., Jackson, A., and Rickwood, P. (2006). Trans-dimensional inverse problems, Model Comparison and the Evidence. *Geophysical Journal International*, 167, 528–42.

Schildgen, T. F., van der Beek, P. A., Sinclair, H. D., and Thiede, R. C. (2018). Spatial correlation bias in late-Cenozoic erosion histories derived from thermochronology. *Nature*, 559(7712), 89–93. https://doi.org/10.1038/s41586-018-0260-6.

Schoene, B. (2014). U-Th-Pb Geochronology. In H. D. Holland and K. K. Turekian, eds., *Treatise on Geochemistry*, 2nd ed. Amsterdam: Elsevier, pp. 341–78.

Shuster, D. L, Flowers, R. M., and Farley, K. A. (2006). The influence of natural radiation damage on helium diffusion kinetics in apatite. *Earth and Planetary Science Letters*. https://doi.org/10.1016/J.EPSL.2006.07.028.

Stephenson, J., Gallagher, K., and Holmes, C. (2006). Low temperature thermochronology and modelling strategies for multiple samples 2: Partition modeling for 2D and 3D distributions with discontinuities. *Earth and Planetary Science Letters*, 241, 557–70.

Tagami, T., Galbraith, R., Yamada, R., and Laslett, G. (1998). Revised annealing kinetics of fission tracks in zircon and geological implications. In P. Van den Haute and F. De Corte, eds., *Advances in Fission-Track Geochronology*. Dordrecht: Kluwer, pp. 99–112.

Titterington, D. M., Smith, A. F. M., and Makov, H. E. (1985). *Statistical Analysis of Finite Mixture Distributions*. Chichester: Wiley.

Trotta, R. (2008). Bayes in the sky: Bayesian inference and model selection in cosmology. *Contemporary Physics*, 49, 71–104

Willett, C. D., Fox, M., and Shuster, D.L. (2017). A helium-based model for the effects of radiation damage annealing on helium diffusion kinetics in apatite. *Earth Planetary Science Letters*, 477, 195–204

Willett, S. D., Herman, F., Fox, M. et al. (2021). Bias and error in modelling thermochronometric data: Resolving a potential increase in Plio-Pleistocene erosion rate. *Earth Surface Dynamics*, 9, 1153–221. https://doi.org/10.5194/esurf-9-1153-2021.

Yamada, R., Murakami, M., and Tagami, T. (2007). Statistical modelling of annealing kinetics of fission tracks in zircon; Reassessment of laboratory experiments. *Chemical Geology* 236, 75–91.

12

Inverse Problems in Lava Dynamics

Alik Ismail-Zadeh, Alexander Korotkii, Oleg Melnik, Ilya Starodubtsev, Yulia Starodubtseva, Igor Tsepelev, and Natalya Zeinalova

Abstract: Lava flow and lava dome growth are two main manifestations of effusive volcanic eruptions. Less-viscous lava tends to flow long distances depending on slope topography, heat exchange with the surroundings, eruption rate, and the erupted magma rheology. When magma is highly viscous, its eruption on the surface results in a lava dome formation, and an occasional collapse of the dome may lead to a pyroclastic flow. In this chapter, we consider two models of lava dynamics: a lava flow model to determine the internal thermal state of the flow from its surface thermal observations, and a lava dome growth model to determine magma viscosity from the observed lava dome morphological shape. Both models belong to a set of inverse problems. In the first model, the lava thermal conditions at the surface (at the interface between lava and the air) are known from observations, but its internal thermal state is unknown. A variational (adjoint) assimilation method is used to propagate the temperature and heat flow inferred from surface measurements into the interior of the lava flow. In the second model, the lava dome viscosity is estimated based on a comparison between the observed and simulated morphological shapes of lava dome shapes using computer vision techniques.

12.1 Introduction: Effusive Eruptions and Lava Dynamics

Volcanism takes place mainly along plate margins in areas of subduction and spreading with exceptions such as hotspot volcanism (Turcotte and Schubert, 2002). Athanasius Kircher (1602–80), a German scholar and polymath, was perhaps the first who could assimilate knowledge about volcanoes and their eruptions into the Earth's deep interior, illustrating it in his *Mundus Subterraneus* of 1664 as 'a big fire' in the Earth's core interconnected with 'smaller fires' in the mantle. Today, our knowledge about origin, dynamics, and manifestations of volcanoes has reached the stage where volcanic eruptions can be predicted with a high probability (Poland and Anderson, 2020), and relevant measures can be taken to reduce disaster risks (e.g. Ismail-Zadeh and Takeuchi, 2007).

Molten rock (magma) flows through volcanic conduits in the crust towards the surface, leading to explosive or effusive (non-explosive) eruptions. The magma's viscosity depends on its composition (mainly on silica [SiO_2] content), crystal content, and temperature; viscosity increases with increasing silica or crystal content and decreases with temperature elevation. Silica-poor magma (under 50 wt% SiO_2) feeds basaltic volcanoes, and silica-rich magma (over 50 wt% SiO_2) forms felsic (andesitic to rhyolitic) volcanoes. The basaltic-to-rhyolitic magmas have pre-eruptive viscosities ranging from 10^1 to 10^8 pascals (Pa) (Takeuchi, 2011). High gas and water contents make felsic volcanoes more explosive compared to basaltic volcanoes (Bryant, 2005).

Volcanic eruptions produce a variety of lava flow morphologies depending on the chemical composition and temperature of the erupting material, and the surface topography over which the lava flows (e.g. Griffiths, 2000; Rumpf et al., 2018). The effusion rate (the volume of magma generated over a certain time) controls lava flow dynamics; the higher the effusion rates, the more rapidly and longer lava flows advance (e.g. Walker, 1973; Harris et al., 1998; Castruccio and Contreras, 2016).

Non-explosive volcanic eruptions are associated with lava dome growth and lava flow on the Earth's surface. Lava domes grow by magma extrusion from a volcanic conduit, and the lavas having low average eruption rates and high viscosities are associated with high groundmass crystallinity and substantial yield strength (e.g. Lavallée et al., 2012; Calder et al., 2015; Sheldrake et al., 2016; Tsepelev et al., 2020, 2021; Zeinalova et al., 2021). Non-linear dynamics of lava dome growth result from crystallisation and outgassing of the highly viscous lava in the volcanic conduit (Melnik and Sparks, 1999). Through the intermittent build-up of gas pressure, growing domes can often experience episodes of explosive activity (e.g. Voight and Ellsworth, 2000; Heap et al., 2019). Lava dome destruction can generate pyroclastic flows which can devastate building constructions and infrastructure and become a major threat to the surrounding population (Kelfoun et al., 2021). Although effusive eruptions are least hazardous and permit for evacuation, flows of lava can seriously damage cities, as happened in and around Catania

(Italy) during the Mt. Etna eruption in 1669 (Branca et al., 2013). The lava flow hazard is not negligible as hot lava kills vegetation, destroys infrastructure, and may trigger floods due to melting of snow/ice (e.g. Papale, 2014).

Mathematical/numerical modelling plays an essential role in understanding lava dynamics (e.g. Costa and Macedonio, 2005; Cordonnier et al., 2015; Tsepelev et al., 2019; and references herein). A mathematical model links the basic characteristics of lava dynamics (e.g. temperature, viscosity, and velocity) with its observed/measured properties (e.g. temperature at the interface between the lava flow and the air, morphology, and the volume of erupted lava). The aim of a direct problem is to determine the basic dynamic characteristics of the model for a given set of initial and boundary conditions and for known physical properties of the magma. An inverse model problem is considered when there is a lack of information on the basic characteristics or initial and boundary conditions, but some information on lava thermal and dynamic properties exists (Ismail-Zadeh et al., 2016).

There are several approaches to solving inverse problems of lava dynamics. One approach is associated with data assimilation and the use of a variational method to minimise the difference between the model solution and observations (e.g. Korotkii et al., 2016). This method involves an analytical derivation of the gradient of a cost function, and hence it is restricted to simple cases. Another approach is based on the use of image processing and computer vision techniques to determine the physical properties of a lava dome (e.g. its viscosity) by analysing images of the observed and modelled morphological shapes of the dome at each time of lava dome growth (Starodubtseva et al., 2021). And the third approach is based on the replacement of an inverse problem by the direct problem conjugated with the inverse problem. The direct problem can be solved then by varying model parameters to fit observations at each time step (Zeinalova et al., 2021). Although this approach is simpler than the previous two approaches, it requires a significant number of numerical experiments to find the best fit to the observations.

In Sections 12.2 and 12.3, we consider two inverse problems related to lava flow (model 1) and lava dome growth (model 2). Model 1 uses variational data assimilation to determine the temperature and the velocity of a lava flow from its thermal observations at the interface between the lava and the air. Model 2 determines the lava dome viscosity by comparing morphological shapes of observed and modelled domes and employing computer vision techniques.

12.2 Model 1: Reconstructing the Thermal State of a Lava Flow

The rapid development of ground-based thermal cameras, drones, and satellite data allows collection of repeated thermal images of the surface of active lava flows during a single lava flow eruption (Calvari et al., 2005; Wright et al., 2010; Kelfoun and Vargas, 2015). For example, remote sensing technologies (e.g. air-borne or space-borne infrared sensors) allow for detecting the absolute temperature at the Earth's surface (e.g. Flynn et al., 2001). The Stefan–Boltzmann law relates the total energy radiated per unit surface area of a body across all wavelengths per unit time to the fourth power of the absolute temperature of the body. Hence the absolute temperature can be determined from the measurements by remote sensors (e.g. Harris et al., 2004), and the heat flow could be then inferred from the Stefan–Boltzmann law using the temperature. Is it possible to use the surface thermal data obtained this way to constrain the thermal and dynamic conditions of lava flow below the surface? Following Korotkii et al. (2016), in this chapter we present a quantitative approach to the problem of reconstructing temperature and velocity in a steady-state lava flow when the temperature and the heat flow are known on the lava interface with the air.

12.2.1 Mathematical Statement

To state a lava flow problem mathematically, we need to consider the following components of a model: a computational domain (e.g. topography on which lava flows or slope of the inclined surface or channel as well as vent geometry); equations governing the lava flow; boundary conditions on flow velocity (e.g. effusion rate, free-slip or no-slip or free surface conditions) and on temperature (e.g. eruption temperature as well as radiative, convective, or conductive heat flow at the interface with the air and with the underlying surface); and physical properties of the lava (i.e. viscosity, density, thermal conductivity, etc.).

The model domain $\Omega \subset \mathbb{R}^2$ (Fig. 12.1) covers part of a lava flow at some distance from the volcanic vent and the lava flow front. The boundary of the model domain consists of the following parts: Γ_1 is a line segment connecting points A and D; Γ_2 is a circular arc connecting points A and B; Γ_3 is a line segment connecting points B and C; and Γ_4 is a circular arc connecting points C and D. Although the lava flow rate depends on the effusion rate, a steady-state lava flow is assumed in the modelling for the sake of simplicity (a justification to this assumption is provided by Korotkii et al., 2016).

In a two-dimensional model domain Ω, the Stokes, incompressibility, and heat equations are employed to determine the velocity and temperature of the incompressible heterogeneous fluid under gravity in the Boussinesq approximation (Chandrasekhar, 1961; Ismail-Zadeh and Tackley, 2010):

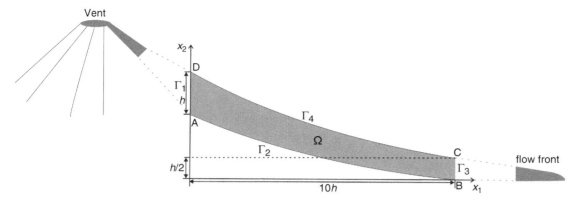

Figure 12.1 Model 1: Geometry of a lava flow. After Tsepelev et al. (2019).

$$\nabla p = \nabla \cdot \left(\eta(T) \left(\nabla \mathbf{u} + \nabla \mathbf{u}^{\mathrm{T}} \right) \right) + RaT\, \mathbf{e}_2, \tag{12.1}$$

$$\nabla \cdot \mathbf{u} = 0, \tag{12.2}$$

$$\nabla \cdot \left(\kappa(T)\, \nabla T \right) = \langle \mathbf{u}, \nabla T \rangle, \tag{12.3}$$

where $\mathbf{x} = (x_1, x_2) \in \Omega$ are the Cartesian coordinates; $\mathbf{u} = (u_1(\mathbf{x}), u_2(\mathbf{x}))$ is the vector velocity; $p = p(\mathbf{x})$ is the pressure; $T = T(\mathbf{x})$ is the temperature; $\eta = \eta(T)$ is the viscosity; $\kappa(T) = k(T)/(\rho_m c_p)$ is the thermal diffusivity; $k = k(T)$ is the heat conductivity; ρ_m is the lava density at the typical lava melting temperature T_m; and c_p is the specific heat capacity. The Rayleigh number is defined as $Ra = \beta g \rho_m \Delta T h^3 \eta_*^{-1} \kappa_*^{-1}$, where β is the thermal expansivity; g is the acceleration due to gravity; η_* and κ_* are the typical lava viscosity and thermal diffusivity, respectively; $\Delta T = T_m - T_s$ is the temperature contrast; T_s is the solidus temperature; h is the typical lava thickness; $\mathbf{e}_2 = (0, -1)$ is the unit vector; ∇, ^{T}, and $\langle \cdot, \cdot \rangle$ denote the gradient vector, the transposed matrix, and the scalar product of vectors, respectively.

Although the lava rheology can be more complicated, we assume that the lava behaves as a Newtonian fluid with a temperature- and volume-fraction-of-crystals-dependent viscosity and temperature-dependent density and thermal conductivity (e.g. Dragoni, 1989; Giordano and Dingwell, 2003; Costa et al., 2009; Tsepelev et al., 2019). It was shown that crystallisation is responsible for the increase of the viscosity during emplacement of mafic lava flows (Chevrel et al., 2013). The following lava viscosity is considered in the modelling:

$$\eta(T) = \eta_* \eta_{12}(T), \quad \eta_{12}(T) = \begin{cases} \eta_1 \eta_2, & \eta_1 \eta_2 < \eta_0, \\ \eta_0, & \eta_1 \eta_2 > \eta_0, \end{cases} \tag{12.4}$$

where η_0 $(= 10^3$ Pa s) introduces a restriction on the exponential growth of the viscosity with temperature. Here η_1 is the temperature-dependent viscosity (Dragoni, 1989)

$$\eta_1(T) = \exp\left(n(T_m - T) \right), \tag{12.5}$$

where $n = 4 \times 10^{-2}$ K^{-1}; and η_2 is the volume-fraction-of-crystals-dependent viscosity (Marsh, 1981; Costa et al., 2009)

$$\eta_2(T) = (1 + \psi^\delta) \left[1 - (1 - \zeta) \cdot \mathrm{erf} \left(\frac{\sqrt{\pi}}{2(1 - \zeta)} \psi(1 + \psi^\gamma) \right) \right]^{-B\phi_*}, \tag{12.6}$$

where $\psi = \phi/\phi_*$; $\phi = 0.5[1 - \mathrm{erf}(b_1[(T - T_s)/\Delta T - 0.5]/b_2)]$ is the volume fraction of crystals; ϕ_* is the specific volume fraction of crystals; the coefficient B is determined from the Einstein equation (Mardles, 1940) and varies from 1.5 to 5 (Jeffrey and Acrivos, 1976); $\delta = 13 - \gamma$, ζ, and γ are the rheological parameters (see Table 12.1; Lejeune and Richet, 1995; Costa et al., 2009); $b_1 = \sqrt{30}$ (Marsh, 1981); $b_2 = 1.5$ (Wright and Okamura, 1977); and $\mathrm{erf}(\cdot)$ is the error function.

The thermal conductivity (Hidaka et al., 2005) and the density (Kilburn, 2000) are represented in the form:

$$k(T) = \begin{cases} 1.15 + 5.9 \cdot 10^{-7}(T - \widetilde{T})^2, & T < \widetilde{T}, \\ 1.15 + 9.7 \cdot 10^{-6}(T - \widetilde{T})^2, & T > \widetilde{T}, \end{cases} \tag{12.7}$$

$$\rho = \rho_m \left(1 - \phi(T) \right) + (\rho_c - \delta\rho)\phi(T), \tag{12.8}$$

where $\widetilde{T} = 1{,}473$ K is the specific temperature at which the thermal conductivity reaches its minimum value (Buttner et al., 1998); ρ_c is the typical crystal density corresponding to the crystals composed of 50% olivine and 50% plagioclase. As the effective density of the lava crust with about 20% volume of vesicles is estimated to be about 2,200 kg m^{-3} (Kilburn, 2000), we decrease the crystal density by $\delta\rho$ (where $\delta\rho = 750$ kg m^{-3}) to permit crustal pieces drifting with and not sinking into the lava. The model parameters are specified in Table 12.1.

The following conditions for temperature T and velocity $\mathbf{u}$ are assumed at the model boundary:

Table 12.1 *Notations, parameters, and their values in modelling*

Notation	parameter, unit	Model 1 Lava flow	Model 2 Lava dome
B	Theoretical value of the Einstein coefficient	2.5	
c_p	Specific heat capacity of lava, J kg^{-1} K^{-1}	1,000	–
h	Typical thickness, m	2	100
g	Acceleration due to gravity, m s^{-2}	9.81	
T_a	Temperature of air, K	300	–
T_m	Lava melting temperature, K	1,333	–
T_s	Lava solidus temperature, K	1,053	–
β	Thermal expansivity of lava, K^{-1}	10^{-5}	–
ϕ_{in}	Initial volume fraction of crystals	0.4	
ϕ_{eq}	Volume fraction of crystals at equilibrium	0.8	
ϕ_*	Specific volume fraction of crystals	0.384 (1)	0.591 (2)
γ	Rheological parameter in Eq. (12.5)	7.701	5.76
η_a	Air viscosity, Pa s	–	10^{-4}
η_*	Typical lava viscosity, Pa s	10^6	–
κ_*	Typical lava thermal diffusivity, m^2 s^{-1}	10^{-6}	–
ρ_a	Air density, kg m^{-3}	–	1
ρ_l	Typical lava density, kg m^{-3}	–	2,500
ρ_c	Crystal density, kg m^{-3}	2,950	–
ρ_m	Lava density at $T = T_m$, kg m^{-3}	2,750	–
ζ	Rheological parameter in Eq. (12.5)	2×10^{-4}	4.63×10^{-4}

Notes: (1) Lejeune and Richet (1995); (2) Costa et al. (2009)

$$\Gamma_1 : \quad T = T_1, \ \mathbf{u} = \mathbf{u}_1, \tag{12.9}$$

$$\Gamma_2 : \quad \mathbf{u} = 0, \tag{12.10}$$

$$\Gamma_3 : \quad T = T_3, \ \sigma \mathbf{n} = 0, \ p = 0, \tag{12.11}$$

$$\Gamma_4 : \quad T = T_4, \ -k\langle \nabla T, \mathbf{n} \rangle = \varphi, \ \langle \mathbf{u}, \mathbf{n} \rangle = 0,$$
$$\sigma \mathbf{n} - \langle \sigma \mathbf{n}, \mathbf{n} \rangle \mathbf{n} = 0, \tag{12.12}$$

where $\sigma = \eta(\nabla \mathbf{u} + \nabla \mathbf{u}^T)$ is the deviatoric stress tensor, and $\mathbf{n}$ is the outward unit normal vector at a point on the model boundary. The *principal problem* is to find the solution to Eqs. (12.1)–(12.8) with the boundary conditions (12.9)–(12.12), and hence to determine the velocity $\mathbf{u} = \mathbf{u}(\mathbf{x})$, the pressure $p = p(\mathbf{x})$, and the temperature $T = T(\mathbf{x})$ in the model domain Ω when temperature T_4 and heat flow $\varphi = k \, \partial T/\partial \mathbf{n}$ are known at boundary Γ_4. We note that Tsepelev et al. (2019) obtained the synthetic temperature and the heat flow at the interface of lava flow and the air (boundary Γ_4) by solving a direct problem of lava flow with relevant boundary conditions. The temperature and the heat flow so obtained were used as the conditions at boundary Γ_4 in the principal problem.

In addition to the principal problem, an *auxiliary problem* is defined as a set of Eqs. (12.1)–(12.8) with the following boundary conditions:

$$\Gamma_1 : \quad T = T_1, \ \mathbf{u} = \mathbf{u}_1, \tag{12.13}$$

$$\Gamma_2 : \quad T = T_2, \ \mathbf{u} = 0, \tag{12.14}$$

$$\Gamma_3 : \quad T = T_3, \ \sigma \mathbf{n} = 0, \ p = 0, \tag{12.15}$$

$$\Gamma_4 : \quad T = T_4, \ \langle \mathbf{u}, \mathbf{n} \rangle = 0 \ , \ \sigma \mathbf{n} - \langle \sigma \mathbf{n}, \mathbf{n} \rangle \mathbf{n} = 0. \tag{12.16}$$

Equations (12.1)–(12.16) are transformed to a dimensionless form assuming that length, temperature, viscosity, and heat conductivity are normalised by h, T_a, η_*, and k_* (see Table 12.1).

The auxiliary problem (12.1)–(12.8), (12.13)–(12.16) is a direct problem compared to the problem (12.1)–(12.12), which is an inverse problem. The conditions at Γ_1 and Γ_3 are the same in the direct and inverse problems. Temperature T_2 at Γ_2 and temperature T_4 (but no heat flow) at Γ_4 are prescribed in the auxiliary problem compared to the inverse problem (12.1)–(12.12). The well- and ill-posedness of the similar problems has been studied by Ladyzhenskaya (1969), Lions (1971), Temam (1977), Korotkii and Kovtunov (2006), and Korotkii and Starodubtseva (2014).

12.2.2 Variational Data Assimilation Method

Consider 'guess' temperature $T_2 = \xi$ at model boundary Γ_2. Solving the auxiliary problem (12.1)–(12.8), (12.13)–(12.16), we can determine the heat flow at model boundary Γ_4 and compare it to the heat flow φ (the known synthetic data) at the same boundary. The following cost functional for admissible functions ξ determined at Γ_2 is to be then assessed:

$$J(\xi) = \int_{\Gamma_4} \left(k(T_\xi) \frac{\partial T_\xi}{\partial \mathbf{n}} - \varphi \right)^2 d\Gamma, \tag{12.17}$$

where $k(T_\xi) \, \partial T_\xi/\partial \mathbf{n}$ is the heat flow at Γ_4 corresponding to temperature $T_2 = \xi$ at Γ_2; and T_ξ is the temperature determined by solving the auxiliary problem. Therefore, we reduce the inverse problem to a minimisation of the functional (12.17) or to a variation of the function ξ at Γ_2, so that the model heat flow at Γ_4 becomes closer to the 'observations' (known heat flow φ) at Γ_4.

The cost functional (12.17) is minimised using the Polak–Ribière conjugate-gradient method (Polak and Ribière, 1969). Korotkii et al. (2016) showed that the gradient of the cost functional can be found as the solution to the adjoint problem. The solution of the minimisation problem is then reduced to solutions of series of well-posed (auxiliary and adjoint) problems. The algorithm for solving the variational data assimilation problem is based on solving iteratively the auxiliary and adjoint problems and on the assessment $J(\xi^{(i)})$, where i is the iteration number. The numerical method and algorithm are not discussed here and can be found in Korotkii et al. (2016).

The performance of the algorithm is evaluated in terms of the number of iterations n required to achieve a prescribed relative reduction of the cost functional (Fig. 12.2). Also, Tsepelev et al. (2019) checked the quality of the gradient of the cost functional with respect to the control variable using the χ-test by Navon et al. (1992). We note that the Polak–Ribière stable iterative conjugate gradient method provides a rapid convergence to the solution to the adjoint and auxiliary problems on the condition that the Rayleigh number is small, and lava viscosity is high (Tsepelev et al., 2019). Meanwhile, the convergence of iterations depends on a choice of the 'guess' temperature: the closer the 'guess' temperature to the 'true' solution, the more rapid convergence of the iteration process. An iterative convergence slows at high Rayleigh numbers, and the iterations diverge at Rayleigh numbers greater than 10^6. At lava viscosity ranging from 10^3 Pa s to 10^6 Pa s and lava flow thickness from 1 m to 10 m, the Rayleigh number takes values from 10^3 to 10^6. As the injection rate of lava into the model domain increases, the minimisation process slows down (Fig. 12.2). A rapid injection results in advection of high temperature with flow and, hence, in a decrease of lava viscosity, and in slowing convergence of iterations and their divergence.

12.2.3 Results

We present here the results of a reconstruction of the thermal state of a lava flow developed by Tsepelev et al. (2019). The model domain covers a portion of the lava flow selected on a two-dimensional profile along the lava flow (the area marked by red in Fig. 12.3a). The lava thickness at the left and right sides of the model domain are 14 m and 31 m, respectively, and the length of the model lava flow is 515 m. The Rayleigh number is about 6×10^5. A lava is injected from the left boundary of the model domain with a prescribed effusion rate. The model problem (12.1)–(12.12) is solved by the variational data assimilation method in the selected domain to determine the lava's temperature, velocity, and viscosity based on the known thermal data on its interface with the air

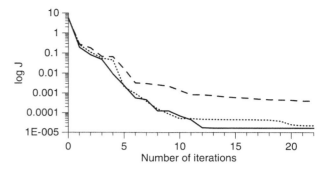

Figure 12.2 Relative reduction of the cost functional J with the number of iterations at three dimensionless rates of lava injection: $\vartheta = 10$ (solid line), $\vartheta = 20$ (dotted line), and $\vartheta = 40$ (dashed line). After Tsepelev et al. (2019).

(Tsepelev et al., 2019). Figure 12.3b presents the target viscosity, temperature, and velocity (the solution of the direct problem presented by Tsepelev et al. (2019), which generated the synthetic thermal conditions at the interface between the lava and the air), and Fig. 12.3c their residuals after the 1st and the 31st iterations. The lava's physical parameters are recovered well enough from the surface thermal data after 30 iterations. A thin crust developing at the left end of the model domain becomes thicker towards its right end, and the flow velocity drops by a factor of about 3 with the lava advancement.

If surface temperature and heat flow data are of high resolution and radiometric accuracy, the temperature and velocity in the lava's interior can be determined properly from measured data using the data assimilation approach. Meanwhile, the spatial resolution of many satellites is too coarse to allow for high-resolution monitoring and precise measurements, and this gives rise to uncertainties in thermal measurement as well as in the inferred parameters (e.g. Zakšek et al., 2015). Hence, if the measured temperature and heat flow data are biased, this information can be improperly assimilated into the lava flow models.

The presented model describes steady-state flow, although lava flows are non-stationary. As measurements of the absolute temperature are discrete in time in most cases (e.g. depending on the location of Landsat satellites), a problem of non-steady-state flow can be reduced to a series of steady-state flow problems with varying model domain and boundary conditions assimilating thermal data available at the discrete-in-time measurements. Also, airborne and space measurements of absolute temperature at the lava interface with the air, being almost instantaneous compared to the duration of the lava flow, allow searching for thermal conditions at the bottom of the lava flow using the cost functional (12.17). Once the boundary conditions at the lava bottom are determined, the steady-state problem can be replaced by a non-steady-state problem, and the lava flow can be modelled forward in time to determine its extent, lava's temperature, and flow rate, as well as backward in

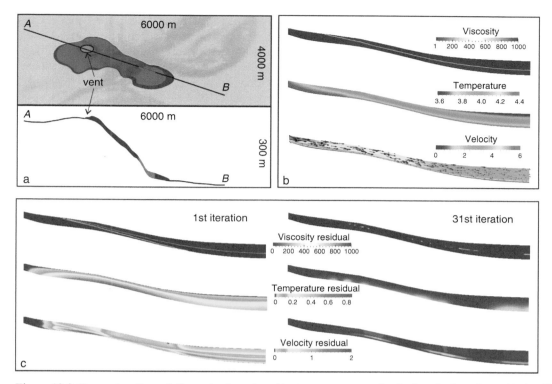

Figure 12.3 Reconstruction of dimensionless viscosity, temperature, and velocity of a lava flow. (a) A relief map (6000 m × 4000 m) of three-dimensional lava flow pattern, view from the top (upper panel) and the cross-section AB (lower panel) along the line indicated in the upper panel; the flow was computed by Tsepelev et al. (2016). The red area marks the numerical model domain. (b) Target viscosity, temperature, and velocity. (c) The relevant residuals after the 1st iteration (the left panel) and after 31st iteration (the right panel). After Tsepelev et al. (2019).

time using variational (Korotkii et al., 2016) or quasi-reversibility (Ismail-Zadeh et al., 2007) methods to search for the initial temperature of the lava flow and for the evolution of the effusion rate.

12.2.4 Sensitivity Analysis

The results of numerical modelling of the problem (12.1)–(12.12) show that the optimisation works effectively: the residuals of temperature, viscosity, and velocity already become small after a few dozen iterations within almost the entire model domain. Meanwhile, measurements (observations) are polluted by errors. For example, in the case of lava temperature measurements at the surface, the accuracy of the calibration curve of remote sensors and the noise of the sensors can influence measurements and contribute to the measurement errors (Short and Stuart, 1983). Korotkii et al. (2016) performed numerical experiments introducing a noise on the 'measured' heat flow data and analysed the sensitivity of the model to the noise. They showed that the temperature and velocity residuals get larger with increase of the noise of the input data but are still acceptable at the level of noise (errors in measurements) of up to 10%. It should be noted

that the error analysis of subpixel temperature retrieval from satellite infrared data showed that errors in measurements of the radiant heat flux are within about 5% to 10%, and can be reduced (Lombardo et al., 2012).

The cost functional related to the inverse (optimisation) problem with noisy measurements at the lava flow interface with the air can be written in the following form:

$$J_\delta(\xi) = \int_{\Gamma_4} \left(k(T_\xi) \frac{\partial T_\xi}{\partial \mathbf{n}} - \varphi_\delta \right)^2 d\Gamma = J(\xi)$$

$$+ 2\delta \int_{\Gamma_4} \left(k(T_\xi) \frac{\partial T_\xi}{\partial \mathbf{n}} - \varphi \right) v d\Gamma + \delta^2 \| v \|^2, \quad (12.18)$$

where $\varphi_\delta = \varphi - \delta v$, δ is the magnitude of the noise, and $v(\cdot)$ is the function generating numbers that are uniformly distributed over the interval [–1, 1]. Substituting the solution ξ^* to Eq. (12.17) into Eq. (12.18), we obtain $J_\delta(\xi^*) \sim \delta^2$, because the first and second terms of the right-hand side of Eq. (12.18) turn to zero at ξ^*. Therefore, at the minimisation of the functional J_δ, the functional will be approaching the non-vanishing value equal to the square of the noise's magnitude (δ^2). In the case of synthetic thermal data prescribed at the upper model boundary (instead of real measurements), the 'plateau' in the curves illustrating the minimisation of the functional (see Fig. 12.2) is likely to be associated with numerical errors. Forcing the solution to the functional to attain zero may lead to an unstable (or erroneous) solution. Moreover, some *a priori* information assists in solving the problem. For example, the temperature inside the model domain cannot be higher than that at the left boundary of the model domain (where the lava is injected into the model domain). This can serve as a control parameter for the computed temperature in the minimisation problem.

Rather accurate reconstructions of the model temperature, viscosity, and flow velocity in this study rely on the chosen method for minimisation of the cost functional (12.18). In the general case, a Tikhonov regularisation term should be introduced in the cost functional as:

$$J_\delta(\xi) = \int_{\Gamma_4} \left(k(T_\xi) \frac{\partial T_\xi}{\partial \mathbf{n}} - \varphi_\delta \right)^2 d\Gamma$$

$$+ \ \omega \int_{\Gamma_2} \Lambda(\xi - \xi_0) d\Gamma, \tag{12.19}$$

where ω is a small positive regularisation parameter, $\Lambda > 0$ is the operator accounting for *a priori* information on the problem's solution (e.g. its monotony property, maximum and minimum values, and the total variation diminishing), and ξ_0 is *a priori* known function close to the solution of the problem. The introduction of the regularisation term in the cost functional makes the minimisation problem more stable and less dependent on measurement errors. For a suitable regularisation parameter $\omega = \omega(\delta)$, the minimum of the regularised cost functional will tend to the minimum of the functional (12.17) at $\delta \to 0$ (Tikhonov and Arsenin, 1977). The choice of the regularisation parameter is a challenging issue as it depends on several factors, for example, on errors of measured data (e.g. Kabanikhin, 2011). Meanwhile, if there is a lack of information on the solution of the problem, a better strategy is to minimise the functional (12.17) using a stable minimisation method (e.g. the Polak–Ribière method).

12.3 Model 2: Lava Viscosity Inferred from Lava Dome Morphology

Lava domes form because of the extrusion of highly viscous magma. The domes develop a solid surface layer remaining mobile and undergoing deformations for days or even months. Several types of lava dome morphology are distinguished. In the endogenous regime, magma intrudes inside the dome without extrusion of fresh magma on the surface. In the exogenous regime, fresh lava pours out over the surface forming various forms of domes, such as obelisks,

lobes, pancake-shaped structures, and some others. Lava dome collapse can cause explosive eruptions, pyroclastic flows, and lahars, and, therefore, studies of the conditions of lava dome growth are important for hazard assessment and risk reduction.

Lava dome growths have been monitored at several volcanoes (e.g. Swanson et al., 1987; Daag et al., 1996; Nakada et al., 1999; Watts et al., 2002; Harris et al., 2003; Wadge et al., 2014; Zobin et al., 2015; Nakada et al., 2019). Monitoring allows mapping the spatial and temporal development of lava domes and determining the morphological changes during the growth as well as the changes in the lava volume over time. The morphology of lava domes is influenced by the rheology of magma and lava discharge rate (DR). Magma viscosity depends on temperature and the volume fraction of crystals, which is determined by crystallisation kinetics, namely, the characteristic time of crystal content growth, CCGT (Tsepelev et al., 2020). At small CCGT values (i.e. fast lava crystallisation), obelisk-type structures develop at lower DR and pancake-like structures at higher DR; at high CCGT values, the domes form either lava lobes or pancake-like structures.

It was shown that cooling does not play a significant role in the development of the lava dome. If the crystal content is controlled only by the cooling, then the lava viscosity increases in the near-surface layer of the dome, and the thickness of the temperature boundary layer remains small compared to the dome height (Tsepelev et al., 2020). In the dome body, a significant increase in the viscosity occurs due to crystallisation caused by a loss of volatiles. Thus, the evolution of the lava dome can be modelled using the rheology depending on CCGT and DR.

How do we determine lava dome viscosity (e.g. the rheological properties of the lava within the dome), if the lava dome morphological shape and the DR are known? Here we present a computer vision approach to solving this inverse problem based on minimising the deviation between the observed and simulated lava dome shapes (see details of the approach in Starodubtseva et al., 2021). Lava domes are modelled numerically at different values of CCGT, DR, and the conduit radius r. Using numerical experiments, a database of morphological shapes of modelled domes is developed for specified extrusion durations. The results of the experiments (the elements of the database) and an observed dome are analysed in the form of two-dimensional images. To estimate the viscosity of the observed lava dome, the difference between the observed and simulated dome shapes is estimated using three different functionals, which are used in computer vision and image processing theory. The viscosity of the observed lava dome is then assessed based on the parameters of the modelled lava dome which shape best fits the shape of the observed dome.

12.3.1 Mathematical Statement

A two-dimensional model of two-phase immiscible incompressible fluid is considered to approximate the lava (one phase) and the air (another phase). The two phases are separated by a moving interface – the lava dome surface. The influence of the air phase on lava dome growth is insignificant due to a large ratio between densities/viscosities of the air and the lava. Meanwhile, numerical schemes are usually inaccurate at the interfaces, where model parameters (e.g. density and viscosity) are discontinuous; it leads to a smearing of the parameters along the interface (e.g. Christensen, 1992; Naimark and Ismail-Zadeh, 1995; Naimark et al., 1998). In this modelling, the viscosity jump at the lava–air interface is significant, and a finer mesh is introduced around the interface to reduce the smearing. In the model domain (Fig. 12.4), the lava motion is described by the following set of equations supplemented by the initial and boundary conditions (Ismail-Zadeh and Tackley, 2010; Tsepelev et al., 2019; 2020).

The Navier–Stokes equations with the initial condition $\mathbf{u}(t = 0, \mathbf{x}) = 0$,

$$\frac{\partial(\rho\mathbf{u})}{\partial t} + \langle\mathbf{u}, \nabla\rangle(\rho\mathbf{u}) - \nabla \cdot \left(\eta(\nabla\mathbf{u} + \nabla\mathbf{u}^T)\right) = -\nabla p - \rho\mathbf{g},$$

(12.20)

and the continuity equation (Eq. 12.2) are employed to describe the lava dynamics, where $t \in [0, \vartheta]$ is the time; ϑ is the duration of the model experiments; ρ is the density; η is the viscosity; and $\mathbf{g} = (0, g)$, g is the acceleration due to gravity. The temperature dependence of the physical parameters and the surface tension forces are neglected in the modelling. Model density and viscosity are represented as $\rho = \rho_l\alpha(t, \mathbf{x}) + \rho_a(1 - \alpha(t, \mathbf{x}))$ and $\eta = \eta_l\alpha(t, \mathbf{x}) + \eta_a(1 - \alpha(t, \mathbf{x}))$, respectively, where ρ_a is the air density, ρ_l is the typical lava density, η_a is the air viscosity, and η_l is the lava viscosity. The

function $\alpha(t, \mathbf{x})$ equals 1 for the lava and 0 for the air at each point $\mathbf{x}$ and at time t, and this function is transported with the velocity $\mathbf{u}$ according to the advection equation

$$\frac{\partial\alpha}{\partial t} + \nabla \cdot (\alpha\mathbf{u}) = 0,$$

(12.21)

with the initial condition $\alpha(t = 0, \mathbf{x}) = 0$, which means that the entire model domain is filled with the air at the initial time.

We assume that lava viscosity depends on the volume fraction of crystals ϕ (Eq. 12.5), which is determined from the evolutionary equation describing the simplified kinetics of crystal growth during crystallisation due to magma degassing:

$$\frac{\partial\phi}{\partial t} + \nabla \cdot (\phi\mathbf{u}) = -\frac{\phi - \phi_{eq}}{\tau},$$

(12.22)

with the initial condition $\phi(t = 0, \mathbf{x}) = 0$. Here ϕ_{eq} is the volume fraction of crystals at the equilibrium; τ is the CCGT. The smaller the CCGT, the faster the crystallisation process converges to its equilibrium state. Note that although the viscosity depends also on the petrological (chemical) composition of the lava and the volatile content of the lava (its water saturation), these viscosity dependencies are not considered here.

The following conditions are set on the boundary $\Gamma = \Gamma_1 \cup \Gamma_2 \cup \Gamma_3 \cup \Gamma_4 \cup \Gamma_5 \cup \Gamma_6$ of the model domain (see Fig. 12.4). At Γ_1, impenetrability condition $<\mathbf{u}, \mathbf{n}> = 0$ and the free slip condition $(\nabla\mathbf{u} + \nabla\mathbf{u}^T)\mathbf{n} - \langle(\nabla\mathbf{u} + \nabla\mathbf{u}^T)\mathbf{n}, \mathbf{n}\rangle\mathbf{n} = 0$ are assumed. Lava enters the model domain through the boundary Γ_2 at the given DR $Q = 0.7$ m³ s⁻¹. At the boundaries Γ_3, Γ_4 and Γ_5, no-slip condition $\mathbf{u} = 0$ is assumed. The outflow conditions are determined at Γ_6 by removing the air from the model domain proportional to the given lava DR and to guarantee the condition of incompressibility. It is assumed that the volume fraction of crystals is equal $\phi = \phi_{in}$ at the boundary Γ_2 and $\phi = 0$ at Γ_6.

12.3.2 Morphological Shapes of Modelled Lava Domes

Initially, the model of lava dome growth (Eqs. 12.2, 12.5, 12.20–12.22, with the boundary and initial conditions) is solved numerically for model parameters specified in Table 12.1, and for the parameter τ (CCGT), the vent's radius r, and the time intervals specified in Table 12.2. The numerical method for solving the model is discussed by Starodubtseva et al. (2021). Morphological shapes $F_k = F(\tau, Q, r, t = t_k)$ at time t_k of the modelled lava domes (Fig. 12.5a) are stored in the database. Here the stored shapes have been generated by a few values of parameters τ, Q, and r; and the database can be extended by varying the parameter values to ensure that different possible morphological shapes are present in the database. A natural lava dome considered here is approximated by Starodubtseva et al. (2021) as a model dome taken from the

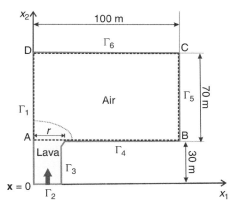

Figure 12.4 Model 2: Geometry of a lava dome. The large black arrow shows the part of the boundary, through which magma enters the model domain. The interface between the lava and the air is indicated by a dotted line. The solid dashed line indicates the border of binary images (Ω_{ABCD}) stored in the model database. Modified after Starodubtseva et al. (2021).

Table 12.2 *Parameters of lava domes in model 2*

Dome number k	τ, s	r, m	t_k, s
1–22	1.8×10^4	15	$\{3 \times 10^4 + k \times 10^3\}$, $k = 0, 1, 2, ..., 21$
23–44	5×10^4	15	$\{3 \times 10^4 + k \times 10^3\}$, $k = 0, 1, 2, ..., 21$
45–92	5×10^4	5	$\{4 \times 10^3 + k \times 10^3\}$, $k = 0, 1, 2, ..., 47$
93–114	5×10^5	15	$\{3 \times 10^4 + k \times 10^3\}$, $k = 0, 1, 2, ..., 21$
115–136	6×10^4	15	$\{3 \times 10^4 + k \times 10^3\}$, $k = 0, 1, 2, ..., 21$
137–181	6×10^4	5	$\{4 \times 10^3 + k \times 10^3\}$, $k = 0, 1, 2, ..., 44$
182–190	7×10^4	15	$\{3 \times 10^4 + k \times 10^3\}$, $k = 0, 1, 2, ..., 8$
191–238	7×10^4	5	$\{4 \times 10^3 + k \times 10^3\}$, $k = 0, 1, 2, ..., 47$
239–260	8×10^4	15	$\{3 \times 10^4 + k \times 10^3\}$, $k = 0, 1, 2, ..., 21$
261–308	8×10^4	5	$\{4 \times 10^3 + k \times 10^3\}$, $k = 0, 1, 2, ..., 47$

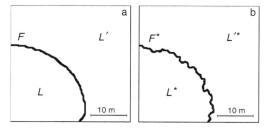

Figure 12.5 Morphological shapes of the modelled dome F (a) and the synthetic dome F^* (b). L and L^* are lava sub-domains, and L' and L'^* are air sub-domains. After Starobubtseva et al. (2021).

database, which morphological shape was distorted by a random perturbation (see Fig. 12.5b). The developed synthetic dome $L^* \cup F^*$ allows for approximating real distortions caused by the growth of a natural lava dome and/or its partial collapse, and/or by errors in measurements of the morphological shape of a lava dome.

To compare the shape of the synthetic dome F^* with the shapes of modelled lava domes F_k from the database, lava domes are presented by binary images. Namely, a uniform rectangular partitioning of the area Ω_{ABCD} into cells $I \times J$ as $\Omega_{ij}\left(\Omega_{ABCD} = \cup_{i=0, j=0}^{I-1, J-1}\Omega_{ij}\right)$ is introduced. A rectangular matrix $P(F) = \{p_{ij}\}_{i=0, j=0}^{I-1, J-1}$ of size $I \times J$ is assigned to each shape of the dome F, where the matrix element p_{ij} equals to 0, if the corresponding cell contains more than 50% of the air, and equals to 1 in all other cases.

12.3.3 Computer Vision Methods for Evaluation of Closeness of Lava Morphology Images

A closeness of the synthetic dome (with shape F^*) to the arbitrary modelled dome (with shape F_k) from the database is assessed by the methods used in the theory of computer

vision and image processing (e.g. Salomon, 2007). Namely, the following functionals are considered.

1. The functional based on the *symmetric difference*:

$$J_1(F^*, F) = k_1 \cdot S\left((L^* \cup L) \setminus (L^* \cap L)\right), \quad (12.23)$$

where $S(\cdot)$ (m²) is the area of the region and k_1 (m⁻²) is the scaling multiplier. Sub-domains L^* and L are presented in Fig. 12.5.

2. The functional based on *peak signal-to-noise ratio measure* (Salomon, 2007):

$$J_2(F^*, F) = k_2\left(k_3 + 10\log_{10}\left[\sum_{i=0, j=0}^{I-1, J-1} (p_{ij} - p_{ij}^*)^2 / (IJ)\right]\right), \quad (12.24)$$

where k_2 is the scaling factor, and k_3 is a positive constant. In numerical implementations, if $P(F)$ and $P^* = P(F^*) = \{p_{ij}^*\}_{i=0, j=0}^{I-1, J-1}$ match completely, the user receives a message containing the number of the modelled dome, where the condition of the complete match between the two matrices is reached.

3. The functional based on the *structure similarity index measure* (SSIM) (Wang et al., 2004):

$$J_3(F^*, F) = \frac{k_4\left(2\mu(P)\mu(P^*) + c_1\right)\left(\sigma(P, P^*) + c_2\right)}{\left(\mu^2(P) + \mu^2(P^*)\right)\left(\sigma^2(P) + \sigma^2(P^*) + c_2\right)}, \quad (12.25)$$

where $\mu(\cdot)$ is the mathematical expectation, $\sigma(\cdot, \cdot)$ is the covariance, $\sigma^2(\cdot)$ is the dispersion, k_4 is the scaling multiplier, $c_1 = 0.01$, and $c_3 = 0.03$. Here we consider a probabilistic model of image representation, namely, the image is considered as a field of random variables, and the value at each point of this field is a realisation of a random variable.

12.3.4 Results

The values of the functionals (Eqs. (12.23)–(12.25)) are calculated for each element of the database and plotted in descending order. Figure 12.6 shows the values of the three functionals on several elements from the database; three functionals reach their minimum at modelled dome image #135. Note that functionals J_1, J_2, and J_3 estimate the quantitative deviation of the modelled and synthetic domes, while the functional J_3 also estimates the structural features of the morphological shapes of the domes, although it leads to time-consuming computations. To consider qualitative and quantitative closeness of the synthetic and modelled domes simultaneously, a linear combination of the described functionals can be employed (Starodubtseva et al., 2021). The time of a lava dome formation, the

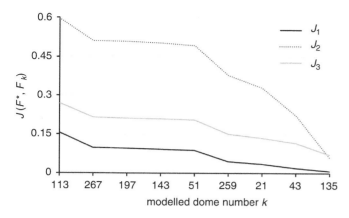

Figure 12.6 The values of functionals J_1, J_2, and J_3 related to selected modelled domes. Modified after Starobubtseva et al. (2021).

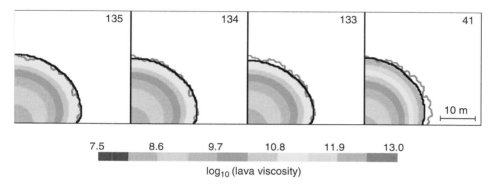

Figure 12.7 Comparison of the shape of the synthetic dome F^* (grey curve) and the shapes of four modelled domes 135, 134, 133, and 41 (black curves). Colours indicate the distribution of lava dome viscosity. Modified after Starodubtseva et al. (2021).

Although the approach presented here is two-dimensional, it can be extended to the three-dimensional case and used to reconstruct the growth conditions of natural lava domes (Starodubtseva et al., 2021).

12.4 Concluding Remarks

If direct problems present models dealing with understanding of phenomena, inverse models are related to finding either initial and boundary conditions or model characteristics (e.g. temperature, viscosity) based on observations and measurements. Data assimilation and artificial intelligence methods become important tools in studies of natural lava flow and lava dome dynamics. Observations and permanent

discharge rate, and the vent's size can help with a practical selection of the closest modelled dome.

Using the functionals (Eqs. (12.23)–(12.25)), the shape of the synthetic dome F^* is compared to dome shapes F_k from the database of morphological shapes. Considering a set of elements F_k, on which the smallest values of the considered functionals are achieved, we see that modelled dome shapes F_{41}, F_{133}, F_{134}, and F_{135} are the closest shapes to the synthetic dome shape F^*. Figure 12.7, presenting the synthetic dome and the four closest modelled domes from the database, shows a distribution of the lava dome viscosity. The lava viscosity can be determined from observations of the morphological shapes of the dome by solving inverse problems using computer vision methods.

Here the inverse problem is related to a search for lava dome viscosity distribution, which is presented as a certain function with a small number of model parameters. This search is reduced to determine CCGT, on which the viscosity depends, and this has been performed by minimising the differences between the morphological shapes of the observed (synthetic, in this case) and modelled lava domes. Natural lava domes are three-dimensional objects.

monitoring of lava flow and dome growth processes facilitate research and data-driven modelling. The presented models employ synthetic data (instead of real observations) as it is an important step to ensure that the data assimilation methods are applicable for problems in lava dynamics. The models and methods discussed here can be used to analyse effusive eruptions, lava flow, and lava dome growth by (i) assimilating of thermal measurements to obtain the information about a thermal state of the lava, and (ii) history matching of dome growth by nudging model forecasts to observations (i.e. minimising misfits between the modelled and observed morphological shapes of domes).

Modelling of direct and inverse problems provides knowledge of the thermal and dynamic characteristics of lava advancement, and it becomes important for lava flow hazard and disaster risk assessments (e.g. Papale, 2014; Cutter et al., 2015; Loughlin et al., 2015). Potentially hazardous volcanic eruptions should be accompanied by their virtual numerical model that is constantly tuned by available new observations and history matching to allow for short- and long-term forecasts of the eruption dynamics and associated hazards (Zeinalova et al., 2021).

Acknowledgements. We thank Donald Dingwell, Andrew Harris, Augusto Neri, Frank Schilling, and Slava Zobin for discussions related to lava dynamics, viscosity, and mineralogy. Also, we are grateful to Catherine Meriaux and Sabrina Sanchez for their constructive comments, which improved the initial version of the manuscript. The research described here was supported in part by the DFG grant IS203/14–1 and the RSF grant 19–17–00027.

References

Branca, S., De Beni, E., and Proietti, C. (2013). The large and destructive 1669 AD eruption at Etna volcano: Reconstruction of the lava flow field evolution and effusion rate trend. *Bulletin of Volcanology*, 75, 694. https://doi.org/10.1007/s00445-013-0694-5.

Bryant, E. (2005). *Natural Hazards*. Cambridge: Cambridge University Press.

Buttner, R., Zimanowski, B., Blumm, J., and Hagemann L. (1998). Thermal conductivity of a volcanic rock material (olivine-melilinite) in the temperature range between 288 and 1470 K. *Journal of Volcanology and Geothermal Research*, 80, 293–302.

Calder, E. S., Lavallée, Y., Kendrick, J. E., and Bernstein, M. (2015). Lava dome eruptions. In H. Sigurdsson, B. Houghton, H. Rymer, J. Stix, and S. McNutt, eds., *Encyclopedia of Volcanoes*, 2nd ed. San Diego, CA: Academic Press, pp. 343–62.

Calvari, S., Spampinato, L., Lodato, L. et al. (2005). Chronology and complex volcanic processes during the 2002–2003 flank eruption at Stromboli volcano (Italy) reconstructed from direct observations and surveys with a handheld thermal camera. *Journal of Geophysical Research*, 110, B02201. https://doi.org/10.1029/2004JB003129.

Castruccio, A., and Contreras, M. A. (2016). The influence of effusion rate and rheology on lava flow dynamics and morphology: A case study from the 1971 and 1988–1990 eruptions at Villarica and Lonquimay volcanoes, Southern Andes of Chile. *Journal of Volcanology and Geothermal Research*, 327, 469–83.

Chandrasekhar, S. (1961). *Hydrodynamic and Hydromagnetic Stability*. Oxford: Oxford University Press.

Chevrel, M. O., Platz, T., Hauber, E., Baratoux, D., Lavallée, Y., and Dingwell, D. B. (2013). Lava flow rheology: A comparison of morphological and petrological methods. *Earth and Planetary Science Letters*, 384, 109–20.

Christensen, U. R. (1992). An Eulerian technique for thermomechanical modeling of lithospheric extension. *Journal of Geophysical Research*, 97, 2015–36.

Cordonnier, B., Lev, E., and Garel, F. (2015). Benchmarking lava-flow models. In A. J. L. Harris, T. De Groeve, F. Garel, and S. A. Carn, eds., *Detecting, Modelling and Responding to Effusive Eruptions*, Special Publications, 426. London: Geological Society, pp. 425–45.

Costa, A., and Macedonio, G. (2005). Computational modeling of lava flows: A review. In M. Manga and G. Ventura, eds., *Kinematics and Dynamics of Lava Flows*, Special Paper 396, Boulder, CO: Geological Society of America, pp. 209–18.

Costa, A., Caricchi, L., and Bagdassarov, N. (2009). A model for the rheology of particle-bearing suspensions and partially molten rocks. *Geochemistry, Geophysics, Geosystems*, 10, Q03010. https://doi.org/10.1029/2008Gc002138.

Cutter, S., Ismail-Zadeh, A., Alcántara-Ayala, I. et al. (2015). Global risk: Pool knowledge to stem losses from disasters. *Nature*, 522, 277–9.

Daag, A. S., Dolan, M. T., Laguerta, E. et al. (1996). Growth of a postclimactic lava dome at Pinatubo Volcano, July–October 1992. In C. Newhall and R. Punongbayan, eds., *Fire and Mud: Eruptions and Lahars of Mount Pinatubo, Philippines*. Seattle: University of Washington Press, pp. 647–64.

Dragoni, M. A. (1989). A dynamical model of lava flows cooling by radiation. *Bulletin of Volcanology*, 51, 88–95.

Flynn, L. P., Harris, A. J. L., and Wright, R. (2001). Improved identification of volcanic features using Landsat 7 ETM+. *Remote Sensing of Environment*, 78, 180–93.

Giordano, D., and Dingwell, D. B. (2003). Viscosity of hydrous Etna basalt: Implications for Plinian-style basaltic eruptions. *Bulletin of Volcanology*, 65, 8–14.

Griffiths, R. W. (2000). The dynamics of lava flows. *Annual Review of Fluid Mechanics*, 32, 477–518.

Harris, A. J. L., Flynn, L. P., Keszthelyi, L. et al. (1998). Calculation of lava effusion rates from Landsat TM data. *Bulletin of Volcanology*, 60, 52–71.

Harris, A. J. L., Flynn, L. P., Matias, O., Rose, W. I., and Cornejo, J. (2004). The evolution of an active silicic lava flow field: An ETM+ perspective. *Journal of Volcanology and Geothermal Research*, 135, 147–68.

Harris, A. J. L., Rose, W. I., and Flynn, L. P. (2003). Temporal trends in lava dome extrusion at Santiaguito 1922–2000. *Bulletin of Volcanology*, 65, 77–89.

Heap, M. J., Troll, V., Kushnir, A. R. L. et al. (2019). Hydrothermal alteration of andesitic lava domes can lead to explosive volcanic behaviour. *Nature Communications*, 10, 5063. https://doi.org/10.1038/s41467-019-13102-8.

Hidaka, M., Goto, A., Umino, S., and Fujita, E. (2005). VTFS project: Development of the lava flow simulation code LavaSIM with a model for three-dimensional convection, spreading, and solidification. *Geochemistry, Geophysics, Geosystems*, 6, Q07008. https://doi.org/10.1029/2004GC000869.

Ismail-Zadeh, A., and Tackley, P. (2010). *Computational Methods for Geodynamics*. Cambridge: Cambridge University Press.

Ismail-Zadeh, A., and Takeuchi, K. (2007). Preventive disaster management of extreme natural events. *Natural Hazards*, 42, 459–67.

Ismail-Zadeh, A., Korotkii, A., Schubert, G., and Tsepelev, I. (2007). Quasi-reversibility method for data assimilation in models of mantle dynamics. *Geophysical Journal International*, 170, 1381–98.

Ismail-Zadeh, A., Korotkii, A., and Tsepelev, I. (2016). *Data-Driven Numerical Modeling in Geodynamics: Methods and Applications*. Heidelberg: Springer.

Jeffrey, D., and Acrivos, A. (1976). The rheological properties of suspensions of rigid particles. *AIChE Journal*, 22, 417–32.

Kabanikhin, S. I. (2011). *Inverse and Ill-Posed Problems: Theory and Applications*. Berlin: De Gruyter.

Kelfoun, K., and Vargas, S. V. (2015). VolcFlow capabilities and potential development for the simulation of lava flows. In A. J. L. Harris, T. De Groeve, F. Farel, and S. A. Carn, eds., *Detecting, Modelling and Responding to Effusive*

Eruptions, Special Publications 426. London: Geological Society, pp. 337–43.

Kelfoun, K., Santoso, A. B., Latchimy, T. et al. (2021). Growth and collapse of the 2018–2019 lava dome of Merapi volcano. *Bulletin of Volcanology*, 83, 8. https://doi.org/10.1007/s00445-020-01428-x.

Kilburn, C. R. J. (2000). Lava flow and flow fields. In H. Sigurdsson, ed., *Encyclopedia of Volcanoes*. San Diego, CA: Academic Press, pp.291–305.

Korotkii, A. I., and Kovtunov, D. A. (2006). Reconstruction of boundary regimes in an inverse problem of thermal convection of a high viscous fluid. *Proceedings of the Institute of Mathematics and Mechanics, Ural Branch of the Russian Academy of Sciences*, 12(2), 88–97.

Korotkii, A. I., and Starodubtseva, Y. V. (2014). Direct and inverse problems for models of stationary reactive-convective-diffusive flow. *Proceedings of the Institute of Mathematics and Mechanics, Ural Branch of the Russian Academy of Sciences*, 20(3), 98–113.

Korotkii, A., Kovtunov, D., Ismail-Zadeh, A., Tsepelev, I., and Melnik, O. (2016). Quantitative reconstruction of thermal and dynamic characteristics of lava from surface thermal measurements. *Geophysical Journal International*, 205, 1767–79.

Ladyzhenskaya, O. A. (1969). *The Mathematical Theory of Viscous Incompressible Flow*. New York: Gordon and Breach.

Lavallée, Y., Varley, N. R., Alatorre-Ibarguengoitia, M. A. et al. (2012). Magmatic architecture of dome building eruptions at Volcán de Colima, Mexico. *Bulletin of Volcanology*, 74, 249–60.

Lejeune, A., and Richet, P. (1995). Rheology of crystal-bearing silicate melts: An experimental study at high viscosity. *Journal of Geophysical Research*, 100, 4215–29.

Lions, J. L. (1971). *Optimal Control of Systems Governed by Partial Differential Equations*. Berlin: Springer.

Lombardo, V., Musacchio, M., and Buongiorno, M. F. (2012). Error analysis of subpixel lava temperature measurements using infrared remotely sensed data. *Geophysical Journal International*, 191, 112–25.

Loughlin, S. C., Sparks, S., Brown, S. K., Jenkins, S. F., and Vye-Brown, C., eds. (2015). *Global Volcanic Hazards and Risk*. Cambridge: Cambridge University Press.

Mardles, E. (1940). Viscosity of suspensions and the Einstein equation. *Nature*, 145, 970. https://doi.org/10.1038/145970a0.

Marsh, B. D. (1981). On the crystallinity, probability of occurrence, and rheology of lava and magma. *Contributions to Mineralogy and Petrology*, 78, 85–98.

Melnik, O., and Sparks, R. S. J. (1999). Nonlinear dynamics of lava dome extrusion. *Nature*, 402, 37–41.

Naimark, B. M., and Ismail-Zadeh, A. T. (1995). Numerical models of subsidence mechanism in intracratonic basin: Application to North American basins. *Geophysical Journal International*, 123, 149–60.

Naimark, B. M., Ismail-Zadeh, A. T., and Jacoby, W. R. (1998). Numerical approach to problems of gravitational instability of geostructures with advected material boundaries. *Geophysical Journal International*, 134, 473–83.

Nakada, S., Shimizu, H., and Ohta, K. (1999). Overview of the 1990–1995 eruption at Unzen Volcano. *Journal of Volcanology and Geothermal Research*, 89, 1–22.

Nakada, S., Zaennudin, A., Yoshimoto, M. et al. (2019). Growth process of the lava dome/flow complex at Sinabung volcano during 2013–2016. *Journal of Volcanology and Geothermal Research*, 382, 120–36.

Navon, I. M., Zou, X., Derber, J. and Sela, J. (1992). Variational data assimilation with an adiabatic version of the NMC spectral model. *Monthly Weather Review*, 120, 1433–46.

Papale, P., ed. (2014). *Volcanic Hazards, Risks and Disasters*. Amsterdam: Elsevier.

Patankar, S. V., and Spalding, D. B. (1972). A calculation procedure for heat and mass transfer in three-dimensional parabolic flows. *International Journal of Heat and Mass Transfer*, 15, 1787–806.

Poland, M. P., and Anderson, K. R. (2020). Partly cloudy with a chance of lava flows: Forecasting volcanic eruptions in the twenty-first century. *Journal of Geophysical Research*, 125, e2018JB016974. https://doi.org/10.1029/2018JB016974.

Polak, E., and Ribière, G. (1969). Note on the convergence of methods of conjugate directions. *Revue Francaise d'Informatique et de Recherche Operationnelle*, 3(16), 35–43.

Rumpf, M. E., Lev, E. and Wysocki, R. (2018). The influence of topographic roughness on lava flow emplacement. *Bulletin of Volcanology*, 80, 63. https://doi.org/10.1007/s00445-018-1238-9.

Salomon, D. (2007). *Data Compression: The Complete Reference*. London: Springer.

Sheldrake, T. E., Sparks, R. S. J., Cashman, K. V., Wadge, G., and Aspinall, W. P. (2016). Similarities and differences in the historical records of lava dome-building volcanoes: Implications for understanding magmatic processes and eruption forecasting. *Earth-Science Reviews*, 160, 240–63.

Short, N. M., and Stuart, L. M. (1983). *The Heat Capacity Mapping Mission (HCMM) Anthology*. Washington, DC: NASA Scientific and Technical Information Branch.

Starodubtseva, Y., Starodubtsev, I., Ismail-Zadeh, A. et al. (2021). A method for magma viscosity assessment by lava dome morphology. *Journal of Volcanology and Seismology*, 15, 159–68. https://doi.org/10.1134/S0742046321030064.

Swanson, D. A., Dzurisin, D., Holcomb, R. T. et al. (1987). Growth of the lava dome at Mount St Helens, Washington, (USA) 1981–1983. In J. H. Fink, ed., *The Emplacement of Silicic Domes and Lava Flows*, Special Paper 212. Boulder, CO: Geological Society of America, pp. 1–16.

Takeuchi, S. (2011). Preeruptive magma viscosity: An important measure of magma eruptibility. *Journal of Geophysical Research*, 116, B10201. https://doi.org/10.1029/2011JB008243.

Temam, R. (1977). *Navier–Stokes Equations: Theory and Numerical Analysis*. Amsterdam: North-Holland.

Tikhonov, A. N. (1963). Solution of incorrectly formulated problems and the regularization method, *Soviet Mathematics Doklady* 4, 1035–8.

Tikhonov, A. N., and Arsenin, V. Y. (1977). *Solution of Ill-Posed Problems*. Washington, DC: Winston.

Tikhonov, A. N., and Samarskii, A. A. (1990). *Equations of Mathematical Physics*. New York: Dover Publications.

Tsepelev, I., Ismail-Zadeh, A., Melnik, O., and Korotkii, A. (2016). Numerical modelling of fluid flow with rafts: An application to lava flows. *Journal of Geodynamics*, 97, 31–41.

Tsepelev, I., Ismail-Zadeh, A., Starodubtseva, Y., Korotkii, A., and Melnik, O. (2019). Crust development inferred from numerical models of lava flow and its surface thermal measurements. *Annals of Geophysics*, 62(2), VO226. https://doi.org/10.4401/ag-7745.

Tsepelev, I., Ismail-Zadeh, A., and Melnik, O. (2020). Lava dome morphology inferred from numerical modelling. *Geophysical Journal International*, 223(3), 1597–609.

Tsepelev, I., Ismail-Zadeh, A., and Melnik, O. (2021). Lava dome evolution at Volcán de Colima, México during 2013: Insights from numerical modeling. *Journal of Volcanology and Seismology*, 15, 491–501. https://doi.org/10.1134/S0742046321060117.

Turcotte, D. L., and Schubert, G. (2002). *Geodynamics*, 2nd ed. Cambridge: Cambridge University Press.

Voight, B., and Elsworth, D. (2000). Instability and collapse of hazardous gas-pressurized lava domes. *Geophysical Research Letters*, 27(1), 1–4.

Wadge, G., Robertson, R. E. A., and Voight, B., eds. (2014). *The Eruption of Soufrière Hills Volcano, Montserrat, from 2000 to 2010*. GSL Memoirs, vol. 39. London: Geological Society. https://doi.org/10.1144/M39.

Wang, Z., Bovik, A. C., Sheikh, H. R., and Simoncelli, E. P. (2004). Image quality assessment: From error visibility to structural similarity. *IEEE Transactions on Image Processing*, 13(4), 600–12.

Walker, G. P. L. (1973). Lengths of lava flows. *Philosophical Transactions of the Royal Society*, 274, 107–18.

Watts, R. B., Herd, R. A., Sparks, R. S. J., and Young, S. R. (2002). Growth patterns and emplacement of the andesitic lava dome at Soufrière Hills Volcano, Montserrat. In T. H. Druitt, and B. P. Kokelaar, eds., *The Eruption of Soufrière Hills Volcano, Montserrat, from 1995 to 1999*. GLS Memoirs, vol. 21. London: Geological Society, pp. 115–52.

Wright, R., Garbeil H., and Davies, A. G. (2010). Cooling rate of some active lavas determined using an orbital imaging spectrometer. *Journal of Geophysical Research*, 115, B06205. https://doi.org/10.1029/2009JB006536.

Wright, T. L., and Okamura, R. T. (1977). Cooling and crystallization of tholeiitic basalt, 1965 Makaopuhi lava lake, Hawaii. US Geological Survey Professional Paper 1004.

Zakšek, K., Hort, M., and Lorenz, E. (2015). Satellite and ground based thermal observation of the 2014 effusive eruption at Stromboli volcano. *Remote Sensing*, 7, 17190–211.

Zeinalova, N., Ismail-Zadeh, A., Melnik, O. E., Tsepelev, I., and Zobin, V. M. (2021). Lava dome morphology and viscosity inferred from data-driven numerical modeling of dome growth at Volcán de Colima, Mexico during 2007–2009. *Frontiers in Earth Science*, 9, 735914. https://doi.org/10.3389/feart.2021.735914.

Zobin, V. M., Arámbula, R., Bretón, M. et al. (2015). Dynamics of the January 2013–June 2014 explosive-effusive episode in the eruption of Volcán de Colima, México: Insights from seismic and video monitoring. *Bulletin of Volcanology*, 77, 31. https://doi.org/10.1007/s00445-015-0917-z.

13

Data Assimilation for Real-Time Shake-Mapping and Prediction of Ground Shaking in Earthquake Early Warning

Mitsuyuki Hoshiba

Abstract: Earthquake early warning (EEW) systems aim to provide advance warning of impending strong ground shaking, in which earthquake ground shaking is predicted in real-time or near real-time. Many EEW systems are based on a strategy which first quickly determines the earthquake hypocentre and magnitude, and then predicts the strength of ground shaking at various locations using the hypocentre distance and magnitude. Recently, however, a new strategy was proposed in which the current seismic wavefield is rapidly estimated by using data assimilation, and then the future wavefield is predicted on the basis of the physics of wave propagation. This technique for real-time prediction of ground shaking in EEW does not necessarily require the earthquake hypocentre and magnitude. In this paper, I review real-time shake-mapping and data assimilation for precise estimation of ongoing ground shaking, and prediction of future shaking in EEW.

13.1 Introduction

Real-time prediction of earthquake ground shaking is a strong tool for disaster prevention and mitigation of earthquake disaster, and it has been applied for earthquake early warning (EEW). In recent decades, EEW systems have been deployed to issue warnings to the general public in Mexico, Japan, Taiwan, and the west coast of the United States (Hoshiba et al., 2008; Cuellar et al., 2014; Chen et al., 2015; Cochran et al., 2019), and their possible use has been investigated in the European Union, Turkey, South Korea, China, Israel, and other regions (Erdik et al., 2003; Peng et al., 2011; Gasparini and Manfredi, 2014; Sheen et al., 2017; Kurzon et al., 2020). Many current EEW systems operate by quickly determining the earthquake source parameters, such as hypocentre location and magnitude, and then predict the strength of ground motions at various sites by applying a ground-motion prediction equation (GMPE) that uses the hypocentral distance and magnitude. This approach is referred as 'the source-based method'.

For example, a typical form of $\log y_{ij} = a\, M_i - b \log r_{ij} + c$ is used, where y_{ij} is the index of peak ground motion, such as peak ground acceleration (PGA) of the seismic waveform or peak ground velocity (PGV) from the ith event at site j, M_i is the magnitude of event i, and r_{ij} is the distance (hypocentral distance, epicentral distance, or fault distance) between event i and site j. Coefficients a, b, and c are usually empirically estimated in advance. The 2011 off the Pacific Coast of Tohoku earthquake (hereinafter, the Tohoku earthquake, Mw(moment magnitude): 9.0) and its quite active aftershocks, however, revealed important technical issues with the approach: the EEW system of the Japan Meteorological Agency (JMA) underpredicted ground motion at distant sites because of the large extent of the fault rupture, and it sometimes overpredicted ground motion because the system was confused by multiple simultaneous aftershocks (Hoshiba et al., 2011; Hoshiba and Ozaki, 2014).

To address the issue of the extent of rupture during large earthquakes, Böse et al. (2018) proposed a method to estimate rupture extents in real-time. To address the issue of multiple simultaneous events, Tamaribuchi et al. (2014) proposed a method to use amplitude data for hypocentre determination. Their approaches are extensions of the source-based method that still relies on rapid estimation of the source parameters. Recently, however, an alternative algorithm that skips the process of the source estimation has been intensively investigated; this algorithm does not necessarily require source parameters to predict the strength of ground motion. Instead, future ground motions are predicted directly from observed ground motion (Hoshiba, 2013a; Hoshiba and Aoki, 2015; Furumura et al., 2019). The wavefield-based method or ground-motion-based method (hereinafter wavefield-based method, for simplicity) first estimates the current wavefield, and then predicts the future wavefield based on the physics of wave propagation. Because source parameters are not estimated, this method avoids the issues of source-based methods (rupture extent, and simultaneous multiple earthquakes). Hoshiba (2021) has reviewed the wavefield-based method for EEW in detail.

In the wavefield-based method, the precise estimation of the current wavefield is key for precise prediction of the future wavefield. Data assimilation (Kalnay, 2003; Awaji et al., 2009) is a powerful technique for estimating the current wavefield. In the data assimilation procedure, the spatial distribution of the wavefield is estimated from not only actual observations but also the simulation of wave propagation based on wave propagation physics, which leads to a precise estimation of the current wavefield. That is, data assimilation incorporates actual observations into the simulation of wave propagation.

The strength of ground motion is measured by PGA, PGV, and/or seismic intensity observed by seismometers, yielding data that constitute a point image of information. At present, except for special observations, seismic networks deploy seismometers at intervals of ~10 to ~100 km; the resulting distribution is too sparse for precise estimation of spread of strong ground motion and potential areas of damage. Map images of the distribution of ground motion, or shake-maps, are a useful approximation. Some organisations publish such images online soon after large earthquakes: examples include the US Geological Survey (USGS) ShakeMap (e.g. Wald et al., 2005), and the JMA Contour-map of Estimated Seismic Intensity.[1] The JMA map images are derived by interpolation of the point image after taking into account site amplification at individual sites (e.g. large amplification at sites in basins and small amplification at sites on hard rock), and the USGS images are made by a combination of the interpolation and estimation by GMPE. These shake-mapping images are estimates of the eventual observations of ground shaking, such as PGA and PGV, and they do not depict the propagation of shaking. Data assimilation makes it possible to obtain an ongoing shake-mapping image in real-time at each time step, because the technique estimates current wavefields. Successive wavefields can be combined into an animation that visualises the evolving wavefield, and shows the propagation of the shaking. Such animations can be helpful for understanding characteristics of seismic wave propagation.

In this chapter, I explain the use of data assimilation for real-time estimation of seismic wavefields (i.e., real-time shake-maps), and then its application for prediction of ground shaking in EEW. I present examples of real-time shake-mapping and prediction, using the 2011 Tohoku earthquake (Mw9.0), the deep 2015 off Bonin islands earthquake (Mw7.9), three inland earthquakes (Mw6.2–6.4) in Japan, and the 2016 Kumamoto earthquake (Mw7.1) which was followed by M~6 triggered earthquake. These examples indicate advantages of data assimilation for precise

estimation of ongoing wavefields and ground-shaking prediction in EEW.

13.2 Seismic Wave Propagation and Its Simulation

Because seismic waves are controlled by the physics of wave propagation, future wavefields can be predicted by using wave propagation theory. In this section, I explain the theoretical background of the physics of seismic wave propagation and its simulation, following Hoshiba (2021). I use a scalar wave expression for simplicity, although seismic waves are vector waves.

13.2.1 Finite Difference Method

Wave propagation is expressed by the wave equation:

$$\frac{1}{c(\boldsymbol{x})^2}\ddot{u}(\boldsymbol{x}, t) = \nabla^2 u(\boldsymbol{x}, t),\qquad(13.1)$$

where $u(\boldsymbol{x}, t)$ is the wave amplitude at location $\boldsymbol{x}$ and time t, c is the phase velocity, ∇^2 is the Laplacian, and $\ddot{u}$ is the second order differential of u with respect to t (i.e. $\partial^2 u/\partial t^2$). This equation implies that the time-evolution of a wave's amplitude, $\ddot{u}$, is determined by its spatial distribution ($\nabla^2 u$). Therefore, future wavefields can be predicted from the spatial distribution of a known wavefield when the velocity structure, $c(\boldsymbol{x})$, is known. Equation (13.1) is approximated as

$$u(\boldsymbol{x}, t + \Delta t) \approx 2u(\boldsymbol{x}, t) - u(\boldsymbol{x}, t - \Delta t) + \Delta t^2 \cdot c(\boldsymbol{x})^2 \cdot \nabla^2 u(\boldsymbol{x}, t).$$
$$(13.2)$$

The wavefield one time step Δt in the future, $u(\boldsymbol{x}, t + \Delta t)$, can be estimated from the current wavefield, $u(\boldsymbol{x}, t)$, and that one time step prior, $u(\boldsymbol{x}, t-\Delta t)$. Then, $u(\boldsymbol{x}, t + 2\Delta t)$ is computed from $u(\boldsymbol{x}, t + \Delta t)$ and $u(\boldsymbol{x}, t)$ as,

$$u(\boldsymbol{x}, t + 2\Delta t) \approx 2u(\boldsymbol{x}, t + \Delta t) - u(\boldsymbol{x}, t)$$
$$+ \Delta t^2 \cdot c(\boldsymbol{x})^2 \cdot \nabla^2 u(\boldsymbol{x}, t + \Delta t).\qquad(13.3)$$

Thus, the wavefield at any future time can be obtained by repeating this procedure.

Furumura et al. (2019) and Oba et al. (2020) applied the finite difference approach to estimate the current wavefield using data assimilation, and then predicted long-period (> 3–10 s) ground motions. Furumura and Maeda (2021) applied a time-reversal propagation technique in an attempt to estimate source images from the current wavefield. At present, however, the finite difference method is not useful for computing short-period ground motions (< 1 s) because the very fine mesh size required exceeds modern computing capabilities, and the very precise velocity structure required

[1] www.data.jma.go.jp/svd/eew/data/suikei/kaisetsu.html, in Japanese.

to simulate wave propagation is not easily obtained by current survey techniques.

13.2.2 Radiative Transfer Theory

To simulate high-frequency wave propagation, the ray theory approach is valid, and radiative transfer theory (RTT) is a powerful tool for representing it, in which scattering, attenuation, and reflection are easily treated, although it is difficult to include refraction. Radiative transfer theory calculates the propagation of energy instead of the propagation of the wave itself. Many authors obtain the time history of energy, $F(x, t)$, from the running average of the squared amplitude of the band-pass filtered waveform, $u(x, t)$, at location x and time t: $F(x, t) = |u(x, t)|^2$. Radiative transfer theory has been widely used to represent the envelope shape of seismic waveforms with frequency greater than 1 Hz (Sato et al., 2012).

When isotropic scattering is assumed, RTT is expressed as:

$$\dot{f}(x, t : q) + c(x)q\nabla f(x, t : q) = -g_s(x) \cdot c(x) \cdot f(x, t : q)$$
$$+ \frac{c(x)}{4\pi} \int g_s(x)f(x, t : q')dq',$$
(13.4)

where f is the energy density at location x and time t travelling in direction q (here q is the unit vector), $c(x)$ is the velocity of the seismic wave at x, and $g_s(x)$ is the strength of scattering at x (Sato et al., 2012). Here it is assumed that scattering does not cause wave conversion between P and S waves, such that the propagation of P waves and S waves can be calculated independently. The time history of energy, $F(x, t)$, is the sum of $f(x, t: q)$ in all directions:

$$F(x, t) = \int f(x, t : q)dq.$$
(13.5)

Here, for simplicity, I assume that both the velocity and scattering structures are homogeneous; thus, velocity and scattering strength are independent of x: $c(x) = c_0$ and $g_s(x) = g_0$. With this assumption Eq. (13.4) is expressed as,

$$\dot{f}(x, t : q) + c_0 q\nabla f(x, t : q) = -g_0 c_0 f(x, t : q)$$
$$+ \frac{c_0}{4\pi} \int g_0 f(x, t : q')dq'. \quad (13.6)$$

The left-hand side of this equation represents advection, and the two terms on the right-hand side represent scattering attenuation and scattering from direction q' to q, respectively. Because the first term on the left-hand side is the differential of f with respect to time, Eq. (13.6) means that it is possible to predict future f provided that the current spatial and directional distributions of f are known. Equation (13.6) is approximated as:

$$f(x, t + \Delta t : q) \approx f(x, t : q) + \Delta t\bigg(- c_0 q\nabla f(x, t : q)$$
$$- g_0 c_0 f(x, t : q) + \frac{c_0}{4\pi} \int g_0 f(x, t : q')dq'\bigg).$$
(13.7)

Repeating this process makes it possible to predict f at any future time.

In actual applications of RTT, a particle method based on the Monte-Carlo technique has been widely used for efficient calculation. Gusev and Abubakirov (1987), Hoshiba (1991), Yoshimoto (2000) and others have used particle methods for simulating high-frequency seismic wave propagation to explain the envelopes of seismic coda.

Hoshiba and Aoki (2015), Wang et al. (2017a,b) and Ogiso et al. (2018) have applied RTT to make predictions of the strength of seismic ground motion for EEW. They have called the method 'numerical shake prediction', because of its analogy to numerical weather prediction in meteorology, in which physical processes are simulated from a precise estimate of present conditions made by data assimilation. Whereas the finite difference method is a good approach for calculating low-frequency waveforms but not high-frequency waveforms, RTT is valid for high-frequency but not necessarily valid for low-frequency because RTT expresses energy transportation based on ray theoretical approach in which refraction is not well included.

13.3 Data Assimilation

The first step in the wavefield-based method is to estimate the current wavefield. Data assimilation is a powerful technique for estimating current conditions that is widely used for this purpose in numerical weather prediction, oceanography, and rocket control (Kalnay, 2003; Awaji et al., 2009). The technique is applied to precise estimation of current seismic wavefield which correspond to ongoing shake-maps in terminology in seismology. In data assimilation, the spatial distribution of the wavefield is estimated from not only actual observations but also the simulation of wave propagation based on wave propagation physics, leading to a precise estimation of the current wavefield. Therefore, data assimilation incorporates actual observations into the simulation of wave propagation. Figure 13.1 schematically indicates the process from data assimilation to prediction. The explanation in this section follows Hoshiba and Aoki (2015) and Hoshiba (2021).

Let u_n indicate the wavefield in the model space at time $t_n = n\Delta t$, in which $u_n = [u(x, n\Delta t), u(x, (n-1)\Delta t)]$ in the finite difference method, or $u_n = [f(x, n\Delta t: q)]$ in RTT, where $[\cdot]$ means the elements of the vector. When the three-dimensional space is discretised as 0 to $L_x\Delta x$, 0 to $L_y\Delta y$ and 0 to $L_z\Delta z$, the number of elements of u_n is $I = 2 \cdot L_x \cdot L_y \cdot L_z$ in the finite difference method, and when the azimuth is discretised

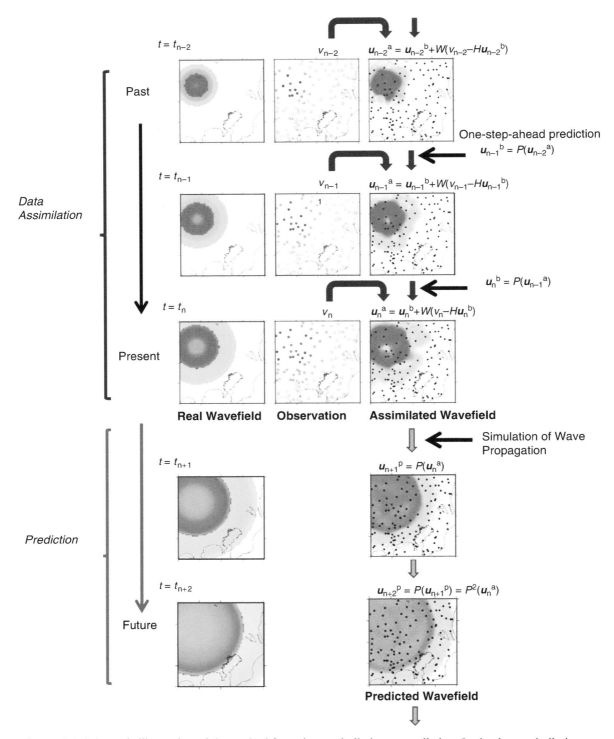

Figure 13.1 Schematic illustration of the method from data assimilation to prediction. In the data assimilation process, one-step ahead prediction, $u^b_n = P(u^a_{n-1})$, is combined with the actual observation, v_n, to estimate the present situation, $u^a_n = u^b_n + W(v_n - Hu^b_n)$. In the prediction process, one-step ahead prediction is repeatedly applied to predict the future situation. In the assimilated and predicted wavefields, small dots indicate the locations of KiK-net stations of the National Research Institute for Earth Science and Disaster Prevention (NIED), and stations of the Japan Meteorological Agency (JMA) network in the Kanto district, Japan. Source: Hoshiba and Aoki, 2015.

as 0 to $L_q\Delta q$, the number is $I = L_x \cdot L_y \cdot L_z \cdot L_q$ in RTT. When u_{n-1} is given, we can predict u_n by simulating the propagation of the wave; this prediction for one time-step-ahead is expressed as $u_n = P(u_{n-1})$, where P is the operator of Eq. (13.2) or (13.7). To discriminate between u_n before and after being combined with the actual observations, the wavefields before and after are denoted by u_n^b and u_n^a, respectively. P is applied to the wavefield after the combination at one time step before (i.e. t_{n-1}); thus, the one-step-ahead prediction is expressed as

$$u_n^b = P(u_{n-1}^a). \tag{13.8}$$

Let $v_n = (v_{n1}, v_{n2}, v_{n3}, \ldots v_{nj}, \ldots v_{nJ})^T$ be the actual observation in observational space at time t_n, in which v_{nj} means the observed data at the j-th element. Let the total number of observation elements be J. Usually I (the number of grids in the model space) is much larger than J (number of observation elements). The data assimilation is expresses as

$$u_n^a = u_n^b + W(v_n - Hu_n^b). \tag{13.9}$$

Here H is the $J \times I$ matrix called the observation matrix, representing the interpolation of grid points onto the location of the observation point, and then $(v_n - Hu_n^b)$ is the difference between the one-step-ahead prediction and the actual observation at time t_n. W is the $I \times J$ matrix called the weight matrix, and $W(v_n - Hu_n^b)$ indicates the correction of the one-step-ahead prediction in the simulation of wave propagation. From u_n^a, u_{n+1}^b is obtained from Eq. (13.8). Iterative application of Eqs. (13.8) and (13.9) produces time-evolution of estimated wavefield, thus constructing ongoing shake-maps.

The parameter setting of W is important in data assimilation, and several techniques have been proposed. The simplest is the optimal interpolation method, in which W is constant irrespective of time n, although in the Kalman filter method, W changes with increasing n. In the optimal interpolation method, matrix W is expressed in relation to the errors in the one-step-ahead prediction (background error, σ_b), and in the observations (observational error, σ_o). When the correlation distance of the background error and the ratio σ_o/σ_b are assumed, matrix W is obtained (for details, see Kalnay, 2003; Awaji et al., 2009; Hoshiba and Aoki, 2015). When the correlation distance is large, the $W(v_n - Hu_n^b)$ correction is applied to a wide area around each observation point, and when the distance is small, the correction is restricted to a small area. When the observational errors are assumed to be much larger than the background errors, $\sigma_o/\sigma_b \gg 1$, $W \approx 0$ and thus $u_n^a \approx u_n^b$. Iterative application of Eqs. (13.8) and (13.9), therefore, results in the simulation of wave propagation alone, because the observation have no effect. In contrast, $\sigma_o/\sigma_b \approx 0$ corresponds to the case where the contours of the actual observations are drawn independently at each time step, because the one-step-ahead prediction has little effect in Eq. (13.9).

Although the seismic wavefield is observable at the ground surface when stations are densely deployed at the surface (i.e. in 2-D space), the underground wavefield at depths of more than a few kilometres cannot be observed because many borehole observations deeper than a few kilometres are not realistic at present. Because actual seismic wavefields are expressed in three-dimensional space, assumptions are required to apply data assimilation to estimate the three-dimensional wavefield. Handling the difference between the two- and three-dimensional spaces is an important subject for future advancement of the data assimilation technique in seismology.

13.4 Shake-Mapping and Prediction

In the process of the data assimilation, the current wavefield, u_n^a, can be estimated in real-time when current observations, v_n, is obtained in real-time, as shown in Eq. (13.9). This real-time estimation corresponds to real-time shake-mapping. Animation of u_n^a (for $1, \ldots, n-3, n-2, n-1, n$) indicates time-evolving wavefields; that is, an ongoing shake-map.

Once the present wavefield, u_n^a, has been precisely estimated by the data assimilation technique, the future wavefield, u^P, is predicted from the current wavefield, u_n^a,

$$u_{n+1}^P = P(u_n^a), \tag{13.10}$$

and u_{n+2}^P is forecast from u_{n+1}^P; that is, $u_{n+2}^P = P(u_{n+1}^P) = P^2(u_n^a)$. Repeating this process

$$\begin{aligned} u_{n+k}^P &= P(u_{n+k-1}^P) = P^2(u_{n+k-2}^P) \\ &= \ldots = P^{k-1}(u_{n+1}^P) = P^k(u_n^a). \end{aligned} \tag{13.11}$$

Future wavefield at any time can be predicted from current wavefield.

13.5 Application Examples

This section presents examples of the use of data assimilation to estimate precise seismic wavefields based on RTT. The first example involves a numerical experiment with multiple simultaneous earthquakes, and then several examples are presented that involve actual earthquakes.

For the latter, given that the amplitude of high-frequency seismic waves depends strongly on subsurface geological conditions around the observation site (i.e. large amplification on sediment sites and small amplification on hard rock sites, and the amplification is frequency-dependent), corrections for site amplification are required, especially for high-frequency seismic waves, to isolate propagation phenomena. Hoshiba (2013b) proposed a method to correct frequency-dependent site amplification in real-time in the time domain, and Ogiso et al. (2016) evaluated the site amplification factors of more than 2,200 stations in Japan to apply this method. In the following examples, site amplification is corrected before data assimilation.

13.5.1 Numerical Experiment with Multiple Simultaneous Earthquakes

As explained in Section 13.1, for a couple of weeks after the 2011 Tohoku Earthquake (Mw9.0), the JMA operational EEW system sometimes overpredicted the strength of ground shaking because it misinterpreted multiple simultaneous aftershocks as a single large event.

Figure 13.2 shows an example of a numerical experiment involving four simultaneous earthquakes, showing (A) true wavefields, (B) observation by seismometers, (C) estimated wavefields (shake-maps) with data assimilation, and (D) estimated wavefields without data assimilation. Physics of wave propagation is considered in panel C by data assimilation, but is not taken into account in panel D where contours are drawn each time step independently from other time step. It is not easy to recognise the four earthquakes from panels B and D, but possible to do it from panel C. This example shows the merit of data assimilation for estimating wavefields and shake-mapping.

13.5.2 Examples of the 2011 Tohoku Earthquake

The 2011 Tohoku earthquake (M_w9.0) occurred off the pacific coast of Japan on 11 March 2011, and generated

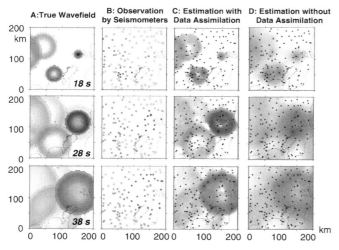

Figure 13.2 Numerical experiment of four simultaneous earthquakes, showing (A) true wavefields, (B) observation by seismometers, (C) estimated wavefields (shake-maps) with data assimilation technique, and (D) shake-maps without data assimilation. Strength of shaking is shown by grey shading. Small dots indicate location of seismometers (K-NET and KiK-net sites).

[2] Fire Disaster Management Agency, Japan, report on 7 March 2014.

a huge tsunami that killed 18,958 people and left 2,655 people missing.[2] Strong motion was observed over a wide area of northeastern Japan. Accelerations exceeding 1,000 cm/s² were recorded at some stations in the Kanto district, more than 300 km from the epicentre (Hoshiba et al., 2011). To explain the wide area of strong ground motion, some authors have proposed source models composed of multiple strong-motion-generation areas (SMGAs): for example, Asano and Iwata (2012) proposed a source model with four SMGAs, and Kurahashi and Irikura (2013) proposed models with five SMGAs. All of these studies identified at least two major SMGAs off Miyagi Prefecture in the Tohoku district and up to several SMGAs off Fukushima Prefecture southwest of the hypocentre.

Figure 13.3 shows the estimated wavefields (current shake-maps) of the earthquake at several elapsed times. All seismic waveforms are corrected according to the frequency-dependent site amplification of Ohtemachi station in Tokyo (Hoshiba and Aoki, 2015). Here P and S waves from the four SMGAs (G1, 2, 3, and 4 identified by Asano and Iwata, 2012) can be recognised: shaking at Sendai was brought by strong motions from G1 and G2, whereas shaking at Tokyo was by those from G3 and G4. Therefore, the source locations of the strong motions are different between Sendai and Tokyo. Data assimilation enables us to identify strong-motion propagation from different SMGAs. Hoshiba and Aoki (2015) used current shake-maps to demonstrate real-time prediction of strong shaking at Tokyo once strong shaking was recognised to approaching to Tokyo (see the 130 s panel in Fig. 13.3).

13.5.3 Example of a Very Deep and Distant Earthquake

On 30 May 2015, a Mw7.9 earthquake occurred off the Bonin Islands with a focal depth of 682 km and an epicentre about 870 km from Tokyo. It was the deepest event (even including small earthquakes) in and around Japan in the recent 100 years of modern seismic observation. Though it was a distant and very deep event, 13 persons were slightly injured in the Tokyo metropolitan area. Hypocentral locations of such distant and deep events are difficult to estimate in EEW, and moreover there are no valid GMPEs for such deep events for predicting the strength of ground shaking. The JMA EEW system did not issue a warning because the focal depth was out of range of the GMPE.

Figure 13.4 shows wavefields that were estimated using data assimilation. The S wave arrived at Tokyo 100 s after the P wave. Shaking due to P wave is visible at 15 s in Fig. 13.4, and propagated northward, after which shaking was weak in the Tokyo metropolitan area until the S wave arrival at 115 s. Strong shaking did not propagate to the west. Shaking in the Tohoku region was stronger on the east coast than the west

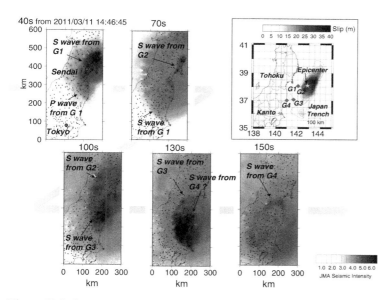

Figure 13.3 Current shake-maps made by using data assimilation showing estimated wavefields of the 2011 Tohoku earthquake (Mw9.0). Upper right map shows four strong-motion-generation areas (SMGAs: G1, 2, 3, and 4) identified by Asano and Iwata (2012) at locations outside the area of large slip (Japan Meteorological Agency, 2012; Hoshiba, 2020). *P* and *S* waves from the SMGAs can be recognised on these shake-maps.

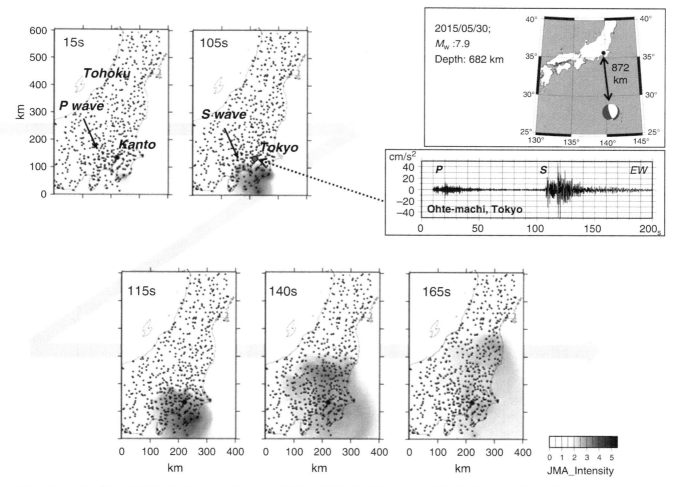

Figure 13.4 Example of the 2015 Bonin Island earthquake (30 May 2015, Mw7.9, depth: 682 km). This event is the deepest one in the recent 100 years of modern seismic observation in and around Japan. *P-S* time was 100 s in Tokyo.

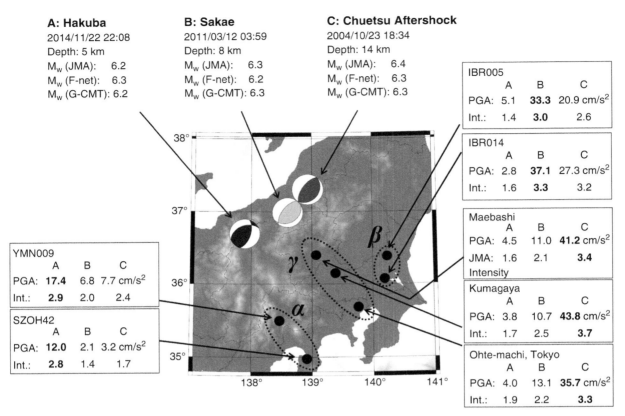

Figure 13.5 Map of central Japan showing locations and focal mechanisms of three island earthquakes – (A) Hakuba, (B) Sakae, and (C) Chuetsu-aftershock – with similar Mw and distances from Tokyo. Note that Mw estimation is slightly different depending on the organisations: F-net, JMA and global CMT. Observation of peak ground acceleration (PGA) and seismic intensity on JMA scale from the three earthquakes are shown.

coast. These noteworthy phenomena are easily recognised in the ongoing shake-maps, which thereby offer geophysical insights.

13.5.4 Examples of Three Inland Earthquakes

Figure 13.5 shows the locations of three Mw6.2–6.4 earthquakes that occurred in central Japan: the 2014 Hakuba earthquake (A), the 2011 Sakae earthquake of 12 March, one day after the 2011 Tohoku earthquake (B), and the largest aftershock of the 2004 Chuetsu earthquake (C). Tokyo is almost equidistant from the three earthquakes, but the strength of ground motion there differed among these events. The PGA during the 2004 Chuetsu aftershock (C) exceeds that of the 2014 Hakuba earthquake (A) by a factor of 9 in Tokyo and Maebashi, and in Kumagaya by a factor greater than 10. For the sites in group γ in Fig. 13.5, the ground motion from the 2004 Chuetsu aftershock (C) was the strongest of the three events. However, for the sites in group α the 2014 Hakuba earthquake (A) was the strongest, and for the sites in group β the 2011 Sakae earthquake (B) was the strongest. This comparison indicates that the source

location and magnitude alone are not enough to reproduce precise distributions (i.e. shake-maps), and that precisely predicting the strength of ground motion is difficult.

Figure 13.6 (left panel) shows the propagation of seismic ground motion (i.e. ongoing shake-map) of the three earthquakes, as estimated by the data assimilation technique. For the 2004 Chuetsu aftershock (C) the strong ground motion propagated towards Tokyo, whereas for the 2011 Sakae earthquake (B) it travelled towards Tsukuba. For the 2014 Hakuba earthquake (A), it propagated south. Data assimilation makes it easy to trace the propagation of ground motion and observe how azimuthal differences in propagation result in large differences in observed ground motion. The non-concentric propagation phenomena were often observed from the other earthquakes occurred around the three earthquakes. The examples of the prediction of ground motion from some time points were indicated in Fig. 13.6 (centre panel). For the 2011 Sakae earthquake (B) at time t = 45 s relatively strong motion was observed towards Tsukuba, and the strong motion was arrived at Tsukuba 10 s later (i.e. t = 55 s). Examples for the 2004 Chuetsu aftershock (C) were indicated in Hoshiba (2021), in

Current wavefield using data assimilation

Predicted wavefield of 10 s later

Estimated wavefield using data assimilation

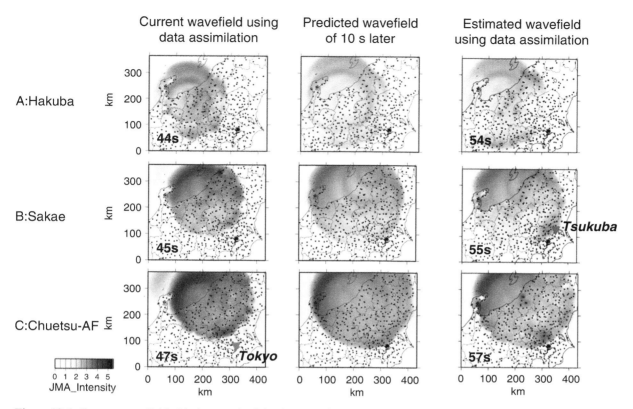

Figure 13.6 Current wavefields (shake-maps) of the three earthquakes estimated from data assimilation (left panel). Wavefields 10 s later predicted from the current wavefield (centre panel). Estimated wavefields 10 s later using data assimilation (right panel).

which strong motion was predicted at Tokyo once strong motion towards Tokyo was recognised.

13.5.5 Examples of the 2016 Kumamoto Earthquake (Mw7.1) and Triggered Event

Figure 13.7 illustrates the example of the 2016 Kumamoto earthquake (Mw7.1; 16 April 2016, focal depth 12 km). The earthquake triggered M~6 earthquake that occurred about 40 s later about 70 km from the hypocentre (e.g. Suzuki et al., 2017). Strong motion from the triggered event, exceeding the seismic motion from the Mw7.1 mainshock, was observed at Yufu City, but the operational JMA EEW system did not identify the triggered earthquake, and therefore underpredicted the strength of ground motion in that area. By using data assimilation to plot ongoing shake-maps, it is easy to recognise the occurrence of the triggered earthquake and predict the ground motion resulting from the earthquake.

13.6 Summary

Data assimilation has been used mainly in meteorology and oceanography in geophysics, and applied to weather prediction. Recently the technique is also used in solid-earth

geophysics. Ongoing shake-mapping and estimation of current wavefields are examples of how the technique enables us to more accurately model seismic wave propagation in detail and recognise new earthquakes that are masked by preceding large earthquakes. From precise knowledge of the current wavefield, we can confidently predict future wavefields. This prediction technique is applicable to EEW for real-time prediction of ground shaking that avoids the problems posed by large rupture events and simultaneous multiple events.

Acknowledgements. I appreciate T. Kagawa's careful reading of the manuscript and his constructive comments. This study used waveform data from K–NET and KiK-net (NIED), and from JMA. I thank NIED and JMA for their efforts. Site-correction was performed by M. Ogiso, and discussions with M. Ogiso, Y. Kodera, S. Aoki, and N. Hayashimoto were useful in this research. M. Kamachi, Y. Fujii, N. Usui, and T. Toyoda helped me learn the data assimilation technique. Figures were produced using Generic Mapping Tools (Wessel and Smith, 1995). This research was partially supported by JSPS KAKENHI Grant Number 17H02064. Any opinions, findings, and conclusions described in this chapter are solely those of the author and do not necessarily reflect the view of JMA.

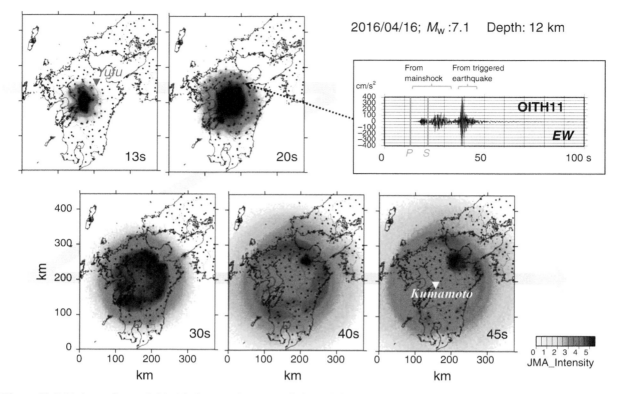

Figure 13.7 Estimated wavefields (shake-maps) at several elapsed times after the 2016 Kumamoto earthquake (16 April 2016, Mw7.1, depth 12 km) using data assimilation, which distinguished triggered M~6 earthquake near Yufu City at about 40 s. Shaking from the triggered earthquake was stronger than that from the M7.1 mainshock at OITH11.

References

Asano, K., and Iwata, T. (2012). Source model for strong ground motion generation in the frequency range 0.1–10 Hz during the 2011 Tohoku earthquake. *Earth Planets Space*, 64, 1111–23. https://doi.org/10.5047/eps.2012.05.003.

Awaji T., Kamachi, M., Ikeda, M., and Ishikawa, Y. (2009). Data Assimilation: Innovation Combining Observation and Model (in Japanese). Kyoto: Kyoto University Press.

Böse, M., Smith, D. E., Felizardo, C. et al. (2018). FinDer v.2: Improved real-time ground-motion predictions forM2–M9 with seismic finite-source characterization. *Geophysical Journal International*, 212, 725–42. https://doi.org/10.1093/gji/ggx430.

Chen, D. Y., Hsiao, N. C., and Wu, Y. M. (2015). The earthworm-based earthquake alarm reporting system in Taiwan. *Bulletin of the Seismological Society of America*, 105, 568–79. https://doi.org/10.1785/0120140147.

Cochran, E. S., Bunn, J., Minson, S. E. et al. (2019). Event detection performance of the PLUM earthquake early warning algorithm in southern California. *Bulletin of the Seismological Society of America*, 109, 1524–41. https://doi.org/10.1785/0120180326.

Cuellar, A., Espinosa-Aranda, J. M., Suárez, G. (2014). The Mexican Seismic Alert System (SASMEX): Its alert signals, broadcast results and performance during the M 7.4 Punta Maldonado earthquake of March 20th, 2012. In F. Wenzel and J. Zschau, eds., *Early Warning for Geological Disasters*. Berlin: Springer, pp. 71–87.

Erdik, M., Fahjan, Y., Ozel, O. et al. (2003). Istanbul earthquake rapid response and the early warning system, *Bulletin of Earthquake Engineering*, 1, 157–63. https://doi.org/10.1023/A:1024813612271.

Furumura, T., and Maeda, T. (2021). High-resolution source imaging based on time-reversal wave propagation simulations using assimilated dense seismic records. *Geophysical Journal International*, 225, 140–57. https://doi.org/10.1093/gji/ggaa586.

Furumura, T., Maeda, T., and Oba, A. (2019). Early forecast of long-period ground motions via data assimilation of observed ground motions and wave propagation simulations. *Geophysical Research Letters*, 46(1), 139–47. https://doi.org/10.1029/2018GL081163.

Gasparini, P., and Manfredi, G. (2014). Development of earthquake early warning systems in the European Union. In J. Zschau and F. Wenzel, eds., *Early Warning for Geological Disasters: Scientific Methods and Current Practice*, Berlin: Springer, pp. 89–101. https://doi.org/10.1007/978-3-642-12233-0_5.

Gusev, A. A., and Abubakirov, I. R. (1987). Monte-Carlo simulation of record envelope of a near earthquake. *Physics of the Earth and Planetary Interiors*, 49, 30–6.

Hoshiba, M. (1991). Simulation of multiple scattered coda wave excitation based on the energy conservation law. *Physics of the Earth and Planetary Interiors*, 67, 123–6.

Hoshiba, M. (2013a). Real-time prediction of ground motion by Kirchhoff–Fresnel boundary integral equation method:

Extended front detection method for earthquake early warning. *Journal of Geophysical Research: Solid Earth*, 118, 1038–50. https://doi.org/10.1002/jgrb.50119.

Hoshiba, M. (2013b). Real-time correction of frequency-dependent site amplification factors for application to earthquake early warning. *Bulletin of the Seismological Society of America*, 103, 3179–88. https://doi.org/10.1785/0120130060.

Hoshiba, M. (2020). Too-late warnings by estimating Mw: earthquake early warning in the near-fault region. *Bulletin of the Seismological Society of America*, 110, 1276–88. https://doi.org/10.1785/0120190306.

Hoshiba, M. (2021), Real-time prediction of impending ground shaking: Review of wavefield-based (ground-motion-based) method for earthquake early warning. *Frontier in Earth Sciences*, 22. https://doi.org/10.3389/feart.2021.722784.

Hoshiba, M., and Aoki, S. (2015). Numerical shake prediction for earthquake early warning: Data assimilation, real-time shake mapping, and simulation of wave propagation. *Bulletin of the Seismological Society of America*, 105, 1324–38. https://doi.org/10.1785/0120140280.

Hoshiba, M., and Ozaki, T. (2014). Earthquake early warning and tsunami warning of the Japan Meteorological Agency, and their performance in the 2011 off the Pacific Coast of Tohoku Earthquake (Mw9.0). In F. Wenzel and J. Zschau, eds., *Early Warning for Geological Disasters: Scientific Methods and Current Practice*. Berlin: Springer, pp. 1–28. https://doi.org/10.1007/978-3-642-12233-0_1.

Hoshiba, M., Iwakiri, K., Hayashimoto, N. et al. (2011). Outline of the 2011 off the Pacific Coast of Tohoku Earthquake (Mw 9.0): Earthquake Early Warning and observed seismic intensity. *Earth Planets Space*, 63, 547–51. https://doi.org/10.5047/eps.2011.05.031.

Hoshiba, M., Kamigaichi, O., Saito, M., Tsukada, S. Y., and Hamada, N. (2008). Earthquake early warning starts nationwide in Japan. *Eos, Transactions, American Geophysical Union*, 89, 73–4. https://doi.org/10.1029/2008EO080001.

Japan Meteorological Agency. (2012). Report of the 2011 off the Pacific Coast of Tohoku earthquake by Japan Meteorological Agency (in Japanese), Japan Meteorological Agency.

Kalnay, E. (2003). *Atmospheric Modeling, Data Assimilation and Predictability*. Cambridge: Cambridge University Press.

Kurahashi, S. and Irikura, K. (2013). Short-period source model of the 2011 Mw 9.0 Off the Pacific Coast of Tohoku earthquake. *Bulletin of the Seismological Society of America*, 103, 1373–93. https://doi.org/10.1785/0120120157.

Kurzon, I., Nof, R. N., Laporte, M. et al. (2020). The 'TRUAA' seismic network: Upgrading the Israel seismic network – Toward national earthquake early warning system. *Seismological Research Letters*, 91, 3236–3255. https://doi.org/10.1785/0220200169.

Oba A., Furumura, T., and Maeda, T. (2020). Data assimilation-based early forecasting of long-period ground motions for large earthquakes along the Nankai Trough. *Journal of Geophysical Research: Solid Earth*, 125(6), e2019JB019047. https://doi.org/10.1029/2019JB019047.

Ogiso, M., Aoki, S., and Hoshiba, M. (2016). Real-time seismic intensity prediction using frequency-dependent site amplification factors. *Earth Planets Space* 68, 83. https://doi.org/10.1186/s40623-016-0467-4.

Ogiso, M., Hoshiba, M., Shito, A., and Matsumoto, S.(2018). Numerical shake prediction for earthquake early warning incorporating heterogeneous attenuation structure: The case of the 2016 Kumamoto earthquake. *Bulletin of the Seismological Society of America*, 108, 3457–68. https://doi.org/10.1785/0120180063.

Peng, H., Wu, Z., Wu, Y. M. et al. (2011). Developing a prototype earthquake early warning system in the Beijing Capital Region. *Seismological Research Letters*, 82, 394–403. https://doi.org/10.1785/gssrl.82.3.394.

Sato, H., Fehler, M. C., and Maeda, T. (2012). *Seismic Wave Propagation and Scattering in the Heterogeneous Earth*, 2nd ed. Berlin: Springer. https://doi.org/10.1007/978-3-642-23029-5.

Sheen, D. H., Park, J. H., Chi, H. C. et al. (2017). The first stage of an earthquake early warning system in South Korea. *Seismological Research Letters*, 88, 1491–98. https://doi.org/https://doi.org/10.1785/0220170062

Suzuki, W., Aoi, S., Kunugi, T. et al. (2017). Strong motions observed by K-NET and KiK-net during the 2016 Kumamoto earthquake sequence. *Earth Planets Space*, 69, 19. https://doi.org/10.1186/s40623-017-0604-8.

Tamaribuchi, K., Yamada, M., and Wu S. (2014). A new approach to identify multiple concurrent events for improvements of earthquake early warning. *Zisin*, 67(2), 41–55. https://doi.org/10.4294/zisin67.41 (in Japanese with English abstract).

Wald, D. J., Worden, B. C., Quitoriano, V., and Pankow, K. L. (2005). ShakeMap manual: Technical manual, user's guide, and software guide, Techniques and Methods 12-A1. https://doi.org/10.3133/tm12A1.

Wang, T., Jin, X., Wei, Y. and Huang, Y. (2017a). Real-time numerical shake prediction and updating for earthquake early warning. *Earthquake Science*, 30, 251–67. https://doi.org/10.1007/s11589-017-0195-2

Wang, T., Jin, X., Huang, Y. and Wei, Y. (2017b). Real-time three-dimensional space numerical shake prediction for earthquake early warning. *Earthquake Science*, 30, 269–81. https://doi.org/10.1007/s11589-017-0196-1.

Wessel, P., and Smith, W. H. F. (1995). New version of the generic mapping tool released. *Eos, Transactions, American Geophysical Union*, 76, 329.

Yoshimoto, K. (2000). Monte Carlo simulation of seismogram envelopes in scattering medium. *Journal of Geophysical Research: Solid Earth*, 105, 6153–61. https://doi.org/10.1029/1999JB900437.

Specific Terms

EEW: Earthquake early warning. Warning of strong shaking before its arrival.

GMPE: Ground-motion prediction equation. Strength of ground motion is empirically estimated from the equation, in which earthquake magnitude and distance (hypocentral distance, epicentral distance, or fault distance) are usually used.

JMA: Japan Meteorological Agency. A national governmental organization in Japan.

K-NET, KiK-net: Observation networks of strong ground motion operated by National Research Institute for Earth Science and Disaster Resilience (NIED) in Japan.

RTT: Radiative transfer theory. A model of wave propagation based on ray theoretical approach, in which scattering and attenuation are included.

14

Global Seismic Tomography Using Time Domain Waveform Inversion

Barbara Romanowicz

Abstract: In this chapter, I present an overview of waveform tomography, in the context of imaging of the Earth's whole mantle at the global scale. In this context, waveform tomography is defined utilising entire wide-band filtered records of the seismic wavefield, generated by natural earthquakes and observed at broadband receivers located at teleseismic distances. This is in contrast to imaging methodologies that first extract secondary observables, such as, most commonly, travel times of the most prominent energy arrivals (i.e. seismic phases), that can be easily identified and isolated in the records. Waveform tomography is a non-linear process that requires the ability to compute the predicted wavefield in a given three-dimensional Earth model and compare it to the observed wavefield. One of its main challenges, is the computational cost involved. I first review the history of methodological developments, specifically focusing on the global, whole mantle Earth imaging problem. I then discuss and contrast the two recent methodologies that have led to the development of the first three-dimensional elastic global shear velocity models that have been published to-date using numerical integration of the wave equation, specifically, using the spectral element method. I discuss how the forward problem is addressed, the data selection approaches, definitions of the misfit function, and computation of kernels for the inverse step of the imaging procedure, as well as the choice of the optimisation method. I also discuss model parametrisation, and, in particular, the important topic of how the strongly heterogeneous crust is modelled. In the final parts of this chapter, I discuss efforts towards resolving the difficult problem of model evaluation and present my views on promising directions and remaining challenges in this rapidly evolving field, aiming at further improving resolution of deep mantle elastic structure with the goal of informing our understanding of the dynamics of our planet.

14.1 Introduction: Some History

The first seismic tomographic models of the Earth's mantle were developed in the late 1970s in the framework of teleseismic travel time tomography, in pioneering work, at the global scale for the lower mantle (Dziewonski et al., 1977), and at the local scale for the upper mantle and crust (Aki et al., 1977). In both cases, advantage was taken of the simple infinite frequency approximation, in which the travel time of a first-arriving teleseismic P or S wave through the Earth could be calculated in a reference, spherically symmetric Earth model (hereafter one-dimensional (1-D) reference model) by invoking Fermat's principle. This also assumed that three-dimensional (3-D) elastic perturbations from the 1-D reference model could be considered small.

Over the following decades, building upon these initial studies, many generations of global and regional mantle models have been developed, supported by the expansion of high-quality digital broadband seismic networks, and relying on travel time measurements of first-arriving P and S waves, sometimes complemented by travel times of other phases that can be well separated on the seismogram (and therefore accurately measured), such as PP, SS, ScS (e.g. see reviews by Romanowicz, 2003, 2020; Thurber and Ritsema, 2015; Ritsema and Lekic, 2020). This has been particularly successful in subduction zone regions, where illumination of the target region from different directions is readily available – a necessary condition for model quality in tomography. Remarkably, these travel time–based models have led to the intriguing observation that subducted slabs change direction as they sink into the mantle, and extend for thousands of kilometres horizontally at the top of the transition zone, or, in several regions, around 1,000 km depth (e.g. Fukao and Obayashi, 2013). Such global seismic images of slabs sinking deep into the mantle have been used to help reconstruct past motions of tectonic plates in the last 200+ Ma (e.g. Richards and Engebretsen, 1992; Replumaz et al., 2004; Sigloch et al., 2008; van der Meer et al., 2010).

Resolving the deep structure beneath ocean basins has been more challenging, given the paucity of recording stations, mostly limited to islands. The addition of surface wave dispersion data helped improve lateral resolution in the upper mantle (e.g. Ritsema et al., 1999, 2011). However, addressing such questions as the presence of narrow mantle plumes beneath hotspot volcanoes, and the depth extent of their roots, required yet higher resolution, and motivated two types of efforts: on the one hand, the deployment of temporary arrays of ocean bottom broadband stations, such as offshore Hawaii, to increase the aperture of the land-based arrays (e.g. Laske et al., 2009), or in Polynesia, to sample regions with few land-based stations (e.g. Suetsugu et al., 2009; Obayashi et al., 2016). On the other hand, theoretical improvements were proposed, such as iterative approaches to account for ray-path perturbations in 3-D structure (e.g. Thurber and Ritsema, 2015), or the introduction of finite-frequency kernels (e.g. Montelli et al., 2004) that helped address wavefront healing effects (e.g. Nolet and Dahlen, 2000) which can make narrow, low-velocity bodies poorly visible when the arrival time of a body wave is picked at the first onset.

Still, the limitation in sampling by direct P and S waves, resulting from too few oceanic stations remained (e.g. van der Hilst and de Hoop, 2005; Boschi et al., 2006). One proposed solution has been to send floats into the oceans, equipped with hydrophones, that detect and record tele-seismic P waveforms and send back the signal through satellites (e.g. Hello et al., 2011). These MERMAIDS (Mobile Earthquake Recorder in Marine Areas by Independent Divers) drift with ocean currents eventually covering large areas of the ocean. It has been shown that, if data were available from 1,000 MERMAIDs providing uniform coverage of the ocean basins, resolution of deep structure could significantly be improved, using only the recording of P waves (Sukhovich et al., 2015). At present, MERMAIDs are being launched in different parts of the ocean, and are starting to provide improved images of target objects, such as the deep structure beneath the Galapagos hotspot (e.g. Nolet et al., 2019). Still, currently, they only record compressional waves at body wave frequencies (1–10 s period; Pipatprathanporn and Simons, 2022).

Meanwhile, a seismic record contains much more information about global Earth structure than that provided by first-arriving waves, even when combined with travel time data from a few additional seismic phases that are well separated from others on the records. Indeed, a seismic record contains information from seismic waves that have bounced around the Earth's interior, reflecting not only off major '1-D' discontinuities such as the Earth's surface or the core-mantle boundary (CMB), but also interacting with smaller-scale structure. This 'scattered' wavefield helps improve illumination, while at the same time providing information about small-scale structures of geophysical interest. This has led to the idea of full-waveform tomography, also referred to as full waveform inversion (FWI), in which entire seismograms – appropriately band-pass filtered – are considered as data in inverse tomographic imaging. In practice, seismograms are also windowed either to remove problematic parts of data or choose waveforms with high correlation to synthetics to reduce non-linearities, as explained in more detail in the following.

Some authors restrict the definition of FWI to a specific class of methods, where the inverse step makes use of adjoint-state kernels (e.g. Fichtner et al., 2010), arguing that in this case, no approximations are made either in the forward or inverse step. This is however not quite correct, as the inverse problem is solved in the framework of the Born approximation, and conditioning is also necessary to obtain a stable solution. I will refer to that class of approaches as 'adjoint tomography'. In contrast, I define full waveform inversion (FWI), as any method that retrieves 3-D seismic structure exclusively from the analysis of seismic waveforms, that is, without first introducing any secondary observables.

In the context of geophysical exploration, the concept of FWI was initially proposed by Tarantola (1984). In the context of earthquake seismology, waveform tomography was first proposed and applied by Woodhouse and Dziewonski (1984) (hereafter referred to as WD84) and led to the development of the first global shear velocity model of the upper mantle based on three component long period waveforms comprised of fundamental mode and overtone surface waves. Here, I will only discuss methodologies and progress in FWI in the context earthquake seismology, focusing primarily on the global scale, with reference to continental scale studies as appropriate.

I will also not discuss – and therefore not give justice to – the many generations of global tomographic studies of the Earth's mantle based on secondary observables, such as travel times of body waves, surface wave dispersion and/or normal mode splitting functions, that have enormously contributed, and continue to contribute, to our knowledge of 3-D global structure. For a more complete overview, the reader is referred to reviews by Romanowicz (2003) and, more recently, Ritsema and Lekic (2020).

Seismic FWI is a significantly non-linear problem which involves an iterative process. I first present the 'forward problem', that is, how the predicted seismic wavefield is computed and compared to observed records at each iteration. I discuss different types of misfit functionals chosen by different groups, and then discuss different approaches to the 'inverse problem', that is, the different methods used to iteratively update the model. Next, I address the challenging question of model assessment and present comparisons of two recent global radially anisotropic shear velocity models obtained by FWI, using different approaches (SEMUCB_WM1, French and Romanowicz, 2014, and GLAD-M25, Lei et al., 2020). Finally, I discuss remaining challenges and possible ways ahead for deep Earth's interior imaging.

I note that Fichtner (2011) describes many technical aspects of adjoint-based FWI in much more detail than is presented here. For more information about progress and current approaches in exploration geophysics, see, for example, reviews by Virieux and Operto (2009) and

Virieux et al. (2014), and for a recent comparison of FWI approaches at scales ranging from exploration geophysics to the global scale, see the review by Tromp (2020).

14.2 The Forward Problem and Sensitivity Kernels in Full Waveform Tomography

Travel time tomography involves relatively rapid computations of the predicted travel times along ray paths, for individual seismic phases, under the approximation of infinite frequency, or, in more recent work, the 'finite-frequency' (i.e. Born) approximation, which results in the famous 'banana-doughnut' sensitivity kernels (e.g. Dahlen et al., 2000). In contrast, when working with full waveforms, the concept of seismic phase is no longer relevant, and the problem is significantly non-linear. In particular, one needs to have a way to compute the full teleseismic wavefield as accurately as possible, in the current iteration 3-D model, in order to compare it to the observed records and compute a misfit function.

In earthquake seismology, until relatively recently, the only practical method for the computation of the teleseismic wavefield has been normal mode summation (e.g. Woodhouse and Deuss, 2015), which is accurate and fast in 1-D Earth models down to ~10 s period, less efficient than other 1-D codes such as the DSM (Direct Solution Method; Geller and Ohminato, 1994) at shorter periods, but has the advantage that it can be extended to 3-D models using normal mode perturbation theory, under various levels of approximation, that I will describe briefly here.

Starting with normal modes computed in a 1-D reference Earth model, and in order to account for the effect of small 3-D elastic perturbations, WD84 introduced minor and great circle average frequency shifts for each mode and each source-station path, corresponding to a zeroth-order asymptotic approximation (high frequency approximation) to first-order normal mode perturbation theory. This approach was later shown to be equivalent to the path average approximation (PAVA), on which most surface wave dispersion studies are based (Mochizuki, 1986; Park, 1987; Romanowicz, 1987). In fact, a 'travelling wave' equivalent to WD84 was already proposed by Lerner-Lam and Jordan (1983), although it involved a two-step procedure, that is, extracting phase velocities first, then inverting the obtained dispersion curves at depth, as opposed to a one-step inversion for 3-D structure as in WD84 and in our definition of FWI.

The underlying assumption in PAVA is that seismic waves are sensitive to the 1-D average structure in the great circle plane containing the source and the receiver, and the corresponding sensitivity kernels are 1-D (i.e. they depend only on depth, not location along a wavepath). Tanimoto (1987) extended this type of FWI to retrieve

structure down to ~1000 km using surface wave overtone waveforms. The WD84 waveform method was later applied, combined with body wave travel times, in the construction of several generations of whole mantle global shear velocity models (e.g. Su et al., 1994; Kustowski et al., 2008). A similar approach, named 'partitioned waveform inversion', was proposed in a propagating wave, rather than standing mode framework (Nolet, 1990), to image upper mantle structure using fundamental mode and overtone surface waveforms. This led to several generations of continental scale (e.g. van der Lee and Nolet, 1997; Schaeffer and Lebedev, 2014), and global scale (Schaeffer and Lebedev, 2013) upper mantle shear velocity models.

PAVA accounts asymptotically for along-branch mode coupling due to 3-D heterogeneity. While it is a good approximation for fundamental mode surface waves, it does not accurately represent the sensitivity of body waves, which are concentrated around the infinitesimal ray-path. To better represent the sensitivity of body waves and overtone surface waves, it is necessary to consider across-branch mode coupling, as shown by Li and Tanimoto (1993) using standing modes, and Marquering et al. (1998) in a propagating wave framework. The formalism of Li and Tanimoto (1993) was combined with PAVA into the non-linear asymptotic coupling theory (NACT; Li and Romanowicz, 1995) and used to develop the first global shear velocity models of the Earth's mantle based entirely on time domain, transverse component waveforms (*SAW12D*, Li and Romanowicz, 1996; *SAW24B16*, Mégnin and Romanowicz, 2000). This approach was later extended to three components and radially anisotropic models (Gung et al., 2003; Panning and Romanowicz, 2006). It is also possible to include a higher order asymptotic approximation, which allows to approximately model off-path effects (e.g. Gung and Romanowicz, 2004). The relative merits of the different normal mode approximations are illustrated and discussed in Romanowicz et al. (2008).

The use of first-order normal mode perturbation theory for the computation of the teleseismic wavefield has many advantages: relatively fast forward calculations and ability to compute the corresponding Fréchet derivatives, and therefore a physics-based Hessian, subsequently leading to fast converging inversion using Gauss-Newton optimisation. It also provides for accurate computation of the effect of perturbations to the gravitational potential, which is important at periods longer than 200 s. However, there are restrictions on its applicability: the lateral variations in the Earth model are assumed to be smooth (i.e. relatively long wavelength) and weak enough for first-order perturbation theory to apply (several percent). In particular, it is no longer appropriate for accurately computing the effects of 3-D structure at scales much shorter than ~2,000 km, and cannot handle the strong lateral variations in Moho depth around the globe. Still, in Fig. 14.1, I illustrate how a model based entirely on waveforms from a relatively small number of three component seismograms (Mégnin and

a) S362ANI   b) S40RTS

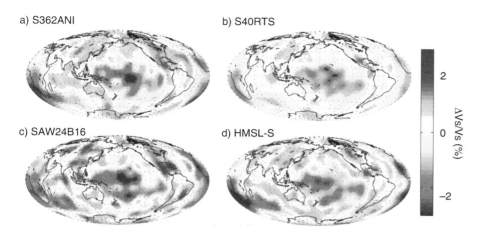

c) SAW24B16   d) HMSL-S

Figure 14.1 Maps of lateral variations in Vs at 2,800 km depth in four whole mantle tomographic studies (A: Kustowski et al., 2008; B: Ritsema et al., 2011; C: Mégnin and Romanowicz, 2000; D: Houser et al., 2008). Model C was constructed entirely using FWI (a total of about 20,000 records), while Model A used a combination of body wave travel times for the lower mantle and waveforms for the upper mantle. Model B was constructed using Rayleigh wave phase velocities, normal mode splitting functions and body wave travel times. Model D was constructed using surface wave phase velocity data and body wave travel times. Models A, B, and D are based on the analysis of hundreds of thousands/millions of travel times.

Romanowicz, 2000), using NACT (Li and Romanowicz, 1995) for both forward and inverse steps, is able to retrieve long-wavelength features in the lower mantle, arguably as well as models based on hundreds of thousands, to millions of travel times of a small number of seismic phases.

To make further progress in resolution, accurate numerical methods based on the direct integration of the wave equation are necessary. Finite difference (FD) methods (e.g. Moczo et al., 2004) can be applied, but present challenges when applied in the global seismology, spherical Earth context (e.g. Fichtner, 2011; Tromp, 2015). In particular, modelling surface waves accurately using FD presents a challenge, because of how the boundary conditions are implemented (strong form of the wave equation). The introduction to global seismology of the spectral element method (SEM) represents a major advance, as it offers the possibility, among others, to accurately compute the teleseismic wavefield in arbitrary 3-D Earth models with theoretically no restrictions on the scale and size of heterogeneities, and at relatively accessible computational cost. The SEM is an FEM-based method where the equation of motion is solved, in its weak form, using Gauss–Lobatto–Legendre (GLL) integration, which results in a diagonal mass-matrix (Komatitsch and Vilotte, 1998; Komatitsch and Tromp, 2002a,b). The main downside of the current SEM implementations is that they rely on the Cowling approximation: the perturbation to the gravitational potential is not included so that Poisson's equation is not solved, which makes it inappropriate for modelling long paths in a 3-D Earth at periods longer than ~200–250 s (e.g. Chaljub and Valette, 2004). The main computational advantage of the SEM is due to the diagonal mass matrix, which is lost when Poisson's equation needs to be solved, significantly slowing down the computations. Several

groups are presently working on ways to include the perturbation to the gravitational potential efficiently, with some recent progress, at least in a 1-D Earth (e.g. van Driel et al., 2021). Note that solutions to Poisson's equation in the context of SEM have been applied in geodesy (Gharti et al., 2018).

The first SEM-based models were developed at local (Chen et al., 2007; Tape et al., 2010) and regional (Fichtner et al., 2009; Lee et al., 2014) crust and uppermost mantle scale, followed by continental scale upper mantle models (e.g. Rickers et al., 2013; Zhu et al., 2015). This list is not meant to be exhaustive and up to date, as in this chapter, I focus on global mantle modelling.

The first global upper mantle, radially anisotropic shear velocity models constructed using FWI and the SEM are SEMum (Lekic and Romanowicz, 2011) and SEMum2 (French et al., 2013), followed by whole mantle radially anisotropic shear velocity models: SEMUCB_WM1 (French and Romanowicz, 2014), GLAD-M16 (Bozdag et al., 2016), and most recently GLAD-M25 (Lei et al., 2020). These models, and the different approaches chosen for their construction, will be discussed in more detail in the following.

In what follows, I restrict further discussion to models that were constructed based exclusively on waveforms at the global, whole mantle scale, and using the SEM for wavefield computations. I will contrast and discuss the pros and cons of the two approaches that have so far led to published global mantle seismic models based on FWI: one in which the SEM is used both in the forward and inverse parts of model construction, the so-called adjoint-based full waveform inversion (hereafter A-FWI), and the other, a 'hybrid full-waveform inversion' approach, where the forward computations are based on SEM, while the inverse part is based on normal mode perturbation theory (hereafter H-FWI).

14.2.1 Data Selection and Misfit Function Computation

In global FWI, teleseismic waveforms are selected for events that typically range in Mw between ~6.0 and ~7.0, in order to ensure a high signal-to-noise ratio, while avoiding complexities due to extended source-time functions and source directivity. Data selection and misfit computations rely on the comparison of observed time-domain earthquake records and 'synthetic seismograms' computed initially in the starting model, and subsequently in the 3-D model obtained from the previous iteration.

The accuracy of the predicted wavefield computation is important in that it controls the shape of the misfit surface, and therefore the location of the misfit minimum towards which the inversion process should proceed (Tarantola, 2005). This is where using the SEM represents a very significant improvement compared to the methodology based on normal mode perturbation theory, on which the first FWI global models had to rely (e.g. Li and Romanowicz, 1996; Mégnin and Romanowicz, 2000).

Usually, the first step in data selection is to reject entire records that are noisy or that have incorrect metadata due to errors in the instrument response function or time-shifts related to clock errors. Comparison with synthetics computed in SEM in the global 3-D model of the current iteration is very helpful at this stage. There are different approaches to additional steps for data selection, generally related to the choice of misfit function, which itself depends on the modelling goal (e.g. elastic versus anelastic structure).

A common approach is to divide each record into time-windows or 'wavepackets', identifying those parts of the record where seismic energy is strong because they contain identifiable seismic 'phases' or a combination thereof:

- The development of the first global A-FWI whole mantle shear velocity models followed the automated FLEXWIN methodology (Maggi et al., 2009) which comprises several stages. The first stage considers the envelope of the synthetic signal and an LTA/STA criterion to identify portions of the record where local maxima of energy are present. In the second stage, a 'water level' is defined and used together with the results of stage 1 to define a set of candidate time windows. Time windows are then rejected if they are too short or fall below the water level. In the last stage, observed and synthetic traces are compared within each remaining window, and different time windows are accepted or rejected according to similarity of the observed and synthetic waveforms. This similarity assessment includes several criteria: signal-to-noise ratio, cross-correlation coefficient and corresponding time-lag, as well as amplitude ratio. The method is described in detail in Maggi et al. (2009). This approach has been used in the development of the first global adjoint-based whole mantle models (Bozdag et al., 2016; Lei et al., 2020) and is illustrated in Fig. 14.2. Note that similar methodologies have been developed by other groups that work with waveforms,

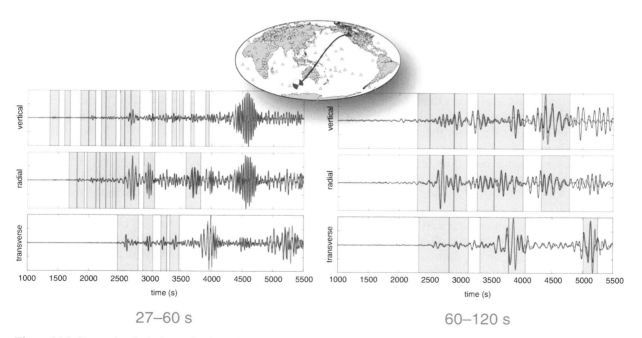

Figure 14.2 Example of window selections (blue) obtained using FLEXWIN (Maggi et al., 2009) for two of the four bandpass categories used in the development of model GLAD-M15 (Bozdag et al., 2016). Observed three component waveforms are in black, and synthetics in red, for an event on the southeast Indian Ridge observed at station LLB (Canada). Courtesy of Ebru Bozdag.

not necessarily in the context of adjoint-based tomography.

- The H-FWI modelling follows the methodology first implemented in Li and Romanowicz (1996) for transverse component records, and later extended to three component records and fully automated (e.g. Panning and Romanowicz, 2006). In this method, contiguous and sometimes overlapping time windows are selected, designed to separate physically meaningful seismic phases, as predicted from travel time curves, both in the case of surface waves and body waves. The windows are precomputed depending on the depth of the event and the source-station distance, for two frequency ranges, one for surface waves (400–250–100–80 s in Li and Romanowicz, 1996; Mégnin and Romanowicz, 2000: extended to 400–250–80-60 s more recently in Lekic and Romanowicz, 2011; French et al., 2013; French and Romanowicz, 2014), and a second one for body waves (300–180–37–32s). Surface wavepackets are separated into fundamental mode, overtone, and 'mixed' wavepackets, as well as first, second and in the case of overtones, third orbit wavetrains. Body wavepackets can

contain one or several body wave phases (e.g. Fig. 14.3). This approach has the advantage that the selected windows can be indexed by the combination of seismic phases they contain, which is later useful to keep track of redundancy in path coverage in a 3-D sense. In each time window, waveforms are compared in the time domain to synthetics computed in the current 3-D model. The initial boundaries of each time window are adjusted in order to take into account shifts in the arrival times of energy due to yet unmodelled 3-D structure, when necessary. Wavepackets are then accepted or rejected, similarly to the FlexWin approach, according to several criteria, including variance reduction and correlation coefficient (see Panning and Romanowicz, 2006, appendix B for details). While synthetics were computed using NACT in the earlier models, substituting them by more accurate SEM-based synthetics in more recent models was straightforward (Lekic and Romanowicz, 2011).

A central issue in FWI is to mitigate cycle skipping, which can occur if the model in which the synthetic seismograms are computed is not close enough to the real Earth and

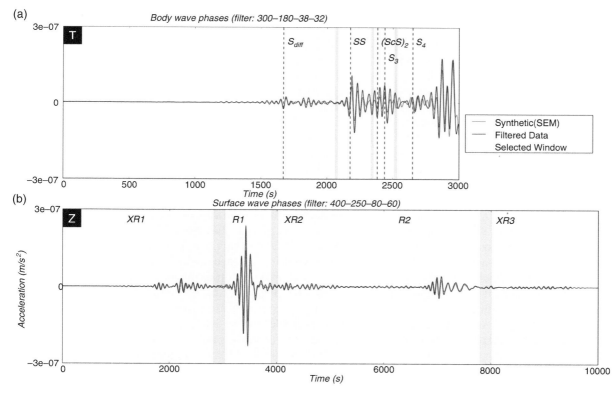

Figure 14.3 Example of window selection used in the development of model SEMUCB-WM1 (French and Romanowicz, 2014), shown for the vertical component in the 'surface wave' bandpass and on the transverse component for the 'body wave' bandpass. The records are from a M6.9 earthquake in Hawaii (4 May 2018) observed at station GNI, at a distance of 117 deg. Windows are marked in blue and can overlap. Synthetics (black) are computed in model SEMUCB-WM1. Windows are labelled according to the seismic phases they contain. Vertical broken lines indicate predicted arrival times of the labelled phases in a 1-D reference model. The second energy arrival in the Sdiff window is SP/PS. Note that all three components were used for both body waves and surface waves in the actual model construction. Courtesy of Heng-Yi Su.

causes large phase shifts (more than half a cycle) on some paths. As is well-known from 'traditional' surface wave phase velocity dispersion measurements, cycle skipping is not a problem at long periods (periods longer than ~80 s), because the observed dispersion curves fall close to those predicted by well-established reference 1-D models. Long wavelength 3-D models (at least to degree 20 is a spherical harmonics expansion laterally) are currently well-enough constrained that they can be used to compute synthetic waveforms down to 80 s period without worrying about cycle-skipping. The waveform dataset can be filtered in different frequency bands, and data in the longer period bands are inverted first, adding shorter periods after some iterations.

In the most recent H-FWI, the lower frequency band-passed data (cut off at 60 s) were first inverted for upper mantle structure (Lekic and Romanowicz, 2011; French et al., 2013) and later shorter-period body waveforms (cut off at 30 s) were added to extend the model to the whole mantle (French and Romanowicz, 2014). The length of record considered for the 'surface wave' bandpass is 10,000 s, starting from earthquake origin time, in order to include both first and second orbit fundamental mode surface waves. The length of record for the 'body wave' bandpass is 5,000 s and includes up to the fundamental Rayleigh wave at a distance of 165° (Fig. 14.3). In the case of A-FWI (Bozdag et al., 2016; Lei et al., 2020) four different 'categories' of waveforms are considered, involving different bandpass filters: one with corners at 250.90 s for surface waves, two with corners at 100.40 s, separately for surface waves and body waves, and the last one with corners at 40.17 s for body waves (Fig. 14.2). The total length of record considered is 10,000 s for the surface wave categories, and 5,000 s for the body wave categories. These authors progressively include data from the shorter period bands as the inversion iterations proceed.

14.2.1.1 *Choice of Misfit Measure*
Because waveforms can be decomposed into phase and amplitude components, and since most studies aim at imaging elastic velocity structure, which is well captured in the phase, some authors favour a time and/or time-frequency dependent phase misfit function, in order to avoid the numerous additional sources of uncertainty in the amplitude. The phase can be computed for entire seismograms (e.g. Fichtner et al., 2008), or, in the case of A-FWI, after selection of specific time windows where energy is strong, distinguishing a time and frequency dependent travel-time misfit computed for the two surface wave windows using multiple tapers, and a travel-time misfit computed for the two body wave windows by cross-correlation (Bozdag et al., 2016). In contrast, as mentioned previously, in H-FWI, point-by-point variance reduction within each wavepacket is measured in the time-domain

(i.e. including amplitude information), as well as cross-correlation coefficient between observed and predicted waveforms. Other misfit functions can be chosen, such as instantaneous phase and/or envelope misfits (e.g. Bozdag et al., 2011) and have been applied in regional scale FWI.

In the case of attenuation tomography based on FWI, which relies on amplitude information, other choices of measure of fit can be implemented. Zhu et al. (2015) considered amplitude ratios between observed and synthetic wavepackets. Karaoglu and Romanowicz (2017) compared misfit definitions based on time-domain matching of waveforms, signal envelope, and amplitude ratios (within each wavepacket), and recommended amplitude ratios in the context of global H-FWI, which they then applied for the construction of the first global upper mantle shear attenuation model based on SEM (Karaoglu and Romanowicz, 2018).

14.2.1.2 *Data Weighting*
Once data selection is completed for the current iteration, an important step is to assign appropriate weights to each datum in order to define the misfit function that will be minimised in the inverse part of the process. These weights are designed to take into account data uncertainty, background noise, and, importantly, path redundancy. In the A-FWI case, the path weighting scheme takes into account source and station proximity, and the number of wavepackets in each of several categories defined as a function of frequency band, and for each component, that is, twelve categories in total for four frequency bands (e.g. Ruan et al., 2019), but only nine categories were used in the development of GLAD-M25. Also, each measurement window is normalised by its own energy, in order to equalise amplitudes among windows.

The Berkeley weighting scheme has more granularity. This is because the windowing scheme of Li and Romanowicz (1996) allows an indexing of windows by the seismic phases they contain. These different phases sample different parts of the mantle, and not all the same windows are necessarily accepted in the data selection for source-station paths that are close to one another. Thus, the proximity of the paths of seismic phases by which wavepackets are indexed are taken into account. For example, in the surface wave frequency band considered in currently published Berkeley models (cut offs at 400 and 60 s), different weights are calculated for fundamental mode, overtone, and mixed wavepackets, also distinguishing first and second orbit arrivals. In the body wave passband (cut offs at 300s and 32s), weights are computed for a total of about 20 different types of wavepackets (per component), corresponding to individual or mixed body wave arrivals that differ according to source–station distance (see table 2 in Mégnin and Romanowicz, 2000, for an earlier version of this labelling scheme). In this case, the misfit function includes not only travel time but also amplitude information, so this weighting scheme additionally considers differences in amplitude due to differences in seismic moment, which can be large. It also takes into account the relative amplitude of different wavepackets within a single record, so as to balance the contribution of strong phases (i.e. fundamental mode surface waves, or SS) compared to smaller amplitude overtones or body waves such as Pdiff or Sdiff, as well as length of the time window corresponding to each wavepacket.

14.2.2 Physical Model Parametrisation

When using three-component waveforms for whole mantle modelling, it is important to include radial anisotropy (i.e. VTI, vertical transverse isotropy) in the parametrisation of the model, at least in the upper mantle, although evidence for the presence of anisotropy in the lowermost mantle is also abundant (e.g. review by Romanowicz and Wenk, 2017). In A-FWI, radial anisotropy has so far been considered only for the upper mantle, due to concerns about data coverage in the lower mantle. In H-FWI, the early whole mantle models (SAW12D, SAW24B16) were developed using only transverse component data, but the subsequent generations of models include VTI in the whole mantle. The same density to Vs scaling is used in both groups (dln ρ/dln Vs = 0.33, Montagner and Anderson, 1989). Parametrisation of VTI is different: in A-FWI, anisotropy is assumed to be significant only for S waves, and the global models are parametrised in terms of bulk sound speed, Vsv, Vsh, and the anisotropic parameter $\eta = F/(A-2L)$. In contrast, H-FWI models are parametrised in terms of Voigt average isotropic velocity Vs_{iso}, and the shear anisotropy parameter $\xi = (Vsh/Vsv)^2$, while the parameters related to P velocity – Vp_{iso}, α, and ϕ – are scaled to Vs using the expressions of Montagner and Anderson (1989), which technically are valid only for upper mantle rocks, and for lack of better geological constraints.

So far, SEM based global elastic models have not included topography of discontinuities, although some synthetic experiments are underway on how to implement adjoint inversion for upper mantle discontinuities (e.g. Koroni and Trampert, 2021). In current generation elastic models, the Q model considered is 1-D and is not updated during the inversion. Both A-FWI models and H-FWI models rely on model QL6 (Durek and Ekström, 1996), which has been shown to predict amplitudes in better agreement with observations for higher orbit surface wave data than the reference PREM model (Dziewonski and Anderson, 1981). Because A-FWI only uses phase information, this distinction is likely not significant, but it does appear in dispersion corrections due to attenuation.

14.2.3 Geographical Parametrisation

The first global model constructed by FWI, model M84C (WD84) was parametrised using globally defined basis functions: spherical harmonics laterally, and Legendre polynomials vertically. This was also the parametrisation chosen for the development of FWI-based whole mantle model SAW12D (Li and Romanowicz, 1996). In SAW24B16 (Mégnin and Romanowicz, 2000), as in other whole mantle shear velocity models of that generation (e.g. Ritsema et al., 2004), a more flexible vertical parametrisation using B-splines was introduced. Later models replaced the spherical harmonics basis by locally defined spherical splines (e.g. Wang and Dahlen, 1995). This type of parametrisation (B-splines vertically and spherical splines horizontally) continued to be used in the development of the SEM-based models SEMum, SEMum2, and SEMUCB_WM1, as well as SEMUCB-WMQ.

For A-FWI, a parametrisation that is anchored on the SEM mesh has been adopted, and Gaussian-function smoothing is applied to balance data coverage and remove numerical noise from kernels (e.g. Ruan et al., 2019). Smoothing and pre-conditioning are also used to speed up convergence (e.g. Modrak and Tromp, 2016).

14.2.4 Implementation of the Earth's Crust

Long period waveforms are sensitive to lateral variations in crustal structure, and especially the large variations in surface topography and Moho depth, but unless short-enough periods are included, it is not possible to constrain this shallow structure accurately by inversion. Yet, properly accounting for strongly heterogeneous crustal structure is important, because it results in strongly non-linear effects and can influence the structure retrieved deeper in the mantle, and particularly so for the anisotropic part of the model (e.g. Bozdag and Trampert, 2008; Ferreira et al., 2010; Lekic et al., 2010; Chang et al., 2014).

Before the introduction of the SEM to seismic tomography, crustal effects were usually dealt with by means of approximate crustal corrections which were computed by combining tectonic regionalisation (to account for large variations in Moho depth), with normal mode perturbations within each region (e.g. Woodhouse and Dziewonski, 1984; Marone and Romanowicz, 2007).

Since the SEM can accurately account for complex crustal structure, it is now possible to include a realistic 3-D crustal model in the seismic wavefield computation and invert for perturbations of the model both in the crust and in the mantle, given appropriate data. However, there are several challenges. First, the presence of thin low-velocity layers results in slowing down the SEM computations, because of the CFL (Courant–Friedrichs–Levy) or Courant condition imposed. Second, at the global scale, fundamental mode surface waveforms have not been considered below ~30 s, at least until now, because of the strong crustal multipathing effects that are observed at shorter periods. This implies that sensitivity of the considered surface wave data to crustal structure is weak, although still strong enough to affect waveforms. Meanwhile, body wave sampling of the crust is sparse. Third, existing 3-D crustal models are not perfect, and, in many poorly sampled regions of the globe, the model is extrapolated from well-sampled regions, based on an a priori tectonic regionalisation. It has been shown that, at least on some paths, predictions from these models do not fit waveform data well (Meier et al., 2007; Pasyanos and Nyblade, 2007).

The A-FWI and H-FWI approaches differ when it comes to accounting for the effect of the crust. In published A-FWI models, a recent global crustal model, Crust2.0 (Bassin et al., 2000), has been implemented in the starting model of this series of models (Bozdag et al., 2016). This model was combined with S362ANI, which required some manipulation given that S362ANI is defined bounded by a spherically symmetric Moho (Tromp et al., 2010; Bozdag et al., 2016). The SEM mesh honours the Moho in regions where the crust is thinner than 15 km or thicker than 35 km, while mesh elements straddle across transition regions. The meshing strategy is described in detail in Zhu et al. (2015). In this approach, the crustal structure is updated at every iteration of the inversion.

A different strategy has been chosen in H-FWI. In order to mitigate some of the issues described when using existing crustal models in SEM, it is possible to use the concept of homogenisation (i.e. the replacement of a complex medium by a smooth medium that is equivalent within a particular target frequency band), from the point of view of seismic wave propagation. Construction of effective media in 3-D is a topic of active research, but in 1-D, it is possible to replace a finely layered velocity model by a smooth radially anisotropic model (e.g. Backus, 1962; Capdeville and Marigo, 2007) that is locally constrained by short-period surface wave dispersion data. Such a model can be conveniently parametrised to match the parametrisation of the SEM mesh in the crust.

Aiming at improving the computational cost of SEM in a model with a 3-D crust, Fichtner and Igel (2008) used a simulated annealing (SA) algorithm to develop a family of smooth, radially anisotropic crustal models designed to fit a synthetic surface wave dispersion dataset derived from model Crust2.0. This was applied, for example, to develop FWI, adjoint-based models of the Australasian continent (Fichtner et al., 2009, 2010).

Instead of homogenising an existing layered crustal model, the published H-FWI models build a smooth crustal model by locally inverting group velocity surface wave dispersion data available globally (Shapiro and Ritzwoller, 2002), while making sure that the SEM mesh honours the Moho discontinuity. In model SEMum (Lekic and Romanowicz, 2011), after performing a series of tests, the apparent crustal thickness was chosen to be uniformly

60 km, while in subsequent models (SEMum2, French et al., 2013; SEMUCB_WM1, French and Romanowicz, 2014), the Moho depth follows that of Crust2.0 (Bassin et al., 2000) in regions where crustal thickness is larger than 30 km, while keeping it 'saturated' at 30 km where it is thinner, thus avoiding the slow-down of the SEM computations due to thin crustal layers. The methodology is described in detail in French and Romanowicz (2014). Because the shortest period considered in the H-FWI models is 30 s, dispersion data down to 25 s, which are well constrained globally, were used. In the future, as the period range of the FWI extends to shorter periods, it is possible to include dispersion data extending to shorter periods, for example in regions where such data are available from ambient noise tomography.

More generally, there are several advantages to working with a smooth crustal model in SEM. First, it avoids thin low-velocity layers, which slow down wavefield computations. Second, because of the non-linearity involved in introducing a sharp fixed Moho that conforms with the SEM mesh, any errors introduced in the starting crustal model will be propagated through model iterations, while a smooth equivalent model is more linearly related to the data. Moreover, the complexity of the Earth's 3-D crustal structure leads to surface wave multipathing at periods of 20 s or shorter, and the corresponding parts of the record are not considered, since they cannot be modelled with presently available approaches, because of theoretical limitations either in an adjoint formalism or in normal-mode perturbation-based approaches.

14.3 The Inverse Step: Choice of Optimisation Approach

FWI is a non-linear inverse problem that needs to be addressed through an iterative process of model updating.

Before the introduction of SEM, synthetic waveforms were computed using asymptotic approximations to normal mode perturbation theory (e.g. PAVA, NACT), which allowed the efficient computation of Fréchet derivatives based on the same theory, and of an approximate, but full and physically meaningful Hessian at relatively small computational cost. This led to fast, quadratically converging, Gauss–Newton optimisation approaches, according to an equation of the form (e.g. Tarantola and Valette, 1982):

$$\delta m_k = (G_k^T C_D^{-1} G_k + C_M^{-1})^{-1} \times \{G_k^T C_D^{-1} [d - g(m_k)] \\ - C_M^{-1} (m_k - m_0)\},$$

where G_k is the matrix of Frechet derivatives of the wavefield with respect to model parameters, computed at iteration k, and $H_k = G_k^T C_D^{-1} G_k$ is the corresponding approximate Hessian, weighted by the data covariance matrix C_D. Here d is the data vector, m_k is the model vector at the k'th iteration, m_0 the starting model, and C_M the covariance matrix in model space, which contains regularisation factors (i.e. norm or roughness damping, correlation lengths and such).

In H-FWI, the computation of the seismic wavefield $g(m_k)$ (i.e. forward problem) is performed using SEM, and the gradient $-G_k^T C_D^{-1}[d-g(m_k)]$ and approximate Hessian H_k are computed using NACT (and attenuation is taken into account in the gradient and Hessian computation). In contrast, A-FWI computes the gradient numerically from the zero-lag correlation of the forward and adjoint wavefields, using SEM (e.g. Tromp et al., 2005). In this case, the computation of H_k is prohibitively costly, so a linear optimisation scheme based on the gradient, such as the conjugate gradient method, is employed, combined with smoothing of the gradient (Bozdag et al., 2016). In the latest iterations of the A-FWI published global model (Lei et al., 2020), conditioning using the quasi-Newton numerical method L-BFGS (e.g. Fichtner et al., 2010; Nocedal and Wright, 2017) was applied.

There are pros and cons to both approaches. The adjoint approach uses an 'exact' gradient which points in the direction of steepest descent, at the cost of requiring a larger number of iterations for convergence, and at least two SEM computations per event. This is only possible with access to considerable HPC computational resources. The methodology relies on the computation of vector-matrix products to avoid handling of huge matrices. There are no limitations on the parametrisation of the 3-D model, which can be based on the SEM mesh, but pre-conditioning of the gradient is applied to speed up convergence, while smoothing by 3-D Gaussian filters is applied to compensate for imperfection in the coverage provided by the available data, which eventually limits resolution (e.g. Zhu et al., 2015; Bozdag et al., 2016). Notably, the numerical conditioning (i.e. L-BFGS) approximates the Hessian in a different way than does H-FWI.

On the other hand, the hybrid H-FWI approach capitalises on the fact that it is more important to compute the misfit function accurately (hence the SEM in the forward computation) than the gradient (Tarantola, 1984; Lekic and Romanowicz, 2011), provided the physics-based theory used in the computation of the gradient and Hessian is a good enough approximation (Valentine and Trampert, 2016). This is the case for NACT, at least at the wavelengths considered so far for the global scale (e.g. Romanowicz et al., 2008). The gradient is indeed approximate, but multiple forward scattering in the current 3-D model is included in the PAVA term, both in the gradient and the Hessian computation. The dimension of the corresponding full Hessian matrix can be very large, depending on the parametrisation of the model, and therefore the target nominal resolution – however an efficient method for parallel assembly of large matrices – has been implemented, using the UPC++ library, an addition to C++ developed at UC Berkeley (French et al., 2015). The

advantage is the relatively fast convergence: starting from a 1-D model in the upper mantle and model SAW24B16 in the lower mantle (e.g. Lekic and Romanowicz, 2011), a total of 14 iterations were needed to construct model SEMUCB_WM1 (French and Romanowicz, 2014), with only one SEM computation per event at each iteration, compared to 25 iterations for GLAD-M25 starting from the whole mantle 3-D model S362ANI (Lei et al., 2020), with two SEM computations per event, albeit accelerated using GPUs, and reaching shorter periods (17 s in GLAD-M25 compared to 30 s in SEMUCB_WM1). The main disadvantage of H-FWI compared to an adjoint inversion, which becomes a more important problem as higher spatial resolution is sought (i.e. at shorter periods), is that, at each iteration, the computations of the gradient and Hessian are tied to a 1-D average model, and even with the path-dependent adjustments that are included through the PAVA term, it does not lend itself conveniently to the implementation of significant discontinuity topographies. An approach currently being investigated in the Berkeley group, is to combine the 'exact' numerical gradient computed using adjoints with the physics-based approximate Hessian.

Another difference between the modelling approaches of the H-FWI and A-FWI is that the period range of one of the A-FWI body wave categories extends down to 17 s (instead of 30 s in H-FWI), and one of the surface wave categories extends to 45 s (instead of 60s). The latter provides somewhat more sensitivity to crust as well uppermost mantle structure. In contrast, for the construction of the latest global H-FWI model, SEMUCB-WM1, there was not enough information in the waveforms down to 60 s to fully constrain the uppermost part of the mantle, so that the waveform inversion for uppermost mantle structure (and crustal updates) was combined with dispersion data in the period range 25 to 150 s. See French and Romanowicz (2014) for details.

14.4 Model Evaluation

As is well-known, model evaluation is generally quite challenging in seismic tomography. In 'traditional' global mantle travel time tomography, it is performed by computing the resolution matrix R, and examining how a series of synthetic input models are recovered or distorted by the inversion process. The resolution matrix is a concept that is theoretically valid only for linear inverse problems (Tarantola, 2005) and therefore does not strictly apply in the case of FWI. Still, when the full Hessian is computed using normal mode perturbation theory, as done in H-FWI, it is possible to compute the resolution matrix at each iteration of the model, and in particular at the last iteration (e.g. French and Romanowicz, 2014). While such tests, and in particular the famous 'checkerboard test', have limited scope (e.g. Levêque et al., 1993), they give some idea of model recovery at different wavelengths and in different locations. They can help assess the robustness of features of geophysical interest, such as, for example, whether periodic low-velocity fingers seen in the upper mantle and aligned with absolute plate motion are robust features (French et al., 2013), or whether the inversion can distinguish a bundle of mantle plumes from a compact large low shear velocity province (e.g. Davaille and Romanowicz, 2020). However, they do not address important aspects of the problem such as the dependence of the final model on the starting model, or the theoretical approximations in either the forward or inverse part of the FWI.

It is not possible, in practice, to compute a full resolution matrix in A-FWI. Stochastic probing of the Hessian or resolution matrix and computation and recovery of point-spread functions and similar methods have been proposed (e.g. review by Rawlinson et al., 2014), and new, ever more efficient methods are the subject of active research both in the applied geophysics and in the global seismology community

(e.g. Liu et al., 2022, and references therein). Notably, in the construction of GLAD-M25, Lei et al. (2020) used the point-spread function approach of Fichtner and Trampert (2011a, b) to probe the model at selected locations of interest.

A more rigorous approach would require starting with a series of relevant test 3-D models, and, for each of them, computing synthetic waveform data matching the distribution of real data, and redo the entire FWI process with its many iterations. This is clearly out of reach due to the computational time required. It is also practically impossible to fully test the dependence of the final model on the starting model. Still, this has been attempted once in H-FWI, as illustrated in Fig. 14.4, to test the reliability of the deep mantle low-velocity plume conduits found in SEMUCB_WM1.

It is likewise challenging to assign reliable uncertainties to model values obtained by FWI. A possible approach is statistical resampling of the data using Bootstrap (e.g. Efron and Tibishirani, 1991), or deleted jackknife (Efron and Stein,

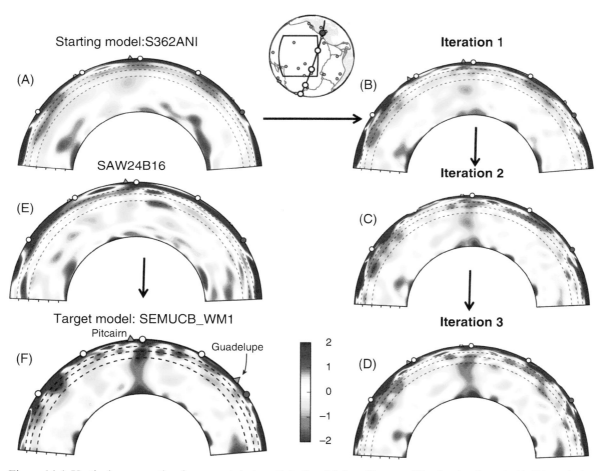

Figure 14.4 Vertical cross-section for a great circle path in the vicinity of hotspot Pitcairn (see insert). (A–D) evolution of the whole mantle model starting from model S362ANI (Kustowski et al., 2008), shown in (A), and performing successively three iterations (B,C,D), using the same waveform dataset as used in the development of SEMUCB_WM1 (French and Romanowicz, 2014). While small-scale details remain different, the large low-velocity conduit extending from the core–mantle boundary through the mantle under Pitcairn progressively appears. For comparison, the same cross section is shown in (E) in model SAW24B16 (Mégnin and Romanowicz, 2000), which was the starting model in the lower mantle for the development of SEMUCB_WM1, while (F) shows model SEMUCB_WM1.

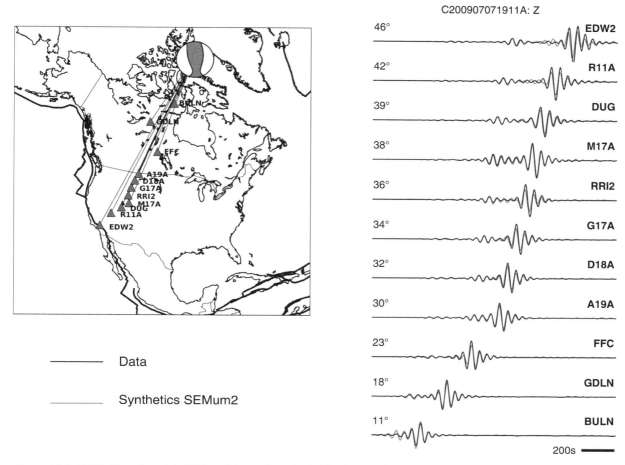

Figure 14.5 Validation of model SEMum2 (French et al., 2013). Comparison of vertical component records for an event in Baffin Island (07/07/2009) observed at stations across North America. Paths and CMT solution are shown on the map on the left side. Observed waveforms are in black, and have not been used in the construction of SEMum2. SEM synthetics computed in SEMum2 using CSEM (Capdeville et al., 2003) are shown in red. Data and synthetics are filtered down to a cut-off period of 40 s, compared to the 60 s cut-off used in the development of model SEMum2.

1981), that is, performing an inversion based on a series of realisations of data from which some waveforms (say 10%) have been randomly removed. This is only possible when the full Hessian is available, and, for that case, details can be found in French and Romanowicz (2014). Note that only the last iteration of the inversion can be tested this way, which addresses the issue only partly.

Another semi-quantitative way to evaluate the model is to compare data with synthetics for a subset of events and source–station paths, that have not been used in the inversion. One can also evaluate how the model performs on selected paths, when extending the period range to higher frequencies than were used in the construction of the model (e.g. Figs. 14.5 and 14.6).

Finally, in lieu of a better approach, comparison of FWI models obtained by different groups with different datasets and methodologies can inform on the robustness of features of geophysical interest. Figure 14.7 shows a comparison between the isotropic part of SEMUCB_WM1 and GLAD-

M25 on two cross sections: one samples the models in the vicinity of the Samoa hotspot, with very similar structure in both models (Fig. 14.7a,b). The other one (Fig. 14.7c,d) samples them in the vicinity of three hotspot volcanoes and one subducting slab. The two models show consistent features: the strong low-velocity conduit beneath Pitcairn (stronger in SEMUCB-WM1), a low-velocity conduit in the vicinity of San Felix, which appears to be better resolved in GLADM25, and a conduit under Easter Island, which is more out of plane for this particular cross-section, in GLADM25 than in SEMUCB-WM1. Both models show the subducted Nazca Plate ponding between 660 and 1,000 km depth, and evidence for older portions of the slab in the lower mantle. The slab is likely better resolved in GLADM25 in the upper mantle and down to the mid-mantle. Combining the information from the two models, one is tempted to infer that individual plume conduits likely exist beneath each of the three hotspots, extending from the core–mantle boundary to the upper mantle.

SEMum2 validation using RegSEM

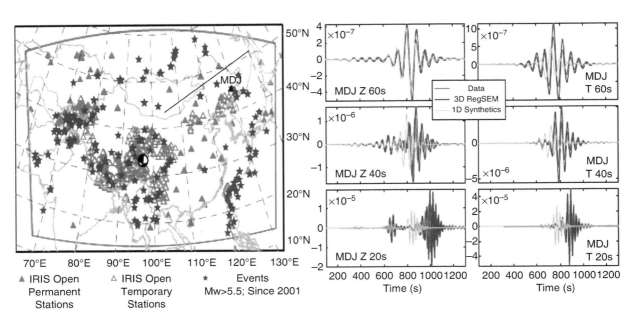

Figure 14.6 Comparison of vertical component observed (red) and synthetic waveforms computed in 1-D (grey) and in model SEMum2 (red) using the regional SEM code RegSEM (Cupillard et al., 2012), for an event in Tibet observed at station MDJ. The waveform comparison is shown in three period bands, with cut-offs at 60 s (top), 40 s (middle), and 20 s (bottom). The cut-off used in the development of SEMum2 was 60 s. The fit, compared to 1-D, for the SEMum2 predictions is satisfactory at 60 s, especially in phase, and still quite good at 40 s and 20 s, although some larger phase differences are visible at 20 s, especially in the Rayleigh wave coda.

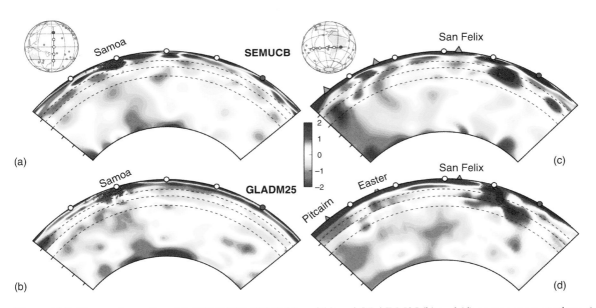

Figure 14.7 Comparison of models SEMUCB_WM1 (a) and (c) and GLADM25 (b) and (d) on two cross-sections. (a) and (b) north–south cross section in the vicinity of the Samoa hotspot. (c) and (d) east–west cross sections at latitude of 20°S. Broken lines indicate depths of 400, 660, and 1,000 km, respectively. The colour scale indicates relative perturbations in Vs referenced to the global average in each model. The scale has been saturated to +/– 2% to better emphasise lower mantle features.

14.5 The Future of Global Mantle FWI

In its quest to further improve resolution of deep mantle elastic structure, global FWI whole mantle modelling in the era of numerical wavefield computations faces many challenges.

First is the computational challenge: increasing resolution implies extending computations to shorter periods and extracting more information from the scattered coda of the main seismic phases, while computational time in SEM increases like the fourth power of frequency, and the choice of the misfit function becomes more critical at short periods, where cycle skipping is a threat.

Second, accounting accurately for the effects of 3-D crustal structure in the full frequency range used for FWI, and notably in parts of the world where current crustal models are not accurate, is important to avoid contamination of deeper structure, while trying to avoid slowing down computations (the thin layer problem). Current approaches (both A-FWI and H-FWI) update the crust in successive iterations, but there is still progress to be made to exploit information contained in multi-pathing surface waves at periods shorter than 25 s.

Third, even with exploiting as much as possible of the information contained in waveforms, resolution is limited by the non-uniform distribution of high-quality broadband stations and sources around the world. Finally, the focus so far has been on inversion for radially anisotropic structure, while FWI applied to imaging azimuthal anisotropy (e.g. Bozdag et al., 2021) and 3-D anelastic attenuation (e.g. Karaoglu and Romanowicz, 2018) at the global scale, is emerging. Meanwhile, resolution of 3-D density structure remains in its infancy.

Efforts towards addressing each of these challenges are underway. Here I briefly discuss some of them.

To address the computational challenge, source stacking, or source encoding, has been proposed. In this approach, the seismic wavefield is computed for a large number of sources simultaneously, by first aligning them on the earthquake origin time. This was first proposed by Capdeville et al. (2005), who showed that they could recover a long wavelength global whole mantle shear velocity model (up to degree 6 in a spherical harmonics expansion, total number of model parameters 274) using stacked waveforms from 84 events filtered at periods longer than 160 s, recorded at 174 stations, and adjoint kernels. While the concept is attractive, as it can potentially lead to orders of magnitude fewer wavefield computations, it presents several drawbacks. First, it is no longer possible to weigh the data so as to balance contributions from waves sampling different paths and different parts of the mantle. To overcome this, a promising approach, at least in H-FWI, is to combine stacking with correlation of the stacked traces, which allows path weighting and, to some extent, windowing and

weighting down of large amplitude fundamental mode surface waves (Romanowicz et al., 2013, 2019).

When using A-FWI, another complication is the presence of cross-talk in the computation of the kernels. To minimise the effect of such cross-talk, random source encoding is used in exploration geophysics (e.g. Krebs et al., 2009). This still leaves the challenge of application to real data at the global scale, which is plagued by missing data: indeed, when stacking N sources, the synthetics are computed automatically at all stations, regardless of whether usable data are available at any particular station. To address the missing data issue, Zhang et al. (2018) proposed a source encoding approach based on Fourier decomposition of the wavefield, which allows the synthetic wavefield for each event to be extracted from the synthetic stacked record at each station. Application of this approach to global scale A-FWI appears promising (Tromp and Bachmann, 2019).

Another way of extending the frequency range to shorter periods in global scale FWI, while keeping computational time manageable, is to try and restrict the volume in which the numerical wavefield computations and inversion iterations are performed. For many years, hybrid methods have been designed and applied to forward model structures in specific areas of the deep mantle near the core–mantle boundary. The wavefield is computed by coupling a fast '1-D' method outside the target region and a numerical scheme, either in two-dimensional, such as finite differences (e.g. Wen and Helmberger, 1998), or in 3-D (Capdeville et al., 2003b). Recently, 1-D synthetics outside the target region computed using DSM (Geller and Ohminato, 1994) have been combined with 3-D FWI inside it using the Born approximation (Kawai et al., 2014) to image structures in D″ in the northern and western Pacific (Suzuki et al., 2016, 2020) and in the transition zone in central America (Borgeaud et al., 2019).

Full waveform inversion using numerical wavefield computations inside the target region, coupled with DSM outside of it has been developed and applied for regional upper mantle tomography (Monteiller et al., 2015; Lin et al., 2019). In such a case, the iterative 3-D wavefield computations need only be performed inside the small target region, considerably saving computational time. It is desirable to be able to accurately account for 3-D structure outside the target region, which requires coupling a global 3-D solver and a regional 3-D solver. In global seismology, one such approach has been coined 'box tomography' (Masson et al., 2014; Masson and Romanowicz, 2017a, 2017b) and has so far been applied at continental scale in North America (using different numerical solvers outside and inside the region), for the specific case when sources are outside the target region and stations inside (Clouzet et al., 2018). A forward modelling approach, in the case where both sources and stations are outside the target region (as would be for targets in the mid and lower mantle), has been demonstrated by Pienkowska et al. (2020), coupling the 1D instaseis code (van Driel et al., 2015) outside the target region with SPECFEM3D Cartesian (Tromp et al., 2005) inside it. Recently, Adourian et al. (2023) coupled SPECFEM3D_globe outside with a regional 3D SEM code, RegSEM (Cupillard et al., 2012) inside a target region.

These types of approaches hold promise for global tomography, not only because they can save considerable computation time, but also because the distribution of sources and stations is limited on the globe, so that attaining equally high resolution everywhere is not possible. Application to the global scale of the mini-batch approach combined with a multi-scale approach (van Herwaarden et al., 2020) or using wavefield adapted meshes (e.g. Thrastarson et al., 2020) are other ways that are being explored to save computational time while progressively improving resolution.

In many parts of the world, significantly improving resolution will not be possible until the ocean floor can be populated with high-quality broadband observatories. Much effort has been devoted internationally to the development of such ocean floor observatories for decades (e.g. COSOD-II, 1987; Romanowicz and Suyehiro, 2001), but the logistics and costs have thwarted progress. Initiatives for the deployment of large aperture broadband ocean floor arrays, coordinated internationally, such as PacificArray are much needed and represent a critical target for future progress in seismic imaging towards improving our understanding of deep mantle dynamics.[1]

As for applying some form of FWI to resolve deep mantle lateral variations in density at the global scale, it is at present only possible using first-order normal mode perturbation theory (e.g. Yang and Tromp, 2015; Al-Attar et al., 2012) – although higher-order developments exist (e.g. Clévédé and Lognonné, 1996) – while awaiting the full development of numerical methods that do not require Cowling's approximation, but with reasonable computational requirements.

Finally, beyond the SEM, more efficient numerical solvers of the wave equation in a spherical Earth, with complex topography, may become available in the years to come (e.g. Masson, 2023). This topic is beyond the scope of this chapter.

14.6 Conclusions

Full waveform tomography of the Earth's deep mantle has benefited in the last decade from the availability of an efficient numerical solver of the wave equation in its weak form, the spectral element method. I have discussed the contrasting approaches used by the two groups that so far have published global whole mantle elastic models that rely on the SEM. The resulting models have revealed or confirmed features within our planet that were previously not resolved, or poorly so. In order to improve resolution of the Earth's mantle structure going forward, including

3-D imaging of attenuation, anisotropy, and density structure, it is necessary to increase efforts at instrumenting ocean basins with seismic sensors – and in particular three-component broadband ones – and to continue developing promising approaches aiming at improving computational efficiency, so that higher frequencies can be reached at reasonable cost.

Acknowledgements. The author thanks Andreas Fichtner and an anonymous reviewer for their constructive suggestions that improved the quality of the. BR acknowledges support from NSF grant EAR-1758198.

References

Adourian, S., Lyu, C., Masson, Y., Munch, F., and Romanowicz, B. (2023). Combining different 3-D global and regional seismic wave propagation solvers towards box tomography in the deep Earth. *Geophysical Journal International*, 232(2), 1340–56.

Aki, K., Christofferson, A., and Husebye, E. S. (1977). Determination of the three-dimensional seismic structure of the lithosphere. *Journal of Geophysical Research*, 82, 277–96.

Al-Attar, D., Woodhouse, J. H., and Deuss, A. (2012). Calculation of normal mode spectra in laterally heterogeneous earth models using an iterative direct solution method. *Geophysical Journal International*, 189, 1038–46.

Backus, G. (1962). Long-wave elastic anisotropy produced by horizontal layering. *Journal of Geophysical Research*, 67(11), 4427–40.

Bassin, C., Laske, G., and Masters, G. (2000). The current limits of resolution for surface wave tomography in North America. *EOS Transactions American Geophysical Union*, 81, 1351–75.

Borgeaud, A. F. E., Kawai, K., and Geller, R. J. (2019). Three-dimensional S velocity structure of the mantle transition zone beneath Central America and the Gulf of Mexico inferred using waveform inversion. *Journal of Geophysical Research: Solid Earth*, 124, 9664–81.

Boschi, L., Becker, T. W., Soldati, G., and Dziewonski, A. M. (2006). On the relevance of Born theory in global seismic tomography. *Geophysical Research Letters*, 33, L06302.

Bozdag, E., and Trampert, J. (2008). On crustal corrections in surface wave tomography. *Geophysical Journal International*, 172, 1066–82.

Bozdag, E., Trampert, J., and Tromp, J. (2011). Misfit functions for full waveform inversion based on instantaneous phase and envelope measurements. *Geophysical Journal International*, 185, 845–70.

Bozdag, E., Peter, D., Lefebvre, M., et al. (2016). Global adjoint tomography: First generation model. *Geophysical Journal International*, 207(3), 1739–66.

Bozdag, E., Orsvuran, R., Ciardelli, C., Peter, D., and Wang, Y. (2021). Upper mantle anisotropy from global adjoint tomography. AGU Fall Meeting 2021, New Orleans, LA, 13–17 December 2021, abstract DI42A-08.

Capdeville, Y., and Marigo, J. J. (2007). Second order homogenization of the elastic wave equation for non-periodic layered media. *Geophysical Journal International*, 170, 823–38.

[1] http://eri-ndc.eri.u-tokyo.ac.jp/PacificArray.

Capdeville, Y., Chaljub, E., Vilotte, J. P., and Montagner, J. P. (2003a). Coupling the spectral element method with a modal solution for elastic wave propagation in global Earth models. *Geophysical Journal International*, 152, 34–66.

Capdeville, Y., Gung, Y., and Romanowicz, B. (2005). Towards global Earth tomography using the spectral element method: A technique based on source stacking. *Geophysical Journal International*, 162, 541–54.

Capdeville, Y., To, A., and Romanowicz, B. A. (2003b). Coupling spectral elements and modes in a spherical Earth: An extension to the 'sandwich' case. *Geophysical Journal International*, *154*, 44–57.

Chaljub, E., and B. Valette, B. (2004). Spectral element modelling of three-dimensional wave propagation in a self-gravitating Earth with an arbitrarily stratified outer core, *Geophysical Journal International*, 158, 131–41.

Chang, S.-J., Ferreira, A. M., Ritsema, J., van Heijst, H. J., and Woodhouse, J. H. (2014). Global radially anisotropic mantle structure from multiple datasets: a review, current challenges, and outlook. *Tectonophysics*, 617, 1–19.

Chen, P., Zhao, L., and Jordan, T. H. (2007). Full three-dimensional tomography: A comparison between the scattering-integral and adjoint-wavefield methods. *Bulletin of the Seismological Society of America*, 97, 1094–120.

Clévédé, E., and Lognonné, P. (1996). Fréchet derivatives of coupled seismograms with respect to an anelastic rotating Earth. *Geophysical Journal International*, 124, 456–82.

Clouzet, P., Masson, Y., and Romanowicz, B. (2018). Box tomography: First application to the imaging of upper mantle shear velocity and radial anisotropy structure beneath the North American continent. *Geophysical Journal International*, 213, 1849–75. doi: 10.1093/gji/ggy078

COSOD-II (1987). Report of the Second Conference on Scientific Ocean Drilling (Cosod II): Strasbourg, 6–8 July, 1987, Joint Oceanographic Institutions for Deep Earth Sampling. https://archives.eui.eu/en/fonds/476326?item=ESF-1224.

Cupillard, P., Delavaud, E., Burgos, G. et al. (2012). RegSEM: a versatile code based on the spectral element method to compute seismic wave propagation at the regional scale. *Geophysical Journal International*, 188, 1203–20.

Dahlen, F. A., Hung, S.-H., and Nolet, G. (2000). Fréchet kernels for finite-frequency traveltimes – I. Theory. *Geophysical Journal International*, 141, 157–74.

Davaille, A., and Romanowicz, B. (2020). Deflating the LLSVPs: Bundles of mantle thermochemical plumes, rather than thick 'stagnant' piles. *Tectonics*, 39, e2020TC006265. https://doi.org/10.1029/2020TC006265.

Durek, J. J., and Ekström, G. (1996). A radial model of anelasticity consistent with long-period surface-wave attenuation. *Bulletin of the Seismological Society of America*, 86, 144–58.

Dziewonski, A., and Anderson, D. (1981). Preliminary reference Earth model. *Physics of the Earth and Planetary Interiors*, 25, 297–356.

Dziewonski, A., Hager, B., and O'Connell, R. (1977). Large-scale heterogeneities in the lower mantle. *Journal of Geophysical Research*, 82, 239–55.

Efron, B., and Stein, C. (1981). The jackknife estimate of variance. *Annals of Statistics*, 9(3), 586–96.

Efron, B., and Tibishirani, R. J. (1991). *An Introduction to the Bootstrap*. Boca Raton, FL: Chapman and Hall.

Ferreira, A. M., Woodhouse, J. H., Visser, K., and Trampert, J. (2010). On the robustness of global radially anisotropic surface wave tomography. *Journal of Geophysical Research*, 115, B04313. https://doi.org/:10.1029/2009JB006716.

Fichtner, A. (2011). *Full Seismic Waveform Modelling and Inversion*. Cham: Springer.

Fichtner, A., and Igel, H. (2008). Efficient numerical surface wave propagation through the optimization of discrete crustal models: A technique based on non-linear dispersion curve matching (DCM). *Geophysical Journal International*, 173, 519–33.

Fichtner, A., and Trampert, J. (2011a). Hessian kernels of seismic data functionals based upon adjoint techniques. *Geophysical Journal International*, 185, 775–98.

Fichtner, A., and Trampert, J. (2011b). Resolution analysis in full waveform inversion. *Geophysical Journal International*, 187, 1604–24. https://doi.org/10.1111/j.1365-246X.2011.05218.x.

Fichtner, A., Kennett, B. L. N., Igel, H., and Bunge, H.-P. (2008). Theoretical background for continental- and global-scale full-waveform inversion in the time-frequency domain. *Geophysical Journal International*, 175, 665–85.

Fichtner, A., Kennett, B. L. N., Igel, H., and Bunge, H. P. (2009). Full seismic waveform tomography for upper-mantle structure in the Australasian region using adjoint methods. *Geophysical Journal International*, 179, 1703–25.

Fichtner, A., Kennett, B. L. N., Igel, H., and Bunge, H.-P. (2010). Full waveform tomography for radially anisotropic structure: New insights into present and past states of the Australasian upper mantle. *Earth and Planetary Science Letters*, 290, 270–80.

French, S., Lekic, V., and Romanowicz, B. (2013). Waveform tomography reveals channeled flow at the base of the oceanic asthenosphere. *Science*, 342, 227–30.

French, S. W. & Romanowicz, B. (2014). Whole-mantle radially anisotropic shear velocity structure from spectral-element waveform tomography. *Geophysical Journal International*, 199(3), 1303–27.

French, S. W., Zheng, Y., Romanowicz, B., and Yelick, K. (2015). Parallel Hessian assembly for Seismic Waveform inversion using Global updates, *Proceedings of the 29th IEEE International Parallel and Distributed Processing Symposium* (2015). https://doi.org/10.1109/IPDPS.2015.58.

Fukao, Y., and Obayashi, M. (2013). Subducted slabs stagnant above, penetrating through, and trapped below the 660 km discontinuity. *Journal of Geophysical Research*, 118(11), 5920–38.

Gharti, H. N., Tromp, J., and Zampini, S. (2018). Spectral-infinite-element simulations of gravity anomalies, *Geophysical Journal International*, 215, 1098–117. https://doi.org/10.1093/gji/ggy324.

Geller R. J., and Ohminato T (1994). Computation of synthetic seismograms and their partial derivatives for heterogeneous media with arbitrary natural boundary conditions using the direct solution method. *Geophysical Journal International*, 116, 421–46.

Gung, Y., and Romanowicz, B. (2004). Q tomography of the upper mantle using three-component long-period waveforms. *Geophysical Journal International*, 157(2), 813–30.

Gung, Y., Panning, M., and Romanowicz, B. (2003). Global anisotropy and the thickness of continents. *Nature*, 422 (6933), 707–11.

Hello, Y., Ogé, A. Sukhovich, A., and Nolet, G. (2011). Modern mermaids: New floats image the Deep Earth. *EOS Transactions American Geophysical Union*, 92, 337–48.

Houser, C., Masters, G., Shearer, P., and Laske, G. (2008). Shear and compressional velocity models of the mantle from cluster analysis of long period waveforms. *Geophysical Journal International*, 174, 195–212.

Karaoglu, H., and Romanowicz, B. (2017). Global seismic attenuation imaging using full-waveform inversion: a comparative assessment of different choices of misfit functionals. *Geophysical Journal International*, 212, 807–26.

Karaoglu, H., and Romanowicz, B. (2018). Inferring global upper-mantle shear attenuation structure by waveform tomography using the spectral element method. *Geophysical Journal International*, 213, 1536–58.

Kawai, K., Konishi, K., Geller, R. J., and Fuji, N. (2014). Methods for inversion of body-wave waveforms for localized three-dimensional seismic structure and an application to D″ structure beneath Central America. *Geophysical Journal International*, 197(1), 495–524. https://doi.org/10.1093/gji/ggt520.

Komatitsch, D., and Tromp, J. (2002a). Spectral-element simulations of global seismic wave propagation – I. Validation. *Geophysical Journal International*, 149, 390–412.

Komatitsch, D., and Tromp, J. (2002b). Spectral-element simulations of global seismic wave propagation – II. Three-dimensional models, oceans, rotation and self-gravitation. *Geophysical Journal International*, 150, 303–18.

Komatitsch, D., and Vilotte, J. P. (1998).The spectral element method: an efficient tool to simulate the seismic response of 2D and 3D geological structures. *Bulletin of the Seismological Society of America*, 88, 368–92.

Koroni, M., and Trampert, J. (2021). Imaging global mantle discontinuities: a test using full-waveforms and adjoint kernels. *Geophysical Journal International*, 226, 1498–15.

Krebs, J., Anderson, J., Hinkley, D. et al. (2009). Fast full-wavefield seismic inversion using encoded sources. *Geophysics*, 74, WCC177–WCC188.

Kustowski, B., Ekström, G., and Dziewonski, A.M. (2008). Anisotropic shear-wave velocity structure of the Earth's mantle: A global model. *Journal of Geophysical Research: Solid Earth*, 113(B6), 2156–202.

Laske, G., Collins, J. A., Wolfe, C. J. et al. (2009). Probing the Hawaiian hotspot with new broadband ocean bottom instruments, *Eos, Transactions, American Geophysical. Union*, 90(41), 362–636.

Lee, E.-J., Chen, P., Jordan, T. H. et al. (2014). Full-3-D tomography for crustal structure in southern California based on the scattering-integral and the adjoint-wavefield methods. *Journal of Geophysical Research*, 119, 6421–51.

Lei, W., Ruan, Y., Bozdag, E. et al. (2020). Global adjoint tomography: model GLAD-M25. *Geophysical Journal International* 223, 1–21.

Lekic, V., and Romanowicz, B. (2011). Inferring upper-mantle structure by full waveform tomography with the spectral

element method. *Geophysical Journal International*, 185, 799–831.

Lekic, V., Panning, M., and Romanowicz, B. (2010). A simple method for improving crustal corrections in waveform tomography. *Geophysical Journal International*, 182, 265–78.

Lerner-Lam, A.L., and, Jordan, T. H. (1983). Earth structure from fundamental and higher mode waveform analysis. *Geophysical Journal of the Royal Astronomical Society*, 75 (3), 759–97.

Lévêque, J. J., Rivera, L., and Wittlinger, G. (1993). On the use o the checker-board test to assess the resolution of tomographic inversions. *Geophysical Journal International*, 115, 313–18

Li, X.-D., and Romanowicz, B. (1995). Comparison of global waveform inversions with and without considering cross-branch modal coupling. *Geophysical Journal International*, 121(3), 695–709.

Li, X.-D., and Romanowicz, B. (1996). Global mantle shear velocity model developed using non-linear asymptotic coupling theory. *Journal of Geophysical Research*, 101, 22245–72.

Li, X.-D., and Tanimoto, T. (1993). Waveforms of long-period body waves in a slightly aspherical Earth model. *Geophysical Journal International*, 112, 92–102.

Lin, C., Monteiller, V., Wang, K., Liu, T., Tong, P., and Liu, Q. (2019). High-frequency seismic wave modelling of the Deep Earth based on hybrid methods and spectral-element simulations: a conceptual study. *Geophysical Journal International*, 219, 1948–69.

Liu, Q., Beller, S., Lei, W., Peter, D., and Tromp, J. (2022). Pre-conditioned BFGS-based uncertainty quantification in elastic full-waveform inversion. *Geophysical Journal International*, 228(2), 796–815.

Maggi, A., Tape, C., Chen, M., Chao, D., and Tromp, J. (2009). An automated time-window selection algorithm for seismic tomography. *Geophysical Journal International*, 178, 257–81.

Marquering, H., Nolet, G., and Dahlen, F. A. (1998). Three-dimensional waveform sensitivity kernels. *Geophysical Journal International*, 132, 521–34.

Marone, F., and Romanowicz, B. (2007). Non-linear crustal corrections in high-resolution regional waveform seismic tomography. *Geophysical Journal International*, 170, 460–67.

Masson, Y. (2023). Distributed finited difference modelling of seismic waves, *Geophysical Journal International* , 233, 264–96.

Masson, Y., Cupillard, P., Capdeville, Y., and Romanowicz, B. (2014). On the numerical implementation of time reversal mirrors for tomographic imaging. *Geophysical Journal International*, 196, 1580–99. https://doi.org/10.1093/gji/ggt459.

Masson, Y., and Romanowicz, B. (2017a). Fast computation of synthetic seismograms within a medium containing remote localized perturbations: a numerical solution to the scattering problem. *Geophysical Journal International*, 208(2), 674–92.

Masson, Y., and Romanowicz, B. (2017b). Box tomography: localized imaging of remote targets buried in an unknown medium, a step forward for understanding key structures in the deep Earth. *Geophysical Journal International*, 211(1), 141–63.

Mégnin, C., and Romanowicz, B. (2000). The three-dimensional shear velocity structure of the mantle from the inversion of body, surface and highermode waveforms. *Geophysical Journal International*, 143, 709–28.

Meier, U., Curtis, A., and Trampert, J. (2007). Fully nonlinear inversion of fundamental mode surface waves for a global crustal model. *Geophysical Research Letters*, 34, L16304. https://doi.org/10.1029/2007GL030989.

Mochizuki, E. (1986). Free oscillations and surface waves of an aspherical Earth. *Geophysical Research Letters*, 13, 1478–81.

Moczo, P., Kristek, J., and Halada, L. (2004). *The Finite-Difference Method for Seismologists*. Bratislava: Comenius University. www.spice-rtn.org.

Modrak, R., and Tromp, J. (2016). Seismic waveform inversion best practices: Regional, global and exploration test cases. *Geophysical Journal International*, 206, 1864–89.

Montagner, J., and Anderson, D. (1989). Petrological constraints on seismic anisotropy. *Physics of the Earth and Planetary Interiors*, 54, 82–105.

Monteiller, V., Chevrot, S., Komatitsch, D., and Wang, Y. (2015). Three-dimensional full waveform inversion of short period teleseismic wavefields based upon the SEM-DSM hybrid method. *Geophysical Journal International*, 202, 811–27.

Montelli, R., Nolet, G., Dahlen, F.A. et al. (2004a). Finite-frequency tomography reveals a variety of plumes in the mantle. *Science*, 303(5656), 338–43.

Nocedal, J., and Wright, S. J. (2017). *Numerical Optimization*, 2nd ed. New York: Springer.

Nolet, G. (1990). Partitioned waveform inversion and two-dimensional structure under the network of autonomously recording seismograph. *Journal of Geophysical Research*, 95, 8499–8512.

Nolet, G., and Dahlen, F. A. (2000). Wave front healing and the evolution of seismic delay times. *Journal of Geophysical Research*, 105, 19043–54.

Nolet, G., Hello, Y., van der Lee, S. et al. (2019). Imaging the Galapagos mantle plume with an unconventional application of floating seismometers. *Scientific Reports*, 9, 1326.

Panning, M., and Romanowicz, B. (2006). A three-dimensional radially anisotropic model of shear velocity in the whole mantle. *Geophysical Journal International*, 167, 361–79.

Park, J. (1987). Asymptotic coupled-mode expressions for multiplet amplitude anomalies and frequency shifts on an aspherical earth. *Geophysical Journal of the Royal Astronomical Society*, 90(1), 129–69.

Pasyanos, M., and Nyblade, A. (2007). A top to bottom lithospheric study of Africa and Arabia. *Tectonophysics*, 444(1–4), 27–44.

Pienkowska, M., Monteiller, V., and Nissen-Meyer, T. (2020). High-frequency global wavefields for local 3-D structures by wavefield injection and extrapolation. *Geophysical Journal International*, 225, 1782–98.

Pipatprathanporn, P., and Simons. F. J. (2022). One year of sound recorded by a MERMAID float in the Pacific: hydroacoustic earthquake signals and infrasonic ambient noise. *Geophysical Journal International*, 228(1), 193–212.

Obayashi, M., Yoshimitsu, J., Sugioka, H. et al. (2016). Mantle plumes beneath the South Pacific Superswell revealed by finite frequency P tomography using regional seafloor and island data. *Geophysical Research Letters*, 43(11), 628–11634.

Rawlinson, N., Fichtner, A., Sambridge, M., and Young, M. K. (2014). Seismic tomography and the assessment of uncertainty. *Advances in Geophysics*, 55, 1–76.

Replumaz, A., Karason, H., van der Hilst, R. D., Besse, J., and Tapponnier, P. (2004). 4-D evolution of SE Asia's mantle from geological reconstructions and seismic tomography. *Earth and Planetary Science Letters*, 221, 103–15.

Richards, M. A., and Engebretson, D. C. (1992). Large scale mantle convection and the history of subduction. *Nature*, 355, 437–40.

Rickers, F., Fichtner, A., and Trampert, J. (2013). The Iceland–Jan Mayen plume system and its impact on mantle dynamics in the North Atlantic region: Evidence from full-waveform inversion. *Earth and Planetary Science Letters*, 367, 39–51.

Ritsema, J., and Lekic, V. (2020). Heterogeneity of seismic wave velocity in Earth's mantle. *Annual Review of Earth and Planetary Sciences*, 48, 377–401

Ritsema, J., Deuss, A., van Heijst, H. J., and Woodhouse, J. H. (2011). S40RTS: A degree-40 shear-velocity model for the mantle from new Rayleigh wave dispersion, teleseismic traveltime and normal-mode splitting function measurements. *Geophysical Journal International*, 184(3), 1223–36.

Ritsema, J., van Heijst, H. J., and Woodhouse, J. H. (1999). Complex shear velocity structure imaged beneath Africa and Iceland. *Science*, 286, 1925–8.

Ritsema, J., van Heijst, H.J., and Woodhouse, J.H. (2004). Global transition zone tomography. *Journal of Geophysical Research*, 109. https://doi.org/10.1029/2003JB002610.

Romanowicz, B. (1987). Multiplet-multiplet coupling due to lateral heterogeneity: Asymptotic effects on the amplitude and frequency of the Earth's normal modes. *Geophysical Journal of the Royal Astronomical Society*, 90(1), 75–100.

Romanowicz, B. (2003). Global mantle tomography: Progress status in the last 10 years. *Annual Review of Earth and Planetary Sciences*, 31(1), 303–28.

Romanowicz, B. (2020). Seismic tomography of the Earth's mantle. In D. Alderton and S. A. Elias, eds., *Encyclopedia of Geology*, vol. 1, 2nd ed. Cambridge, MA: Academic Press, pp. 587–609.

Romanowicz, B., and Suyehiro, K. (2001). History of the International Ocean Network. http://eri-ndc.eri.u-tokyo.ac.jp/OHP-sympo2/report/index.html.

Romanowicz, B., and Wenk, R. (2017). Anisotropy in the Deep Earth. *Physics of the Earth and Planetary Interiors*, 269, 58–90. https://doiorg/10.1016/j.pepi.2017.05.005.

Romanowicz, B., Chen, L.-W., and French, S. W. (2019). Accelerating full waveform inversion via source stacking and cross-correlations. *Geophysical Journal International*, 220(1), 308–22.

Romanowicz, B., Panning, M., Gung, Y., and Capdeville, Y. (2008). On the computation of long period seismograms in a 3-D Earth using normal mode based approximations. *Geophysical Journal International*, 175, 520–36.

Romanowicz, B. A., French, S. W., Rickers, F., and Yuan, H. (2013). Source stacking for numerical wavefield computations: Application to continental and global scale seismic mantle tomography, in *American Geophysical Union*, Fall Meeting 2013, abstract S21E-05.

Ruan, Y., Lei, W., Modrak, R. et al. (2019). Balancing unevenly distributed data in seismic tomography: a global adjoint tomography example. *Geophysical Journal International*, 219(2), 1225–36

Schaeffer, A. J., and Lebedev, S. (2013). Global shear speed structure of the upper mantle and transition zone. *Geophysical Journal International*, 194(1), 417–49.

Schaeffer, A. J., and Lebedev, S. (2014). Imaging the North American continent using waveform inversion of global and USArray data. *Earth and Planetary Science Letters*, 402, 26–41.

Shapiro, N., and Ritzwoller, M. (2002). Monte-Carlo inversion for a global shear-velocity model of the crust and upper mantle. *Geophysical Journal International*, 151, 88–105.

Sigloch, K., McQuarrie, N., and Nolet, G. (2008). Two-stage subduction history under North America inferred from multiple-frequency tomography. *Nature Geoscience*, 1, 458–63.

Su, W.-J., Woodward, R. L., and Dziewonski, A. M. (1994). Degree-12 model of shear velocity heterogeneity in the mantle. *Journal of Geophysical Research*, 99(4), 4945–80.

Suetsugu, D., Isse, T., Tanaka, S. et al. (2009). South Pacific mantle plumes imaged by seismic observation on islands and seafloor. *G-Cubed*, 10, Q11014.

Sukhovich, A., Bonnieux, S., Hello, Y. et al. (2015). Seismic monitoring in the oceans by autonomous floats. *Nature Communications*, 6, 8027–33.

Suzuki, Y., Kawai, K., Geller, R. J., Borgeaud, A. F. E., and Konishi, K. (2016). Waveform inversion for 3-D S-velocity structure of D″ beneath the Northern Pacific: Possible evidence for a remnant slab and a passive plume. *Earth, Planets and Space*, 68(1). https://doi.org/10.1186/s40623-016-0576-0.

Suzuki, Y., Kawai, K., Geller, R. J. et al. (2020). High-resolution 3-D S-velocity structure in the D' region at the western margin of the Pacific LLSVP: Evidence for small-scale plumes and paleoslabs. *Physics of the Earth and Planetary Interiors*, 307, 106544.

Tanimoto, T. (1987). The three-dimensional shear wave structure in the mantle by overtone waveform inversion – I. Radial seismogram inversion. *Geophysical Journal of the Royal Astronomical Society*, 89(2), 713–40.

Tape, C., Liu, Q., Maggi, A., and Tromp, J. (2010). Seismic tomography of the southern California crust based upon spectral-element and adjoint methods. *Geophysical Journal International*, 180, 433–62.

Tarantola, A. (1984). Inversion of seismic reflection data in the acoustic approximation. *Geophysics*, 49, 1259–66.

Tarantola, A. (2005). *Inverse Problem Theory and Methods for Model Parameter Estimation*. Philadelphia, PA: Society for Industrial and Applied Mathematics.

Tarantola, A., and Valette, B. (1982). Generalized nonlinear inverse problems solved using the least squares criterion. *Reviews of Geophysics*, 20(2), 219–232.

Thrastarson, S., van Driel, M., Krischer, L. et al. (2020). Accelerating numerical wave propagation by wavefield adapted meshes. Part II: Full-waveform inversion. *Geophysical Journal International*, 221, 1591–604.

Thurber, C., and Ritsema, J. (2015). Theory and Observations- Seismic Tomography and Inverse Methods. In G. Schubert, ed., *Treatise on Geophysics*, vol. 1. Amsterdam: Elsevier, pp. 307–37.

Tromp, J. (2015). 1.07 – Theory and Observations: Forward Modeling and Synthetic Seismograms, 3D Numerical Methods. In G. Schubert, ed., *Treatise on Geophysics*. Amsterdam: Elsevier, pp. 231–51.

Tromp, J. (2020). Seismic wavefield imaging of earth's interior across scales. *Nature Reviews Earth and Environment*, 1, 40–53.

Tromp, J., and Bachmann, E. (2019). Source encoding for adjoint tomography. *Geophysical Journal International*, 218, 2019–44.

Tromp, J., Tape, C. & Liu, Q. Y. (2005). Seismic tomography, adjoint methods, time reversal and banana-doughnut kernels. *Geophysical Journal International*, 160, 195–216.

Tromp, J., Komatitsch, D., Hjörleifsdóttir, V. et al. (2010). Near real-time simulations of global CMT earthquakes. *Geophysical Journal International*, 183(1), 381–9.

Valentine, A. P., and Trampert, J. (2016). The impact of approximations and arbitrary choices on geophysical images. *Geophysical Journal International*, 204, 59–73.

van der Hilst, R., and de Hoop, M. V. (2005). Banana-doughnut kernels and mantle tomography. *Geophysical Journal International*, 163, 956–61.

van der Lee, S., and Nolet, G. (1997). Upper mantle S velocity structure of North America. *Journal of Geophysical Research: Solid Earth*, 102, 22815–38.

van der Meer, D. G., Spakman, W., van Hinsbergen, D. J. J, Amaru, M. L., and Torsvik, T. H. (2010). Towards absolute plate motions constrained by lower-mantle slab remnants. *Nature Geoscience*, 3, 36–40.

van Driel, M., Krischer, L., Stähler, S. C., Hosseini, K., and Nissen-Meyer, T. (2015). Instaseis: instant global seismograms based on a broadband waveform database. *Solid Earth*, 6(2), 701–17.

van Driel, M., Kemper, J., and Boehm, C. (2021). On the modelling of self-gravitation for full 3-D global seismic wave propagation. *Geophysical Journal International*, 227, 632–43.

van Herwaarden, D. P., Boehm, C., Afanasiev, M. et al. (2020). Accelerated full-waveform inversion using dynamic mini-batches. *Geophysical Journal International*, 221, 1427–38.

Virieux, J., and Operto, S. (2009). An overview of full-waveform inversion in exploration geophysics. *Geophysics*, 74, WCC127–WCC152.

Virieux, J., Asnaashari, A., Brossier, R. et al. (2014). 6. An introduction to full waveform inversion. In *Encyclopedia of Exploration Geophysics*, Geophysical References Series: R1-1-R1-40. Tulsa, OK: Society of Exploration Geophysicists.

Wang, Z., and Dahlen, F.A. (1995). Spherical-spline parameterization of three-dimensional Earth models. *Geophysical Research Letters*, 22, 3099–102.

Wen, L., and Helmberger, D.V. (1998). A two-dimensional P-SV hybrid method and its application to modeling localized structures near the core-mantle boundary. *Journal of Geophysical Research*, 103, 17901–18.

Woodhouse, J. H., and Dziewonski, A. M. (1984). Mapping the upper mantle: Three dimensional modeling of Earth structure by inversion of seismic waveforms. *Journal of Geophysical Research*, 89, 5953–86.

Woodhouse, J. H., and Deuss, A. (2015). 1.03 – Theory and Observations: Earth's Free Oscillations. In G. Schubert, ed., *Treatise on Geophysics*. Amsterdam: Elsevier, pp. 31–65.

Yang, H., and Tromp, J. (2015). Synthetic free-oscillation spectra: an appraisal of various mode-coupling methods. *Geophysical Journal International*, 203, 1179–92.

Zhang, Q., Mao, W., Zhou, H., Zhang, H., and Chen, Y. (2018). Hybrid-domain simultaneous-source full waveform inversion without crosstalk noise. *Geophysical Journal International*, 215(3), 1659–81.

Zhu, H., Bozdag, E., and Tromp, J. (2015). Seismic structure of the European upper mantle based on adjoint tomography. *Geophysical Journal International*, 201(1), 18–52.

15

Solving Larger Seismic Inverse Problems with Smarter Methods

Lars Gebraad, Dirk-Philip van Herwaarden, Solvi Thrastarson, and Andreas Fichtner

Abstract: The continuously increasing quantity and quality of seismic waveform data carry the potential to provide images of the Earth's internal structure with unprecedented detail. Harnessing this rapidly growing wealth of information, however, constitutes a formidable challenge. While the emergence of faster supercomputers helps to accelerate existing algorithms, the daunting scaling properties of seismic inverse problems still demand the development of more efficient solutions. The diversity of seismic inverse problems – in terms of scientific scope, spatial scale, nature of the data, and available resources – precludes the existence of a silver bullet. Instead, efficiency derives from problem adaptation. Within this context, this chapter describes a collection of methods that are smart in the sense of exploiting specific properties of seismic inverse problems, thereby increasing computational efficiency and usable data volumes, sometimes by orders of magnitude. These methods improve different aspects of a seismic inverse problem, for instance, by harnessing data redundancies, adapting numerical simulation meshes to prior knowledge of wavefield geometry, or permitting long-distance moves through model space for Monte Carlo sampling.

15.1 Introduction

Recordings of the seismic wavefield are an enormously rich source of information that may be used to constrain the structure and evolution of the Earth, to infer the locations and dynamics of earthquakes and other sources (e.g. underground explosions, ocean dynamics, urban noise), or to produce estimates of future ground motion for hazard assessment.

Much of the history of seismology revolves around efforts to harness this wealth of information in a more comprehensive fashion. A key element of these efforts is the development of theories and methods that allow us to simulate the seismic wavefield more completely and more accurately. Numerical wave propagation, enabled by modern supercomputers and sophisticated spatiotemporal

discretisation schemes (e.g. Komatitsch and Vilotte, 1998; Moczo et al., 2002; de la Puente et al., 2007; Afanasiev et al., 2019) may be considered the preliminary culmination of this trend. Combined with adjoint techniques, or variants thereof (e.g. Tarantola, 1988; Tromp et al., 2005; Fichtner et al., 2006; Plessix 2006; Chen et al., 2007a), numerical wave propagation may be combined into a class of inverse problem solutions, loosely and somewhat exuberantly referred to as full-waveform inversion (FWI). Conceived already in the late 1970s and early 1980s (e.g. Bamberger et al., 1977, 1982), FWI for realistic 3-D problems is a more recent achievement that has produced detailed images of Earth structure from local to global scales (e.g. Chen et al., 2007b; Fichtner et al., 2009; Sirgue et al., 2010; Tape et al., 2010; Bozdag et al., 2016; Fichtner et al., 2018).

The basic mechanics of FWI have been described in several review papers and books (e.g. Virieux and Operto, 2009; Fichtner, 2010; Liu and Gu 2012), and will be repeated here only in a very condensed form that provides some context for the following sections. In a narrow sense, FWI can be regarded as an optimisation problem, aiming to find an Earth model that explains observed seismograms to within their uncertainties. Starting from an initial model, often obtained from ray-based travel-time tomography, the first step consists in the calculation of synthetic seismograms using a suitable (i.e. problem-adapted), numerical method. The pairwise comparison of observed and synthetic seismograms yields a misfit value that we wish to minimise. Adjoint techniques provide the gradient of the misfit with respect to the Earth model parameters of interest (e.g. P- and S-wave speeds). The misfit gradient can then be used in a gradient-based descent method, which iteratively repeats the sequence of (1) synthetic seismogram calculation, (2) misfit evaluation, (3) gradient computation, and (4) model updating.

This basic FWI theme has numerous variations, depending on the numerical wave propagation method, the misfit functional used to compare observed and synthetic seismograms, the iterative optimisation and regularisation scheme, and the choice of model parameters that are in- or excluded.

Furthermore, and in a broader sense, any inversion involves both some sort of optimisation and an analysis of alternative models that may explain the observations equally well (Backus and Gilbert, 1968). Clearly, the search for alternative models, closely related to uncertainty quantification, adds yet another level of complexity, that we will consider more closely in Section 15.4.

Though waveform inversion methods based on simplified forward problem solutions date back, at least, to the pioneering work of Woodhouse and Dziewonski (1984), FWI in the current and admittedly diffuse interpretation, may naively be considered the silver bullet for the solution of seismic, or similar wave-based, inverse problems. Despite being computationally expensive, quasi-continuous streams of success stories about ever-growing supercomputers should make us hopeful that an assimilation of the full observable bandwidth and volume of seismic data is within reach. Shouldn't it?

A more critical look quickly reveals that the scaling of seismic inverse problems puts the increase of computational power into a different perspective. In fact, the cost of a forward numerical wavefield simulation in 3-D scales with frequency f as f^4 (e.g. Fichtner, 2010; Igel, 2016). Increasing frequency by a certain factor, shrinks the volume of a Fresnel zone by the same factor, meaning that, loosely speaking, higher frequencies 'see' less of the Earth, though in more detail. Using more data compensates for the shrinking of Fresnel zones but also modifies the frequency scaling from f^4 to roughly f^5. Even an optimistic interpretation of Moore's law then implies that growing computational power allows us to increase the maximum frequency in FWI by only about 7% per year.

An alternative to waiting for several decades until supercomputers are finally fast enough is suggested by the No-Free-Lunch theorem (Wolpert and Macready, 1997; Mosegaard, 2012). In simple words, it states that the efficiency of any algorithm does not rest within the algorithm itself, but instead derives from its specialisation or tuning towards a specific problem. In this sense, the No-Free-Lunch theorem is a mathematical corroboration of our intuition and experience that problem-adapted solutions may greatly outperform a one-size-fits-all approach.

Problem adaptation in the context of FWI is the central theme of this chapter. Its purpose is to describe and highlight a collection of recent developments that improve the performance of existing inversions, sometimes by orders of magnitude. This leads not only to more detailed images of Earth structure, but it also opens up new perspectives, for instance, in uncertainty quantification.

This chapter is organised as follows. In Section 15.2, we consider improvements on the data side of the inverse problem, including source stacking and source encoding approaches, as well as non-linear optimisation schemes that automatically take advantage of data redundancies.

This is followed in Section 15.3 by the description of forward problem solvers that reduce computational requirements through the use of prior knowledge on the geometry of seismic wavefields. Finally, in Section 15.4, we discuss novel approaches to uncertainty quantification based on efficient, approximate solutions to Bayesian inference problems.

This chapter complements other chapters in this book. Chapter 2 on *Emerging Directions in Geophysical Inversion* (Valentine and Sambridge, this volume) summarises promising methodological developments, including approaches based on non-Euclidean metrics, new ensemble- and sampling-based methods, as well as generative models. Chapter 11 on *Global Seismic Tomography Using Time-Domain Waveform Inversion* (Romanowicz, this volume) focuses on FWI strategies for whole-Earth imaging, with an emphasis on data selection, the choice of misfit, model parameterisation, and optimisation schemes.

15.2 Static and Dynamic Optimisation of Wavefield Sources and Receiver Configurations

This section is dedicated to an overview of techniques that reduce the computational requirements of FWI by producing more useful data in the first place, or by using already available data more efficiently. In this context, we will discuss three main directions. The first is optimal experimental design, which attempts the balancing act of positioning sources and receivers statically, prior to the experiment, such that their numbers can be minimised while maximising the information content of the data. The second is source stacking and source encoding. With this family of techniques, it becomes possible to simultaneously simulate wavefields of multiple sources, which reduces the total number of wavefield simulations. Finally, we discuss the use of mini-batches, which can be interpreted as a dynamic version of optimal experimental design that adjusts automatically to the current state of an interactive FWI. With mini-batches, model updates are computed from variable subsets of the complete dataset, which are chosen in a quasi-random fashion that exploits redundancies.

15.2.1 Optimal Experimental Design

Optimal experimental design (OED), as the name suggests, revolves around the design of an optimal experiment, where the amount of exploitable information can be quantified and maximised. Though different from today's common understanding of OED, the foundations of geophysical inverse theory laid by Backus and Gilbert (1968, 1970) may be considered the pioneering contributions in this field. Their approach aims to construct observables that maximise the resolution of Earth structure. Seismological research on OED, as it is now mostly interpreted, started with the goal to design optimal seismic receiver networks (e.g. Kijko, 1977; Rabinowitz and Steinberg, 1990; Hardt and Scherbaum, 1994). These early works had the goal to improve the resolution power of seismic networks for the determination of the epicentre and focal depth of earthquakes. Later, OED was applied to select optimal source and receiver locations to maximise image quality in the context of travel-time tomography (e.g. Curtis, 1999; Ajo-Franklin, 2009). Within the context of FWI, OED may be used to reduce the computational cost by maximising data information content for the smallest number of source and receiver locations (Krampe et al., 2021). In contrast to OED for travel-time tomography, this, however, requires the computation of sensitivity kernels for representative subsurface models (e.g. Djikpesse et al., 2012; Maurer et al., 2017). While these techniques may be of great use in exploration geophysics, their applicability is more limited in the context of earthquake seismology, where the locations of earthquakes cannot be chosen, and where the geometry of networks is strongly influenced by topography, oceans, and political circumstances. Nevertheless, OED techniques provide suggestions for optimal new station locations. Variations of the OED theme within the FWI context, which are still in their infancy, are the design of optimal model parameterisations (Sieminski et al., 2009) and of misfit functionals that are particularly (in)sensitive to selected model parameters (Bernauer et al., 2014).

15.2.2 Source Encoding

In a conventional FWI, sources are simulated individually. This means that the simulation cost scales linearly with the number of sources present in the dataset. Since the wave equation is linear with respect to the source input, the idea arose to simulate multiple sources simultaneously. This approach, commonly referred to as source stacking, was first proposed in seismology by Capdeville et al. (2005). The method, however, suffers from the problem that all sources are required to be recorded by all stations, which is hardly the case in seismology. In seismic exploration, a variant of source stacking, termed source encoding, was proposed (e.g. Krebs et al., 2009; Choi and Alkhalifah, 2011; Schiemenz and Igel, 2013). The challenge here was that physically unrelated forward and adjoint components have the chance to interact, leading to a noisy 'crosstalk' component in the gradients, which may then reduce the final image quality. To mitigate this issue, several strategies can be applied, such as limiting the number of sources in a stack (e.g. Romero et al., 2000), or randomly changing the source-encoding functions between iterations (e.g. Krebs et al., 2009). Eventually, this helps to average out the noisy component over the course of many iterations.

More recently, crosstalk-free methods have been developed to mitigate these earlier issues completely. Krebs et al. (2018) and Huang and Schuster (2018) employed narrow-band filtering to ensure that source and receiver wavefields

are orthogonal. Essentially, more sources can be simulated at the cost of a loss of frequency content in each source. However, as Sirgue and Pratt (2004) showed, reconstructions can be made successfully from only a limited number of all frequencies. Zhang et al. (2018) introduced a crosstalk-free method where sources are given a frequency band, and forward and adjoint wavefields are simulated until they reach a steady state. At this point, the steady-state wavefields are decoded by integrating over time to obtain the contributions of each individual source. The introduction of these crosstalk-free techniques also meant that the earlier requirement that all sources be recorded by all stations was removed.

Tromp and Bachmann (2019) further extended this approach and showed how it may be used in combination with explicit time-domain solvers and in combination with a variety of misfit functions. Using this technique, the overall computational cost may be reduced to just two simulations per iteration when adjoint sources are computed based on measurements of the source-encoded Fourier coefficients of the observed and synthetic seismograms. This theoretically leads to a very large performance improvement for a large number of sources. However, when standard measurements based on specific time windows are used, a separate simulation for each source is still required. Only two extra simulations are required to compute the gradient, but since each forward wavefield needs to be computed individually, the computational gain is upper-bounded by a factor of three under ideal conditions.

15.2.3 Mini-Batches

A third option to reduce the computational cost of large-scale FWI by adaptations on the data side is the use of mini-batches (i.e. subsets of the complete dataset), which represents a form of stochastic optimisation. In contrast to standard FWI without source encoding, where the cost per iteration scales linearly with the total number of sources, the cost of a mini-batch iteration depends only on the number of sources contained in a mini-batch. Misfits and gradients of individual events are sample-averaged rather than summed, which enables the approximation of the sample average of the large dataset using a smaller number of samples.

Van Leeuwen and Herrmann (2013) explored this idea, giving it the name 'fast waveform inversion without source encoding'. They argued that randomised source encoding, developed to reduce crosstalk (e.g. Krebs et al., 2009), is equivalent to stochastic approaches that also rely on multiple iterations with approximate versions of gradients and misfits. Hence, there is no need to use explicit source encoding to reap the benefits of the fast convergence that stochastic optimisation techniques have in the early phase of optimisation. Van Leeuwen and Herrmann (2013) proposed a hybrid approach, using gradually growing batches that simultaneously bring the benefits of both stochastic and conventional optimisation, while performing equally well as source encoding techniques. This was illustrated with several synthetic examples of a reflection seismology setup.

Van Herwaarden et al. (2020) further developed these ideas to make them suitable for FWI on large scales using earthquake data. The use of earthquake data, rather than active source data, poses specific extra challenges. Data from earthquakes tend to be heterogeneous. Within a particular frequency band, certain earthquakes may create high-quality waveform data, while others do not. Furthermore, earthquakes tend to be recorded by temporarily existing arrays. These two factors result in the fact that some earthquakes have many high-quality recordings, while others only have few.

This, in turn, may cause a seismologist to select many more measurement windows for one event than for the other, which makes the misfit and gradient of the complete dataset much more sensitive to these particular high-quality events. This, combined with the observation by van Leeuwen and Hermann (2013) that the misfit and model convergence per iteration depends on the quality of the gradient approximation, led van Herwaarden et al. (2020) to propose the use of dynamic mini-batches.

This technique differs from earlier mini-batch approaches in several ways. First, it uses control-group events. These are a subset of the mini-batch that is used again in the next iteration, in order to check if the misfit indeed reduces between iterations without requiring any extra simulations. Second, the dynamic mini-batches seek to become a reasonable approximation of the full gradient. This is achieved through a gradient approximation algorithm that tests how the mini-batch sample average gradient direction changes as events are removed from the mini-batch. If the removal of events does not significantly alter the sample average, it is seen as an indicator that the mini-batch approximation is good, and that the number of samples in the batch is appropriate. On the other hand, if the gradient direction changes significantly when events are removed, it is seen as an indicator that the mini-batch may give a poor approximation of the full dataset, after which the number of sources in the next mini-batch is increased.

In addition to the dynamic sizing of the batches, the dynamic mini-batch method also keeps track of which events had a significant contribution to the sample average gradient. Events with a small contribution in earlier iterations have a smaller chance of being randomly selected again. In short, it seeks to iterate more often with the most informative events, while making sure that the mini-batch approximation is of sufficient quality to ensure rapid convergence.

Figure 15.1 shows an example of a synthetic inversion that aims to recover a known input model using an actual distribution of earthquakes and seismic stations on and around the African continent. Though the recovered models

are visually indistinguishable and equally similar to the input model in terms of L_2 distance, the dynamic mini-batch approach requires only about 20% of the computational resources of a 'traditional' FWI that uses all sources in each iteration. Similar results were achieved with a real-data example presented in van Herwaarden et al. (2021).

15.3 Goal-Oriented Wavefield Simulations

With the large number of wavefield simulations required for FWI, it is worth considering the most computationally feasible way to perform the simulations. In studies of regional Earth structure using teleseismic data, the domain for expensive numerical calculations may be limited by wavefield decomposition and extrapolation (e.g. Masson and Romanowicz, 2017), leading to the concept of 'box tomography' (Clouzet et al., 2018).

In addition to the domain, the geometry and type of relevant waves may also be exploited to optimise simulation algorithms. For example, exploration-scale studies focus mostly on reflected and refracted waves off interfaces in the shallow subsurface and ignore surface waves. In contrast, regional- to global-scale studies are dominated by surface waves and transmitted body waves. Although these two problems have the same underlying physical laws, the way the wavefield interacts with the medium is quite different, suggesting that they should not be solved in the same way. In fact, there is room to adapt the numerical solution of the wave equation to the respective problems.

Here we focus on large-scale problems and demonstrate how the wave equation can be solved numerically in a way that properly accounts for the physics most relevant to the problem, while ignoring less relevant aspects.

15.3.1 Wavefield-Adapted Meshes

When examining a rock, it can be viewed at multiple scales. From far away, the eye is only sensitive to the bulk structure; when looking closer, however, it becomes apparent that the rock is composed of multiple minerals, and looking even closer into the minerals, they have an atomic structure. The Earth can, in a similar way, be seen as a multi-scale medium. The way in which a seismic wave experiences the medium depends on the wavelength of the wave. Seismic waves are not directly sensitive to individual heterogeneities much smaller than their minimum wavelength but rather to some kind of average over all heterogeneities, that is, to their ensemble action, which can be described in terms of an effective medium (Capdeville et al., 2010).

Another important feature of most wavefields is that their complexity depends on the direction of inspection. For example, wavefields in smooth media exhibit approximate spherical symmetry with rapid oscillations in propagation (radial) direction, while changing only slowly in the

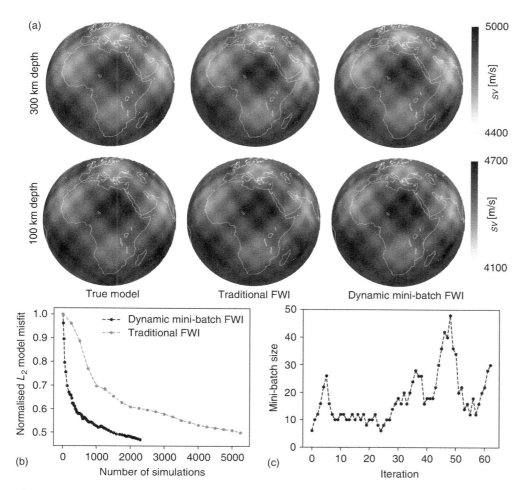

Figure 15.1 Overview of results of the dynamic mini-batch method for a synthetic inversion with a 'true' input model. (a) Starting from a 1D Earth model, both traditional FWI and dynamic mini-batch FWI were used to invert for the true model shown on the left. The panels in the middle and right show spherical slices of the recovery at 100 and 300 km depth for traditional FWI and dynamic mini-batch FWI, respectively. (b) Convergence in terms of the L_2 distance between the input model and the recovered model for both methods. Dynamic mini-batch FWI recovers the structure at a significantly lower cost. (c) Number of events in each mini-batch iteration. There is a general tendency to use more events as the inversion progresses.

perpendicular (azimuthal) direction. This observation was the inspiration behind AxiSEM3D by Leng et al. (2016), where the radial and latitudinal dimensions are solved using the spectral-element method, and the azimuthal one is solved using the pseudo-spectral method. In favourable cases, such as simulating global wave propagation through smooth tomographic models, the required simulation cost can be reduced by over an order of magnitude, compared to the standard approach of using cubed-sphere meshes (Ronchi et al., 1996) and the spectral-element method in all three dimensions.

With the two presented features in mind, van Driel et al. (2020) developed an approach to adapt the computational mesh to the expected wavefield propagating from the source. As the number of required elements in a mesh is controlled by the complexity of the wavefield, fewer elements are needed to cover the azimuthal dimension, compared

to the other dimensions (see Figs. 15.2c and 15.3a). A wavefield travelling through laterally smooth media can thus be modelled using fewer elements than traditionally. The novel meshing approach can result in an order of magnitude reduction in compute cost, while no modifications are needed in the wave propagation algorithm.

15.3.2 Use Case: Full-Waveform Inversion
Being able to simulate a wavefield propagating through a given medium can already be useful in data analysis. Furthermore, it can be used to infer the structure of that medium through FWI. Doing so requires observed data, multiple wavefield simulations, and a method to compute the gradient of the misfit between observed and simulated data with respect to the model parameters.

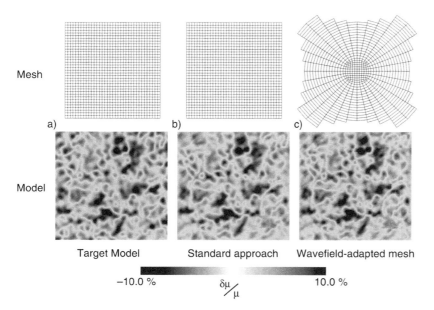

Figure 15.2 A comparison of two inversions. The mesh is in the top panel with its corresponding resulting model in the bottom panel. The models show percentage deviations of the shear modulus from the mean value. (a) The target model that the synthetic inversion aims to recover. Artificial data were created using a rectilinear grid. (b) The results obtained by using the standard approach to FWI, with a rectilinear grid as a mesh. (c) The results obtained with the wavefield-adapted meshes. The displayed mesh is for a source in the middle, but in the complete FWI each modelled source needs its own mesh. The mesh has a Cartesian rectilinear grid at the source location, but then expands away using cylindrical coordinates. This result only used 12% of the computational time required to reach the result in (b). Such wavefield-adapted meshes are an ideal way to solve large-scale simulations in laterally smooth media. As the medium becomes more complex (e.g. in the course of an iterative FWI), more elements in azimuthal direction may need to be added. For a strongly scattering medium, the mesh would eventually converge towards a standard mesh, and the method loses its benefits.

With the need for multiple wavefield simulations, being able to compute them efficiently becomes critical. Using wavefield-adapted meshes for the inversion can therefore dramatically reduce the computational cost of the whole inversion. In contrast to standard FWI workflows, however, wavefield adapted meshes require a different mesh for each source, thereby requiring slightly more book-keeping.

Thrastarson et al. (2020) introduced an FWI workflow that exploits wavefield-adapted meshes and demonstrated its computational benefits. In a 2D experiment, they compared the FWI results from a standard approach to their developed workflow and showed that they could achieve the same quality of results at only 12% of the previous compute cost. The experiment aimed to reconstruct a random target medium (Fig. 15.2a), using nine sources and eighty receivers. As a reference, the medium was reconstructed using a rectilinear grid (Fig. 15.2b). The same reconstruction was then attempted using wavefield-adapted meshes, achieving the same quality result (Fig. 15.2c).

The synthetic proof of concept serves as the methodological basis for a real-data, global-scale FWI (Thrastarson et al., 2021, 2022). For this, wavefield-adapted meshes are constructed in 3D, as shown in Fig. 15.3a, and combined with the dynamic mini-batch method of van Herwaarden et al. (2020), described before in Section 15.2.

Starting from the spherically symmetric Earth model PREM (Dziewonski and Anderson, 1981), data from 1,200 earthquakes were inverted, using waveforms in the 100–200 s period range. The latest model, obtained after 70 iterations, is shown in Fig. 15.3b. The velocities are plotted as percentage deviations from the lateral mean velocity of the depth level.

The cost of the 70 iterations was comparable to only 70% of a single iteration using a standard FWI approach with a cubed-sphere mesh. Mini-batch iterations and full iterations are not directly comparable, however, being able to construct a geologically plausible model using fewer computations than a single model update, showcases the benefit of the methods.

15.3.3 Discussion

The presented methods for accelerating numerical wave propagation, wavefield-adapted meshes and AxiSEM3D, in both forward and adjoint mode, can add to the benefits of other approaches described in this chapter. As they rely on single source simulations, they do, however, not function in combination with source stacking or source encoding methods. Yet, wavefield-adapted meshes and AxiSEM3D harmonise well with the dynamic mini-batch approach, as presented in the previous paragraph, and

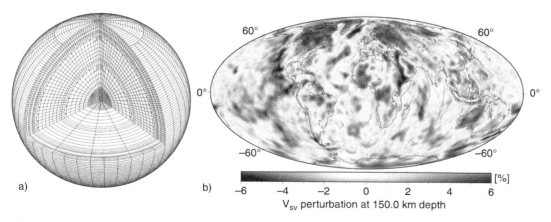

Figure 15.3 Global velocity model inferred by using wavefield-adapted meshes. (a) A 3D wavefield-adapted mesh for a source at the North Pole. The elements are stretched along the equatorial direction (azimuthal direction) compared to the other directions. (b) A slice through the model recovered at 150 km depth. Displayed are percentage variations in vertically polarised shear wave velocities from the lateral mean velocity.

they may enable the solution of Bayesian inference problems for uncertainty quantification, covered in the following paragraphs.

15.4 Tomographic Uncertainty Quantification

Improved algorithms combined with modern HPC capabilities have not only satisfied the quest for solving larger inverse problems in seismology, but also enabled more complete and less biased uncertainty quantification. Model uncertainties arise from measurement errors and a lack of sensitivity, which, in turn, is related to data coverage in space and frequency. The uneven source–receiver distribution controls which parts of the wavefield can be recorded and which regions of the Earth can be constrained. Additionally, seismic waves do not carry the same amount of information on every Earth model parameter of interest. The canonical example is density, which is notoriously difficult to resolve using seismic data.

Inversions utilising gradient-based deterministic algorithms do not provide information on either of these aspects of uncertainty. Furthermore, in cases where information on a specific parameter or part of the Earth is particularly weak, gradient-based methods rely heavily on regularisation, the effect of which is hardly distinguishable from actual data constraints. Hence, when interpreting tomographic models, one might over-interpret features that were actually imposed by regularisation.

Throughout this section, we focus on how the latest algorithms can tackle uncertainty quantification in seismic tomography. Before delving into the details of concrete approaches to uncertainty quantification in a tomographic context, we

provide a condensed review of Bayesian inference. This will be followed by the fundamentals of Hamiltonian Monte Carlo (HMC) sampling. We will discuss a case of HMC applied to a synthetic FWI experiment, and an alternative to HMC sampling applied to the same test case.

The two methods share the probabilistic interpretation of parameter estimation problems, which is drastically different from deterministic approaches. Importantly, both methods also employ gradients typically used in deterministic approaches.

15.4.1 Bayesian Interpretation of Inverse Problems and Its Use

Deterministic inversion aims to produce a single model that explains observations within their errors, starting from a hopefully suitable initial model that prevents trapping in a local minimum. The probabilistic interpretation, in contrast, aims to construct a posterior probability distribution of the model parameters, taking data errors and prior information into account (e.g. Tarantola 2005; Fichtner 2021). The posterior distribution is considered the solution of the inverse problem, which contains all information one can possibly obtain from the data. This wealth of information includes, but is not limited to, the maximum-likelihood model, other plausible models, uncertainties of specific parameters, and trade-offs between pairs of parameters.

A prior distribution $p(\mathbf{m})$ is a probability density that encodes information about model parameters $\mathbf{m}$ that is available prior to the experiment. Prior information may constrain the approximate value of a specific parameter or correlations between parameters. The classical example is a Gaussian distribution as prior with mean zero,

$$p(\mathbf{m}) \; \alpha \; exp\left(-\tfrac{1}{2}\, \mathbf{m}^T \mathbf{M}^{-1} \mathbf{m}\right), \qquad (15.1)$$

where ^{T} denotes a transpose, and $\mathbf{M}^{-1}$ is the inverse of the prior model covariance matrix. More sophisticated prior distributions may be derived from physical laws that place bounds on some parameters, from smoothness or sparsity constraints, from requirements of invariance and symmetry, or from the statistics of laboratory samples (e.g. Calvetti and Somersalo 2017; Fichtner 2021). Frequently, priors are also built on the basis of physical intuition, which adds an explicit subjective component to an inverse problem.

The probability distribution related to measurement errors, referred to as the likelihood function, is denoted by $p(\mathbf{d}|\mathbf{m})$. It describes the probability that a model $\mathbf{m}$ generates the observed data $\mathbf{d}$. The numerical value of $p(\mathbf{d}|\mathbf{m})$ can be computed by solving the forward problem $\mathbf{d}_{syn} = F(\mathbf{m})$ with $\mathbf{m}$ as input and then comparing the computed data $\mathbf{d}_{syn}$ to the observed data $\mathbf{d}$. Again, the classical example is that of Gaussian observational errors with covariance matrix $\mathbf{D}$, which induce the probability density,

$$p(\mathbf{d}|\mathbf{m}) \; \alpha \; exp\left[-\tfrac{1}{2}\, (\mathbf{d} - \mathbf{d}_{syn})^T \mathbf{D}^{-1} (\mathbf{d} - \mathbf{d}_{syn})\right]. \qquad (15.2)$$

Although this equation is Gaussian in $\mathbf{d}$ and $\mathbf{d}_{syn}$, the appearance of a potentially non-linear forward problem $F(\mathbf{m})$ may cause $p(\mathbf{d}|\mathbf{m})$ to be non-Gaussian in $\mathbf{m}$. Different observational error statistics produce different expressions for $p(\mathbf{d}|\mathbf{m})$.

Bayes' theorem allows us to combine the prior probability densities $p(\mathbf{m})$ and $p(\mathbf{d}|\mathbf{m})$ into the posterior distribution $p(\mathbf{m}|\mathbf{d})$, which captures all information one can possibly infer about $\mathbf{m}$, (Bayes, 1764):

$$p(\mathbf{m}|\mathbf{d}) = p(\mathbf{d}|\mathbf{m})\, p(\mathbf{m})/p(\mathbf{d}). \qquad (15.3)$$

The denominator $p(\mathbf{d})$, called the evidence, scales the posterior $p(\mathbf{m}|\mathbf{d})$ so that its integral over $\mathbf{m}$ equals 1. It describes the probability that the assumptions made in the numerics and parameterisation can explain the observations at all, and therefore plays an important role in trans-dimensional inference (e.g. Green, 1995; Malinverno, 2002; Sambridge et al., 2006; Bodin et al., 2012; Sambridge et al., 2013). Since the evidence does not affect the relative probability of different models, this section is limited to the proportionality of $p(\mathbf{m}|\mathbf{d})$ to the product $p(\mathbf{d}|\mathbf{m})\, p(\mathbf{m})$, which, in the case of Gaussian priors, takes the form

$$p(\mathbf{m}|\mathbf{d}) \; \alpha \; exp\left[-\tfrac{1}{2}\, (\mathbf{d} - \mathbf{d}_{syn})^T \mathbf{D}^{-1} (\mathbf{d} - \mathbf{d}_{syn})\right].$$
$$exp\left(-\tfrac{1}{2}\, \mathbf{m}^T \mathbf{M}^{-1} \mathbf{m}\right). \qquad (15.4)$$

The posterior distribution is closely linked to the misfit function $X(\mathbf{m})$ of deterministic inversions. In fact, one may find the maximum-likelihood model by minimising a misfit defined as the negative logarithm of the posterior, which, again in the Gaussian case, produces the familiar least-squares misfit

$$X(\mathbf{m}) = -\log p(\mathbf{m}|\mathbf{d}) = \tfrac{1}{2}\, (\mathbf{d} - \mathbf{d}_{syn})^T \mathbf{D}^{-1} (\mathbf{d} - \mathbf{d}_{syn})$$
$$+ \tfrac{1}{2}\, \mathbf{m}^T \mathbf{M}^{-1} \mathbf{m} + \text{const.} \qquad (15.5)$$

This equation reveals that the prior $p(\mathbf{m})$ in probabilistic inversion and the regularisation term $\tfrac{1}{2}\, \mathbf{m}^T \mathbf{M}^{-1} \mathbf{m}$ in deterministic inversion play comparable roles.

15.4.2 Hamiltonian Monte Carlo

A class of methods often considered to be the staple of Bayesian inference is Markov chain Monte Carlo (MCMC) sampling, which aims to draw a sequence of test models, called samples in this context, with a density that is proportional to the posterior distribution. This means that the number of samples falling within an arbitrary (and potentially infinitesimally small) subvolume of model space V equals $const \int_V p(\mathbf{m})d\mathbf{m}$. Hence, relative probabilities and probability densities can be quantified simply by counting samples within model space subvolumes. MCMC achieves this goal by randomly drawing samples, evaluating the distribution at that point, and deciding whether or not this is acceptable. As this decision only depends on the previous sample in the algorithm, it becomes a Markov chain, that is, a memoryless stochastic process.

MCMC theory does not prescribe how the proposal for a new sample should be made, and there are more or less efficient ways of doing this. Applying a suitable variation of the general MCMC concept to a specific problem is just another instance of the No-Free-Lunch theorem.

With this in mind, we discuss HMC (Duane et al., 1987; Neal, 2011; Betancourt, 2017) for probabilistic seismic tomography. HMC integrates the randomness of MCMC with the gradients used in deterministic optimisation. For problems of sufficient dimensionality n, traditional MCMC techniques such as Metropolis–Hastings (MH) sampling (e.g. Chib and Greenberg, 1995) quickly become too computationally expensive. This is because the amount of forward problem solutions required to obtain an independent sample grows unfavourably with dimensions, for example, with $O(n^2)$ (Neal, 2011) for the MH algorithm. Using HMC relaxes this to $O(n^{5/4})$ (Neal, 2011), making it particularly well-suited for large-scale problems, including FWI.

Hamiltonian Monte Carlo works by interpreting the misfit surface as a gravitational potential that controls the movement of a model through model space. Models with relatively low misfit correspond to potential lows. Gravitational forces, proportional to the gradient of the potential, point towards these lows. The movement of the model is determined by Hamilton's equations, the solutions of which are trajectories that roughly orbit local minima. This recipe only allows for the movement in model space giving some starting location, but it does not provide the randomness required from MCMC methods. Therefore, the movement of this particle is stopped after some time, and its momentum is randomly refreshed.

The gradients used in deterministic inversions for the optimisation of the misfit find their place in HMC in the integration of Hamilton's equations. In fact, the same gradient that guides gradient-descent methods towards the models of lower misfit, guides the HMC particle in its gravitational potential.

15.4.3 Test Case: Probabilistic FWI Using HMC

To demonstrate the strength of uncertainty quantification using HMC for tomographic problems, we show how the algorithm can be applied to a synthetic FWI. The high dimensionality of the model space paired with the computational costs of the forward problem of FWI has so far limited its applicability to low-dimensional special cases (Käufl et al., 2013; Afanasiev et al., 2014; Hunziker et al., 2019; Visser et al., 2019; Kotsi et al., 2020). For the same reasons, resolution and uncertainty analysis in FWI is still mostly based on making the assumption of a Gaussian posterior centred near a hopefully meaningful approximation

of the maximum-likelihood model (Fichtner and Trampert 2011; Bui-Thanh et al., 2013; Fichtner and van Leeuwen 2015; Liu et al., 2019a, 2019b), that is, the result of a gradient-descent type algorithm.

Gebraad et al. (2020) investigated P-SV wave propagation in a synthetic transmission-type setting, shown in Fig. 15.4. The free parameters for this example are P-wave velocity, S-wave velocity, and density. For each of these, there are values at 180×60 grid points to estimate, bringing the total number of free parameters to 32,400. The sources are randomly oriented moment tensor sources with 50 Hz dominant frequency. At this frequency, the structure of the target is partially sub-wavelength. The waveforms are recorded at the top of the medium, and the wave equation is solved using a finite-difference scheme (Virieux, 1986).

Using HMC, it was possible to reliably recover important aspects of the posterior. Shown in Fig. 15.4 are two posterior properties: the posterior means and standard deviations. Although these quantities fail to describe the non-linear aspects of the posterior fully, they do in this case provide useful first-order information, such as the approximate best model, and model resolution. For example, it can be seen

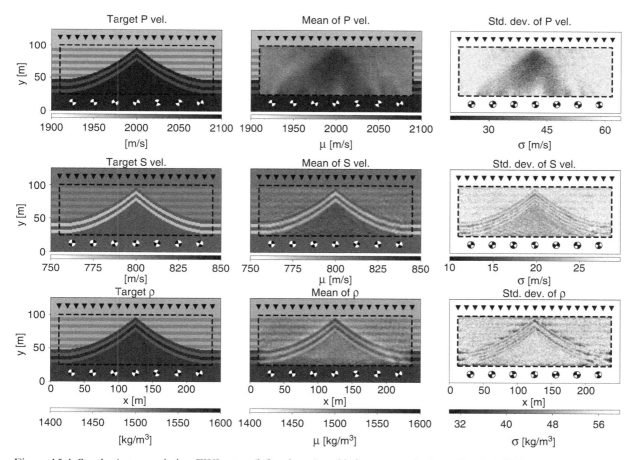

Figure 15.4 Synthetic transmission FWI setup (left column) and inference results from HMC (middle and right column). All columns contain, from top to bottom, P-wave velocity, S-wave velocity, and density. The posterior statistics are computed using 30,000 samples. Adapted from Gebraad et al. (2020).

that S-wave velocity is much better constrained than P-wave velocity. This is due to the fact that S-waves have larger amplitudes, thereby having a larger weighting in the likelihood function. Additionally, the experimental geometry also shapes the posterior. Depending on the strength of material variations, the posterior can be significantly non-Gaussian, as found in several P-wave and density marginals. These, along with additional details, can be found in Gebraad et al. (2020).

A particularly significant result from this example is the appraisal of density, which recently received new attention with the emergence of FWI (e.g. Blom et al., 2017). Known to be a notoriously ill-resolved parameter, density has typically been strongly regularised in seismic tomography. Deterministic results may therefore fail to quantify if the resulting density images are controlled by regularisation or required by the data (e.g. Kuo and Romanowcz 2002; Resovsky and Trampert, 2002). Using HMC allows us to circumvent the regularisation effect. The resulting density image is rather poor, showing only regions where density contrast is strong. However, one can assert that this is due to the data, which provides more confidence in the images and their uncertainties.

15.4.4 Variational Inference

A second approach to approximate the posterior distribution $p(\mathbf{m}|\mathbf{d})$ is variational inference. It rests on the design of a convenient analytical function, for instance, a superposition of Gaussians, that closely matches $p(\mathbf{m}|\mathbf{d})$. Since it is impossible to compute both $p(\mathbf{m}|\mathbf{d})$ and its analytical approximation over the entire model space to compare them, they are evaluated only for specific models $\mathbf{m}$. Following such an evaluation, the analytical approximation is updated to better fit $p(\mathbf{m}|\mathbf{d})$. Specifically, the aim of variational inference is to minimise the Kullback–Leibler (KL; Kullback and Leibler, 1951) divergence, which measures the discrepancy between two probability distributions , e.g. the posterior and its approximation.

The specific functional form of the analytical approximation can, in principle, be chosen freely. It controls the amount of detail that can be represented, as well as computational cost. A very basic approximation would be to fit a Gaussian to an unknown posterior $p(\mathbf{m}|\mathbf{d})$, but its ability to capture non-linear effects, including local optima, is obviously limited. The real power of variational inference lies in using more advanced functions to approximate the posterior.

Much work has been done on finding optimal approximating functions in many different settings. For seismology, and especially probabilistic tomography, readily accessible gradient information already points in the direction of variational methods that employ gradients. One such method, Stein Variational Gradient Descent (SVGD; Liu and Wang, 2016), works by transporting an ensemble

of models through model space until their distribution density matches that of the posterior distribution $p(\mathbf{m}|\mathbf{d})$. Since computations for the individual models in the ensemble are independent, parallelisation on HPC systems is straightforward.

The SVGD is given special mention, as a study by Zhang and Curtis (2020) applied the method to the same posterior as shown in Fig. 15.4. They found posterior statistics (the means and standard deviations) very similar to those found by the study from Gebraad et al. (2020), which increases confidence in both methods.

15.4.5 Discussion

In the previous paragraphs, we described two modern algorithms for uncertainty quantification in seismic inverse problems. Although both methods were applied to a synthetic FWI, their success in this computationally expensive problem with many free parameters indicates a potential for future developments towards large-scale uncertainty quantification in seismology.

Although HMC is shown to compute reliable posterior quantities, the runtimes required for the algorithm are much higher than for deterministic methods. The simulation of physics (e.g. the wave equation), can typically be parallelised on HPC systems. However, parallelising MCMC by running multiple copies of it, typically does not scale performance linearly with increasing computational power.

The SVGD lends itself better to parallelisation on modern HPC systems, and the analytical approximation of the posterior facilitates the calculation of statistical properties. These advantages are balanced by the inherent simplifications of the variational approach, which limit the ability to explore more complex posterior distributions. These simplifications include the prescribed functional shape of the distribution and the number of ensemble members used to actually compute an approximation. Hence, it is impossible to obtain arbitrarily precise posteriors by running the algorithm for more iterations. This is in contrast to MCMC methods, which often exhibit slower convergence but higher precision.

These trade-offs between methods show how both are specialised tools. The common denominator between these methods is that they exploit all readily available information as much as possible, which, in the case of many seismological problems, is misfit and gradient information. The decision between the discussed and possibly other methods, including those presented in Sambridge (this volume), needs to be carefully made, depending on the aim of a study and the inverse problem considered. Streamlining this selection process will require the development of concepts and tools for the quantitative comparison of algorithms for a usefully broad range of target problems. This topic is a focus of current research.

15.5 Concluding Remarks

The past decade of seismological research has been marked by an astonishing diversification of forward and inverse modelling methods. This trend is largely driven by the need to achieve higher resolution and lower uncertainties in tomographic images or earthquake source models. Hence, we act in accord with the No-Free-Lunch theorem, often without knowing it explicitly.

Ultimately, however, diversification is bound by the complexity of the algorithms that we can use without misusing them, and that we can maintain with realistic funding and human resources. To some extent, we can push the boundary of complexity by adopting a better programming style, where codes are properly commented, documented, continuously tested, and complemented by reproducible and educationally valuable examples.

The number of recently proposed methods may seem daunting, even for experienced scientists. Identifying the methods that ultimately move our science forward requires community-driven trial and error. Similar to natural selection, this process maps out the niches where certain approaches are fitter than others. Considering the omnipresence of exponential distributions in nature, science, and technology, it may be suspected that the vast majority of niches will be very small, whereas few others will dominate the ecosystem of seismological applications.

On average, newly proposed methods seem to become more complicated, building on increasingly sophisticated mathematical concepts. Hence, the application of a specific method and the community-driven trial and error require proper education, to which this chapter aims to contribute.

References

Afanasiev, M., Boehm, C., van Driel, M. et al. (2019). Modular and flexible spectral-element waveform modelling in two and three dimensions. *Geophysical Journal International*, 216, 1675–92.

Afanasiev, M. V., Pratt, R. G., Kamei, R., and McDowell, G. (2014). Waveform-based simulated annealing of crosshole transmission data: A semi-global method for estimating seismic anisotropy. *Geophysical Journal International*, 199, 1586–607.

Ajo-Franklin, J. B. (2009). Optimal experiment design for time-lapse traveltime tomography. *Geophysics*, 74, Q27–Q40.

Backus, G. E., and Gilbert, F. (1968). The resolving power of gross Earth data. *Geophysical Journal of the Royal Astronomical Society*, 16, 169–205.

Backus, G. E., and Gilbert, F. (1970). Uniqueness in the inversion of inaccurate gross Earth data. *Philosophical Transactions of the Royal Society A*, 266(1173), 123–92.

Bamberger, A., Chavent, G., and Lailly, P. (1977). Une application de la théorie du contrôle à un problème inverse sismique. *Annales Geophysicae*, 33, 183–200.

Bamberger, A., Chavent, G., Hemons, C., and Lailly, P. (1982). Inversion of normal incidence seismograms. *Geophysics*, 47, 757–70.

Bayes, T. (1764). An essay toward solving a problem in the doctrine of chances. *Philosophical Transactions of the Royal Society of London*, 53, 370-418.

Bernauer, M., Fichtner, A., and Igel, H. (2014). Optimal observables for multi-parameter seismic tomography. *Geophysical Journal International*, 198, 1241–54.

Betancourt, M. (2017). A conceptual introduction to Hamiltonian Monte Carlo. arXiv:1701.02434 [stat.ME].

Blom, N., Boehm, C., and Fichtner, A. (2017). Synthetic inversion for density using seismic and gravity data. *Geophysical Journal International*, 209, 1204–20.

Bodin, T., Sambridge, M., Rawlinson, N., and Arroucau, P. (2012). Transdimensional tomography with unknown data noise. *Geophysical Journal International*, 189, 1536–56.

Bozdag, E., Peter, D., Lefebvre, M. et al. (2016). Global adjoint tomography: First-generation model. *Geophysical Journal International*, 207, 1739–66.

Bui-Thanh, T., Ghattas, O., Martin, J., and Stadler, G. (2013). A computational framework for infinite-dimensional Bayesian inverse problems. Part I: The linearized case, with application to global seismic inversion. *SIAM Journal on Scientific Computing*, 35, A2494–A2523.

Calvetti, D., and Somersalo, E. (2017). Inverse problems: From regularization to Bayesian inference. *Computational Statistics*, 10(3). http://doi.org/10.1002/wics.1427.

Capdeville, Y., Guillot, L., and Marigo, J. J. (2010). 2-D non-periodic homogenization to upscale elastic media for P–SV waves. *Geophysical Journal International*, 182, 903–22.

Capdeville, Y., Gung, Y., and Romanowicz, B. (2005). Towards global Earth tomography using the spectral element method: A technique based on source stacking. *Geophysical Journal International*, 162, 541–54.

Chen, P., Jordan, T. H., and Zhao, L. (2007a). Full 3D waveform tomography: A comparison between the scattering-integral and adjoint-wavefield methods. *Geophysical Journal International*, 170, 175–81.

Chen, P., Zhao, L., and Jordan, T. H. (2007b). Full 3D tomography for the crustal structure of the Los Angeles region. *Bulletin of the Seismological Society of America*, 97, 1094–120.

Chib, S., and Greenberg, E. (1995). Understanding the Metropolis-Hastings algorithm. *American Statistician*, 49, 327–35.

Choi, Y., and Alkhalifah, T. (2011). Source-independent time-domain waveform inversion using convolved wavefields: Application to the encoded multisource waveform inversion. *Geophysics*, 76, R125–R134.

Clouzet, P., Masson, Y., and Romanowicz, B. (2018). Box tomography: First application to the imaging of upper mantle shear velocity and radial anisotropy structure beneath the north American continent. *Geophysical Journal International*, 213, 1849–75.

Curtis, A. (1999). Optimal experiment design: Cross-borehole tomographic examples. *Geophysical Journal International*, 136, 637–50.

de la Puente, J., Dumbser, M., Käser, M., and Igel, H. (2007). An arbitrary high-order discontinuous Galerkin method for elastic waves on unstructured methods. IV. Anisotropy. *Geophysical Journal International*, 169, 1210–28.

Djikpesse, H. A., Khodja, M. R., Prange, M. D., Duchenne, S., and Menkiti, H. (2012). Bayesian survey design to optimize resolution in waveform inversion. *Geophysics*, 77, R81–R93.

Duane, S., Kennedy, A. D., Pendleton, B. J., and Roweth, D. (1987). Hybrid Monte Carlo. *Physics Letters B*, 195, 216–22.

Dziewonski, A. M., and Anderson, D. L. (1981). Preliminary reference Earth model. *Physics of the Earth and Planetary Interiors*, 25, 297–356.

Fichtner, A., Bunge, H.-P., and Igel, H. (2006). The adjoint method in seismology – I. Theory. *Physics of the Earth and Planetary Interiors*, 157, 105–23.

Fichtner, A., Kennett, B. L. N., Igel, H., and Bunge, H.-P. (2009). Full seismic waveform tomography for upper-mantle structure in the Australasian region using adjoint methods. *Geophysical Journal International*, 179, 1703–25.

Fichtner, A. (2010). *Full seismic waveform modeling and inversion*. Heidelberg: Springer.

Fichtner, A., and Trampert, J. (2011). Resolution analysis in full waveform inversion. *Geophysical Journal International*, 187, 1604–24.

Fichtner, A. (2021). *Lecture Notes on Inverse Theory*. Cambridge: Cambridge Open Engage, http://doi.org/10.33774/coe-2021-qpq2j.

Fichtner, A., van Herwaarden, D.-P., Afanasiev, M. et al. (2018). The Collaborative Seismic Earth Model: Generation I. *Geophysical Research Letters*, 45, 4007–16.

Fichtner, A., and van Leeuwen, T. (2015). Resolution analysis by random probing. *Journal of Geophysical Research: Solid Earth*, 120, 5549–73.

Gebraad, L., Boehm, C., and Fichtner, A. (2020). Bayesian elastic full-waveform inversion using Hamiltonian Monte Carlo. *Journal of Geophysical Research: Solid Earth*, 125, e2019JB018428.

Green, P. J. (1995). Reversible jump Markov chain Monte Carlo computation and Bayesian model determination. *Biometrika*, 82, 711–32.

Hardt, M., and Scherbaum, F. (1994). The design of optimum networks for aftershock recordings. *Geophysical Journal International*, 117, 716–26.

Huang, Y., and Schuster, G. T. (2018). Full-waveform inversion with multisource frequency selection of marine streamer data. *Geophysical Prospecting*, 66, 1243–57.

Hunziker, J., Laloy, E., and Linde, N. (2019). Bayesian full-waveform tomography with application to crosshole ground penetrating radar data. *Geophysical Journal International*, 218, 913–31.

Igel, H. (2016). *Computational Seismology: A Practical Introduction*. Cambridge: Cambridge University Press.

Kijko, A. (1977). An algorithm for the optimum distribution of a regional seismic network – I. *Pure and Applied Geophysics*, 115, 999–1009.

Komatitsch, D., and Vilotte, J.-P. (1998). The spectral element method: An effective tool to simulate the seismic response of 2D and 3D geological structures. *Bulletin of the Seismological Society of America*, 88, 368–92.

Käufl, P., Fichtner, A., and Igel, H. (2013). Probabilistic full waveform inversion based on tectonic regionalisation:

Development and application to the Australian upper mantle. *Geophysical Journal International*, 193, 437–51.

Kotsi, P., Malcolm, A., and Ely, G. (2020). Time-lapse full-waveform inversion using Hamiltonian Monte Carlo: A proof of concept. *SEG Technical Program Expanded Abstracts*, 845–49.

Krampe, V., Edme, P., and Maurer, H. (2021). Optimized experimental design for seismic full waveform inversion: A computationally efficient method including a flexible implementation of acquisition costs. *Geophysical Prospecting*, 69, 152–66.

Krebs, J. R., Anderson, J. E., Hinkley, D. et al. (2009). Fast full-wavefield seismic inversion using encoded sources. *Geophysics*, 74, WCC177–WCC188.

Krebs, J. R., Cha, Y. H., Lee, S. et al. ExxonMobil Upstream Research Co (2018). *Orthogonal Source and Receiver Encoding*. U.S. Patent 10,012,745.

Kullback, S., and Leibler, R. A. (1951). On information and sufficiency. *Annals of Mathematical Statistics*, 22, 79–86.

Kuo, C., and Romanowicz, B. (2002). On the resolution of density anomalies in the Earth's mantle using spectral fitting of normal mode data. *Geophysical Journal International*, 150, 162–79.

Leng, K., Nissen-Meyer, T., and van Driel, M. (2016). Efficient global wave propagation adapted to 3-D structural complexity: A pseudospectral/spectral-element approach. *Geophysical Journal International*, 207, 1700–21.

Liu, Q., and Gu, Y. (2012). Seismic imaging: From classical to adjoint tomography. *Tectonophysics*, 566–567, 31–66.

Liu, Q., and Peter, D. (2019). Square-root variable metric based elastic full-waveform inversion. Part 2: Uncertainty estimation. *Geophysical Journal International*, 218(2), 1100–20.

Liu, Q., and Wang, D. (2016). Stein variational gradient descent: A general purpose Bayesian inference algorithm. *Advances in Neural Information Processing Systems*, 2378–86. arXiv:1608.04471.

Liu, Q., Peter, D., and Tape, C. (2019). Square-root variable metric based elastic full-waveform inversion. Part 1: Theory and validation. *Geophysical Journal International*, 218(2), 1121–35.

Masson, Y., and Romanowicz, B. (2017). Fast computation of synthetic seismograms within a medium containing remote localized perturbations: A numerical solution to the scattering problem. *Geophysical Journal International*, 218, 674–92.

Maurer, H., Nuber, A., Martiartu, N. K. et al. (2017). Optimized experimental design in the context of seismic full waveform inversion and seismic waveform imaging. In L. Nielsen, ed., *Advances in Geophysics*, vol. 58. Cambridge, MA: Academic Press, pp. 1–45.

Malinverno, A. (2002). Parsimonious Bayesian Markov chain Monte Carlo inversion in a nonlinear geophysical problem. *Geophysical Journal International*, 151, 675–88.

Moczo, P., Kristek, J., Vavrycuk, V., Archuleta, R., and Halada, L. (2002). 3D heterogeneous staggered-grid finite-difference modeling of seismic motion with volume harmonic and arithmetic averaging of elastic moduli. *Bulletin of the Seismological Society of America*, 92, 3042–66.

Mosegaard, K. (2012). Limits to nonlinear inversion. In K. Jónasson, ed., *Applied Parallel and Scientific Computing*, Berlin: Springer, pp. 11–21.

Neal, R. M. (2011). MCMC using Hamiltonian dynamics. In S. Brooks, A. Gelman, G. Jones, and X.-L. Meng, eds., *Handbook of Markov chain Monte Carlo*. New York: Chapman and Hall, pp. 113–62.

Plessix, R.-E. (2006). A review of the adjoint-state method for computing the gradient of a functional with geophysical applications. *Geophysical Journal International*, 167, 495–503.

Rabinowitz, N., and Steinberg, D. M. (1990). Optimal configuration of a seismographic network: A statistical approach. *Bulletin of the Seismological Society of America*, 80, 187–96.

Resovsky, J., and Trampert, J. (2002). Reliable mantle density error bars: An application of the Neighbourhood Algorithm to normal-mode and surface wave data. *Geophysical Journal International*, 150, 665–72.

Romero, L. A., Ghiglia, D. C., Ober, C. C., and Morton, S. A. (2000). Phase encoding of shot records in prestack migration. *Geophysics*, 65, 426–36.

Ronchi, C., Iacono, R., and Paolucci, P. S. (1996). The 'cubed sphere': A new method for the solution of partial differential equations in spherical geometry. *Journal of Computational Physics*, 124, 93–114.

Sambridge, M. S., Bodin, T., Gallagher, K., and Tkalcic, H. (2013). Transdimensional inference in the geosciences. *Philosophical Transactions of the Royal Society A*, 371, 20110547.

Sambridge, M. S., Gallagher, K., Jackson, A., and Rickwood, P. (2006). Trans-dimensional inverse problems, model comparison, and the evidence. *Geophysical Journal International*, 167, 528–42.

Schiemenz, A., and Igel, H. (2013). Accelerated 3-D full-waveform inversion using simultaneously encoded sources in the time domain: Application to Valhall ocean-bottom cable data. *Geophysical Journal International*, 195, 1970–88.

Sieminski, A., Trampert, J., and Tromp, J. (2009). Principal component analysis of anisotropic finite-frequency kernels. *Geophysical Journal International*, 179, 1186-98.

Sirgue, L., and Pratt, R. G. (2004). Efficient waveform inversion and imaging: A strategy for selecting temporal frequencies. *Geophysics*, 69, 231–48.

Sirgue, L., Barkved, O. I., Dellinger, J. et al, (2010). Full-waveform inversion: The next leap forward in imaging at Valhall. *First Break*, 28, 65–70.

Tape, C., Liu, Q., Maggi, A., and Tromp, J. (2010). Seismic tomography of the southern California crust based upon spectral-element and adjoint methods. *Geophysical Journal International*, 180, 433–62.

Tarantola, A. (1988). Theoretical background for the inversion of seismic waveforms, including elasticity and attenuation. *Pure and Applied Geophysics*, 128, 365–99.

Tarantola, A. (2005). *Inverse Problem Theory and Methods for Model Parameter Estimation*, 2nd ed. Philadelphia, PA: Society for Industrial and Applied Mathematics.

Thrastarson, S., van Driel, M., Krischer, L. et al. (2020). Accelerating numerical wave propagation by wavefield adapted meshes. Part II: full-waveform inversion. *Geophysical Journal International*, 221, 1591–604.

Thrastarson, S., van Herwaarden, D.-P. and Fichtner, A. (2021). Inversionson: Fully Automated Seismic Waveform Inversions. *EarthArXiv*, http://doi.org/10.31223/X5F31V.

Thrastarson, S., van Herwaarden, D.-P., Krischer, L. et al. (2022). Data-adaptive global full-waveform inversion. *Geophysical Journal International*, 230, 1374–93, https://doi.org/10.1093/gji/ggac122.

Tromp, J., Tape, C., and Liu, Q. (2005). Seismic tomography, adjoint methods, time reversal and banana-doughnut kernels. *Geophysical Journal International*, 160, 195–216.

Tromp, J., and Bachmann, E. (2019). Source encoding for adjoint tomography. *Geophysical Journal International*, 218, 2019–44.

van Driel, M., Boehm, C., Krischer, L., and Afanasiev, M. (2020). Accelerating numerical wave propagation using wavefield adapted meshes. Part I: forward and adjoint modelling. *Geophysical Journal International*, 221, 1580–90.

van Herwaarden, D.-P., Boehm, C., Afanasiev, M. et al. (2020). Accelerated full-waveform inversion using dynamic mini-batches. *Geophysical Journal International*, 221, 1427–38.

van Herwaarden, D.-P., Afanasiev, M., Thrastarson, S., and Fichtner, A. (2021). Evolutionary full-waveform inversion. *Geophysical Journal International*, 224, 306–11.

van Leeuwen, T., and Herrmann, F. J. (2013). Fast waveform inversion without source-encoding. *Geophysical Prospecting*, 61, 10–19.

Virieux, J. (1986). P-SV wave propagation in heterogeneous media: Velocity-stress finite difference method. *Geophysics*, 51, 889–901.

Virieux, J., and Operto, S. (2009). An overview of full waveform inversion in exploration geophysics. *Geophysics*, 74, WCC127–WCC152.

Visser, G., Guo, P., and Saygin, E. (2019). Bayesian transdimensional seismic full-waveform inversion with a dipping layer parameterization. *Geophysical Journal International*, 84(6), R845–R858.

Woodhouse, J. H., and A. M. Dziewonski (1984). Mapping the upper mantle: Three-dimensional modeling of Earth structure by inversion of seismic waveforms. *Journal of Geophysical Research: Solid Earth*, 89, 5953–86.

Wolpert, D. H., and Macready, W. G. (1997). No Free Lunch theorems for optimization. *IEEE Transactions on Evolutionary Computation*, 1, 67–88.

Zhang, Q., Mao, W., Zhou, H., Zhang, H., and Chen, Y. (2018). Hybrid-domain simultaneous-source full waveform inversion without crosstalk noise. *Geophysical Journal International*, 215, 1659–81.

Zhang, X., and Curtis, A. (2020). Variational full-waveform inversion. *Geophysical Journal International*, 222, 406–11.

16

Joint and Constrained Inversion as Hypothesis Testing Tools

Max Moorkamp

Abstract: In this chapter, I discuss an alternative perspective on interpreting the results of joint and constrained inversions of geophysical data. Typically such inversions are performed based on inductive reasoning (i.e. we fit a limited set of observations and conclude that the resulting model is representative of the Earth). While this has seen many successes, it is less useful when, for example, the specified relationship between different physical parameters is violated in parts of the inversion domain. I argue that in these cases a hypothesis testing perspective can help to learn more about the properties of the Earth. I present joint and constrained inversion examples that show how we can use violations of the assumptions specified in the inversion to study the subsurface. In particular I focus on the combination of gravity and magnetic data with seismic constraints in the western United States. There I see that high velocity structures in the crust are associated with relatively low density anomalies, a possible indication of the presence of melt in a strong rock matrix. The concepts, however, can be applied to other types of data and other regions and offer an extra dimension of analysis to interpret the results of geophysical inversion algorithms.

16.1 Introduction

Within the Earth Sciences and, more specifically, solid-Earth geophysics the term joint inversion typically describes approaches where datasets obtained through different measurement techniques are inverted within a single inversion algorithm. This can be data sensitive to one physical parameter (e.g. seismic shear wave velocity), but sensing the Earth in fundamentally different ways (e.g. receiver functions and surface wave dispersion measurements; Julia et al., 2000), or datasets sensitive to different physical parameters (e.g. seismic velocity and electrical resistivity; Gallardo and Meju, 2003). In joint inversion, all data under consideration are input into the algorithm and the combined parameter model is adjusted until all data are fit to a satisfactory level (e.g. Lines et al., 1986). In contrast, constrained inversion in the context of this discussion relates to methods that utilise a fixed model as a constraint in the inversion of a single

dataset (e.g. Mackie et al., 2020; Harmon et al., 2021; Martin et al., 2021). Other types of constrained inversion exist, for example, those based on fitting physical properties to *a priori* defined clusters (e.g. Carter-McAuslan et al., 2015), possibly with some updates to the cluster centres (e.g. Sun and Li, 2015b; Astic and Oldenburg, 2019) or specifying different parameter ranges in different parts of the model (e.g. Darijani et al., 2021). I focus on model-constrained inversions, i.e. where spatial similarity to a known model is enforced (e.g. Zhou et al., 2014; Franz et al., 2021), as they are very similar to joint inversion methods. However, many of the concepts described here can extended to other types of joint and constrained inversion.

There have been a variety of reviews of joint inversion approaches in recent years (Gallardo and Meju, 2011; Haber and Holtzman Gazit, 2013; Moorkamp et al., 2016b; Moorkamp, 2017; Spichak, 2020) which describe both the theory behind joint inversion and give examples of applications for Earth imaging. I will therefore keep the description relatively brief and focus on issues relevant to using inversion as a hypothesis testing tool. Similar ideas have been expressed previously (Bosch and McGaughey, 2001; Sun and Li, 2015a; Kamm et al., 2015), but, in my view, have not received the attention they deserve. I will also mention new developments that have arisen in the last couple of years.

16.2 Basic Principles of Joint and Constrained Inversion

In order to simplify the mathematical notation, I will discuss the joint inversion and model constrained inversion in the context of gravity and surface wave dispersion data which I also use in some of the examples here. The principles extend to any other coupled inversion with two methods. More general mathematical treatments and extensions to more than two methods are given in Moorkamp et al. (2011) and Bosch (2016), for example. Considering measurements of gravitational acceleration $\mathbf{g}_z$ and Rayleigh wave dispersion $\mathbf{d}_r$ we have the corresponding physical parameters density ρ and shear wave velocity $\mathbf{v}_s$. Rayleigh wave dispersion also has a minor sensitivity to the density structure of the Earth. However, in practice, this sensitivity is often ignored (e.g. Fishwick, 2010; Weise, 2021) and so I will do the same here. To solve the joint inverse problem for these two methods we have to define an objective function Φ, viz.

$$\Phi(\rho,\mathbf{v}_s) = \Phi_{grav,d}(\rho) + \Phi_{dis,d}(\mathbf{v}_s) + \lambda_{grav}\Phi_{grav,r}(\rho)$$
$$+ \lambda_{dis}\Phi_{dis,r}(\mathbf{v}_s) + \nu\Phi_{coupling}(\rho,\mathbf{v}_s).$$

Here $\Phi_{grav,d}(\rho)$ is the data misfit term for the gravity data, $\Phi_{dis,d}(\mathbf{v}_s)$ the misfit for the dispersion measurements, $\Phi_{grav,r}(\rho)$ the regularisation for density, $\Phi_{dis,r}(\mathbf{v}_s)$ the regularisation for shear wave velocity, and $\Phi_{coupling}(\rho,\mathbf{v}_s)$ the coupling between density and shear wave velocity; λ_{grav} and λ_{dis} are the Lagrange parameters for the regularisation and ν the weight for the coupling term.

Note that only the coupling term contains both physical parameters, all other terms only depend on one of the two parameters. Thus the coupling term is crucial for the joint inversion problem and I will give some examples of popular approaches. Using the same notation, the objective function for a density inversion constrained by a seismic velocity model can be written as

$$\Phi_{\mathbf{v}_s}(\rho) = \Phi_{grav,d}(\rho) + \lambda_{grav}\Phi_{grav,r}(\rho) + \nu\Phi_{coupling}(\rho,\mathbf{v}_s).$$

Here the notation $\Phi_{\mathbf{v}_s}$ reflects the fact that the velocity model is provided by the user and needed for the coupling calculation, but does not change during the inversion. The example of the model-constrained inversion particularly highlights the importance of the coupling term, as it is the only thing that distinguishes it from a regular inversion of a single dataset. The goal of the inversion algorithm is then to find a model that fits the observed data and has a small coupling constraint, corresponding to high similarity or correlation between the models. The regularisation largely serves to stabilise the inversion and model roughness considerations are often secondary for joint inversion (e.g. Moorkamp et al., 2016a).

To couple different parameters, a variety of methods have been proposed (e.g. Haber and Oldenburg, 1997; Gallardo and Meju, 2003; Haber and Holtzman Gazit, 2013; Zhdanov et al., 2012; Sun and Li, 2016; Heincke et al., 2017). I will focus on three methods: (i) direct functional parameter relationships, (ii) cross-gradient coupling, and (iii) variation of information coupling. A more extensive discussion of different coupling methods can be found in Meju and Gallardo (2016) and Colombo and Rovetta (2018).

Direct functional relationships are conceptually and mathematically among the simplest ways to couple two different physical parameters. One defines a function $\rho = f(\mathbf{v}_s)$ and assumes that it describes the connection

between the different parameters in the inversion domain. Such a relationship can be linear (e.g. Tiberi et al., 2003; Moorkamp et al., 2011) or more complex (e.g. Maceira and Ammon, 2009; Panzner et al., 2016; Heincke et al., 2017), and in some cases the relationship is allowed to vary (e.g. O'Donnell et al., 2011). Direct parameter relationships establish a strong link between the physical parameters and thus a strong coupling between the methods (Moorkamp, 2017). However, they can be difficult to estimate from independent information and it is often questionable to which degree an estimated relationship is representative of the region of interest. Even where high-quality borehole-logs exist and the geology can be assumed to be relatively homogeneous, e.g. in sub-basalt hydrocarbon exploration, different researchers can estimate different relationships with significant impact on the final results (Panzner et al., 2016; Heincke et al., 2017; Moorkamp, 2017). Therefore direct parameter relationships are not very common in joint inversion with the exception of seismic-gravity inversions (e.g. Tiberi et al., 2003; Maceira and Ammon, 2009; O'Donnell et al., 2011; Blom et al., 2017; Zhao et al., 2020), where a linear relationship is often deemed appropriate (Birch, 1961; Gardner et al., 1974; Barton, 1986).

The most commonly applied coupling constraint in joint inversion, the cross-gradient constraint (Gallardo and Meju, 2003), side-steps issues with unknown direct relationships by focusing on structural similarity (Meju and Gallardo, 2016). The underlying assumption is that there should be a correlation between parameter changes if all methods sense the same geological structures. Compared to previous structural coupling constraints (Haber and Oldenburg, 1997), the cross-gradient even permits cases where one parameter changes and the other stays constant. These very loose assumptions suggest that cross-gradient coupling should be appropriate in the majority of geological scenarios and has led to its general success (e.g. Gallardo and Meju, 2007; Lelièvre et al., 2016; Linde and Doetsch, 2016; Bennington et al., 2015; Shi et al., 2017; Gross, 2019). Furthermore, we can use the results from a cross-gradient–based joint inversion to estimate parameter relationships (e.g. Linde et al., 2006; Moorkamp et al., 2013) and further analyse these relationships for geological classification (e.g. Sun et al., 2020; Li and Sun, 2022) or estimate petrophysical properties (e.g. Meju et al., 2018). This data-driven analysis can complement approaches that use petrophysical relationships directly in the joint inversion (e.g. Wagner et al., 2019; Afonso et al., 2019; Manassero et al., 2021). However, there are some situations where cross-gradient coupling is too weak to result in any meaningful changes in the inversion output (Franz et al., 2021; Weise, 2021), and this appears to be particularly the case when two methods with moderate resolution are combined.

Recently I presented a new coupling constraint based on the concept of variation of information (VI) (Moorkamp,

2021, 2022). Variation of information is closely related to mutual information, an information theoretical quantity used in machine learning and medical imaging (e.g. Pluim et al., 2003). Using mutual information for geophysical joint inversion was first proposed by Haber and Holtzman Gazit (2013), but they reported severe issues with convergence and did not show any practical results. Later Mandolesi and Jones (2014) used mutual information for 1D constrained inversion of MT data and seismic models. Until recently no successful examples of a full joint inversion based on VI or mutual information have been shown. The core idea of using VI for joint inversion is that it measures the amount of information that is lost when transforming from one variable to the other (Meilă, 2003). In other words, how much does knowing variable a (e.g. seismic velocity), tell us about variable b (e.g. density). Variation of information decreases the smaller the amount of information loss and thus a joint inversion approach that minimises VI as part of the objective function seeks a one-to-one correspondence between the physical parameters without specifying a preferred form (e.g. clusters, functional), between the quantities. Thus, VI might be a reasonable middle ground between the strong assumption of a specified relationship and the loose coupling provided by the cross-gradient.

16.3 Joint Inversion and Hypothesis Testing

The common idea behind joint inversion expressed in most publications is that we want to obtain models that are better in some loosely defined sense (e.g. Moorkamp et al., 2013; Giraud et al., 2017; Astic et al., 2021; Ghalenoei et al., 2021; Moorkamp, 2021; Tu and Zhdanov, 2021). More importantly though, a model that is constructed from a variety of observations is likely to be a better representation of the Earth. However, this view point also leads to a variety of issues that frequently arise when judging joint inversion results and are controversially discussed in the joint inversion community:

1. Does it make sense to combine methods with different resolution capabilities (e.g. full-waveform seismic and electromagnetic data) or are we just transferring structures from the high-resolution method to the low-resolution method?
2. To which degree do the regions of sensitivity have to overlap, particularly when combining methods sensitive to different parameters? Is it enough when there is some overlap in sensitivity or do they have to largely agree?
3. Which coupling approach is appropriate?
4. What does it mean when a joint inversion fails, that is, we cannot find a model that fits all observations satisfactorily and has a low coupling constraint?

5. How do we reconcile different results produced by joint inversions with different types of coupling?

These are valid and important questions when trying to understand the value of a joint inversion result for interpretation. It is also related to the fact that we often use inversions in an explorative manner, meaning, we do not have a pre-conceived idea, but try to image unknown structures in the Earth. This exploration is often based on inductive reasoning (Elsasser, 1950; Fullagar and Oldenburg, 1984; Hong and Sen, 2009). Based on the misfit at a limited number of measurement locations, we conclude how representative the Earth model is in the whole region of investigation. In addition, we often perform selected synthetic tests and infer from the success of these tests that the inversion algorithm also produces reasonable Earth models with the real observations. However, when using inversion for hypothesis testing many of the aforementioned issues can be resolved.

The first thing to realise is that joint inversions operate largely in the null-space of the individual methods. Early hopes that, due to the additional data and different sensitivities, joint inversion leads to better-behaved inverse problems and thus potentially to solutions that match the observations better than individual inversions have not been realised. In fact, many current joint inversion approaches need more iterations and are more difficult to lead to converge than individual inversions (Moorkamp et al., 2011, 2016a; Heincke et al., 2017). Thus the typical goal for a joint inversion is to reach a comparable misfit to the individual inversion results for each dataset under consideration (e.g. Kamm et al., 2015; Paulatto et al., 2019). Thus when comparing the models from joint inversion and individual inversion from the perspective of a single method, we have two different models with the same data misfit and thus by definition the models are connected through the model null-space. Thus when looking at the joint inversion results for each of the methods, the coupling in the joint inversion is akin to regularisation. Instead of seeking only a smooth model, the inversion seeks a model that is similar to the current model from the other methods under the coupling constraint.

I illustrate this idea in Fig. 16.1. Within the nullspace for each method we have the individual inversion results ρ_{ind} and v_{ind} with a specific regularisation term and two sets of joint inversion results for two different types of coupling. Each of those joint inversion results has its own null-space under the respective coupling constraint. This null-space is a sub-space of the null-space of each method individually and might or might not overlap with the null-space from the other joint inversion results. When the joint inversion literature discusses improved resolution or decreased ambiguity of the results, this is what is typically meant: Under a given coupling constraint the null-space is a sub-region of the nulls-space for each method individually. However, two

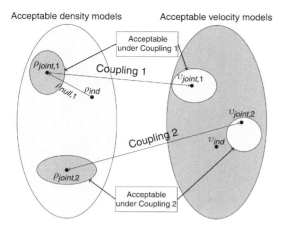

Figure 16.1 Illustration of the relationship between the results of individual inversions (ρ_{ind}, v_{ind}) and the results of two joint inversions with different coupling approaches, ($\rho_{joint,1}$, $v_{joint,1}$) and ($\rho_{joint,2}$, $v_{joint,2}$), within the null-space of each method.

different coupling methods will result in different results that are typically not reconcilable in the sense that the null-spaces do not overlap. Thus when comparing the results from two different joint inversions, we might end up with differing results that fit all observations to the same level and where the models are strongly linked to the respective coupling constraint. Accordingly, it becomes difficult to decide which of these results should be preferred and used for further interpretation.

In my view, the situation becomes conceptually easier and intellectually better to handle when we switch from the explorative application of inverse methods ('show me what is there?') to a hypothesis testing viewpoint. If we use inversion with a single method for hypothesis testing, the hypothesis to be tested is: 'No distribution of physical properties with the chosen discretisation and smoothness level exists that can produce data that fit the observations.' If the inversion finds a model that fits the data to a satisfactory level we can refute this hypothesis. For joint inversion, the hypothesis to be tested is modified to: 'No distributions of physical properties with the chosen discretisation and smoothness level that also adhere to the coupling constraints exists that can produce data that fit all the observations.'

For the inversion of an individual dataset, the hypothesis testing approach might appear relatively dull and not particularly interesting. After all, we know that, for example, a density distribution must exist that produced the gravity observations. So the hypothesis test is mainly about technical details such as discretisation and level of smoothness. Once we have found appropriate values for those and other parameters for the inversion algorithm, we should be able to find a model. It is different for the joint inversion problem though. If we perform individual inversions first, we have demonstrated that we can fit the observations with some parameter distribution for each method separately. Thus the hypothesis test is now about

the appropriateness of the coupling approach. This leaves us with two interesting possible outcomes for our joint inversion experiment: (i) We can fit the data to the same degree as the individual inversions and the coupling criterion is adhered to throughout the modelling domain. We have therefore demonstrated that this coupling constraint is a potential candidate to connect those two physical properties and can analyse the connection to learn more about the Earth. (ii) We cannot fit the observations to the same level or have regions of the model that violate the coupling constraint. Thus the coupling constraint is not a feasible candidate or it is only applicable in parts of the inverted region. I find the last scenario (i.e. a result that suggests that the relationship is applicable in large parts of the model but not everywhere), particularly interesting. Of course, a large number of coupling constraints exist that are obviously nonsensical. For example, we could force electrical resistivity and seismic velocity in each cell to have identical numerical values. The fact that the used coupling works in parts of the model suggests that it is a reasonable assumption. However, the existence of a region that does not conform with this assumption indicates that something unexpected is happening there. To say it in the famous quote ascribed to Isaac Asimov: 'The most exciting phrase to hear in science, the one that heralds new discoveries, is not "Eureka!" but "That's funny."' Furthermore, the existence of this region must be mandated by the data, as the coupling constraint works to reduce the deviation from the assumptions. In particular, all data need to have significant sensitivity to the properties of this structure as otherwise the minimisation of the objective function would produce a model that does not include the constraint violation.

To my knowledge, the first example of using joint inversion in this way was presented in Moorkamp et al. (2007) and, in more detail, in Moorkamp (2007). There I analyse magnetotelluric and receiver function data from the Slave Craton, Canada with a 1D joint inversion approach. Individual inversions of both datasets indicate a conductive structure that approximately coincides with the depth of the Moho based on the receiver functions. Given the uncertainty in each dataset, it is conceivable that the top of this conductor is located just above the Moho in the lowermost crust or just below the Moho (i.e. in the uppermost mantle). I therefore designed two joint inversion experiments based on coincident layer interfaces: one where the conductor is forced to be located below the Moho and one where the conductor is allowed to start above the Moho. The results of these experiments are shown in Fig. 16.2. When the conductor is forced to be located just below the Moho, the joint inversion cannot fit both datasets simultaneously as indicated by the strong trade-off in fitting the two datasets (red squares in Fig. 16.2a). When the receiver function data are fit, the conductor is too deep to explain the MT measurements. Conversely, when the MT data are

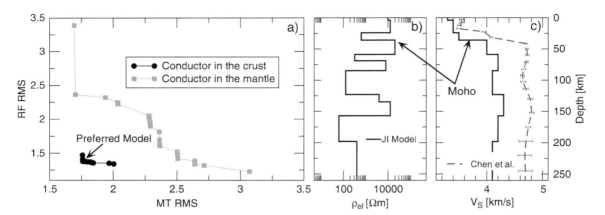

Figure 16.2 Joint inversion results for magnetotelluric and receiver function data for station BOXN located on the Slave Craton, Canada (redrawn after Moorkamp et al. (2007)). (a) Trade-off between fitting the MT and receiver function observations for different assumptions about the shallow conductor. (b) Preferred electrical resistivity model. (c) Preferred velocity model together with the surface wave dispersion based model of Chen et al. (2007).

fit, the Moho is shallower than permitted by the receiver functions. As soon as the restriction of the conductor to the mantle is lifted, the issue of not fitting all data vanishes (black circles in Fig. 16.2a) and a model can be found that fits all observations. The combined experiments clearly demonstrate that, under the assumption of a layered Earth, the conductive material is located in the crust and not in the mantle.

This first example demonstrates the power of negative or unsuccessful joint inversion results when seen through the perspective of hypothesis testing. It might become apparent at this point that I am using the term hypothesis testing in a qualitative sense, even though it is a well-developed field of statistics with clear procedures. While this is feasible in the context of assessing error-weighted data misfits under the assumption of Gaussian errors, the statistical distributions for coupling constraints such as the cross-gradient or variation of information are currently not known. Thus a rigorous statistical significance test is currently not possible and the assessment of what constitutes a significant deviation or a violation of the hypothesis is left to the researcher. Despite these difficulties the concept is useful as the first example illustrates.

16.4 Examples from the Western United States

To demonstrate these concepts further, I use data from the western United States around the Yellowstone Hotspot and the Snake River plain (SRP; Fig. 16.3). This is a region of high interest due to the location of the hotspot, the adjacent flood basalts, and the transition from active tectonic extension in the Basin and Range province in the south to old cratonic structures of the Wyoming Craton in the north (e.g. Kelbert et al., 2012; Meqbel et al., 2014). Thanks to the

USArray program, the region is well covered with MT stations (de Groot-Hedlin et al., 2003–2004; Schultz et al., 2006–2018) and passive seismic measurements (IRIS, 2003). As a result, a variety of velocity and resistivity models have been created to investigate the structure of the lithosphere (e.g. Schmandt and Humphreys, 2011; Kelbert et al., 2012; Bedrosian and Feucht, 2014; Meqbel et al., 2014; Liu and Gao, 2018; Gao and Shen, 2014). In addition, the European Space Agency's GRACE and GOCE missions provide global coverage of gravity data (e.g. Pail et al., 2018; Zingerle et al., 2020) and magnetic measurements are available through the North America magnetic database (Bankey et al., 2002). Thus it is an ideal setting to demonstrate the capabilities of joint inversion and the idea of hypothesis testing.

The study region is shown as a blue rectangle in Fig. 16.3. For all inversion experiments, I discretise the Earth into $64 \times 64 \times 30$ rectilinear cells with horizontal dimensions of 10 km $\times$ 10 km. In vertical direction, the cell size increases from 1 km at the surface to 16 km at depth and the model covers the upper 160 km of the Earth. I extract Bouguer-corrected gravity data with a spacing of 0.1 degree from the XGM2019 global gravity model (Zingerle et al., 2020) resulting in 3,350 data points. Magnetic data with identical spacing are taken from the North America magnetic database (Bankey et al., 2002). To reduce the influence of small-scale near-surface features, I upward continue the full datasets to a height of 10,000 m before filtering and downsampling. As a seismic reference model, I use the shear wave velocity model of Gao and Shen (2014). It is based on full-waveform ambient-noise tomography and covers the upper 200 km of the lithosphere. Its resolution characteristics are comparable with the lateral resolution that can be expected from potential field inversions and thus is a suitable candidate for a constrained inversion.

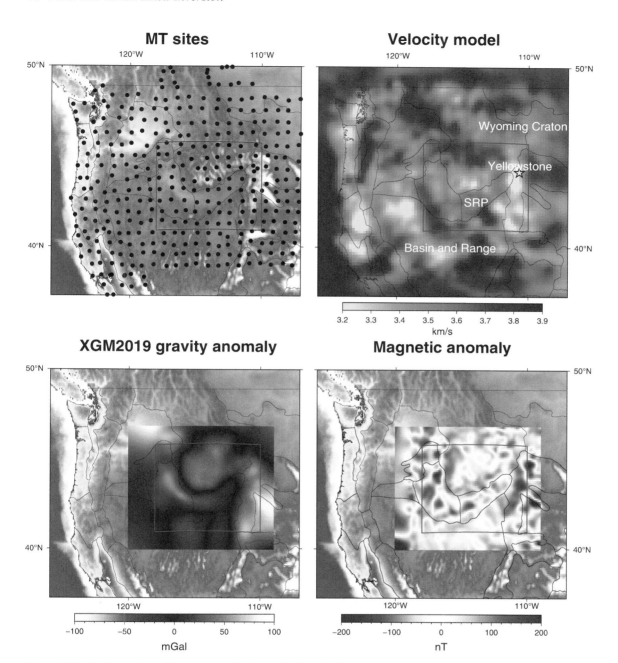

MT sites

Velocity model

XGM2019 gravity anomaly

Magnetic anomaly

Figure 16.3 Study region with magnetotelluric (MT) site distribution from the USArray MT initiative (top left) (de Groot-Hedlin et al., 2003–2004, Kelbert et al., 2011). I also plot a horizontal slice through the reference velocity model of Gao and Shen (2014) at a depth of 15 krn (top right), the Bouguer gravity anomaly from XGM2019 (bottom left), and the magnetic anomaly from the North America magnetic database (Bankey et al., 2002) (bottom right). The blue lines mark the extent of the inversion domain.

As a first example, I perform a constrained inversion of total field magnetic data and the seismic model coupled by a VI constraint. Strictly speaking, I am testing two hypotheses simultaneously: (a) The velocity model of Gao and Shen (2014) is a reasonable representation of crustal structure. (b) There is a one-to-one correspondence between seismic velocity in this model and magnetic susceptibility. Given that the seismic velocity model has been

widely used in other studies, for the purpose of this experiment I will assume (a) to be true. While certainly the model is not a perfect representation of the Earth and some structures are due to regularisation choices and inversion strategy, assumption (b) is more likely to be problematic. First of all, the regions of sensitivity are not identical. Gao and Shen (2014) give the depth range of 40–150 km as the region of highest resolution and structures above 20 km depth are likely to be poorly resolved. In contrast, magnetic measurements are typically only sensitive to crustal structures where Earth materials are above the Curie temperature which is estimated to be reached at a depth of 23–30 km in the study area (Bouligand et al., 2009). Thus there should be some overlapping sensitivity, but there are certainly some discrepancies in resolution at different depth levels.

A further complication is that I model the data under the assumption of an inducing field with magnitude and direction corresponding to the current magnetic field according to the WMM model (Chulliat et al., 2020) and thus neglect remnant magnetisation. Given that flood basalts are present throughout the Snake River plane, this is a simplification. However, measurements of magnetisation directions on basalts from the SRP show that these coincide with the current spin axis of the Earth (Tauxe et al., 2004). Thus assuming a single source of magnetisation will result in anomalies with the correct shape, but will attribute the amplitude of the anomaly only to changes in susceptibility. Other magnetic studies in the region have also assumed coincident directions of induced and remanent magnetic directions (Finn and Morgan, 2002; Bouligand et al., 2014).

For the inversion, I start with a high coupling weight v, trying to force as much coupling as possible. When the inversion stops progressing, I reduce the coupling weight v by a factor of 10. In all inversions performed here, the algorithm progresses again after the reduction of the coupling weight, indicating that the coupling is the main obstacle preventing an adequate data fit. For the constrained magnetic inversion, the initial starting value of $v = 10^6$ is reduced once to a final value of $v = 10^5$. I show horizontal slices through the inversion results at depths of 15 km and 30 km as well as the recovered velocity susceptibility relationship in Fig. 16.4. On average, the magnetic observations are fit within ±1 nT, which is at the lower end of the reported data uncertainty (Bankey et al., 2002). Comparing the susceptibility model with the velocity model used as a constraint at a depth of 15 km (upper row in Fig. 16.4) reveals some similarities, but also significant differences between the two models. For example, there appears to be a spatial correlation between low velocity and low susceptibility features along the eastern edge of the modelling domain. In contrast, high velocities are associated with high susceptibility in some regions, e.g. south of the SRP, and average susceptibility elsewhere. In addition, some of the high susceptibility anomalies within the SRP show different shapes compared to the features in the seismic velocity model.

The situation is different at a depth of 30 km (middle row in Fig. 16.4). Here we can see good correspondences between most structures in the susceptibility model and the velocity model. Particularly in the western half of the region, the two models vary in a similar manner. The impression of disparity between the models at shallow depths is confirmed

by the parameter cross-plots for the two models (bottom row in Fig. 16.4). Here I plot susceptibility versus velocity in each model cell for the entire inversion domain (left) and the depth slices at 15 km and 30 km depth (right), respectively. When looking at the relationship for the whole inversion domain, we can see strong variability of susceptibility corresponding to velocities between 2.5 and 4.0 km/s. Despite the inclusion of the variation of information constraint, no discernible relationship can be identified. As can be seen from the colours and suggested by the relatively low seismic velocity, these scattered estimates correspond largely to the crust. The constrained inversion indicates that throughout the entire crust no relationship between the seismic model and the susceptibility model can be established.

This observation changes abruptly when looking at susceptibilities corresponding to shear wave velocities > 4.0 km/s, i.e. the upper mantle. Here a well-defined relationship emerges. As discussed, this is somewhat suspicious as rocks at this depth should be significantly above the Curie temperature and thus lose their magnetic properties (Pasquale, 2011). In addition, due to the decay of the sensitivity kernel with depth, magnetic data are most sensitive to near-surface structures (e.g. Hinze et al., 2013). In this example, tests confirm that the data have little to no sensitivity to structures below 30 km depth. Thus this relationship is an artefact of the constrained inversion approach. As the data have no sensitivity at this depth range, the inversion algorithm is free to set the susceptibility values and thus the coupling constraint forces a unique relationship, as it is designed to do. The depth slice at 30 km is an example of the transition between a region with significant sensitivity and low sensitivity at depth. Large parts of the model are transferred from the seismic velocity model without much influence on the data fit. However, the regions that show differences in structures and where the parameter relationship is nonunique, correspond to aspects of the model that are required by the magnetic data in direct contradiction to the coupling constraint. Therefore, these regions are probably the most interesting aspects of the constrained model.

Based on the constrained inversion, we can refute the hypothesis that there is a susceptibility model with a one-to-one relationship with the seismic velocity model in the crust. Two possible conclusions can be drawn from this: (a) The velocity model is not a good representation of structures in the crust, or (b) Susceptibility and velocity sense different structures above 30 km depth. Based on the resolution analysis of the seismic model discussed above, it is likely that some velocity structures are missing or poorly represented. As the following experiment with constrained inversion of gravity shows, the overall structure of the crust appears to be reasonably represented though. Thus my conclusion is that there is a difference in causative formations for the two physical properties in the Earth. A full joint inversion could shed further light in this question, as there might be alternative models that are compatible with the

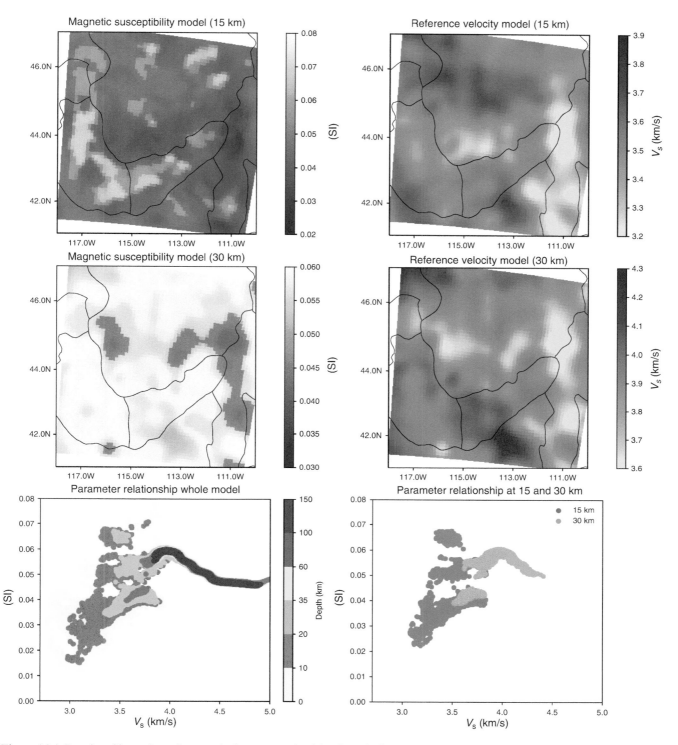

Figure 16.4 Results of inversion of magnetic data constrained by the velocity model of Gao and Shen (2014). I show horizontal slices through the susceptibility model (left column) and the reference velocity model (right column) at a depth of 20 km (top row) and 30 km (middle row). The bottom row shows the velocity–susceptibility relationship for the whole model (left) and for the depth slice at 30 km (right). Note that the colourbars are different between the upper two rows to improve the visibility of features.

seismological observations. This is, however, beyond the scope of this discussion.

Instead, I want to discuss the results of a constrained inversion of gravity data with the same seismic model to demonstrate some of the differences. The general inversion approach is identical to the inversion of magnetic data. The mean is subtracted from the data and the inversion is formulated in terms of density anomaly. I start with a large coupling constraint that is gradually reduced when the inversion stops making progress (from $v = 10^6$ to $v = 10^5$). The final misfit is 1–2 mGal on average and is compatible with the uncertainty of the gravity data. A slice through the resulting model at a depth of 30 km and parameter relationships are shown in Fig. 16.5. For this inversion we see a strong correspondence between density anomaly and

velocity. Boundaries and shapes of structures in the velocity model are mirrored in the density anomaly results. The parameter relationship for the whole model shows a well-defined shape at all velocities below 4.4 km/s. It appears to rise linearly between 3.0 km/s and 4.0 km/s and then changes slope and exhibits a more complex behaviour between 4.0 and 4.4 km/s. At the highest velocities, the relationship for the whole model starts to scatter strongly but different branches can still be identified. In this context, it is also important to note that these branches are not associated with different depths, but density values at similar depth show strong variability. Thus this effect cannot be explained by the fact that I invert for density anomaly instead of absolute density.

When only looking at the relationship for the 30 km depth slice in Fig. 16.5, a more well-defined relationship is evident.

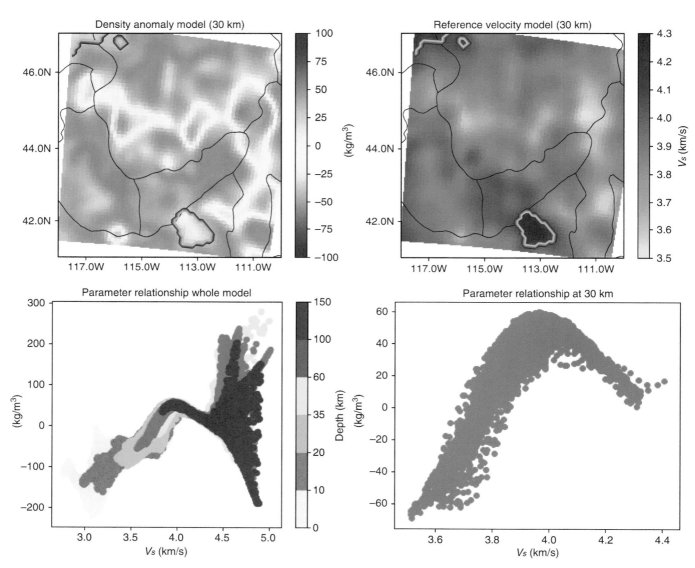

Figure 16.5 Result of gravity inversion constrained by the velocity model of Gao and Shen (2014). Horizontal slices through the inversion result and reference model are shown at a depth of 30 km (top row). Bottom row shows the parameter relationship for the whole model (left). The outlined regions in the horizontal slices correspond to the regions with seismic velocities greater than 4.2 km/s.

Each velocity corresponds to a density range of approximately ± 25 kg/m^3. The non-linear behaviour of this relationship is interesting and somewhat surprising. Instead of a monotonously increasing density anomaly with increasing velocity as expected from typical velocity-density relationships (e.g. Nafe and Drake, 1957; Birch, 1961; Barton, 1986), the density anomaly corresponding to the highest seismic velocities is close to zero. I outline the regions of highest seismic velocities in Fig. 16.5. Apart from some small regions in the north-western corner of the model which could be due to boundary effects, there is a high-velocity region at the southern boundary of the Snake River plain. It coincides with the Albion–Raft River–Grouse Creek metamorphic core complex (Konstantinou et al., 2012) where a lower crustal mush zone has been invoked to explain the development of the complex. While more focused investigations are necessary, it is conceivable that the density model identifies a low-density zone caused by a mix of fluids and melts while the seismic velocity is dominated by high velocities of the surrounding rocks. This is compatible with the interpretation based on the joint inversion of resistivity and density for the same region (Moorkamp, 2022).

On a broader level, the well-defined relationship at shear wave velocities below 4.5 km/s and the good fit to the gravity data of the constrained model indicate that both methods image compatible structures. The complexities of the relationship suggest that simple linear velocity–density relationships might not be appropriate as previously discussed by Barton (1986). More work is needed to understand the origin of this relationship and its connection to petrophysics of the crust and mantle (Afonso et al., 2016). Here the combination with resistivity from magnetotellurics and a full joint inversion with seismic data can provide further insights.

16.5 Discussion and Conclusions

The examples collected here demonstrate how unexpected results and deviations from the assumed parameter relationships can be used to gain insights into Earth structure if considered in the framework of hypothesis testing. The core idea is that such departures from the constraints imposed during the inversion are caused by features in the data and only evolve because the data demand it. Conversely when the parameters adhere to the assumed relationship, it can be seen as evidence supporting the assumption but also due to a lack of resolution of one of the datasets as shown in the magnetic data inversion. In that example, structures at depth are clearly an expression of the null-space of magnetic susceptibility inversions and not a reflection of true Earth structure. The constrained gravity inversion shows how seismology and gravimetry, for the most part, sense the Earth in a similar manner as can be expected from

theoretical considerations. Still, the recovered relationship exhibits complexities that do not adhere to the often-assumed linear density velocity relationships.

Regardless whether the results confirm or refute our assumptions, it is essential to explore different model parametrisations, regularisation parameters, and other inversion parameters. If not performed with care, inadequate model discretisation etcetera can lead to spurious inversion results and therefore incorrect conclusions. This is even more critical in joint and constrained inversions than in inversions of a single dataset, as the assumptions of all involved methods have to be considered. However, when used correctly, we can utilise both the successes and failures of integrated inversion approaches to learn about the Earth.

Acknowledgements. This work was funded by the German Research Foundation, DFG, under grant MO 2265/6-1 and by the European Space Agency ESA as part of the Support to Science Element 3D Earth. Insightful review by J. Kamm and an anonymous reviewer helped to clarify the ideas presented here.

All model and data files used for the two constrained inversion examples including Python scripts to plot them can be downloaded from https://doi.org/10.5281/zenodo.6552879. The gravity data have been extracted from the global XGM2019 gravity model (Zingerle et al., 2020), which is accessible at http://doi.org/10.5880/icgem.2017.003. Grids for the study region can be obtained through the ICGEM calculation service http://icgem.gfz-potsdam.de/tom_longtime by selecting XGM2019 (model 176) and calculating the Bouguer anomaly (gravity anomaly bg). Magnetic data can be downloaded from https://mrdata.usgs.gov/magnetic/. The seismic velocity model used as a constraint can be downloaded from https://doi.org/10.17611/DP/10009352.

Information about the inversion codes used for this study can be found at: https://sourceforge.net/projects/jif3d/.

Bibliography

Afonso, J. C., Moorkamp, M., and Fullea. J. (2016). Imaging the lithosphere and upper mantle: Where we are at and where we are go. In M. Moorkamp, P. G. Lelièvre, N. Linde, and A. Khan (eds.) *Integrated Imaging of the Earth: Theory and Applications*, AGU Geophysical Monograph 218. Hoboken, NJ: John Wiley & Sons, pp. 191–218.

Afonso, J. C., Salajegheh, F., Szwillus, W., Ebbing, J., and Gaina C. (2019). A global reference model of the lithosphere and upper mantle from joint inversion and analysis of multiple data sets. *Geophysical Journal International*, 217(3), 1602–28.

Astic, T., and Oldenburg, D. W. (2019). A framework for petrophysically and geologically guided geophysical inversion using a dynamic Gaussian mixture model prior. *Geophysical Journal International*, 219(3), 1989–2012.

Astic, T., Heagy, L. J., and Oldenburg, D. W. (2021). Petrophysically and geologically guided multi-physics inversion using a dynamic Gaussian mixture model. *Geophysical Journal International*, 224 (1), 40–68.

Bankey, V., Cuevas, A., Daniels, D. et al. (2002). Digital data grids for the magnetic anomaly map of North America. USGS Open-File Report 02-414. https://pubs.usgs.gov/of/2002/ofr-02-414/.

Barton, P. J. (1986). The relationship between seismic velocity and density in the continental crust: A useful constraint? *Geophysical Journal International*, 87(1), 195–208.

Bedrosian, P. A., and Feucht, D. W. (2014). Structure and tectonics of the northwestern United States from EarthScope USArray magnetotelluric data. *Earth and Planetary Science Letters*, 402, 275–89.

Bennington, N. L., Zhang, H., Thurber, C. H., and Bedrosian, P. A. (2015). Joint inversion of seismic and magnetotelluric data in the Parkfield region of California using the normalized cross-gradient constraint. *Pure and Applied Geophysics*, 172(5), 1033–52.

Birch, F. (1961). The velocity of compressional waves in rocks to 10 kilobars: Part 2. *Journal of Geophysical Research*, 66(7), 2199–224.

Blom, N., Boehm, C., and Fichtner, A. (2017). Synthetic inversions for density using seismic and gravity data. *Geophysical Journal International*, 209(2), 1204–20.

Bosch. M. (2016). Inference networks in Earth models with multiple components and data. In M. Moorkamp, P. G. Lelièvre, N. Linde, and A. Khan (eds.) *Integrated Imaging of the Earth: Theory and Applications*, AGU Geophysical Monograph 218. Hoboken, NJ: John Wiley & Sons, pp. 29–47.

Bosch M., and McGaughey. J. (2001). Joint inversion of gravity and magnetic data under lithologic constraints. *The Leading Edge*, 20(8), 877–81.

Bouligand, C., Glen, J. M. G., and Blakely, R. J. (2009). Mapping Curie temperature depth in the western United States with a fractal model for crustal magnetization. *Journal of Geophysical Research:* Solid Earth, 114(B11).

Bouligand, C., Glen, J. M. G., and Blakely, R. J. (2014). Distribution of buried hydrothermal alteration deduced from high-resolution magnetic surveys in Yellowstone National Park. *Journal of Geophysical Research: Solid Earth*, 119(4), 2595–630.

Carter-McAuslan, A., Lelièvre, P. G. and Farquharson, C. G. (2015). A study of fuzzy c-means coupling for joint inversion, using seismic tomography and gravity data test scenarios. *GEOPHYSICS*, 80(1), W1–W15.

Chen, C. W., Rondenay, S., Weeraratne, D. S., and Snyder D. B. (2007).New constraints on the upper mantle structure of the Slave Craton from Rayleigh wave inversion. *Geophysical Research Letters*, 34, L10301. https://doi.org/10.1029/2007 GL029535

Chulliat, A., Brown, W., Alken, P. et al. (2020). The US/UK world magnetic model for 2020–2025: Technical Report. National Centers for Environmental Information (U.S.); British Geological Survey. https://doi.org/10.25923/ytk1-yx35

Colombo, D., and Rovetta, D. (2018). Coupling strategies in multiparameter geophysical joint inversion. *Geophysical Journal International*, 215(2), 1171–84.

Darijani, M., Farquharson, C. G., and Lelièvre. P. G. (2021). Joint and constrained inversion of magnetic and gravity data: A case history from the McArthur River area, Canada. *Geophysics*, 86(2), B79–B95.

de Groot-Hedlin, C., Constable, S., and Weitemeyer, K. (2003–4). Transfer functions for deep magnetotelluric sounding along the Yellowstone-Snake River hotspot track. https://doi.org/10.17611/DP/EMTF/YSRP/2004.

Elsasser, W. M. (1950). The Earth's interior and geomagnetism. *Reviews of Modern Physics*, 22(1), 1–35.

Finn, C. A., and Morgan, L. A. (2002). High-resolution aeromagnetic mapping of volcanic terrain, Yellowstone National Park. *Journal of Volcanology and Geothermal Research*, 115 (1-2), 207–31.

Fishwick, S. (2010). Surface wave tomography: Imaging of the lithosphere-asthenosphere boundary beneath central and southern Africa? *Lithos*, 120(1–2), 63–73.

Franz, G., Moorkamp, M., Jegen, M., Berndt, C., and Rabbel, W. (2021). Comparison of different coupling methods for joint inversion of geophysical data: A case study for the Namibian continental margin. *Journal of Geophysical Research: Solid Earth*, 126 (12), e2021JB022092.

Fullagar, P. K., and Oldenburg, D. W. (1984). Inversion of horizontal loop electromagnetic frequency soundings. *Geophysics*, 49(2), 150–64.

Gallardo, L. A., and Meju, M. A. (2003). Characterization of heterogeneous near-surface materials by joint 2D inversion of dc resistivity and seismic data. *Geophysical Research Letters*, 30(13), 1658.

Gallardo, L. A., and Meju, M. A. (2007). Joint two-dimensional cross-gradient imaging of magnetotelluric and seismic travel-time data for structural and lithological classification. *Geophysical Journal International*, 169, 1261–72.

Gallardo, L. A., and Meju, M. A. (2011). Structure-coupled multiphysics imaging in geophysical sciences. *Reviews of Geophysics*, 49(1).

Gao, H., and Shen, Y. (2014). Upper mantle structure of the Cascades from full-wave ambient noise tomography: Evidence for 3D mantle upwelling in the back-arc. *Earth and Planetary Science Letters*, 390, 222–33.

Gardner, G. H. F., Gardner, L. W., and Gregory, A. R. (1974). Formation velocity and density: The diagnostic basics for stratigraphic traps. *Geophysics*, 39(6), 770–80.

Ghalenoei, E., Dettmer, J., Ali, M. Y., and Kim, J. W. (2021). Gravity and magnetic joint inversion for basement and salt structures with the reversible-jump algorithm. *Geophysical Journal International*, 227(2), 746–58.

Giraud, J., Pakyuz-Charrier, E., Jessell, M. et al. (2017). Uncertainty reduction through geologically conditioned petrophysical constraints in joint inversion. *Geophysics*, 82 (6), ID19–ID34.

Gross, L. (2019). Weighted cross-gradient function for joint inversion with the application to regional 3-D gravity and magnetic anomalies. *Geophysical Journal International*, 217(3), 2035–46.

Haber, E., and Holtzman Gazit, M. (2013). Model fusion and joint inversion. *Surveys in Geophysics*, 34(5), 675–95.

Haber, E., and Oldenburg, D. W. (1997). Joint inversion: A structural approach. Inverse *Problems*, 13(1), 63–77.

Harmon, N., Wang, S., Rychert, C. A. Constable, S., and Kendall, J. M. (2021). Shear velocity inversion guided by resistivity structure from the pi-lab experiment for integrated estimates of partial melt in the mantle. *Journal of Geophysical Research: Solid Earth*, e2021JB022202.

Heincke, B., M., Jegen, M., Moorkamp, R. W., Hobbs, and J. Chen (2017). An adaptive coupling strategy for joint inversions that use petrophysical information as constraints. *Journal of Applied Geophysics*, 136, 279–97.

Hinze, W. J., Von Frese, R. R. B., and Saad, A. H. (2013). *Gravity and magnetic exploration: Principles, practices, and applications*. Cambridge: Cambridge University Press.

Hong, T., and Sen, M. K. (2009). A new MCMC algorithm for seismic waveform inversion and corresponding uncertainty analysis. *Geophysical Journal International*, 177 (1), 14–32.

IRIS. USArray Transportable Array. (2003). https://doi.org/10.7914/SN/TA.

Julia, J., Ammon, C. J., Herrmann, R. B,. and Correig, A. M. (2000). Joint inversion of receiver function and surface wave dispersion observations. *Geophysical Journal International*,143 (1), 99–112.

Kamm, J., Lundin, I. A., Bastani, M., Sadeghi, M., and Pedersen, L. B. (2015). Joint inversion of gravity, magnetic, and petrophysical data: A case study from a gabbro intrusion in Boden, Sweden. *Geophysics*, 80(5), B131–B152.

Kelbert, A., Egbert, G. D., and Schultz, A. (2011). IRIS DMC data services products: EMTF, the magnetotelluric transfer functions. https://doi.org/10.17611/DP/EMTF.1.

Kelbert, A., Egbert, G. D., and deGroot-Hedlin, C. (2012). Crust and upper mantle electrical conductivity beneath the Yellowstone Hotspot Track. *Geology*, 40(5), 447–50.

Konstantinou, A., Strickland, A., Miller, E. L., and Wooden, J. P. (2012). Multistage Cenozoic extension of the Albion–Raft River–Grouse Creek metamorphic core complex: Geochronologic and stratigraphic constraints. *Geosphere*, 8 (6), 1429–66.

Lelièvre, P. G., Bijani, R., and Farquharson, C. G. (2016). Joint inversion using multi-objective global optimization methods. In *78th EAGE Conference and Exhibition 2016*. Houten: European Association of Geoscientists & Engineers. https://doi.org/10.3997/2214-4609.201601655.

Li, X., and Sun, J. (2022). Towards a better understanding of the recoverability of physical property relationships from geophysical inversions of multiple potential-field data sets. *Geophysical Journal International*, 230(3), 1489–507.

Linde, N., Binley, A., Tryggvason, A., Pedersen, L. B., and Revil, A. (2006). Improved hydrogeophysical characterization using joint inversion of cross-hole electrical resistance and ground-penetrating radar traveltime data. *Water Resources Research*, 42: 12404.

Linde, N., and Doetsch, J. (2016). Joint Inversion in Hydrogeophysics and Near-Surface Geophysics. In M. Moorkamp, P. G. Lelièvre, N. Linde, and A. Khan, eds, *Integrated Imaging of the Earth: Theory and Applications*. Hoboken, NJ: John Wiley & Sons, pp. 117–35.

Lines, L. R., Schultz, A. K., and Treitel, S. (1986). Cooperative inversion of geophysical data. *Geophysics*, 53(1), 8–20.

Liu, L., and Gao, S. S. (2018). Lithospheric layering beneath the contiguous United States constrained by S-to-P receiver functions. *Earth and Planetary Science Letters*, 495, 79–86.

Maceira, M., and Ammon, C. J. (2009). Joint inversion of surface wave velocity and gravity observations and its application to central Asian basins shear velocity structure. *Journal of Geophysical Research: Solid Earth*, 114(B2), B02314.

Mackie, R. L., Meju, M. A., Miorelli, F. et al. (2020). Seismic image-guided 3D inversion of marine controlled-source electromagnetic and magnetotelluric data. *Interpretation*, 8(4), SS1–SS13.

Manassero, M. C., Afonso, J. C., Zyserman, F I. et al. (2021). A reduced order approach for probabilistic inversions of 3D magnetotelluric data II: Joint inversion of MT and surface-wave data. *Journal of Geophysical Research: Solid Earth*, 126 (12), e2021JB021962.

Mandolesi, E., and Jones, A. G. (2014). Magnetotelluric inversion based on mutual information. *Geophysical Journal International*, 199(1), 242–52.

Martin, R., Giraud, J., Ogarko, V. et al. (2021). Three-dimensional gravity anomaly data inversion in the Pyrenees using compressional seismic velocity model as structural similarity constraints. *Geophysical Journal International*, 225(2), 1063–85.

Meilă, M. (2003). Comparing clusterings by the variation of information. In B. Schölkopf and M. K. Warmuth, eds., *Learning Theory and Kernel Machines: Lecture Notes in Computer Science*, vol. 2777. Berlin: Springer, pp. 173–87. https://doi.org/10.1007/978-3-540-45167-9_14.

Meju, M., and Gallardo, L. A. (2016). Structural Coupling Approaches in Integrated Geophysical Imaging. In M. Moorkamp, P. G. Lelièvre, N. Linde, and A. Khan, eds., *Integrated Imaging of the Earth: Theory and Applications*. Hoboken, NJ: . John Wiley & Sons, pp. 49–67.

Meju, M., Saleh, A. S., Mackie, R. L. et al. (2018). Workflow for improvement of 3D anisotropic CSEM resistivity inversion and integration with seismic using cross-gradient constraint to reduce exploration risk in a complex fold-thrust belt in offshore northwest Borneo. *Interpretation*, 6(3), SG49–SG57.

Meqbel, N. M., Egbert G. D., Wannamaker, P. E., Kelbert, A., and Schultz, A. (2014). Deep electrical resistivity structure of the northwestern US derived from 3-D inversion of USArray magnetotelluric data. *Earth and Planetary Science Letters*, 402,290–304.

Moorkamp. M. (2007). Joint inversion of MT and receiver-function data. PhD thesis, National University of Ireland, Galway.

Moorkamp. M. (2017). Integrating electromagnetic data with other geophysical observations for enhanced imaging of the Earth: A tutorial and review. *Surveys in Geophysics*, 38(5), 935–62.

Moorkamp. M. (2021). Joint inversion of gravity and magnetotelluric data from the Ernest Henry IOCG deposit with a variation of information constraint. In B. Swinford and A. Abubakar *First International Meeting for Applied Geoscience & Energy*. Houston, TX: Society of Exploration Geophysicists, pp. 1711–15.

Moorkamp. M. (2022). Deciphering the state of the lower crust and upper mantle with multi-physics inversion. *Geophysical Research Letters*, 49 (9), e2021GL096336.

Moorkamp, M., Jones, A. G., and Eaton, D. W. (2007). Joint inversion of teleseismic receiver functions and magnetotelluric data using a genetic algorithm: Are seismic velocities and electrical conductivities compatible? *Geophysical Research Letters*, 34(16), L16311.

Moorkamp, M., Roberts, A. W., Jegen, M., Heincke, B., and Hobbs, R. W. (2013). Verification of velocity-resistivity relationships derived from structural joint inversion with borehole data. *Geophysical Research Letters*, 40(14), 3596–601.

Moorkamp, M., Heincke, B., Jegen, M., Roberts, A. W., and Hobbs, R. W. (2011). A framework for 3-D joint inversion of MT, gravity and seismic refraction data. *Geophysical Journal International*, 184, 477–93.

Moorkamp, M., Heincke, B., Jegen, M., Roberts, A. W., and Hobbs, R. W. (2016a). Joint Inversion in Hydrocarbon Exploration. In M. Moorkamp, P. G. Lelièvre, N. Linde, and A. Khan, eds., *Integrated Imaging of the Earth: Theory and Applications*. Hoboken, NJ: John Wiley & Sons, pp. 167–189.

Moorkamp, M., Lelièvre, P. G., Linde, N., and Khan, A., eds. (2016b). *Integrated Imaging of the Earth*. Hoboken, NJ: John Wiley & Sons.

Nafe, J. E., and Drake, C. L. (1957). Variation with depth in shallow and deep water marine sediments of porosity, density and the velocities of compressional and shear waves. *Geophysics*, 22(3), 523–52.

O'Donnell, J. P., Daly, E., Tiberi, C. et al. (2011). Lithosphere-asthenosphere interaction beneath Ireland from joint inversion of teleseismic p-wave delay times and grace gravity. *Geophysical Journal International*, 184(3), 1379–96.

Pail, R., Fecher, T., Barnes, D. et al. (2018). Short note: The experimental geopotential model XGM2016. *Journal of Geodesy*, 92(4), 443–51.

Panzner, M., Morten, J. P., Weibull, W. W., and Arntsen, B. (2016). Integrated seismic and electromagnetic model building applied to improve subbasalt depth imaging in the Faroe-Shetland basin. *Geophysics*, 81(1), E57–E68.

Pasquale, V. (2011). Curie temperature. In H. K. Gupta, ed., *Encyclopedia of Solid Earth Geophysics*. Dordrecht: Springer, pp. 89–90. https://doi.org/10.1007/978-90-481-8702-7_109.

Paulatto, M., Moorkamp, M., Hautmann, S. et al. (2019). Vertically extensive magma reservoir revealed from joint inversion and quantitative interpretation of seismic and gravity data. *Journal of Geophysical Research: Solid Earth*, 124 (11), 11170–91.

Pluim, J. P. W. Maintz, J. B. A. and Viergever, M. A. (2003). Mutual-information-based registration of medical images: A survey. *IEEE Transactions on Medical Imaging*, 22(8), 986–1004.

Schmandt, B., and Humphreys, E. (2011). Seismically imaged relict slab from the 55 Ma Siletzia accretion to the northwest United States. *Geology*, 39(2), 175–8.

Schultz, A., Egbert, G. D., Kelbert, A. et al., and staff of the National Geoelectromagnetic Facility and their contractors. (2006–8). USArray TA magnetotelluric transfer functions. https://doi.org/10.17611/DP/EMTF/USARRAY/TA.

Shi, Z., Hobbs, R. W., Moorkamp, M., Tian, G., and Jiang, L. (2017). 3-D cross-gradient joint inversion of seismic refraction and dc resistivity data. *Journal of Applied Geophysics*, 141, 54–67.

Spichak, V. V. (2020). Modern methods for joint analysis and inversion of geophysical data. *Russian Geology and Geophysics*, 61(3), 341–57.

Sun, J., and Li, Y. (2015a). Advancing the understanding of petrophysical data through joint clustering inversion: A sulfide deposit example from Bathurst mining camp. In R. V. Schneider, ed., *SEG Technical Program Expanded Abstracts 2015*. Houston, TX: Society of Exploration Geophysicists, pp. 2017–21.

Sun, J., and Li, Y. (2015b). Multidomain petrophysically constrained inversion and geology differentiation using guided fuzzy c-means clustering. *Geophysics*, 80(4), ID1–ID18.

Sun, J., and Li, Y. (2016). Joint inversion of multiple geophysical data using guided fuzzy c-means clustering. *Geophysics*, 81(3), ID37–ID57.

Sun, J., Melo A. T., Kim, J. D., and Wei, X. (2017). Unveiling the 3D undercover structure of a Precambrian intrusive complex by integrating airborne magnetic and gravity gradient data into 3D quasi-geology model building. *Interpretation*, 8(4), SS15–SS29.

Tauxe, L., Luskin, C., Selkin, P., Gans, P., and Calvert, A. (2004). Paleomagnetic results from the Snake River Plain: Contribution to the time-averaged field global database. *Geochemistry, Geophysics, Geosystems*, 5(8), Q08H13. https://doi.org/10.1029/2003GC000661.

Tiberi, C., Diament, M., Déverchère, J. et al. (2003). Deep structure of the Baikal rift zone revealed by joint inversion of gravity and seismology. *Journal of Geophysical Research: Solid Earth*, 108 (B3), 2133. https://doi.org/10.1029/2002JB001880.

Tu, X., and Zhdanov, M. S. (2021). Joint Gramian inversion of geophysical data with different resolution capabilities: Case study in Yellowstone. *Geophysical Journal International*, 226 (2), 1058–85.

Wagner F. M., Mollaret, C., Günther, T., Kemna, A., and Hauck, C. (2019). Quantitative imaging of water, ice and air in permafrost systems through petrophysical joint inversion of seismic refraction and electrical resistivity data. *Geophysical Journal International* 219(3), 1866–75.

Weise, B. (2021). Joint Inversion of magnetotelluric, seismic and gravity data. PhD thesis, University of Leicester.

Zhao, Y., Guo, L., Guo, Z. et al. (2020). High resolution crustal model of SE Tibet from joint inversion of seismic p-wave travel times and Bouguer gravity anomalies and its implication for the crustal channel flow. *Tectonophysics*, 792, 228580.

Zhdanov, M. S., Gribenko, A., and Wilson, G. (2012). Generalized joint inversion of multimodal geophysical data using Gramian constraints. *Geophysical Research Letters*, 39(9).

Zhou, J. Revil, A. Karaoulis, M. et al. (2014). Image-guided inversion of electrical resistivity data. *Geophysical Journal International*, 197 (1), 292–309.

Zingerle, P., Pail, R., Gruber, T., and Oikonomidou, X. (2020). The combined global gravity field model xgm2019e. *Journal of Geodesy*, 94(7), 1–12.

17

Crustal Structure and Moho Depth in the Tibetan Plateau from Inverse Modelling of Gravity Data

Shuanggen Jin and Songbai Xuan

Abstract: Although many geophysical observations and models are available for the Tibetan Plateau (TP) and its surroundings regions, our knowledge and understanding of the uplift and deformation in the TP caused by the India–Asia collision is still incomplete. Due to the environmental complexity, the gravity method is indispensable to investigate the evolution of the TP. This study concentrates on the Moho depth and crustal density structure in the TP from gravity inversion of Bouguer anomalies. The results show Moho deeper than 60 km in the regions of the TP, suggesting thickening of the crust. Two sinking Moho belts in the southern and northern plateau regions and the linearly increasing Moho depth from the Indian Plate (IP) to the Indus-Yalu suture can be used to infer the crustal fold that has resulted from the India–Asia collision. On the other hand, the density structures show the lower density is commonly found in the crust and the underlying lithospheric mantle beneath the TP, contrasting with the high density in the surrounding blocks. Notably, the high density of the IP is observed underneath the Himalayas, suggesting that the Indian lithosphere extends northward, at least reaching the Indus-Yalu suture. Corresponding to the sinking Moho belts, the crustal densities in these regions present relatively low, which may be evidence for the absence of the eclogites in the lower crust beneath the Himalayas and the Lhasa terrane. In contrast, the relatively high densities underneath the Bangong-Nujiang suture are potentially contributions to the interpretation of the eclogitised lower crust.

17.1 Introduction

The India–Asia collision began at ~50 Ma and resulted in the uplift of the Himalayas and the Tibetan Plateau (hereafter TP; Molnar and Tapponnier, 1975; Yin and Harrison, 2000). With an average elevation of 4–5 km and ~70 km thick crust (Li et al., 2006; Teng et al., 2013, 2020; Li et al., 2014), the TP is an excellent region for studying the mechanics of continental deformation (Jin and Park, 2006; Jin et al., 2007). Over the last few decades, several geophysical investigations have been conducted and many tectonic activities beneath the TP have

been investigated (Owens and Zandt, 1997; Wang et al., 2003; Huang and Zhao, 2006; Kumar et al., 2006; Hetényi et al., 2007; Shin et al., 2007, 2009, 2015; Zheng et al., 2007; Jiménez-Munt et al., 2008; Li et al., 2008; Nábělek et al., 2009; He et al., 2010; Bai et al., 2013; Zhao et al., 2013; Zhang et al., 2014; Bao et al., 2015; Liang et al., 2016; Wang et al., 2019; Teng et al., 2020; Zhao et al., 2020). However, the long-standing question on the tectonic mechanism responsible for the crustal thickening and uplift in the TP caused by the India–Asia collision is still debated. Two contrasting models to explain the rheology and structure of the lithosphere are the so-called jelly sandwich and crème brûlée models (Searle et al., 2011). Several remarkable models, including internal deformation (Houseman and England, 1986), block extrusion (Tapponnier et al., 1982), and lower crustal flow (Royden et al., 1997; Clark and Royden, 2000), have been proposed to interpret the India–Asia collision.

The Tibetan Plateau, generated by the India–Asia collision, is bounded by the Tarim Basin and Qaidam Basin to the north, and the Himalaya, Karakoram, and Pamir mountain chains to its south and west. As shown in Fig. 17.1, following Yin and Harrison (2000), four bordering sutures – the Indus-Yalu suture (IYS), the Bangong-Nujiang suture (BNS), the Jinshajiang suture (JS), and the Anyimaqen-Kunlun-Mutztagh suture (AKMS) – separate the TP into the Himalaya mountains (HM), the Lhasa terrane (LT), the Qiangtang terrane (QT), the Songpan-Ganzi terrane (ST), and the Kunlun-Qilian terrane (KT). West–east variations in the distances of the under thrusting Indian lithosphere along the HM and LT have been inferred in previous studies (e.g. Liang et al., 2016). As the main belt responded to Indian–Eurasian convergence and collision, the crustal structures underneath the HM and LT provide important clues for understanding the tectonic environments and dynamic process of continent-continent collision, such as the thrusting distance of the Indian lithosphere (Searle et al., 2011; Liang et al., 2016) and the eclogitised lower crust (Hetényi et al., 2007; Bai et al., 2013).

Gravity modelling for investigations of the crustal structure has a distinct advantage in the inaccessible regions of the

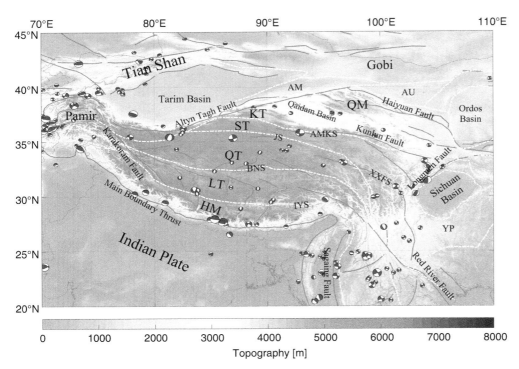

Figure 17.1 Topographic map of the TP region showing the major faults (solid lines), geo-block boundaries (white dotted lines), and earthquakes with $M_w \geq 6$ since 1976 (www.globalcmt.org). Abbreviations are as follows: *HM*, Himalayan Mountain Range; *LT*, Lhasa Terrane; *QT*, Qiangtang Terrane; *ST*, Songpan-Ganzi Terrane; *KT*, Kunlun Terrane; *QM*, Qilian Mountain Range; *AM*, Altyn Tagh Mountains; *AU*, Alashan Uplift; *YP*, Yungui Plateau; *IYS*, Indus-Yarlung Suture; *BNS*, Bangong-Nujiang Suture; *JS*, Jinsha River Suture; *AKMS*, Anyimaqen-Kunlun-Mutztagh Suture; *XXFS*, Xianshuihe-Xiaojiang Fault System.

TP. Although previous studies have reported the Moho undulation (Braitenberg et al., 2000; Shin et al., 2007, 2009, 2015; Tenzer et al., 2015; Xu et al., 2017; Zhao et al., 2020) and crustal density (Hetényi et al., 2007; Jiménez-Munt et al., 2008; Bai et al., 2013) in the TP, there are only few studies of them together. In this chapter, we present the Moho depth and crustal density structure in the TP from inverse modelling of gravity data. The results allow us to understand the interactions between the TP and the adjacent tectonic blocks, as well as the responses of the tectonic blocks to the lateral variations beneath the central part of the TP.

17.2 Data and Methods

17.2.1 Gravity Data and Its Multi-scale Analysis

The Bouguer gravity anomaly data shown in Fig. 17.2 is calculated by the Bureau Gravimetrique International on the basis of the EGM2008 spherical harmonic coefficients (Pavlis et al., 2008). Overall, the gravity anomalies in the western and central areas are lower in magnitude than in other areas. The medium to large wavelength negative gravity anomalies which are related to crustal thickening due to isostatic compensation are better visualised in the Bouguer anomaly map. In addition, the gravity anomaly data

demonstrates an excellent correlation with the topography, wherein a lower gravity anomaly value corresponds to a higher terrain. The high mountains (with heights of over 6 km) in the central region correspond to low Bouguer gravity anomalies (approximately −500 mGal), while the Tarim Basin and Sichuan Basin, which are less than 1.5 km, have high anomaly (−100 mGal). The altitude of the Qaidam Basin is at least 2 km, which is lower than that of the surrounding mountains, and its gravity anomaly values are approximately −200 mGal. Moreover, large-scale tectonic features can be revealed directly from the Bouguer gravity anomaly data. Negative gravity anomalies in the central area indicate present–day mass deficits beneath the region. The positive–negative alternating zones and high–low gravity gradient belts correspond to the boundaries of major tectonic blocks. The significant feature of this Bouguer gravity map is that the gradient belts are consistent with the major faults around the TP and Pamir Plateau, such as the Main Boundary Thrust to the south, the Altyn Tagh Fault to the north, and the Longmenshan to the east. Positive anomaly values over 100 mGal cover the Indian Plate. The high negative anomaly values are distributed across the TP and Pamir Plateau and correspond to a higher terrain and. In particular, the −250 mGal contour is nearly coincident with the margin of the plateaus. In the

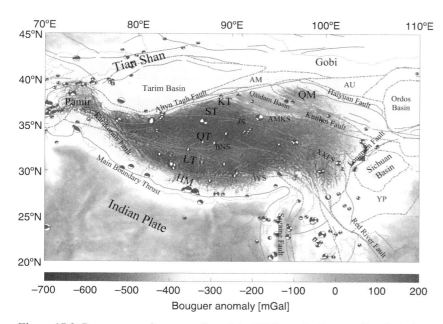

Figure 17.2 Bouguer gravity anomalies of the TP from the Bureau Gravimétrique International (http://bgi.omp.obs-mip.fr). Abbreviations are as Fig. 17.1.

Tian Shan region, the anomaly values are approximately – 200 mGal.

Although the Bouguer gravity anomaly data can reveal rich tectonic and geophysical information, it must be noted that it reflects the integrated effects of heterogeneous materials sourced from different scales and depths within the Earth's interior. To extract the signal caused exclusively by Moho undulations, the Bouguer gravity anomaly data must be further decomposed and analysed accordingly. The discrete wavelet transform method was applied for the multi-scale analysis of the Bouguer gravity anomaly data in the TP (Jiang et al., 2012; Xuan et al., 2016; Xu et al., 2017). Xu et al. (2017) decomposed the Bouguer gravity anomaly data in the TP using the base function of Coif3. Following Xu et al. (2017), the sum of the first-, second-, and third-order wavelet details (Fig. 17.3a) primarily reflected the tectonic structure of the upper to middle crust. Although interference from covered sediments is obvious within the first through the third order wavelet details, the major tectonic units and boundaries, such as HM, Pamir, Longmenshan Fault (LMS) and Altyn Tagh Fault, can still be identified. The sixth-order wavelet detail (Fig. 17.3c) reflects the material distribution at the bottom of the lithosphere, as argued by Xu et al. (2017). The sixth-order wavelet detail (Fig. 17.3c) with respective average source depths of 130 km suggested an attenuating lateral density inhomogeneity within the upper mantle.

Xuan et al. (2016) presented the fourth- and fifth-order wavelet details in the south-eastern TP (Fig. 17.4). Gravity anomalies are complex, reflecting the complex density structures in the middle crust. It is evident that a high-gravity anomaly existed beneath the JS near the western boundary of the Chuan-Dian block (CDB) and a low-gravity anomaly

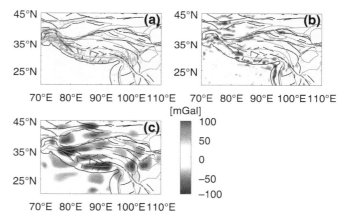

Figure 17.3 Decomposed gravity anomalies in the TP following Xu et al. (2017). (a) Sum of the first-, second, and third-order wavelet details. (b) Sum of the fourth- and fifth-order wavelet details. (c) Sixth-order wavelet detail.

existed beneath the Jiali fault. The parallel anomalies in the southwestern region of the study area implied strong deformation in the lower crust, induced by eastward extrusion of the Burmese block and clockwise rotation of the CDB around the Eastern Himalayan syntaxis. The fifth-order detail map (Fig. 17.4b) reveals the density structure of the lower crust. The local anomalies and the fifth-order details mainly result from inhomogeneity in the density distribution of the whole and lower crust, respectively. In the Indo–China block and Eastern Himalayan syntaxis, eastward arcs of lower and high anomalies indicate that subduction of the Burmese block occurs in the

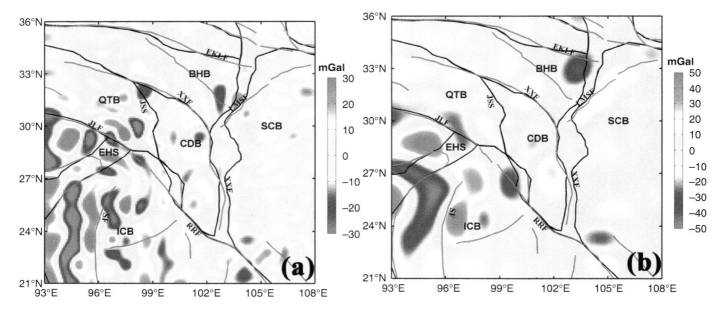

Figure 17.4 Decomposed gravity anomalies of the Chuan-Dian region (Xuan et al., 2016). (a) Fourth-order wavelet detail. (b) Fifth-order wavelet detail. Abbreviations as follows: *ICB*, Indo-China block; *QTB*, Qiangtang block; *CDB*, Chuan-Dian block; *BHB*, Bayan Har block, *SCB*, Sichuan Basin, *EHS*, Eastern Himalayan syntaxis; *JSS*, Jinshajiang Suture; *EKLF*, East Kunlun Fault; *XXF*, Xianshuihe–Xiaojiang Fault System; *LMS*, Longmenshan Fault; *RRF*, Red River Fault; *SF*, Sagaing Fault.

lower crust, a phenomenon that is not obvious in the middle crust (Fig. 17.4a). In the Longmenshan region, high anomalies are found along the LMS, and low anomalies exist in the north-eastern area of the LMS, suggesting that the effects of the eastward extrusion of the Bayan Har block exist in the lower crust.

17.2.2 Moho Inversion
The gravity anomaly, $\Delta g(x, y)$, is related to the undulation of the density interface, $h(x, y)$, around the reference level, z_0, in the Fourier domain described by Parker (1973):

$$F[\Delta g(x,y)] = -2\pi G \Delta \rho e^{(-kz_0)} \sum_{n=1}^{\infty} \frac{k^{n-1}}{n!} F[h^n(x,y)], \qquad (17.1)$$

where $F[\ ]$ represents the Fourier transform, G is the Newton's gravitational constant, k is the wave vector of the transformed function, and $\Delta \rho$ is the crust–mantle density contrast. Oldenburg (1974) presented an iterative algorithm via modification of Eq. (17.1):

$$F[h(x,y)] = -\frac{F[\Delta g(x,y)]e^{(-kz_0)}}{2\pi G \Delta \rho} - \sum_{n=2}^{\infty} \frac{k^{n-1}}{n!} F[h^n(x,y)], \qquad (17.2)$$

and the gravity-derived Moho depth, z_{cg}, is obtained as follows:

$$z_{cg} = z_0 + h(x, y). \qquad (17.3)$$

Oldenburg (1974) defined a filter $B(k)$ to restrict parts of the observations with high frequencies via the frequency parameters WH and SH as follow:

$$B(k) = \begin{cases} 1 & |k/2\pi| < WH, \\ \frac{1}{2}\left[1 + \cos\left(\frac{k - 2\pi WH}{2(SH - WH)}\right)\right] & WH \le |k/2\pi| \le SH, \\ 0 & |k/2\pi| > SH. \end{cases} \qquad (17.4)$$

Given the reference level z_0, the crust–mantle density contrast $\Delta \rho$, and the cut-off frequencies (WH and SH) in Eq. (17.4), it is possible to compute the Moho relief $h(x, y)$, versus reference depth using Eq. (17.2) iteratively, and then obtain the gravity Moho depth z_{cg} using Eq. (17.3).

17.2.3 Density Inversion
To solve the non-unique and ill-conditioned linear inverse problem $\boldsymbol{g} = \boldsymbol{G}\boldsymbol{\rho}$, the density transformed from V_p is considered as the priori density $\boldsymbol{\rho}_{prior}$, and the least-squares criterion is used to determine the density model $\widetilde{\boldsymbol{\rho}}$ by minimising the objective function (Barnoud et al., 2016):

$$\phi(\widetilde{\boldsymbol{\rho}}) = (\boldsymbol{g} - \boldsymbol{G}\widetilde{\boldsymbol{\rho}})^{\mathrm{T}} \boldsymbol{C}_{\boldsymbol{g}}^{-1} (\boldsymbol{g} - \boldsymbol{G}\widetilde{\boldsymbol{\rho}}) + (\widetilde{\boldsymbol{\rho}} - \boldsymbol{\rho}_{prior}) \boldsymbol{C}_{\rho}^{-1} (\widetilde{\boldsymbol{\rho}} - \boldsymbol{\rho}_{prior}), \qquad (17.5)$$

where $\boldsymbol{C}_{\boldsymbol{g}}$ and $\boldsymbol{C}_{\rho}$ are the data and priori model covariance matrices, the superscript 'T' denotes the transposition. The second term in Eq. (17.5) is a regularisation term. The solution of Eq. (17.5), solved in the data space, can be written as

$$\widetilde{\rho} = \rho_{prior} + (G^T C_g^{-1} G + C_\rho^{-1})^{-1} G^T C_g^{-1} (g - G\rho_{prior}). \quad (17.6)$$

The data can be considered to be independent, therefore, the C_g is a diagonal matrix and its diagonal values σ_g^2 are estimated errors of the input data. The covariance matrix of the priori model C_ρ includes a variance on density and a spatial correlation. The details can be found in Barnoud et al. (2016).

17.3 Gravity-Inverted Moho Depth beneath the Tibetan Plateau

Several results of the Moho depth from the gravity inversion using the Parker–Oldenburg method beneath the TP have been published (Braitenberg et al., 2000; Shin et al., 2007; Shin et al., 2009; Steffen et al., 2011; Bagherbandi, 2012; Shin et al., 2015; Xuan et al., 2015; Zhang et al., 2015; Chen, 2017; Chen and Tenzer, 2017; Xu et al., 2017; Zhao et al., 2020), although the values and locations of the maximum depths were different among these models. In particular, two depressed belts were found in the LT and QT using the Moho maps (Braitenberg et al., 2000; Shin et al., 2007; Shin et al., 2009; Shin et al., 2015; Xuan et al., 2015; Xu et al., 2017; Zhao et al., 2020).

Figure 17.5 showed the gravity-inverted Moho depth modified from that of Xuan et al. (2015). The reference Moho depth of 40 km and crust–density contrast of 0.4 kg/m^3 are used during the gravity inversion. The main features of this Moho depth model were that the deeper

Moho (> 60 km) covers the TP and Pamir Plateau, while the maximum depth of ~74 km occurs in the southern TP. The Moho depths of the tectonic units in the surroundings of the TP are not consistent with those of the plateau, that is, a depth of ~40 km under the Indian Plate, ~50 km beneath the Tarim Basin, and about 40 km beneath the Sichuan Basin. The Pamir region, northwest of the TP, has a deeper Moho with a maximum depth of ~65 km; and the Tian Shan region, north of the Tarim Basin, also has a deeper Moho depth of ~60 km. It should also be noted that a Moho uplift of ~5 km occurs under the central part of the Tarim Basin.

For comparison, four results of the Moho depth from CRUST1.0 (Laske et al., 2013), seismic study (Li et al., 2014), gravity inversion (Zhao et al., 2020), and local isostasy are shown in Fig. 17.6. Moho depth from CRUST1.0 (Fig. 17.6a) shows the deepest Moho (~70 km) is observed in the middle LT, and the shallower Moho (50–60 km) in the northwestern TP and Pamir region. Moho depth of Xuan et al. (2015) (Fig. 17.5) was deeper than that from CRUST1.0 (> 10 km) in the north-western Tibet region, eastern Tarim Basin, and Pamir region. The seismic Moho depth shown in Fig. 17.6b shows that the deepest Moho exceeding 80 km was found in the eastern TP. As shown in Fig. 17.6c, the deepest Moho was observed in the Pamir region reported by Zhao et al. (2020). Moho depth deeper than that of Zhao et al. (2020) was found in the eastern Tarim Basin, LT, and QT along the BNS in western Tibet, as well as the western HM. It is worth noting that the three deep belts in the TP reported by Shin et al. (2007), Xuan et al. (2015), and Zhao et al. (2020) were not found from the seismic-derived Moho (Fig. 17.6b) and the local isostatic Moho (Fig. 17.6d).

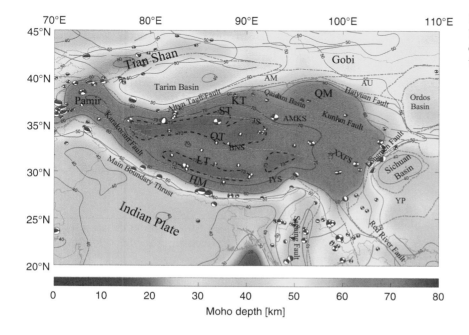

Figure 17.5 Gravity-inverted Moho depth beneath the TP following Xuan et al. (2015). The contours of 73 km (black dotted lines) outline the two depressed belts at the base of the crust.

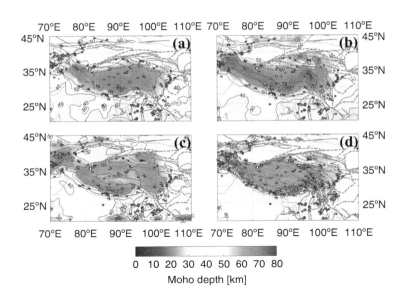

Figure 17.6 Maps of Moho depth from (a) CRUST1.0 (Laske et al., 2013), (b) seismic study (Li et al., 2014), (c) gravity inversion (Zhao et al., 2020), and (d) local isostasy.

The Moho is deeper than 70 km in the western LT and QT; however, it was ~60 km throughout the eastern part of the TP. Along the Indian subduction zone, the Moho depth exhibited a narrow gradient belt and increases linearly from ~40 km to exceeding 70 km with different angles from west to east (Fig. 17.7), and the relatively wider gradient belt was consistent with the smaller angle underneath the western and eastern ends of the Himalayan Mountains. Although Moho depth derived from Vening Meinesz–Moritz isostatic hypothesis are in better agreement with the CRUST2.0 Moho depth (Bagherbandi, 2012; Tenzer and Bagherbandi, 2012), the deepest Moho of ~61.5 km is shallower than that of ~70 km presented by the most of the previous studies (e.g. Shin et al., 2007, 2009; Li et al., 2014; Chen et al., 2017; Xu et al., 2017; Zhao et al., 2020), suggesting that the isostatic gravity disturbance is more suitable for determining isostatic Moho depth than Bouguer gravity anomaly (Vaníček et al., 2004; Tenzer and Bagherbandi, 2012; Sjöberg et al., 2013).

Xuan et al. (2016) presented the 3D view of the topography and the Moho depth around the Eastern Himalayan Syntaxis region (Fig. 17.8), where the crustal thickness decreased from northwest to southeast beneath the CDB. There was a significant change in the depth of the Moho undulation between the TP and the surrounding regions. The obvious steep belt of deep Moho appeared in both the LMS and the Eastern Himalayan syntaxis. Fu and She (2017) suggested the Moho depth increased linearly from south to north with an angle of 12° across the Eastern Himalayan syntaxis, however this larger dip angle was not found in Figs. 17.5 and 17.8, potentially resulting from the space resolution of the different gravity data sets. The Moho dip angle of Fu and She (2017) is larger than 8–9°

in the middle thrusting belt of the Indian Plate (Bai et al., 2013). The depth of the Moho on both sides of the LMS decreased from nearly 70 km to less than 50 km from northwest to southeast. There were abnormal belts parallel to the LMS at a depth of 65 km in the northwest and 30–44 km in the Sichuan Basin to the southeast. The depth of the Moho was approximately 60 km in the northern portion of the CDB and approximately 50 km in the southern portion, reaching a depth of approximately 70 km in the Qiantang block (QTB). On both the west and east side of the Sagaing Fault, the depth of the Moho is less than 40 km. In the south segment of the Xianshuihe–Xiaojiang Fault system, the depth increased from 50 to 55 km in the west. From the west, near the Eastern Himalayan syntaxis across the middle of the CDB, to the east (the LMS region), the Moho had a Y-shaped relief. In conjunction with the topography belt, this suggested that the CDB was a transitional zone of the TP, the South China block and the Indo-China block.

17.4 Crustal Density Structure beneath the Tibetan Plateau

17.4.1 Initial Density

The density ρ from surface to 80 km depth can be derived from *P*-wave velocity V_p according to the velocity-density relationship (Brocher, 2005):

$$\rho = 1.6612 V_P - 0.4721 V_P^2 + 0.0671 V_P^3 - 0.0043 V_P^4 + 0.000106 V_P^5. \tag{17.7}$$

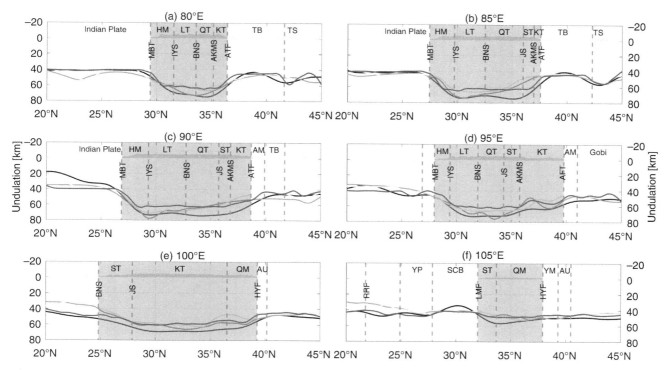

Figure 17.7 North-south trending sections of the topography (in green) and Moho depth across the TP. The black, magenta, cyanine, and blue lines denote Moho models from gravity inversion (Xuan et al., 2015), seismic model (Li et al., 2014), CRUST1.0 (Laske et al., 2013), and local isostasy. The range of the TP is shaded in grey. The dotted red lines respected the boundaries of the tectonic units. Abbreviations are as follows: *MBT*, Main Boundary Thrust; *TB*, Tarim Basin; *SCB*, Sichuan Basin; *TS*, Tian Shan; *ATF*; Altyn Tagh Fault; *RRF*, Red River Fault; Other abbreviations are as Figure 17.1.

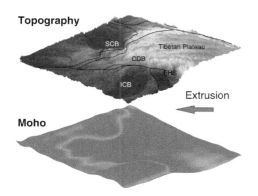

Figure 17.8 3D map view of the regional topography and underlying Moho relief.

The layered V_p maps from surface to 80 km in the region 80–108°E and 23–42°N shown in Fig. 17.9 (Zhang et al., 2011).[1] The horizontal spatial resolution of this dataset is 0.25° × 0.25°. The reference densities of 2,750 kg/m³, 2,900 kg/m³, and 3,300 kg/m³ were used for 0–20 km, 20–60 km, and 60–80 km, respectively. Subsequently, the initial density shown

[1] www.researchgate.net/profile/Yangfan-Deng

in Fig. 17.10 was transformed from the V_p maps (Fig. 17.9) using the Eq. (17.7). The calculated anomalies induced by the initial density and the residual anomalies after removal of the calculated anomalies from the Bouguer anomalies are presented in Fig. 17.11. The residual anomalies were used to determine the density model.

17.4.2 Density Structure from Modelling of Gravity Data

We modelled density from the surface to 80 km in the TP using the inversion method described above and the initial density transformed from V_p (Fig. 17.11). The layered and profile densities were shown in Figs. 17.12a–h and 17.13. The average and standard deviation (STD) of residuals are –2.44 and 0 mGal (Fig. 17.12i), respectively, suggesting the inversion result is reasonable.

Densities of ~2.5–2.8 g/cm³ present at 0–30 km depth in the whole region (Fig. 17.12a –c), and the lateral heterogeneity were not apparent. By contrast, the high densities in the rigid blocks around the TP at a depth of 30–80 km are visible (Fig. 17.12d–h), such as for the Indian Plate, the Sichuan Basin and Tarim Basin, especially at the depth of 50–80 km (Figs. 17.12f–h and 17.13a–d). Significantly, the high densities (~3.0–3.4 g/cm³) in the region from the Indian Plate to the HM, clearly observed in the profile images (Figs. 17.13a–c), suggest the lithosphere of the Indian

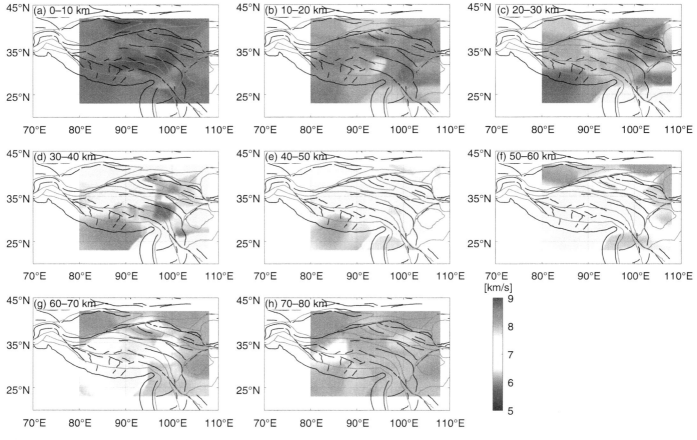

Figure 17.9 Layered seismic velocity (V_p) at depth of 0–80 km in the TP.

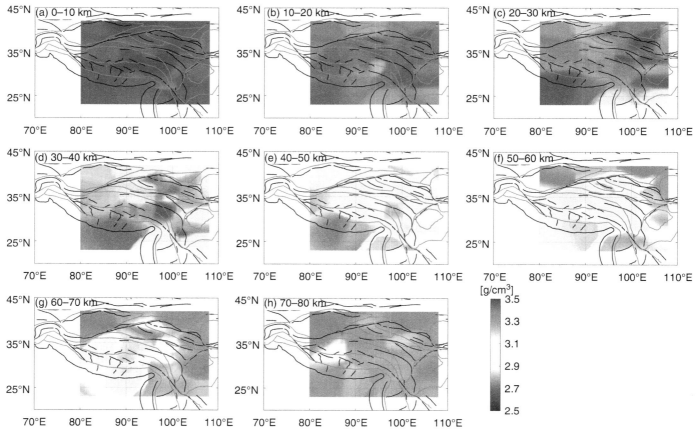

Figure 17.10 Initial density estimated from seismic velocity (V_p) at depth of 0–80 km in the TP.

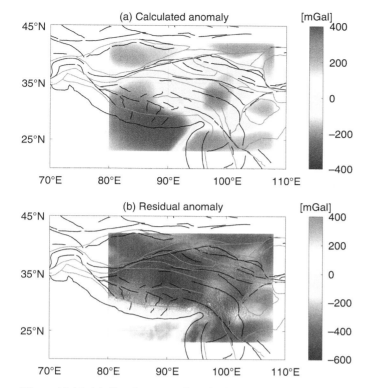

Figure 17.11 (a) Gravity anomaly calculated from initial density (Fig. 17.10) and (b) residual anomaly after removing the calculated anomaly from the Bouguer gravity anomaly values (Fig. 17.2).

Plate is being subducted northwardly under the HM. The subducting Indian lithosphere reached the IYS at least inferred by the high-density distribution (Figs. 17.13a–c). Although the density in the lithospheric mantle of the Tarim Basin was less than that in the Indian lithospheric mantle, it was greater than that in the TP, reaching ~3.2 g/cm^3 (Figs. 17.12f–h and 17.13a,b) suggesting the rigid block of the Qaidam Basin.

Interestingly, there was a relative high-density belt with density of ~2.9 g/cm^3 beneath the BNS along the profiles of 80°E and 90°E (Fig. 17.13a,b), but the relative low densities occurred along the IYS and JS, where there are negative belts of the gravity anomaly (Fig. 17.3c) and two belts of the Moho subsidence (Fig. 17.5). The low-density in the crust of the HM and LT was corresponding to low velocity (Monsalve et al., 2008; He et al., 2010; Basuyau et al., 2013) and low density (Basuyau et al., 2013). However, Tiwari et al. (2006) argued that the lower crust presents high density inferred from the gravity modelling. In contrast, the density at depth of 30–70 km beneath the BNS is relatively high, reaching ~3.0 g/cm^3, which is probably related to the eclogites in the lower crust (Bai et al., 2013).

17.5 Discussion and Conclusions

Following the India–Asia collision, the continuing northward movement and subducting under the

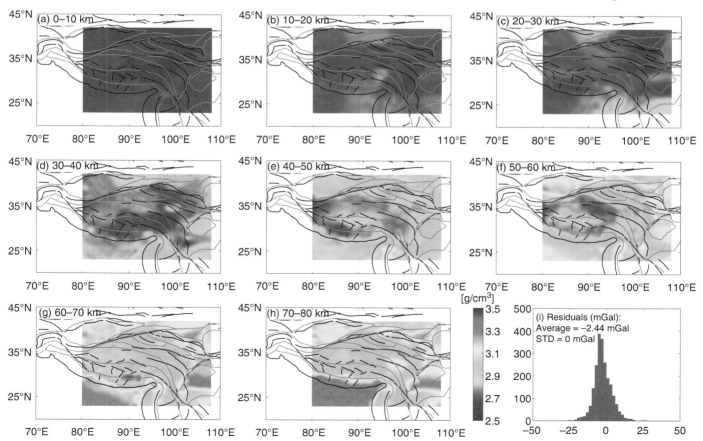

Figure 17.12 (a–h) Layered density structure derived from gravity inversion in the TP and (i) statistics of residuals.

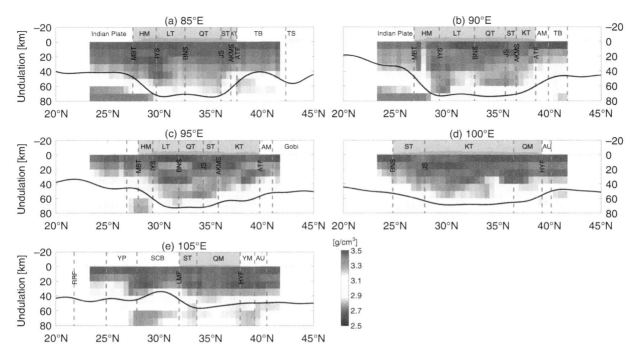

Figure 17.13 North–south trending sections for densities of 0–80 km depth from Figure 17.12 with Moho depth shown in Figure 17.5 from Xuan et al. (2015) across the TP.

Eurasian Plate has resulted in continued and accelerated crustal thickening and uplift of the TP. The Himalayan Mountains (HM) and Lhasa Terrane (LT) have acted as the main belts of convergence and collision between the Indian and Eurasian plates. Their crustal structures can be used to understand the dynamic process of continent–continent collision and the thrusting earthquakes that occur in the subduction zone. Correspondingly, the Moho depth linearly increased from south to north (Figs. 17.5 and 17.7) and the high-density was noted beneath the HM. However, lower density can be found in the lower crust beneath LT in the southern TP (Figs. 17.12f–h; Figs. 17.13a,b).

The Indian lithosphere subducted under the TP with small angle and thickened the relatively weak Tibetan crust resulting from the resistance of the rigid blocks (e.g. Tarim Basin). The dense lithosphere of the Indian Plate has sunk at the subduction zone due to its negative buoyancy (Stern, 2007; Chen et al., 2020), which allowed for the delamination of the subducting lithospheric mantle from the crust. The suction induced by the subducting slab has resulted in mantle convection on both sides of the slab (Conrad and Lithgow-Bertelloni, 2004). If the downwelling of the Indian lithospheric mantle with large angle occurs beneath the HM, the thin lithospheric mantle caused by the upwelling of the asthenosphere will occur to the north of the subducting lithospheric mantle, which was in agreement with the low-velocity anomaly reported by seismic studies (Zhang et al., 2014; Liang et al., 2016). Furthermore, the

less-dense lithospheric mantle under the LT and Qiangtang terrane (QT) (Fig. 17.13a,b) is likely to be associated with the partial melting due to the upwelling of the asthenospheric material (Tilmann et al., 2003; Zhang et al., 2014). To accommodate the convergence of the subducting Indian and Asian lithosphere, the Tibetan crust beneath Indus-Yarlung suture (IYS) and the Jinsha suture (JS) are folded and the Moho depth exceed 70 km (Figs. 17.5 and 17.7b,c). The increasing Moho depth from south to north is probably due to the negative buoyancy of the under-thrusting Indian Plate, which would induce the local mantle convection under the southern and northern parts of the under-thrusting plate (Tilmann et al., 2003). The upwelling asthenospheric materials are considered to be the main reason for the low density in the lithospheric mantle of the TP (Fig. 17.13a,b).

The lateral variations in the density in the TP (Figs. 17.12; 17.13a,b) may infer the probable presence and absence of eclogites in the lower crust beneath the HM, LT, and QT. The presence of eclogites has been widely used in support of the theory of mass transfer from the lower crust to the underlying mantle (Tiwari et al., 2006; Hetényi et al., 2007). Bai et al. (2013) suggested that the lower crust beneath HM and the southern LT was not eclogitised based on the lower density from gravity modelling, which was in keep with our inversing density model (Figs 17.12 and 17.13). The lower crust beneath the Bangong-Nujiang suture with relatively high density (Figs 17.12e–g; Figs. 17.13a,b) aids in interpreting the presence of the eclogites.

The eastward extruded Tibetan lithosphere was expected to change direction towards the south and southwest because of resistance from the Sichuan Basin within the NE trending margin. Eastward arcs of negative and positive anomalies (Fig. 17.4) and 40–50 km Moho relief (Figs. 17.5 and 17.8) near the Eastern Himalayan syntaxis suggest that the effects of subduction of the Burmese block in the lower crust (Socquet and Pubellier, 2005; Wang et al., 2007) spread to the CDB. Thus, the movement towards the south and southwest would cause the CDB to rotate clockwise around the Eastern Himalayan syntaxis along the strike-slip faults (Royden et al., 1997; Wang et al., 2008), corresponding to the multiple-order *en echelon* patterns of positive and negative anomalies across the Indo-China block from west to east (Fig. 17.4).

Acknowledgements. We thank Chris Rizos and an anonymous reviewer for their valuable comments of this manuscript. This work was supported by the National Key Research and Development Program of China Project (Grant No. 2018YFC0603502) and the National Natural Science Foundation of China (Grant No. 42074090).

References

Bagherbandi, M. (2012). A comparison of three gravity inversion methods for crustal thickness modelling in Tibet plateau. *Journal of Asian Earth Sciences*, 43(1), 89–97.

Bai, Z. M., Zhang, S. F., and Braitenberg, C. (2013). Crustal density structure from 3D gravity modeling beneath Himalaya and Lhasa blocks, Tibet. *Journal of Asian Earth Sciences*, 78, 301–17.

Bao, X. W., Song, X. D., and Li, J. T. (2015). High-resolution lithospheric structure beneath Mainland China from ambient noise and earthquake surface-wave tomography. *Earth and Planetary Science Letters*, 417, 132–41.

Barnoud, A., Coutant, O., Bouligand, C., Gunawan, H., and Deroussi, S. (2016). 3-D linear inversion of gravity data: Method and application to Basse-Terre volcanic island, Guadeloupe, Lesser Antilles. *Geophysical Journal International*, 205(1), 562–74.

Basuyau, C., Diament, M., Tiberi, C. et al. (2013). Joint inversion of teleseismic and GOCE gravity data: Application to the Himalayas. *Geophysical Journal International*, 193(1), 149–60.

Braitenberg, C., Zadro, M., Fang, J., Wang, Y., and Hsu, H. T. (2000). Gravity inversion in Qinghai-Tibet plateau. *Physics and Chemistry of the Earth Part A: Solid Earth and Geodesy*, 25(4), 381–36.

Brocher, T. A. (2005). Empirical relations between elastic wavespeeds and density in the earth's crust. *Bulletin of the Seismological Society of America*, 95(6), 2081–92.

Chen, L., Wang, X., Liang, X. F., Wan, B., and Liu, L. J. (2020). Subduction tectonics vs. Plume tectonics: Discussion on driving forces for plate motion. *Science China Earth Sciences*, 63 (3), 315–28.

Chen, W. J. (2017). Determination of crustal thickness under Tibet from gravity-gradient data. *Journal of Asian Earth Sciences*, 143, 315–25.

Chen, W. J., and Tenzer, R. (2017). Moho modeling in spatial domain: A case study under Tibet. *Advances in Space Research*, 59(12), 2855–69.

Clark, M. K., and Royden, L. H. (2000). Topographic ooze: Building the eastern margin of Tibet by lower crustal flow. *Geology*, 28, 703–6.

Conrad, C. P., and Lithgow-Bertelloni, C. (2004). The temporal evolution of plate driving forces: Importance of 'slab suction' versus 'slab pull' during the Cenozoic. *Journal of Geophysical Research: Solid Earth*, 109 (B10407). https://doi.rog/10.1029/2004jb002991.

Fu, G. Y., and She, Y. W. (2017). Gravity anomalies and isostasy deduced from new dense gravimetry around the Tsangpo Gorge, Tibet. *Geophysical Research Letters*, 44(20), 10233–9.

He, R. Z., Zhao, D. P., Gao, R., and Zheng, H.W. (2010). Tracing the Indian lithospheric mantle beneath central Tibetan Plateau using teleseismic tomography. *Tectonophysics*, 491 (1–4), 230–43.

Hetényi, G., Cattin, R., Brunet, F. et al. (2007). Density distribution of the India plate beneath the Tibetan plateau: Geophysical and petrological constraints on the kinetics of lower-crustal eclogitization. *Earth and Planetary Science Letters*, 264(1-2), 226–44.

Houseman, G., and England, P. (1986). Finite strain calculations of continental deformation 1. Method and general results for convergent zones. *Journal of Geophysical Research*, 91(B3), 3651–63.

Huang, J. L., and Zhao, D. P. (2006). High-resolution mantle tomography of China and surrounding regions. *Journal of Geophysical Research: Solid Earth*, 111, B09305. https://doi.org/10.1029/2005JB004066.

Jiang, W. L., Zhang, J. F., Tian, T., and Wang, X. (2012). Crustal structure of Chuan-Dian region derived from gravity data and its tectonic implications. *Physics of the Earth and Planetary Interiors*, 212, 76–87.

Jiménez-Munt, I., Fernàndez, M., Vergés, J., and Platt, J. P. (2008). Lithosphere structure underneath the Tibetan Plateau inferred from elevation, gravity and geoid anomalies. *Earth and Planetary Science Letters*, 267, 276–89.

Jin, S. G., and Park, P. (2006). Strain accumulation in South Korea inferred from GPS measurements. *Earth Planets Space*, 58(5), 529–34.

Jin, S. G., Park, P., and Zhu, W. (2007). Micro-plate tectonics and kinematics in northeast Asia inferred from a dense set of GPS observations. *Earth and Planetary Science Letters*, 257(3–4), 486–96.

Kumar, P., Yuan, X. H., Kind, R., and Ni, J. (2006). Imaging the colliding Indian and Asian lithospheric plates beneath Tibet. *Journal of Geophysical Research: Solid Earth*, 111, B06308. https://doi.org/10.1029/2005JB003930.

Laske, G., Masters., G., Ma, Z., and Pasyanos, M. (2013). Update on CRUST1.0: A 1-degree global model of Earth's crust. *Geophysical Research Abstracts*, 15, EGU2013-2658.

Li, C., Van der Hilst, R .D., Meltzer, A. S., and Engdahl, E. R. (2008). Subduction of the Indian lithosphere beneath the

Tibetan Plateau and Burma. *Earth and Planetary Science Letters*, 274(1–2), 157–68.

Li, S. L., Mooney, W. D., and Fan, J. C. (2006). Crustal structure of mainland China from deep seismic sounding data. *Tectonophysics*, 420(1–2), 239–52.

Li, Y. H., Gao, M. T., and Wu, Q. J. (2014). Crustal thickness map of the Chinese mainland from teleseismic receiver functions. *Tectonophysics*, 611, 51–60.

Liang, X. F., Chen, Y., Tian, X. B. et al. (2016). 3D imaging of subducting and fragmenting Indian continental lithosphere beneath southern and central Tibet using body-wave finite-frequency tomography. *Earth and Planetary Science Letters*, 443, 162–75.

Molnar, P., and Tapponnier, P. (1975). Cenozoic tectonics of Asia: Effects of a continental collision. *Science*, 189, 419–26.

Monsalve, G., Sheehan, A., Rowe, C., and Rajaure, S. (2008). Seismic structure of the crust and the upper mantle beneath the Himalayas: Evidence for eclogitization of lower crustal rocks in the Indian Plate. *Journal of Geophysical Research: Solid Earth*, 113(B8). https://doi.rog/10.1029/2007jb005424.

Nábělek, J., Hetényi, G., Vergne, J. et al. (2009). Underplating in the Himalaya-Tibet collision zone revealed by the Hi-CLIMB experiment. *Science*, 325, 1371–74.

Oldenburg, D. W. (1974). The inversion and interpretation of gravity anomalies. *Geophysics*, 39(4), 526–36.

Owens, T. J., and Zandt, G. (1997). Implications of crustal property variations for models of Tibetan plateau evolution. *Nature*, 387, 37–43.

Parker, R. L. (1973). The rapid calculation of potential anomalies. *Geophysical Journal of the Royal Astronomical Society*, 31, 447–55.

Pavlis, N. K., Holmes, S. A., Kenyon, S. C., and Factor, J. K. (2008). *An Earth gravitational model to degree 2160: EGM2008, EGU General Assembly*. Vienna: European Geosciences Union.

Royden, L. H., Burchfiel, B. C., King, R. W. et al. (1997). Surface deformation and lower crustal flow in eastern Tibet. *Science*, 276(5313), 788–90.

Searle, M. P., Elliott, J. R., Phillips, R. J., and Chung, S. L. (2011). Crustal-lithospheric structure and continental extrusion of Tibet. *Journal of the Geological Society*, 168(3), 633–72.

Shin, Y. H., Xu, H., Braitenberg, C., Fang, J., and Wang, Y. (2007). Moho undulations beneath Tibet from GRACE-integrated gravity data. *Geophysical Journal International*, 170(3), 971–85.

Shin, Y. H., Shum, C. K., Braitenberg, C. et al. (2009. Three-dimensional fold structure of the Tibetan Moho from GRACE gravity data. *Geophysical Research Letters*, 36. https://doi.rog/10.1029/2008gl036068.

Shin, Y. H., Shum, C. K., Braitenberg, C. et al. (2015). Moho topography, ranges and folds of Tibet by analysis of global gravity models and GOCE data. *Scientific Reports*, 5. https://doi.rog/10.1038/srep11681.

Sjöberg, L .E. (2013). On the isostatic gravity anomaly and disturbance and their applications to Vening Meinesz-Moritz gravimetric inverse problem. *Geophysical Journal International*, 193(3), 1277–82.

Socquet, A., and Pubellier, M. (2005). Cenozoic deformation in western Yunnan (China–Myanmar border). *Journal of Asian Earth Sciences*, 24(4), 495–515.

Steffen, R., Steffen, H., and Jentzsch, G. (2011). A three-dimensional Moho depth model for the Tien Shan from EGM2008 gravity data. *Tectonics*, 30(TC5019). https://doi.rog/10.1029/2011tc002886.

Stern, R. J. (2007). When and how did plate tectonics begin? Theoretical and empirical considerations. *Chinese Science Bulletin*, 52(5), 578–1.

Tapponnier, P., Peltzer, G., Le Dain, A. Y., Armijo, R., and Cobbold, P. (1982). Propagating extrusion tectonics in Asia: New insights from simple experiments with plasticine. *Geology*, 10: 611–16.

Teng, J. W., Zhang, Z. J., Zhang, X. K. et al. (2013). Investigation of the Moho discontinuity beneath the Chinese mainland using deep seismic sounding profiles. *Tectonophysics*, 609, 202–16.

Teng, J. W., Yang, D. H., Tian, X. B. et al. (2020). Geophysical investigation progresses of the Qinghai-Tibetan Plateau in the past 70 years. *Scientia Sinica Terrae*, 49, 1546–64.

Tenzer, R., and Bagherbandi, M. (2012). Reformulation of the Vening-Meinesz Moritz inverse problem of isostasy for isostatic gravity disturbances. *International Journal of Geosciences*, 3: 918–29.

Tenzer, R., Chen, W., and Jin, S. G. (2015). Effect of the upper mantle density structure on the Moho geometry. *Pure and Applied Geophysics*, 172(6), 1563–83.

Tilmann, F., Ni, J., and Team, I. I. S. (2003). Seismic imaging of the downwelling Indian lithosphere beneath central Tibet. *Science*, 300(5624), 1424–7.

Tiwari, V. M., Rao, M. B. S. V., and Singh, M. B. (2006). Crustal structure across Sikkim, NE Himalaya from new gravity and magnetic data. *Earth and Planetary Science Letters*, 247, 61–9.

Vaníček, P., Tenzer, R., Sjöberg, L. E., Martinec, Z., and Featherstone, W. E. (2004). New views of the spherical Bouguer gravity anomaly. *Geophysical Journal International*, 159, 460–72.

Wang, C. Y., Chan, W. W., and Mooney, W. D. (2003). Three-dimensional velocity structure of crust and upper mantle in southwestern China and its tectonic implications. *Journal of Geophysical Research: Solid Earth*, 108(B9). https://doi.org/10.1029/2002JB001973.

Wang, C. Y., Han, W. B., Wu, J. P., Lou, H., and Chan, W. W. (2007). Crustal structure beneath the eastern margin of the Tibetan Plateau and its tectonic implications. *Journal of Geophysical Research: Solid Earth*, 112(B7). https://doi.rog/10.1029/2005jb003873.

Wang, C. Y., Lou, H., Lue, Z. Y. et al. (2008). S-wave crustal and upper mantle's velocity structure in the eastern Tibetan Plateau: Deep environment of lower crustal flow. *Science in China Series D: Earth Sciences*, 51(2), 263–74.

Wang, Z. W., Zhao, D. P., Gao, R., and Hua, Y. Y. (2019). Complex subduction beneath the Tibetan plateau: A slab warping model. *Physics of the Earth and Planetary Interiors*, 292, 42–54.

Xu, C., Liu, Z. W., Luo, Z. C., Wu, Y. H., and Wang, H. H. (2017). Moho topography of the Tibetan Plateau using multi-scale gravity analysis and its tectonic implications. *Journal of Asian Earth Sciences*, 138: 378–86.

Xuan, S., Shen, C. Y., Li, H., and Tan, H. B. (2016). Structural interpretation of the Chuan-Dian block and surrounding regions using discrete wavelet transform. *International Journal of Earth Sciences*, 105(5), 1591–602.

Xuan, S., Shen, C. Y., Tan, H. B., and Li, H. (2015). Inversion of Moho depth in China mainland from EGM2008 gravity data. *Journal of Geodesy and Geodynamics*, 35 (2), 309–11, 317.

Yin, A., and Harrison, T. M. (2000). Geologic Evolution of the Himalayan-Tibetan Orogen. *Annual Review of Earth and Planetary Sciences*, 28: 211–80.

Zhang, Z. J., Deng, Y. F., Teng, J. W. et al. (2011). An overview of the crustal structure of the Tibetan plateau after 35 years of deep seismic soundings. *Journal of Asian Earth Sciences*, 40(4), 977–89.

Zhang, Z. J., Teng, J. W., Romanelli, F. (2014). Geophysical constraints on the link between cratonization and orogeny: Evidence from the Tibetan Plateau and the North China Craton. *Earth-Science Reviews*, 130, 1–48.

Zhang, C., Huang, D. N., Wu, G. C. et al. (2015). Calculation of Moho depth by gravity anomalies in Qinghai-Tibet plateau based on an improved iteration of Parker-Oldenburg inversion. *Pure and Applied Geophysics*, 172(10), 2657–68.

Zhao, G. D., Liu, J. X., Chen, B., Kaban, M. K., and Zheng, X. Y. (2020). Moho beneath Tibet based on a joint analysis of gravity and seismic data. *Geochemistry, Geophysics, Geosystem*, 21(2), e2019GC008849.

Zhao, L. F., Xie, X. B., He, J. K., Tian, X. B., and Yao, Z. X. (2013). Crustal flow pattern beneath the Tibetan plateau constrained by regional Lg-wave Q tomography. *Earth and Planetary Science Letters*, 383, 113–22.

Zheng, H. W., Li, T. D., Gao, R., Zhao, D. P., and He, R. Z. (2007). Teleseismic P-wave tomography evidence for the Indian lithospheric mantle subducting northward beneath the Qiangtang terrane. *Chinese Journal of Geophysics*, 50(5), 1223–32.

18

Geodetic Inversions
and Applications in Geodynamics

Grigory M. Steblov and Irina S. Vladimirova

Abstract: The primary observables of the Global Positioning
System (GPS) ground tracking sites for geodynamics are the
Earth's surface motions, and their geophysical interpretation
is based on the numerical models of various tectonic processes. The key issues for geophysical interpretation of the
GPS observations are adequate mechanical models of brittle
and ductile rock behaviour used to predict surface motions
related to various tectonic processes, and the corresponding
inversion techniques which allow separation of the processes,
and evaluation of their parameters. For large-scale heterogeneous processes, the inversion of the GPS observations
requires regularisation because it implies evaluation of some
complicated distributed underground motions from their discrete manifestation at the surface. One of the fastest growing
applications of the satellite geodetic observations is investigation of the seismotectonic deformation associated with
great earthquakes worldwide at all stages of the seismic
cycle – inter-seismic, co-seismic, post-seismic. The inversion
techniques based on dislocation models in elastic or viscoelastic medium is one of the approaches that may be widely
used for GPS-based studies of various seismotectonic
deformations.

18.1 Introduction

Satellite geodetic observations based on Global Navigation
Satellite Systems (GPS, GLONASS, etc.) systems have
made significant progress since their initial deployment
worldwide in early 1990s. Originally designed for navigation
purposes, they reached an accuracy level suitable for tracking tectonic motions of the Earth's surface, thereby becoming suitable for geodynamic research (Dixon, 1991).
Occasional observations of selective active tectonic belts
were among of the first well-known applications of the
satellite geodetic investigations for geodynamics, and, as
the precision of the measurements and their processing technique evolved, global-scale tectonic plate motions (Argus
and Gordon, 1991; Argus and Heflin, 1995; Kogan and
Steblov, 2008) also become the subject of such research.
As the tracking network expanded and densified, turning
into a set of permanent GPS sites well distributed globally,
a lot of irregular time-varying phenomena added to the

topics addressed by space geodesy. The whole scope of
geodynamic processes investigated by means of GPS has
been continuously growing, now including global scale
plate motions, plate boundary deformation, mantle and
asthenosphere flow, seismotectonic deformation, volcanology, postglacial isostatic rebound, ice flow, water mass
flow, and so on. All of them are characterised by various
spatial and time scales, from global to regional, from
instantaneous to centennial.

One set of applications of satellite geodesy, actively developed for more than two decades, involves observations of
seismotectonic deformations at all stages of the seismic
cycle, including inter-seismic, co-seismic, post-seismic. The
most significant factor of such geodetic applications is their
ability to directly resolve slow aseismic motions, both inter-
seismic and post-seismic, as well as co-seismic static offsets.
All these processes are usually beyond the sensitivity of
seismic instrumentation, which show only indirect effects
of those motions, such as foreshocks, aftershocks, and
main shocks. Seismic instrumentation is also subject to
limited frequency response, especially at low band, and
magnitude saturation at close proximity of the great earthquakes. Analysis of such observations substantially
improved understanding of seismotectonic deformations
backgrounds: distributed asperities associated with locking
along the seismogenic fault, slip distribution for the finite
fault seismic source model, mechanisms of post-seismic
transient response.

Using satellite geodesy data, initially only quite simple
models were implemented for inversion for the first-order
features, such as a single plane uniform co-seismic slip to
evaluate the earthquake rupture size, a uniform interplate
locking to evaluate its depth, and so on. Appropriate models
were developed to predict surface motions based on dislocation model in various media: single rectangular patch in
homogenous half-space (Okada, 1992), point sources in
layered half-space (He et al., 2003) or multiple rectangular
patches in stratified spherical media (Rundle, 1980;
Piersanti et al., 1995; Pollitz, 1996, 1997). This approach
was used for numerous investigations of inter-seismic coupling and asperity distribution in many subduction zones, for
example, Kamchatka (Bürgmann et al., 2005) as well as for

co-seismic slip and afterslip evaluation after great earthquakes like 1997 Kronotsky Kamchatka earthquake (Bürgmann et al., 2001) and 2000 Uglegorsk earthquake (Kogan et al., 2003). Subduction zones release $\sim$90% of the total global seismic moment, and thus constitute much of the Earth's seismic potential. To better determine the seismic potential of a specific subduction zone, the spatial and temporal distribution of elastic strain accumulation and release along the plate boundary underthrust must be understood (Bürgmann et al., 2005).

While the observation network density increased, the more detailed patterns were sought for inversion of the surface displacements. The spatial-temporal variations of the fault locking and spatially distributed slip became resolvable, which raised the issue of stability of the inversion technique. For large-scale spatially heterogeneous processes the inversion of the GPS observations requires regularisation because it implies evaluation of some complicated distributed underground motions from their discrete manifestation at the surface. Thus, the key issues for geophysical interpretation of the GPS observations are adequate rheological models used to predict surface motions related to various tectonic processes and the corresponding inversion technique which allows to separate the processes and to evaluate their parameters.

18.2 Dislocation Model

The most common approach to evaluate surface displacements due to active faults motion is based on the dislocation model in various media, like an elastic half space or a spherical layered elastic or viscoelastic media. Such models are widely used for investigation of either static co-seismic offsets or aseismic motions, the latter significantly contributing to the strain inter-seismic build-up and its release during post-seismic creep and viscoelastic relaxation in the asthenosphere.

In general, while solving linear equations of static or quasi-static equilibrium with a point source, the surface response to any deep dislocation may be expressed through convolution with a Green's function:

$$\mathbf{u}(\mathbf{r}, t) = \iint\limits_S \mathbf{G}(\mathbf{r}, \mathbf{r}_s, t, t_s) \mathbf{U}(\mathbf{r}_s, t_s) dS, \qquad (18.1)$$

where $\mathbf{U}(\mathbf{r}_s, t_s)$ is the source dislocation at the point $\mathbf{r}_s$ at the moment t_s, S is the dislocation surface, $\mathbf{u}(\mathbf{r}, t)$ is the surface displacement at the point $\mathbf{r}$ at the moment t, and $\mathbf{G}(\mathbf{r}, \mathbf{r}_s, t, t_s)$ is the Green's function determined by the medium's rheology.

Then the inversion of the surface motion for the deep dislocation assuming the noise in the observable $\mathbf{u}(\mathbf{r}, t)$ is equivalent to finding the $\mathbf{U}(\mathbf{r}_s, t_s)$ which minimises the residuals of the observation equation (Eq. 18.1):

$$\mathbf{U}(r_s, t_s) = \arg \min_{\mathbf{U}(r_s, t_s)} \left\{ \left\| \mathbf{u}(r, t) - \iint\limits_S \mathbf{G}(r, r_s, t, t_s) \right. \right.$$
$$\left. \mathbf{U}(r_s, t_s) dS \right\| + \alpha R[\mathbf{U}(r_s, t_s)] \right\}, \qquad (18.2)$$

where the regularisation term $R[\mathbf{U}(r_s, t_s)]$ is introduced to stabilise and/or smooth the solution with the weighting coefficient α. Various types of regularisation can be applied to minimise not only misfit to the data, but also to minimise/constrain any of the following commonly used forms of $R[\mathbf{U}(r_s, t_s)]$:

- the magnitude of the estimated parameters (with Tikhonov (L^2) or Manhattan norm);
- the second spatial derivative of the parameter vector (Laplacian smoothing);
- the difference between estimated model and *a priori* model.

In practice, the observations are sampled in discrete manner in the spatial and time domain, which means that we have a set of surface displacements $\mathbf{u}_{ij} = \mathbf{u}(\mathbf{r}_i, t_j)$ at a finite number of sites $\mathbf{r}_i$ and moments t_j. Possible significant spatial variations of the Green's function $\mathbf{G}(\mathbf{r}, \mathbf{r}_s, t, t_s)$ necessitate consideration of a non-uniform spatial distribution of $\mathbf{U}(\mathbf{r}_s, t_s)$ for great earthquakes in the convolution (Eq. 18.1). Thus, numerical modelling of the distributed slip for large scale faults requires discretisation of the $\mathbf{U}(\mathbf{r}_s, t_s)$. The common approach for discretisation of slip distribution is the approximation of the fault surface S by a finite number of elements. In the two-dimensional case, triangular or rectangular elements are most often used. Triangular elements are more suitable to approximate areas bounded by curvilinear contours than rectangular. But the latter are easier to apply, because the widely used dislocation modelling approaches are designed for rectangular dislocations (Okada, 1985; Pollitz, 1996, 1997). In general, the finite element method can be performed using irregular grids, but numerical procedures for solving problems with such grids become more complicated. An irregular grid provides better resolution in case of sharp uneven spatial variations of any parameter, but for most geophysical problems in which the modelling domain is defined by simple geometric shapes, it is most appropriate to use a regular grid (Ismail-Zadeh and Tackley, 2010). In this chapter a regular grid is used to discretise the modelling area, and rectangles are chosen as finite elements. Thus, for numerical calculations the solution is sought in discretised form as a piecewise constant vector field: $\mathbf{U}(\mathbf{r}_s, t_s) = \mathbf{U}_k(t_s)$ for $\mathbf{r}_s \in S_k$, where S_k are the elements of the dislocation surface $S = \{S_1, S_2, \ldots, S_m\}$ (Fig. 18.1).

Expansion of each vector $\mathbf{U}_k(t_s)$ into its components in terms of an orthonormal basis $\mathbf{e}_{k1}, \mathbf{e}_{k2}$ for each surface element S_k:

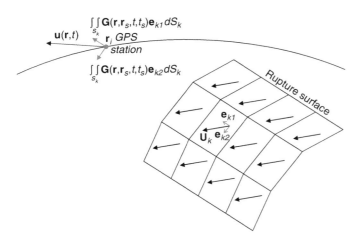

Figure 18.1 Non-uniform dislocation: discretisation.

$$\mathbf{U}_k(t_s) = \mathbf{e}_{k1} U_{k1}(t_s) + \mathbf{e}_{k2} U_{k2}(t_s) \qquad (18.3)$$

yields Eq. (18.1) in a discrete form:

$$\mathbf{u}_{ij} = \sum_{k,l} U_{kl}(t_s) \iint\limits_{S_k} \mathbf{G}(\mathbf{r}_i, \mathbf{r}_s, t_j, t_s) \mathbf{e}_{kl} dS \qquad (18.4)$$

which is a system of linear equations:

$$\mathbf{Lm} = \mathbf{d}. \qquad (18.5)$$

Here $\mathbf{d}$ is the data column composed of the components of the observed displacements vectors $\mathbf{u}_{ij}$, $\mathbf{m}$ is model parameters column vector composed of unknown dislocation components $U_{kl}(t_s)$, $\mathbf{L}$ is the operator predicting the data from the model, composed of the separate integrals over fault elements $\iint\limits_{S_k} \mathbf{G}(\mathbf{r}_i, \mathbf{r}_s, t_j, t_s) \mathbf{e}_{kl} dS$, each representing the surface response to a uniform dislocation $\mathbf{e}_{kl}$ over the surface element S_k. Then the constrained damped least squares can be used to invert the observed surface motions for the dislocation distribution over the fault model. Similar to Eq. (18.1) in discrete form, the inversion minimises the objective function

$$OBJ = \chi_r^2 + \alpha \sum_l m_l^2, \qquad (18.6)$$

where $\chi_r^2 = \frac{1}{N} \sum_{i=1}^{N} \frac{1}{\sigma_i^2} \left(d_i - \sum_{j=1}^{M} L_{ij} m_j \right)^2$ is conventionally called the reduced chi-square of the inverse problem with zeroth order regularisation (Press et al., 2007).

Such a representation allows to use various existing algorithms to calculate the integrals $\iint\limits_{S_k} \mathbf{G}(\mathbf{r}_i, \mathbf{r}_s, t_j, t_s) \mathbf{e}_{kl} dS$ of the Green's functions over each element of the dislocation surface.

The most straightforward is the case when the elements S_k are rectangles. The most widely used are the Okada (Okada, 1992) and Pollitz (Pollitz, 1996, 1997; Wang et al., 2003; Barbot and Fialko, 2010; Tanaka, 2013) approaches. The former approach (Okada, 1992) is designed for a rather simple medium, an elastic homogeneous half-space and it is an explicit finite representation of the strains and displacements over the medium with a uniform rectangular dislocation. The latter approach (Pollitz, 1996, 1997) is rather more complicated, for the uniform rectangular dislocations set in a spherically symmetric layered elastic or viscoelastic medium, and it is based on a series expansion in spherical functions of the solution. The approach of Okada is well applicable at small distances for shallow dislocations when sphericity and elastic stratification are neglectable. For deeper dislocations and large distances, the Pollitz approach is preferable since neglecting sphericity and stratification yields tangible errors (up to 20%). Either approach provides a tool for direct calculation of the surface integrals of a unit dislocation over the elements S_k in Eq. (18.4), thus making the minimisation task (Eq. 18.6) straightforward with use of standard optimisation techniques.

Non-linear programming algorithm implemented by NPSOL software (Gill et al., 1986) is one of such optimisation tools, used in this chapter to minimise non-linear objective function in Eq. (18.2) with linear and nonlinear constraints on the parameter vector. For visualisation purposes, the resulting slip and coupling patterns were further smoothed by means of B-spline interpolation when plotted with the GMT software (Wessel et al., 2019).

This methodology for interpreting the Earth's surface offsets can be used with minor adjustments to analyse various satellite geodetic data in the vicinity of great earthquakes. The results of the analysis, depending on the input data used, are, for example, estimates of the rate and variations in the rate of the elastic stresses accumulation at the fault surface, co-seismic source slip distribution, parameters of post-seismic processes, and also estimates of the rheological parameters of the medium.

18.3 Preliminary Setting of the Modelling

The primary issues of GPS-data interpretation based on dislocation techniques are the choice of an adequate rheological model of the medium and the correct specification of the fault geometry. The PREM Earth model (Dziewoński and Anderson, 1981) used in this chapter can be considered as quite suitable to model elastic deformations in a stratified medium with the viscosity introduced to model viscoelastic behaviour when appropriate (Table 18.1).

The most common methods to determine the linear dimensions of the initial co-seismic fault include on-site investigation methods such as direct mapping of the faults at the surface in the case of strong crustal earthquakes or geodetic measurements of the offsets and tilts of the Earth's surface induced by

Table 18.1 *Lithosphere and mantle rheology*

Depth, km	Maxwell viscosity, Pa·s	Elastic moduli
0–63	$\eta_1 \to \infty$	PREM Earth model
63–220	$\eta_1 = 1 \times 10^{17} \div 1 \times 10^{19}$	(Dziewoński and
220–670	$\eta_1 = 5 \times 10^{20}$	Anderson, 1981)
670–2900	$\eta_1 = 5 \times 10^{21}$	

an earthquake. It is also possible to use theoretical methods based on considerations about the possible tensile strength of the material in the vicinity of the source considering knowledge of such quantities as seismic energy or material elasticity (Riznichenko, 1985).

The following approach for specification of the geometry of the fault zone is suitable for any large earthquake with long fault zone. Many great earthquakes with long fault occur in subduction zones. The main difficulty of the fault geometry evaluation in the case of subduction earthquakes is their typical underwater localisation. That means that investigation of the surface trace or direct offsets measurements in the source zones in most cases is impossible, and thus the source modelling is based on seismological data and empirical patterns instead of direct on-site observations. An approximate fault geometry can be inferred from instrumental data on the epicentral location and depth of the main shock, foreshocks, and aftershocks, and the relative position of the foreshock, main shock, and aftershocks; data on the tsunami source location, approximating seismic source location; an empirical dependence of the fault sizes on the earthquake magnitude (Fedotov, 1965; Riznichenko, 1985).

One of the widely used methods to determine the seismic rupture dimensions is the outlining of the aftershocks cloud, which is expected to coincide approximately with the rupture zone subject to maximum displacements and deformations at the moment of the main shock. And at the same time, the highest concentration of aftershocks should be observed along the edges of the rupture (Fedotov, 1965; Kuchay, 1982; Das and Henry, 2003). The characterisation of aftershock sequences for estimating the linear dimensions of the rupture is based on the statistical or cluster analysis of seismic event catalogues (Molchan and Dmitrieva, 1992).

The resolution of the GPS data inversions for the seismic rupture depends on many factors, including the source signal pattern and the rupture model discretisation, as well as the observing network density. A general idea of the data resolution can be inferred from a checkerboard test, which is applicable to any inversion procedure independent of the internal operator of the inverse problem. The test is based on simulating a hypothesised signal at the GPS sites used for inversion. The source signal pattern is usually composed of spatially alternating patches with opposite signals (source slip, for example) with random noise applied. Usually, the random errors are con-

sidered as Gaussian with zero mathematical expectation and standard deviation equal to the mean data errors. Then the synthesised observations are inverted for the source signal and the recovery of the original source signal pattern is considered as a stability criterion (Lévêque et al., 1993; Zelt et al., 2006).

Inversion of the surface motions for a finite fault model also requires information on its spatial orientation. Attempts to simultaneously estimate the source slip distribution together with the dislocation surface orientation greatly increases the number of degrees of freedom in the inversion, thus necessitating additional *a priori* constraint on the vertical cross section of the faults. For subduction zones, such information can be inferred from the local seismicity as the seismic focal zone profile delineated by the subduction type earthquakes. Then the dip of each fault element S_k in Eq. (18.4) is estimated from piecewise linear approximation of such profiles.

An example of the co-seismic rupture evaluation and discretisation for the Great Tohoku earthquake 2011 is shown in Fig. 18.2.

18.4 Co-seismic Motions

Considering static surface offsets following an earthquake as the result of finite fault rupture, all the variables in Eq. (18.1), including the Green's functions, take static form without time dependence:

$$\mathbf{u}(\mathbf{r}) = \iint\limits_S \mathbf{G}(\mathbf{r}, \mathbf{r}_s)\mathbf{U}(\mathbf{r}_s)dS. \tag{18.7}$$

With this static form, inversion based on Eqs. (18.1)–(18.6) of the static surface offsets for slip distribution provides an estimate which is usually in good agreement with various teleseismic inversions. The advantage of inversion of surface geodetic offsets is its stability for great events in the near field zone, while the seismic inversions usually degrade due to limitations of the seismic equipment frequency response, especially at low frequencies, and amplitude saturation at small epicentral distances. This is especially important for the case of shallow undersea events for correct assessment of their tsunamigenic potential.

Some examples of geodetic co-seismic offsets analysis performed by the authors of this chapter for the earthquake source investigation follow. All GPS time series used in the simulations have been obtained by the authors from the raw GPS data. An example of simulation for the 2006–7 Simushir earthquakes, including the input data, is available through a public repository (Vladimirova et al., 2019). Raw GPS data for the 2011 Tohoku earthquake are available upon request from the Geospatial Information Authority of Japan (GSI).[1]

---

[1] www.gsi.go.jp/ENGLISH/geonet_english.html.

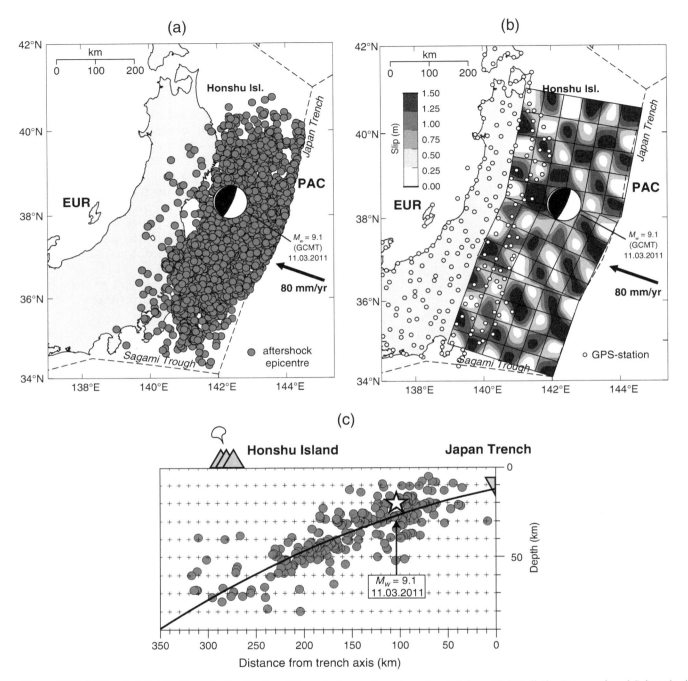

Figure 18.2 (a) Spatial distribution of aftershocks of the Tohoku earthquake extracted from ISC Bulletin (International Seismological Centre, 2022) data, (b) optimal grid spacing for slip distribution obtained by checkerboard test for Tohoku earthquake, and (c) second-degree polynomial regression of the vertical profile of Japan subduction zone, obtained by mapping the depths of the subduction earthquakes from the Centennial Catalog (Engdahl and Villaseñor, 2002).

18.4.1 Mid-Kuril (Simushir) Earthquakes 2006–2007

On 15 November 2006, a thrust event ruptured the subduction interface between the Pacific and North American plates; then on 13 January 2007, an extensional event ruptured the outer rise of the Pacific lithosphere near the Kuril trench. The earthquakes struck at a distance of about 100 km from each other in the Kuril arc segment where such large events had not happened since 1915 (Fedotov, 1965; United States Geological Survey (USGS)[2]). The tsunami runup of the 2006 earthquake reached 20 m on the Kuril Islands

---

[2] http://earthquake.usgs.gov.

(Bourgeois et al., 2007). The detected horizontal co-seismic offsets for these two events captured at the regional satellite geodetic network ranged from several mm to over half a metre at the nearest GPS stations Matua (MATC) and Ketoy (KETC), shown in Fig. 18.3a. The rupture surface S to model each event was chosen to comply with the subduction interface. Its vertical cross section was constrained by the local shallow seismicity in the centre of the Kuril subduction zone by projecting hypocentres of thrust events since 1976 from the region bounded by the modelling surface. The boundaries of the rupture surface model were chosen so as to include the area that comprises the GCMT hypocentre and the aftershocks cloud from National Earthquake Information Center (NEIC) (Masse and Needham, 1989). To compose the linear operator in Eq. (18.5), the contributions from fault elements $\iint_{S_k} \mathbf{G}(\mathbf{r}_i, \mathbf{r}_s)\mathbf{e}_{kl}dS$ were calculated with the (Pollitz, 1996) approach for a spherically symmetrical layered Earth.

For the first Mid-Kuril event (Fig. 18.3a), the following *a priori* constraints were imposed for inversion:

$$rake = 90° \pm 30°, |slip| \leq 30 \text{ m}, \| slip - slip_{averaged} \| \leq 6 \text{ m},$$

which yielded $\chi^2 = 2.6$.

And for the second Mid-Kuril event (Fig. 18.3b), *a priori* constraints

$$|slip| \leq 30 \text{ m}, \| slip - slip_{averaged} \| \leq 5 \text{ m}$$

yielded $\chi^2 = 5.5$.

Comparison of various source slip models (Table 18.2) shows that the distributed slip over finite fault model in the spherically symmetric layered media provides the most adequate models for great earthquakes, compared to a point source or uniform slip model. The distributed slip model agrees better in terms of scalar seismic moment with the updated teleseismic body-wave solution (Lay et al., 2009).

The inversion of the surface offsets for slip distribution shows that the highest slip locations for the 2006 and 2007 earthquakes are adjacent to each other (Fig. 18.3); this correlation suggests (although does not prove) that the 2007 extensional event was triggered by redistribution of stresses following the 2006 thrust event.

Table 18.2 *Scalar moment release (din · cm) and magnitude: Various estimates*

	Event	
Solution	15/11/2006	13/01/2007
Distributed slip	$M_0 = 5.93 \times 10^{28}$, Mw = 8.5	$M_0 = 3.05 \times 10^{28}$, Mw = 8.3
Uniform slip	$M_0 = 1.57 \times 10^{28}$, Mw = 8.1	$M_0 = 0.85 \times 10^{28}$, Mw = 7.9
GCMT	$M_0 = 3.51 \times 10^{28}$, Mw = 8.3	$M_0 = 1.78 \times 10^{28}$, Mw = 8.1
P-waves, body-waves (Lay et al., 2009)	$M_0 = 5.0 \times 10^{28}$ Mw = 8.4	$M_0 = 2.6 \times 10^{28}$ Mw = 8.2

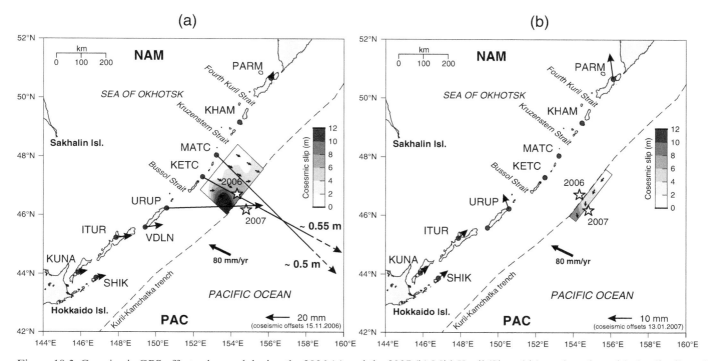

Figure 18.3 Co-seismic GPS offsets observed during the 2006 (a) and the 2007 (b) Mid-Kuril (Simushir) earthquakes with the distributed source slip models for these earthquakes.

18.4.2 Tohoku Earthquake 2011

Another example of inversion of co-seismic surface offsets for distributed source slip is the great Tohoku earthquake. On 11 March 2011, a thrust event ruptured the subduction interface between the Pacific and Eurasian plates where such a large event had not happened since 869 (Namegaya and Satake, 2014). The tsunami runup of the 2011 earthquake was as high as 40 m along the Sanriku coast (Lay et al., 2011; Yamazaki et al., 2018). The detected horizontal co-seismic offsets for this event captured at the GEONET satellite geodetic network ranged from several centimetres to over 5 m, shown in Fig. 18.4. Considerations similar to those for the Mid-Kuril earthquakes 2006–7 were used to choose the rupture surface in compliance with the subduction interface delineated by the shallow thrust events in the north part of the Japan subduction zone. The boundaries of the fault model were chosen to include the GCMT hypocentre and the aftershocks cloud from NEIC. The approach of (Pollitz, 1996) for a spherical layered Earth was used to calculate the convolution fault element integrals constituting the linear operator in Eq. (18.5). Due to the diurnal intervals used to estimate the position time series, the static offsets taken for inversion include not only instantaneous co-seismic jumps associated with the main shock, but also co-seismic jumps caused by its two largest aftershocks with Mw = 7.9 and Mw = 7.6 occurred on the same day. The following *a priori* constraints were imposed for

Table 18.3 *Scalar moment release (din · cm) and magnitude: Various estimates*

Solution	Event 11/03/2011
Distributed slip	$M_0 = 4.12 \times 10^{29}$, Mw = 9.01
Uniform slip (Zhou et al., 2018)	$M_0 = 3.55 \times 10^{29}$, Mw = 8.97
GCMT	$M_0 = 5.31 \times 10^{29}$, Mw = 9.1
P-waves, body-waves (Yokota et al., 2011)	$M_0 = 4.3 \times 10^{29}$, Mw = 9.0

inversion: $rake = 90° \pm 30°, |slip| \leq 50$ m, which yielded $\chi^2 = 25.18$.

Similar to the Mid-Kuril earthquakes 2006–7, a comparison of various source models (Table 18.3) shows that the distributed slip over finite fault model in the spherically symmetric layered media provides the most adequate models for this great earthquake compared to point source model or uniform slip. The distributed slip complies better with the teleseismic body-waves solution (Yokota et al., 2011).

The inversion of the surface offsets for slip distribution shows that the highest slip for the 2011 earthquake was located at shallow depth close to the trench (Fig. 18.4). The revealed large near-trench slip (up to 28 m) is one of the reasons for such a devastating tsunami.

18.5 Inter-seismic Motions at the Plate Boundaries

The analysis of inter-seismic motion is based on a well-known superposition approach (Savage, 1983). The main idea of this approach is that the accumulation of elastic stresses can be described as a superposition of steady state subduction, which doesn't contribute to the deformation of the medium, and a repetitive cycle of slip on the shallow main thrust zone causing elastic deformation observable on the earth's surface. Thus, the surface deformation caused by the accumulation of elastic stresses due to the coupling $\varphi(\mathbf{r}_s, t)$ of the interplate interface and relative plate motion $\mathbf{V}_{conv}(\mathbf{r}_s, t)$ can be modelled by adding a reverse (back) slip at each point of the interface plane, the magnitude of which is determined as:

$$\mathbf{V}_{back}(\mathbf{r}_s, t) = -\varphi(\mathbf{r}_s, t) \cdot \mathbf{V}_{conv}(\mathbf{r}_s, t). \tag{18.8}$$

Differentiation of (18.1) with respect to time yields the relation between the surface velocities and the back slip rate over the fault:

$$\mathbf{v}(\mathbf{r}, t) = \iint_S \mathbf{G}(\mathbf{r}, \mathbf{r}_s) \mathbf{V}_{back}(\mathbf{r}_s, t) dS. \tag{18.9}$$

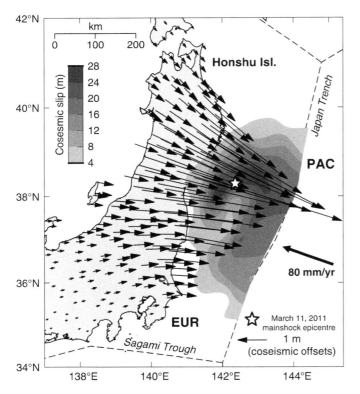

Figure 18.4 Co-seismic GPS offsets observed by GEONET stations during the Tohoku earthquake and model of distributed slip in the source zone of the Tohoku earthquake.

Here inelastic response components are considered neglectable at a given strain rate thus making the Green's function $\mathbf{G}(\mathbf{r}, \mathbf{r}_s)$ independent of time while possible temporal variations of the velocities $\mathbf{V}_{back}(\mathbf{r}_s, t)$ are presumed.

The back slip is interpreted as a result of the plate boundaries motion opposite to the relative plates motion $\mathbf{V}_{conv}(\mathbf{r}_s, t)$ at their interface due to asperities. The coupling ratio φ between the residual displacement at the interface (slip deficit) $\mathbf{V}_{sl.def}(\mathbf{r}_s, t) = \mathbf{V}_{conv}(\mathbf{r}_s, t) + \mathbf{V}_{back}(\mathbf{r}_s, t)$ and the collinear relative rigid plate motion $\mathbf{V}_{conv}(\mathbf{r}_s, t)$ indicates the degree of mechanical locking at their boundaries:

$$\varphi(\mathbf{r}_s, t) = \frac{\mathbf{V}_{sl.def}(\mathbf{r}_s, t)}{\mathbf{V}_{conv}(\mathbf{r}_s, t)}, 0 \le \varphi(\mathbf{r}_s, t) \le 1. \quad (18.10)$$

This ratio reflects the strain build-up rate and it is an analogue of the seismic locking calculated as the ratio of the seismic part of the relative plates motion to their full rate. The latter was commonly used to evaluate the seismogenic potential of various subduction zones before space geodesy. The advantage of the geodetic kinematic estimates of the mechanical locking compared to the seismic locking is due to the better spatial-temporal resolution. The seismic locking estimates are based on the following assumptions: plate convergence rate is estimated from geological data averaged over a few million years (NUVEL-1A (DeMets et al., 1994); MORVEL (DeMets et al., 2010), etc.); the whole seismogenic layer is fully locked between the megathrust earthquakes, which makes the averaging period to be the duration of the seismic cycle of dozens or hundreds of years. In contrast, the geodetic kinematic estimates are based on the current plate convergence rates obtained in a rather short period of GPS observations for several years over stable plates interior, and the seismogenic layer is considered locked irregularly.

The simplest model of subduction thrusts implies an inter-seismically locked portion of the plate boundary interface within a continuous depth interval with upper edge between 5 and 10 km and lower edge between 30 and 70 km, with adjacent updip and downdip segments deforming aseismically (Fig. 18.5) (Savage, 1983). The downdip width of the seismogenic zone differs significantly for any particular

subduction zone as well as between each of them (Pacheco et al., 1993; Tichelaar and Ruff, 1993; Oleskevich et al., 1999).

The possible existence of large portions of aseismically slipping subduction interface follows from comparisons of plate motions and estimated locking widths with observed seismic moment release (Pacheco et al., 1993; Tichelaar and Ruff, 1993; Wang and Dixon, 2004). Comparison of the source areas of large historic subduction earthquakes and non-slipping fault areas deduced from geodetic data suggests that asperities on the subduction interface are persistently locked inter-seismically, slip in large stick-slip events, and are surrounded by areas of stable sliding. Thus, the subduction thrust should be considered a spatially heterogeneous structure (Pacheco et al., 1993; Yamanaka and Kikuchi, 2004) that may potentially evolve over time. A wide variety of possible scenarios for different subduction zones have been revealed by GPS measurements throughout the world: some plate interface faults appear persistently locked (Mazzotti et al., 2000), others are partially locked (Lundgren et al., 1999), and some are steadily aseismically slipping (Freymueller and Beavan, 1999). Such kinematic interpretation is not related to any considerations about the frictional strength of either locked or creeping patches of the fault (Lay and Kanamori, 1981), but only infers their existence from available observations. Modern studies also indicate a discrepancy that occurs in some cases between geodetically inferred asperities and sources of large historical earthquakes (Govers et al., 2017). Such discrepancies can be explained by the limitations of GPS data (a rather small number and limited distribution of GPS stations, a limited observation duration) and possible changes in the structure of the subduction interface over time.

The factors that control the geometry of the seismogenic portion of the subduction interface usually include: temperature-related phenomena (transition in behaviour of clay minerals in unconsolidated accretionary prism, crystal-plastic flow of rocks at the depth with $\sim 350°C$) (Peacock, 1996; Oleskevich et al., 1999), hydration of the forearc mantle material, the plate convergence rate and/or the absolute velocity of the upper plate, the size and composition of the accretionary wedge, and the existence of heterogeneous features (such as seamounts) on the subducting plate. A general global correlation of negative free-air gravity anomalies along subduction zones with the major subduction asperities broken in great earthquakes was noticed by (Wells et al., 2003) and (Song and Simons, 2003), which may indicate either erosion or frictional shear stress on the plate interface.

An example of geodetic analysis of interplate motions applied to the Japan subduction zone is shown in Fig. 18.6. The following *a priori* constraints were imposed for inversion: *rake* = 90° ± 30°, | *slip rate* | ≤ 8 cm/year, $\chi^2 \le 30$, which yielded $\chi^2 = 5.09 \div 16.7$.

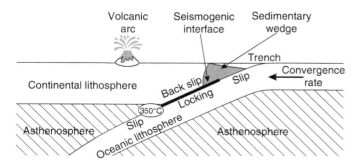

Figure 18.5 Subduction zone: Inter-seismic state general scheme.

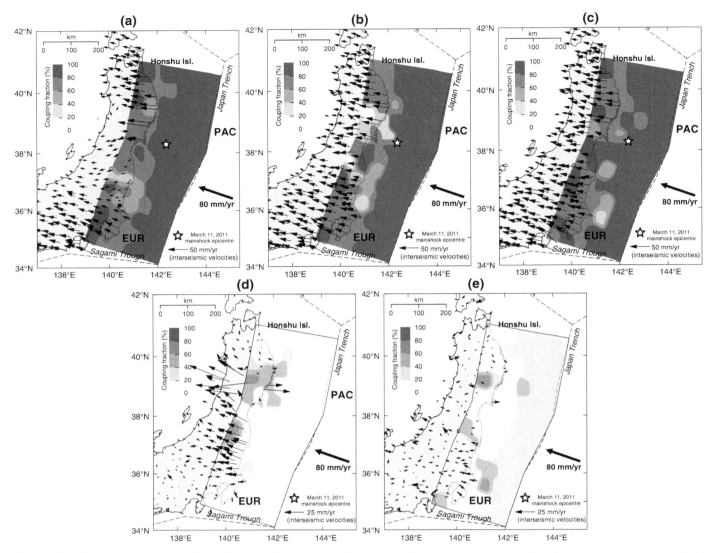

Figure 18.6 Temporal variations of interplate coupling in the northern part of Japan subduction zone based on inter-seismic velocity field estimated over the periods: (a) 11.03.2008–10.03.2009; (b) 11.03.2009–10.03.2010; (c) 11.03.2010–10.03.2011. Subfigures (d) and (e) show differences between (b)–(a) and (c)–(b), respectively.

In general, the calculated yearly locking patterns are characterised by relative stability before the Tohoku earthquake (Fig. 18.6a–c), while the observed downdip locking attenuation agrees with the same decrease of seismicity along the seismic focal zone with depth. At the same time, before the Tohoku 2011 earthquake, an increase in locking near the future source is observed, which is reflected at the surface as a noticeable acceleration of the GPS station displacements (Fig. 18.6d–e). So, the interplate locking ratio immediately prior to the earthquake was relatively high with its maximum near the trench, which could be one of the factors that induced the well-known devastating tsunami after the Tohoku earthquake. The patches of substantially lower

locking are seen at the bottom edge of the whole rupture zone that could be the reason further propagation of the mainshock rupture was prevented. A detailed analysis of the phenomenon of slip acceleration, which has been going on since at least 1996, is given in (Mavrommatis et al., 2015). The analysis performed on the basis of GPS data and data on recurrence intervals of repeating earthquakes suggests that the rupture zone of the 2011 Tohoku earthquake was partly outlined by slip acceleration, implying that a substantial portion of the megathrust experienced accelerating aseismic slip, while most of the rupture area of the Tohoku earthquake was either locked or creeping at a constant rate at least 15 years before its occurrence.

18.6 Post-Seismic Motions

The stress-strain state variations in the vicinity of the seismogenic faults after the great earthquakes typically show a transient post-seismic response in the surface motions. Among the most common mechanisms, the following three are usually proposed: (1) viscoelastic relaxation (Freed et al., 2007; Pollitz et al., 2008; Wang et al., 2012), (2) frictional afterslip (Marone et al., 1991), and (3) poroelastic rebound (Peltzer et al., 1996). Correct assessment of the duration, intensity, spatial scales, and energy characteristics of these processes contributes to understanding the mechanics of faulting and the features of the seismic cycle in a particular subduction zone, as well as to refinement of regional rheology. In practice, distinction of these processes from each other requires assumptions on spatial and temporal features specific for each mechanism, and that would allow independent study of each of them. As a rule, for the first months after an earthquake, afterslip dominates in the near field zone, while after about half a year and in the far zone, the process of viscoelastic relaxation in the asthenosphere and upper mantle usually dominates (Scholz, 2019). At the same time, recent observations of post-seismic motions with seafloor GPS stations after the Tohoku earthquake (Sun et al., 2014; Sun and Wang, 2015) showed that neglecting the short-term viscoelastic relaxation for events with Mw $\geq$ 8.0 leads to an incorrect estimation of the amplitude of the displacements caused by afterslip. This reflects in the underestimation of the displacement values at shallow depths and their overestimation at depths exceeding the lower edge of the rupture.

Rheological modelling of post-seismic processes requires consideration of a number of factors, among which the most important are the rock mechanics dependence on the temperature and pressure variations with depth, as well as the time scale (Scholz, 1988, 2019). At low temperatures and pressures, typical for the upper layers of the Earth's crust, the process of brittle faulting prevails, but in the lower part of the Earth's crust, the elastic-brittle behaviour of rocks turns into plastic. For further deep layers, as the temperature and pressure grow, such rather complex rheological behaviour becomes simpler, similar to non-linear or even Newtonian viscous fluid (Turcotte and Schubert, 2001).

The timing factor implies various response of the medium depending on the rapidness and duration of the stress rising and attenuation. If deformation processes occur at higher temperatures and pressures, then both transient creep (on small time scales) and viscous behaviour of the medium are possible (Savage et al., 2005). Such dual viscoelastic behaviour is typically modelled for the Earth with Burgers rheology, which includes both Maxwell and Kelvin elements. The behaviour of such a bi-viscous medium (Fig. 18.7c) is determined by the initial rapid response of the low-viscosity Kelvin element, while at longer time intervals, the Maxwell element response prevails.

18.6.1 Afterslip Modelling

Rapid frictional afterslip and long aftershock processes both usually follow immediately after great shallow earthquakes. The relationship between these two processes is not quite clear and remains a controversial issue. The total seismic moment released in aftershocks, as a rule, is significantly lower than that released in afterslip. This suggests that the whole post-seismic process is unlikely to be the consequence of aftershock activity only (Helmstetter and Shaw, 2009). The spatial and temporal correlations between those processes revealed by geodetic methods suggest that at least some part of the aftershocks is induced by frictional afterslip (Pritchard and Simons, 2006; Wang, 2010). Simulation results using a simple spring-slider model that captures the main features of the temporal evolution of seismicity and deformation during the seismic cycle (Perfettini and Avouac, 2004; Perfettini et al., 2005), also support the idea that the decay rate of aftershocks may be controlled by reloading due to deep afterslip in the brittle creep fault zone located deeper than the source zone along the plate interface.

The temporal dependence of the afterslip is approximated as rapid initial motion logarithmically decaying with time after the earthquake due to presence of the rate-strengthening patches at the plates interface (Marone et al., 1991; Scholz, 1998), with cumulative post-seismic displacements ranging from decimetres up to metres.

Considering the elastic response of the medium to afterslip with stationary Green's functions, the cumulative amount of slip over the co-seismic rupture and adjacent patches can be expressed as:

$$\mathbf{u}(\mathbf{r}, \mathbf{r}_s, t, t_s) = \iint\limits_{S} \mathbf{G}(\mathbf{r}, \mathbf{r}_s)\mathbf{U}(\mathbf{r}_s, t_s)dS. \qquad (18.11)$$

Then the discretisation in Eqs. (18.3)–(18.4) can be applied for inversion (Eq. 18.6) of the observed surface post-seismic motions for the afterslip distributed model.

18.6.2 Viscoelastic Relaxation Modelling

The intensity and spatial distribution of the process of viscous relaxation in the asthenosphere and upper mantle highly depend on rheological parameters of the medium and on the magnitude of stresses transferred to the asthenosphere at the co-seismic and early post-seismic stages of the seismic cycle, which, in turn, is associated with the spatial distribution and the magnitude of co-seismic slip and afterslip in the source zone of subduction earthquake and its surroundings. The process of afterslip following the main shock is also one of the reasons that explain the possible expansion of the search area for post-seismic slip distribution relative to the co-seismic one.

The first step of the analysis of the viscoelastic relaxation requires consideration of the compressibility and rheological stratification of the upper layers of the Earth. The issue of the

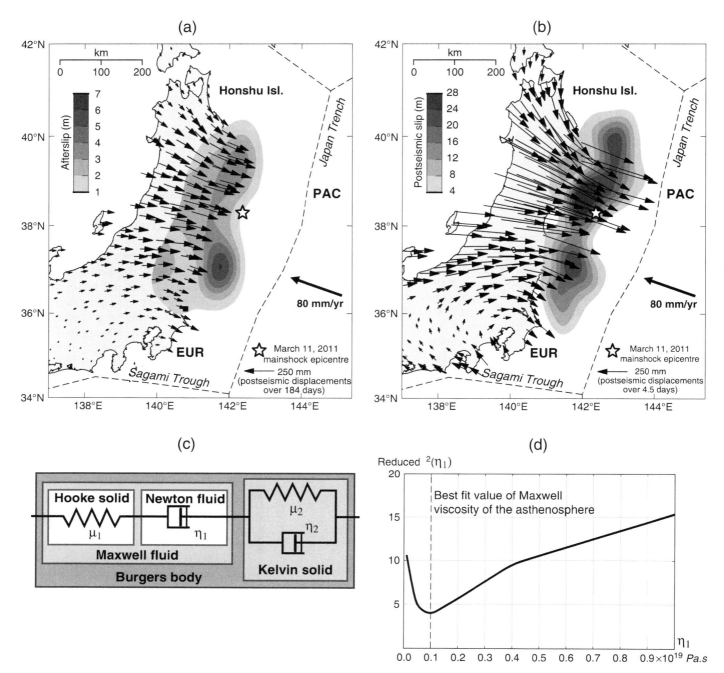

Figure 18.7 (a)Distribution of the afterslip in the first six months after the Tohoku earthquake; (b) model of effective slip distribution causing viscoelastic relaxation; (c) Burgers body which describes the rheology of the asthenosphere; and (d) search for best fit value of Maxwell viscosity of the asthenosphere under Japan islands.

asthenosphere rheological properties is rather debatable; however, the study of post-seismic deformations after the great earthquakes makes it possible to set some constraints on these properties (Muto et al., 2019). Since the Maxwell viscosity of the asthenosphere is one of the parameters determining non-stationary medium response to the static co-seismic slip, the Green's function becomes explicitly dependent on time and viscosity:

$$\mathbf{u}(\mathbf{r}, t) = \iint_S \mathbf{G}(\eta_1, \mathbf{r}, \mathbf{r}_s, t, t_s)\mathbf{U}(\mathbf{r}_s, t_s)dS. \qquad (18.12)$$

Given the observed surface motions $\mathbf{u}(\mathbf{r}, t)$ and initial co-seismic source slip $\mathbf{U}(\mathbf{r}_s, t_s)$, the inversion

$$\eta_1 = \arg\min_{\eta_1}\left\{ \left\| \mathbf{u}(r,t) - \iint_S \mathbf{G}(\eta_1, r, r_s, t, t_s)\mathbf{U}(r_s, t_s)dS \right\| \right\},$$

$$(18.13)$$

after discretisation similar to Eqs. (18.2)–(18.5), provides an estimate of the Maxwell viscosity η_1. Here the co-seismic source slip $\mathbf{U}(\mathbf{r}_s, t_s)$ can be taken either from the inversion of co-seismic surface displacements or from various other inversions like teleseismic USGS, NEIC, etc. The convolution integrals $\iint_{S_k} \mathbf{G}(\eta_1, \mathbf{r}_i, \mathbf{r}_s, t_j, t_s)\mathbf{e}_{kl}dS$ of the unit dislocations over rectangular dislocation elements used in discretisation of Eq. (18.13) can be calculated with the viscoelastic modelling approach of (Pollitz, 1997). To solve Eq. (18.13), various optimisation methods can be used, such as a grid search over viscosity. An example of estimating the best-fit Maxwell viscosity η_1 for the central part of the Japan subduction zone is shown in Fig. 18.7d.

On the second step of the analysis of viscoelastic relaxation, having the constraints set on the viscosity, one can construct another inversion of Eq. (18.12) similar to Eq. (18.2) to solve for the post-seismic (total effective) slip distribution:

$$\mathbf{U}(r_s, t_s) = \arg\min_{\mathbf{U}(r_s, t_s)}\left\{ \left\| \mathbf{u}(r,t) - \iint_S \mathbf{G}(\eta_1, r, r_s, t, t_s)\mathbf{U}(r_s, t_s) \right. \right.$$

$$\left. \left. dS \right\| + \alpha R[\mathbf{U}(r_s, t_s)] \right\}.$$

$$(18.14)$$

While solving Eq. (18.14), one can resolve the total effective slip distribution which has induced the observed viscoelastic response and that may differ from initial co-seismic source slip inversions due to additional contribution of the frictional afterslip at the early post-seismic stage. Examples of modelling post-seismic processes after the great Tohoku 2011 earthquake follow (Fig. 18.7a–b). A priori constraints imposed for inversion for afterslip in the first six months after the Tohoku earthquake (Fig. 18.7a) were: *rake* = 90° ± 30°, | *slip* | ≤ 20 m, which yielded $\chi^2 = 1.48$.

To model the effective slip distribution that induced the process of viscoelastic relaxation in the asthenosphere, its rheology was represented by a bi-viscous Burgers body, comprising both Maxwell and Kelvin viscosities (Fig. 18.7c). A priori constraints imposed for inversion for effective slip (Fig. 18.7b) were: *rake* = 90 ° ± 30°, | *slip* | ≤ 50 m, $\chi^2 \leq 30$, which yielded $\chi^2 = 9.79$.

The slip model distribution (Fig. 18.7b) shows a bilateral pattern that covers a much wider area than co-seismic pattern (Fig. 18.4b), thus indicating the wider spatial scales of the viscoelastic relaxation process. The areas of the largest displacements in the effective source slip are localised near

the lower edge of the co-seismic fault, apparently marking the areas of additional release of residual stresses that had not dropped during the earthquake. The model of cumulative afterslip (Fig. 18.7a) also demonstrates the bilateral expansion of the initial co-seismic source slip pattern, mostly at the lower edge, where the lack of co-seismic slip has been assumed (Fig. 18.4b). The general pattern of afterslip being downdip of the co-seismic slip agrees with independent models (e.g. Muto et al., 2019). This agreement confirms the contribution of the initial stage afterslip into the total effective source of further viscoelastic relaxation. The influence of the additional energy release during the afterslip process and the features of the viscoelastic relaxation process result in the difference between the slip distributions shown in Figs. 18.4 and 18.7b. The estimated effective Maxwell viscosity of the asthenosphere (1×10^{18} Pa·s) is ten times less than the usual mean value for the Earth, but it is typical for the transient viscosity estimated in many post-seismic studies (Pollitz, 2019). According to the prediction based on this model, the viscoelastic relaxation in the asthenosphere will decay in about 30 years.

18.7 Conclusion

Development of rheological models predicting surface motions induced by various tectonic deformations is the key issue for improving interpretation of satellite geodetic observations. Investigation of many active plate boundaries usually reveals a superposition of different simultaneous processes such as rigid block motions, boundary deformations, co-seismic jumps, and post-seismic transient response, which are the mostly known to affect the surface deformations. Their specific spatial and temporal features need quantitative evaluation for better distinction, enabling us to assess their interrelations and predictive potential. Stabilisation of inversions of the observed surface motions for deep underground processes requires not only mathematical constraints usually applied for regularisation but also development of the observation networks. One of the obvious ways to improve the resolution of the satellite geodetic data is expansion and densification of the observation network. The effect of expansion was demonstrated by Japanese GPS network GEONET, when the sea floor GPS sites were installed near the deep ocean trench off the Honshu Island coast. This has led to new details discovered of the co-seismic and post-seismic motions related to the great Tohoku earthquake in 2011, which had not previously been resolved.

Additional knowledge also comes from other kind of observations, such as gravity, seismicity, geology/morphology, and so on. Such data provide information on the fault geometry, the faulting depth, deep masses transport, and so on. Combination of all available data in conjunction with development of each kind of observation substantially improves geodetic inversions thus making understanding of tectonic processes better.

References

Argus, D. F., and Gordon, R. G. (1991). No-net-rotation model of current plate velocities incorporating plate motion model NUVEL-1. *Geophysical Research Letters*, 18(11), 2039–2042. https://doi.org/10.1029/91GL01532.

Argus, D. F., and Heflin, M. (1995). Plate motion and crustal deformation estimated with geode-tic data from the Global Positioning System. *Geophysical Research Letters*, 22(15), 1973–6. https://doi.org/10.1029/95GL02006.

Barbot, S., and Fialko, Y. (2010). Fourier-domain Green's function for an elastic semi-infinite solid under gravity, with applications to earthquake and volcano deformation. *Geophysical Journal International*, 182, 568–82.

Bourgeois, J., Pinegina, T., Razhegaeva, N. et al. (2007). Tsunami runup in the middle Kuril Islands from the great earthquake of 15 Nov 2006. *Eos Transactions of the American Geophysical Union*, 88(52), Abstract S51C-02.

Bürgmann, R., Kogan, M. G., Levin, V. E. et al. (2001). Rapid aseismic moment release following the 5 December 1997 Kronotsky, Kamchatka, Earthquake. *Geophysical Research Letters*, 28, 1331–4. https://doi.org/10.1029/2000GL012350.

Bürgmann, R., Kogan, M. G., Steblov, G. M. et al. (2005). Interseismic coupling and asperity distribution along the Kamchatka subduction zone. *Journal of Geophysical Research*, 110, B07405. https://doi.org/10.1029/2005JB003648.

Das S., and Henry, C. (2003). Spatial relation between main earthquake slip and its aftershock distribution. *Reviews of Geophysics*, 41(3), 1–26.

DeMets, C., Gordon, R. G., Argus, D. F., and Stein, S. (1994). Effect of recent revisions to the geomagnetic reversal time scale on estimates of current plate motions. *Geophysical Research Letters*, 21(20), 2191–4.

DeMets, C., Gordon, R. G., and Argus, D. F. (2010). Geologically current plate motions. *Geophysical Research Letters*, 181, 1–80.

Dixon, T. H. (1991). An introduction to the Global Positioning System and some geological applications. *Reviews of Geophysics* 29(2), 249–76.

Dziewoński, A. M., and Anderson, D. L. (1981). Preliminary reference earth model, *Physics of the Earth and Planetary Interiors* 25, 297–356. https://doi.org/10.1016/0031-9201(81)90046-7.

Engdahl, E. R., and Villaseñor, A. (2002). Global Seismicity: 1900–1999. In W. H. K. Lee et al., eds., *International Handbook of Earthquake Engineering and Seismology*. Amsterdam: Academic Press, pp. 665–90.

Fedotov, S. A. (1965). Regularities of distribution of large earthquakes of Kamchatka, Kuril Islands and North-Eastern Japan. Seismic microzoning. *Transactions of the Institute of Physics of the Earth of the USSR Academy of Sciences*, 36, 66–93 (in Russian).

Freed, A. M., Bürgmann, R., and Herring, T. (2007). Far-reaching transient motions after Mojave earthquakes require broad mantle flow beneath a strong crust. *Geophysical Research Letters*, 34. https://doi.org/10.1029/2007GL030959.

Freymueller, J. T., and Beavan, J. (1999). Absence of strain accumulation in the western Shumagin segment of the Alaska subduction zone. *Geophysical Research Letters*, 26, 3233–6.

Gill, P. E., Murray, W., Saunders, M. A., and Wright, M.H. (1986). User's Guide for NPSOL (Version 4.0): A Fortran Package for Nonlinear Programming, Report SOL 86–2, Department of Operations Research, Stanford University.

Govers, R., Furlong, K. P., van de Wiel, L., Herman, M. W., and Broerse, T. (2018). The geodetic signature of the earthquake cycle at subduction zones: Model constraints on the deep processes. *Reviews of Geophysics*, 56, 6–49. https://doi.org/10.1002/2017RG000586.

He, Y.-M., Wang, W.-M., and Yao, Z.-X. (2003). Static deformation due to shear and tensile faults in a layered half-space. *Bulletin of the Seismological Society of America*, 93, 2253–63.

Helmstetter, A., and Shaw, B. E. (2009). Afterslip and aftershocks in the rate-and-state friction law. *Journal of Geophysical Research*, 114(B01308), 1−24.

International Seismological Centre (2022). On-line Bulletin. https://doi.org/10.31905/D808B830.

Ismail-Zadeh, A., and Tackley, P. (2010). *Computational Methods for Geodynamics*. Cambridge: Cambridge University Press.

Kogan, M. G., and Steblov, G. M. (2008). Current global plate kinematics from GPS (1995–2007) with the plate-consistent reference frame. *Journal of Geophysical Research*, 113 (B04416). https://doi.org/10.1029/2007JB005353.

Kogan, M. G., Bürgmann, R., Vasilenko, N. F. et al. (2003). The 2000 Mw 6.8 Uglegorsk earthquake and regional plate boundary deformation of Sakhalin from geodetic data. *Geophysical Research Letters*, 30(3), 1102. https://doi.org/10.1029/2002GL016399.

Kuchay, O. A. (1982). Spatial Regularities of Aftershock Deformation of the Source Region of a Strong Earthquake. *Izv., Physics of the Solid Earth*, 10, 62–7. (in Russian)

Lay, T., and Kanamori, H. (1981). An asperity model of large earthquake sequences. In D. W. Simpson and P.G. Richards, eds., *Earthquake Prediction*, vol. 4. Washington, DC: American Geophysical Union, pp. 579–92. https://doi.org/10.1029/ME004p0579.

Lay, T., Kanamori, H., Ammon, C. J. et al. (2009). The 2006–2007 Kuril Islands great earthquake sequence. *Journal of Geophysical Research*, 114(B11308), 1–31.

Lay, T., Ammon, C. J., Kanamori, H., Kim, M. J., and Xue, L. (2011). Possible large near-trench slip during the 2011 Mw 9.0 off the Pacific coast of Tohoku Earthquake. *Earth, Planets and Space*, 63, 713–18.

Lévêque, J.-J., Rivera, L., and Wittlinger, G. (1993). On the use of the checker-board test to assess the resolution of tomographic inversions. *Geophysical Journal International*, 115, 313–18.

Lundgren, P., Protti, M., Donnellan, A. et al. (1999). Seismic cycle and plate margin deformation in Costa Rica: GPS observations from 1994 to 1997. *Journal of Geophysical Research*, 104 (28), 915–26.

Marone C. J., Scholz C. H., and Bilham R. G. (1991). On the mechanics of earthquake afterslip. *Journal of Geophysical Research*, 96(B5), 8441–52.

Masse, R. P., and Needham, R. E. (1989). NEIC – the National Earthquake Information Center. *Earthquakes & Volcanoes (USGS)*, 21(1), 4–44.

Mavrommatis, A. P., Segall, P., Uchida, N., and Johnson, K. M. (2015). Long-term acceleration of aseismic slip preceding the Mw 9 Tohoku-oki earthquake: Constraints from repeating earthquakes. *Geophysical Research Letters*, 42, 9717–25. https://doi.org/10.1002/2015GL066069.

Mazzotti, S., Le Pichon, X., and Henry, P. (2000). Full interseismic locking of the Nankai and Japan-west Kurile subduction zones: An analysis of uniform elastic strain accumulation in Japan constrained by permanent GPS. *Journal of Geophysical Research*, 105, 13159–77.

Molchan, G. M., and Dmitrieva, O. E. (1992). Aftershock identification: Methods and new approaches. *Geophysical Journal International*, 109, 501–16. https://doi.org/10.1111/j.1365-246X.1992.tb00113.x.

Muto, J., Moore, J. D. P., Barbot, S. et al. (2019). Coupled afterslip and transient mantle flow after the 2011 Tohoku earthquake. *Science Advances*, 5(9), eaaw1164. https://doi.org/10.1126/sciadv.aaw1164.PMID:31579819;PMCID:PMC6760927.

Namegaya, Y., and Satake, K. (2014). Reexamination of the A.D. 869 Jogan earthquake size from tsunami deposit distribution, simulated flow depth, and velocity. *Geophysical Research Letters*, 41, 2297–303.

Okada, Y. (1985). Surface deformation due to shear and tensile faults in a half-space. *Bulletin of the Seismological Society of America*, 75(4), 1135–54.

Okada, Y. (1992). Internal deformation due to shear and tensile faults in a half-space. *Bulletin of the Seismological Society of America*, 82(2), 1018–40.

Oleskevich, D., Hyndman, R. D., and Wang, K. (1999). The updip and downdip limits of subduction earthquakes: Thermal and structural models of Cascadia, south Alaska, S.W. Japan, and Chile. *Journal of Geophysical Research*, 104(14), 965–91.

Pacheco, J. F., Sykes, L. R., and Scholz, C. H. (1993). Nature of seismic coupling along simple plate boundaries of the subduction type. *Journal of Geophysical Research*, 98, 14,133–59.

Peacock, S. M. (1996). Thermal and petrologic structure of subduction zones. In G. E. Bebout, D. W. Scholl, S. H. Kirby, and J. P. Platt, eds., *Subduction: Top to Bottom. Geophysical Monograph 96*. Washington, DC: American Geophysical Union, pp.119–33. https://doi.org/10.1029/GM096p0119.

Peltzer, G., Rosen, P. A., Rogez, P., and Hudnut, K. (1996). Postseismic rebound in fault step-overs caused by pore fluid flow. *Science*, 273, 1202–4. https://doi.org/10.1126/science.273.5279.1202.

Perfettini, H., and Avouac, J.-P. (2004). Stress transfer and strain rate variations during the seismic cycle. *Journal of Geophysical Research*, 109, B06402. https://doi.org/10.1029/2003JB002917.

Perfettini, H., Avouac, J.-P., and Ruegg, J.-C. (2005). Geodetic displacements and aftershocks following the 2001 Mw = 8.4 Peru earthquake: Implications for the mechanics of the earthquake cycle along subduction zones. *Journal of Geophysical Research*, 110, B09404. https://doi.org/10.1029/2004JB003522.

Piersanti, A., Spada, G., Sabadini, R., and Bonafede, M. (1995). Global postseismic deformation. *Geophysical Journal International*, 120, 544–66.

Pollitz, F. F. (1996). Coseismic deformation from earthquake faulting on a layered spherical Earth. *Geophysical Journal International*, 125, 1–14.

Pollitz, F. F. (1997). Gravitational viscoelastic postseismic relaxation on a layered spherical Earth. *Journal of Geophysical Research*, 102, 17921–41.

Pollitz, F.F. (2019). Lithosphere and shallow asthenosphere rheology from observations of post-earthquake relaxation. *Physics of the Earth and Planetary Interiors*, 293. https://doi.org/10.1016/j.pepi.2019.106271.

Pollitz, F., Banerjee, P., Grijalva, K., Nagarajan, B., and Bürgmann, R. (2008). Effect of 3-D viscoelastic structure on post-seismic relaxation from the 2004 M = 9.2 Sumatra earthquake. *Geophysical Journal International*, 173, 189–204. https://doi.org/10.1111/j.1365-246X.2007.03666.x.

Press, W. H., Teukolsky, S. A., Wetterling, W. T., and Flannery, B. P. (2007). *Numerical Recipes: The Art of Scientific Computing*, 3rd ed. New York: Cambridge University Press.

Pritchard, M. E., and Simons, M. (2006). An aseismic slip pulse in northern Chile and along-strike variations in seismogenic behavior. *Journal of Geophysical Research*, 111, 1–14.

Riznichenko, Y. V. (1985). *Problems of Seismology: Selected Works*. Moscow: Nauka (in Russian).

Rundle, J. B. (1980). Static elastic-gravitational deformation of a layered half-space by point couple sources. *Journal of Geophysical Research*, 85, 5354–63.

Savage, J. C. (1983). A dislocation model of strain accumulation and release at a subduction zone. *Journal of Geophysical Research*, 88, 4984–96. https://doi.org/10.1029/JB088IB06P04984.

Savage, J. C., Svarc, J. L., and Yu, S.-B. (2005). Postseismic relaxation and transient creep. *Journal of Geophysical Research*, 110, B11402. https://doi.org/10.1029/2005JB003687.

Scholz, C. H. (1998). Earthquakes and friction laws. *Nature*, 391, 37–42.

Scholz, C. H. (1988). The brittle-plastic transition and the depth of seismic faulting. *Geologische Rundschau*, 77, 319–28.

Scholz, C. H. (2019). *The Mechanics of Earthquakes and Faulting*, 3rd ed. Cambridge: Cambridge University Press.

Song, T.-R. A., and Simons, M. (2003). Large trench-parallel gravity variations predict seismogenic behaviour in subduction zones. *Science*, 301, 630–3.

Sun T., Wang K., Iinuma T. et al. (2014). Prevalence of viscoelastic relaxation after the 2011 Tohoku-oki earthquake. *Nature*, 514 (7520), 84–87.

Sun, T., and Wang, K. (2015). Viscoelastic relaxation following subduction earthquakes and its effects on afterslip determination. *Journal of Geophysical Research: Solid Earth*, 120(2), 1329–44.

Tanaka, Y. (2013). Theoretical computation of long-term postseismic relaxation due to a great earthquake using a spherically symmetric viscoelastic Earth model. *Journal of the Geodetic Society of Japan*, 59, 1–10.

Tichelaar, B. W., and Ruff, L. J. (1993). Depth of seismic coupling along subduction zones. *Journal of Geophysical Research*, 98, 2017–37.

Turcotte, D. L., and Schubert, D. (2001). *Geodynamics*, 2nd ed. Cambridge: Cambridge University Press.

Vladimirova, I. S., Lobkovsky, L. I., Gabsatarov, Y. V. et al. (2019). Source data for Vladimirova et al. (2020). Patterns of seismic cycle in the Kuril Island arc from GPS observations. *Pure and Applied Geophysics*, figshare, Dataset. https://doi.org/10.6084/m9.figshare.10028582.v2.

Wang, K., and Dixon, T. H. (2004). Coupling semantics and science in earthquake research. *EOS Transactions of the American Geophysical Union*, 85, 180–1.

Wang, K., Hu, Y., and He, J. (2012). Deformation cycles of subduction earthquakes in a viscoelastic Earth. *Nature*, 484, 327–32.

Wang, L. (2010). Analysis of Postseismic Processes: Afterslip, Viscoelastic Relaxation and Aftershocks. Ph.D. thesis, Institute of Geology, Mineralogy and Geophysics. Ruhr University Bochum, Germany.

Wang, R., Martin, F. L., and Roth, F. (2003). Computation of deformation induced by earthquakes in a multi-layered elastic crust: FORTRAN programs EDGRN/EDCMP. *Computers & Geosciences*, 29(2), 195–207.

Wells, R. E., Blakely, R. J., Sugiyama, Y., Scholl, D. W., and Dinterman, P. A. (2003). Basin-centered asperities in great subduction zone earthquakes: A link between slip, subsidence, and subduction erosion? *Journal of Geophysical Research*, 108 (B10), 2507. https://doi.org/10.1029/2002JB002072.

Wessel, P., Luis, J. F., Uieda, L. et al. (2019). The generic mapping tools, version 6. *Geochemistry, Geophysics, Geosystems*, 20, 5556–64. https://doi.org/10.1029/2019GC008515

Zhou, X., Cambiotti, G., Sun, W. K., and Sabadini, R. (2018). Co-seismic slip distribution of the 2011 Tohoku (M_W 9.0) earthquake inverted from GPS and space-borne gravimetric data. *Earth and Planetary Physics*, 2, 120–38. http://doi.org/10.26464/epp2018013.

Yamanaka, Y., and Kikuchi, M. (2004). Asperity map along the subduction zone in northeastern Japan inferred from regional seismic data. *Journal of Geophysical Research*, 109, B07307. https://doi.org/10.1029/2003JB002683.

Yamazaki, Y., Cheung, K. F., and Lay, T. (2018). A self-consistent fault slip model for the 2011 Tohoku earthquake and tsunami. *Journal of Geophysical Research: Solid Earth*, 123, 1435–8. https://doi.org/10.1002/2017JB014749.

Yokota, Y., Koketsu, K., Fujii, Y. et al. (2011). Joint inversion of strong motion, teleseismic, geodetic, and tsunami datasets for the rupture process of the 2011 Tohoku earthquake. *Geophysical Research Letters*, 38, L00G21. https://doi.org/10.1029/2011GL050098.

Zelt, C. A., Azaria, A., and Levander, A. (2006). 3D seismic refraction traveltime tomography at a ground water contamination site. *Geophysics*, 71(5), H67–H78.

19

Data Assimilation in Geodynamics: Methods and Applications

Alik Ismail-Zadeh, Igor Tsepelev, and Alexander Korotkii

Abstract: In this chapter, we review basic methods for data assimilation used in geodynamic modelling: backward advection (BAD), variational/adjoint (VAR), and quasi-reversibility (QRV). The VAR method is based on a search for model parameters (e.g. mantle temperature and flow velocity in the past) by minimising the differences between present observations of the relevant physical parameters (e.g. temperature derived from seismic tomography, geodetic measurements) and those predicted by forward models for an initial guess temperature. The QRV method is based on introduction of the additional term involving the product of a small regularisation parameter and a higher-order temperature derivative into the backward heat equation. The data assimilation in this case is based on a search of the best fit between the forecast model state and the observations by minimising the regularisation parameter. To demonstrate the applicability of the considered data assimilation methods, a numerical model of the evolution of mantle plumes is considered. Also, we present an application of the data assimilation to dynamic restoration of the thermal state of the mantle beneath the Japanese islands and their surroundings. The geodynamic restoration for the last 40 million years is based on the assimilation of the present temperature inferred from seismic tomography, and constrained by the present plate movement derived from geodetic observations, and paleogeographic and paleomagnetic plate reconstructions. Finally, we discuss some challenges, advantages, and disadvantages of the data assimilation methods.

19.1 Introduction

Geodynamics deals with physical and chemical processes in the Earth's interior and their surface manifestations to provide a better understanding of the thermal convection, hotspots and plumes in the mantle; lithosphere dynamics including plate tectonics, spreading, and subduction; and orogeny, sedimentary basins evolution, volcanism, and seismicity (e.g. Turcotte and Schubert, 2002). Many geodynamic problems can be described by mathematical models, i.e. by a set of partial differential equations and boundary and/or initial conditions defined in a specific domain. A mathematical model links the causal characteristics of a geodynamic process with its effects. The causal characteristics of the model process include, for example, parameters of the initial and boundary conditions, coefficients of the differential equations, and geometrical parameters of a model domain (e.g. Ismail-Zadeh et al., 2016).

A mathematical model of a geodynamic problem can be transformed into a numerical model to determine quantitatively the effects of geophysical processes related to the geodynamic problem based on the knowledge of its causes. This type of geodynamic modelling is associated with solving *direct* problems. An *inverse* problem is the opposite of a direct problem. An inverse problem is considered when there is a lack of information on the model's causal characteristics, but an information on the effects of the relevant geodynamic processes exists, and it comes from observations and measurements of physical and chemical parameters involved in the process. This information presents a set of geophysical, geodetic, geochemical, and other data, and utilising the data to estimate the causal characteristics of the model (e.g. temperature, velocity, rheology) and to study the evolving geodynamic process is the principal goal of *data assimilation* (e.g. Ismail-Zadeh et al., 2016).

Data assimilation in geodynamical models can be defined as a combination of theory related to geodynamic processes and mathematical/numerical modelling with available (observed and/or measured) geo-data. The incorporation of data and initial/boundary conditions into an explicit dynamic model will provide time continuity and coupling among the physical characteristics of a geodynamic problem. In this chapter, we use a classical approach to data assimilation considering a mathematical model and observations as a true model and true data with measurement errors, respectively.

There are several data assimilation techniques used in geodynamics (e.g. Ismail-Zadeh et al., 2016). The principal mathematical difficulty in assimilating geo-data into geodynamic models (e.g. thermal convective circulations in the mantle) in the presence of the heat diffusion is the ill-posedness of the backward heat problem (e.g. Kirsh, 1996). One of the simplest methods in data assimilation in geodynamics is the backward advection (BAD) method which suggests neglecting the thermal diffusion term in the heat equation. The resulting heat advection equation can then be solved backward in time. In the case of advection-dominated flows with an insignificant diffusion, this approach is valid. Both direct (forward in time) and inverse (backward in time) advection problems are well-posed. This is because the time-dependent advection equation has the same form of characteristics for the direct and inverse velocity field: the vector velocity reverses its direction when time is reversed. Therefore, numerical algorithms used to solve the direct problem can also be used in studies of the time-reverse problems by replacing positive time steps with negative ones.

The BAD method has been applied to restore diapiric structures to their earlier stages (e.g. Ismail-Zadeh et al., 2001a; Kaus and Podladchikov, 2001; Ismail-Zadeh et al., 2004b; Massimi et al., 2007; Schuh-Senlis et al., 2020) as well as to reconstruct a mantle flow in the past from present-day mantle density heterogeneities (e.g. Forte and Mitrovica, 1997; Steinberger and O'Connell, 1997, 1998; Conrad and Gurnis, 2003; Moucha and Forte, 2011; Glišović et al., 2014). For example, using the BAD method, Steinberger and O'Connell (1998) studied the motion of hotspots relative to the deep mantle. They combined the advection of plumes, which are thought to cause the hotspots on the Earth's surface, with a large-scale mantle flow field and constrained the viscosity structure of the Earth's mantle. Conrad and Gurnis (2003) modelled the history of mantle flow using a tomographic image of the mantle beneath southern Africa as an input (initial) condition for the backward mantle advection model while reversing the direction of flow. If the resulting model of the evolution of thermal structures obtained by the BAD method is used as a starting point for a forward mantle convection model, present mantle structures can be reconstructed if the time of assimilation does not exceed 50–75 million years (Myr). Moucha and Forte (2011) simulated mantle convection using the BAD method to reconstruct the evolution of dynamic topography of Africa over the past 30 Myr.

In sequential data assimilation, a model is computed forward in time, and model predictions are updated each time where observations are available. A sequential data assimilation was used to compute mantle circulation models (Bunge et al., 2002). It was shown that the sequential data assimilation is well adapted to mantle circulation studies. However, the sequential assimilation requires the information of mantle states at earlier geological times, which are poorly known, and hence the updates are rare or uncertain.

The variational (or adjoint) method of data assimilation is based on the optimisation of a given criterion (for example, minimisation of a difference between model forecasts and observations). It was pioneered by meteorologists and used very successfully to improve operational weather forecasts (e.g. Kalnay, 2003). The adjoint method has also been widely used in oceanography (e.g. Bennett, 1992) and

in hydrological studies (e.g. McLaughlin, 2002). The use of variational method of data assimilation in models of mantle dynamics to estimate mantle temperature and flow in the geological past has been put forward by Bunge et al. (2003) and Ismail-Zadeh et al. (2003a,b). The major differences between the two approaches are that Bunge et al. (2003) applied this method to the coupled Stokes, continuity, and heat equations (generalised inverse), whereas Ismail-Zadeh et al. (2003a) applied it only to the heat equation. The variational approach by Ismail-Zadeh et al. (2003a) is computationally less expensive because it does not involve the Stokes equation into the iterations between the direct and adjoint problems.

Another method used in assimilation of data in geodynamic models is a quasi-reversibility (QRV) method (Lattes and Lions, 1969). This method was introduced in geodynamic modelling by Ismail-Zadeh et al. (2007) and employed to assimilate data in models of lithosphere/mantle dynamics beneath Carpathians and Japanese Islands (Ismail-Zadeh et al., 2008, 2013). A QRV method was used to reconstruct the global mantle dynamics focusing on the regions of the Pacific plate, Indian plate, North Atlantic plate, and the North Atlantic region (Glišović and Forte, 2014, 2016, 2017, 2019).

In the following subsections, we describe two essential methods for data assimilation in geodynamics – variational and quasi-reversibility – and present their performance in the case of reconstructions of mantle plume dynamics.

19.2 Variational (VAR) Method

The VAR data assimilation is based on a search of the best fit between the forecast model state and the observations by minimising an objective functional (a normalised residual between the target model and observed variables) over space and time. To minimise the objective functional over time, an assimilation time interval is defined and an adjoint model is typically used to find the derivatives of the objective functional with respect to the model states. The VAR method (sometimes referred to as the adjoint method) can be formulated with a weak constraint (so-called, a generalised inverse), where errors in the model formulation are taken into account (Bunge et al., 2003), or with a strong constraint where the model is assumed to be perfect except for the errors associated with the initial conditions (Ismail-Zadeh et al., 2003a,b). The generalised inverse of mantle convection considers model errors, data misfit and the misfit of parameters as control variables. As the generalised inverse presents a computational challenge, Bunge et al. (2003) considered a simplified generalised inverse imposing a strong constraint on errors, ignoring all errors except for the initial condition errors. Therefore, the strong constraint makes the problem computationally tractable.

The variational data assimilation method was employed for numerical restoration of models of present prominent mantle plumes to their past stages (Ismail-Zadeh et al., 2004a; Hier-Majumder et al., 2005). Effects of thermal diffusion and temperature-dependent viscosity on the evolution of mantle plumes was studied by Ismail-Zadeh et al. (2006) to recover the structure of mantle plumes prominent in the past from that of present plumes weakened by thermal diffusion. Liu and Gurnis (2008) simultaneously inverted mantle properties and initial conditions and applied this method to reconstruct the evolution of the Farallon Plate subduction (Liu et al., 2008, 2010; Spasojevic et al., 2009) and northern South America (Shephard et al., 2010). Horbach et al. (2014) demonstrated the practicality of the method for use in a high-resolution mantle circulation model by restoring a representation of present mantle heterogeneity derived from the global seismic shear wave study backward in time for the past 40 million years. Worthen et al. (2014) used an adjoint method to infer mantle rheological parameters from surface velocity observations and instantaneous mantle flow models. Ratnaswamy et al. (2015) developed an adjoint-based approach to infer plate boundary strength and rheological parameters in models of mantle flow from surface velocity observations, although, compared to Worthen et al. (2014), they formulated the inverse problem in a Bayesian inference framework. An adjoint method was derived and used for a compressible mantle convection model (Ghelichkhan and Bunge, 2016) and thermochemical convection model (Ghelichkhan and Bunge, 2018). An adjoint method was also used to recover simultaneously initial temperature conditions and viscosity parameters in time-dependent mantle convection models from the current mantle temperature and historic plate motion (Li et al., 2017). Application of the variational data assimilation to spherical mantle circulation models are discussed by Bunge et al. (2022, this volume).

In what follows below we use a strong constraint in variational data assimilation assuming that the model is perfect except for errors associated with the initial conditions.

19.2.1 Mathematical Statement

We consider that the Earth's mantle behaves as a Newtonian incompressible fluid with a temperature-dependent viscosity. The viscous flow in the mantle is described by heat, motion, and continuity equations (Chandrasekhar, 1961). To simplify the governing equations, the Boussinesq approximation (Boussinesq, 1903) is used by keeping the density constant everywhere except for buoyancy term in the equation of motion. In the model domain $\Omega = [0, x_1 = h_1] \times [0, x_2 = h_2] \times [0, x_3 = h_3]$, where $\mathbf{x} = (x_1, x_2, x_3)$ are the Cartesian coordinates and h_1, h_2, and h_3 are the domain's dimensions, we consider

two coupled mathematical problems: (1) the *boundary value problem for the flow velocity* including the Stokes equations (Eq. 19.1) and the incompressibility equation (Eq. 19.2), subject to impenetrability and perfect slip conditions at the model boundary (19.3)

$$\nabla p = \nabla \cdot [\eta(\nabla \mathbf{u} + \nabla \mathbf{u}^T)] + Ra T \mathbf{e}, \quad \mathbf{x} \in \Omega, \quad (19.1)$$

$$\nabla \cdot \mathbf{u} = 0, \quad \mathbf{x} \in \Omega, \quad (19.2)$$

$$\langle \mathbf{u}, \mathbf{n} \rangle = 0, \ \tau\mathbf{n} - \langle \tau\mathbf{n}, \mathbf{n} \rangle \mathbf{n} = 0, \quad \mathbf{x} \in \partial\Omega, \quad (19.3)$$

and (2) the *initial-boundary value problem for temperature* including the heat equation (Eq. 19.4), subject to appropriate boundary (19.5) and initial (19.6) conditions

$$\partial T/\partial t + \langle \mathbf{u}, \nabla T \rangle = \nabla^2 T, \quad \mathbf{x} \in \Omega, t \in (0, \vartheta), \quad (19.4)$$

$$\sigma_1 T + \sigma_2 \partial T/\partial \mathbf{n} = 0, \quad \mathbf{x} \in \partial\Omega, \quad (19.5)$$

$$T(t = 0, \mathbf{x}) = T_0(\mathbf{x}), \quad \mathbf{x} \in \Omega. \quad (19.6)$$

Here t, $\mathbf{u} = (u_1, u_2, u_3)$, T, p, and η are dimensionless time, velocity, temperature, pressure, and viscosity, respectively; $\mathbf{e} = (0,0,1)$ is the unit vector; $\mathbf{n}$ is the outward unit normal vector at a point on the model boundary; and $\tau = \eta(\nabla\mathbf{u} + \nabla\mathbf{u}^T)$ is the stress tensor. In the dimensionless equation (Eqs. 19.1–19.6), the length, temperature, and time are normalised by h, ΔT, and $h^2\kappa^{-1}$, respectively. The Rayleigh number is defined as $Ra = \alpha g \rho_{ref} \Delta T h^3 \eta_{ref}^{-1} \kappa^{-1}$, where α is the thermal expansivity, g is the acceleration due to gravity, ρ_{ref} and η_{ref} are the reference typical density and viscosity, respectively; ΔT is the temperature contrast between the lower and upper boundaries of the model domain; and κ is the thermal diffusivity. Here ∇, $\nabla\cdot$, ^{T}, and $\langle \cdot, \cdot \rangle$ denote the gradient operator, the divergence operator, the transposed matrix, and the scalar product of vectors, respectively. The σ_1 and σ_2 are piecewise smooth functions or constants such that $\sigma_1^2 + \sigma_2^2 \neq 0$; choosing σ_1 and σ_2 in a proper way we can specify temperature or heat flux at the model boundaries; $T_0(\mathbf{x})$ is the temperature at time $t = 0$. The boundary value problem for the flow velocity (Eqs. 19.1–19.3) together with the initial-boundary value problem for temperature (Eqs. 19.4–19.6) describe a thermal convective viscous flow (circulation) in the mantle.

19.2.2 Objective Functional

Consider the following objective (cost) functional at $t \in [0, \vartheta]$,

$$J(\varphi) = \| T(\vartheta, \cdot; \varphi) - \chi(\cdot) \|^2, \quad (19.7)$$

where $\| \cdot \|$ denotes the norm in the space $L_2(\Omega)$, the Hilbert space with the norm defined as $\| y \| = \left[\int_\Omega y^2(\mathbf{x}) d\mathbf{x} \right]^{1/2}$. Here $T(\vartheta, \cdot; \varphi)$ is the solution of the initial-boundary value problem for temperature (Eqs. 19.4–19.6) at the final time ϑ, which corresponds to some (unknown as yet) initial temperature distribution $\varphi(\mathbf{x})$; $\chi(\mathbf{x}) = T(\vartheta, \mathbf{x}; T_0)$ is the known temperature (e.g. the temperature inferred from measurements or observations) at the final time, which corresponds to the initial temperature $T_0(\cdot)$. The functional has its unique global minimum at value $\varphi \equiv T_0$ and $J(T_0) \equiv 0$, $\nabla J(T_0) \equiv 0$ (Vasiliev, 2002).

The gradient method is employed to find the minimum of the functional (7) ($k = 0, \ldots, j, \ldots$):

$$\varphi_{k+1} = \varphi_k - \beta_k \nabla J(\varphi_k), \quad \varphi_0 = T_*, \tag{19.8}$$

$$\beta_k = \begin{cases} J(\varphi_k)/\|\nabla J(\varphi_k)\|^2, & 0 \leq k \leq k_*, \\ k^{-1}, & k > k_* \end{cases}, \tag{19.9}$$

where T_* is an initial temperature guess. The minimisation method belongs to a class of limited-memory quasi-Newton methods (Zou et al., 1993), where approximations to the inverse Hessian matrices are chosen to be the identity matrix. Equation (19.9) is used to maintain the stability of the iteration scheme (19.8).

Let us consider that the gradient of the objective functional $\nabla J(\varphi_k)$ is computed with an error $\|\nabla J_\delta(\varphi_k) - \nabla J(\varphi_k)\| < \delta$, where $\nabla J_\delta(\varphi_k)$ is the computed value of the gradient.

Introducing the function $\varphi^\infty = \varphi_0 - \sum_{k=1}^{\infty} \beta_k \nabla J(\varphi_k)$ (and assuming that the infinite sum exists) and the function $\varphi_\delta^\infty = \varphi_0 - \sum_{k=1}^{\infty} \beta_k \nabla J_\delta(\varphi_k)$ (as the computed value of φ^∞), the following inequality should be held for stability of the iteration method (Eq. 19.8):

$$\left\| \varphi_\delta^\infty - \varphi^\infty \right\| = \left\| \sum_{k=1}^{\infty} \beta_k (\nabla J_\delta(u_k) - \nabla J(u_k)) \right\| \leq \sum_{k=1}^{\infty} \beta_k$$

$$\left\| \nabla J_\delta(\varphi_k) - \nabla J(\varphi_k) \right\| \leq \delta \sum_{k=1}^{\infty} \beta_k. \tag{19.10}$$

The sum $\sum_{k=1}^{\infty} \beta_k$ is finite, if $\beta_k = 1/k^p$, $p > 1$. If $p = 1$, but the number of iterations is limited, the iteration method is conditionally stable, although the convergence rate of these iterations is low. Meanwhile the gradient of the objective functional $\nabla J(\varphi_k)$ decreases steadily with the number of iterations providing the convergence, although the absolute value of $J(\varphi_k)/\|\nabla J(\varphi_k)\|^2$ increases with the number of iterations, and it can result in instability of the iteration process (Samarskii and Vabischevich, 2007).

19.2.3 Adjoint Problem
The minimisation algorithm requires the calculation of the gradient of the objective functional, ∇J. This can be done

using the *adjoint problem* for the problem (Eqs. 19.4–19.6), which can be represented in the following form:

$$\partial \Psi / \partial t + \langle \mathbf{u}, \nabla \Psi \rangle + \nabla^2 \Psi = 0, \quad \mathbf{x} \in \Omega, \ t \in (0, \vartheta), \tag{19.11}$$

$$\sigma_1 \Psi + \sigma_2 \partial \Psi / \partial \mathbf{n} = 0, \quad \mathbf{x} \in \Gamma, \ t \in (0, \vartheta), \tag{19.12}$$

$$\Psi(\vartheta, \mathbf{x}) = 2(T(\vartheta, \mathbf{x}; \varphi) - \chi(\mathbf{x})), \quad \mathbf{x} \in \Omega. \tag{19.13}$$

Ismail-Zadeh et al. (2004a) proved that the solution Ψ to the adjoint problem (Eqs. 19.11–19.13) is the gradient of the objective functional (19.7), that is $\nabla J(\varphi) = \Psi(t = 0, x; \varphi)$. Therefore, the solution of the backward heat problem is reduced to iterations between two forward problems, which are known to be well-posed (Tikhonov and Samarskii, 1990): the heat problem (Eqs. 19.4–19.6) and the adjoint problem (Eqs. 19.11–19.13).

19.2.4 Solution Method
Here we present a method for solving the coupled problem (Eqs. 19.1–19.6) backward in time. A uniform partition of the time axis is defined at points $t_n = \vartheta - \delta t\, n$, where δt is the time step, and n successively takes integer values from 0 to some natural number $m = \vartheta/\delta t$. At each subinterval of time $[t_{n+1}, t_n]$, the search of temperature T and flow velocity $\mathbf{u}$ at $t = t_{n+1}$ consists of the following basic steps.

Step 1 Given the temperature $T = T(t_n, \mathbf{x})$ at $t = t_n$, solve the boundary value problem (19.1)–(19.3) to determine the velocity $\mathbf{u}$.

Step 2 The 'advective' temperature $T_{adv} = T_{adv}(t_{n+1}, \mathbf{x})$ is then determined by solving the initial-boundary value problem (Eqs. 19.4–19.6), where the diffusion term in the heat equation (19.4) is neglected and $T = T(t_n, \mathbf{x})$ is considered as the initial temperature in (19.6). As in the case of the BAD method, the solution of the problem backward in time is found by replacing the positive time step with a negative one. Given temperature $T = T_{adv}$ at $t = t_{n+1}$, steps 1 and 2 are then repeated to find the velocity $\mathbf{u}_{adv} = \mathbf{u}(t_{n+1}, \mathbf{x}; T_{adv})$.

Step 3 The problem (Eqs. 19.4– 19.6) is solved with the initial condition (guess temperature) $\varphi_k(\mathbf{x}) = T_{adv}(t_{n+1}, \mathbf{x})$, $k = 0, 1, 2, \ldots, m, \ldots$ forward in time using velocity $\mathbf{u}_{adv}$ to find $T(t_n, \mathbf{x}; \varphi_k)$.

Step 4 The adjoint problem (Eqs. 19.11–19.13) is then solved backward in time with the initial condition (13) as $\Psi(t_n, \mathbf{x}) = 2\Big(T(t_n, \mathbf{x}; \varphi_k) - \chi(\mathbf{x})\Big)$ and $\mathbf{u} = \mathbf{u}_{adv}$ to determine $\nabla J(\varphi_k) = \Psi(t_{n+1}, \mathbf{x}; \varphi_k)$.

Step 5 The coefficient β_k is determined from (19.9), and the temperature is updated (i.e. φ_{k+1} is determined) from (19.8).

Steps 3 to 5 are repeated until

$$\delta\varphi_n = J(\varphi_n) + \| \nabla J(\varphi_n) \|^2 < \varepsilon, \qquad (19.14)$$

where ε is a small constant. Temperature φ_k is then considered to be the approximation to the target value of the initial temperature $T(t_{n+1}, \mathbf{x})$. And finally,

Step 6 The boundary value problem (Eqs. 19.1–19.3) is solved at $T = T(t_{n+1}, \mathbf{x})$ to determine the flow velocity $\mathbf{u}(t_{n+1}, \mathbf{x}; T(t_{n+1}, \mathbf{x}))$.

Note that Step 2 introduces a pre-conditioner to accelerate the convergence of temperature iterations in Steps 1 to 3 at a higher Rayleigh number. At lower Ra, Step 2 is omitted and $\mathbf{u}_{adv}$ is replaced by $\mathbf{u}$. After these algorithmic steps, temperature $T = T(t_n, \mathbf{x})$ and flow velocity $\mathbf{u} = \mathbf{u}(t_n, \mathbf{x})$ (corresponding to $t = t_n$, $n = 0, \ldots, m$) are obtained. Based on the obtained results, and when required, interpolations can be used to reconstruct the process on the time interval $[0, \vartheta]$ in more detail.

Hence, at each subinterval of time: (i) the VAR method is applied to the heat equation only; (ii) the direct and conjugate problems for the heat equation are solved iteratively to find temperature; and (iii) the backward flow is determined from the Stokes and continuity equations twice (for 'advective' and 'true' temperatures). Compared to the VAR generalised inverse approach (Bunge et al., 2003), the described numerical approach is computationally less expensive, because the Stokes equation is not involved in the iterations between the direct and conjugate problems (note that the numerical solution of the Stokes equation is the most time-consuming).

19.2.5 Challenges in VAR Data Assimilation

The VAR data assimilation can theoretically be applied to many geodynamic problems, although a practical implementation of the technique for modelling of real geodynamic processes backward in time is not a simple task. The mathematical model of mantle dynamics (Eqs. 19.1–19.6) is simple, and many complications are omitted. For example, the mantle rheology is more complex (e.g. Karato, 2010); the adiabatic heating/cooling affects mantle temperature, especially near the thermal boundary layer; no phase transformations are considered, although they can influence the thermal convection pattern. To consider these and other complications in the VAR data assimilation, the adjoint equations should be derived each time when the set of the equations is changed. The cost to be paid is in software development since an adjoint model should be developed.

The solution to the heat problem (Eqs. 19.4–19.6) is a sufficiently smooth function. The temperature derived from the seismic tomography should be smooth as well; actually they are, being the solution to the relevant seismic velocity–temperature optimisation problem. The requirement of smoothness is important, because if the initial temperature is not a smooth function of space variables, recovery of this

temperature using the VAR method is not effective as the iterations converge very slowly to the target temperature. Ismail-Zadeh et al. (2006) illustrated the issue of convergence in the cases of a smooth, piecewise smooth, and discontinuous target function. It was shown that iterations converge rapidly for the sufficiently smooth target function (only a few iterations required), and a large number of iterations (more than 500) is required in the cases of non-smooth functions.

If the initial temperature guess φ_0 is a smooth function, all successive temperature iterations φ_k in scheme (19.8) should be smooth functions too, because the gradient of the objective functional ∇J is a smooth function as the solution to the adjoint problem (Eqs. 19.11–19.13). However, the temperature iterations φ_k are polluted by small errors, which are inherent in numerical experiments, and these perturbations grow with time. Samarskii et al. (1997) applied a VAR method to a 1-D backward heat diffusion problem and showed that the solution to this problem becomes noisy, if the initial temperature guess is slightly perturbed, and the amplitude of this noise increases with the initial perturbations of the temperature guess. To reduce the noise, they suggested to use a filter based on the replacement of iterations (19.8) by $\mathbf{B}(\varphi_{k+1} - \varphi_k) = -\beta_k \nabla J(\varphi_k)$, where $\mathbf{B}y = y - \nabla^2 y$ (Tsepelev, 2011). An employment of this filter increases the number of iterations to obtain the target temperature, and it becomes quite expensive computationally, especially when the model is three-dimensional. Another way to reduce the noise is to employ high-order adjoint (Alekseev and Navon, 2001) or regularisation (e.g. Tikhonov, 1963; Lattes and Lions, 1969; Samarskii and Vabischevich, 2007) techniques.

19.3 Quasi-Reversibility (QRV) Method

The mathematical idea of the QRV method is based on the transformation of an ill-posed problem into a well-posed problem (Lattes and Lions, 1969). In the case of the backward heat equation, this implies an introduction of an additional term into the equation, which involves the product of a small regularisation parameter and a higher-order temperature derivative. The additional term should be sufficiently small compared to other terms of the heat equation and allow for simple additional boundary conditions. The data assimilation in this case is based on a search of the best fit between the forecast model state and the observations by minimising the regularisation parameter. The QRV method is proven to be well suited for smooth and non-smooth input data (Lattes and Lions, 1969; Samarskii and Vabishchevich, 2007). Note that any regularisation has its advantages and disadvantages. A regularising operator is used in a mathematical problem to accelerate a convergence, to fulfil the physical laws (e.g. maximum principal, conversation of energy) in discrete equations, to suppress a noise in input data and in numerical computations, and to account for *a priori* information about an unknown solution and hence to improve a quality of computations. The major drawback of regularisation is that the accuracy of the solution to a regularised problem is always lower than that to a non-regularised problem.

The transformation to the regularised backward heat problem is not only a mathematical approach to solving ill-posed backward heat problems, but has some physical meaning: it can be explained on the basis of the concept of relaxing heat flux for heat conduction (e.g. Vernotte 1958). The classical Fourier heat conduction theory provides the infinite velocity of heat propagation in a region. The instantaneous heat propagation is unrealistic, because the heat is a result of the vibration of atoms and the vibration propagates in a finite speed (Morse and Feshbach, 1953). To accommodate the finite velocity of heat propagation, a modified heat flux model was proposed by Vernotte (1958) and Cattaneo (1958). The modified Fourier constitutive equation is expressed as $\vec{Q} = -k\nabla T - \tau\, \partial \vec{Q}/\partial t$, where $\vec{Q}$ is the heat flux, and k is the coefficient of thermal conductivity. The thermal relaxation time $\tau = k/(\rho c_p v^2)$ is usually recognised to be a small parameter (Yu et al., 2004), where ρ is the density, c_p is the specific heat, and v is the heat propagation velocity. The situation for $\tau \to 0$ leads to instantaneous diffusion at infinite propagation speed, which coincides with the classical thermal diffusion theory. The heat conduction equation $\partial T/\partial t = \nabla^2 T + \tau\, \partial^2 T/\partial t^2$ based on non-Fourier heat flux can be considered as a regularised heat equation. If the Fourier law is modified further by an addition of the second derivative of heat flux (e.g. $\vec{Q} = -k\nabla T + \beta \frac{\partial^2 \vec{Q}}{\partial t^2}$), where small $\beta > 0$ is the relaxation parameter of heat flux (Bubnov, 1976, 1981), the heat conduction equation can be transformed into a higher-order regularised heat equation.

19.3.1 Mathematical Statement
We consider a mathematical model of thermo-convective flow in the mantle and search for the velocity $\mathbf{u} = \mathbf{u}(t, \mathbf{x})$, the pressure $p = p(t, \mathbf{x})$, and the temperature $T = T(t, \mathbf{x})$ satisfying the boundary value problem (Eqs. 19.1–19.3) and the initial-boundary value problem (Eq. 19.4–19.6). The inverse problem can be formulated in this case as follows: find the velocity, pressure, and temperature satisfying the boundary value problem (Eqs. 19.1–19.3) and the final-boundary value problem including Eq. (19.4), Eq. (19.5), and the final condition:

$$T(t = \vartheta, \mathbf{x}) = T_\vartheta(\mathbf{x}), \qquad \mathbf{x} \in \Omega, \tag{19.15}$$

where T_ϑ is the temperature at time $t = \vartheta$.

To solve the inverse problem by the QRV method, Ismail-Zadeh et al. (2007) considered the following regularised backward heat problem to define temperature in the past from the known present temperature $T_\vartheta(\mathbf{x})$:

$$\partial T_\beta/\partial t - \langle \mathbf{u}_\beta, \nabla T_\beta \rangle = \nabla^2 T_\beta - \beta\,\Lambda\,(\partial T_\beta/\partial t), \quad t \in [0, \vartheta],$$
$$\mathbf{x} \in \Omega, \tag{19.16}$$

$$\sigma_1 T_\beta + \sigma_2 \partial T_\beta/\partial \mathbf{n} = T_*, \qquad t \in (0, \vartheta), \mathbf{x} \in \partial\Omega, \tag{19.17}$$

$$\sigma_1 \partial^2 T_\beta/\partial \mathbf{n}^2 + \sigma_2 \partial^3 T_\beta/\partial \mathbf{n}^3 = 0, \qquad t \in (0, \vartheta), \mathbf{x} \in \partial\Omega, \tag{19.18}$$

$$T_\beta(\vartheta, \mathbf{x}) = T_\vartheta(\mathbf{x}), \qquad \mathbf{x} \in \Omega, \tag{19.19}$$

where $\Lambda(T) = \partial^4 T/\partial x_1^4 + \partial^4 T/\partial x_2^4 + \partial^4 T/\partial x_3^4$, and the boundary value problem to determine the fluid flow:

$$\nabla P_\beta = -\nabla \cdot [\eta(T_\beta)(\nabla \mathbf{u}_\beta + \nabla \mathbf{u}_\beta{}^T)] + Ra T_\beta \mathbf{e}, \quad \mathbf{x} \in \Omega, \tag{19.20}$$

$$\nabla \cdot \mathbf{u}_\beta = 0, \qquad \mathbf{x} \in \Omega, \tag{19.21}$$

$$\langle \mathbf{u}_\beta, \mathbf{n} \rangle = 0, \qquad \sigma_\beta \mathbf{n} - \langle \sigma_\beta\, \mathbf{n}, \mathbf{n} \rangle \mathbf{n} = 0, \qquad \mathbf{x} \in \partial\Omega, \tag{19.22}$$

where $\sigma_\beta = \eta(\nabla \mathbf{u}_\beta + \nabla \mathbf{u}_\beta{}^T)$. The sign of the velocity field is changed ($\mathbf{u}_\beta$ by $-\mathbf{u}_\beta$) in Eqs. (19.16) and (19.20) to simplify the application of the total variation diminishing (TVD) method (Ismail-Zadeh and Tackley, 2010) for solving equations (19.16)–(19.19). Hereinafter temperature T_ϑ is referred to as the input temperature for the problem (Eqs. 19.16–19.22). The core of the transformation of the heat equation is the addition of a high-order differential expression $\Lambda(\partial T_\beta/\partial t)$ multiplied by a small parameter $\beta > 0$. Note that Eq. (19.18) is added to the boundary conditions to properly define the regularised backward heat problem. The solution to the regularised backward heat problem is stable for $\beta > 0$, and the approximate solution to Eqs. (19.16)–(19.22) converges to the solution of (19.1)–(19.5), and (19.15) in some spaces, where the conditions of well-posedness are met (Samarskii and Vabischevich, 2007). Thus, the inverse problem of thermo-convective mantle flow is reduced to determination of the velocity $\mathbf{u}_\beta = \mathbf{u}_\beta(t, \mathbf{x})$, the pressure $P_\beta = P_\beta(t, \mathbf{x})$, and the temperature $T_\beta = T_\beta(t, \mathbf{x})$, satisfying Eqs. (19.16)–(19.22).

19.3.2 Optimisation Problem
A maximum of the following functional is sought with respect to the regularisation parameter β:

$$\delta - \| T(t = \vartheta, \cdot; T_{\beta_k}(t = 0, \cdot)) - \varphi(\cdot) \| \to \max_k, \tag{19.23}$$

$$\beta_k = \beta_0 q^{k-1}, \qquad k = 1, 2, ..., \Re, \tag{19.24}$$

where sign $\| \cdot \|$ denotes the norm in the space $L_2(\Omega)$. Here $T_k = T_{\beta_k}(t = 0, \cdot)$ is the solution to the regularised backward heat problem (Eqs. 19.16–19.19) at $t = 0$;

$T(t = \vartheta, \cdot; T_k)$ is the solution to the heat problem (Eqs. 19.4–19.6) at the initial condition $T(t = 0, \cdot) = T_k$ at time $t = \vartheta$; φ is the known temperature at $t = \vartheta$ (the input data, i.e. the present temperature); $\beta_0 > 0$, $0 < q < 1$, and $\delta > 0$ is a given accuracy. When q tends to unity, the computational cost becomes large; and when q tends to zero, the optimal solution can be missed.

The prescribed accuracy δ is composed from the accuracy of the initial data and the accuracy of computations. When the input noise decreases and the accuracy of computations increases, the regularisation parameter is expected to decrease. However, estimates of the initial data errors are usually inaccurate. Estimates of the computation accuracy are not always known, and when they are available, the estimates are coarse. In practical computations, it is more convenient to minimise the following functional with respect to Eq. (19.24)

$$\| T_{\beta_{k+1}}(t = 0, \cdot) - T_{\beta_k}(t = 0, \cdot) \| \to \min_{k}, \quad (19.25)$$

where misfit between temperatures obtained at two adjacent iterations must be compared. To implement the minimisation of temperature residual (19.23), the inverse problem (Eqs. 19.16–19.22) must be solved on the entire time interval as well as the direct problem (Eqs. 19.1–19.6) on the same time interval. This at least doubles the number of computations. The minimisation of functional (19.25) has a lower computational cost, but it does not rely on *a priori* information.

19.3.3 Numerical Approach
The numerical algorithm for solving the inverse problem of thermo-convective mantle flow using the QRV method can be described as follows. Consider a uniform temporal partition $t_n = \vartheta - \delta t \, n$ (as defined in Section 19.2.4) and prescribe some values to parameters β_0, q, and $\Re$ (e.g. $\beta_0 = 10^{-3}$, $q = 0.1$, and $\Re = 10$). According to (19.24), a sequence of the values of the regularisation parameter $\{\beta_k\}$ is defined. For each value $\beta = \beta_k$, model temperature and velocity are determined in the following way.

Step 1 Given the temperature $T_\beta = T_\beta(t, \cdot)$ at $t = t_n$, the velocity $\mathbf{u}_\beta = \mathbf{u}_\beta(t_n, \cdot)$ is found by solving problem (Eqs. 19.20–19.22). This velocity is assumed to be constant on the time interval $[t_{n+1}, t_n]$.

Step 2 Given the velocity $\mathbf{u}_\beta = \mathbf{u}_\beta(t_n, \cdot)$, the new temperature $T_\beta = T_\beta(t, \cdot)$ at $t = t_{n+1}$ is found on the time interval $[t_{n+1}, t_n]$ subject to the final condition $T_\beta = T_\beta(t_n, \cdot)$ by solving the regularised problem (Eqs. 19.16–19.19) backward in time.

Step 3 Upon the completion of Steps 1 and 2 for all $n = 0, 1, \ldots, m$, the temperature $T_\beta = T_\beta(t_n, \cdot)$ and the velocity $\mathbf{u}_\beta = \mathbf{u}_\beta(t_n, \cdot)$ are obtained at each $t = t_n$. Based on the computed solution, find the temperature and flow velocity at each point of time interval $[0, \vartheta]$ using interpolation.

Step 4a The direct problem (Eqs. 19.4–19.6) is solved assuming that the initial temperature is given as $T_\beta = T_\beta(t = 0, \cdot)$, and the temperature residual (19.23) is found. If the residual does not exceed the predefined accuracy, the calculations are terminated, and the results obtained at Step 3 are considered as the final ones. Otherwise, parameters β_0, q, and $\Re$ entering Eq. (19.24) are modified, and the calculations are continued from Step 1 for new set $\{\beta_k\}$.

Step 4b The functional (19.25) is calculated. If the residual between the solutions obtained for two adjacent regularisation parameters satisfies a predefined criterion (the criterion should be defined by a user, because no *a priori* data are used at this step), the calculation is terminated, and the results obtained at Step 3 are considered as the final ones. Otherwise, parameters β_0, q, and $\Re$ entering Eq. (19.24) are modified, and the calculations are continued from Step 1 for new set $\{\beta_k\}$.

In a particular implementation, either Step 4a or Step 4b is used to terminate the computation. This numerical algorithm allows organising a certain number of independent computational modules for various values of the regularised parameter β_k that find the solution to the regularised problem using Steps 1–3 as well as determining an acceptable result *a posteriori* according to Step 4a or 4b.

19.4 Restoration of Mantle Plumes

A plume is hot, narrow mantle upwelling that is invoked to explain hotspot volcanism. In a fluid with the temperature-dependent viscosity as in the case of the mantle viscosity model, a plume is characterised by a mushroom-shaped head and a thin tail. Mantle plumes evolve in three distinguishing stages: *immature* – an origin and initial rise of the plumes; *mature* – plume-lithosphere interaction, gravity spreading of plume head and development of overhangs beneath the bottom of the lithosphere, and partial melting of the plume material (e.g. Ribe and Christensen, 1994; Moore et al., 1998); and *overmature* – slowing-down of the plume rise and fading of the mantle plumes due to thermal diffusion (Davaille and Vatteville, 2005; Ismail-Zadeh et al., 2006). The ascent and evolution of mantle plumes depend on the properties of the source region (i.e. the thermal boundary layer) and the viscosity and thermal diffusivity of the ambient mantle. The properties of the source region determine temperature and viscosity of the mantle plumes. Structure, flow rate, and heat flux of the plumes are controlled by the properties of the mantle through which the plumes rise. While properties of the lower mantle (e.g. viscosity, thermal conductivity) are relatively constant during about 150 Myr lifetime of most plumes, source region properties can vary substantially with time as the thermal basal boundary layer feeding the plume is depleted of hot material (Schubert et al., 2001). Complete local depletion of this boundary layer cuts the plume off from its source.

A mantle plume is a well-established structure in computer modelling and laboratory experiments. Numerical experiments on dynamics of mantle plumes (Trompert and Hansen, 1998; Zhong, 2005) showed that the number of plumes increases and the rising plumes become thinner with an increase in Rayleigh number. Disconnected thermal plume structures appear in thermal convection at *Ra* greater than 10^7 (e.g. Hansen et al., 1990). At high *Ra* (in the hard turbulence regime), thermal plumes are torn off the boundary layer by the large-scale circulation or by non-linear interactions between plumes (Malevsky and Yuen, 1993). Plume tails can also be disconnected when the plumes are tilted by plate scale flow (e.g. Olson and Singer, 1985). Ismail-Zadeh et al. (2006) presented an alternative explanation for the disconnected mantle plume heads and tails, which is based on thermal diffusion of mantle plumes.

19.4.1 Model Setup and Methods

To model the evolution of mantle plumes, Ismail-Zadeh et al. (2006) used Eqs. (19.1)–(19.6) with impenetrability and perfect slip conditions at the model boundary Ω. A temperature-dependent viscosity $\eta(T) = \exp(M/[T + G] - M/[0.5 + G])$ is employed, where $M = [225/\ln(r)] - 0.25 \ln(r)$, $G = 15/\ln(r) - 0.5$ and r is the viscosity ratio between the upper and lower boundaries of the model domain (Busse et al., 1993). The model domain is divided into $37 \times 37 \times 29$ rectangular finite elements to approximate the vector velocity potential by tricubic splines, and a uniform grid $112 \times 112 \times 88$ is employed for approximation of temperature, velocity, and viscosity (Ismail-Zadeh et al., 2001b, 2006). Temperature in the heat equation (19.4) is approximated by finite differences and determined by the semi-Lagrangian method (Ismail-Zadeh and Tackley, 2010). A numerical solution to the Stokes and incompressibility equations (19.1) and (19.2) is based on the introduction of a two-component vector velocity potential and on the application of the Eulerian finite-element method with a tricubic-spline basis for computing the potential (Ismail-Zadeh et al., 2001b; Ismail-Zadeh and Tackley, 2010). Such a procedure results in a set of linear algebraic equations solved by the conjugate gradient method (e.g. Ismail-Zadeh and Tackley, 2010).

19.4.2 Forward Modelling

We present here the evolution of mature mantle plumes modelled forward in time. Considering the following model parameters, $\alpha = 3 \times 10^{-5}$ K^{-1}, $\rho_{ref} = 4000$ kg m^{-3}, $\Delta T = 3000$ K, $h = 2{,}800$ km, $\eta_{ref} = 8 \times 10^{22}$ Pa s, and $\kappa = 10^{-6}$ m^{-2} s^{-1}, the Rayleigh number is estimated to be $Ra = 9.5 \times 10^5$. While plumes evolve in the convecting heterogeneous mantle, at the initial time, it is assumed that the plumes develop in a laterally homogeneous temperature field, and hence the initial mantle temperature is considered to increase linearly with depth.

Mantle plumes are generated by random temperature perturbations at the top of the thermal source layer associated with the core–mantle boundary (Fig. 19.1a). The mantle material in the basal source layer flows horizontally towards the plumes. The reduced viscosity in this basal layer promotes the flow of the material to the plumes. Vertical upwelling of hot mantle material is concentrated in low-viscosity conduits near the centrelines of the emerging plumes (Fig. 19.1b,c). The plumes move upward through the model domain, gradually forming structures with well-developed heads and tails. Colder material overlying the source layer (e.g. portions of lithospheric slabs subducted to the core–mantle boundary) replaces hot material at the locations where the source material is fed into mantle plumes. Some time is required to recover the volume of source material depleted due to plume feeding (Howard, 1966). Because the volume of upwelling material is comparable to the volume of the thermal source layer feeding the mantle plumes, hot material could eventually be exhausted, and mantle plumes would be starved thereafter.

Ismail-Zadeh et al. (2006) showed that the plumes diminish in size with time (Fig. 19.1d), and the plume tails disappear before the plume heads (Fig. 19.1e,f). Therefore, plumes start disappearing from bottom up and fade away by thermal diffusion. At different stages in the plume decay, one sees quite isolated plume heads, plume heads with short tails, and plumes with nearly pinched off tails. Different amounts of time are required for different mantle plumes to vanish into the ambient mantle, the required time depending on the geometry of the plume tails.

19.4.3 Performance of the VAR Data Assimilation

To restore the prominent state of the plumes (Fig. 19.1d) in the past from their 'present' weak state (Fig. 19.1f), Ismail-Zadeh et al. (2006) employed the VAR data assimilation. Figure 19.2 illustrates the restored states of the plumes (panel b) and the temperature residuals δT (panel e) between the temperature $\widetilde{T}(\mathbf{x})$ predicted by the forward model and the temperature $\widetilde{T}(\mathbf{x})$ reconstructed to the same age.

The performance of the VAR data assimilation for various Rayleigh numbers (*Ra*) and the mantle viscosity ratio (*r*) is evaluated in terms of the number of iterations n required to achieve a prescribed relative reduction of $\delta\varphi_n$ (inequality 19.14). Figure 19.3 presents the evolution of the objective functional $J(\varphi_n)$ versus the number of iterations at time $\sim 0.5\theta$. The objective functional show a quite rapid decrease after about seven iterations for $Ra = 9.5 \times 10^5$ and $r = 20$ (curves 1). As *Ra* decreases and thermal diffusion increases (curves 2–4) the performance of the algorithm becomes poor: more iterations are needed to achieve the prescribed ε.

Despite its simplicity, the minimisation algorithm (19.8) provides for a rapid convergence and good quality of

(a) 335 Myr

(b) 195 Myr

(c) 155 Myr

(d) 120 Myr

(e) 80 Myr

(f) present

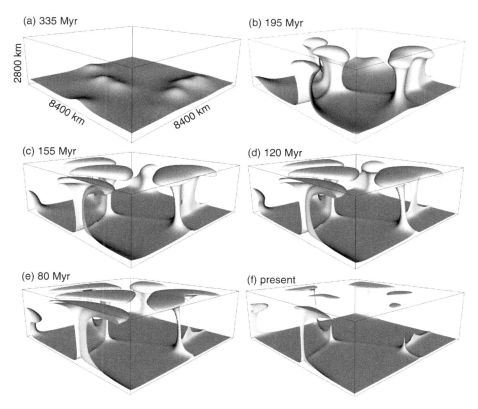

Figure 19.1 Mantle plumes in the forward modelling at successive times: from 335 Myr ago (a) to the 'present' state of the plumes (f). The plumes are represented here and in Fig. 19.2 by isothermal surfaces at 3000 K. After Ismail-Zadeh et al. (2006).

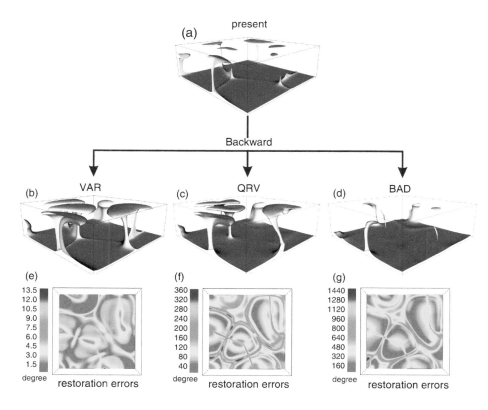

Figure 19.2 Assimilation of the mantle temperature and flow to the time of 100 Myr ago and temperature residuals between the present temperature model (a) and the temperature assimilated to the same age, using the VAR (b and e), QRV (c and f; $\beta = 10^{-7}$), and BAD (d and g) methods, respectively. After Ismail-Zadeh et al. (2007).

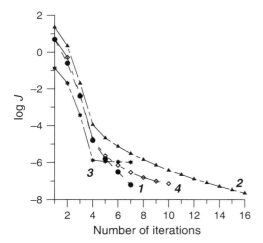

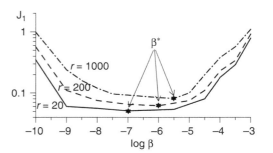

Figure 19.4 Temperature misfit J_1 as the function of the regularisation parameter β. The minimum of the temperature misfit is achieved at β^*, an optimal regularisation parameter. Solid curves: $r = 20$; dashed curves: $r = 200$; and dash-dotted curves: $r = 1000$. After Ismail-Zadeh et al. (2007).

Figure 19.3 Relative reductions of the objective functional J as functions of the number of iterations. Curves: 1, $r = 20$, $Ra = 9.5 \times 10^5$; 2, $r = 20$, $Ra = 9.5 \times 10^2$; 3, $r = 200$, $Ra = 9.5 \times 10^3$; 4, $r = 200$, $Ra = 9.5 \times 10^2$. After Ismail-Zadeh et al. (2006).

optimisation at high Rayleigh numbers (Ismail-Zadeh et al., 2006). The convergence rate and the quality of optimisation become worse with the decreasing Rayleigh number. The use of the limited-memory quasi-Newton algorithm L-BFGS (Liu and Nocedal, 1989) might provide for a better convergence rate and quality of optimisation (Zou et al., 1993). Although an improvement of the convergence rate by using another minimisation algorithm (e.g. L-BFGS) will reduce the computational expense associated with the solving of the problem under question, this reduction would be not significant, because the largest portion (about 70%) of the computer time is spent to solve the Stokes equations.

19.4.4 Performance of the QRV Data Assimilation

To demonstrate the performance of the QRV data assimilation and to compare the results with those obtained by the VAR and BAD methods, the same forward model for mantle plume evolution as in Section 19.4.2 is used (Ismail-Zadeh et al., 2007). Figure 19.2 (panels b, c, and d) illustrate the restored state of mantle plumes with the temperature residuals δT (panels e, f, and g) obtained by application of the VAR, QRV, and BAD assimilation, respectively. Ismail-Zadeh et al. (2007) showed that the VAR method provides the best performance for the diffused plume restoration. The BAD method cannot restore the diffused parts of the plumes because temperature is only advected backward in time. The QRV method restores the diffused thermal plumes, meanwhile the restoration results are not so perfect as in the case of VAR method. Namely, the temperature residuals obtained in the case of the VAR method do not exceed 15 degrees (Fig. 19.2e). Meanwhile, the residuals range from tens to a few hundred degrees in the case of the QRV method

(Fig. 19.2f) and from a few hundred to about 1,000 degrees in the case of the BAD method (Fig. 19.2g). Although the accuracy of the QRV data assimilation is lower compared to the VAR data assimilation, the QRV method does not require any additional smoothing of the input data and filtering of temperature noise as the VAR method does.

Figure 19.4 presents the residual $J_1(\beta) = \| T_0(\cdot) - T_\beta(t = t_0, \cdot; T_\vartheta) \|$ between the target temperature T_0 and the restored temperature obtained by the QRV data assimilation with the input temperature T_ϑ. The optimal accuracy is attained at $\beta^* = \arg\min\{J(\beta) : \beta = \beta_k, \ k = 1, 2, ..., 10\} \approx 10^{-7}$ in the case of $r = 20$, and at $\beta^* \approx 10^{-6}$ and $\beta^* \approx 10^{-5.5}$ in the cases of the viscosity ratio $r = 200$ and $r = 1000$, respectively. Comparison of the temperature residuals for three values of the viscosity ratio r indicates that the residuals become larger as the viscosity ratio increases. The numerical experiments show that the algorithm for solving the inverse problem performs well when the regularisation parameter is in the range $10^{-8} \leq \beta \leq 10^{-6}$. For greater values, the solution of the inverse problem retains stability but is less accurate. The numerical procedure becomes unstable at $\beta < 10^{-9}$, and the computations must be stopped.

19.5 Reconstruction of Plate Subduction

In this section, we consider the application of the QRV data assimilation to plate subduction beneath the Japanese islands (Ismail-Zadeh et al., 2013; 2016). An interaction of the Pacific, Okhotsk, Eurasian, and Philippine Sea lithosphere plates with the deeper mantle around the Japanese islands is complicated by active subduction of the plates (Fukao et al., 2001; Furumura and Kennett, 2005) and back-arc spreading (Jolivet et al., 1994), which cannot be understood by the plate kinematics only (Fig. 19.5). The Pacific plate subducts under the Okhotsk and the Philippine Sea plates with the relative speed of about 9 cm yr^{-1} and 5 cm yr^{-1}, respectively, whereas the Philippine Sea plate subducts under the Eurasian plate with the relative

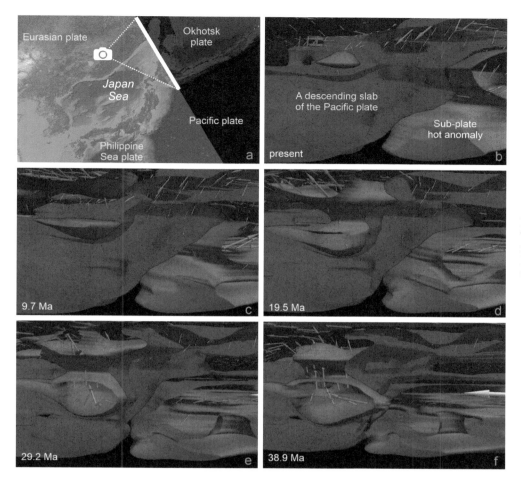

Figure 19.5 Reconstruction of mantle dynamics beneath the Japanese islands to the Middle Eocene times (~40 Myr ago (Ma); Ismail-Zadeh et al., 2013). (a) Topography map of the Japanese Islands and surroundings. The bold white line marks the surface projection of the vertical 'window', through which the model domain is seen from SW. (b)–(f) Snapshots of the evolution of the lithospheric slab (blue) and the mantle upwelling (red) seen through the 'window': (b) present state; (c) 9.7 Ma; (d) 19.5 Ma; €̃ 29.2 Ma; and (f) 38.9 Ma. Arrows indicate the mantle flow velocities. Images courtesy A. Helfrich-Schkarbanenko.

speed of about 5 cm yr⁻¹ (Drewes, 2009). Back arcs of these subduction zones are also known as the site of active spreading in the past and recent as inferred from both the geophysical and geological surveys (Jolivet et al., 1994).

P-wave seismic tomography of the mantle beneath the subducting Pacific plate near the Japanese islands revealed a low-velocity region extending oceanward at depths around the 410-km seismic discontinuity, and this low-velocity anomaly region was interpreted as a zone with an excess temperature of 200 K and the associated fractional melt of less than 1% (Obayashi et al., 2006). To clarify the origin of the hot temperature anomaly beneath the Pacific plate and its implication for back-arc basin evolution, Ismail-Zadeh et al. (2013) studied the mantle evolution beneath the Japanese islands and their surroundings based on the assimilation of temperature inferred from seismic tomography, the present movements derived from geodetic observations, and the past plate motion inferred from paleogeographic and paleomagnetic plate reconstructions.

19.5.1 Mathematical Statement and Model Setup

In the domain $\overline{\Omega} = [0, x_1 = l_1 = 4000 \text{ km}] \times [0, x_2 = l_2 = 4000 \text{ km}] \times [0, x_3 = h = 800 \text{ km}]$, and for time interval, the regularised Stokes, the incompressibility, and the backward heat balance equations are solved with relevant boundary and initial conditions (presented here) using the QRV method and the extended Boussinesq approximation (Christensen and Yuen, 1985):

$$-\nabla P + \nabla \cdot (\eta[\nabla \mathbf{u} + (\nabla \mathbf{u})^T]) = (\mathbf{E} + \varsigma\nabla^2)^{-1}$$
$$[RaT - a_1 La\,\Phi_1(\pi_1) - a_2 La\,\Phi_2(\pi_2)]\mathbf{e}, \quad (19.26)$$

$$\nabla \cdot \mathbf{u} = 0, \quad (19.27)$$

$$\frac{\partial}{\partial t}(\mathbf{E} + \beta\nabla^2)^2 T - \mathbf{u}\cdot\nabla T - A^{-1}B\ Di^* Ra\ u_3\ T$$
$$= -A^{-1}\left(-\nabla^2 T + Di^*\eta\sum_{i,j=1}^{3}(e_{ij})^2\right), \quad (19.28)$$

where

Table 19.1 *Parameters of the model by Ismail-Zadeh et al. (2013) used in Section 19.5*

PARAMETER	SYMBOL	VALUE
Dimensionless density jump at the 410-km phase boundary	a_1	0.05
Dimensionless density jump at the 660-km phase boundary	a_2	0.09
Thermal conductivity	c	1,250 W m^{-1} K^{-1}
Activation energy	E_a	3×10^5 J mol^{-1}
Acceleration due to gravity	g	9.8 m s^{-2}
Depth	h	800 km
Length (in x-direction)	l_1	4,000 km
Length (in y-direction)	l_2	4,000 km
Universal gas constant	R	8.3144 J mol^{-1} K^{-1}
Difference between the temperatures at the lower (T_l) and upper (T_u) model boundaries	T^*	1,594 K
Dimensionless temperature at the upper model boundary	T_u	290 / T^*
Dimensionless temperature at the lower model boundary	T_l	1,884 / T^*
Dimensionless temperature at the 410-km phase boundary	T_1	1,790 / T^*
Dimensionless temperature at the 660-km phase boundary	T_2	1,891 / T^*
Activation volume	V_a	4×10^{-6} m^3 mol^{-1}
Dimensionless width of the 410-km phase transition	w_1	10 km / h
Dimensionless width of the 660-km phase transition	w_2	10 km / h
Dimensionless depth of the 410-km phase boundary	z_1	390 km / h
Dimensionless depth of the 660-km phase boundary	z_2	140 km / h
Thermal expansivity	α	3×10^{-5} K^{-1}
QRV regularisation parameter	β	0.00001
Dimensionless Clapeyron (pressure-temperature) slope at the 410-km phase boundary	$\bar{\gamma}_1$	4×10^6 Pa K^{-1} $\times T^*(\rho^*gh)^{-1}$
Dimensionless Clapeyron slope at the 660-km phase boundary	$\bar{\gamma}_2$	-2×10^6 Pa K$^{-1} \times T^*(\rho^*gh)^{-1}$
Reference viscosity	η^*	10^{21} Pa s
Thermal diffusivity	κ	10^{-6} m^{2}s^{-1}
Reference density	ρ^*	3,400 kg m^{-3}
Phase regularisation parameter	ς	0.0001

$$A = \left[1 + \left(a_1 \frac{2}{w_1}(\Phi_1 - \Phi_1{}^2)\bar{\gamma}_1^2 + a_2 \frac{2}{w_2}(\Phi_2 - \Phi_2{}^2)\bar{\gamma}_2^2 \right) Di^* \, La \, T \right] > 0,$$

$$B = \left[1 + \frac{La}{Ra}\left(a_1 \frac{2}{w_1}(\Phi_1 - \Phi_1{}^2)\bar{\gamma}_1 + a_2 \frac{2}{w_2}(\Phi_2 - \Phi_2{}^2)\bar{\gamma}_2 \right) \right],$$

$$\Phi_i = \frac{1}{2}\left[1 + \tanh\frac{\pi_i}{w_i} \right], \pi_i = z_i - x_3 - \bar{\gamma}_i(T - T_i), \quad i = 1, 2.$$

Here ϑ is the present time; $e_{ij}(\mathbf{u}) = \{\partial u_i/\partial x_j + \partial u_j/\partial x_i\}$ is the strain rate tensor; and $\mathbf{E}$ is the unit operator. With regard to the phase changes around 410 km and 660 km, respectively, π_1 and π_2 are the dimensionless excess pressures; Φ_1 and Φ_2 are the phase functions describing the relative fraction of the heavier phase, respectively, and varying between 0 and 1 as a function of depth and temperature. The Rayleigh (Ra), Laplace (La), and modified dissipation (Di^*) dimensionless numbers are defined as $Ra = \alpha g \rho^* T^* h^3$

$(\eta^*\kappa)^{-1}$, $La = \rho^*gh^3(\eta^*\kappa)^{-1}$, and $Di^* = \eta^*\kappa(\rho^*ch^2T^*)^{-1}$, respectively. The operator $(\mathbf{E} + \varsigma\nabla^2)^{-1}$, $\varsigma = \text{const} > 0$ is applied to the right-hand side of the Stokes equations (19.26) to smooth temperature jumps at the phase boundaries and to enhance the stability of our computations. According to the QRV method, the higher dissipation term, whose magnitude is controlled by the small (regularisation) parameter β, is introduced to regularise the heat balance equation (19.28). The physical parameters used in this study are listed in Table 19.1.

Although the mantle dynamics is coupled to the lithosphere dynamics, the coupling can be weak or strong depending on the viscosity contrast between the lithosphere and the underlying mantle (Doglioni et al., 2011). There is still a debate about driving forces of plate tectonics: whether a mantle–lithosphere interaction is driven by a slab pull or by mantle upwellings (e.g. Coltice et al., 2019). Because of uncertainties in the driving forces, kinematic conditions (the direction and rate of plate motion) are normally prescribed to the plates in numerical modelling of the mantle–lithosphere dynamics. Current kinematic conditions can be estimated from geodetic measurements. However, the past kinematic conditions are less reliable due to uncertainties in plate tectonic reconstructions (e.g. Rowley, 2008).

To constrain the horizontal lithosphere motion, Ismail-Zadeh et al. (2013) prescribed plate motion velocities at the upper boundary of the model domain. The present plate motion velocities were determined from the Actual Plate Kinematic and Deformation Model (APKIM2005) derived from geodetic data (Drewes, 2009) and from the PB2002 model for the Philippine Sea and Okinawa Plates (Bird, 2003). To determine mantle dynamics in the geological past, the past plate velocities were derived from paleogeographic reconstructions of the Philippine Sea (Seno and Maruyama, 1984) and Japanese Islands (Maruyama et al., 1997), Philippine Sea plate motion from paleomagnetic studies (Yamazaki et al., 2010), relative motion of the Pacific plate (Northrup et al., 1995), and Cenozoic plate tectonic evolution of the south-eastern Asia (Hall, 2002).

The input temperature for data assimilation was obtained from seismic tomography (*P*-wave velocity anomalies) data beneath the Japanese islands (Fukao et al., 2001; Obayashi et al., 2006, 2009) using a non-linear inversion method (Ismail-Zadeh et al., 2005) and considering the effects of mantle composition, anelasticity, and partial melting on seismic velocities as well as surface heat flow data to constrain the crustal temperature (Wang et al., 1995; Yamano et al., 2003). The seismic thermal state of the back-arc region (Ismail-Zadeh et al., 2013) is characterised by shallow hot anomalies reflecting the remnants of the back-arc spreading (Jolivet et al., 1994) and deep cold anomalies related to the stagnation of the lithospheric slabs (Fukao et al., 2001). The mantle beneath the Pacific plate is characterised by the shallow cold anomalies reflecting the existence of the old oceanic Pacific plate and the deep broad hot anomaly of unknown origin. This temperature model is used as the initial condition (input temperature) for a restoration model.

In the numerical modelling, Ismail-Zadeh et al. (2013) assumed that the Earth's mantle behaves as a temperature-dependent Newtonian fluid $\eta(T(\mathbf{x}), x_3) = \eta_0 \exp[(E_a + \rho_* g x_3 V_a)/(RT)]$, where η_0 is determined so that it will give 2.905×10^{20} Pa s at the depth of 290 km and temperature of 1,698 K; the activation energy is $E_a = 3 \times 10^5$ J mol^{-1}, and the activation volume of $V_a = 4 \times 10^{-6}$ m^3 mol^{-1}. Other parameters of the rheological law are listed in Table 19.1. The upper limit of the viscosity is set to be ~10^{22} Pa s, which results in the viscosity increase from the upper to the lower mantle by about two orders of magnitude.

19.5.2 Results

In backward sense, the high-temperature patchy anomaly beneath the back-arc Japan Sea basin splits into two prominent anomalies showing two small-scale upwellings beneath

the northern part (presented in Fig. 19.5) and the southwestern part of the Japan Sea. The present hot anomalies in the back-arc region move down eastward, while the broad hot anomaly under the Pacific plate moves slowly down westward (Fig. 19.5). The hot anomalies tend to merge at 38.9 Myr ago (Ma). The model shows the link between hot anomaly in the back-arc region and that in the sub-slab mantle at depths of about 440–560 km in the Middle to Late Eocene time. The upwellings are likely to be generated in the sub-slab hot mantle and penetrated through breaches/tears of the subducting Pacific plate into the mantle wedge. Hence, Ismail-Zadeh et al. (2013) proposed that the present hot anomalies in the back-arc and sub-slab mantle had a single origin located in the sub-lithospheric mantle. These small-scale upwellings beneath the northern part of the Japan Sea predicted by the assimilation of geophysical, geodetic, and geological data could be a source contributing to the rifting and back-arc basin opening.

Ismail-Zadeh et al. (2013) performed a sensitivity analysis to understand how stable the numerical solution was with respect to small perturbations of input data (the seismic temperature model). The seismic temperature model was perturbed randomly by 0.5–1% and then assimilated to the past to find the initial temperature. A misfit between the initial temperatures related to the perturbed and unperturbed present temperature is about 3–5%, which proves the stability of the solution.

A test related to the QRV reconstruction accuracy can be also performed. This test considers the reconstructed temperature as an initial temperature for the forward numerical model. The model is then solved with this initial temperature forward in time (from the past to the present). The 'present' temperature so obtained is compared to the input data of the QRV model. However, the misfit between the 'present' temperature and the input data may become significant depending on the time interval of the input data assimilation, the regularisation parameter β, and the errors in the input data (Samarskii and Vabishchevich, 2007; Ismail-Zadeh et al., 2007). This misfit can be amplified by the errors associated with the numerical solutions backward and forward in time.

For a given temperature, mantle dynamics depends on various factors including rheology (Billen, 2008), phase changes (Liu et al., 1991; Honda et al., 1993), and boundary conditions. Ismail-Zadeh et al. (2013) conducted a search over the ranges of uncertain parameters in the temperature- and pressure-dependent viscosity (activation energy and activation volume) to achieve 'plate-like' behaviour of the colder material. Also, they tested the influence of phase changes, model depth variations and boundary conditions on the model results. In numerical models of mantle dynamics, the choice of boundary conditions and the size of the model domain influence the pattern of flow and slab dynamics. If the depth of the model domain is significantly smaller than the horizontal dimensions of the domain, the thermo-

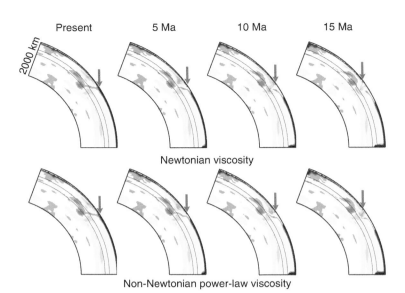

Figure 19.6 Cross-section view of the reconstruction of mantle dynamics beneath the Japanese islands to the Middle Miocene times (~15 Ma) along the latitude 53°N from the longitude 105°E to 175°E (unpublished work by Honda and Ismail-Zadeh). Snapshots of the evolution of the lithospheric slab (bluish) and the hotter mantle (reddish) in the case of the Newtonian rheology (upper panels) and the non-Newtonian rheology (lower panels). Numbers at the top are the age measured from the present in million years. Red arrows show the positions of the plate boundaries.

convective flow in the model with a low-viscosity upper mantle and higher-viscosity lower mantle will generate the return flow focused in the upper mantle. Increasing the model domain's depth removes the artificial lateral return flow in the upper mantle. The sensitivity analysis related to the presence of phase transformations and to changes in boundary conditions show that the model is robust, and the principal results of the model do not change (Ismail-Zadeh et al., 2013).

Numerical experiments using non-Newtonian power-law rheology showed that the dynamic features of the past plate subduction look rather similar to the Newtonian rheology (Fig. 19.6). The Newtonian viscosity was modelled as $\eta_l = \eta_r \eta_{zp}[\exp(E_1/(RT) - E_1/(RT_r))]$, and non-Newtonian power-law viscosity as the combination of both diffusion and dislocation creeps $(\eta_n)^{-1} = (\eta_r \eta_{zp})^{-1}[\exp(E_1/(RT) - E_1/(RT_r)) + \exp(E_2/(RT) - E_2/(RT_r)) (\tau/\tau_r)^{n-1}]$, where η_{zp} is the depth- and phase-dependent viscosity; η_r (= 10^{21} Pa s) is the reference viscosity at reference mantle temperature $T = T_r$ (= 1330°C); E_1 (= 300 kJ mol^{-1}), and E_2 (= 540 kJ mol^{-1}) are the activation energy of diffusion creep and diffusion/dislocation creep, respectively; R is the universal gas constant; T is the temperature; n (= 3.5) is the power-law index of the non-Newtonian rheology; τ is the second invariant of stress tensor, and τ_r (= 10^6 Pa) is the transitional stress at $T = T_r$, where the linear and effective non-linear viscosities give the same value η_r. There exist some differences in the style of subduction between the two rheological models; for example, the necking of the subducting lithosphere near the oceanic side of subduction and the link between the mantle wedge and sub-slab mantle become clearer. Also, we can observe a significant migration of the trench of the descending slab. Using data assimilation, Peng and Liu (2022) showed that slabs can migrate laterally at significant distances due to horizontal mantle

flows (trench migration) and surface plate motions influence the horizontal migration of slab.

19.6 Discussion

In this chapter we have presented two basic methods for data assimilation in geodynamic models and illustrated their applicability to analyse the mantle plume dynamics and lithosphere subduction in the geological past. Each method has its advantages and disadvantages. Table 19.2 summaries the differences in the methodology of data assimilation. The VAR data assimilation assumes that the direct and adjoint problems are constructed and solved iteratively forward in time. The structure of the adjoint problem is identical to the structure of the original problem, which considerably simplifies the numerical implementation. However, the VAR method imposes some requirements for the mathematical model (i.e. a derivation of the adjoint problem). Moreover, for an efficient numerical implementation of the VAR method, the error level of the computations must be adjusted to the parameters of the algorithm, and this complicates computations.

The QRV method allows employing sophisticated mathematical models (because it does not require derivation of an adjoint problem as in the VAR data assimilation) and hence expands the scope for applications in geodynamics (e.g. thermo-chemical convection, phase transformations in the mantle). It does not require that the desired accuracy of computations be directly related to the parameters of the numerical algorithm. However, the regularising operators usually used in the QRV method enhance the order of the system of differential equations to be solved.

The BAD method does not require any additional work (neither analytical nor computational). The major difference between the BAD method and two other methods

Table 19.2 *Comparison of data assimilation methods (after Ismail-Zadeh et al., 2007)*

	QRV method	VAR method	BAD method
Method	Solving the regularised backward heat problem with respect to parameter β	Iterative sequential solving of the direct and adjoint heat problems	Solving of heat advection equation backward in time
Solution's stability	Stable for parameter β to numerical errors[1] and conditionally stable for parameter β to arbitrarily assigned initial conditions (numerically)	Conditionally stable to numerical errors depending on the number of iterations (theoretically[2]) and unstable to arbitrarily assigned initial conditions (numerically[3])	Stable theoretically and numerically
Solution's convergence	Numerical solution to the regularised backward heat problem converges to the solution of the backward heat problem in the special class of admissible solutions[4]	Numerical solution converges to the exact solution in the Hilbert space[5]	Not applied
Solution's accuracy	Acceptable accuracy for both synthetic and geophysical data	High accuracy for synthetic data.	Low accuracy for both synthetic and geophysical data in conduction-dominated mantle flow
Time interval for data assimilation	Limited by the characteristic thermal diffusion time	Limited by the characteristic thermal diffusion time and the accuracy of the numerical solution	No specific time limitation; depends on mantle flow intensity
Analytical work	Choice of the regularising operator	Derivation of the adjoint problem	No additional analytical work
Algorithmic work	New solver for the regularised equation should be developed	No new solver should be developed	Solver for the advection equation is to be used

[1] Lattes and Lions, 1969; [2] Ismail-Zadeh et al., 2004a; [3] Ismail-Zadeh et al., 2006; [4] Tikhonov and Arsenin, 1977; [5] Tikhonov and Samarskii, 1990.

(VAR and QRV methods) is that the BAD method is by design expected to work (and hence can be used) *only* in advection-dominated heat flow. In the regions of high temperature/low mantle viscosity, where heat is transferred mainly by convective flow, the use of the BAD method is justified, and the results of numerical reconstructions can be satisfactory. Otherwise, in the regions of conduction-dominated heat flow (due to either high mantle viscosity or high conductivity of mantle rocks), the use of the BAD method cannot even guarantee any similarity of reconstructed structures. If mantle structures are diffused significantly, the remaining features of the structures can be only backward advected with the flow.

Data assimilation is a useful tool for improving our understanding of the thermal and dynamic evolution of the Earth's structures. The geometry of the thermal structures in the mantle changes with time due to heat advection, which deforms the structures, and heat conduction, which smooths the complex shapes of the structures. This creates difficulties in understanding the evolution of the mantle structures in the past. A quantitative assimilation of the present mantle temperature and flow into the geological past provides a tool for restoration of thermal shapes of prominent structures in the past from their diffusive shapes at present. An assimilation of geophysical, geodetic, and geological data and plate tectonic constraints allows us to reconstruct prominent features of hot upwelling or cold downwelling in the mantle.

We have presented here the VAR and QRV data assimilation methods and their realisations with the aim to restore the evolution of the thermal structures. The VAR and QRV methods have been compared to the BAD method. It is shown that the BAD method can be employed only in models of advection-dominated mantle flow, that is, in the regions where the Rayleigh number is high enough (e.g. $>10^7$), whereas the VAR and QRV methods are suitable for the use in models of conduction-dominated flow (lower Rayleigh numbers). The VAR method provides a higher accuracy of model restoration compared to the QRV method, meanwhile the latter method can be applied to assimilate both smooth and non-smooth data. Depending on a geodynamic problem one of the three methods can be employed in data-assimilation modelling.

Acknowledgements. We are very thankful to Gerald Schubert and Satoru Honda for their discussion on the methods for data assimilation and invaluable cooperation in the studies of geodynamic problems using the methods.

References

Alekseev, A. K., and Navon, I. M. (2001). The analysis of an ill-posed problem using multiscale resolution and second order adjoint techniques. *Computer Methods in Applied Mechanics and Engineering*, 190, 1937–53.

Bennett, A. F. (1992). *Inverse Methods in Physical Oceanography*. Cambridge: Cambridge University Press.

Billen, M. I. (2008). Modeling the dynamics of subducting slabs. *Annual Review of Earth and Planetary Sciences*, 36, 325–56.

Bird, P. (2003). An updated digital model of plate boundaries. *Geochemistry, Geophysics, Geosystems*, 4, 1027. https://doi.org/10.1029/2001GC000252.

Boussinesq, J. (1903). *Theorie Analytique de la Chaleur*, vol. 2. Paris: Gauthier-Villars.

Bubnov, V. A. (1976). Wave concepts in the theory of heat. *International Journal of Heat and Mass Transfer*, 19, 175–84.

Bubnov, V. A. (1981). Remarks on wave solutions of the nonlinear heat-conduction equation. *Journal of Engineering Physics and Thermophysics*, 40(5), 565–71.

Bunge, H.-P., Richards, M. A., and Baumgardner, J. R. (2002). Mantle circulation models with sequential data-assimilation: Inferring present-day mantle structure from plate motion histories. *Philosophical Transactions of the Royal Society A*, 360, 2545–67.

Bunge, H.-P., Hagelberg, C. R., and Travis, B. J. (2003). Mantle circulation models with variational data assimilation: Inferring past mantle flow and structure from plate motion histories and seismic tomography. *Geophysical Journal International*, 152, 280–301.

Busse, F. H., Christensen, U., Clever, R. et al. (1993). 3D convection at infinite Prandtl number in Cartesian geometry: A benchmark comparison. *Geophysical and Astrophysical Fluid Dynamics*, 75, 39–59.

Cattaneo, C. (1958). Sur une forme de l'equation de la chaleur elinant le paradox d'une propagation instantanee. *Comptes Rendus de l'Académie des Sciences*, 247, 431–33.

Chandrasekhar, S. (1961). *Hydrodynamic and Hydromagnetic Stability*. Oxford: Oxford University Press.

Christensen, U. R., and Yuen, D. A. (1985). Layered convection induced by phase transitions. *Journal of Geophysical Research*, 90, 10291–300.

Coltice, N., Husson, L., Faccenna, C., and Arnould, M. (2019). What drives tectonic plates? *Science Advances*, 5, eaax4295.

Conrad, C. P., and Gurnis, M. (2003). Seismic tomography, surface uplift, and the breakup of Gondwanaland: Integrating mantle convection backwards in time. *Geochemistry, Geophysics, Geosystems*, 4(3). https://doi.org/10.1029/2001GC000299.

Davaille, A., and Vatteville, J. (2005). On the transient nature of mantle plumes. *Geophysical Research Letters*, 32, L14309. https://doi.org/10.1029/2005GL023029.

Doglioni C., Ismail-Zadeh A., Panza G., and Riguzzi F. (2011). Lithosphere-asthenosphere viscosity contrast and decoupling. *Physics of the Earth and Planetary Interiors*, 189, 1–8.

Drewes, H. (2009). The Actual Plate Kinematic and Crustal Deformation Model APKIM2005 as basis for a non-rotating ITRF. In H. Drewes, ed., *Geodetic Reference Frames*, IAG Symposia series 134. Berlin: Springer, pp. 95–99.

Forte, A. M., and Mitrovica, J. X. (1997). A resonance in the Earth's obliquity and precession over the past 20 Myr driven by mantle convection. *Nature*, 390, 676–80.

Fukao, Y., Widiyantoro, S., and Obayashi, M. (2001). Stagnant slabs in the upper and lower mantle transition region. *Reviews of Geophysics*, 39, 291–323.

Furumura, T., and Kennett, B. L. N. (2005). Subduction zone guided waves and the heterogeneity structure of the subducted plate: Intensity anomalies in northern Japan. *Journal of Geophysical Research*, 110, B10302. https://doi.org/10.1029/2004JB003486.

Ghelichkhan, S., and Bunge, H.-P. (2016). The compressible adjoint equations in geodynamics: derivation and numerical assessment. *International Journal on Geomathematics*, 7, 1–30.

Ghelichkhan, S., and Bunge, H.-P. (2018). The adjoint equations for thermochemical compressible mantle convection: derivation and verification by twin experiments. *Proceedings of the Royal Society A*, 474, 20180329.

Glišović, P., and Forte, A. M. (2014). Reconstructing the Cenozoic evolution of the mantle: Implications for mantle plume dynamics under the Pacific and Indian plates. *Earth and Planetary Science Letters*, 390, 146–156.

Glišović, P., and Forte, A. M. (2016). A new back-and-forth iterative method for time-reversed convection modeling: Implications for the Cenozoic evolution of 3-D structure and dynamics of the mantle. *Journal of Geophysical Research*, 121, 4067–84.

Glišović, P., and Forte, A. M. (2017). On the deep-mantle origin of the Deccan Traps. *Science*, 355(6325), 613–16.

Glišović, P., and Forte, A M. (2019). Two deep-mantle sources for Paleocene doming and volcanism in the North Atlantic. *Proceedings of the National Academy of Sciences USA*, 116(27), 13227–32.

Hall, R. (2002). Cenozoic geological and plate tectonic evolution of SE Asia and the SW Pacific: computer-based reconstructions, model and animations. *Journal of Asian Earth Sciences*, 20, 353–431.

Hansen, U., Yuen, D. A., and Kroening, S. E. (1990). Transition to hard turbulence in thermal convection at infinite Prandtl number. *Physics of Fluids*, A2(12), 2157–63.

Hier-Majumder, C. A., Belanger, E., DeRosier, S., Yuen, D. A., and Vincent, A. P. (2005). Data assimilation for plume models. *Nonlinear Processes in Geophysics*, 12, 257–67.

Honda, S., Yuen, D. A., Balachandar, S., and Reuteler D. (1993). Three-dimensional instabilities of mantle convection with multiple phase transitions. *Science*, 259, 1308–11.

Horbach, A., Bunge, H.-P., and Oeser, J. (2014). The adjoint method in geodynamics: derivation from a general operator formulation and application to the initial condition problem in a high resolution mantle circulation model. *International Journal on Geomathematics*, 5, 163–94.

Howard, L. N. (1966). Convection at high Rayleigh number. In H. Goertler, and P. Sorger, eds., *Applied Mechanics, Proc. of the 11th Intl Congress of Applied Mechanics, Munich, Germany 1964*. New York: Springer-Verlag, pp. 1109–15.

Ismail-Zadeh, A., and Tackley, P. (2010). *Computational Methods for Geodynamics*. Cambridge: Cambridge University Press.

Ismail-Zadeh, A. T., Talbot, C. J., and Volozh, Y. A. (2001a). Dynamic restoration of profiles across diapiric salt structures: Numerical approach and its applications. *Tectonophysics*, 337, 21–36.

Ismail-Zadeh, A. T., Korotkii, A. I., Naimark, B. M., and Tsepelev, I. A. (2001b). Numerical modelling of three-dimensional viscous flow with gravitational and thermal effects. *Computational Mathematics and Mathematical Physics*, 41 (9), 1331–45.

Ismail-Zadeh, A. T., Korotkii, A. I., and Tsepelev, I. A. (2003a). Numerical approach to solving problems of slow viscous flow backwards in time. In K. J. Bathe, ed., *Computational Fluid and Solid Mechanics*. Amsterdam: Elsevier Science, pp. 938–41.

Ismail-Zadeh, A. T., Korotkii, A. I., Naimark, B. M., and Tsepelev, I. A. (2003b). Three-dimensional numerical simulation of the inverse problem of thermal convection. *Computational Mathematics and Mathematical Physics*, 43(4), 587–99.

Ismail-Zadeh, A., Schubert, G., Tsepelev, I., and Korotkii, A. (2004a). Inverse problem of thermal convection: Numerical approach and application to mantle plume restoration. *Physics of the Earth and Planetary Interiors*, 145, 99–114.

Ismail-Zadeh, A. T., Tsepelev, I. A., Talbot, C. J., and Korotkii, A. I. (2004b). Three-dimensional forward and backward modelling of diapirism: Numerical approach and its applicability to the evolution of salt structures in the Pricaspian basin. *Tectonophysics*, 387, 81–103.

Ismail-Zadeh, A., Mueller, B., and Schubert, G. (2005). Three-dimensional modeling of present-day tectonic stress beneath the earthquake-prone southeastern Carpathians based on integrated analysis of seismic, heat flow, and gravity observations. *Physics of the Earth and Planetary Interiors*, 149, 81–98.

Ismail-Zadeh, A., Schubert, G., Tsepelev, I., and Korotkii, A. (2006). Three-dimensional forward and backward numerical modeling of mantle plume evolution: Effects of thermal diffusion. *Journal of Geophysical Research*, 111, B06401. https://doi.org/10.1029/2005JB003782.

Ismail-Zadeh, A., Korotkii, A., Schubert, G., and Tsepelev, I. (2007). Quasi-reversibility method for data assimilation in models of mantle dynamics. *Geophysical Journal International*, 170, 1381–98.

Ismail-Zadeh, A., Schubert, G., Tsepelev, I., and Korotkii, A. (2008). Thermal evolution and geometry of the descending lithosphere beneath the SE-Carpathians: An insight from the past. *Earth and Planetary Science Letters*, 273, 68–79.

Ismail-Zadeh, A., Honda, S., and Tsepelev, I. (2013). Linking mantle upwelling with the lithosphere descent and the Japan Sea evolution: a hypothesis. *Scientific Reports*, 3, 1137. https://doi.org/10.1038/srep01137.

Ismail-Zadeh, A., Korotkii, A., and Tsepelev, I. (2016). *Data-Driven Numerical Modeling in Geodynamics: Methods and Applications*. Heidelberg: Springer.

Jolivet, L., Tamaki, K., and Fournier, M. (1994). Japan Sea, opening history and mechanism: A synthesis. *Journal of Geophysical Research*, 99, 22232–59.

Kalnay, E. (2003). *Atmospheric Modeling, Data Assimilation and Predictability*. Cambridge: Cambridge University Press.

Karato, S. (2010). Rheology of the Earth's mantle: A historical review. *Gondwana Research*, 18, 17–45.

Kaus, B. J. P., and Podladchikov, Y. Y. (2001). Forward and reverse modeling of the three-dimensional viscous Rayleigh–Taylor instability. *Geophysical Research Letters*, 28, 1095–8.

Kirsch, A. (1996). *An Introduction to the Mathematical Theory of Inverse Problems*, New York: Springer-Verlag.

Lattes, R., and Lions, J. L. (1969). *The Method of Quasi-Reversibility: Applications to Partial Differential Equations*. New York: Elsevier.

Li, D., Gurnis, M., and Stadler, G. (2017). Towards adjoint-based inversion of time-dependent mantle convection with nonlinear viscosity. *Geophysical Journal International*, 209(1), 86–105.

Liu, D. C., and Nocedal, J. (1989). On the limited memory BFGS method for large scale optimization. *Mathematical Programming*, 45, 503–28.

Liu, L., and Gurnis M. (2008). Simultaneous inversion of mantle properties and initial conditions using an adjoint of mantle convection. *Journal of Geophysical Research*, 113, B08405. https://doi.org/10.1029/2008JB005594.

Liu, L., Spasojevic, S., and Gurnis, M. (2008). Reconstructing Farallon plate subduction beneath North America back to the Late Cretaceous. *Science*, 322, 934–38.

Liu, L., Gurnis, M., Seton, M. et al. (2010). The role of oceanic plateau subduction in the Laramide orogeny. *Nature Geoscience*, 3, 353–7.

Liu, M., Yuen, D. A., Zhao, W., and Honda, S. (1991). Development of diapiric structures in the upper mantle due to phase transitions. *Science*, 252, 1836–9.

Malevsky, A. V., and Yuen, D. A. (1993). Plume structures in the hard-turbulent regime of three-dimensional infinite Prandtl number convection. *Geophysical Research Letters*, 20, 383–6.

Maruyama, S., Isozaki, Y., Kimura, G., and Terabayashi, M. (1997). Paleogeographic maps of the Japanese Islands: Plate tectonic synthesis from 750 Ma to the present. *The Island Arc*, 6, 121–42.

Massimi, P., Quarteroni, A., Saleri, F., and Scrofani, G. (2007). Modeling of salt tectonics. *Computational Methods in Applied Mathematics*, 197, 281–93.

McLaughlin, D. (2002). An integrated approach to hydrologic data assimilation: Interpolation, smoothing, and forecasting. *Advances in Water Resources*, 25, 1275–86.

Moore, W. B., Schubert, G., and Tackley, P. (1998). Three-dimensional simulations of plume–lithosphere interaction at the Hawaiian Swell. *Science*, 279, 1008–11.

Morse, P. M., and Feshbach, H. (1953). *Methods of Theoretical Physics*. New York: McGraw-Hill.

Moucha, R., and Forte, A. M. (2011). Changes in African topography driven by mantle convection. *Nature Geoscience*, 4, 707–12.

Northrup, C. J., Royden, L. H., and Burchfiel, B. C. (1995). Motion of the Pacific plate relative to Eurasia and its potential relation to Cenozoic extension along the eastern margin of Eurasia. *Geology*, 23, 719–22.

Obayashi, M., Sugioka, H., Yoshimitsu, J., and Fukao, Y. (2006). High temperature anomalies oceanward of subducting slabs at the 410-km discontinuity. *Earth and Planetary Science Letters*, 243, 149–58.

Obayashi, M., Yoshimitsu, J., and Fukao, Y. (2009). Tearing of stagnant slab. *Science*, 324, 1173–5.

Olson, P., and Singer, H. (1985). Creeping plumes. *Journal of Fluid Mechanics*, 158, 511–31.

Peng, D., and Liu, L. (2022). Quantifying slab sinking rates using global geodynamic models with data-assimilation, *Earth-Science Reviews*, 230, 104039.

Ratnaswamy, V., Stadler, G., and Gurnis, M. (2015). Adjoint-based estimation of plate coupling in a non-linear mantle flow model: Theory and examples. *Geophysical Journal International*, 202, 768–86.

Ribe, N. M., and Christensen, U. (1994). Three-dimensional modeling of plume–lithosphere interaction. *Journal of Geophysical Research*, 99, 669–82.

Rowley, D. B. (2008). Extrapolating oceanic age distributions: Lessons from the Pacific region. *Journal of Geology*, 116, 587–98.

Samarskii, A. A., and Vabishchevich, P. N. (2007). *Numerical Methods for Solving Inverse Problems of Mathematical Physics*. Berlin: De Gruyter.

Samarskii, A. A., Vabishchevich, P. N., and Vasiliev, V. I. (1997). Iterative solution of a retrospective inverse problem of heat conduction. *Mathematical Modeling*, 9, 119–27.

Schubert, G., Turcotte, D. L., and Olson, P. (2001). *Mantle Convection in the Earth and Planets*. Cambridge: Cambridge University Press.

Schuh-Senlis, M., Thieulot, C., Cupillard, P., and Caumon, G. (2020). Towards the application of Stokes flow equations to structural restoration simulations. *Solid Earth*, 11, 1909–30.

Seno, T., and Maruyama, S. (1984). Paleogeographic reconstruction and origin of the Philippine Sea. *Tectonophysics*, 102, 53–84.

Shephard, G., Müller, R., Liu, L. et al. (2010). Miocene drainage reversal of the Amazon River driven by plate–mantle interaction. *Nature Geoscience*, 3, 870–75.

Spasojevic, S., Liu, L., and Gurnis, M. (2009). Adjoint models of mantle convection with seismic, plate motion, and stratigraphic constraints: North America since the Late Cretaceous. *Geochemistry, Geophysics, Geosystems*, 10, Q05W02. https://doi.org/10.1029/2008GC002345.

Steinberger, B., and O'Connell, R.J. (1997). Changes of the Earth's rotation axis owing to advection of mantle density heterogeneities. *Nature*, 387, 169–73.

Steinberger, B., and O'Connell, R. J. (1998). Advection of plumes in mantle flow: implications for hotspot motion, mantle viscosity and plume distribution. *Geophysical Journal International*, 132, 412–34.

Tikhonov, A. N. (1963). Solution of incorrectly formulated problems and the regularization method. *Soviet Mathematics Doklady*, 4, 1035–8.

Tikhonov, A. N., and Arsenin, V. Y. (1977). *Solution of Ill-Posed Problems*. New York: Halsted Press.

Tikhonov, A. N., and Samarskii, A. A. (1990). *Equations of Mathematical Physics*. New York: Dover Publications.

Trompert, R. A., and Hansen, U. (1998). On the Rayleigh number dependence of convection with a strongly temperature-dependent viscosity. *Physics of Fluids*, 10, 351–60.

Tsepelev, I. A. (2011). Iterative algorithm for solving the retrospective problem of thermal convection in a viscous fluid. *Fluid Dynamics*, 46, 835–42.

Turcotte, D. L., and Schubert, G. (2002). *Geodynamics*, 2nd ed. Cambridge: Cambridge University Press.

Vasiliev, F. P. (2002). *Methody optimizatsii*. Moscow: Factorial Press.

Vernotte, P. (1958). Les paradoxes de la theorie continue de l'equation de la chaleur. *Comptes Rendus de l'Académie des Sciences*, 246, 3154–5.

Wang, K., Hyndman, R. D., and Yamano, M. (1995). Thermal regime of the Southwest Japan subduction zone: Effects of age history of the subducting plate. *Tectonophysics*, 248, 53–69.

Worthen, J., Stadler, G., Petra, N., Gurnis, M., and Ghattas, O. (2014). Towards an adjoint-based inversion for rheological parameters in nonlinear viscous mantle flow. *Physics of the Earth and Planetary Interiors*, 234, 23–34.

Yamano, M., Kinoshita, M., Goto, S., and Matsubayashi, O. (2003). Extremely high heat flow anomaly in the middle part of the Nankai Trough. *Physics and Chemistry of the Earth*, 28, 487–97.

Yamazaki, T., Takahashi, M., Iryu, Y. et al. (2010). Philippine Sea Plate motion since the Eocene estimated from paleomagnetism of seafloor drill cores and gravity cores. *Earth Planets Space*, 62, 495–502.

Yu, N., Imatani, S., and Inoue, T. (2004). Characteristics of temperature field due to pulsed heat input calculated by non-Fourier heat conduction hypothesis. *JSME International Journal Series A*, 47(4), 574–80.

Zhong, S. (2005). Dynamics of thermal plumes in three-dimensional isoviscous thermal convection. *Geophysical Journal International*, 162, 289–300.

Zou, X., Navon, I. M., Berger, M. et al. (1993). Numerical experience with limited-memory quasi-Newton and truncated Newton methods. *SIAM Journal of Optimization*, 3(3), 582–608.

20

Geodynamic Data Assimilation: Techniques and Observables to Construct and Constrain Time-Dependent Earth Models

Hans-Peter Bunge, Andre Horbach,
Lorenzo Colli, Siavash Ghelichkhan,
Berta Vilacís, and Jorge N. Hayek

Abstract: Variational data assimilation through the adjoint method is a powerful emerging technique in geodynamics. It allows one to retrodict past states of the Earth's mantle as optimal flow histories relative to the current state, so that poorly known mantle flow parameters such as rheology and composition can be tested explicitly against observations gleaned from the geologic record. By yielding testable time dependent Earth models, the technique links observations from seismology, geology, mineral physics, and paleomagnetism in a dynamically consistent way, greatly enhancing our understanding of the solid Earth system. It motivates three research fronts. The first is computational, because the iterative nature of the technique combined with the need of Earth models for high spatial and temporal resolution classifies the task as a grand challenge problem at the level of exa-scale computing. The second is seismological, because the seismic mantle state estimate provides key input information for retrodictions, but entails substantial uncertainties. This calls for efforts to construct 3D reference and collaborative seismic models, and to account for seismic data uncertainties. The third is geological, because retrodictions necessarily use simplified Earth models and noisy input data. Synthetic tests show that retrodictions always reduce the final state misfit, regardless of model and data error. So the quality of any retrodiction must be assessed by geological constraints on past mantle flow. Horizontal surface velocities are an input rather than an output of the retrodiction problem; but viable retrodiction tests can be linked to estimates of vertical lithosphere motion induced by mantle convective stresses.

20.1 Introduction

Mantle convection is a key element of the Earth system. The relentless deformation inside Earth's mantle from slow, viscous creep has a far greater impact on our planet than might be immediately evident. Continuously reshaping Earth's surface, mantle convection provides the driving forces necessary to support large-scale horizontal motion in the form of plate tectonics and the associated earthquake and mountain building activity.

Mantle convection models have reached an impressive level of sophistication (e.g. Zhong et al., 2015). Their need for high numerical resolution has led to the development of codes based on state-of-the-art numerical techniques suitable for massively parallel architectures (e.g. Burstedde et al., 2013; Heister et al., 2017; Kronbichler et al., 2012; Bauer et al., 2019). However, many model features, such as complex rheologies or thermochemical flow properties, involve ad hoc parameterisations and long-range extrapolations. This calls for growing capabilities to test mantle convection models against observables. The long time scales of mantle flow, on the order of millions of years (Myr), rule out predictions of future system states. But tests of mantle convection models, to resolve uncertainties in model parameters and the assumptions they are based upon, are available by constructing time trajectories of past mantle states, obtained through so-called retrodictions, and comparing them with constraints gleaned from the geologic record.

To this end, our understanding of how to retrodict past mantle states has progressed significantly. Early backward advection schemes (Bunge and Richards, 1992; Steinberger and O'Connell, 1997), where one integrates model heterogeneity back in time by reversing the time step of the energy equation and ignoring thermal diffusion, have given way to a formal inverse problem based on an adjoint approach, with so-called adjoint equations providing sensitivity information in the geodynamic model relative to earlier system states. Adjoint equations have been derived for incompressible (Bunge et al., 2003; Ismail-Zadeh et al., 2004; Horbach et al., 2014), compressible (Ghelichkhan and Bunge, 2016), and thermochemical (Ghelichkhan and Bunge, 2018) mantle flow. There are also reports on savings in computational cost of the adjoint method by optimising the step sizes (Price and Davies, 2018), on using a hybrid forward-adjoint scheme (Zhou and Liu, 2017), on simultaneous recoveries of initial temperature condition and rheology (Li et al., 2017), and on multiphysics adjoint modelling (Reuber and Simons, 2020).

Mantle convection is a chaotic process. This seemingly rules out any construction of robust flow time trajectories (Bello et al., 2014); but the chaotic nature of mantle convection is mitigated if one assimilates the horizontal surface velocity field (Colli et al., 2015). Knowledge of the latter is therefore essential to assure convergence of the inverse problem (Vynnytska and Bunge, 2014). This makes horizontal surface motions in the form of plate motion histories (e.g. Müller et al., 2016) the input of retrodictions rather than their output, implying that viable tests of mantle flow retrodictions should be linked to inferences of vertical lithosphere motion induced by mantle convective stresses. Simply put: it is not possible to construct self-consistent geodynamic models of plate tectonics that are testable against the geologic record, because the horizontal velocity field of past plate motions is an input to the inverse problem of mantle flow retrodictions.

To this end, geodynamicists have long known that convective stresses deflect Earth's surface away from its isostatically compensated state (Pekeris, 1935). Termed *dynamic topography* by Hager et al. (1985), the deflections have received renewed attention (Braun, 2010), for instance in passive margin environments (Bunge and Glasmacher, 2018), where the proximity to a base-level allows one to gauge topographic changes better than at other places. There has been much improvement in the `amount and quality of dynamic topography inferences in recent years. Information on the present-day scale and amplitude of topography in convective support comes from oceanic residual depth surveys (Hoggard et al., 2017). Additional geologic indicators constrain the temporal evolution of dynamic topography. They include studies of river profiles (e.g. Roberts and White, 2010), sediment compaction (Japsen, 2018) and provenance (e.g. Meinhold, 2010, Şengör, 2001), landform analysis (Guillocheau et al., 2018), thermochronological data (e.g. Flowers et al., 2008; Reiners and Brandon, 2006), quantifications of sediment budgets at the scale of continental margins (Guillocheau et al., 2012; Said, Moder, Clark, and Abdelmalak, 2015; Said, Moder, Clark, and Ghorbal, 2015), paleobiological and paleoenvironmental data (Fernandes and Roberts, 2020), or sequence stratigraphy (Czarnota et al., 2013; Hartley et al., 2011). Inferences can also be drawn from geological hiatus maps. The latter yield powerful constraints at interregional to continental scales on past dynamic topography (Friedrich et al., 2018; Vibe et al., 2018; Carena et al., 2019; Hayek et al., 2020,

2021). An effective review of observations of dynamic topography through space and time is given by Hoggard et al. (2021).

From this it is clear that techniques and observations are available to construct and constrain past mantle flow. However, the most severe limitation for geodynamic retrodictions arguably comes from uncertainties in the assumed modelling parameters for Earth's mantle. The rheology is not well known. Information comes from geodynamic studies of glacial isostatic adjustment (e.g. Mitrovica, 1996) and the geoid (e.g. Richards and Hager, 1984). A robust conclusion from this work is that the upper part of the mantle has a lower viscosity than its lower part. The resolving power, however, is limited and the resulting inference involves a strong trade-off between the thickness of the low-viscosity upper layer and its viscosity reduction (e.g. Paulson and Richards, 2009; Schaber et al., 2009). For geoid models, the trade-off is aggravated by our limited knowledge of the loading function (i.e. the assumed mantle density heterogeneity structure). Because geoid models solve an instantaneous Stokes equation, the loading function is necessarily fixed in space and time. Mantle flow retrodictions yield time-varying loading functions (i.e. density anomalies advected by mantle flow). Comparing such time-dependent Earth models with geologic indicators of evolving dynamic topography should help to reduce the trade-off between competing mantle viscosity models.

The mantle thermochemical structure from which one retrodicts past mantle states is also not well known. In principle, one can map seismic heterogeneity to temperature and density through thermodynamically self-consistent mantle mineralogy models (e.g. Piazzoni et al., 2007; Stixrude and Lithgow-Bertelloni, 2011; Chust et al., 2017); but the approach suffers from a well-known trade-off between compositional and thermal variations (Mosca et al., 2012). Some considerations are therefore required. First, mantle density increases by nearly a factor of two (Dziewonski and Anderson, 1981) from the surface to the core–mantle boundary (CMB) due to compression induced by self-gravitation. So compressibility effects should be considered in retrodictions to account for mantle heterogeneity in a dynamically consistent way. For instance, the depthwise heterogeneity increase in the lower mantle revealed by seismic imaging (Ritsema et al., 2011; Simmons et al., 2012; French and Romanowicz, 2014) seems best explained in mantle convection models restricted to an incompressible formulation by invoking compositional variations (McNamara and Zhong, 2005; McNamara, 2019). But compressibility effects and mantle subadiabaticity (Bunge, 2005) raise the excess temperature of mantle plumes with depth (e.g. Schuberth et al., 2009). This makes it plausible to account for deep mantle heterogeneity by temperature alone (Schuberth, Bunge and Ritsema, 2009; Davies et al., 2012; Schuberth et al., 2012). Second, the limited data coverage available for seismic studies makes tomographic inverse problems ill-posed. So an explicit regularisation, usually in the form of damping and/or smoothing of the seismic model, is needed. While regularisation parameters directly impact the size and amplitude of seismic anomalies, their choice is to a large extent subjective (Ritsema et al., 2011). Zaroli et al. (2013) proposed some objective rationales to constrain the regularisation parameters. But their span is still sufficiently large to permit a factor of ≈ 2 uncertainty in the RMS amplitude of seismic anomalies (Zaroli et al., 2013). Damping and filtering effects are illustrated by synthetic studies, where one constructs a tomographic image from a geodynamically plausible mantle convection input structure (e.g. Mégnin et al., 1997; Bunge and Davies, 2001; Schuberth, Bunge and Ritsema, 2009) and finds that dynamically significant features are dampened and either smeared or absent in the imaged output structure. Additional uncertainties arise from theoretical simplifications (e.g. the high-frequency approximation in raypath travel time tomography) and unmodelled effects. Tomographic techniques based on finite-frequency and full-waveform methods employ more complete physics to improve data coverage and reduce modelling errors. Seismic tomographies based on these techniques (Fichtner et al., 2009; Colli et al., 2013) should yield sharper images of seismic anomalies.

In summary, we note that mantle state estimates involve uncertainties in mantle heterogeneity structure and amplitude. State estimate uncertainties combine with model uncertainties. This necessarily restricts our ability to retrodict past mantle states. Still, current global tomographic models agree over length scales of thousands of kilometres (Becker and Boschi, 2002), making it feasible to construct mantle flow retrodictions. In the following, we present forward and adjoint mantle convection equations and report the impact of model inconsistencies and imperfect mantle state estimates on the outcome of the adjoint method. Next we turn to geologic archives that constrain long wavelength vertical motion of the lithosphere. To this end, continent-scale hiatus maps are beginning to yield powerful proxies for mantle-flow-induced dynamic topography. We then briefly illustrate the dynamic topography evolution of four recent global mantle flow retrodictions from the early Cenozoic onward, finding they capture some first-order dynamic topography changes over the last 50 Myr. This bodes well for future modelling studies. We complete the chapter with concluding remarks.

20.2 Geodynamic Data Assimilation with Adjoint Techniques

20.2.1 Forward and Adjoint Equations

Mantle convection is modelled by forward equations that embody the conservation principles for mass, momentum, and energy. To account for compressibility effects, they are solved in the truncated anelastic-liquid approximation (Jarvis and Mckenzie, 1980; Baumgardner, 1985) in a time interval $I := [t_0, t_1]$ within a spherical shell V (i.e. the Earth's mantle) with boundary $\partial V = S \cup C$, where S denotes the Earth's surface and C the CMB:

$$\nabla \cdot (\rho_r v) = 0,$$

$$\nabla \cdot \left[\eta \left(\nabla v + (\nabla v)^T - \frac{2}{3} (\nabla \cdot v \mathbf{1}) \right) \right]$$
$$- \nabla P + \alpha \rho_r \mathbf{g} (T_{av} - T) = 0,$$

$$\partial_t T + \gamma T \nabla \cdot v + v \cdot \nabla T - \frac{1}{\rho_r c_v} \{ \nabla \cdot (k \nabla T) + \tau : \nabla v \}$$
$$+ H = 0,$$

where v is the velocity, ρ_r is the radial density profile, η is the viscosity, P is the pressure, α is the thermal expansivity, $\mathbf{g}$ is the gravitational acceleration, and T is the temperature, while T_{av} is its layer-averaged value, γ is the Grüneisen parameter, c_v is the specific heat capacity at constant volume, k is the thermal conductivity, τ is the deviatoric stress tensor, and H is the rate of radiogenic heat production. At the surface, mantle convection models commonly employ a free-slip (no tangential shear stress) or no-slip (fixed velocity value) boundary condition for the momentum equation. Mantle circulation models (e.g. Bunge et al., 1998) instead impose plate motion histories (e.g. Müller et al., 2016) on S through a time-dependent velocity boundary condition. The latter is a form of sequential data assimilation (Bunge et al., 2002) such that geologic information on the surface velocity history enters the flow. The C is treated as free-slip. The energy equation commonly applies Dirichlet boundary conditions (fixed temperatures) on both boundaries.

The corresponding adjoint equations for compressible mantle flow have been derived by Ghelichkhan and Bunge (2016). They use so-called adjoint variables for three fields, termed adjoint velocity ϕ, pressure λ, and temperature Ψ, in analogy to the forward velocity, pressure, and temperature fields:

$$\nabla \cdot \phi = 0,$$

$$\nabla \cdot \left[\eta \left(\nabla \phi + (\nabla \phi)^T \right) \right] + \Psi \nabla T - \rho_r \nabla \lambda - 2 \nabla \cdot \left(\frac{\Psi}{\rho_r c_v} \tau \right) = 0,$$

$$\partial_t \Psi + v \cdot \nabla \Psi - (\gamma - 1) \Psi \nabla \cdot v + \nabla \cdot \left(k \nabla \left(\frac{\Psi}{\rho_r c_v} \right) \right)$$
$$+ \alpha \rho_r \mathbf{g} \cdot \phi = \partial_{T^F} \hat{\chi} (T^F).$$

Forward and adjoint equations are similar, so that similar computational strategies and methods can be used for their solution. The main difference lies in the adjoint energy equation, where the diffusion term has an opposite sign.

This makes the equation unconditionally stable to integration back in time. The adjoint energy equation also includes a source term related to a *misfit functional* which acts as a final time condition and links to the present-day state estimate. The adjoint equations, too, require a set of boundary conditions in addition to this final time condition. A detailed explanation of the equations, variables, and boundary conditions is given by Ghelichkhan and Bunge (2016).

The *misfit functional* χ quantifies the difference between estimates of the present mantle state T^E and its model prediction T^F. From this, one computes the total (Fréchet) derivative $\mathfrak{D}_{T^I}\chi(T^F)(\Delta T^I)$ that describes the sensitivity of χ relative to changes ΔT^I in the initial condition T^I. The Fréchet derivative is obtained from the solution of the *adjoint equations*. In an iterative procedure (e.g. Bunge et al., 2003), and starting from an arbitrary *first-guess* initial condition T_0^I, one computes a *first-guess* final state T_0^F. The adjoint equations are then solved to provide the necessary information to obtain an improved initial condition, T_1^I. From this, in turn, one computes a new final state T_1^F and the procedure is repeated until a desired level of minimisation in χ is achieved. We give a visual indication (from Ghelichkhan et al., 2021) for the iterative misfit functional reduction in Fig. 20.1. Numerical values for χ as function of the iteration for a number of retrodiction models are shown in Fig. 20.2. They indicate that a significant misfit reduction can be achieved in the first few iterations.

20.2.2 Impact of Model-Inconsistencies and Data Error

Mantle flow retrodictions rely on three information sources: a geodynamic model with forward and corresponding adjoint equations, a present-day mantle state estimate, and a history of the horizontal surface velocity field. The geodynamic model expresses our knowledge of the physical laws governing the flow evolution as described by the conservation equations for mass, momentum, and energy. The present-day mantle heterogeneity state ties the general physical system described by these conservation equations to its one specific realisation in the Earth system (e.g. Carrassi and Vannitsem, 2010), while the surface velocity field counteracts the chaotic nature of mantle convection and assimilates information on the surface velocity history (Vynnytska and Bunge, 2014; Colli et al., 2015). The accuracy of mantle flow restorations depends on the error associated with these three essential information sources. The conservation equations for mantle flow are not in question. But, as mentioned before, significant uncertainty exists in the choice of key geodynamic modelling parameters, such as rheology and composition, and the mantle state estimate obtained from seismic tomography.

Synthetic tests, known as *twin experiments*, provide the means to explore the impact of model and data uncertainties on the accuracy of mantle flow retrodictions. To this end, one generates a reference mantle circulation flow trajectory via numerical modelling. This is called the *reference twin*.

The final state of the reference twin serves as a target state that drives the restoration problem. The reconstructed flow trajectory is then compared against the reference twin trajectory to assess the inversion quality. Early geodynamic twin experiments took the true final state and surface velocity history as input for the inversion, together with the exact same model parameters that were used to compute the reference trajectory. So they assumed perfect knowledge of the physical system and error-free data. For this ideal condition the twin experiments demonstrate that mantle flow can be restored over time scales comparable to a convective transit time and that successive iterations improve the restored flow trajectory (e.g. Bunge et al., 2003). But such ideal conditions constitute an unrealistic scenario. The act of using the same model to generate synthetic data and to invert them has been named an *inverse crime* (Colton and Kress, 1992), as it leads to overly optimistic results.

Colli et al. (2020) explored more realistic twin experiments. By inserting on purpose a mismatch between the geodynamic model used to generate the target final state and the model used for carrying out the inversion, they found that mismatched model parameters do not inhibit misfit reduction: the adjoint method still produces a flow history that optimally fits the target final state. But the recovered initial state can be a poor approximation of the true initial state and deteriorates with increasing iteration number. So, in the presence of model and data error, a complete reduction of the cost function may not be desirable and a limited number of adjoint iterations seems advisable when the goal is a best fit to the initial condition. When the target final state is a noisy low-pass version of the true final state, as implied by the finite resolution of seismic tomography, an appropriate misfit function choice can help to reduce the generation of artefacts in the initial state. Figure 20.3 gives numerical values for the iterative misfit functional reduction for a range of twin experiments reported by Colli et al. (2020).

20.3 Hiatus Maps as Proxies of Past Dynamic Topography

Theoretical considerations suggest to link viable tests of mantle flow retrodictions to inferences of evolving dynamic topography. Such dynamic topography histories are beginning to emerge for the continents, because the transient nature of dynamic topography leaves geologic evidence in sedimentary archives (Şengör, 2001). The approach was pioneered for regions that underwent periods of low dynamic topography, such as the *Cretaceous Interior Seaway* of North America (e.g. Mitrovica et al., 1989; Burgess et al., 1997), because low topography enables the deposition and preservation of sediments. High dynamic topography, instead, creates erosional/non-depositional environments expressed as time gaps, that is hiatuses, in

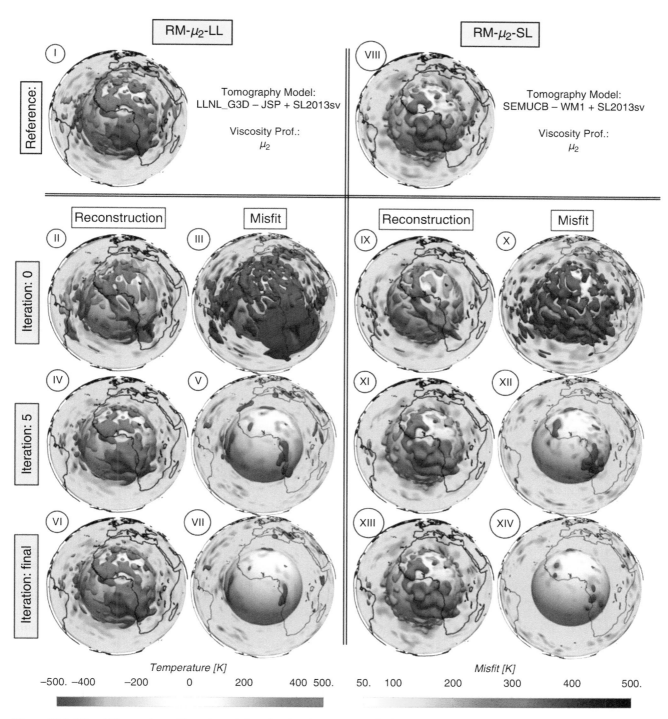

Figure 20.1 Visual illustration of iterative misfit reduction for two retrodiction models from Ghelichkhan et al. (2021). In both panels, the top row shows the reference temperature fields that represent two different assumed present-day mantle state estimates. The latter also serve as first-guess for the unknown initial condition at 50 Ma. Second, third and fourth rows in each panel show the reconstructed final temperature field for the present-day (II, IV, and VI, or IX, XI, and XIII), and the corresponding misfit (III, V, and VI, or X, XII, and XIIV) to the assumed present-day mantle state estimate, after zeroth, fifth, and final (here 13th) iteration. The error isosurface is chosen at 300 K. Taking left panel as an example: the model starts from necessarily incorrect first-guess initial condition. After first forward run (iteration zero) terminal state shows narrow bands of hot material beneath western margin of Africa extending to the southeastern parts of the continent. The pattern differs substantially from reference temperature field (top row, I), as expected, and indicated by large misfit amplitudes, because the incorrect first-guess initial condition yields an incorrect final state. Successive model updates improve the reconstructed final temperature field (IV). The general pattern of the African large low velocity province and the girdling subduction regions are restored, as confirmed by much reduced misfit isosurface (V). Misfit reductions in subsequent iterations are minor and occur on shorter wavelengths, such that misfit isosurface after 13 iterations (VII) is further reduced. Details in Ghelichkhan et al. (2021).

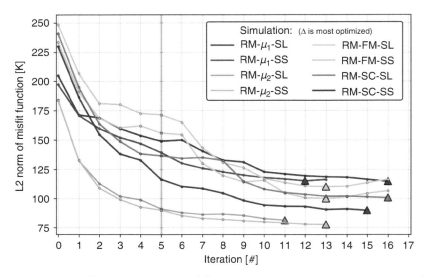

Figure 20.2 Difference between modelled and assumed present-day temperature field as function of iteration for eight retrodiction models from Ghelichkhan et al. (2021). The most minimised iteration is indicated by a colour-filled triangle. A significant reduction of $\hat{\chi}$ is achieved in the first five iterations, and $\hat{\chi}$ is reduced on average by 55% in all retrodictions. Details in Ghelichkhan et al. (2021).

the geologic record. The latter are known as nonconformities and unconformities (see Miall, 2016, for a review). To this end, Friedrich et al. (2018) and Friedrich (2019) introduced a hiatus-area mapping approach, because, at continental scales, what is normally perceived as a lack of data (material eroded or not deposited) becomes part of the dynamic topography signal. The method has been applied to map the spatiotemporal patterns of conformable and unconformable geological contacts across Europe (Vibe et al., 2018), Africa (Carena et al., 2019), and the Atlantic realm and Australia since the Upper Jurassic (Hayek et al., 2020, 2021). An important finding is that significant differences exist in the spatial extent of hiatus area across and between continents at the time scale of geologic series, that is, ten to a few tens of Myr (see definition of series as a unit of chronostratigraphy in the chronostratigraphic chart; Cohen et al., 2013; updated; Ogg et al., 2016). This is considerably smaller than the mantle transit time, which as the convective time scale is about 100–200 Myr (Iaffaldano and Bunge, 2015). It suggests vigorous upper mantle flow, as illustrated by geodynamic kernels (see Colli et al., 2016, for a review).

Figure 20.4 shows hiatus maps from Hayek et al. (2020, 2021), for North and South America, Europe, Africa, and Australia for eight geologic series from the Pleistocene to the Lower Cretaceous. The resolution of geological series is chosen, because this is the most frequently adopted temporal resolution for interregional geologic maps (Friedrich, 2019), while the choice of the Lower Cretaceous as the oldest stratigraphic unit is motivated by the time scale of the mantle transit time, which is about 100–200 Myr, as noted before. Hayek et al. (2020, 2021) give details of the data sources, method and results. Here we summarise key observations. Red/blue colours depict un/conformable (hiatus/no

hiatus) contacts, respectively, indicative of high/low topography in the preceding geological series, while blank regions reveal the absence of the considered geological series and its immediately preceding unit. Such regions may have undergone intense and/or long-lasting erosion or non-deposition, indicative of intense and/or persistent exhumation and surface uplift (Friedrich et al., 2018; Friedrich, 2019; Vibe et al., 2018; Carena et al., 2019; Hayek et al., 2020, 2021). Overall, Fig. 20.4 reveals significant differences in hiatus distribution across and between continents at the time scale of geologic series. **Base of Pleistocene**, Fig. 20.4A, shows North and South America and Africa with a mix of hiatus and no hiatus surfaces. Extensive hiatus surface exists in Australia, except the *Nullarbor Plain*, Alaska, the eastern margin of Greenland, and north-central Africa. No hiatus surface prevails in Europe and the Congo Basin. **Base of Pliocene**, Fig. 20.4B, reveals minor hiatus surface and blank regions. Hiatus is located in central Africa and near the Canaries. Blank regions occur in eastern North and South America. No hiatus signals dominate elsewhere. In **Base of Miocene**, Fig. 20.4C, hiatus surfaces dominate across the continents. Prominent examples include North and South America, parts of Europe, Australia, and Africa. North America shows conformable contacts surrounding the hiatus regions near the Yellowstone hotspot location. Alaska, Patagonia, and Central Europe also show conformable contacts. **Base of Oligocene**, Fig. 20.4D, displays blank regions in Africa. But the foremost occurrence is in South America, where it signals an almost complete absence of Oligocene and Eocene strata throughout the continent. Limited hiatus surface exists in the western part of North America and the Afar region. Europe, northernmost Africa, the Karoo Basin and much of Australia show prominent no hiatus regions.

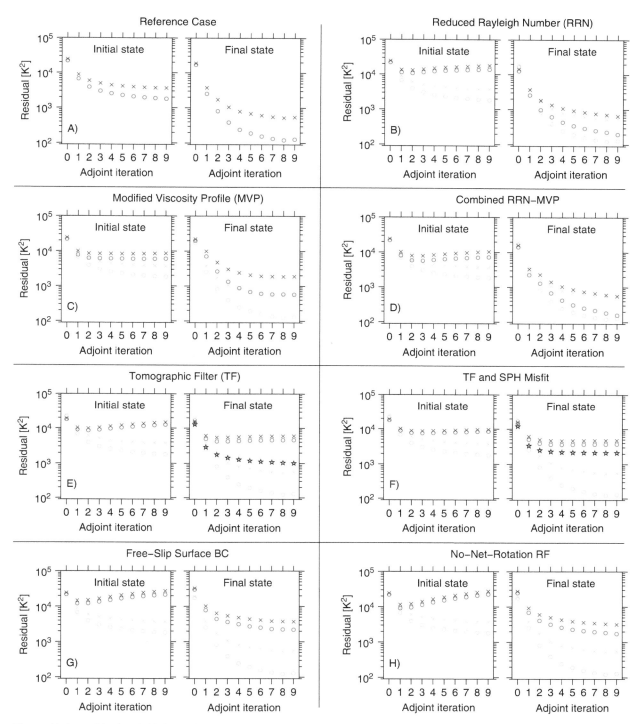

Figure 20.3 Residual at initial and final state over the course of successive adjoint iterations evaluated from twin experiments using a reduced least-squares (blue crosses) and a reduced spherical harmonic misfit (red circles); see Colli et al. (2020). (A) Control case where perfect knowledge of the reference model is assumed (inverse crime, see text). (B) Mismatched case where model viscosity is larger by one order of magnitude compared to reference twin. Residuals of the control case are plotted in light grey for comparison. (C) as (B) but for a mismatched viscosity profile. (D) The viscosity profile is mismatched and the absolute viscosity is larger. (E) Target final state is noisy and low-pass filtered. The actual misfit (i.e. difference between the modelled final state and the filtered target state) is plotted with black stars. (F) as for (E), but the spherical harmonic misfit is used to drive the adjoint inversion. (G) The history of surface velocities is assumed to be unknown (free-slip boundary condition). (H) A mismatched (no-net-rotation) reference frame is assumed for the plate motion history. Note that cases with free slip and mismatched surface velocity fail to restore the initial state despite otherwise identical parameters to reference twin, as expected, because the horizontal surface velocity history is an input rather than an output to retrodiction problem (see text). Note also that all cases minimise the final state residual, which on its own cannot be used to assess the flow restoration quality.

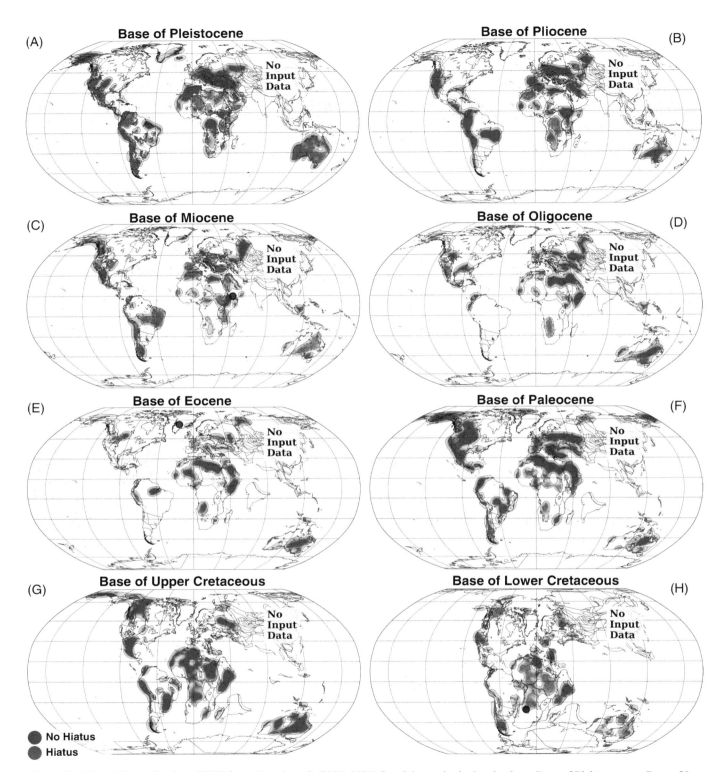

Figure 20.4 Base Hiatus Surface (BHS) from Hayek et al. (2020, 2021) for eight geological series from Base of Pleistocene to Base of Lower Cretaceous (A)–(H) reconstructed paleogeographically with a global Mesozoic-Cenozoic plate motion model (Müller et al., 2016) tied to a reference frame of global moving hotspots and a True Polar Wander (TPW) corrected paleomagnetic reconstruction (Torsvik et al., 2008), with the latter updated by Seton et al. (2012). BHS is placed into a plate tectonic configuration corresponding to the base of each geological series. Red/blue colours represent the hiatus/no hiatus surfaces. Each map serves as a proxy for paleotopography (red=high, blue=low) in the preceding series (see text). Black circles at Base of Miocene (C), Base of Eocene (E), and Base of Lower Cretaceous (H) maps correspond to the location of flood basalts associated with Afar, Iceland, and Tristan hotspots. Blank regions indicate the absence of series and its immediately preceding unit, suggesting long hiatus duration. See Hayek et al. (2020) for further information.

Base of Eocene, Fig. 20.4E, presents hiatus surface and blank regions prominently in two continents: Europe and South America except the Amazon Basin. No hiatus exists in the northernmost part of Africa, the Karoo Basin, and eastern Australia. **Base of Paleocene**, Fig. 20.4F, reveals isolated patches of hiatus surface and blank regions along the east coast of Brazil, in Southern Africa (Karoo Basin), and Australia. No hiatus regions dominate elsewhere throughout the continents. **Base of Upper Cretaceous**, Fig. 20.4G, shows hiatus and blank regions in parts of Europe, North and South America. No hiatus is prominent throughout much of Africa, Australia, and South America in the Paraná region. **Base of Lower Cretaceous**, Fig. 20.4H, shows hiatus surface and blank regions, indicative of high topography in the Upper Jurassic, in much of North and South America, Africa, and Australia. No hiatus, indicative of low Upper Jurassic topography, is prominent in northernmost Africa and Europe.

20.4 Global Mantle Flow Retrodictions

Some prominent topographic changes in the Cenozoic, Earth's current geologic era, are likely mantle flow related. An example is the termination of large-scale marine inundation in North America at the end of the Cretaceous/Paleogene (Fig. 20.4 F/E), for which negative dynamic topography induced by mantle downwellings has long been invoked (e.g. Mitrovica et al., 1989; Burgess et al., 1997), as noted before. Another example includes long wavelength tilting of Australia since the late Cretaceous (e.g. DiCaprio et al., 2009; Sandiford, 2007), which occurred when the continent approached the subduction systems in Southeast Asia on its northward passage. Figures 20.5 and 20.6 show dynamic topography evolutions for four global mantle flow retrodictions (Ghelichkhan et al., 2021). The models, described in detail by Ghelichkhan et al. (2021), employ high numerical resolution and discretise Earth's mantle with ≈670 million finite element nodes. This corresponds to a grid point resolution of 11 km radially, and 14 km tangentially at the surface, decreasing to half that value at the CMB, and allows one to resolve global mantle flow at earth-like convective vigour. The models apply two mantle state estimates, derived for the lower mantle from the whole-mantle tomographic studies by Simmons et al. (2015) and French and Romanowicz (2014), and in the upper mantle from the shear wave speed study of Schaeffer and Lebedev (2013). The two state estimates are combined with two mantle viscosity profiles, μ_1 and μ_2, involving a deep mantle viscosity of $\approx 2 \times 10^{22}$ Pa.s and 10^{23} Pa.s, respectively. This yields the four retrodiction models (RM-μ_1-SL, RM-μ_1-SS, RM-μ_2-SL, and RM-μ_2-SS).

Figure 20.5 shows the models with a view centred on North America. All four models produce a dynamic topography low over the interior of the continent for the earliest retrodicted time – that is, the early Eocene – regardless of the tomographic input structure. This is expected, because a fast seismic velocity structure is robustly imaged in both state estimates of the lower mantle beneath the continent. The low has largely disappeared by mid-Eocene, which is somewhat later than its late-Cretaceous-Paleocene demise inferred by paleoshorelines (Smith et al., 1994). There are additional variations in amplitude, location, and uplift rates of dynamic topography between the models, depending on the chosen state estimate and viscosity profile, with the stiffer lower mantle viscosity profile yielding smaller amplitudes and slower uplift rates, as expected. For instance, at 40 Ma, the North American interior and Central America uplift rates exceed 200 m/Myr in RM-μ_1-SL, while RM-μ_2-SL has an early Cenozoic uplift rate of about half that value. North America thus illustrates the sensitivity of retrodictions to the assumed viscosity profile. There is also early Cenozoic positive dynamic topography along the eastern margin of North America that gives way to subsidence, in agreement with reports (Spasojevic et al., 2008). Importantly, none of the retrodictions yields an Oligo-Miocene dynamic topography growth for the western United States. The latter is indicated by hiatus surfaces (Fig. 20.4C). In other words, the current topographic highstand of the western United States is absent from these models. This is likely owing to the fact that the Yellowstone plume has been imaged only recently in greater detail by seismic models (Nelson and Grand, 2018) and suggests the use of an updated mantle state estimate in future studies.

Figure 20.6 shows the retrodicted dynamic topography over Australia, the western Pacific, and Antarctica for the same four models. We again find similar dynamic topography evolutions for the models, modulated in rate and amplitude by the assumed viscosity profiles, as expected and seen before for North America. By the early to mid-Cenozoic, between 50 and 40 Ma, all four retrodictions develop a regional Cenozoic uplift signal for eastern and southern Australia, in agreement with reports on the spatial and temporal patterns of Australian dynamic topography (Czarnota et al., 2013) and modelling studies (Stotz et al., 2021). The late Neogene, at 10 Ma, sees subsidence in the northernmost part of the continent, in a trend that is consistent with long wavelength northward tilting of the continent, as noted before. Importantly, there is a noticeable difference in the dynamic topography evolution inferred from the two tomographic state estimates. Models RM-μ_1-SS and RM-μ_2-SS (Fig. 20.6 II,IV), combining Schaeffer and Lebedev (2013) and French and Romanowicz (2014) with the μ_1 and μ_2 viscosity profiles, yield a broad dynamic topography low in the early Cenozoic over much of Southeast Asia that gives way to regional uplift. Models RM-μ_1-SL and RM-μ_2-SL (Fig. 20.6 I,III), instead yield a broadly opposite trend. Starting from high dynamic topography in the early Cenozoic over much of Southeast Asia,

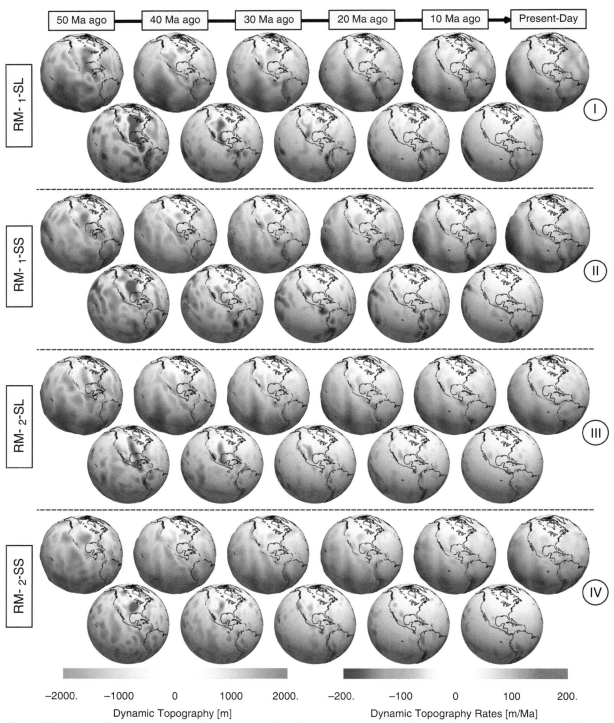

Figure 20.5 Dynamic topography evolution (top row) and its rates (bottom row) over North America computed from four retrodiction models (see text). Paleo coastlines drawn by back-rotation of present-day coastlines using Young et al. (2019). Rates are calculated over 10 Myr time intervals. In all four models, negative dynamic topography in the interior of North America in the early Cenozoic (at 50 Ma) gives way to uplift, such that by about 30 Ma much of the continent-wide negative dynamic topography has disappeared. Models with the stiffer lower mantle viscosity profile μ_2 (RM-μ_2-SL and RM-μ_2-SS) show smaller amplitude of dynamic topography and rates of change, as expected. For example, while RM-μ_1-SL show uplift rates exceeding 200 m/Myr in the North American interior and Central America, RM-μ_2-SL shows an early Cenozoic uplift rate of about half that value. RM-μ_1-SL and RM-μ_2-SL show an early Cenozoic positive dynamic topography along the eastern margin giving way to subsidence at 35 Ma. See text and Ghelichkhan et al. (2021).

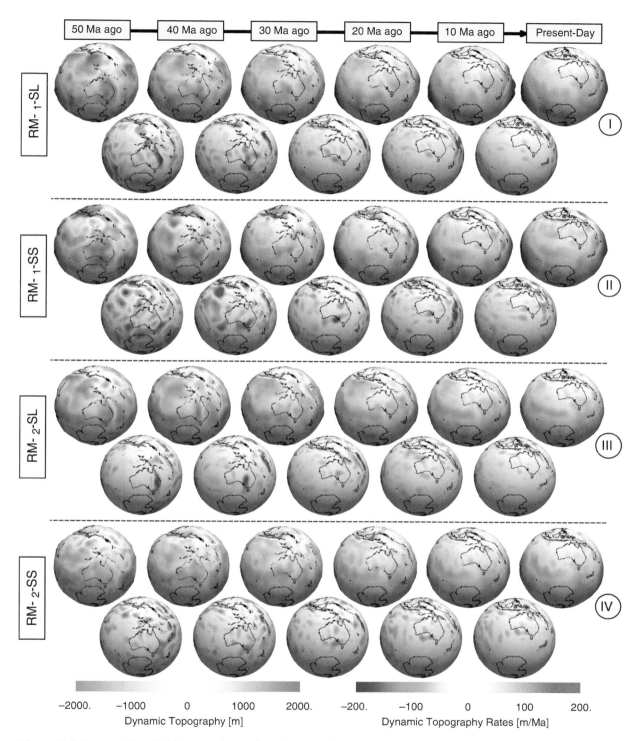

Figure 20.6 Same as Fig. 20.5 but over Australia and Antarctica. Dynamic topography is initially low over much of the Australian continent. Between 50 and 40 Ma, eastern Australia uplifts, while at 30 Ma, uplift moves to the south and southeast of Australia. At 10 Ma, the northernmost part of the continent subsides, in a trend that is more marked in models RM-μ_1-SL and RM-μ_2-SL. While RM-μ_1-SS and RM-μ_2-SS yield a broad dynamic topography low in the early Cenozoic over much of Southeast Asia that gives way to regional uplift, RM-μ_1-SL and RM-μ_2-SL instead yield a broadly opposite trend. See text and Ghelichkhan et al. (2021).

they give way to regional subsidence over the course of the Cenozoic, consistent with reports on a Late Cretaceous-Eocene hiatus in Sundaland (Zahirovic et al., 2016). The Australasian region thus illustrates the sensitivity of retrodictions to the assumed state estimate.

20.5 Concluding Remarks

Mantle flow retrodictions are a powerful tool. They allow one to identify geodynamically relevant observables from the geologic record, such as hiatus area, that are sensitive to mantle convection. These underutilised observations in turn allow one to better constrain key parameters of dynamic Earth models, such as rheology.

The construction of mantle flow trajectories is enabled by growing computational capabilities. But the iterative nature of the adjoint technique poses a large computational cost. One iteration in the high-resolution mantle flow retrodictions of Colli et al. (2018) and Ghelichkhan et al. (2021), requires $\approx$ 40,000 CPUh, translating to $\approx$ 400,000 CPUh for 10 adjoint-forward iterations needed to reduce the misfit function. This limits current abilities to perform broad parameter surveys or to extend the flow trajectories beyond the Cenozoic. It motivates the development of geodynamic codes suitable for peta- and emerging exa-scale supercomputers (Burstedde et al., 2013; Bauer et al., 2019).

Seismic mantle state estimates provide key input information for mantle flow retrodictions. But they are subject to substantial uncertainties. While significant advances are underway in our ability to illuminate the mantle, they should be augmented by techniques to account for seismic data uncertainty (Zaroli, 2016; Freissler et al., 2020) and by efforts to construct 3D reference and collaborative seismic models (Fichtner et al., 2018; Moulik et al., 2021). This will strengthen the link between seismic and geodynamic Earth models.

Plate motion histories (e.g. Müller et al., 2016) provide the other indispensable information for retrodictions. They are an input rather than an output of mantle flow trajectories. So one should link viable tests of mantle flow retrodictions to other datasets, such as inferences of vertical lithosphere motion induced by mantle convective stresses.

Studies on the impact of model and data error on mantle flow retrodictions (Colli et al., 2020) are insightful, because, in solving retrodiction problems for the real Earth, we are bound to use simplified Earth models and noisy input data. The studies demonstrate that retrodictions always achieve a final state misfit reduction. So a reduction of the latter on its own is insufficient to judge the quality of the reconstructed flow history. The quality of any retrodiction must be assessed by geological constraints. Collaborations across the Earth Sciences will advance our understanding of the surface expressions of paleo mantle flow so that dynamic Earth models can be improved.

References

Bauer, S., Huber, M., Ghelichkhan, S. et al. (2019). Large-scale simulation of mantle convection based on a new matrix-free approach. *Journal of Computational Science*, 31, 60–76.

Baumgardner, J. R. (1985). Three-dimensional treatment of convective flow in the Earth's mantle. *Journal of Statistical Physics*, 39(5/6).

Becker, T. W., and Boschi, L. (2002). A comparison of tomographic and geodynamic mantle models. *Geochemistry, Geophysics, Geosystems*, 3(1).

Bello, L., Coltice, N., Rolf, T., and Tackley, P. J. (2014). On the predictability limit of convection models of the Earth's mantle. *Geochemistry, Geophysics, Geosystems*, 15, 2319–28.

Braun, J. (2010). The many surface expressions of mantle dynamics. *Nature Geoscience*, 3(12), 825–33.

Bunge, H.-P. (2005). Low plume excess temperature and high core heat flux inferred from non-adiabatic geotherms in internally heated mantle circulation models. *Physics of the Earth and Planetary Interiors*, 153(1–3), 3–10.

Bunge, H.-P., and Davies, J. H. (2001). Tomographic images of a mantle circulation model. *Geophysical Research Letters*, 28(1), 77–80.

Bunge, H.-P., and Glasmacher, U. (2018). Models and observations of vertical motion (MoveOn) associated with rifting to passive margins: Preface. *Gondwana Research*, 53, 1–8.

Bunge, H. P., Hagelberg, C. R., and Travis, B. J. (2003). Mantle circulation models with variational data assimilation: Inferring past mantle flow and structure from plate motion histories and seismic tomography. *Geophysical Journal International*, 152(2), 280–301.

Bunge, H.-P., and Richards, M. A. (1992). The backward-problem of plate tectonics and mantle convection (abstract). *Eos, Transactions, American Geophysical Union*, 73(14), 281.

Bunge, H.-P., Richards, M. A., and Baumgardner, J. R. (2002). Mantle-circulation models with sequential data assimilation: Inferring present-day mantle structure from plate-motion histories. *Philosophical Transactions. Series A, Mathematical, Physical, and Engineering Sciences*, 360 (1800), 2545–67.

Bunge, H.-P., Richards, M. A., Lithgow-Bertelloni, C. et al. (1998). Time scales and heterogeneous structure in geodynamic Earth models. *Science*, 280(5360), 91–5.

Burgess, P. M., Gurnis, M., and Moresi, L. (1997). Formation of sequences in the cratonic interior of North America by interaction between mantle, eustatic, and stratigraphic processes. *Geological Society of America Bulletin*, 109(12), 1515–35.

Burstedde, C., Stadler, G., Alisic, L. et al. (2013). Large-scale adaptive mantle convection simulation. *Geophysical Journal International*, 192(3), 889–906.

Carena, S., Bunge, H.-P., and Friedrich, A. M. (2019). Analysis of geological hiatus surfaces across Africa in the Cenozoic and implications for the timescales of convectively-maintained topography. *Canadian Journal of Earth Sciences*, 56(12), 1333–46.

Carrassi, A., and Vannitsem, S. (2010). Accounting for model error in variational data assimilation: A deterministic formulation. *Monthly Weather Review*, 138(9), 3369–86.

Chust, T. C., Steinle-Neumann, G., Dolejš, D., Schuberth, B. S. A., and Bunge, H. P. (2017). MMA-EoS: A computational framework for mineralogical thermodynamics. *Journal of Geophysical Research: Solid Earth*, 122(12), 9881–920.

Cohen, K. M., Finney, S., Gibbard, P. L., and Fan, J.-X. (2013). The ICS International Chronostratigraphic Chart. *Episodes*, 36(3), 199–204.

Colli, L., Bunge, H.-P., and Oeser, J. (2020). Impact of model inconsistencies on reconstructions of past mantle flow obtained using the adjoint method. *Geophysical Journal International*, 221(1), 617–39.

Colli, L., Bunge, H.-P., and Schuberth, B. S. A. (2015). On retrodictions of global mantle flow with assimilated surface velocities. *Geophysical Research Letters*, 42(20), 8341–8.

Colli, L., Fichtner, A., and Bunge, H.-P. (2013). Full waveform tomography of the upper mantle in the South Atlantic region: Imaging a westward fluxing shallow asthenosphere? *Tectonophysics*, 604, 26–40.

Colli, L., Ghelichkhan, S., and Bunge, H.-P. (2016). On the ratio of dynamic topography and gravity anomalies in a dynamic Earth. *Geophysical Research Letters*, 43(6), 2510–16.

Colli, L., Ghelichkhan, S., Bunge, H.-P., and Oeser, J. (2018). Retrodictions of Mid Paleogene mantle flow and dynamic topography in the Atlantic region from compressible high resolution adjoint mantle convection models: Sensitivity to deep mantle viscosity and tomographic input model. *Gondwana Research*, 53, 252–72.

Colton, D., and Kress, R. (1992). *Inverse Acoustic and Electromagnetic Scattering Theory*. Berlin: Springer Verlag.

Czarnota, K., Hoggard, M., White, N., and Winterbourne, J. (2013). Spatial and temporal patterns of Cenozoic dynamic topography around Australia. *Geochemistry, Geophysics, Geosystems*, 14(3), 634–58.

Davies, D. R., Goes, S., Davies, J. H. et al. (2012). Reconciling dynamic and seismic models of Earth's lower mantle: The dominant role of thermal heterogeneity. *Earth and Planetary Science Letters*, 353–4(0), 253–69.

DiCaprio, L., Gurnis, M., and Müller, R. D. (2009). Long-wavelength tilting of the Australian continent since the Late Cretaceous. *Earth and Planetary Science Letters*, 278(3–4), 175–85.

Dziewonski, A. M., and Anderson, D. L. (1981). Preliminary reference Earth model. *Physics of the Earth and Planetary Interiors*, 25(4), 297–356.

Fernandes, V. M., and Roberts, G. G. (2020). Cretaceous to recent net continental uplift from paleobiological data: Insights into sub-plate support. *GSA Bulletin*, 133(5–6), 1217–36.

Fichtner, A., Kennett, B. L. N., Igel, H., and Bunge, H.-P. (2009). Full seismic wave-form tomography for upper-mantle structure in the Australasian region using adjoint methods. *Geophysical Journal International*, 179(3), 1703–25.

Fichtner, A., van Herwaarden, D.-P., Afanasiev, M. et al. (2018). The collaborative seismic Earth model: Generation 1. *Geophysical Research Letters*, 45(9), 4007–16.

Flowers, R., Wernicke, B., and Farley, K. (2008). Unroofing, incision, and uplift history of the southwestern Colorado Plateau from apatite (U-Th)/He thermochronometry. *GSA Bulletin*, 120(5–6), 571–87.

Freissler, R., Zaroli, C., Lambotte, S., and Schuberth, B. S. (2020). Tomographic filtering via the generalized inverse: A way to account for seismic data uncertainty. *Geophysical Journal International*, 223(1), 254–69.

French, S. W., and Romanowicz, B. A. (2014). Whole-mantle radially anisotropic shear velocity structure from spectral-element waveform tomography. *Geophysical Journal International*, 199(3), 1303–27.

Friedrich, A. M. (2019). Palaeogeological hiatus surface mapping: A tool to visualize vertical motion of the continents. *Geological Magazine*, 156(2), 308–19.

Friedrich, A. M., Bunge, H.-P., Rieger, S. M. et al. (2018). Stratigraphic framework for the plume mode of mantle convection and the analysis of interregional unconformities on geological maps. *Gondwana Research*, 53, 159–88.

Ghelichkhan, S., and Bunge, H.-P. (2016). The compressible adjoint equations in geodynamics: Derivation and numerical assessment. *GEM – International Journal on Geomathematics*, 7(1), 1–30.

Ghelichkhan, S., and Bunge, H.-P. (2018). The adjoint equations for thermochemical compressible mantle convection: Derivation and verification by twin experiments. *Proceedings of the Royal Society A: Mathematical, Physical and Engineering Sciences*, 474(2220), 20180329.

Ghelichkhan, S., Bunge, H.-P., and Oeser, J. (2021), Global mantle flow retrodictions for the early Cenozoic using an adjoint method: Evolving dynamic topographies, deep mantle structures, flow trajectories and sublithospheric stresses. *Geophysical Journal International*, 226(2), 1432–60.

Guillocheau, F., Rouby, D., Robin, C. et al. (2012). Quantification and causes of the terrigeneous sediment budget at the scale of a continental margin: A new method applied to the Namibia-South Africa margin. *Basin Research*, 24(1), 3–30.

Guillocheau, F., Simon, B., Baby, G. et al. (2018). Planation surfaces as a record of mantle dynamics: The case example of Africa. *Gondwana Research*, 53, 82–98.

Hager, B. H., Clayton, R. W., Richards, M. A., Comer, R. P., and Dziewonski, A. M. (1985). Lower mantle heterogeneity, dynamic topography and the geoid. *Nature*, 313(6003), 541–5.

Hartley, R. A., Roberts, G. G., White, N. J., and Richardson, C. (2011). Transient convective uplift of an ancient buried landscape. *Nature Geoscience*, 4(8), 562–5.

Hayek, J. N., Vilacís, B., Bunge, H.-P. et al. (2020). Continent-scale hiatus maps for the Atlantic Realm and Australia since the Upper Jurassic and links to mantle flow induced dynamic topography. *Proceedings of the Royal Society A: Mathematical, Physical and Engineering Sciences*, 476(2242), 20200390.

Hayek, J. N., Vilacʹıs, B., Bunge, H.-P. et al. (2021). Correction: Continent-scale hiatus maps for the Atlantic Realm and Australia since the Upper Jurassic and links to mantle flow-induced dynamic topography. *Proceedings of the Royal Society A: Mathematical, Physical and Engineering Sciences*, 477(2251), 20210437.

Heister, T., Dannberg, J., Gassmöller, R., and Bangerth, W. (2017). High accuracy mantle convection simulation through modern numerical methods. II: Realistic models and problems. *Geophysical Journal International*, 210(2), 833–51.

Hoggard, M. J., Austerman, J., Randel, C., and Stephenson, S. (2021). Observational estimates of dynamic topography through space and time. In H. Marquardt, S. Ballmer, M. adn Cottaar, and J. Konter, eds., *Mantle Convection and Surface Expressions*. Washington DC: American Geophysical Union (AGU), pp. 371–411.

Hoggard, M. J., Winterbourne, J., Czarnota, K., and White, N. (2017). Oceanic residual depth measurements, the plate cooling model, and global dynamic topography. *Journal of Geophysical Research: Solid Earth*, 122(3), 2328–72.

Horbach, A., Bunge, H. P., and Oeser, J. (2014). The adjoint method in geodynamics: Derivation from a general operator formulation and application to the initial condition problem in a high resolution mantle circulation model. *GEM – International Journal on Geomathematics*, 5(2), 163–94.

Iaffaldano, G., and Bunge, H.-P. (2015). Rapid plate motion variations through geological time: Observations serving geodynamic interpretation. *Annual Review of Earth and Planetary Sciences* 43, 571–92.

Ismail-Zadeh, A., Schubert, G., Tsepelev, I., and Korotkii, A. (2004). Inverse problem of thermal convection: numerical approach and application to mantle plume restoration. *Physics of the Earth and Planetary Interiors*, 145(1–4), 99–114.

Japsen, P. (2018). Sonic velocity of chalk, sandstone and marine shale controlled by effective stress: Velocity-depth anomalies as a proxy for vertical movements. *Gondwana Research*, 53, 145–58.

Jarvis, G. T., and Mckenzie, D. P. (1980). Convection in a compressible fluid with infinite Prandtl number. *Journal of Fluid Mechanics*, 96(03), 515–83.

Kronbichler, M., Heister, T., and Bangerth, W. (2012). High accuracy mantle convection simulation through modern numerical methods. *Geophysical Journal International*, 191, 12–29.

Li, D., Gurnis, M., and Stadler, G. (2017). Towards adjoint-based inversion of time-dependent mantle convection with nonlinear viscosity. *Geophysical Journal International*, 209(1), 86–105.

McNamara, A. K. (2019). A review of large low shear velocity provinces and ultra low velocity zones. *Tectonophysics*, 760, 199–220.

McNamara, A. K., and Zhong, S. (2005). Thermochemical structures beneath Africa and the Pacific Ocean. *Nature*, 437 (7062), 1136–9.

Mégnin, C., Bunge, H.-P., Romanowicz, B., and Richards, M. A. (1997). Imaging 3-D spherical convection models: What can seismic tomography tell us about mantle dynamics? *Geophysical Research Letters*, 24(11), 1299–302.

Meinhold, G. (2010). Rutile and its applications in Earth sciences. *Earth–Science Reviews*, 102(1), 1–28.

Miall, A. D. (2016). The valuation of unconformities. *Earth-Science Reviews*, 163, 22–71.

Mitrovica, J. X. (1996). Haskell [1935] revisited. *Journal of Geophysical Research*, 101(B1), 555.

Mitrovica, J. X., Beaumont, C., and Jarvis, G. T. (1989). Tilting of continental interiors by the dynamical effects of subduction. *Tectonics*, 8(5), 1079–94.

Mosca, I., Cobden, L., Deuss, A., Ritsema, J., and Trampert, J. (2012). Seismic and mineralogical structures of the lower mantle from probabilistic tomography. *Journal of Geophysical Research*, 117(B6).

Moulik, P., Lekic, V., Romanowicz, B. et al. (2021). Global reference seismological datasets: Multi-mode surface wave dispersion. *Geophysical Journal International*, 228(3).

Müller, R. D., Seton, M., Zahirovic, S. et al. (2016). Ocean basin evolution and global-scale plate reorganization events since Pangea breakup. *Annual Review of Earth and Planetary Sciences*, 44, 107–38.

Nelson, P. L., and Grand, S. P. (2018). Lower-mantle plume beneath the Yellowstone hotspot revealed by core waves. *Nature Geoscience*, 11(4), 280–4.

Ogg, J. G., Ogg, G. M., and Gradstein, F. M. eds. (2016). Introduction. In J. G. Ogg, G. M. Ogg, and F. M. Gradstein, eds., *A Concise Geologic Time Scale*. Amsterdam: Elsevier, pp. 1–8.

Paulson, A., and Richards, M. A. (2009). On the resolution of radial viscosity structure in modelling long-wavelength postglacial rebound data. *Geophysical Journal International*, 179 (3), 1516–26.

Pekeris, C. L. (1935). Thermal convection in the interior of the Earth. *Geophysical Journal International*, 3(8), 343–67.

Piazzoni, A. S., Steinle-Neumann, G., Bunge, H., and Dolejš, D. (2007). A mineralogical model for density and elasticity of the Earth's mantle. *Geochemistry, Geophysics, Geosystems*, 8(11).

Price, M. G., and Davies, J. H. (2018). Profiling the robustness, efficiency and limits of the forward-adjoint method for 3D mantle convection modelling. *Geophysical Journal International*, 212(2), 1450–62.

Reiners, P. W., and Brandon, M. T. (2006). Using thermochronology to understand orogenic erosion. *Annual Review of Earth and Planetary Sciences*, 34(1), 419–66.

Reuber, G. S., and Simons, F. J. (2020). Multi-physics adjoint modeling of Earth structure: Combining gravimetric, seismic, and geodynamic inversions. *GEM – International Journal on Geomathematics*, 11, 30. https://doi.org/10.1007/s13137-020-00166-8.

Richards, M. A., and Hager, B. H. (1984). Geoid anomalies in a dynamic Earth. *Journal of Geophysical Research*, 89(B7), 5987–6002.

Ritsema, J., Deuss, A., Van Heijst, H.-J., and Woodhouse, J. H. (2011). S40RTS: A degree-40 shear-velocity model for the mantle from new Rayleigh wave dispersion, teleseismic traveltime and normal-mode splitting function measurements. *Geophysical Journal International*, 184(3), 1223–36.

Roberts, G. G., and White, N. (2010). Estimating uplift rate histories from river profiles using African examples. *Journal of Geophysical Research: Solid Earth* 115(B2), B02406.

Said, A., Moder, C., Clark, S., and Abdelmalak, M. M. (2015). Sedimentary budgets of the Tanzania coastal basin and implications for uplift history of the East African rift system. *Journal of African Earth Sciences*, 111, 288–95.

Said, A., Moder, C., Clark, S., and Ghorbal, B. (2015). Cretaceous-Cenozoic sedimentary budgets of the Southern Mozambique Basin: Implications for uplift history of the South African Plateau. *Journal of African Earth Sciences*, 109, 1–10.

Sandiford, M. (2007). The tilting continent: A new constraint on the dynamic topographic field from Australia. *Earth and Planetary Science Letters*, 261(1-2), 152–63.

Schaber, K., Bunge, H.-P., Schuberth, B., Malservisi, R., and Horbach, A. (2009). Stability of the rotation axis in high-resolution mantle circulation models: Weak polar wander despite strong core heating. *Geochemistry, Geophysics, Geosystems*, 10, Q11W04. https://doi.org/10.1029/2009GC002541.

Schaeffer, A. J., and Lebedev, S. (2013). Global shear speed structure of the upper mantle and transition zone. *Geophysical Journal International*, 194(1), 417–49.

Schuberth, B. S. A., Bunge, H.-P., and Ritsema, J. (2009). Tomographic filtering of high-resolution mantle circulation models: Can seismic heterogeneity be explained by temperature alone? *Geochemistry, Geophysics, Geosystems*, 10(5).

Schuberth, B. S. A., Bunge, H.-P., Steinle-Neumann, G., Moder, C., and Oeser, J. (2009). Thermal versus elastic heterogeneity in high-resolution mantle circulation models with pyrolite composition: High plume excess temperatures in the lowermost mantle. *Geochemistry, Geophysics, Geosystems*, 10 (1).

Schuberth, B. S. A., Zaroli, C., and Nolet, G. (2012). Synthetic seismograms for a synthetic Earth: Long-period Pand S-wave traveltime variations can be explained by temperature alone. *Geophysical Journal International*, 188(3), 1393–412.

Sengör, A. M. C. (2001). Elevation as indicator of mantle-plume activity. *Mantle Plumes: Their identification through Time*, 352, 183–245.

Seton, M., Müller, R. D., Zahirovic, S. et al. (2012). Global continental and ocean basin reconstructions since 200 Ma. *Earth-Science Reviews*, 113(3–4), 212–70.

Simmons, N. A., Myers, S. C., Johannesson, G., and Matzel, E. (2012). LLNL-G3Dv3: Global P-wave tomography model for improved regional and teleseismic travel time prediction. *Journal of Geophysical Research: Solid Earth*, 117(10), 1–28.

Simmons, N. A., Myers, S. C., Johannesson, G., Matzel, E., and Grand, S. P. (2015). Evidence for long-lived subduction of an ancient tectonic plate beneath the southern Indian Ocean. *Geophysical Research Letters*, 42(21), 9270–8.

Smith, A., Smith, D., and Funnel, B. (1994). *Atlas of Mesozoic and Cenozoic landmasses*. Cambridge: Cambridge University Press.

Spasojevic, S., Liu, L., Gurnis, M., and Müller, R. D. (2008). The case for dynamic subsidence of the U.S. east coast since the Eocene. *Geophysical Research Letters*, 35(8).

Steinberger, B., and O'Connell, R. J. (1997). Changes of the Earth's rotation axis owing to advection of mantle density heterogeneities. *Nature*, 387(6629), 169–73.

Stixrude, L., and Lithgow-Bertelloni, C. (2011). Thermodynamics of mantle minerals II. Phase equilibria. *Geophysical Journal International*, 184(3), 1180–213

Stotz, I. L., Tassara, A., and Iaffaldano, G. (2021). Pressure-driven Poiseuille flow inherited from Mesozoic mantle circulation led to the Eocene separation of Australia and Antarctica. *Journal of Geophysical Research: Solid Earth*, 126(4), e2020JB019945.

Torsvik, T. H., Müller, R. D., Van Der Voo, R., Steinberger, B., and Gaina, C. (2008). Global plate motion frames: Toward a unified model. *Reviews of Geophysics*, 46(3), RG3004.

Vibe, Y., Friedrich, A. M., Bunge, H.-P., and Clark, S. R. (2018). Correlations of oceanic spreading rates and hiatus surface area in the North Atlantic realm. *Lithosphere*, 10(5), 677–84.

Vynnytska, L., and Bunge, H. (2014). Restoring past mantle convection structure through fluid dynamic inverse theory: Regularisation through surface velocity boundary conditions. *GEM – International Journal on Geomathematics*, 6(1), 83–100.

Young, A., Flament, N., Maloney, K. et al. (2019). Global kinematics of tectonic plates and subduction zones since the late Paleozoic Era. *Geoscience Frontiers*, 10(3), 989–1013.

Zahirovic, S., Flament, N., Dietmar Müller, R., Seton, M., and Gurnis, M. (2016). Large fluctuations of shallow seas in low-lying Southeast Asia driven by mantle flow. *Geochemistry, Geophysics, Geosystems*, 17(9), 3589–607.

Zaroli, C. (2016). Global seismic tomography using Backus-Gilbert inversion. *Geophysical Journal International*, 207(2), 876–88.

Zaroli, C., Sambridge, M., Le´ve^que, J.-J., Debayle, E., and Nolet, G. (2013). An objective rationale for the choice of regularisation parameter with application to global multiple-frequency S-wave tomography. *Solid Earth*, 4(2), 357–71.

Zhong, S. J., Yuen, D. A., Moresi, L. N., and Knepley, M. G. (2015). Numerical methods for mantle convection, in D. Bercovici, ed., *Treatise on Geophysics. Vol. 7: Mantle Dynamics*, 2nd ed. Amsterdam: Elsevier, pp. 197–222.

Zhou, Q., and Liu, L. (2017). A hybrid approach to data assimilation for reconstructing the evolution of mantle dynamics. *Geochemistry, Geophysics, Geosystems*, 18(11), 3854–68.

21

Understanding and Predicting Geomagnetic Secular Variation via Data Assimilation

Weijia Kuang, Kyle Gwirtz, Andrew Tangborn, and Matthias Morzfeld

Abstract: Geomagnetic data assimilation is a recently established research discipline in geomagnetism. It aims to optimally combine geomagnetic observations and numerical geodynamo models to better estimate the dynamic state of the Earth's outer core, and to predict geomagnetic secular variation. Over the past decade, rapid advances have been made in geomagnetic data assimilation on various fronts by several research groups around the globe, such as using geomagnetic data assimilation to understand and interpret the observed geomagnetic secular variation, estimating part of the core state that is not observable on the Earth's surface, and making geomagnetic forecasts on multi-year time scales. In parallel, efforts have also been made on proxy systems for understanding fundamental statistical properties of geomagnetic data assimilation, and for developing algorithms tailored specifically for geomagnetic data assimilation. In this chapter, we provide a comprehensive overview of these advances, as well as some of the immediate challenges of geomagnetic data assimilation, and possible solutions and pathways to move forward.

21.1 Introduction

The observed Earth's magnetic field, measured on the ground and in orbits, is the sum of contributions from various magnetic sources, both within the Earth and external to it. Among them is the field originating in the Earth's fluid outer core, which accounts for more than 99% of the observed magnetic energy as described by the Mausberger–Lowes power spectra (Langel and Estes, 1982). This part of the field, called the 'core field' and the 'geomagnetic field' interchangeably in this chapter, is generated and maintained by turbulent convection in the outer core via dynamo action (Larmor, 1919), and was first modelled numerically a quarter century ago (e.g. Glatzmaier and Roberts, 1995; Kageyama and Sato, 1997; Kuang and Bloxham, 1997). Thus, geomagnetic observations and geodynamo simulations are powerful tools for understanding the dynamical processes in the Earth's outer core, the thermo-chemical properties in the deep Earth, and the interactions between the outer core and other components of the Earth system.

Both geomagnetic observations and geodynamo models can provide independent glimpses of the core dynamic state. Surface and orbital measurements can determine the geomagnetic field up to degree $L_{obs} \leq 14$ in spherical harmonic expansion (Langel and Estes, 1982), and its slow-time variation, called the secular variation (SV), at higher degrees with the data collected from the current Swarm satellite constellations (e.g. Finlay et al., 2020; Sabaka et al., 2020). The observed SV can be used to infer core flow beneath the core–mantle boundary (CMB) via the 'frozen flux' approximation (Roberts and Scott, 1965) and additional constraints on the core flow properties (e.g. Holme, 2007). The observed field and the inferred flow provide (observational) pieces of the core dynamics puzzle (Schaeffer et al., 2016; Aubert and Finlay, 2019; Kloss and Finlay, 2019), but the dominant part of the core state remains opaque.

On the other hand, numerical geodynamo simulations can provide self-consistent approximations of the core dynamic state, by numerically solving the dynamo equations with given boundary and initial conditions. These dynamo equations are the non-linear partial differential equations derived from first principles with various simplifications (Braginsky and Roberts, 1995). Due to computational constraints, numerical geodynamo simulations could not be made with arbitrarily high spatial temporal resolutions, and are thus limited to the parameter regimes far from those appropriate to the Earth's outer core (e.g. Roberts and King, 2013; Wicht and Sanchez, 2019). But estimates of the core state are still attempted by the asymptotic limits (scaling laws) derived from numerical simulations with (computationally permitted) broad ranges of parameter values (e.g. Christensen, 2010; Yadav et al., 2013; Aubert et al., 2017; Kuang et al., 2017). Still the differences between the 'true' core state and the numerical asymptotic limits remain uncertain.

The capabilities and limitations of geomagnetic observations and of geodynamo simulations led to the birth and growth of geomagnetic data assimilation (GDA) for making optimal estimates of the core dynamic state, and interpreting and predicting SV, using available geomagnetic data and numerical geodynamo models (Fournier et al., 2010). The first 'proof-of-concept' studies were carried out using simplified magnetohydrodynamic (MHD) systems (Fournier et al., 2007; Sun et al., 2007; Morzfeld and Chorin, 2012), or numerical dynamo models (Liu et al., 2007). These ignited subsequent efforts in GDA for understanding observational constraints on core dynamics (e.g. Kuang et al., 2009; Aubert and Fournier, 2011; Fournier et al., 2011; Aubert, 2014; Kuang and Tangborn, 2015); GDA error statistics and model developments (e.g. Kuang et al., 2008; Canet et al., 2009; Hulot et al., 2010; Fournier et al., 2011; Li et al., 2011, 2014; Sun and Kuang, 2015; Tangborn and Kuang, 2015, 2018; Sanchez et al., 2016, 2019; Gwirtz et al., 2021); and geomagnetic predictions (Kuang et al., 2010; Fournier et al., 2015, 2021b; Morzfeld et al., 2017; Minami et al., 2020; Sanchez et al., 2020; Tangborn et al., 2021). Despite these advancements, many fundamental questions still remain on geodynamic approximations, assimilation algorithms and model/observation error statistics, such as model and observation bias corrections, forecast covariance matrix convergences, and non-linear assimilation algorithms. However, it is expected that GDA will continue to rapidly advance and will be broadly used in geodynamo and geomagnetic field modelling, and in studies of the Earth's deep interior.

The goal of this chapter is to present the reader with an overview of GDA which is easy to comprehend and can provide a first step into geomagnetic data assimilation research. This chapter is organised as follows: the geomagnetic field and geodynamo modelling are reviewed in Section 21.2; the assimilation algorithm is given in Section 21.3. The current research results are presented in Section 21.4, followed by the challenges and future development in Section 21.5. Conclusions and Discussions are in Section 21.6.

21.2 Geomagnetic Field and Geodynamo Modelling

The world (global) magnetic maps can be constructed from observatory and satellite magnetic measurements, historical navigation data, and archeo- and paleomagnetic data (e.g. Jackson et al., 2000; Lesur et al., 2010; Panovska et al., 2019; Finlay et al., 2020; Huder et al., 2020; Sabaka et al., 2020; Alken et al., 2021b; Brown et al., 2021). For details of geomagnetic observations and field models, see, for example, Mandea and Korte (2011) and Sanchez (this volume). Despite subtle differences in the algorithms utilised to produce these field models, they all share the same objective: optimally separating the different sources that contribute to the magnetic measurements. Among those contributions is the magnetic field originated from the Earth's core, called the Earth's intrinsic magnetic field or simply the core field. In this section, this field is also called the geomagnetic field, and the corresponding models are called geomagnetic field models. These models provide descriptions of the spatial and temporal variations of the modern field, as well as lower-accuracy descriptions going as far as 100k years back in time. In a geomagnetic field model, the observed geomagnetic field $\mathbf{B}^{(o)}$ is approximated as a potential field and is described mathematically by the following spherical harmonic expansion

$$\mathbf{B}^{(o)} = -\nabla V,$$

$$V = a \sum_{0 \leq m \leq l}^{L_{obs}} \left(\frac{a}{r}\right)^{l+1} (g_l^m \cos m\phi + h_l^m \sin m\phi) P_l^m(\theta), \quad (21.1)$$

where (g_l^m, h_l^m) are called the Gauss coefficients; P_l^m are the Schmidt normalised associated Legendre polynomials of degree l and order m; L_{obs} is the highest degree resolved with the data; a is the mean radius of the Earth's surface; θ and φ are the co-latitude and longitude, respectively. The highest degree L_{obs} in (21.1) depends on the quality of the data available, and thus varies over time. By (21.1), $\mathbf{B}^{(o)}$ can be continued downward from the surface to any location r^* in the interior, as long as the region $r^* \leq r \leq a$ is electrically insulating. In a GDA system, r^* is typically the mean radius r_c of the CMB if the entire mantle is assumed electrically insulating in the geodynamo model; or, if there is an electrically conducting D″-layer at the base of the mantle, it is the mean radius r_d of the top of the layer. For example, the top row of Fig. 21.1 are the mean observed radial component B_r and its SV $\dot{B}_r$ B_r in 2010–15 at $r_d = 3{,}520$ km (i.e. assuming a 20 km thick D″-layer). Notice that B_r reverses its sign in the areas around the tip of South America and south of Africa, which coincide with the South Atlantic Anomaly (SAA), a region with exceptionally low field intensity (see Fig. 21.2).

The working of the geodynamo can be described simply as follows. The secular cooling and differentiation through the Earth's evolution have provided the buoyancy force which drives convection in the outer core. Since the core fluid is an iron-rich liquid alloy and is therefore highly electrically conducting (e.g. Nimmo, 2007; Hirose et al., 2013), an additional magnetic field is generated by the core convection given any seed (background) magnetic field. A self-consistent dynamo is achieved if the generated magnetic field can be maintained without the presence of the seed field (Larmor, 1919). Braginsky and Roberts (1995) provide the full set of the partial differential equations for the geodynamo. But the earliest dynamo models, first by Glatzmaier and Roberts (1995), and later by Kageyama and Sato (1997) and Kuang and Bloxham (1997), were

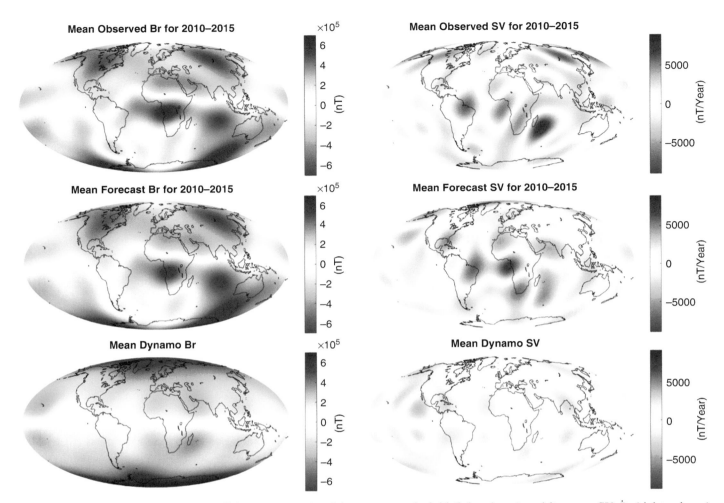

Figure 21.1 Snapshots of the mean radial component B_r of the geomagnetic field (left column) and its mean SV $\dot{B}_r$ (right column) at the top of the D″-layer for 2010–15. The top row are downward continued from that of CM6 field model at the mean surface (Sabaka et al., 2020), the centre row are from NASA GEMS assimilation solutions (Sun and Kuang, 2015; Tangborn et al., 2021), and the bottom row are from MoSST geodynamo simulation solutions (Kuang and Chao, 2003; Jiang and Kuang, 2008).

developed with simplified versions of the Braginsky–Roberts equations. Since then, many more dynamo models have been developed with different physics and/or numerical algorithms implemented. Details of these dynamo models can be found in the past community dynamo benchmark efforts (Christensen et al., 2001; Jones et al., 2011; Matsui et al., 2016).

To better explain the core state defined in geodynamo models and its correlation to geomagnetic observations, we use the MoSST core dynamics model (Kuang and Bloxham, 1999; Kuang and Chao, 2003; Jiang and Kuang, 2008) as an example. Formulations can be easily adapted to other dynamo models. In MoSST, the core fluid is assumed Boussinesq; and the core state is described by the velocity field **v**, the magnetic field **B**, and the temperature anomaly Θ (from the background conducting state). Since **v** and **B** are

solenoidal (divergence-free), they are decomposed into the poloidal and the toroidal components, for example,

$$\mathbf{B} = \nabla \times (T_B \hat{\mathbf{r}}) + \nabla \times \nabla \times (P_B \hat{\mathbf{r}}), \qquad (21.2)$$

where $\hat{\mathbf{r}}$ is the unit radial vector, and T_B and P_B are the toroidal and poloidal scalars, respectively. In MoSST, all scalar fields are described by spherical harmonic expansions, with the spectral coefficients discretised in radius, for example,

$$P_B = \sum_{0 \le m \le l}^{L_d} b_l^m(r_i) Y_l^m(\theta, \phi) + C.C., \text{ for } i = 0, 1, \dots N_r,$$

$$(21.3)$$

where L_d is the highest degree, N_r is the number of the radial grid points r_i, Y_l^m are the orthonormal spherical harmonic

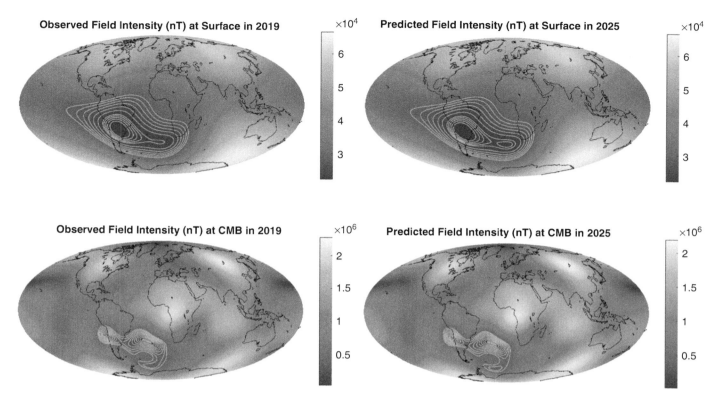

Figure 21.2 The South Atlantic Anomaly observed in 2019 (left column) and predicted in 2025 (right column). The top are the magnetic intensity at the Earth's mean surface, and the bottom are those at the top of the D''-layer. The intensity levels decrease from the outermost contour (30,000 nT on the surface) to the innermost contour (22,800 nT on the surface). The observations are from CM6 (Sabaka et al., 2020), and the forecasts are made by Tangborn et al. (2021).

functions, and C.C. stands for the complex conjugate. With these definitions, the core state vector is

$$\mathbf{x} = [\{b_l^m(r_i)\}, \{j_l^m(r_i)\}, \{v_l^m(r_i)\}, \{\omega_l^m(r_i)\}, \{\zeta_l^m(r_i)\}]^T,$$

$$(21.4)$$

where j_l^m, v_l^m, w_l^m and ζ_l^m are the complex spectral coefficients of the toroidal magnetic scalar, the poloidal and toroidal velocity scalars, and the temperature anomaly, respectively. Thus the dimension of $\mathbf{x}$ is $N_x = 5(N_r + 1)L_d(L_d + 2)$, which can be very large, for example, $N_x \approx 10^6$ if $N_r, L_d \sim 100$. The time evolution of $\mathbf{x}$ is evaluated via given discretised schemes, which are of the form

$$\mathbf{x}^{k+1} = \mathbf{L}\mathbf{x}^k + \mathbf{f}(\mathbf{x}^k),$$

$$(21.5)$$

where $\mathbf{L}$ is the linear matrix describing, for example, the dissipative effects; $\mathbf{f}$ is the vector describing the quadratic interactions among the fields, for example, motional induction; and the superscripts k and $k + 1$ indicate the state vector at the time steps t_k and t_{k+1}, respectively. In MoSST, the entire dynamo system is non-dimensionalised, and is therefore defined with a set of non-dimensional parameters, namely the Rayleigh number R_{th} (for the buoyancy force), the magnetic Rossby number R_o (for the fluid inertia), the Ekman number E (for the fluid viscosity), and

the modified Prandtl number q_κ (ratio of the thermal conductivity to the magnetic diffusivity). They are referred to as the dynamo parameters in this chapter, and are embedded in $\mathbf{L}$ and $\mathbf{f}$ of (21.5). It should be pointed out that the mathematical description (21.5) applies to all dynamo models, though the state vector and the dynamo parameters may differ. It should also be noted that most of the currently available dynamo models are non-dimensional (e.g. Christensen et al., 2001; Jones et al., 2011; Matsui et al., 2016), thus requiring appropriate rescaling to compare with the observed geomagnetic field.

Geodynamo simulations, which are also called the 'free-running models' in data assimilation as they are not constrained by observations, can provide axial-dipolar–dominant magnetic fields that change in both space and time. But they are also expected to differ from observations, mainly because the dynamo parameter values used in numerical simulation are far from those appropriate for the Earth's core (see Section 21.5 for more discussions). For example, in the bottom row of Fig. 21.1, are the typical mean B_r and mean SV at the top of the D''-layer over a 100-year period from MoSST dynamo simulation results with the parameters $R_o = E = 1.25 \times 10^{-6}$ and $q_\kappa = 1$. Compared to the observations (the top row in Fig. 21.1), the axial dipole component of the simulated field is too strong, and the simulated SV is too weak. These significant differences show that geodynamo simulations cannot provide accurate estimates of the core state on their own.

Aiming at improving core state estimates, one may attempt to constrain geodynamo simulations with observations, and data assimilation is an optimal choice. The observational constraints can be made through connections between the Gauss coefficients defined in (21.1), and the poloidal field spectral coefficients defined in (21.3). In MoSST, an electrically conducting D''-layer at the base of the mantle is implemented in the model. With this feature, the connection is made at the top of the D''-layer $r = r_d$:

$$b_l^m = b_l^{m(o)} \equiv \frac{(-1)^m}{l\mathcal{B}} \left(\frac{a}{r_d}\right)^{l+2} \sqrt{\frac{2\pi(1+\delta_{m0})}{2l+1}} (g_l^m - ih_l^m) \tag{21.6}$$

for $l \le L_{obs}$. In (21.6), δ_{m0} is the Kronecker delta ($\delta_{m0} = 1$ if $m = 0$, and 0 otherwise), and $\mathcal{B}$ is a magnetic scaling factor used to match the dimensional Gauss coefficients (from geomagnetic field models) and the non-dimensional poloidal spectral coefficients (from geodynamo models). The matching condition (21.6) shows also that only part of the poloidal magnetic field can be observed, leaving the rest of the core state unobserved. It also defines the projection of $\mathbf{x}$ onto the observational subspace:

$$\mathbf{y} = \mathbf{Hx}, \tag{21.7}$$

where $\mathbf{y}$ [of the dimension $N_y = L_{obs}(L_{obs} + 2)$] is the observed part of the core state, and $\mathbf{H}$ is an $N_y \times N_x$ matrix, called the observation operator in data assimilation (see Section 21.3). Since the observation $\mathbf{y}$ is defined only at r_d (the top of the D''-layer), $\mathbf{H}$ is very simple: each row of $\mathbf{H}$ contains a single non-zero entry of 1 corresponding to the observed $b_l^{m(o)}$. The relations (21.6) and (21.7) are needed for GDA. It should be noted that the methodology is applicable to other geodynamo models, though the state vector $\mathbf{x}$ and therefore the relation (21.6) may need to be modified accordingly.

21.3 Mathematics of Data Assimilation

In this section, we provide a brief review of data assimilation (DA) while highlighting details relevant to geomagnetism. We begin with a short introduction to the fundamental framework of DA. Specific methods of DA are then outlined and the computational limitations they are subject to in practice are briefly discussed.

21.3.1 The Basic Framework of Data Assimilation

Data assimilation merges a computational model of a process with observations, in order to produce an improved estimate of the state of a system. Methods of DA are typically constructed within a Bayesian framework as follows. Let $\mathbf{x}^t$ be a vector of N_x elements representing the true state of a system at a particular time (the geodynamo in GDA). Knowledge of the system state is recorded in $\mathbf{y}$, a vector of N_y observations which is related to $\mathbf{x}^t$ according to

$$\mathbf{y} = \mathbf{Hx}^t + \boldsymbol{\varepsilon}, \tag{21.8}$$

where $\mathbf{H}$ is the observation operator and $\boldsymbol{\varepsilon}$ is the observation noise which is frequently assumed to be Gaussian with mean zero and covariance $\mathbf{R}$. In GDA, $\mathbf{x}^t$ and $\mathbf{y}$ may consist of spherical harmonic coefficients defining the state of the geodynamo and knowledge of the poloidal magnetic field near the CMB, respectively (see Section 21.2), in which case the observation operator is the $N_y \times N_x$ matrix $\mathbf{H}$. Equation (21.8) defines the likelihood $p(\mathbf{y}|\mathbf{x}^t)$, that is, the probability distribution of the observations given a system state. In sequential DA systems, numerical simulations can be used to define a prior distribution $p_0(\mathbf{x}^t)$(see Section 21.3.2 for details) and by Bayes' rule, a posterior distribution

$$p(\mathbf{x}^t|\mathbf{y}) \propto p_0(\mathbf{x}^t)p(\mathbf{y}|\mathbf{x}^t), \tag{21.9}$$

is defined by the product of the likelihood and the prior. The ultimate objective of various approaches to DA is the approximation of this posterior distribution.

21.3.2 The Ensemble Kalman Filter

The ensemble Kalman filter (EnKF), is a widely used method that has been employed in multiple GDA systems (see, e.g., Fournier et al., 2013; Sun and Kuang, 2015; Sanchez et al., 2020; Tangborn et al., 2021). It approximates the posterior distribution of (21.9) by combining a Monte Carlo approach with the Kalman filter (see, e.g., Evensen, 2006). Specifically, the distributions in Section 21.3.1 are estimated by sampling them through multiple, simultaneous runs of a numerical model. In an EnKF, a *forecast ensemble* of N_e unique forecasts $\mathbf{X}^f = \{\mathbf{x}_1^f, ..., \mathbf{x}_{N_e}^f\}$ is produced at a time when observations are to be assimilated. This ensemble is taken to be a sample of the prior distribution $p_0(\mathbf{x}^t)$. The purpose of the EnKF is to adjust these forecasts by merging them with information contained in the observations $\mathbf{y}$. This collection of 'adjusted' forecasts forms an *analysis ensemble* which is taken to be the desired sampling of the posterior distribution $p(\mathbf{x}^t|\mathbf{y})$. Typically, the mean of the analysis ensemble is used as an estimate of the true state of the system, with the ensemble variance indicating the estimate's uncertainty.

An EnKF can be implemented in the following way. The initial collection of forecasts are used to determine the *forecast covariance*

$$\mathbf{P}^f = \frac{1}{N_e - 1} \sum_{i=1}^{N_e} (\mathbf{x}_i^f - \bar{\mathbf{x}})(\mathbf{x}_i^f - \bar{\mathbf{x}})^T, \tag{21.10}$$

where $\bar{\mathbf{x}} = (1/N_e)\Sigma_{i=1}^{N_e} \mathbf{x}_i^f$. An *analysis ensemble* can then be determined by

$$\mathbf{x}_i^a = \mathbf{x}_i^f + \mathbf{K}[\mathbf{y} - (\mathbf{H}\mathbf{x}_i^f + \boldsymbol{\varepsilon}_i)], \tag{21.11}$$

for $i = 1, \ldots N_e$, where $\boldsymbol{\varepsilon}_i \sim \mathcal{N}(\mathbf{0}, \mathbf{R})$, and

$$\mathbf{K} = \mathbf{P}^f \mathbf{H}^T (\mathbf{H}\mathbf{P}^f \mathbf{H}^T + \mathbf{R})^{-1} \tag{21.12}$$

is the estimate of the Kalman gain. Under appropriate conditions, the analysis ensemble is a sampling of the posterior and thus provides the desired approximation of (21.9). The analysis ensemble members can then be propagated forward in time by the numerical model, to the next instance when observations are available for assimilation and the process is repeated. The particular EnKF algorithm outlined here is known as the *stochastic EnKF*. Other implementations exist (see, e.g., Tippett et al., 2003; Hunt et al., 2007; Buehner et al., 2017) and differ in their details; however, all rely on a Monte Carlo approximation of the Kalman gain and are designed such that, under certain conditions, the analysis ensemble they produce is distributed according to the posterior.

21.3.3 Variational and Hybrid Methods

Variational methods produce an estimate of the system state by seeking the maximum of the posterior distribution of (21.9). This approach effectively transforms the assimilation of observations to an optimisation problem. Assume that $\mathbf{m}$ is the numerical model for advancing the state vector in time

$$\mathbf{x}(t_k) = \mathbf{m}[\mathbf{x}(t_{k-1})], \tag{21.13}$$

where $\mathbf{x}(t_k)$ is the state vector of a numerical simulation at time t_k. Assume also that at t_k, the observations $\mathbf{y}(t_k)$ are made with error covariance $\mathbf{R}$. For simplicity, we denote $\mathbf{x}_k = \mathbf{x}(t_k)$ and $\mathbf{y}_k = \mathbf{y}(t_k)$ in the rest of the discussion. With this notation, the posterior distribution of the true initial state $\mathbf{x}_0^t$ at time t_0, given observations $\mathbf{y}_1$ at time t_1 is

$$p(\mathbf{x}_0^t | \mathbf{y}_1) \propto p(\mathbf{x}_0^t) p(\mathbf{y}_1 | \mathbf{x}_0^t). \tag{21.14}$$

Under the assumption that the prior and likelihood are Gaussian, maximising (21.14) is equivalent to minimising

$$J(\mathbf{x}_0) = (\mathbf{x}_0 - \boldsymbol{\mu})^T \mathbf{B}^{-1} (\mathbf{x}_0 - \boldsymbol{\mu}) + [\mathbf{Hm}(\mathbf{x}_0) - \mathbf{y}_1]^T \mathbf{R}^{-1} \\ [\mathbf{Hm}(\mathbf{x}_0) - \mathbf{y}_1], \tag{21.15}$$

where $\boldsymbol{\mu}$ and $\mathbf{B}$ are the mean and background covariance of the prior distribution, respectively. This particular approach to variational DA is referred to as 4D-Var (see, e.g., Courtier, 1997). Determining the minimiser of the cost function $J(\mathbf{x}_0)$ is an iterative process that can be computationally challenging, in part because optimization requires computation of the gradient of the cost function. Note the dependence of Eq. (21.15) on $\mathbf{m}(\mathbf{x}_0)$, indicating that evaluations of $J(\mathbf{x}_0)$ require runs of the numerical model. Methods for improving the efficiency of the optimisation process are known, however, many require code for

a tangent linear model $\mathbf{M}$ of the full numerical model $\mathbf{m}$ (see, e.g., Talagrand and Courtier, 1987), that is, approximating $\mathbf{x}_1$ as $\mathbf{M}\mathbf{x}_0$. But constructing $\mathbf{M}$ for large, non-linear, numerical models can be a significant challenge.

Hybrid techniques have been developed which combine variational and ensemble-based approaches to DA. For example, one may run an ensemble of 4D-Var systems (Bonavita et al., 2012) or couple an EnKF to a 4D-Var setup (Zhang and Zhang, 2012), and use the ensemble mean and covariance to define $\boldsymbol{\mu}$ and $\mathbf{B}$, the statistics of the prior distribution appearing in Eq. (21.15). Recently, for the first time, a hybrid method (4DEnVar) was employed in GDA to propose an SV candidate model for IGRF-13 (Minami et al., 2020).

21.3.4 The Role of Ensemble Size

Under appropriate conditions, approximations of the posterior distribution produced by EnKF implementations or hybrid methods will converge as $N_e \to \infty$. But in practice, one aims to use sufficiently large, but finite ensemble sizes. However, what constitutes a 'sufficiently large' ensemble depends on individual models. It depends, among other things, on the dimension N_x of the state space $\mathbf{x}$, and the quality and extent of the observations $\mathbf{y}$ (Chorin and Morzfeld, 2013). Typically, the larger the state space and the sparser the observations, the larger the ensemble size required. In GDA, a typical 3-D geodynamo model with modest numerical resolutions has a dimension $N_x \sim 10^6$, while observations are limited to $N_y \sim 10^2$ on a 2-D spherical surface (e.g. the outer boundary of a geodynamo model). In many DA applications, the computational expense of the numerical model limits the ensemble size with which it is practical to implement EnKF/hybrid methods. This is the case in GDA where the computational demands of numerical geodynamos have typically restricted ensemble sizes to the hundreds. Efforts towards making GDA with ensembles of limited size more effective are discussed in Sections 21.4.3 and 21.5.

21.4 Geomagnetic Data Assimilation: Current Results

The geomagnetic field varies on timescales ranging from sub-annual to geological time scales, as found from paleomagnetic data (e.g. Panovska et al., 2019), historical magnetic navigation data (e.g. Jackson et al., 2000), and ground observatory and satellite magnetic measurements (e.g. Finlay et al., 2020; Huder et al., 2020; Sabaka et al., 2020). Over the past century, details of small-scale (high spherical harmonic degree) changes in the geomagnetic field morphology have been discovered with observatory and satellite magnetic measurements, such as persistent localised magnetic fluxes (e.g. Jackson, 2003; Finlay and Jackson, 2003); geomagnetic acceleration and geomagnetic jerks (e.g. Mandea et al., 2010; Chulliat and Maus, 2014); high-degree geomagnetic variations (e.g. Hulot et al., 2002; Olsen and Mandea, 2008; Kloss and Finlay, 2019); and the South Atlantic Anomaly (SAA), a localised region of extra low magnetic field intensity at the Earth's mean surface (e.g. Finlay et al., 2020). Historical, archaeo- and paleomagnetic data can provide information on large-scale (low spherical harmonic degree) geomagnetic field variations, such as the strong decay of the axial dipole moment in the past century (e.g. Brown et al., 2018; Jackson et al., 2000), persistent westward drift in the northern hemisphere over millennial time scales (e.g. Nilsson et al., 2020), and the well-known polarity reversals of the geomagnetic field over the last 150 Myr (Cande and Kent, 1995; Lowrie and Kent, 2004; Ogg, 2012).

These observations provide the data used in GDA and highlight the primary purposes of GDA: forecasting, hind-casting, and interpreting geomagnetic SV. The GDA-based SV studies provide one of the few windows we have into the dynamics of Earth's deep interior. Additionally, geomagnetic forecasts have broad scientific and societal applications, such as their contributions to time-varying global geomagnetic models which are widely used in various scientific communities (e.g. Doglioni et al., 2016), for navigation and survey applications (e.g. Kaji et al., 2019), and for space exploration (e.g. Heirtzler et al., 2002).

In this section, we provide a brief overview of some of the current results in using GDA to understand and predict SV. The discussion is organised according to the types of the dynamic models used in the assimilation system. We begin with a review of GDA relying on self-consistent 3-D dynamo models, followed by a discussion of alternative physical and statistical models, and finally, simplified models used in, for example, GDA algorithm development and the prediction of very long-term geomagnetic variations.

21.4.1 GDA with Self-consistent Dynamo Models

The first attempt to predict SV through GDA was carried out a little more than a decade ago (Kuang et al., 2010) and made use of a 3-D numerical geodynamo model. The resulting forecast contributed as a candidate SV model to the 11th generation International Geomagnetic Reference Field (IGRF-11, Finlay et al., 2010). While this demonstrated clearly the value of GDA using Gauss coefficients from geomagnetic field models (not direct geomagnetic measurements) and geodynamo models (with non 'Earth-like' dynamo parameters, see Section 21.5), it was also limited in estimating model uncertainties and biases. For example, the forecast errors in the study by Kuang et al. (2010) are approximated by a simple, time-invariant mathematical description (analogous to the Optimal Interpolation scheme in data assimilation), not by the

covariance matrix of a forecast ensemble (as in an EnKF scheme). Fournier et al. (2015) continued the effort to make GDA-based SV forecasts as a candidate model for IGRF-12, with a major improvement in utilising the dynamo solution covariances for the model error statistics. However, the covariances in their assimilation were based only on free-running models, not updates from assimilations (Aubert, 2014).

In the most recent IGRF release (IGRF-13, Alken et al., 2021a), of the fourteen candidate SV models included, four are products of GDA systems using 3-D numerical geodynamos (Minami et al., 2020; Sanchez et al., 2020; Fournier et al., 2021b; Tangborn et al., 2021). Each of these systems make use of an ensemble-based method for assimilation, including the first-ever use of a hybrid variational method (Minami et al., 2020). In addition, observations over the past decades (Minami et al., 2020) and longer (Sanchez et al., 2020; Tangborn et al., 2021) were assimilated. These help produce assimilation solutions which are dynamically consistent over time, as suggested by earlier work with sequential GDA systems (Tangborn and Kuang, 2018; Sanchez et al., 2019). For a more recent description of SV forecasts made by GDA using full dynamo models, we refer the reader to, for example, Fournier et al. (2021a).

The quality of geomagnetic data decreases rapidly when looking back in time, and thus may present the ultimate limitation to the forecast accuracy and dynamic consistency of assimilation solutions from sequential data assimilation systems. But the variational approach by Minami et al. (2020) has the potential to reduce such limitations, as it provides an opportunity to improve earlier geomagnetic data with much more accurate satellite magnetic measurements. All GDA systems face astronomical computational expense if they are to assimilate all available geomagnetic and paleomagnetic data (see Section 21.5 for further discussion).

GDA with self-consistent geodynamo models presents an opportunity to predict the future geomagnetic field over several decades – much longer than the five-year IGRF period. For example, Aubert (2015) showed that forecasts using the dynamic models of GDA systems can outperform linear extrapolations on decadal time scales. This is possible because the estimated 'unobserved' part of the core state is utilised by the dynamic models to predict the future magnetic field. Such forecasts currently predict a continuation of the weakening of the axial dipole, and the expansion and weakening of the SAA over the coming 50–100 years (Aubert, 2015; Sanchez et al., 2020). In particular, as shown in Fig. 21.2, the forecasts predict that a second minimum will grow to split the SAA region in 2025. This highlights a significant contribution of GDA besides simply forecasting the magnetic field: GDA systems are powerful tools for obtaining insight into Earth's deep interior and the origins of SV.

21.4.2 GDA with Alternative Dynamic Models

While the use of full numerical dynamo models in GDA has grown over the last decade, approaches employing alternative models based on various physical, mathematical, and statistical descriptions of the magnetic field continue to be successfully developed and employed. These models are, in general, much simpler than the full dynamo models in both mathematical formulation and numerical simulation. However, they have unique advantages in, for example, allowing for assimilation runs on million-year time scales that would be prohibitively expensive with GDA systems using a full dynamo model. Alternative models can also be useful in testing and validating specific physical and/or mathematical approximations which are applicable to geodynamo models. Such models can be particularly useful as testbeds for advancing GDA methods.

The first such approach to GDA with an alternative model is the geomagnetic forecasts made with magnetic induction via core surface flows (e.g. Maus et al., 2008; Beggan and Whaler, 2009, 2010). The flows used for the induction are also derived from geomagnetic observations. It has been shown that a steady core flow model is capable of producing accurate SV forecasts over five-year periods, and are, in principle, improved 'linear extrapolations' of the SV from current and past geomagnetic observations. These results agree also with studies of Aubert (2015).

Improvement could be made if time-varying core flows are considered, such as the quasi-geostrophic flows in the studies of Pais and Jault (2008), Canet et al. (2009) and Aubert (2015), and the 'inertia-free' core flow of Li et al. (2014). In the former approach, the full dynamo model is replaced by a quasi-geostrophic flow which can be determined without the Lorentz force (i.e. the magnetic field in the outer core). In the latter, the momentum equation becomes 'diagnostic', making it equally numerically stable for both forecast and hindcast, a welcome simplification in, for example, a variational data assimilation approach. In summary, these models allow one to avoid solving the full dynamo equations for GDA, greatly reducing computational demands.

Another approach relies on purely statistical models, such as those of Barrois et al. (2018) and Bärenzung et al. (2020). While these models may not capture the detailed physics of a numerical dynamo, it makes the consideration of multiple magnetic sources outside of the core (e.g. the lithosphere and magnetosphere), computationally tractable, which, in turn, allows for the direct assimilation of geomagnetic measurements (as opposed to geomagnetic field models). Computationally less demanding dynamic models can also make ensemble sizes attainable which would otherwise be impractical with numerical dynamos, resulting in more reliable uncertainty estimates. For example, the statistical model of Bärenzung et al. (2018) permits an EnKF with an ensemble size of $N_e = 40,000$, nearly two orders of magnitude larger than those used in EnKFs with numerical dynamos.

From long-term archeo- and paleomagnetic data, one can find some persistent global scale magnetic features over very long time periods (e.g. Amit et al., 2011; Constable et al., 2016). Thus, an interesting approach is to use low-dimensional models based on either stochastic PDEs (see, e.g., Morzfeld and Buffett, 2019; Pétrélis et al., 2009) or deterministic ODEs (Gissinger, 2012) for long-term SV, such as the behaviour of the axial dipole component of Earth's magnetic field, and for the occurrence of reversals. These models have been used to investigate the predictability of reversals (see, e.g., Gwirtz et al., 2020) including an effort involving the assimilation of paleomagnetic data (Morzfeld et al., 2017). Results from the latter work indicate that assimilations with simplified models of the axial dipole may be useful for anticipating reversals within a window of a few millennia. Limited paleomagnetic data however, makes the validation of reversal prediction strategies a challenge. But it is expected that, as more paleomagnetic data and better low-degree models become available, GDA will also become a powerful tool for predicting very slow secular variations and geomagnetic reversals.

21.4.3 Proxy Models for GDA Development

Simplified models have also played a role in the development of GDA outside of directly being used to make predictions about the Earth. The earliest works concerning GDA involved demonstrating its viability through observing system simulation experiments (OSSEs) with simplified MHD systems (Fournier et al., 2007; Sun et al., 2007; Morzfeld and Chorin, 2012). The dynamic models of those works consisted of 1-D scalar fields intended to represent the magnetic field and fluid velocity of the outer core. While these 'proxy models' were significantly simpler than the geodynamo, they enabled extensive numerical studies of some of the challenges of GDA. This approach of using proxy models has been widely used in the successful development of DA in other applications, such as numerical weather prediction and oceanography. Surprisingly, the pursuit of proxy models for GDA was discontinued until a two-dimensional geomagnetic proxy system (TGPS) recently developed by Gwirtz et al. (2021) was used to study assimilation strategies for use with numerical dynamos.

The TGPS is a magnetoconvection system consisting of 2-D magnetic and velocity fields, on either a plane or spherical surface, which are non-linearly coupled. The right side of Fig. 21.3 shows a snapshot of a solution to the TGPS in a spherical geometry.

The magnetic field and the velocity field are defined by the scalar fields A and ω, respectively (shading in the image of the fields), permitting a complete description of the system state in spherical harmonics, similar to geodynamo models such as MoSST (see Section 21.2). The TGPS was designed to mimic

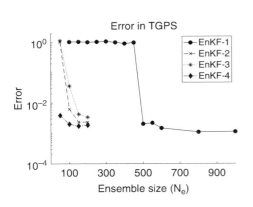

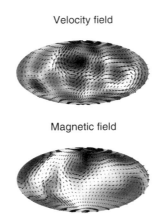

Velocity field

Magnetic field

Figure 21.3 A 2-D proxy model for studying GDA. *On the left* are average forecast errors during OSSEs as a function of ensemble size, for the TGPS in a square geometry when using EnKFs with various modifications (labelled 1–4). The errors are scaled by the average forecast error of a free-running model assimilating no data. *On the top right* is a snapshot of the 2-D velocity field (vectors) and the normal component of vorticity (ω, shading) for the TGPS on a sphere. *On the bottom right* is a snapshot of the 2-D magnetic field (vectors) and the magnetic scalar (A, shading) for the TGPS on a sphere.

the scenario where spherical harmonics, determined from geomagnetic field models, are assimilated into numerical geodynamos. This is accomplished through OSSEs in which only noisy 'observations' of the large length-scale spectral coefficients of A, which defines the magnetic field of the TGPS, are assimilated. As with proxy models for other DA applications, a major advantage of the TGPS is that its computational simplicity allows for a large number of OSSEs which would otherwise be impractical with numerical geodynamos.

Extensive OSSEs with the TGPS have been used to explore and propose assimilation strategies for improving accuracy and reducing the computational demands of operational GDA systems. The left panel of Fig. 21.3 shows the average forecast errors with the TGPS during various OSSEs, as a fraction of the average forecast error of a free-running model using no assimilations. The OSSEs differ only in the particular details of the EnKFs (labelled 1–4) and the size of the ensemble (horizontal axis). The curve determined by the black circles (EnKF-1) relies on the standard stochastic EnKF described in Section 21.3.2, while the others (EnKF 2–4) use various modifications generally known as *localisation* and *inflation* (see, e.g., Kotsuki et al., 2017; Shlyaeva et al., 2019). It can be seen that while the unmodified EnKF-1 requires an ensemble of $N_e = 500$ to reduce forecast errors, similar results can be achieved at reduced ensemble sizes when the modified EnKFs (2–4) are used. These findings support the recent implementation of 'localised' EnKFs in GDA (Sanchez et al., 2019, 2020) and suggest additional EnKF modifications which might be useful in reducing the necessary ensemble size, and therefore computational demands, of GDA systems.

21.5 Geomagnetic Data Assimilation: Challenges and Developments

GDA has advanced greatly in the past decade by utilising knowledge accumulated in other Earth sciences, in particular in numerical weather prediction (NWP), and has been

recognised as a unique tool for geomagnetic forecasting and core-state estimation. But many challenges still remain in areas ranging from observations and physics to mathematical and computational techniques. Future progress in GDA will rely on overcoming these hurdles. Some could be addressed by leveraging knowledge from, for example, NWP; but many others are unique to geomagnetism, the geodynamo, and core dynamics. Among these challenges are GDA system spin-up given the limited availability of high resolution geomagnetic observations from the past; the astronomical computational requirements of GDA systems; and the systematic errors (model biases) arising from large gaps between the dynamo parameter values used in numerical dynamo simulations and those appropriate to the Earth's core.

A major challenge in GDA is the differences between the observed geomagnetic field and the magnetic field from dynamo simulations. EnKF-type assimilation algorithms such as (21.11) assume that forecast errors are random with the zero mean, that is, no systematic error (bias). It also requires that the observations $\mathbf{y}$ and the forecasts $\mathbf{x}^f$ in (21.11) are defined in the same units. But both are difficult to implement in current GDA systems. First, model biases exist because of the large parameter gaps between the parameter values used in numerical dynamo simulation and those appropriate for the Earth's outer core. The dynamo parameters described in Section 21.2, such as the magnetic Rossby number R_o, the Ekman number E, and the modified Prandtl number q_κ, are very small: $R_o \sim 10^{-9}$, $E \sim 10^{-15}$, and $q_\kappa \sim 10^{-6}$ in the outer core if the molecular fluid viscosity, magnetic diffusivity and thermal conductivity are used (e.g. Braginsky and Roberts, 1995). In numerical simulations, they are at least two orders of magnitude larger (e.g. Wicht and Sanchez, 2019), simply due to computational limitations. Numerical dynamo solutions with such large parameter gaps certainly differ from the (unknown) true core dynamic state. This is particularly significant in dynamical processes that are directly related to these small parameters, such as torsional oscillations (waves) in the outer core. These waves are excited to leading order by the balance between the fluid inertia and the Lorentz force (e.g. Braginsky, 1970; Wicht and Christensen, 2010). Thus their typical frequencies are $\sim \mathcal{O}(R_o^{-1/2})$, implying that two orders of magnitude differences in R_o result in one order of magnitude differences in the wave frequencies (and thus the time scales). Since these waves are conjectured to play a major role in explaining the observed sub-decadal geomagnetic SV (e.g. Bloxham et al., 2002; Cox et al., 2016; Aubert and Finlay, 2019), the impacts of the large parameter gaps must be properly addressed for accurate geomagnetic forecasts. Although continuous efforts are made to narrow the parameter gap (Aubert et al., 2017; Aubert and Gillet, 2021), numerical dynamos which are practical for GDA will continue to use parameters which differ significantly from those representative of the Earth in near future.

Model biases also arise from uncertainties in the thermochemical properties of the deep Earth, such as the adiabatic and the total heat fluxes across the CMB (e.g. Nimmo, 2007; Nakagawa, 2020), and the heterogeneity in the inner core and the lower mantle (e.g. Garnero, 2000; Deuss, 2014). The former directly affects the Rayleigh numbers R_{th}, and the latter affects the boundary conditions at the CMB and the inner core boundary (ICB) for numerical dynamo models. It is expected that the parameter gaps will be narrowed and the thermo-chemical uncertainties will be reduced in the coming decades, but will not vanish. The model biases in GDA systems will therefore remain in the foreseeable future.

One possible approach is to estimate the model biases with the asymptotic limits (scaling laws) derived from systematic numerical dynamo simulations with wide ranges of parameter values (e.g. Christensen, 2010; Yadav et al., 2013; Kuang et al., 2017; Petitdemange, 2018). But this could be very difficult, since the numerical asymptotes may not agree with the core state, and since the computational needs for acquiring such asymptotes could be comparable or even higher than those for GDA runs with large ensembles.

Another approach is to rescale $\mathbf{y}$ (or equivalently $\mathbf{x}^f$) based on the properties of the observed field and of the dynamo model used in GDA (e.g. Kuang et al., 2010; Aubert, 2014; Fournier et al., 2015, 2021b; Tangborn et al., 2021). This is perhaps more pragmatic since such rescaling is needed if non-dimensional numerical dynamo models are used in GDA, and since the canonical scaling rules employed in dynamo modelling are inappropriate due to non-'Earth-like' dynamo parameters in simulation. In the approach of Kuang et al. (2010) and Tangborn et al. (2021), the numerical and the observed axial dipole moments are used for the magnetic field rescaling, but the time rescaling remains the same as the canonical time scale of the numerical dynamo model. In the approach of Aubert (2014) and Fournier et al. (2015), the typical time scales of the numerical dynamo solutions and of the observed SV are used for the time rescaling, but the magnetic rescaling relies on the

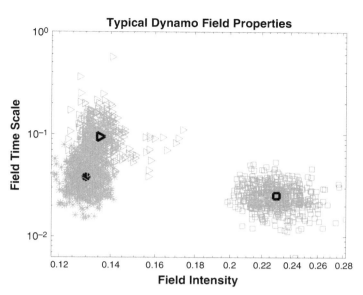

Figure 21.4 The typical intensity (x-axis) and the typical time scale (y-axis) of the ensembles of numerical poloidal magnetic fields with (1) $R_{th} = 1811$, $R_o = E = 1.25 \times 10^{-6}$ (squares); (2) $R_{th} = 1811$, $R_o = E = 6.25 \times 10^{-7}$ (stars); (3) $R_{th} = 905$, $R_o = E = 6.25 \times 10^{-7}$ (triangles). The dark bold-face symbols are the ensemble mean values. The intensities are for the poloidal magnetic field at the top of D″-layer for spherical harmonic degrees $l \leq 13$. Each of the three ensembles consist of 512 snapshots of numerical dynamos selected from large pools of well-developed free-running dynamo solutions obtained with the MoSST core dynamics model (e.g. Jiang and Kuang, 2008).

asymptotic properties derived from independent numerical dynamo simulations. As such, both approaches may lead to inconsistencies between the magnetic rescaling and the time rescaling since, as shown in Fig. 21.4, the typical time scales and the typical intensities of the numerical magnetic fields vary with the dynamo parameters. An immediate development could be to select both the magnetic rescaling and the time rescaling to match the observed and the modelled field intensities and time scales. Consistencies of the scalings and their potential improvements may be tested and validated with various OSSEs.

The GDA forecast spin-up has always been a concern because, as discussed in Section 21.2, the geomagnetic observations are very sparse (the dimension of $\mathbf{y}$ is more than four orders of magnitudes less than that of $\mathbf{x}^f$), and are not 'in-situ' as in many other Earth systems ($\mathbf{y}$ is determined by downward continuing the surface observations to the outer boundary of the dynamo system). The spin-up can be measured by the time evolution of the forecast accuracy $(\text{O-F})^B \equiv \mathbf{y} - \mathbf{H}\mathbf{x}^f$: for an EnKF GDA system, it is expected to decrease in time as more data are assimilated, until it reaches some minimum level (thus the system is fully spun-up). Reaching the minimum level is critical for minimising the SV forecast error which, by definition, can be determined by the field forecast errors at different times:

$$(\text{O-F})^{SV} \equiv \dot{\mathbf{y}} - \mathbf{H}\dot{\mathbf{x}}^f = \frac{1}{\Delta t}\left[(\text{O-F})^B_{t+\Delta t} - (\text{O-F})^B_t\right]. \qquad (21.16)$$

However, previous studies suggest that $(\text{O-F})^B$ may not always decay monotonically in time (e.g. Tangborn and Kuang, 2015, 2018). Though SV forecasts can be improved by some additional calibration of $(\text{O-F})^B$ (Kuang et al., 2010; Tangborn et al., 2021), or by careful selection of the initial ensembles of the inverse dynamo solutions matching earlier geomagnetic observations (Aubert, 2014; Fournier et al., 2015), better solutions could be found by improved assimilation algorithms, such as the rescaling approach discussed in the previous paragraph, and by assimilating other geodynamic observables, such as the length of day variation on decadal time scales which is likely due to angular momentum exchange between the solid Earth and the fluid core (e.g. Jault et al., 1988). It should be noted that the angular momentum $\mathbf{M}_{oc}$ of the outer core (relative to the solid mantle) is the volume integral $\mathbf{M}_{oc} = \int \rho_{oc}\mathbf{r} \times \mathbf{v}dV$ (where ρ_{oc} is the core fluid density).

Development and validation of new assimilation algorithms needed to address the issues outlined here will add to the already very challenging computational needs of GDA. Take a Boussinesq dynamo model for an illustration. As shown in Section 21.2, there are five independent scalar fields in the state vector $\mathbf{x}$ of the dimension $5N_1N_2N_3$ (where $N_{1,2,3}$ are the numerical resolutions in a 3-D dynamo domain). In the MoSST core dynamics model (e.g. Kuang and Bloxham, 1999; Jiang and Kuang, 2008), for example, the resolution is defined by the radial grid points N_x and the truncation order L_x of the spherical harmonic expansions. Thus, in one time step, there are $\sim \mathcal{O}(50 N_x L_x^3 \ln L_x)$ floating point operations (with spherical harmonic transforms). For a modest resolution $N_x = L_x = \mathcal{O}(10^2)$ and time step $\Delta t \sim 10^{-6}$ (typical for dynamo simulations with $R_o, E \sim 10^{-6}$) a total of $\sim 10^{16}$ floating point operations are needed for a dynamo simulation over a magnetic free-decay time. This amounts to ~ 3 h on a tera floating point operations per second (teraflops) computing system (excluding the communication time across computing nodes). The computing needs will be N_e-fold more for GDA runs using ensembles of size N_e. Therefore, there is a need for more efficient assimilation algorithms to make the GDA computing needs bearable, such as those aiming at reducing the necessary ensemble size for EnKF based GDA systems (e.g. Sanchez et al., 2019; Gwirtz et al., 2021). Development of proxy models, such as those of Canet et al. (2009) and Gwirtz et al. (2021), are of particular importance for advancing GDA, as they can provide dynamically complex, but computationally economical platforms for at least early stage proof-of-concept studies of assimilation algorithms and physical approximations.

21.6 Discussion

In this chapter, we have provided an overview of geomagnetic data assimilation (GDA), including some basics of geomagnetic observations, geodynamo models, and assimilation methodologies. We have also presented a wide range of GDA results in understanding core dynamical processes, interpreting observed SV, and geomagnetic forecasting. In addition, we have elaborated on some of the challenges in GDA and possible pathways to move forward. As such, this chapter serves as a quick and comprehensive introduction for those who wish to learn GDA and/or work on GDA-related research and applications.

We would like to point out a particular useful application of the proxy models described in Section 21.4. Since these models are mathematically simple and computationally affordable, they are very handy for teaching/learning GDA. Compared to any full geodynamo model, these models are easy to analyse. In particular, simulation and assimilation (with these models) can be completed quickly on desktops and laptops, thus making them ideal in, for example, student projects.

While we have made an effort to include representative GDA results and developments, this chapter does not cover all GDA activities, in part due to the page limit and the rapid development of GDA in recent years. For example, the description of variational geomagnetic data assimilation is very brief in this chapter, and we refer the reader to relevant references for more details. Regardless, this should not affect the main purpose of this chapter, which is to provide a comprehensive understanding of geomagnetic data assimilation.

References

Alken, P., Thébault, E., Beggan, C. D. et al. (2021a). International Geomagnetic Reference Field: The thirteenth generation. *Earth, Planets and Space*, 73, 49. https://doi.org/10.1186/s40623-020-01288-x.

Alken, P., Chulliat, A., and Nair, M. (2021b). NOAA/NCEI and University of Colorado candidate models for IGRF-13. *Earth, Planets and Space*, 73, 44. https://doi.org/10.1186/s40623-020-01313-z.

Amit, H., Korte, M., Aubert, J., Constable, C., and Hulot, G. (2011). The time-dependence of intense archeomagnetic flux patches. *Journal of Geophysical Research: Solid Earth*, 116, B12106. https://doi.org/10.1029/2011JB008538.

Aubert, J. (2014). Earth's core internal dynamics 1840–2010 imaged by inverse geodynamo modelling. *Geophysical Journal International*, 197, 1321–34.

Aubert, J. (2015). Geomagnetic forecasts driven by thermal wind dynamics in the Earth's core. *Geophysical Journal International*, 203(3), 1738–51.

Aubert, J., and Finlay, C. C. (2019). Geomagnetic jerks and rapid hydromagnetic waves focusing at Earth's core surface. *Nature Geoscience*, 12, 393–8.

Aubert, J., and Fournier, A. (2011). Inferring internal properties of Earth's core dynamics and their evolution from surface observations and a numerical geodynamo model. *Nonlinear Processes in Geophysics*, 18, 657–74.

Aubert, J., and Gillet, N. (2021). The interplay of fast waves and slow convection in geodynamo simulations nearing Earth's core conditions. *Geophysical Journal International*, 225(3), 1854–73.

Aubert, J., Gastine, T., and Fournier, A. (2017). Spherical convective dynamos in the rapidly rotating asymptotic regime. *Journal of Fluid Mechanics*, 813, 558–93.

Bärenzung, J., Holschneider, M., Wicht, J., Lesur, V., and Sanchez, S. (2020). The Kalmag model as a candidate for IGRF-13. *Earth, Planets and Space*, 72(163).

Bärenzung, J., Holschneider, M., Wicht, J. Sanchez, S., and Lesur, V. (2018). Modeling and predicting the short-term evolution of the geomagnetic field. *Journal of Geophysical Research: Solid Earth*, 123(6), 4539–60.

Barrois, O., Hammer, M. D., Finlay, C. C., Martin, Y., and Gillet, N. (2018). Assimilation of ground and satellite magnetic measurements: inference of core surface magnetic and velocity field changes. *Geophysical Journal International*, 215(1), 695–712.

Beggan, C. D., and Whaler, K. A. (2009). Forecasting change of the magnetic field using core surface flows and ensemble Kalman filtering. *Geophysical Research Letters*, 36, L18303. https://doi.org/10.1029/2009GL039927.

Beggan, C. D., and Whaler, K. A. (2010). Forecasting secular variation using core flows. *Earth, Planets and Space*, 62, 821–28.

Bloxham, J, Zatman, S., and Dumburry, M. (2002). The origin of geomagnetic jerks. *Nature*, 420, 65–68.

Bonavita, M., Isaksen, L., and Hólm, E. (2012). On the use of EDA background error variances in the ECMWF 4D-Var. *Quarterly Journal of the Royal Meteorological Society*, 138(667), 1540–59.

Braginsky, S. I. (1970). Torsional magnetohydrodynamic vibrations in the Earth's core and variation in day length. *Geomagnetism and Aeronomy*, 10, 1–8.

Braginsky, S. I., and Roberts, P. H. (1995). Equations governing convection in Earth's core and the geodynamo. *Geophysical and Astrophysical Fluid Dynamics*, 79, 1–97.

Brown, M., Korte, M., Holme, R., Wardinski, I., and Gunnarson, S. (2018). Earth's magnetic field is probably not reversing. *PNAS*, 115, 5111–16.

Brown, W. J., Beggan, C. D., Cox, G. A., and Macmillan, S. (2021). The BGS candidate models for IGRF-13 with a retrospective analysis of IGRF-12 secular variation forecasts. *Earth, Planets and Space*, 73 (42). https://doi.org/10.1186/s40623-020-01301-3.

Buehner, M., McTaggart-Cowan, R., and Heilliette, S. (2017). An Ensemble Kalman filter for numerical weather prediction based on variational data assimilation: VarEnKF. *Monthly Weather Review*, 145(2), 617–35.

Cande, S. C., and Kent, D. V. (1995). Revised calibration of the geomagnetic polarity timescale for the late Cretaceous and Cenozoic. *Journal of Geophysical Research*, 100, 6093–6095.

Canet, E., Fournier, A., and Jault, D. (2009). Forward and adjoint quasigeostrophyic models of geomagnetic secular variations. *Journal of Geophysical Research*, 114, B11101.

Chorin, A. J., and Morzfeld, M. (2013). Conditions for successful data assimilation. *Journal of Geophysical Research: Atmospheres*, 118 (20), 11522–33.

Christensen, U. R. (2010). Dynamo scaling laws and applications to the planets. *Space, Science, Reviews*, 152, 565–90.

Christensen, U.R., Aubert, J., Cardin, P. et al. (2001). A numerical dynamo benchmark. *Physics of the Earth and Planetary Interiors*, 128(1), 25–34.

Chulliat, A., and Maus, S. (2014). Geomagnetic secular acceleration, jerks, and localized standing wave at the core surface from 2000 to 2010. *Journal of Geophysical Research: Solid Earth*, 119, 1531–43.

Constable, C., Korte, M., and Panovska, S. (2016). Persistent high paleosecular variation activity in southern hemisphere for at least 10000 years. *Earth and Planetary Science Letters*, 453, 78–86.

Courtier, P. (1997). Variational methods. *Journal of the Meteorological Society of Japan*, 75(1B), 211–18.

Cox, G. A., Livermore, P. W., and Mound, J. E. (2016). The observational signature of modelled torsional waves and comparison to geomagnetic jerks. *Physics of the Earth and Planetary Interiors*, 255, 50–65.

Deuss, A. (2014). Heterogeneity and anisotropy of Earth's inner core. *Annual Review of Earth and Planetary Sciences*, 42, 103–26.

Doglioni, C., Pignatti, J., and Coleman, M. (2016). Why did life develop on the surface of the Earth in the Cambrian? *Geoscience Frontiers*, 7, 865–75.

Evensen, G. (2006). *Data assimilation: The ensemble Kalman filter*. Springer.

Finlay, C. C., and Jackson, A. (2003). Equatorially dominated magnetic field change at the surface of the Earth's core. *Science*, 300, 2084–6.

Finlay, C. C., Maus, S., Beggan, C. D. et al. (2010). International Geomagnetic Reference Field: The eleventh generation. *Geophysical Journal International*, 183(3), 1216–30.

Finlay, C. C., Kloss, C., Olsen, N. et al. (2020). The CHAOS-7 geomagnetic field model and observed changes in the South Atlantic Anomaly. *Earth, Planets and Space*, 72, 156. https://doi.org/10.1186/s40623-020-01252-9.

Fournier, A., Eymin, C., and Alboussier, T. (2007). A case for variational geomagnetic data assimilation: Insights from a one-dimensional, nonlinear, and sparsely observed MHD system. *Nonlinear Processes in Geophysics*, 14, 163–80.

Fournier, A., Aubert, J., and Thébault, E. (2011). Inference on core surface flow from observations and 3-D dynamo modelling. *Geophysical Journal International*, 186, 118–36.

Fournier, A., Nerger, L., and Aubert, J. (2013). An ensemble Kalman filter for the time-dependent analysis of the geomagnetic field. *Geochemistry, Geophysics, Geosystems*, 14(10), 4035–43. https://doi.org/10.1002/ggge.20252.

Fournier, A., Aubert, J., and Thébaut, E. (2015). A candidate secular variation model for IGRF-12 based on Swarm data and inverse geodynamo modeling. *Earth, Planets and Space*, 67. https://doi.org/10.1186/s40623-015-0245-8.

Fournier, A., Aubert, J., Lesur, V., and Thébault, E. (2021a). Physics-based secular variation candidate models for the IGRF. *Earth, Planets and Space*. https://doi.org/10.1186/s40623-021-01507-z.

Fournier, A., Aubert, J., Lesur, V., and Ropp, G. (2021b). A secular variation candidate model for IGRF-13 based on Swarm data and ensemble inverse geodynamo modeling. *Earth, Planets and Space*. https://doi.org/10.1186/s40623-020-01309-9.

Fournier, A. G., Hulot, G., Jault, D. et al. (2010). An introduction to data assimilation and predictability in geomagnetism. *Space Science Reviews*. https://doi.org/10.1007/s11214-010-9669-4.

Garnero, E. J. (2000). Heterogeneity of the lowermost mantle. *Annual Review of Earth and Planetary Sciences*, 28, 509–37.

Gissinger, C. (2012). A new deterministic model for chaotic reversals. *European Physical Journal B*, 85, 137.

Glatzmaier, G. A., and Roberts, P. H. (1995). A three-dimensional convective dynamo solution with rotating and finitely conducting inner core and mantle. *Physics of the Earth and Planetary Interiors*, 91, 63–75.

Gwirtz, K, Morzfeld, M, Fournier, A, and Hulot, G. (2020). Can one use Earth's magnetic axial dipole field intensity to predict reversals? *Geophysical Journal International*, 225(1), 277–97.

Gwirtz, K., Morzfeld, M., Kuang, W., and Tangborn, A. (2021). A testbed for geomagnetic data assimilation. *Geophysical Journal International*, 227, 2180–203.

Heirtzler, J. R., Allen, J. H., and Wilkinson, D. C. (2002). Everpresent South Atlantic Anomaly damages spacecraft. *Eos, Transactions American Geophysical Union*, 83(15), 165–9.

Hirose, K., Labrosse, S., and Hernlund, J. (2013). Composition and state of the core. *Annual Review of Earth and Planetary Sciences*, 41(1), 657–91.

Holme, R. (2007). Large-scale flow in the core. In P. Olson, ed., *Treatise on Geophysics: Vol. 8. Core Dynamics*. Amsterdam: Elsevier, pp. 107–30.

Huder, L., Gillet, N., Finlay, C. C., Hammer, M. D., and Tchoungui, H. (2020). COV-OBS.x2: 180 years of geomagnetic field evolution from ground-based and satellite observations. *Earth, Planets and Space*, 72(160). https://doi.org/10.1186/s40623-020-01194-2.

Hulot, G., Eymin, C., Langlais, B., Mandea, M., and Olsen, N. (2002). Smallscale structure of the geodynamo inferred from Oested and Magsat satellite data. *Nature*, 416, 620–3. https://doi.org/10.1038/416620a.

Hulot, G., Lhuillier, F., and Aubert, J. (2010). Earth's dynamo limit of predictability. *Geophysical Research Letters*, 37(6). https://doi.org/10.1029/2009GL041869.

Hunt, B. R., Kostelich, E. J., and Szunyogh, I. (2007). Efficient data assimilation for spatiotemporal chaos: A local ensemble transform Kalman filter. *Physica D*, 230(1), 112–26.

Jackson, A. (2003). Intense equatorial flux spots on the surface of the Earth's core. *Nature*, 424, 760–63.

Jackson, A., Jonkers, A. R. T., and Walker, M. R. (2000). Four centuries of geomagnetic secular variation from historical records. *Philosophical Transactions of the Royal Society of London. Series A: Mathematical, Physical and Engineering Sciences*, 358(1768), 957–90.

Jault, D., Gire, C., and LeMouël, J.-L. (1988). Westward drift, core motions and exchanges of angular momentum between core and mantle. *Nature*, 333, 353–6.

Jiang, W., and Kuang, W. (2008). An MPI-based MoSST core dynamics model. *Physics of the Earth and Planetary Interiors*, 170(1), 46–51.

Jones, C. A., Boronski, P., Brun, A. et al. (2011). Anelastic convection-driven dynamo benchmarks. *Icarus*, 216, 120–35.

Kageyama, A., and Sato, T. (1997). Generation mechanism of a dipole field by a magnetohydrodynamic dynamo. *Physical Review E*, 55, 4617–26.

Kaji, C. V., Hoover, R. C., and Ragi, S. (2019). Underwater navigation using geomagnetic field variations. *2019 IEEE International Conference on Electro Information Technology*. https://doi.org/10.1109/EIT.2019.8834192 of:

Kloss, C, and Finlay, C. C. (2019). Time-dependent low-latitude core flow and geomagnetic field acceleration pulses. *Geophysical Journal International*, 217(1), 140–68.

Kotsuki, S., Ota, Y., and Miyoshi, T. (2017). Adaptive covariance relaxation methods for ensemble data assimilation: experiments in the real atmosphere. *Quarterly Journal of the Royal Meteorological Society*, 143(705), 2001–15.

Kuang, W., and Bloxham, J. (1997). An Earth-like numerical dynamo model. *Nature*, 389, 371–4.

Kuang, W., and Bloxham, J. (1999). Numerical Modeling of Magnetohydrodynamic Convection in a Rapidly Rotating Spherical Shell: Weak and Strong Field Dynamo Action. *J. Comput. Phys.*, 153(1), 51–81.

Kuang, W., and Chao, B. F. (2003). Geodynamo Modeling and Core-Mantle Interactions. In V. Dehant, K. Creager, S. Karato, and S. Zatman, eds., *Earth's Core: Dynamics, Structure, Rotation, Geodynamics Series 31*. Washington, DC: American Geophysical Union (AGU), pp. 193–212.

Kuang, W., and Tangborn, A. (2015). Dynamic responses of the Earth's outer core to assimilation of observed geomagnetic secular variation. *Progress in Earth and Planetary Science*, 2. https://doi.org/10.1186/s40645-015-0071-4.

Kuang, W., Tangborn, A., Jiang, W. (2008). MoSST– DAS: The first generation geomagnetic data assimilation framework. *Communications in Computational Physics*, 3, 85–108.

Kuang, W., Tangborn, A., Wei, Z., and Sabaka, T. J. (2009). Constraining a numerical geodynamo model with 100 years of surface observations. *Geophysical Journal International*, 179(3), 1458–68, https://doi.org/10.1111/j.1365–246X.2009.04376.x.

Kuang, W., Wei, Z., Holme, R., and Tangborn, A. (2010). Prediction of geomagnetic field with data assimilation: a candidate secular variation model for IGRF-11. *Earth, Planets and Space*, 62, 775–85.

Kuang, W., Chao, B. F., and Chen, J. (2017). Decadal polar motion of the Earth excited by the convective outer core from geodynamo simulations. *Journal of Geophysical Research: Solid Earth*, 122(10), 8459–73.

Langel, R. A., and Estes, R. H. (1982). A geomagnetic field spectrum. *Geophysical Research Letters*, 9, 250–3.

Larmor, J. (1919). How could a rotating body such as the Sun become a magnet? *Reports of the British Association*, 87, 159–60.

Lesur, V., Wardinski, I., Hamoudi, M., and Rother, M. (2010). The second generation of the GFZ internal magnetic model: GRIMM-2. *Earth, Planets and Space*, 62, 765–73.

Li, K., Jackson, A., and Livermore, P. W. (2011). Variational data assimilation for the initial-value dynamo problem. *Physical Review E*, 84. https://doi.org/10.1103/PhysRevE.84.056321.

Li, K., Jackson, A., and Livermore, P. W. (2014). Variational data assimilation for a forced, inertia-free magnetohydrodynamic dynamo model. *Geophysical Journal International*, 199, 1662–76.

Liu, D., Tangborn, A., and Kuang, W. (2007). Observing system simulation experiments in geomagnetic data assimilation. *Journal of Geophysical Research*, 112. https://doi.org/10.1029/2006JB004691.

Lowrie, W., and Kent, D. V. (2004). Geomagnetic polarity time scale and reversal frequency regimes. *Timescales of the paleomagnetic field*, 145, 117–29.

Mandea, M., and Korte, M., eds. (2011). *Geomagnetic Observations and Models*. Dordrecht: Springer.

Mandea, M., Holme, R., Pais, A. et al. (2010). Geomagnetic jerks: Rapid core field variations and core dynamics. *Space Science Reviews*, 155, 147–75.

Matsui, H., Heien, E., Aubert, J. (2016). Performance benchmarks for a next generation numerical dynamo model. *Geochemistry, Geophysics, Geosystems*, 17(5), 1586–607.

Maus, S., Silva, L., and Hulot, G. (2008). Can core-surface flow models be used to improve the forecast of the Earth's main magnetic field? *Journal of Geophysical Research*, 113, B08102. https://doi.org/10.1029/2007JB005199.

Minami, T., Nakano, S., Lesur, V. et al. (2020). A candidate secular variation model for IGRF-13 based on MHD dynamo simulation and 4DEnVar data assimilation. *Earth, Planets and Space*, 72, 136. https://doi.org/10.1186/s40623–020–01253–8.

Morzfeld, M., and Buffett, B. A. (2019). A comprehensive model for the kyr and Myr timescales of Earth's axial magnetic dipole field. *Nonlinear Processes in Geophysics*, 26(3), 123–42.

Morzfeld, M., and Chorin, A. J. (2012). Implicit particle filtering for models with partial noise, and an application to geomagnetic data assimilation. *Nonlinear Processes in Geophysics*, 19 (3), 365–82.

Morzfeld, M., Fournier, A., and Hulot, G. (2017). Coarse predictions of dipole reversals by low-dimensional modeling and data assimilation. *Physics of the Earth and Planetary Interiors*, 262, 8–27.

Nakagawa, T. (2020). A coupled core-mantle evolution: review and future prospects. *Progress in Earth and Planetary Science*, 7. https://doi.org/10.1186/s40645–020–00374–8.

Nilsson, A., Suttie, N., Korte, M., Holme, R., and Hill, M. (2020). Persistent westward drift of the geomagnetic field at the core-mantle boundary linked to recurrent high-latitude weak/reverse flux patches. *Geophysical Journal International*, 222, 1423–32.

Nimmo, F. (2007). Energetics of the core. In P. Olson, ed., *Treatise on Geophysics: Vol. 8. Core Dynamics*. Amsterdam: Elsevier, pp. 31–66.

Ogg, J. G. (2012). Geomagnetic polarity time scale. In F. M. Gradstein, J. G. Ogg, M. Schmitz, and G. Ogg, eds., *The Geologic Time Scale 2012*. Amsterdam: Elsevier Science, pp. 85–113.

Olsen, N., and Mandea, M. (2008). Rapidly changing flows in the Earth's core. *Nature Geoscience*, 1, 390–94.

Pais, M. A., and Jault, D. (2008). Quasi-geostrophyic flows responsible for the secular variation of the Earth's magnetic field. *Geophysical Journal International*, 173, 421–43.

Panovska, S., Korte, M., and Constable, C. G. (2019). One hundred thousand years of geomagnetic field evolution. *Reviews of Geophysics*, 57(4), 1289–337.

Petitdemange, L. (2018). Systematic parameter study of dynamo bifurcations in geodynamo simulations. *Physics of the Earth and Planetary Interiors*, 277, 113–32.

Pétrélis, F., Fauve, S., Dormy, E., and Valet, J.-P. (2009). Simple mechanism for reversals of Earth's magnetic field. *Physical Review Letters*, 102, 144503.

Roberts, P. H., and Scott, S. (1965). On analysis of the secular variation. *Journal of Geomagnetism and Geoelectricity*, 17, 137–51.

Roberts, P. H, and King, E. M. (2013). On the genesis of the Earth's magnetism. *Reports on Progress in Physics*, 76(9), 096801.

Sabaka, T. J., Tøffner-Clausen, L., Olsen, N., and Finlay, C. C. (2020). CM6: A comprehensive geomagnetic field model derived from both CHAMP and Swarm satellite observations. *Earth, Planets and Space*, 72, 80.

Sanchez, S., Fournier, A., Aubert, J., Cosme, E., and Gallet, Y. (2016). Modeling the archaeomagnetic field under spatial constraints from dynamo simulations: A resolution analysis. *Geophysical Journal International*, 207, 983–1002.

Sanchez, S., Wicht, J., Bärenzung, J., and Holschneider, M. (2019). Sequential assimilation of geomagnetic observations: Perspectives for the reconstruction and prediction of core dynamics. *Geophysical Journal International*, 217, 1434–50.

Sanchez, S., Wicht, J., Bärenzung, J., and Holschneider, M. (2020). Predictions of the geomagnetic secular variation based on the ensemble sequential assimilation of geomagnetic field models by dynamo simulations. *Earth, Planets and Space*, 72, 157. https://doi.org/10.1186/s40623–020–01279–y.

Schaeffer, N., Lora Silva, E., and Pais, M. A. (2016). Can core flows inferred from geomagnetic field models explain the Earth's dynamo? *Geophysical Journal International*, 204(2), 868–77.

Shlyaeva, A., Whitaker, J. S., and Snyder, C. (2019). Model-space localization in serial ensemble filters. *Journal of Advances in Modeling Earth Systems*, 11(6), 1627–36.

Sun, Z., and Kuang, W. (2015). An ensemble algorithm based component for geomagnetic data assimilation. *Terrestrial Atmospheric and Oceanic Sciences*, 26, 53–61.

Sun, Z., Tangborn, A., and Kuang, W. (2007). Data assimilation in a sparsely observed one-dimensional modeled MHD system. *Nonlinear Processes in Geophysics*, 14, 181–92.

Talagrand, O., and Courtier, P. (1987). Variational assimilation of meteorological observations with the adjoint vorticity equation. I: Theory. *Quarterly Journal of the Royal Meteorological Society*, 113(478), 1311–28.

Tangborn, A., and Kuang, W. (2015). Geodynamo model and error parameter estimation using geomagnetic data assimilation. *Geophysical Journal International*, 200, 664–75.

Tangborn, A., and Kuang, W. (2018). Impact of archeomagnetic field model data on modern era geomagnetic forecasts. *Physics of the Earth and Planetary Interiors*, 276, 2–9.

Tangborn, A., Kuang, W., Sabaka, T. J., and Yi, C. (2021). Geomagnetic secular variation forecast using the NASA GEMS ensemble Kalman filter: A candidate SV model for IGRF-13. *Earth, Planets and Space*, 73, 47. https://doi.org/10.1186/s40623–020–01324–w.

Tippett, M. K., Anderson, J. L., Bishop, C. H., Hamill, T. M., and Whitaker, J. S. (2003). Ensemble square root filters. *Monthly Weather Review*, 131, 1485–90.

Wicht, J., and Christensen, U. R. (2010). Torsional oscillations in dynamo simulations. *Geophysical Journal International*, 181, 1367–80.

Wicht, J., and Sanchez, S. (2019). Advances in geodynamo modelling. *Geophysical & Astrophysical Fluid Dynamics*, 113(1–2), 2–50.

Yadav, R. K., Gastine, T., and Christensen, U. R. (2013). Scaling laws in spherical shell dynamos with free-slip boundaries. *Icarus*, 225, 185–93.

Zhang, M., and Zhang, F. (2012). E4DVar: Coupling an ensemble Kalman filter with four-dimensional variational data assimilation in a limited-area weather prediction model. *Monthly Weather Review*, 140(2), 587–600.

22

Pointwise and Spectral Observations in Geomagnetic Data Assimilation: The Importance of Localization

Sabrina Sanchez

Abstract: Geomagnetic data assimilation aims at constraining the state of the geodynamo working at the Earth's deep interior by sparse magnetic observations at and above the Earth's surface. Due to difficulty separating the different magnetic field sources in the observations, spectral models of the geomagnetic field are generally used as inputs for data assimilation. However, the assimilation of raw pointwise observations can be relevant within certain configurations, specifically with paleomagnetic and historical geomagnetic data. Covariance localisation, which is a key ingredient to the assimilation performance in an ensemble framework, is relatively unexplored, and differs with respect to spectral and pointwise observations. This chapter introduces the main characteristics of geomagnetic data and magnetic field models, and explores the role of model and observation covariances and localisation in typical assimilation set-ups, focusing on the use of 3D dynamo simulations as the background model.

22.1 Introduction

The Earth's magnetic field has displayed many intriguing features throughout its past. During the last decade, the North magnetic pole has rapidly migrated towards Siberia (Livermore et al., 2020), unmatched by its Southern counterpart. The South Atlantic Anomaly (SAA, a region where the surface magnetic field is abnormally weak) has been drifting westwards, deepening and increasing in complexity (Amit et al., 2021). Moreover, the axial dipole (the largest scale of the magnetic field) has been continuously decreasing though the past century (Finlay, 2008) – from –32 μT in year 1850 to about –29 μT in 2020. The understanding of these variations can be greatly improved by geomagnetic data assimilation (GDA), which has been thriving during the past decade. Not only can GDA provide reanalyses of shallow and deep processes in the Earth's interior, it also offers the potential to increase the quality of predictions of the magnetic field. An

overview of GDA can be found in Kuang et al. (this volume), as well as in Fournier et al. (2010).

Unlike many other applications of DA in geosciences, in geomagnetism, observations are only available at a great distance from its main source, the Earth's core. The core is located beneath the Earth's mantle, at 2,886 km below the surface, as shown in Fig. 22.1. In its outer part, among a liquid alloy of iron and nickel at extremely high temperatures and pressures, the geomagnetic field is induced by the interaction with the electrically conducting flow, via a self-sustained dynamo process. The geodynamo's magnetic field evolves within a variety of timescales: rapid magnetic variations with interannual frequencies connected to wave dynamics atop the core (Gillet et al., 2021), decadal variations associated to flow convection (like the changes in the North pole and SAA mentioned before), polarity reversals within many thousand years, and variations in this frequency through the Earth's history (Amit et al., 2010; Valet and Fournier, 2016). Such rich temporal behaviour can be witnessed through different observations of the magnetic field.

The geomagnetic field is currently observed at a high level of accuracy by satellite missions (notably by *Swarm*, Friis-Christensen et al., 2006), with a high spatial and frequency coverage. In addition, ground observatories provide close to continuous time series of the magnetic field, some reaching back 150 years. Previous to that, only historical records are available, beginning around 1450s. Beyond the historical period is the domain of archeo- and paleomagnetism, which consider the measuring of the magnetic field locked in rocks and artefacts from the time of their formation. The observations, of course, get sparse in space and time through the past, and the errors and uncertainties affecting them grow accordingly. Given the heterogeneity of the data sets, often inverse spectral models are used to compare field characteristics through time; for instance, the CHAOS model (Finlay et al., 2020) is used as reference for the satellite epoch, and the CALS family of models (Korte et al., 2011) provides a reference for the paleomagnetic era. The drastic changes in magnetic data have prevented the construction of global magnetic field models bridging different epochs (with a few exceptions, e.g. Arneitz et al., 2017).

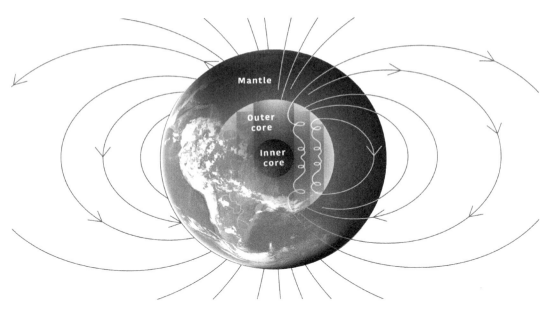

Figure 22.1 Basic structure of the Earth's interior superimposed with the magnetic field lines originating from the geodynamo in the outer core.

Although the natural path for GDA would be to employ 3D magneto-hydrodynamic (MHD) dynamo simulations as background models, many successful attempts have instead used reduced stochastic models of induction at the core–mantle boundary (CMB). This choice allows for a more portable set-up and simplification of the model dynamics. For instance, Barrois et al. (2018), Baerenzung et al. (2020), and Ropp et al. (2020) have used ensemble Kalman filters (EnKF) to assimilate pointwise observatory and satellite data with stochastic models. The main goal has been to provide a reanalysis of field and/or flow at the CMB, as well as short-term field forecasts. On the other side of the spectrum, simpler stochastic models have also been used in paleomagnetism. Morzfeld et al. (2017) employed particle filters to assimilate virtual axial dipole moment (reconstruction of the first harmonic degree of the magnetic field by paleomagnetic data) models in order to attempt predicting field reversals and excursions.

Nonetheless, 3D dynamo simulations might be advantageous within the gap between the data-sparse paleomagnetic period and the data-dense satellite era. In this case, data availability might be enough to properly propagate information from the surface observations to the deep core interior, at least to the larger scales of the system. Although efforts have been made in this direction (e.g. Kuang et al., 2009; Minami et al., 2020), most studies are related to testing and therefore working with synthetic set-ups (e.g. Liu et al., 2007; Fournier et al., 2013). Moreover, in most cases of data assimilation with dynamo simulations spectral field models are conveniently employed instead of

pointwise observations. However, field models only show a smooth picture of the geomagnetic field, and might also bear potential biases, which the assimilation algorithm should account for (Tangborn and Kuang, 2015). The possibility of assimilating pointwise observations with a dynamo model background has been scarcely explored, but showed promising results (Sanchez, 2016).

In this chapter, we describe the main characteristics of the magnetic field observations through the past millennium (Section 22.2), dynamo simulations and their relevant aspects for DA (Section 22.3), and present a synthetic GDA exercise using an EnKF (Section 22.4). In the latter, particular attention is given to the difference in assimilating spectral and pointwise observations, as well as the importance of covariance localisation.

22.2 Geomagnetic Data

Observations of the magnetic field are either given in terms of a complete set or individual components of the magnetic field. As shown in Fig. 22.2a, the North, East and downwards components are traditionally known as X, Y, and Z, respectively. The field is also often given in terms of the following nonlinear components: F the total field intensity, H the horizontal field intensity, D the declination, and I the inclination. In the rest of this section, we list and describe the main subcategories of magnetic field data, but the interested reader can find more in-depth discussions in Matzka et al. (2010).

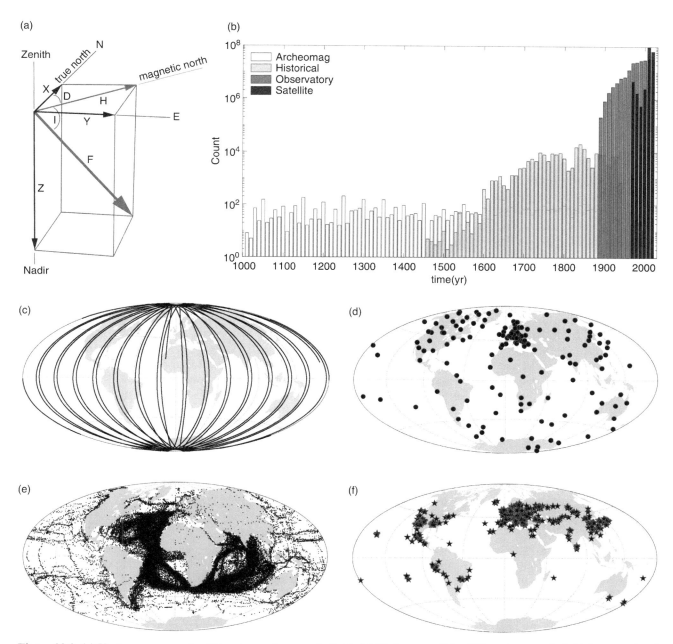

Figure 22.2 (a) Vector components of the observed magnetic field. (b) Histogram of the different magnetic field data sets for the past millennium separated in bins of ten years (observatory and satellite data from hourly-means). (c) Coverage of the Swarm Alpha satellite over the period of one day. (d) Network of geomagnetic observatories from INTERMAGNET. (e) Distribution of historical data between 1450 AD and 1800 AD from HISTMAG (Arneitz et al., 2017). (f) Geographical distribution of the paleo- and archeomagnetic data sets for the period 1000 AD to 2000 AD from Geomagia (Brown et al., 2015). (c)–(f) are shown in Hammer-Aitoff projection.

22.2.1 Modern Era

The geomagnetic field has been monitored from space as early as 1957 by Sputnik, and since then, five main satellite missions have undertaken the task: POGO (1965–71), Magsat (1979–80), Ørsted (1999–2005), CHAMP (2000–10), SAC-C (2001–4). Figure 22.2b shows the temporal distribution of these data. Since the end of 2013, the Swarm mission uses three satellites to cover the low orbit of the Earth, measuring the geomagnetic field at a 1 Hz frequency with an unprecedented accuracy of 2 nT for vector components and 45 pT for field intensity (Fratter et al., 2016). Although satellite data are the most precise and have the best spatial coverage among the available geomagnetic data catalogs, as seen in Fig. 22.2c, they are also the most subjected to space weather. Space weather refers to external phenomena linked to the interaction of the solar activity and the Earth's magnetic field creating geomagnetic disturbances.

The external field is mainly generated by electric currents flowing through and in between the magnetosphere and ionosphere. 'Quiet' days can show disturbances of a few nT, depending on local time and latitude, while geomagnetic 'storms' can easily reach variations of several hundreds, sometimes thousands, of nT (Mandea and Chambodut, 2020). These fields can also induce electric currents at and under the Earth's surface, generating internal fields of external origin. The separation of the magnetic field sources contributing to the observed signal becomes extremely important, even more so due to the orbital emplacement of the satellites between the ionosphere and magnetosphere.

At the Earth's surface, an extensive network of magnetic observatories is responsible for constantly recording the geomagnetic field. Some observatories reach very long time series, with durations of about a few centuries. The INTERMAGNET database offers past and real time data of high standard geomagnetic observatories.[1] The observatories provide fair coverage of the Earth's surface, despite a clustering over Europe and obvious lower density at the oceans, as shown in Fig. 22.2d. Currently, observatory data are given every second, with a typical precision level of about 5 nT (Reda et al., 2011). To compensate for the uneven distribution of observatories, the database is extended by 'repeat stations', where every few years the field at a given location is measured under similar conditions. Finally, regional data from land and sea magnetic surveys are also available, but are mostly used to study the smaller scale crustal magnetic field.

Although less subjected to space-weather phenomena, ground-based observations are also strongly affected by the external field. When interested in decreasing external field sources, core field modellers often perform strict data selections, like discarding day-time data (very affected by the ionospheric signal), data at times of high geomagnetic indexes (related to magnetic storms), and data from above a certain latitude (to decrease the influence of aurora-related interference). Such a strong data exclusion could be potentially avoided by developing better external field models, notably through space-weather DA. Shprits et al. (this volume), explores, for instance, DA in the near-Earth space environment.

22.2.2 Historical Period

Scattered measurements of the geomagnetic field already existed previous to the sixteenth century, but the Age of Discovery was a turning point (Jonkers, 2003). During the Great Navigations, the magnetic field was of utter importance for geolocalization, hence the number of observations increased exponentially until the nineteenth century, as seen in Fig. 22.2b. The boom in data coincided with diverse mathematical breakthroughs, when science quickly shifted from qualitative to more quantitative. However, magnetic field data was often recorded incomplete, and mostly in term of its directional components. At the beginning, only the magnetic declination (D) provided by the compass was recorded, followed then by the dip of the magnetic field (inclination, I) within the early seventeenth century. It was only in 1832 that a method for measuring the magnetic field absolute intensity (F) was first developed.

Although land surveys were also recorded, most of the historical measurements were performed by mariners, so the data distribution is highly concentrated at sea, mostly over popular routes around the Atlantic and Indian oceans, as seen in Fig. 22.2e. The database HISTMAG provides an easily accessible compilation of the magnetic data from the historical era (Arneitz et al., 2017).[2] Uncertainties are rarely provided within the historical data, but estimates suggest that they can reach reasonably high measurement precision, of about 1.0° for the angles D and I and 0.2 μT for F (Arneitz et al., 2017). A major source of uncertainties within the navigational data corresponds to the position. Although latitudes were accurately calculated astronomically, the longitude was calculated through 'dead reckoning', which could accumulate to substantial errors on the course of the enterprise (Jackson et al., 2000; Arneitz et al., 2017).

22.2.3 Paleomagnetism

Going further back into the past greatly limits the observability of the geodynamo. However, valuable information can be gained by indirect observations of the ancient magnetic field. Certain rocks and human artefacts with rich iron content, such as lava flows and bricks, can record, to some extent, the ambient magnetic field at the time of their formation/fabrication. The process through which these objects can lock in a magnetic field is known as thermal remnant magnetisation (TRM). Thermal remnant magnetisation can be achieved as the magnetic moments of a sample, which are scattered at high temperatures, align with the ambient field when lowered beneath a critical point, the Curie temperature (Merril et al., 1996). If the sample position at the time of cooling down is known, the field directions can be retrieved. Otherwise only the intensity can be estimated. Observations of TRM are examples of paleomagnetic (or archeomagnetic when only artefacts are considered) data, and can be very useful to geomagnetic field modelling when accompanied by reliable dating of the

---

[1] https://intermagnet.github.io/.   [2] https://cobs.zamg.ac.at/data/index.php/en/models-and-databases/histmag.

samples. Other rocks can also bear magnetisation, like sediments and stalagmites, although through a different process known as detrital remnant magnetisation.

Paleomagnetic data distribution in space and time is very uneven and sparse, respectively. As shown in Fig. 22.2f, for observations from volcanic rocks and artefacts, there is a strong clustering around Europe, and in general very little data is available in the Southern Hemisphere. Also, as seen in Fig. 22.2b, the availability of data from the mentioned data set reaches about 6 observations per year on average through the past millennium. Previous to that, data count decreases to about 4 data per year from 0 AD to 1000 AD and to roughly 1 datum every 15 years from 1000 BC to 0 AD. However, there is a continuous effort of the paleomagnetic community to increase the database (Panovska et al., 2019).

Typical paleomagnetic data uncertainties fall to about 2.5°, 2°, and 4 μT for declination, inclination, and intensity, respectively. Although many data are provided with their respective uncertainty, disagreement between protocols and mismatch with neighbouring data often make modellers suspect the uncertainties to be untrustworthy. In that case, data with small uncertainties are often penalised during the modelling strategy (Korte et al., 2009; Licht et al., 2013), which can in some cases mean the loss of valuable accurate data. Lastly, one of the greatest sources of uncertainties in paleomagnetic data is their timing. Age estimations through dating methods (archaeological, radiocarbon, among others) can easily bear uncertainties of many decades, even centuries in some cases.

Although all geomagnetic data are subjected to external and internal field contributions which are not originated from the geodynamo, the relatively large error budget of paleomagnetic data easily accommodates these potential signals. Separating external field sources is therefore, in principle, dispensable, though not impossible.

22.2.4 Spectral Field Models
The Earth's magnetic field can be expressed as the gradient of potential fields of external and internal origins,

$$\mathbf{B}(r, \theta, \phi, t) = -\nabla[V_{int}(r_o, \theta, \phi, t) + V_{ext}(r_o, \theta, \phi, t)], \quad (22.1)$$

where the subscripts refer to the potential V components. Within a region free of field sources, the potentials satisfy Laplace's equation, and can be written in terms of a spherical harmonic (SH) basis. For the internal potential,

$$V_{int}(r, \theta, \phi, t) = r_o \sum_{\ell=1}^{\infty} \sum_{m=0}^{\ell} \left(\frac{r_o}{r}\right)^{\ell+1} [g_\ell^m(t) \cos m\phi$$
$$+ h_\ell^m(t) \sin m\phi] P_\ell^m(\cos\theta), \quad (22.2)$$

where ℓ and m are, respectively, the degree and order of the SH expansion, g and h are the so-called Gauss coefficients, and P_l^m are the Schmidt quasi-normalised associated Legendre functions. A similar equation can be derived for the external field, but with a different radial dependency. In order to model the magnetic field potential, a certain truncation L must be sought. There exist a number of models spanning different epochs, of which the most representative and well-established are the CHAOS models (Finlay et al., 2020) for the satellite era, the COV-OBS models (Huder et al., 2020) for satellite and observatory, the *gufm1* model (Jackson et al., 2000) based on historical and observatory data, and the CALS models (Korte et al., 2011) based on paleomagnetic data. These models share the same underlying characteristics, and mostly differ in their temporal regularisation.

Depending on the data set used, the models reach different resolutions. The temporal resolution of field models ranges from a few decades to yearly variations (Korte and Constable, 2008). In space, paleomagnetic field models' resolution can range from SH degree $\ell = 3$ to 5 through the past three millennia (Sanchez et al., 2016). In the modern era, core field models are usually considered to be resolved up until SH degree 14. In fact, $\ell = 14$ is often taken as the upper bound in geomagnetic field model resolution, not due to the data accuracy and sampling, but to the masking of smaller length scales by the crustal magnetic field, which is also of internal origin. However, recent models currently employing DA have managed to better separate core and crustal field sources, so that the core field truncation can be expanded. For instance, the Kalmag model (Baerenzung et al., 2020) achieves a theoretical resolution of $\ell = 20$. The separation can be better visualised by the field's power spectrum, given by

$$W_{int}(\ell) = (\ell + 1)\left(\frac{r_o}{r}\right)^{2\ell+4} \sum_{m=0}^{\ell} [(g_\ell^m)^2 + (h_\ell^m)^2], \quad (22.3)$$

and shown in Fig. 22.3a. The spectrum of the internal field at the surface decreases linearly at the same time that the crustal field increases up until SH degree 15. Beyond these length scales, the crustal field dominates, and the uncertainties in the core field reach the same magnitude as the signal. Assuming the mantle is electrically insulating, the potential field can be downward continued up until its source region, the CMB. When extrapolated at the CMB, the core spectrum is close to 'white', and the crustal field diverges, as shown in Fig. 22.3b.

The core field morphology is therefore very different at the Earth's surface and at the CMB. As shown in Fig. 22.3c, the radial surface field is smooth, and the most prominent features are the wavy magnetic equator, the high latitude flux lobes, and the region of low intensity underneath the SAA. As seen in Fig. 22.3d, the core field at the CMB is much more complex, particularly regarding the reverse magnetic flux patches underneath the SAA region.

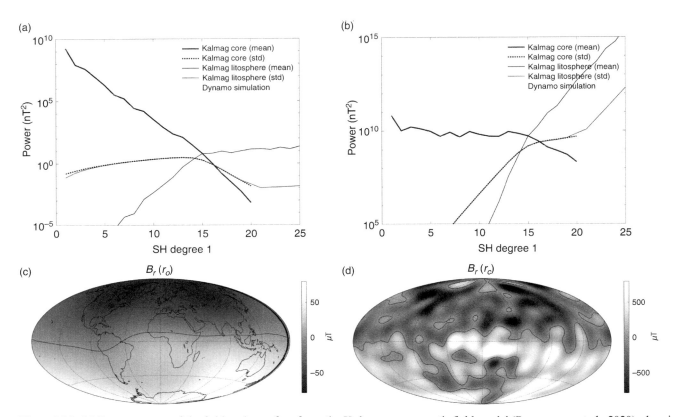

Figure 22.3 (a) Power spectra of the field at the surface from the Kalmag geomagnetic field model (Baerenzung et al., 2020), showing core and litospheric components, as well as the average spectra of a dynamo simulation described in Section 22.3. (b) The same power spectra as in (a), but evaluated at the CMB. (c) Radial component of the core magnetic field from Kalmag at the Earth's surface in 2020. (d) Downward continuation of (c) to the CMB.

As previously mentioned, spectral field models are often used as input data in GDA for simplicity, since the background models are often built on a spectral set up. Also, in such case there is no need to deal with separation of magnetic field sources or potential inconsistencies between data and their errors. However, although field models are nowadays usually provided with uncertainties, many popularly used field models are not (e.g. Jackson et al., 2000; Finlay et al., 2020). Since data uncertainties are key in DA, this can become a major drawback, and hinder the interpretation of results. Moreover, spectral field models can also bear biases issued from regularisation and data selection, but these issues remain relatively unexplored (Johnson and Constable, 1997; Tangborn and Kuang, 2018, to cite a couple of paleomagnetic examples). In principle, assimilating pointwise data with dynamo simulations is feasible, at least within a synthetic framework (Sanchez, 2016). Application to real data remains to be seen.

22.3 Dynamo Simulations

Simulations of the geodynamo solve for the interaction between the magnetic field $\vec{B}$ and flow $\vec{V}$ within a convecting spherical shell of thickness D rotating

with angular velocity Ω around a $\hat{z}$ axis. Convection is driven by gravity acting on density ρ variations, and is described as disturbances around an adiabatic hydrostatic background state. Both temperature and composition are coupled in a single variable C, the codensity (Braginsky and Roberts, 1995). The Boussinesq approximation is generally applied, where variations in density are only considered through the buoyancy force. The equations being solved are then the induction equation

$$\frac{d\vec{B}}{dt} = \nabla \times (\vec{V} \times \vec{B}) + \frac{E}{Pm} \nabla^2 \vec{B}, \qquad (22.4)$$

the Navier-Stokes equation

$$\frac{d\vec{V}}{dt} + \vec{V} \cdot \nabla \vec{V} + 2\hat{z} \times \vec{V} = -\nabla p$$
$$+ \text{Ra} \frac{\vec{r}}{r_c} C + (\nabla \times \vec{B}) \times \vec{B} + E \nabla^2 \vec{V}, \qquad (22.5)$$

the codensity equation

$$\frac{dC}{dt} + V \cdot \nabla C = \frac{E}{Pr} \nabla^2 C + S, \qquad (22.6)$$

and the continuity equations for magnetic and flow fields

$$\nabla \cdot \vec{B} = 0; \; \nabla \cdot \vec{V} = 0, \qquad (22.7)$$

where r_c is the outer shell boundary, p is the non-hydrostatic pressure, and S is a source/sink term representing the slow secular cooling and/or accommodation of light elements emanated from the growing inner core. The equations here have been non-dimensionalised, resulting in the following non-dimensional control parameters: the Ekman number $E = \nu/(\Omega D^2)$, the Prandtl number $Pr = \nu/\kappa$, the magnetic Prandtl number $Pm = \nu/\lambda$, and the modified Rayleigh number $Ra = g_c F/4\pi\rho\Omega^3 D^4$ (Aubert et al., 2009). Further, g_c is the gravity at the CMB, F the mass anomaly flux, and ν, κ, and λ the kinetic, thermal, and magnetic diffusivities, respectively. Boundary conditions can differ depending on the set-up sought, and the interested reader can find more details about dynamo simulations in Roberts and King (2013) and Christensen and Wicht (2015).

Due to the solenoidal nature of flow and field, dynamo simulations usually employ the poloidal-toroidal composition. For the magnetic field this writes

$$\vec{B} = \nabla \times \nabla \times [\mathcal{P}(r)\vec{r}] + \nabla \times [\mathcal{T}(r)\vec{r}], \qquad (22.8)$$

where $\mathcal{P}$ and $\mathcal{T}$ are, respectively, the poloidal and toroidal scalar potentials. Further, the potentials are expanded in fully normalised SH basis, which, for the poloidal scalar, is given as

$$\mathcal{P}(r) = \sum_{\ell=1}^{L_{max}} \sum_{m=0}^{\ell} \mathcal{P}_\ell^m(r) Y_\ell^m(\theta,\phi) + c.c., \qquad (22.9)$$

where $Y_\ell^m(\theta,\phi)$ are the complex-valued, fully normalised SH functions and $c.c.$ refers to the complex conjugate. As previously mentioned, only the surface field can be directly observed, and a downward continuation of the field to the top of the core is possible considering that the mantle is insulating. Here $\mathcal{P}(r_c)$ is, therefore, the sole observable of the system. A transformation from fully normalised Eq. (22.9) to Schmidt normalised coefficients in Eq. (22.2) can be found in Sanchez et al. (2016).

22.3.1 Earth's Core Conditions
The parameters mentioned previously can be written in terms of the typical time scales of the system, most importantly the magnetic diffusion time scale τ_λ, the viscous diffusion time scale τ_ν, the flow advection time τ_V, and the rotation time τ_Ω. The Ekman number can be written as $E = \tau_\Omega/\tau_\nu$ and is extremely small under Earth's core conditions, $E \sim \mathcal{O}(10^{-15})$, due to the rapid rotation and nearly inviscid fluid. The magnetic Prandtl number, $Pr = \tau_\lambda/\tau_\nu \sim \mathcal{O}(10^{-6})$, exposes the dominance of ohmic dissipation in the system, and therefore the scale separation between a small-scale flow and a large-scale magnetic field. This is also reflected in diagnostic parameters, such as the

hydrodynamic and magnetic Reynolds number, respectively $Re = VD/\nu$ and $Rm = VD/\lambda$. Flow inversions close to the CMB suggest $V \approx 14$ km/yr, which results in $Re \sim \mathcal{O}(10^9)$, revealing a high level of turbulence in the core. However, $Rm \sim \mathcal{O}(10^3)$ showing a more moderate level of magnetic turbulence.

22.3.2 'Down-to-Earth' Conditions
The core-like conditions mentioned in the previous section are unreachable in simulations, despite the great computational advances through the past decade. Affordable dynamo simulations tend to work within $E > 10^{-5}, Pr \sim 0.1-1, Pm > 0.1$. However, many of them surprisingly manage to reproduce some of the main characteristics of the geomagnetic field, mostly in terms of its morphology. In fact, simulations with $Rm > 10^2$ are seen to be 'Earth-like', provided that $E/Pm < 10^{-4}$ (Christensen et al., 2010). So to promote a more Earth-like magnetic field, the considerably large E in the simulations can be compensated by increasing Pm. Figure 22.3a,b shows the field mean spectrum from a rescaled dynamo simulation of moderate complexity, considering $E = 3\times10^{-4}$, $Pr = 1$ and $Pm = 5$ and $Ra = 5\times10^{-6}$ (more details can be found in Sanchez et al., 2019). The spectrum follows the general tendency displayed by the data-based models over the observable scales.

Reproducing the observed dynamics of the geomagnetic field is much more challenging. The secular variation (SV) time scale, $\tau_{SV} = B/\dot{B}$, is generally used for the comparison between data and simulations. Although decadal time-scale variations in the simulations are in line with observations (Lhuillier et al., 2011), some specific features and interannual variations are still unmatched. As an example, dynamo simulations do not present, in general, the westward drift suggested by the observed SV during the past century (a notable exception can be found in Aubert et al., 2013). Recently, high-end simulations have shown that faster, interannual dynamics start to manifest when control parameters sufficiently approach core conditions (Schaeffer et al., 2017). The crucial requirement for reaching such conditions seems to be a sufficient energetic separation between the first-order force balance between ageostrophic Coriolis, Lorentz, and buoyancy forces (known as the MAC balance), and viscous and inertial forces (Aubert et al., 2017). Under such configuration, hydromagnetic waves seem to manifest in appropriate time scales, sometimes imprinting jerk-like signals in the magnetic field at the surface (Aubert and Finlay, 2019).

22.4 Geomagnetic Data Assimilation
The dynamo models discussed in the previous section are complex, highly non-linear systems of high

dimensionality. Because of this, ensemble sequential methods have been prioritised in dynamo-based GDA. In particular, the EnKF (Evensen, 1994) has been used in a number of studies (Fournier et al., 2013; Sanchez et al., 2019; Tangborn et al., 2021, to cite a few). An example of an EnKF with synthetic data will be given in this section.

22.4.1 Observations

For the simulation mentioned in Section 22.3, the state vector **x** comprising all the simulation fields has a size of $N_{\mathbf{x}} \approx 2.6 \times 10^6$. Observations, however, are much smaller – typical observation size for spectral field models consist of $N_{\mathbf{y}} \approx 100$. With pointwise data this can change drastically depending on the data type involved (Fig. 22.2b). In a DA formulation, the observations **y** at a given time i can be written as

$$\mathbf{y}_i = \mathcal{H}_i(\mathbf{x}_i^t) + \epsilon_i, \qquad (22.10)$$

where $\mathcal{H}$ is an observation functional and ϵ is a vector of observation errors. The observation error covariance is given by $\mathbf{R}_i$.

As mentioned in Section 22.2, only the large-scale core magnetic field can be observed at the surface, due to concealment by the crustal magnetic field. For spectral observations, the observation operator is linear, **H**, and simply screens out the poloidal field $P(r_c)$ truncated to a certain observable degree L_o (with attention to SH normalisation if conversion is needed for Gauss coefficients). The observed radial field can be expressed in the spatial domain as

$$B_r^o(r_c, \theta, \phi) = \sum_{\ell=1}^{L_o} \sum_{m=0}^{\ell} \frac{\ell(\ell+1)}{r_c} \mathcal{P}_\ell^m(r_c) Y_\ell^m(\theta, \phi). \qquad (22.11)$$

Figure 22.4a shows $B_r^{o(rc)}$ from a snapshot of the chosen dynamo simulation truncated to SH degree $L_o = 10$.

Pointwise measurements of the magnetic field at the Earth's surface can be linked to the full field at the CMB $B_r(r_c)$ through a data kernel. For instance, the Z component can be written as

$$Z(r_o, \theta_o, \phi_o) = \int_S G_Z(r_o, \theta_o, \phi_o | r_c, \theta, \phi) B_r(r_c, \theta, \phi) dS, \qquad (22.12)$$

Where G_Z is the Green function corresponding to a Z observation, and similarly to X and Y measurements (Constable et al., 1993). Directional and intensity data are nonlinear functions of the magnetic field vector, that can be either calculated using the X, Y, and Z field components or by linearised kernels (Sanchez et al., 2016). Figure 22.4b shows the Z component of the field from Fig. 22.4a at the Earth's surface. Superposed is an observation grid with $N_o = (L_o + 1)^2 - 1$ points, which corresponds to the L_o resolution, nearly equally distributed over the sphere.

22.4.2 Ensemble Kalman Filter

The EnKF works with the common forecast-and-analysis steps of the Kalman filter (Kalman, 1960), but within an ensemble framework (Evensen, 1994; Burgers et al., 1998). After a given initialisation, a forecast by the model $\mathcal{M}$ is performed by an ensemble of size N_e, so that for every ensemble member e,

$$\mathbf{x}_{i,e}^f = \mathcal{M}(\mathbf{x}_{i-1,e}^f), \qquad (22.13)$$

where the superscript denotes the forecast. Whenever observations are available (Eq. 22.10), an update is performed:

$$\mathbf{x}_{i,e}^a = \mathbf{x}_{i,e}^f + \mathbf{K}_i(\mathbf{y}_{i,e} - \mathbf{H}_i \mathbf{x}_{i,e}^f), \qquad (22.14)$$

where the superscript denotes the analysis,

$$\mathbf{K}_i = \mathbf{P}_i^f \mathbf{H}_i^\dagger (\mathbf{H}_i \mathbf{P}_i^f \mathbf{H}_i^\dagger + \mathbf{R}_i)^{-1} \qquad (22.15)$$

is the Kalman gain, and $\mathbf{P}^f$ is the forecast error covariance. Due to the ensemble, **P** can be approximated by the sample covariance. Moreover, instead of calculating the error covariance, the EnKF allows the direct calculation of $\mathbf{P}_i^f \mathbf{H}_i^\dagger$ and $\mathbf{H}_i \mathbf{P}_i^f \mathbf{H}_i^\dagger$ in Eq. (22.15), which are much more portable (e.g. Fournier et al., 2013).

Due to the computational cost of dynamo simulations, it is of course of interest to lower the ensemble size to a minimum while maintaining the filter performance. However, the EnKF is known to underperform when the ensemble size is too small in comparison with the model size. In particular, a too-small ensemble tends to introduce spurious correlations, while at the same time underestimating the variances. The former problem greatly impacts the analysis step, since the observable information is carried to the hidden state through the covariance. For instance, if a given correlation between observed and hidden variable is overestimated, the update of the hidden variable will also be overestimated, which will be accompanied by a decrease in the variance of the variable. Apart from degrading the quality of the assimilation, this issue can lead in some situations to a catastrophic filter divergence (Gottwald and Majda, 2013).

22.4.3 Covariances

Given a sequence of well-spaced snapshots, one can easily compute the sample covariance for the observed magnetic field in spectral space. Figure 22.4c shows the absolute normalised covariance (therefore the correlation matrix) of the large-scale poloidal magnetic field at the upper boundary $\mathcal{P}_\ell^m(r_c)$ for an ensemble of size $N_e = 512$. The matrix is organised by sections of the same order m, within which follow the different degrees ℓ. The correlations from the magnetic field are clustered around the diagonal, in blocks

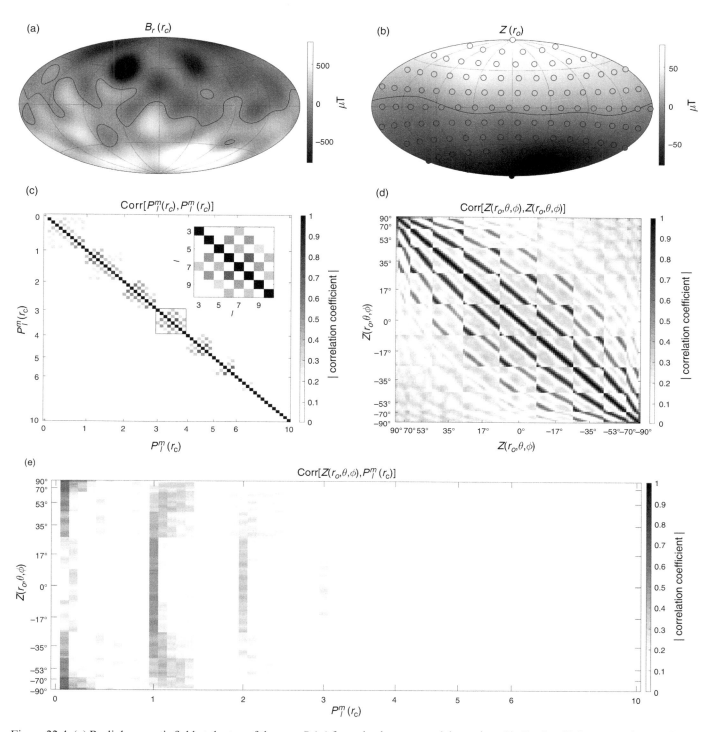

Figure 22.4 (a) Radial magnetic field at the top of the core $B_r(r_c)$ from the dynamo model mentioned in Section 22.3 truncated to SH degree $L_0 = 10$. (b) Vertical component of the magnetic field at the surface $Z(r_o)$ from the same dynamo model, with a uniformly distributed grid of Z observations shown by circles. (c) Spectral correlation matrix, showing the cross-correlations of the poloidal magnetic field atop the core $P_\ell^m(r_c)$ truncated at $L_0 = 10$, where the axes are organised through consecutive blocks of the same wave number m. In the insert, the $m = 3$ block is zoomed in showing correlations between SH degrees $\ell = 3$ to 10. (d) Correlations between the Z magnetic field component of the grid shown in (b), where the axes labels correspond to the grid latitudes. (e) Correlations between the Z pointwise observations and the spectral coefficients $P_\ell^m(r_c)$.

of the same wave number m. Within these blocks, there is a clear checkerboard pattern, as highlighted in the inset of the $m = m' = 3$ block. Similar configurations appear in the cross-covariance between field, flow, or codensity.

The likely reason behind the m-block clustering is that the correlations reflect the interaction between flow and field around an axi-symmetric background magnetic field (corresponding to the average field state around which the covariance is calculated). The nonlinear coupling between flow and field via the induction Eq. (22.4) transfers energy between modes following certain selection rules (Bullard and Gellman, 1954). For instance, when a background axisymmetric field is considered ($m = 0$), a flow with a certain m wave number would induce field within the same m. The checkerboard pattern in Fig. 22.4c stem from symmetry constraints. Equations (22.4) and (22.5) feature 'curl' and 'cross' products, operations that when combined leave the equatorial symmetry unchanged. It follows, for instance, that an equatorially symmetric flow acting on equatorially antisymmetric magnetic field produces more equatorially antisymmetric field (Gubbins and Zhang, 1993). Symmetries can be generalised in terms of $\ell + m$, even or odd contributions from poloidal and toroidal flow and field formalism (Sanchez et al., 2019, 2020).

The correlations in the spatial domain are considerably different. Figure 22.4d shows the correlation matrix of a grid of Z observations at the surface (Fig. 22.4b), from the same previously described ensemble. The pattern reveals strong local correlations between neighbouring sites, but also weaker, large-scale negative correlations elsewhere (not shown). The patterns look considerably different for the X and Y field components. Figure 22.4e shows the correlation between the Z observations and the poloidal field at the CMB $P_\ell^m(r_c)$. It is clear that, near the poles, Z data are quite sensitive to the axisymmetric modes, while, closer to the equator, modes with $m = 1$ and higher play a bigger role. Figure 22.4c,d corresponds to representations of $\mathbf{HP}^f\mathbf{H}^\dagger$, while Fig. 22.4e shows a subset of $\mathbf{HP}^f$ corresponding to correlations of the observations solely with the large scale $P_\ell^m(r_c)$. As we can see through Eq. (22.15), $\mathbf{HP}^f$ is crucial for propagating observational information to the hidden state.

22.4.4 Localisation
One way to mitigate the effects of ensemble size is to apply covariance localisation. As the name suggests, this technique aims at reducing the effects of long-range spurious correlations beyond a certain distance to the observation site (Hamill et al., 2001). Covariance localisation is typically described by

$$\mathbf{P}^{f'} = \mathbf{C} \circ \mathbf{P}^f, \tag{22.16}$$

where $\mathbf{C}$ is a 'mask' containing elements from 0 to 1 and $\circ$ is the element-wise matrix product. As mentioned in Section 22.4.2, the forecast covariance is not directly computed, so, instead, it is practical to apply localisation in the form

$$\mathbf{K}' = [\mathbf{C} \circ (\mathbf{P}^f\mathbf{H}^\dagger)][\mathbf{C} \circ (\mathbf{HP}^f\mathbf{H}^\dagger) + \mathbf{R}]^{-1}. \tag{22.17}$$

The application of such localization is reasonably straightforward in thin-shell systems, but is problematic for the Earth's core. Due to the thickness of the domain and the strong rotation of the Earth, core flows are expected to be nearly geostrophic (show small departures along the rotation axis, as sketched in Fig. 22.1). This, summed with the spectral nature of the simulations and the observations (in the case of spectral field models), motivates to consider covariance localisation in the spectral domain. In the case of spectral observations, one can take advantage of the m-block clustering and checkerboard pattern described in the previous section and shown in Fig. 22.4c. As shown in Sanchez et al. (2020), a suitable spectral localisation matrix can be described through the following relationship between SH degree and order

$$C(\ell, m, k; \ell', m', k') = \delta_m^{m'} \delta_{\mathrm{mod}(m+\ell,2)}^{\mathrm{mod}(m'+\ell',2)}, \tag{22.18}$$

where mod is an operator defining the remainder of the division of the two arguments. Its application for $\mathbf{HP}^f\mathbf{H}^\dagger$ and $\mathbf{P}^f\mathbf{H}^\dagger$ in Eq. (22.17) is straightforward, due to the spectral nature of both observation and state vector. Throughout this chapter, Eq. (22.18) is coined ML- localisation.

For pointwise observations no localisation formalism has been proposed yet. Although a localisation based on the distance between observations would be of easy application to $\mathbf{HP}^f\mathbf{H}^\dagger$ (Fig.22.4d), the connection between pointwise observations and spectral field in $\mathbf{P}^f\mathbf{H}^\dagger$ (Fig.22.4e) is not straightforward, and different for each field component.

22.4.5 Synthetic Experiments
In order to closely assess the impact of the magnetic observations and the previously discussed localisation, we resort to observing system simulation experiments (OSSEs), in particular to twin experiments. An ensemble of N_e snapshots of a long run from the dynamo model discussed in Section 22.3.2 is used as the initial condition for the assimilating ensemble. The observations are constructed from a nature run of the same dynamo model, but from a time interval different than the initial ensemble. In order to build up a synthetic test which serves as a compromise between paleomagnetic and historical periods described in Section 22.2, we prepare observations of the field with a $\Delta t = 10$-yr frequency with a spatial resolution corresponding to SH degree $L_o = 10$.

The observations are given either in spectral form or as the three vector field components within a grid of uniformly distributed sites over the surface sphere. A spatial representation of the field observations at the CMB truncated to $L_o = 10$ is shown in Fig. 22.4a and the Z surface grid data are shown in Fig. 22.4b. The observation window spans a period of 1,000 years. The observation uncertainties are obtained from the standard deviation of the field over twice as long a period, and multiplied by an arbitrary factor to correspond to, on average, 10% of the observed field.

22.4.6 Results

The impact of the assimilation strategies in the OSSEs is accessible by comparing their normalised error. The error ε is calculated as the RMS difference between the analysis and the true state, from which the observations were created, and the normalisation consists of the RMS value of the true state. Figure 22.5a,b shows the normalised error of the radial magnetic field at the CMB, separated into the large scales (filtered to $L = 10$) and the fully resolved field, respectively. Also shown are

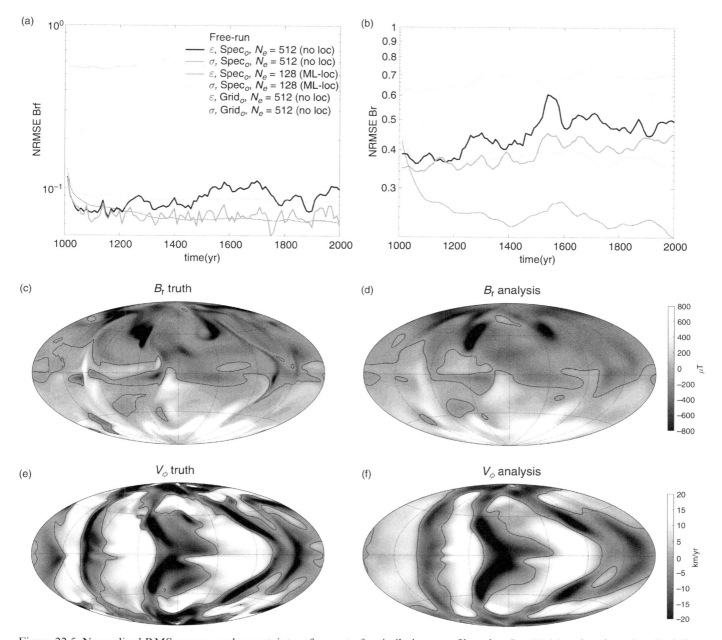

Figure 22.5 Normalised RMS errors ε and uncertainty σ for a set of assimilation tests filtered to $L = 10$ (a) and evaluated at the full truncation $L = 85$ (b). Spec or grid relate to the type of observations assimilated, N_e is the ensemble size and 'loc' refers to the localisation used. (c) and (e) show the true state and (d) and (f) the analysis for the last cycle of the 'Spec$_0$, $N_e = 128$, ML-loc' assimilation test in terms of the radial magnetic field and azimuthal flow at and underneath the CMB, respectively.

the standard deviation σ of the estimates, which work as proxy for the uncertainties. The tests are compared to an error upper bound, corresponding to the free-run case, where no observations are assimilated. The results of the DA test with a large ensemble using spectral observations is shown by the black curve ('Spec$_o$, $N_e = 512$, no loc'). The larger-scale errors remain relatively small, but the full field shows they increase while the uncertainties decrease. This behaviour is known as filter divergence, in this case diagnosed by insufficient ensemble size, creating spurious correlation in the covariance.

The employment of a covariance localisation brings a substantial improvement, as seen by the dark grey curve in Fig. 22.5a,b. In this case, the assimilation uses the ML covariance localisation introduced in Section 22.4.4. Not only is the ensemble reduced to $N_e = 128$, but the error is decreased, roughly matching the uncertainty level. However, the improvement is not perfect and towards the end of the assimilation both begin to slightly diverge. The divergence between error and uncertainty is better seen in hidden variables (such as the flow underneath the CMB, not shown). Although there certainly is still room for improvement, the results are reassuring, as suggested by Fig. 22.5c–f. They compare the estimated flow and field beneath and at the CMB with the true state for the last analysis of the assimilation case 'Spec$_o$, $N_e = 128$, (ML-loc)', for which the corresponding observation is shown in Fig. 22.4a. Both field and flow are well reconstructed, and most differences rely on the smaller length scales. Although observations are large scale (Fig. 22.4a), the analysis displays small-scale field and flow features, which is possible due to the model statistics (covariance between large and small scales) and dynamics.

The assimilation of pointwise observation is shown in Fig. 22.5a,b through the light grey curve, for an ensemble of $N_e = 512$. Since no localisation is used, the case suffers from similar filter divergence issues as the corresponding case with spectral observations. However, a big difference between the two cases is seen with respect to the large-scale field error and uncertainty levels. The difference owns to the fact that the spectral observations perform a direct mapping of the model, and instead, the pointwise observations indirectly map the whole set of spectral coefficients at the CMB (Eqs. 22.11 and 22.12). The pointwise data assimilation updates, in a more distributed way, the length scales relative to the spectral observation case. This points to a great advantage of assimilating pointwise observations in comparison to spectral field models, but also shows how it is more challenging to localise. Unfortunately, no localisation strategy has been developed yet for the assimilation of pointwise observations.

22.5 Discussion

The results shown in the previous section raise the question of whether it is worthwhile to assimilate pointwise observations, when spectral models seem to be simpler to implement and yield better results in synthetic tests (Fig. 22.5). Although there is no definitive answer yet, it is important to point out that spectral observations come with certain disadvantages. The field proposed by the spectral models is usually too smooth in space and time, mostly concerning the paleomagnetic and historical periods (Korte and Constable, 2008), and often not provided with uncertainties which are crucial for a reliable DA scheme. Also, different data selection criteria are employed in the construction of the field models, such as the assignment of lower bounds in the paleomagnetic data uncertainties (e.g. Licht et al., 2013) that might not be required in GDA. Finally, GDA using pointwise observations offers a good opportunity to incorporate data from both paleomagnetic and historical periods, which are rarely combined in the usual inverse framework due to their very different spatial and temporal distribution (one exception is the BIGMUDI model by Arneitz et al., 2017).

As seen in Fig. 22.5a,b, the assimilation of pointwise observations shows signs of filter divergence, associated with insufficient ensemble size. Although this problem can be strongly mitigated by covariance localisation, this venue remains yet unexplored within the pointwise GDA set-up. Though localisation in grid space in $\mathbf{HP}^f\mathbf{H}^\dagger$ might be simple and based on the distance between observations (Fig. 22.4d), the localisation of the spectral-pointwise covariance $\mathbf{P}^f\mathbf{H}^\dagger$ (Fig. 22.4e) remains elusive. Strategies such as 'shrinkage' proposed by Gwirtz et al. (2021) might provide a practical solution, whereby every non-diagonal covariance matrix is damped by a certain factor. Yet, this abrupt localisation might lead to a loss of useful covariance information.

Covariance localisation is indeed essential in ensemble-based DA, as mentioned in Sections 22.4.3 and 22.4.4. In the spectral DA tests, the ensemble size could be decreased by four while diminishing errors and mending, to a large extent, the filter divergence. However, as seen in Fig. 22.5b, error and uncertainty tend to slightly diverge still showing the extra need to deal with ensemble size problems. Further localisation (over depth, for instance) and 'inflation' are therefore needed to address this issue. The latter consists of increasing the ensemble variances by a certain factor, which allows to counterbalance variance underestimation typical of small ensemble sizes (Anderson and Anderson, 1999). However, in typical implementations, the inflation factor has to be tuned depending on each assimilation set-up, as it depends on model and ensemble size and observation type and uncertainties. So a systematic parameter search has to be performed in order to find an efficient inflation for every assimilation set-up (Gwirtz et al., 2021).

22.6 Conclusions

Magnetic field observations provide an invaluable window through which to study the Earth's core dynamics. Covering the past millennium, the very heterogeneous catalog of observations is composed of direct measurements, such as satellite, magnetic observatory data, and early records from the historical period. Indirect observations are also available in the form of paleomagnetic data. Geomagnetic DA offers great potential to model the Earth's magnetic field through this diverse data set. In particular, using dynamo simulations as background models can, in principle, allow the reanalysis of the core state deep beneath the CMB over decadal time scales (Fournier et al., 2013; Sanchez et al., 2019). Most efforts have favoured spectral field models instead of point-wise data as observation input for the DA algorithms, since the simulations themselves inhabit a spectral domain (Kuang et al., 2010; Aubert, 2015; Tangborn et al., 2021).

The spectral covariance shows signatures reminiscent of selection rules between SH degrees and modes, as well as symmetry pairing, from the magnetic induction process. The spectral ML localisation not only improves the GDA performance, but also allows an important decrease in ensemble size since it eliminates many spurious correlations in the covariance (Sanchez et al., 2020). However, further localisation or inflation should still be considered in order to eliminate remaining divergences between error and uncertainties, mostly in the unobserved state.

Despite the simplicity, the assimilation of spectral models might over-smooth field signals due to strong regularisation and biases, mostly over the paleomagnetic period (Korte and Constable, 2008). Assimilating pointwise observations can be particularly fitting to bridge the late paleomagnetic and historical period, even more so since the time window of a few millennia is ideal to investigate decadal to centennial scale convection dynamics. Within such a period, the data error budget surpasses external field variations, making field source separation in principle unnecessary. However, no localisation strategy has been considered as yet for point-wise geomagnetic observations, so that large ensembles have to be used. Toy models such as the one presented in Gwirtz et al. (2021) can provide an excellent test-bed for assimilating pointwise observations into spectral models.

References

Amit, H., Leonhardt, R., and Wicht, J. (2010). Polarity reversals from paleomagnetic observations and numerical dynamo simulations. *Space Science Reviews*, 155(1), 293–335.

Amit, H., Terra-Nova, F., Lézin, M., and Trindade, R. I. (2021). Non-monotonic growth and motion of the south Atlantic anomaly. *Earth, Planets and Space*, 73(1), 1–10.

Anderson, J. L., and Anderson, S. L. (1999). A Monte Carlo implementation of the nonlinear filtering problem to produce ensemble assimilations and forecasts. *Monthly Weather Review*, 127(12), 2741–58.

Arneitz, P., Leonhardt, R., Schnepp, E. et al. (2017). The HISTMAG database: Combining historical, archaeomagnetic and volcanic data. *Geophysical Journal International*, 210(3), 1347–59.

Aubert, J. (2015). Geomagnetic forecasts driven by thermal wind dynamics in the Earth's core. *Geophysical Journal International*, 203(3), 1738–51.

Aubert, J., and Finlay, C. C. (2019). Geomagnetic jerks and rapid hydromagnetic waves focusing at Earth's core surface. *Nature Geoscience*, 12(5), 393–8.

Aubert, J., Finlay, C. C., and Fournier, A. (2013). Bottom-up control of geomagnetic secular variation by the Earth's inner core. *Nature*, 502(7470), 219–23.

Aubert, J., Gastine, T., and Fournier, A. (2017). Spherical convective dynamos in the rapidly rotating asymptotic regime. *Journal of Fluid Mechanics*, 813, 558–93.

Aubert, J., Labrosse, S., and Poitou, C. (2009). Modelling the palaeo-evolution of the geodynamo. *Geophysical Journal International*, 179(3), 1414–28.

Baerenzung, J., Holschneider, M., Wicht, J., Lesur, V., and Sanchez, S. (2020). The Kalmag model as a candidate for IGRF-13. *Earth, Planets and Space*, 72(1), 1–13.

Barrois, O., Hammer, M., Finlay, C., Martin, Y., and Gillet, N. (2018). Assimilation of ground and satellite magnetic measurements: Inference of core surface magnetic and velocity field changes. *Geophysical Journal International*, 215(1), 695–712.

Braginsky, S. I., and Roberts, P. H. (1995). Equations governing convection in Earth's core and the geodynamo. *Geophysical & Astrophysical Fluid Dynamics*, 79(1–4), 1–97.

Brown, M. C., Donadini, F., Korte, M. et al. (2015). Geomagia50. v3: 1. General structure and modifications to the archeological and volcanic database. *Earth, Planets and Space*, 67(1), 1–31.

Bullard, E. C., and Gellman, H. (1954). Homogeneous dynamos and terrestrial magnetism. *Philosophical Transactions of the Royal Society of London A*, 247(928), 213–78.

Burgers, G., Jan van Leeuwen, P., and Evensen, G. (1998). Analysis scheme in the Ensemble Kalman Filter. *Monthly Weather Review*, 126(6), 1719–24.

Christensen, U., and Wicht, J. (2015). Numerical dynamo simulations. In G. Schubert, ed., *Treatise on Geophysics*, 2nd ed. Oxford: Elsevier, pp. 245–77.

Christensen, U. R., Aubert, J., and Hulot, G. (2010). Conditions for Earth-like geodynamo models. *Earth and Planetary Science Letters*, 296(3), 487–96.

Constable, C. G., Parker, R. L., and Stark, P. B. (1993). Geomagnetic field models incorporating frozen-flux constraints. *Geophysical Journal International*, 113(2), 419–33.

Evensen, G. (1994). Sequential data assimilation with a nonlinear quasi-geostrophic model using Monte Carlo methods to forecast error statistics. *Journal of Geophysical Research: Oceans*, 99(C5), 10143–62.

Finlay, C. C. (2008). Historical variation of the geomagnetic axial dipole. *Physics of the Earth and Planetary Interiors*, 170(1–2), 1–14.

Finlay, C. C., Kloss, C., Olsen, N. et al. (2020). The CHAOS-7 geomagnetic field model and observed changes in the South Atlantic anomaly. *Earth, Planets and Space*, 72(1), 1–31.

Fournier, A., Hulot, G., Jault, D. et al. (2010). An introduction to data assimilation and predictability in geomagnetism. *Space Science Reviews*, 155(1–4), 247–91.

Fournier, A., Nerger, L., and Aubert, J. (2013). An ensemble Kalman filter for the time dependent analysis of the geomagnetic field. *Geochemistry, Geophysics, Geosystems*, 14(10), 4035–43.

Fratter, I., Léger, J.-M., Bertrand, F. et al. (2016). Swarm absolute scalar magnetometers first in-orbit results. *Acta Astronautica*, 121, 76–87.

Friis-Christensen, E., Lühr, H., and Hulot, G. (2006). Swarm: A constellation to study the earth's magnetic field. *Earth, Planets and Space*, 58(4), 351–8.

Gillet, N., Gerick, F., Angappan, R., and Jault, D. (2021). A dynamical prospective on interannual geomagnetic field changes. *Surveys in Geophysics*, 43, 71–105.

Gottwald, G. A., and Majda, A. (2013). A mechanism for catastrophic filter divergence in data assimilation for sparse observation networks. *Nonlinear Processes in Geophysics*, 20(5), 705–12.

Gubbins, D., and Zhang, K. (1993). Symmetry properties of the dynamo equations for palaeomagnetism and geomagnetism. *Physics of the Earth and Planetary Interiors*, 75(4), 225–41.

Gwirtz, K., Morzfeld, M., Kuang, W., and Tangborn, A. (2021). A testbed for geomagnetic data assimilation. *Geophysical Journal International*, 227(3), 2180–203.

Hamill, T. M., Whitaker, J. S., and Snyder, C. (2001). Distance-dependent filtering of background error covariance estimates in an ensemble Kalman filter. *Monthly Weather Review*, 129(11), 2776–90.

Huder, L., Gillet, N., Finlay, C. C., Hammer, M. D., and Tchoungui, H. (2020). Cov-obs. x2: 180 years of geomagnetic field evolution from ground-based and satellite observations. *Earth, Planets and Space*, 72(1), 1–18.

Jackson, A., Jonkers, A. R., and Walker, M. R. (2000). Four centuries of geomagnetic secular variation from historical records. Philosophical Transactions of the Royal Society of London A: Mathematical. *Physical and Engineering Sciences*, 358(1768), 957–90.

Johnson, C. L., and Constable, C. G. (1997). The time-averaged geomagnetic field: global and regional biases for 0–5 Ma. *Geophysical Journal International*, 131(3), 643–66.

Jonkers, A. R. T. (2003). *Earth's Magnetism in the Age of Sail*. Baltimore, MD: John Hopkins University Press.

Kalman, R. E. (1960). A new approach to linear filtering and prediction problems. *Journal of basic Engineering*, 82(1), 35–45.

Korte, M., and Constable, C. (2008). Spatial and temporal resolution of millennial scale geomagnetic field models. *Advances in Space Research*, 41(1), 57–69.

Korte, M., Donadini, F., and Constable, C. (2009). Geomagnetic field for 0–3 ka: 2. A new series of time-varying global models. *Geochemistry, Geophysics, Geosystems*, 10(6).

Korte, M., Constable, C., Donadini, F., and Holme, R. (2011). Reconstructing the Holocene geomagnetic field. *Earth and Planetary Science Letters*, 312(3–4), 497–505.

Kuang, W., Tangborn, A., Wei, Z., and Sabaka, T. (2009). Constraining a numerical geodynamo model with 100 years of surface observations. *Geophysical Journal International*, 179(3), 1458–68.

Kuang, W., Wei, Z., Holme, R., and Tangborn, A. (2010). Prediction of geomagnetic field with data assimilation: a candidate secular variation model for igrf-11. *Earth, Planets and Space*, 62(10), 775–85.

Lhuillier, F., Fournier, A., Hulot, G., and Aubert, J. (2011). The geomagnetic secular-variation timescale in observations and numerical dynamo models. *Geophysical Research Letters*, 38, L09306.

Licht, A., Hulot, G., Gallet, Y., and Thébault, E. (2013). Ensembles of low degree archeomagnetic field models for the past three millennia. *Physics of the Earth and Planetary Interiors*, 224, 38–67.

Liu, D., Tangborn, A., and Kuang, W. (2007). Observing system simulation experiments in geomagnetic data assimilation. *Journal of Geophysical Research: Solid Earth*, 112, B08103.

Livermore, P. W., Finlay, C. C., and Bayliff, M. (2020). Recent north magnetic pole acceleration towards Siberia caused by flux lobe elongation. *Nature Geoscience*, 13(5), 387–91.

Mandea, M., and Chambodut, A. (2020). Geomagnetic field processes and their implications for space weather. *Surveys in Geophysics*, 41(6), 1611–27.

Matzka, J., Chulliat, A., Mandea, M., Finlay, C., and Qamili, E. (2010). Geomagnetic observations for main field studies: from ground to space. *Space Science Reviews*, 155(1), 29–64.

Merril, R. T., McElhinny, M., and McFadden, P. L., eds. (1996). *The Magnetic Field of the Earth*, vol. 63. Cambridge, MA: Academic Press.

Minami, T., Nakano, S., Lesur, V. et al. (2020). A candidate secular variation model for IGRF-13 based on MHD dynamo simulation and 4DEnVar data assimilation. *Earth, Planets and Space*, 72, 136.

Morzfeld, M., Fournier, A., and Hulot, G. (2017). Coarse predictions of dipole reversals by lowdimensional modeling and data assimilation. *Physics of the Earth and Planetary Interiors*, 262, 8–27.

Panovska, S., Korte, M., and Constable, C. (2019). One hundred thousand years of geomagnetic field evolution. *Reviews of Geophysics*, 57(4), 1289–337.

Reda, J., Fouassier, D., Isac, A. et al. (2011). Improvements in geomagnetic observatory data quality. In M. Mandea and M. Korte, eds., *Geomagnetic Observations and Models*. Dordrecht: Springer, pp. 127–48.

Roberts, P. H., and King, E. M. (2013). On the genesis of the Earth's magnetism. *Reports on Progress in Physics*, 76(9), 096801.

Ropp, G., Lesur, V., Baerenzung, J., and Holschneider, M. (2020). Sequential modelling of the earth's core magnetic field. *Earth, Planets and Space*, 72(1), 1–15.

Sanchez, S. (2016). Assimilation of geomagnetic data into dynamo models, an archeomagnetic study. PhD thesis, Institut de Physique du Globe de Paris (IPGP), France.

Sanchez, S., Fournier, A., Aubert, J., Cosme, E., and Gallet, Y. (2016). Modelling the archaeomagnetic field under spatial constraints from dynamo simulations: a resolution analysis. *Geophysical Journal International*, 207, 983–1002.

Sanchez, S., Wicht, J., and Bärenzung, J. (2020). Predictions of the geomagnetic secular variation based on the ensemble sequential assimilation of geomagnetic field models by dynamo simulations. *Earth, Planets and Space*, 72(1), 1–20.

Sanchez, S., Wicht, J., Bärenzung, J., and Holschneider, M. (2019). Sequential assimilation of geomagnetic observations: perspectives for the reconstruction and prediction of core dynamics. *Geophysical Journal International*, 217(2), 1434–50.

Schaeffer, N., Jault, D., Nataf, H.-C., and Fournier, A. (2017). Turbulent geodynamo simulations: a leap towards Earth's core. *Geophysical Journal International*, 211(1), 1–29.

Tangborn, A., and Kuang, W. (2015). Geodynamo model and error parameter estimation using geomagnetic data assimilation. *Geophysical Journal International*, 200(1), 664–75.

Tangborn, A., and Kuang, W. (2018). Impact of archeomagnetic field model data on modern era geomagnetic forecasts. *Physics of the Earth and Planetary Interiors*, 276, 2–9.

Tangborn, A., Kuang, W., Sabaka, T. J., and Yi, C. (2021). Geomagnetic secular variation forecast using the NASA GEMS ensemble Kalman filter: A candidate SV model for IGRF-13. *Earth, Planets and Space*, 73(1), 1–14.

Valet, J.-P., and Fournier, A. (2016). Deciphering records of geomagnetic reversals. *Reviews of Geophysics*, 54(2), 410–46.

INDEX